THE NEUROSCIENCES INSTITUTE
of the Neurosciences Research Program

Gerald M. Edelman, *Director*
W. Einar Gall, *Research Director*
W. Maxwell Cowan, *Chairman,*
Scientific Advisory Committee

The Neurosciences Institute was founded in 1981 by the Neurosciences Research Program to promote the study of scientific problems within the broad range of disciplines related to the neurosciences. It provides visiting scientists with facilities to plan and review of experimental and theoretical research with emphasis on understanding the biological basis for higher brain function.

Support for the Neurosciences Research Foundation, Inc., which makes the Institute's programs possible, has come in part from the generous gifts by The Vincent Astor Foundation, Lily Auchincloss, John R. Costantino, The Charles A. Dana Foundation, Francois de Menil, Ford Motor Company, Sibyl & William T. Golden Foundation, Golden Family Foundation, Doris & Ralph E. Hansmann Foundation, Lita Annenberg Hazen, Lita Annenberg Hazen Charitable Trust, Carl B. Hess, The IFF Foundation, Johnson and Johnson, Harvey L. Karp, John D. & Catherine T. MacArthur Foundation, Rockefeller Brothers Fund, Alfred P. Sloan Foundation, Timber Hill, Inc., van Ameringen Foundation, The G. Unger Vetlesen Foundation, The Vollmer Foundation, and several anonymous donors.

The Neurosciences Institute Publications Series

Neurophysiological Approaches to Higher Brain Function
Edward V. Evarts, Yoshikazu Shinoda, and Steven P. Wise

Protein Phosphorylation in the Nervous System
Eric J. Nestler and Paul Greengard

Dynamic Aspects of Neocortical Function
Gerald M. Edelman, W. Einar Gall, and W. Maxwell Cowan, Editors

Molecular Bases of Neural Development
Gerald M. Edelman, W. Einar Gall, and W. Maxwell Cowan, Editors

The Cell in Contact: Adhesions and Junctions as Morphogenetic Determinants
Gerald M. Edelman and Jean Paul Thiery, Editors

Synaptic Function
Gerald M. Edelman, W. Einar Gall, and W. Maxwell Cowan, Editors

Auditory Function: Neurobiological Bases of Hearing
Gerald M. Edelman, W. Einar Gall, and W. Maxwell Cowan, Editors

Morphoregulatory Molecules
Gerald M. Edelman, Bruce A. Cunningham, and Jean Paul Thiery, Editors

MORPHOREGULATORY MOLECULES

MORPHOREGULATORY MOLECULES

Edited by

GERALD M. EDELMAN
The Rockefeller University

BRUCE A. CUNNINGHAM
The Rockefeller University

JEAN PAUL THIERY
Ecole Normale Supérieure, CNRS

A Neurosciences Institute Publication

JOHN WILEY & SONS

New York • Chichester • Brisbane • Toronto • Singapore

Library of Congress Cataloging in Publication Data:

Morphoregulatory molecules / edited by Gerald M. Edelman, Bruce A.
   Cunningham, Jean Paul Thiery.
      p.   cm.—(The Neurosciences Institute publications series)
   Bibliography: p.
   Includes index.
   ISBN 0-471-51261-3
      1. Cell adhesion—Molecular aspects.   2. Cell adhesion molecules.
I. Edelman, Gerald M.   II. Cunningham, Bruce A.   III. Thiery, Jean
Paul.   IV. Series.
QH623.M67   1989   89-9014
574.87—dc20   CIP

Printed in the United States of America

10 9 8 7 6 5 4 3 2 1

Contents

v

MORPHOREGULATORY MOLECULES

Introduction

GERALD M. EDELMAN
BRUCE A. CUNNINGHAM
JEAN PAUL THIERY

A double movement is occurring in developmental biology, yielding conjugate results that promise to clarify the genetic basis of pattern formation and morphogenesis. The first part of this movement is concerned with the analysis of developmentally important regulatory gene cascades, notably in *Drosophila*. The second is concerned with the analysis of the cellular and morphogenetic effects of the expression of adhesion molecules and it is this arena that is the focus of this volume. Extensive searches over the last decade have revealed that at least three families of molecules are involved in one or another aspect of adhesion. The families comprise cell adhesion molecules (CAMs), substrate adhesion molecules (SAMs), and cell junctional molecules (CJMs). In discussing them separately, it is important to note that molecules in each family are structurally and functionally different, appear in development on different schedules and at different sites, and differ in their contributory roles to morphogenetic events.

In a previous volume (Edelman and Thiery, 1985), we touched upon the functional properties of some of these molecules and their families. Since the publication of that volume, considerable progress has been made in detailing the structures of adhesion molecules as well as those of the genes specifying them.

In addition to its obvious mechanical role in development, adhesion is a regulatory process because it constrains cell and tissue sheet movement. The temporal and spatial regulation of the expression of particular adhesion molecules is thus critical to development. Moreover it appears to be controlled by local signals so that a change in form and tissue pattern can be guided at different stages by different contributions of molecules in each of the families (Edelman, 1986). For this reason, we have chosen to call the various molecules belonging to these families *morphoregulatory* molecules.

Adhesion molecules were first defined by assays measuring *in vitro* aggregation and attachment of cells to extracellular matrix components. In

1

contrast, the adhesive functions of specialized junctions were inferred from ultrastructural studies. Each of the three major families of morphoregulatory molecules can be defined in terms of their basic functions and their localization at the cell surface. CAMs link cells together at their surfaces to form collectives even at the earliest stages of development and the borders of such collectives are determined by CAM specificities. CAMs thus can be considered para- digmatic morphoregulatory molecules. By a cell collective, we mean a group of adjacent cells with common phenotypic properties acting as a source of or target for signals; in epithelia, cells are usually linked by CAMs but in mesenchyme they are not. All of the well-characterized CAMs are large cell surface glycoproteins; the evidence suggests that they are also intrinsic membrane proteins (Cunningham, 1988). Primary CAMs (e.g., N-CAM and L- CAM) are present immediately following fertilization and are also expressed in most tissues that are derived later from the primary germ layers together with secondary CAMs. Secondary CAMs (e.g., Ng-CAM, cell-CAM 105) are thus more or less specifically expressed in development in particular cell types. The expression of the different CAMs can be modulated by alternative splicing, by posttranslational mechanisms, and by *cis* and *trans* interactions at the cell surface; in addition, changes in the amount of a CAM at the cell surface (prevalence modulation) can provide fine tuning by altering the strength of adhesion.

SAMs include various extracellular matrix components having adhesive function together with their cell surface receptors or integrins (Yamada, 1983). Certain SAMs (e.g., fibronectins and their receptors, and laminin and its receptors) are expressed very early during development and are maintained in many tissues; other SAMs (e.g., cytotactin or tenascin) have a more restricted developmental distribution.

SAMs have been implicated in many histogenetic processes including stabilization of epithelia and connective tissues, cell migration, wound healing, and regeneration. Different SAM receptors share structural similarities; at the same time each recognizes one or another subset of extracellular matrix components (Hynes, 1987; Ruoslahti and Pierschbacher, 1987). Many of the SAMs are multifunctional glycoproteins capable of interacting with other members of the extracellular matrix as well as with cellular receptors. Different functional domains of particular SAMs have been implicated in cell attachment, spreading, and motility (Dufour et al., 1989). Local and global surface modulation of SAM receptors and the activity of multiple, potentially available adhesion sites in different molecules of the matrix together provide both the specificity and constraints for different cellular behaviors at different locales. In other words, SAMs form a complex modulatory network consisting of molecules interacting with each other and with cell surfaces in a combinatorial fashion. Particular forms of this expression are specific to particular mor- phological regions and according to its variations, this expression alters cell behavior in those regions.

An additional level of cell–cell interaction has been defined by studies of defined junctional complexes (CJMs). These include tight junctions, intermediate junctions, and gap junctions (Bock and Clark, 1987). These complex structures serve to link epithelial sheets controlling their polarity and intercellular communication. Tight junctions, which play a critical role in the establishment and maintenance of cell polarity and physical barriers in epithelia, have still to be defined in molecular terms and only one cytoplasmic protein has been localized so far at the junctional site. In the case of other junctions, however, progress has been more rapid. For example, the structure and function of some of the components involved in desmosome assembly have now been established. Attention has been focused on the submembranous microdomains of these junctions because they are the physical link between the transmembrane proteins acting as intercellular ligands and the intermediate filaments that contribute to the cytoskeleton. Characterization of intermediate junctions has shown that they share at least one common component, plakoglobin, in the dense plaque. Various specialized junctions differ in a number of other aspects; for example, microfilaments interact only with intermediate junctions that are enriched in the transmembrane ligand A-CAM. Recent progress in defining gap junctional structures and in determining their relative specificity in different tissues has been particularly important in evaluating the roles of these junctions in intercellular communication by electrical coupling or by metabolic cooperation. These studies promise a molecular picture of interactions underlying the function of CJMs.

The relation of the structural properties to the function of morphoregulatory molecules underlines a key biological issue. Structure–function studies have led to the concept that cell adhesion molecules are not simply ligands between cells and between cells and the extracellular matrix. Recently, molecular probes for cell adhesion molecules have been used in a series of *in vitro* and *in vivo* perturbation experiments designed to evaluate the consequences of the alteration of each adhesion mode (Thiery et al., 1985; Edelman, 1988a). Such studies have shown that patterns of development are affected irreversibly if one or another of several adhesion molecules is perturbed. Moreover, the dynamic expression and surface modulation of the different adhesion molecules are accompanied by drastic changes in cell behavior; this must imply that the transduction of adhesive signals has pleiotropic effects. Such events affect cell shape through cytoskeletal reorganization and modify gene expression, reflecting the occurrence of global cell surface modulation.

In each morphogenetic event analyzed so far, the concomitant or sequential expression of a particular adhesion molecule appears to follow fixed rules. This has prompted the general notion that, in the overall course of development, CAMs appear first, followed by SAMs and then by CJMs. Although this precedence hypothesis must not be interpreted too strictly, the evidence suggests that particular CAMs are possibly required to act before the formation of certain cellular junctions and it appears highly likely, for example, that gap junctions are formed only after cells are linked by CAMs (Mege et al., 1988).

In the preface to the previous volume (Edelman and Thiery, 1985), several unresolved issues about CAMs, SAMs, and CJMs were raised in an anticipatory mode. Since then, considerable progress has been made in understanding the structure, function, and genetic control of these molecules. The molecules in the three families do not appear structurally or evolutionarily related, but the receptors in the SAM family are related to a wide variety of tissue proteins (Hynes, 1987), and the evolutionary precursors of N-CAM appear to be the origin of the entire immunoglobulin superfamily (Edelman, 1987). We may soon be in a position to develop a solidly based molecular biology.

Despite this progress, much remains to be done to clarify questions related to gene regulation in each of the families of morphoregulatory molecules. Genes regulating the coordinate expression of CAMs, SAMs, and CJMs must act fairly independently of the differentiation program followed by each tissue but at the same time such genes must be responsible for the establishment of an appropriate hierarchy in different adhesion mechanisms in order to create and maintain borders between developing tissues while permitting epithelial–mesenchyme interconversion, remodeling of tissues, various types of cell migrations, and folding of epithelial sheets. As knowledge develops, it will be intriguing to consider the evolutionary role of these processes. Although composite fate maps of adhesion molecules are similar in all the vertebrates studied so far (Edelman, 1986), it is already apparent that heterochrony in CAM gene expression early in development may contribute to variance in morphogenesis while still contributing to the same general body plan.

The outstanding gap in the scheme relating the structure and function of adhesion and functional molecules to development concerns the nature of signals that lead to their expression. Evidence is accumulating to suggest that growth control and differentiation factors (Friedlander et al., 1986; Deuel, 1987; Heine et al., 1987; Massagué, 1987) may govern the pattern of expression of adhesion molecules. Such factors (including hormones) may, in fact, have dual functions, acting as inducers or morphogens and as growth and differentiation factors. By their action, they may actually connect the expression of two relatively independent networks of regulatory genes, those for morphoregulatory molecules and those for the historegulatory molecules that define specific intracellular functions (Edelman, 1988b).

Obviously, one would like to know more about the role of such molecules in the development of those animals such as *Drosophila* in which an intricate network of genes controlling morphogenesis has already been described. Progress on the expression of developmentally important genes in *Drosophila* (Scott and O'Farrell, 1986) should open the way to understanding similar cascades in the regulative development of vertebrates. It is already clear that SAMs are present in such invertebrates and one would expect additional CAMs to be found as well.

We ended the previous volume with a list of forward-looking questions, and in light of the foregoing discussion we may profit by doing the same here:

1. How are the expression sequences of CAMs, SAMs, and CJMs coordinated during development?
2. How are the interacting cascades of genes regulating these molecules related to their expression at particular places and times of development?

3. What kinds of molecular signals lead to the expression of these genes during embryonic induction and histogenesis?
4. How is the genetic control of morphoregulatory molecules related to the control of cytodifferentiation in particular tissues?
5. What is the evolutionary generality of such gene control mechanisms? Do they provide a basis for both heterochrony and the origin of special physiological systems such as regeneration and repair, metamorphosis, blood clotting, and immunity?

Clearly, these questions set many challenging tasks for the future. As we have suggested in this introduction, this volume obviously cannot contain the latest data in the rapidly expanding field of cell interactions during development. Rather, its main aim is to emphasize some key features of morphoregulatory molecules as major molecular determinants of animal form. The matters discussed in the ensuing chapters should serve as a base for those interested in analyzing primary processes in embryogenesis, regeneration, and disease, and they may ultimately help to focus attention on the fascinating puzzle of the molecular basis of morphologic evolution.

SELECTED REFERENCES

Bock, G., and S. Clark, eds. (1987) *Junctional Complexes of Epithelial Cells,* Ciba Foundation Symposium 125, Wiley, Chichester.

Cunningham, B. A. (1989) Structure and function of cell adhesion molecules. *Adv. Cell Biol.* (in press).

Deuel, T. F. (1987) Polypeptide growth factors: Role in normal and abnormal cell growth. *Annu. Rev. Cell Biol.* 3:443–492.

Dufour, S., J.-L. Duband, A. R. Kornblihtt, and J. P. Thiery (1989) Role of fibronectins during embryonic development and cell migration. *Trends Genet.* (in press).

Edelman, G. M. (1986) Cell adhesion molecules in the regulation of animal form and tissue pattern. *Annu. Rev. Cell Biol.* 2:81–116.

Edelman, G. M. (1987) CAMs and Igs: Cell adhesion and the evolutionary origins of immunity. *Immunol. Rev.* 100:9–43.

Edelman, G. M. (1988a) Morphoregulatory molecules. *Biochemistry* 27:3533–3543.

Edelman, G. M. (1988b) *Topobiology: An Introduction to Molecular Embryology,* Basic Books, New York.

Edelman, G. M., and J. P. Thiery, eds. (1985) *The Cell in Contact: Adhesions and Junctions as Morphogenetic Determinants,* Wiley, New York.

Friedlander, D. R., M. Grumet, and G. M. Edelman (1986) Nerve growth factor enhances expression of neuron–glia cell adhesion molecule in PC12 cells. *J. Cell Biol.* 102:413–419.

Heine, U. I., E. F. Munoz, K. C. Flanders, L. R. Ellingsworth, H.-Y. P. Lam, N. L. Thompson, A. B. Roberts, and M. B. Sporn (1987) The role of transforming growth factor-β in the development of the mouse embryo. *J. Cell Biol.* 105:2861–2876.

Hynes, R. O. (1987) Integrins: A family of cell surface receptors. *Cell* 48:549–554.

Massagué, J. (1987) The TGF-β family of growth and differentiation factors. *Cell* 49:437–438.

Mege, R.-M., F. Matsuzaki, W. J. Gallin, J. I. Goldberg, B. A. Cunningham, and G. M. Edelman (1988) Construction of epithelioid sheets by transfection of mouse sarcoma cells with cDNAs for chicken cell adhesion molecules. *Proc. Natl. Acad. Sci. USA* **85**:7274–7278.

Ruoslahti, E., and M. D. Pierschbacher (1987) New perspectives on cell adhesion: RGD and integrins. *Science* **238**:490–497.

Scott, M. P., and P. H. O'Farrell (1986) Spatial programming of gene expression in early *Drosophila* embryo. *Annu. Rev. Cell Biol.* **2**:201–229.

Thiery, J. P., J.-L. Duband, and G. C. Tucker (1985) Cell migration in the vertebrate embryo: Role of cell adhesion on tissue environment in pattern formation. *Annu. Rev. Cell Biol.* **1**:91–113.

Yamada, K. M. (1983) Cell surface interactions with extracellular materials. *Annu. Rev. Biochem.* **52**:761–799.

Section 1

Molecular Biology and Chemistry of Cell Adhesion Molecules

Chapter 1

Structure, Expression, and Cell Surface Modulation of Cell Adhesion Molecules

BRUCE A. CUNNINGHAM
GERALD M. EDELMAN

ABSTRACT

The characterization of the molecular structures, properties, and expression of cell adhesion molecules (CAMs) has suggested that these cell surface glycoproteins, together with the cell–substrate adhesion molecules (SAMs) and molecules found in cellular junctions (CJMs), serve as regulators of morphogenesis and histogenesis. The CAMs are structurally distinct not only from SAMs and CJMs but also from each other. Their expression and activity can be modulated by a variety of mechanisms, including changes in their levels and locations on the cell surface, differential gene splicing, changes in posttranslational modifications, differences in their requirements for specific cations, and differential responses to growth factors.

The neural CAM (N-CAM) and liver CAM (L-CAM) are primary CAMs that are expressed on the earliest embryonic cells and are differentially distributed on cell collectives at sites of embryonic induction. In adult tissues L-CAM is found on most epithelia, whereas N-CAM is prominent in brain, muscle, heart, and kidney. The neuron–glia CAM (Ng-CAM) appears later in development and is restricted to the nervous system and is thus a secondary CAM. Studies with vesicles and transfected cells have established that N-CAM and L-CAM mediate adhesion via homophilic mechanisms; a CAM on one cell binds the same CAM on another. Ng-CAM mediates neuron–neuron adhesion by a homophilic mechanism and neuron–glia adhesion by a heterophilic mechanism.

N-CAM appears as a variety of closely related polypeptides that arise from a single gene by alternative RNA splicing; the polypeptides are differentially expressed in various tissues. All known N-CAM polypeptides contain immunoglobulinlike domains in their binding regions, and covalently bound polysialic acids that influence binding and decrease in amount during development. By contrast, L-CAM appears in only a single form with no polysialic acid; it is the first known example of a family of structurally related molecules that mediates calcium-dependent adhesion in various tissues. A variety of considerations suggests that the precursor of cell adhesion molecules provides the basis for other sophisticated molecular recognition phenomena, including immune regulation.

Cell adhesion has long been recognized as an important process in biology—particularly in development. The study of the differential binding and sorting out of cells in multicellular organisms began with sponges (Wilson, 1907) and coelenterates (DeMorgan and Drew, 1914; Chalkey, 1945). Holtfreter (1939, 1948a,b) showed that when cells from different embryonic amphibian tissues were mixed, they could sort out to form structures characteristic of parent tissues, indicating that there is selective adherence among cells of various types. Moscona (1952, 1962) and his colleagues used this paradigm to show similar phenomena in cells from chicken and mice. All of these studies utilized long-term assays, so it was not possible to distinguish cell type–specific adhesion from differentiation. Various attempts were subsequently made to develop more direct short-term assays of cell–cell adhesion (for a review, see Frazier and Glaser, 1979) and to isolate molecular fractions that might be responsible for differential selectivity or specificity (Balsamo and Lilien, 1974; Merrell et al., 1975; Oppenheimer, 1975; Shur and Roth, 1975; Hausman and Moscona, 1976). Only limited success was achieved, however, and various interpretations of the results reflected different views about the nature of adhesion.

More recently, immunological assays (Gerisch and Malchow, 1976; Brackenbury et al., 1977) have been used to isolate specific cell adhesion molecules (CAMs); analyses of the expression and properties of these glycoproteins have provided a new, more dynamic view of the role of CAMs in development (for reviews, see Edelman, 1983, 1984a, 1985; Damsky et al., 1984). In addition, these studies have provided a clearer discrimination between molecules mediating cell–cell interaction (CAMs), those involved in cell–substrate interaction (SAMs, substrate adhesion molecules), and those appearing at cell junctions (CJMs, cell junctional molecules). The three families have been termed morphoregulatory molecules because it is now apparent that the regulation of the time and place of their expression guides and constrains the primary processes leading to morphogenesis and histogenesis (Edelman and Thiery, 1985; Edelman, 1988). CAMs, SAMs, and CJMs play different roles in the regulation of morphogenetic sequences, as indicated by differences in their chemical structures, locations, and schedules of appearance: CAMs are critical in *initial* boundary formation, embryonic induction and migration, tissue stabilization, and regeneration; SAMs function in relation to migration, stabilization of epithelia, and the development of solid tissues; and CJMs are involved in the formation of specialized cell connections, cell communication (as seen in gap junctions), and the sealing of the surfaces of epithelial sheets (see Edelman and Thiery, 1985, for discussions by various authors).

CAMs were first definitively identified by means of short-term assays in which specific antibodies capable of blocking cell adhesion *in vitro* were used to purify cell surface molecules as putative CAMs (Brackenbury et al., 1977). These are N-CAM (neural cell adhesion molecule; Thiery et al., 1977; Hoffman et al., 1982), L-CAM (liver cell adhesion molecule; Bertolotti et al., 1980; Gallin et al., 1983), and Ng-CAM (neuron–glia cell adhesion molecule; Grumet and Edelman, 1984; Grumet et al., 1984a,b). Their role as CAMs was indicated by the chemical definition of their binding mechanisms, their appearance at the cell surface at specific developmental stages, and the ability of antibodies against them to perturb tissue integrity. These glycoproteins were all identified in tissues from embryonic chicken, but analogous molecules were subsequently identified by a variety of investigators in other species, and some were given different names: D_2 (Jorgensen et al., 1980) and BSP-2 (Hirn et al., 1983) are rodent equivalents of N-CAM; L1 (Faissner et al., 1985) and NILE (Salton et al., 1983a,b; Bock et al., 1985; Friedlander et al., 1986) are rodent equivalents of Ng-CAM; and uvomorulin (Hyafil et al., 1981), E-cadherin (Yoshida and Takeichi, 1982), cell-CAM 120/80 (Damsky et al., 1983), and Arc 1 (Imhof et al., 1983) are mammalian equivalents of L-CAM.

N-CAM and L-CAM appear in the earliest embryonic cells and are also expressed on adult tissues derived from all three germ layers; they are thus designated primary CAMs (Thiery et al., 1982; Edelman et al., 1983a; Thiery et al., 1984; Crossin et al., 1985). Although they were named for the cell types from which they were first isolated, each appears on a variety of tissues: N-CAM appears in brain, heart, kidney, skeletal muscle, and smooth muscle (Crossin et al., 1985); L-CAM appears on most epithelia (Gallin et al., 1983) including, although rarely, neuroepithelia (Levi et al., 1987). CAMs that appear later in development and are more restricted in their tissue distribution are termed secondary CAMs; Ng-CAM, which appears only on postmitotic neurons and Schwann cells (Grumet et al., 1984a,b; Faissner et al., 1985; Daniloff et al., 1986a), was the first example of a secondary CAM.

All known CAMs are large integral membrane proteins and as such share many basic features. They are distinct molecules, however, each with its unique properties. N-CAM has large amounts of sialic acid in α-2,8 homopolymers (Hoffman et al., 1982; Rougon et al., 1982; Finne et al., 1983) attached to as many as three asparagine-linked oligosaccharides (Cunningham et al., 1983; Crossin et al., 1984). The amount of sialic acid is greatest in embryonic brain N-CAM (130 moles/mole protein) and decreases as the embryo matures, so that N-CAM in adult tissues has only about one-third the amount found in embryos (Rothbard et al., 1982). This embryonic (E) to adult (A) conversion occurs in all known forms of N-CAM. In different brain regions the rate of E-to-A conversion generally parallels the maturation of each region (Edelman and Chuong, 1982). The change is due to a decrease in the activity of one or more sialyl transferases (Friedlander et al., 1985) and influences N-CAM binding (Hoffman and Edelman, 1983).

L-CAM does not contain polysialic acid nor does it undergo any form of E-to-A conversion (Gallin et al., 1983; Cunningham et al., 1984). It appears, however, to be the first of a family of closely related proteins that mediate calcium-dependent adhesion. Other members of this family appear later in development than L-CAM and have different but overlapping tissue distributions with L-CAM and with each other. These include N-cadherin (Hatta et al.,

1985; Hatta and Takeichi, 1986; Shirayoshi et al., 1986) and A-CAM (Volk and Geiger, 1984, 1986), found in brain, lens, and heart, and P-cadherin (Nose and Takeichi, 1986; Nose et al., 1987), detected initially in mouse placenta. These molecules differ from L-CAM in size and in the products produced by limited proteolysis (Volk and Geiger, 1986); they are synthesized from different genes, but they are highly similar (~50% identity) to L-CAM and each other in their amino acid sequences (Shirayoshi et al., 1986; Nose et al., 1987; Hatta et al., 1988).

The secondary CAM, Ng-CAM, appears in neurons but mediates both neuron–neuron and neuron–glia adhesion (Grumet et al., 1984b; Grumet and Edelman, 1988). It is particularly important in neurite fasciculation (Stallcup and Beasley, 1985; Fischer et al., 1986; Hoffman et al., 1986). It is equivalent to the NILE glycoprotein (Bock et al., 1985; Friedlander et al., 1986), and hence its expression is enhanced by nerve growth factor.

CAM STRUCTURE

Because of their potential significance in regulating developmental events, N-CAM and L-CAM from chickens were the first CAMs to be studied in detail; the complete amino acid sequences of both molecules have been determined (Cunningham et al., 1987; Gallin et al., 1987), and the structures of the N-CAM gene (Owens et al., 1987) and the L-CAM gene (Sorkin et al., 1988) have been described. More recently, homologous molecules in other species have been characterized (Barthels et al., 1987; Dickson et al., 1987; Nagafuchi et al., 1987; Ringwald et al., 1987; Santoni et al., 1987; Small et al., 1987). Similar studies have been undertaken for Ng-CAM (Grumet et al., 1984a; Faissner et al., 1985; Sorkin et al., 1985; Tacke et al., 1987). Recently, the structure of L1 has been determined, and it appears to be homologous to that of N-CAM (Moos et al., 1988). Similar findings have been obtained for Ng-CAM (M. Grumet, M. P. Burgoon, V. Mauro, G. M. Edelman, and B. A. Cunningham, unpublished data). The results of these studies provide a firm basis for classifying the CAMs and for relating their structures to their various functions. Moreover, the antibodies and molecular biological reagents derived from this work have opened new avenues for analyzing CAM function *in vitro* and for correlating cell adhesion with the other primary processes in development *in vivo*. Some examples using CAM cDNAs to transfect cells are discussed below.

N-CAM

N-CAM is expressed in multiple forms, and the expression of the variants appears to be tissue restricted. Four distinct polypeptides have been well defined (Figure 1), and still others are known to exist. All, however, are derived from a single gene, located on chromosome 9 in mice (D'Eustachio et al., 1985) and chromosome 11 in humans (Nguyen et al., 1985). The gene (Figure 2) in chickens (Owens et al., 1987) is large (>50 kb), with more than 19 exons

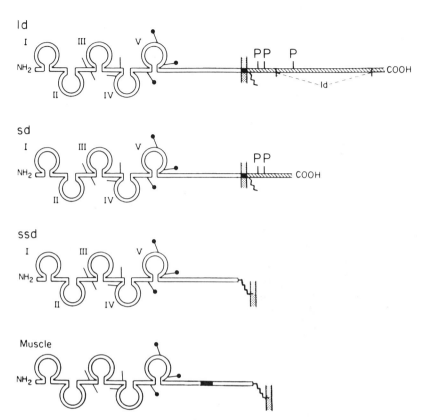

Figure 1. *Schematic drawing of the structure of some known N-CAM polypeptides. Loops* (I–V) denote immunoglobulinlike domains; the *lines* in regions III, IV, and V indicate potential attachment sites for asparagine-linked oligosaccharides; *closed circles* indicate potential sites for the attachment of polysialic acid. The cell membrane is denoted by the *stippled region*; the *dark bar* indicates the membrane-spanning region of the ld and sd chains. The *stair-step symbol attached to the membrane* represents the phosphatidylinositol link for the ssd chain and the muscle-specific form, whereas the *stair-step symbol inside the membrane* indicates fatty acids that are incorporated into the ld and sd chains. P, phosphothreonine and phosphoserine. The region of polypeptide specific to the ld chain is indicated by *vertical lines*. The *shaded area* in the muscle protein denotes its specific inserts in chicken cardiac and skeletal muscle (Prediger et al., 1988) and human skeletal muscle (Dickson et al., 1987).

utilized in the coding regions (Owens et al., 1987); additional noncoding exons are located 20–30 kb 5' to the exon specifying the amino terminus (exon 1). The largest N-CAM polypeptide (ld, large cytoplasmic domain polypeptide) is specified by exons 1–14 plus exons 16–19, whereas the next smaller (sd, small cytoplasmic domain polypeptide) is identical but lacks the sequence specified by exon 18. In addition to the carboxyl termini of the ld and sd chains, exon 19 specifies a large (3.5-kb) 3' untranslated sequence shared by their mRNAs. The smallest (ssd, small surface domain polypeptide) is also specified by exons 1–14 plus exon 15. A form in human muscle (Dickson et al., 1987) appears to be

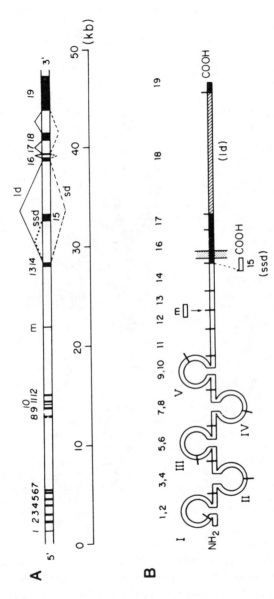

Figure 2. Structure of the N-CAM gene in chickens (A) in relation to the structure of the polypeptides (B). Of exons 1–19 (*dark bars*), 1–14 and 16–19 are used for the ld chain (*solid lines*), 1–14, 16–17, and 19 for the sd chain (*dashed lines*), and 1–15 for the ssd chain (*dotted line*). m denotes the assumed position of the exons unique to muscle-specific N-CAM (Dickson et al., 1987; Prediger et al., 1988); this polypeptide resembles the ssd chain.

like the ssd form but has an additional 37-amino-acid insert. Recent studies (Prediger et al., 1988) have described an ssd form specific to chicken cardiac and skeletal muscle with a similar 31-amino-acid insert; this insert is specified by four exons located in the region between exons 12 and 13. This region in the chicken gene contains still other alternative N-CAM exons (Prediger et al., 1988).

The amino acid sequences of the ld, sd and ssd polypeptides (Hemperly et al., 1986a,b; Cunningham et al., 1987) are shown in Figure 3. The membrane-spanning segment of the ld and sd chains extends from residue 693 to 710, with amino acids 1–692 in the extracellular side of the membrane and 711 to 1090 on the cytoplasmic side. The carboxyl terminus shown here differs from that described in earlier reports in that the carboxy-terminal 7 amino acids are replaced with a new carboxyl terminus of 25 amino acids (Hemperly et al., 1988). The original clone (pEC208) lacked a cytosine present in other clones, which accounts for the sequence reported earlier. Sequences of clones other than pEC208 all agree with the revised sequence, as do the sequences of human (Dickson et al., 1987), mouse (Santoni et al., 1987), and rat (Small et al., 1987) N-CAM.

From the amino terminus to residue 682, the ld, sd, and ssd N-CAM polypeptides are identical (Cunningham et al., 1987). Each polypeptide contains five contiguous segments of about 100 amino acids each that resemble one another and are strikingly similar to other proteins, now collectively termed the immunoglobulin superfamily (for reviews, see Edelman, 1987; Williams, 1987); each segment also contains a pair of cysteines like those conserved in immunoglobulins and related molecules. This portion of N-CAM includes the binding region, which probably involves two or more of the Ig-like domains. The three potential sites for attachment of the polysialic acid (Crossin et al., 1984) are in domain V; each is located at a position that would be on the molecular surface if this N-CAM homology region is folded like an immunoglobulin domain (J. W. Becker, B. A. Cunningham, and G. M. Edelman, unpublished data).

From residue ~483 to the membrane, the polypeptide lacks the defined elements of the Ig-like segments. This region, however, is homologous (J. W. Becker, B. A. Cunningham, and G. M. Edelman, unpublished data) to type III repeats in fibronectin (Petersen et al., 1983), and residues 543 to 549 (Hemperly et al., 1986a) are identical to a sequence in fibronectin that precedes the fibronectin RGD recognition sequence (Pierschbacher and Ruoslahti, 1984). N-CAM, however, contains no RGD sequence (Hemperly et al., 1986a).

The membrane-spanning segments of the ld and sd chains resemble those seen in other membrane proteins, but lack the basic amino acids typically clustered on the cytoplasmic side. Instead, N-CAM has a series of closely spaced cysteine residues, the sulfur atoms of which probably form covalent bonds with one another or with other molecules such as fatty acids; fatty acids can be incorporated covalently into N-CAM molecules (Sorkin et al., 1985). The cytoplasmic region contains several phosphorylation sites on as yet

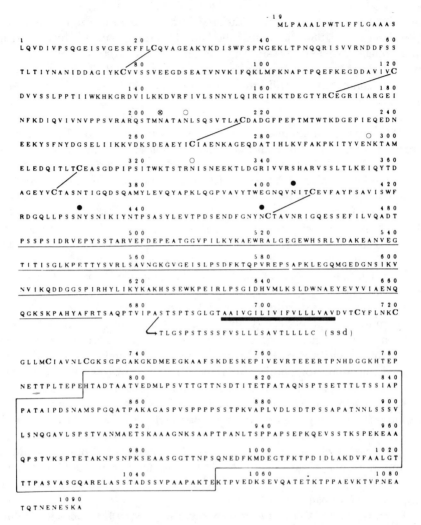

- 1 9
M L P A A A L P W T L F F L G A A A S

1   2 0   4 0   6 0
L Q V D I V P S Q G E I S V G E S K F F L C Q V A G E A K Y K D I S W F S P N G E K L T P N Q Q R I S V V R N D D F S S

                               8 0   1 0 0   1 2 0
T L T I Y N A N I D D A G I Y K C V V S S V E E G D S E A T V N V K I F Q K L M F K N A P T P Q E F K E G D D A V I V C

1 4 0   1 6 0   1 8 0
D V V S S L P P T I I W K H K G R D V I L K K D V R F I V L S N N Y L Q I R G I K K T D E G T Y R C E G R I L A R G E I

                               2 0 0   ⊗   ○   2 2 0   2 4 0
N F K D I Q V I V N V P P S V R A R Q S T M N A T A N L S Q S V T L A C D A D G F P E P T M T W T K D G E P I E Q E D N

                               2 6 0   2 8 0   ○   3 0 0
E E K Y S F N Y D G S E L I I K K V D K S D E A E Y I C I A E N K A G E Q D A T I H L K V F A K P K I T Y V E N K T A M

                               3 2 0   ○   3 4 0   3 6 0
E L E D Q I T L T C E A S G D P I P S I T W K T S T R N I S N E E K T L D G R I V V R S H A R V S S L T L K E I Q Y T D

                               3 8 0   4 0 0   ●   4 2 0
A G E Y V C T A S N T I G Q D S Q A M Y L E V Q Y A P K L Q G P V A V Y T W E G N Q V N I T C E V F A Y P S A V I S W F

         ●   4 4 0   ●   4 8 0
R D G Q L L P S S N Y S N I K I Y N T P S A S Y L E V T P D S E N D F G N Y N C T A V N R I G Q E S S E F I L V Q A D T

                               5 0 0   5 2 0   5 4 0
P S S P S I D R V E P Y S S T A R V E F D E P E A T G G V P I L K Y K A F W R A L G E G E W H S R L Y D A K E A N V E G

                               5 6 0   5 8 0   6 0 0
T I T I S G L K P F T T Y S V R L S A V N G K G V G E I S L P S D F K T Q P V R E P S A P K L E G Q M G E D G N S I K V

                               6 2 0   6 4 0   6 6 0
N V I K Q D D G G S P I R H Y L I K Y K A K H S S E W K P E I R L P S G I D H V M L K S L D W N A E Y E V Y V I A E N Q

                               6 8 0   7 0 0   7 2 0
Q G K S K P A H Y A F R T S A Q P T V I P A S T S P T S G L G T A A I V G I L I V I F V L L L V A V D V T C Y F L N K C

         ↳ T L G S P S T S S S F V S L L L S A V T L L L L C   (s s d)

                               7 4 0   7 6 0   7 8 0
G L L M C I A V N L C G K S G P G A K G K D M E E G K A A F S K D E S K E P I V E V R T E E E R T P N H D G G K H T E P

                               8 0 0   8 2 0   8 4 0
N E T T P L T E P E H T A D T A A T V E D M L P S V T T G T T N S D T I T E T F A T A Q N S P T S E T T T L T S S I A P

                               8 6 0   8 8 0   9 0 0
P A T A I P D S N A M S P G Q A T P A K A G A S P V S P P P P S S T P K V A P L V D L S D T P S S A P A T N N L S S S V

                               9 2 0   9 4 0   9 6 0
L S N Q G A V L S P S T V A N M A E T S K A A A G N K S A A P T P A N L T S P P A P S E P K Q E V S S T K S P E K E A A

                               9 8 0   1 0 0 0   1 0 2 0
Q P S T V K S P T E T A K N P S N P K S E A A S G G T T N P S Q N E D F K M D E G T F K T P D I D L A K D V F A A L G T

                               1 0 4 0   1 0 6 0   1 0 8 0
T T P A S V A S G Q A R E L A S S T A D S S V P A A P A K T E K T P V E D K S E V Q A T E T K T P P A E V K T V P N E A

         1 0 9 0
T Q T N E N E S K A

Figure 3. *Amino acid sequences of the ld, sd, and ssd chains of chicken N-CAM.* Included are the signal sequence (−19 to 0), the ssd-specific region (following the *bent arrow*), and the ld-specific insert (*large box*). *Slanted lines* indicate cysteines assumed to be connected by intrachain disulfide bonds; *circles* are potential sites for attachment of asparagine-linked oligosaccharides, with the X indicating a position known to have carbohydrates and the *closed circles* being the potential sites for attachment of polysialic acid. The *dark bar underline* denotes the membrane-spanning region common to the ld and sd chains. The *thin underline* denotes the region where N-CAM has sequence similarities with fibronectin. The *bent arrow* indicates the beginning of a newly assigned carboxy-terminal sequence for the sd and ld chains; the earlier sequence (Hemperly et al., 1986a) was deduced from a clone that had apparently lost a cytosine as the result of a cloning artifact.

16

unidentified serine and threonine residues (Sorkin et al., 1984). Most of the phosphothreonines and phosphoserines are common to both chains; there are additional sites in the ld chain. The unique sequence in the ld chain (261 amino acids) is rich in serine, threonine, alanine, and proline and deficient in aromatic amino acids (Hemperly et al., 1986a), suggesting that this chain has a specialized structure probably adapted for interacting with specific cytoplasmic elements in neurons (Pollerberg et al., 1986).

The ssd chain lacks the cytoplasmic domain and ends in a sequence resembling those of proteins such as the Thy-1 antigen that are attached to the membrane by a special phosphatidylinositol anchor (Nybroe et al., 1985; He et al., 1986; Hemperly et al., 1986b). In accord with this observation, the ssd chain can be released from cells by treatment with phosphatidylinositol-specific phospholipase C.

Sequences for the ssd chain in mouse (Barthels et al., 1987), the majority of the sequence of an ssd-like chain in human muscle (Dickson et al., 1987), the sd chain in mouse (Santoni et al., 1987), and the sd chain in rat (Small et al., 1987) have all recently been reported and are all closely related to one another and to the chicken N-CAM chains. Overall, the sequences are highly conserved (>80% identity), and some regions are practically invariant; an example for chicken and mouse ssd chains is shown in Figure 4. These findings are consistent with earlier observations (Hoffman et al., 1984) that N-CAM chains from different species bind to one another.

When viewed in the electron microscope (Edelman et al., 1983b; Hall and Rutishauser, 1987), N-CAM ld and sd molecules appear as bent rods clustered about central hubs in arrays of two, three, four, or more. The hubs are largest in the absence of detergent and, we propose that they correspond to elements created by interactions of the hydrophobic portions of the chains. Others (Hall and Rutishauser, 1987), however, have concluded that they may result from N-CAM interacting with itself via the binding region. Recent studies (Becker et al., 1989) definitively exclude this interpretation and define the lengths of the hinged segments and the location of the polysialic acid. The ssd form also appears as a bent rod, but it does not form aggregates.

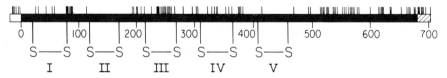

Figure 4. *Comparison of the amino acid sequences of the ssd polypeptides of chicken (Hemperly et al., 1986a,b; Cunningham, 1987) and mouse (Barthels, 1987) N-CAM. Each vertical line represents a position where the sequences differ. The open bar indicates the signal sequence, the dark bar the coding region believed to be common to the ld, sd, and ssd chains, and the hatched bar the sequence unique to the ssd polypeptide. The numbered (I–V) S–S segments are the immunoglobulinlike domains.*

L-CAM

In contrast to N-CAM, L-CAM has been detected in only one molecular form (Figure 5) in all tissues in which it is expressed (Gallin et al., 1983; Cunningham et al., 1984; Thiery et al., 1984), and it appears to be the product of a single, much smaller (~10 kb) gene (Sorkin et al., 1988). The structural gene (Figure 6A) contains 16 exons with an average size of 222 nucleotides. The exon boundaries do not correspond to known structural features of the protein.

The polypeptide (Figure 6B) is initially made as a large precursor (135 kD) that is trimmed at the amino-terminal end to a smaller (124-kD) component before expression on the cell surface (Peyrieras et al., 1983; Gallin et al., 1987). It is an integral membrane protein (Gallin et al., 1983, 1987; Cunningham et al., 1984), with the amino-terminal portion (amino acids 1–544) on the external side of the membrane and the carboxy-terminal end (amino acids 576–727) on the cytoplasmic side. Immediately following the membrane-spanning region

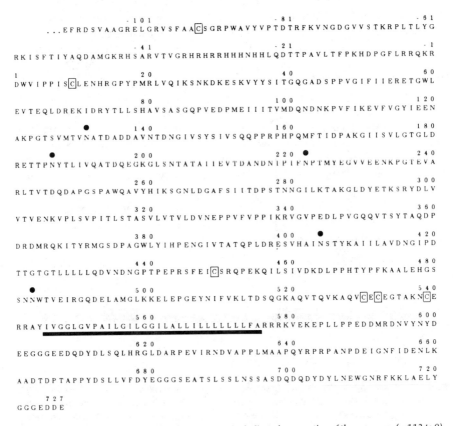

Figure 5. *Amino acid sequences of chicken L-CAM, including a large portion of the precursor (−113 to 0).* Cysteines are enclosed in *boxes*; potential asparagine glycosylation sites are indicated by *closed circles*; the putative membrane-spanning region is denoted by the *thick underline.*

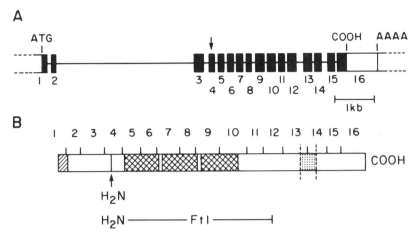

Figure 6. *Schematic drawing of the L-CAM gene and protein. A:* The gene contains 16 exons *(solid vertical bars)* that encode the protein; 3′ and 5′ untranslated regions are denoted by the open portions of exons 1 and 16. The amino terminus of the protein, as it is expressed on the cell surface, is indicated by the *arrow*; a *scale bar* (1 kb) is given below. *B:* The protein has a typical signal sequence *(slanted lines)* followed by a segment of 134 amino acids *(from signal sequence to arrow)* that is removed prior to expression of the molecule on the cell surface. The region *(dotted segments)* that spans the cell membrane *(dashed vertical lines)* and the three contiguous segments that are similar to each other *(cross-hatching)*, as well as the amino (H₂N) and carboxyl (COOH) termini of the protein are indicated. *Vertical lines* mark the boundaries of exons *(numbered above)*. Ft 1 is a fragment of L-CAM released from cell membranes by digestion with trypsin in the presence of calcium.

there is a cluster of basic amino acids on the cytoplasmic side. There is a cluster of three cysteines (amino acids 530, 532, 539) near the membrane-spanning region, but they are on the external side rather than the cytoplasmic side as in N-CAM. Overall, the proteins from chicken and mouse (Nagafuchi et al., 1987; Ringwald et al., 1987) are highly homologous to each other (70% identical), especially in their cytoplasmic segments (90% identical).

L-CAM contains no segments that resemble the Ig superfamily, but residues 30–374 can be divided into three contiguous segments of about 110 amino acids that are 20–40% identical to one another, suggesting that the evolution of the molecule involved one or more duplication events (Gallin et al., 1987). L-CAM contains no polysialic acid but has five potential asparagine glycosylation sites (Gallin et al., 1987), four of which are known to be occupied by three complex and one high-mannose oligosaccharides (Cunningham et al., 1984). Calcium is important both to the L-CAM structure and to its binding, and the molecule binds calcium (Ringwald et al., 1987). Computer searches, however, did not reveal distinct calcium-binding sites analogous to those in other calcium-binding proteins such as the calmodulins. Recent studies (Hatta et al., 1988) indicate that N-cadherin and E-cadherin have structures similar to that of L-CAM.

In the electron microscope, a proteolytic fragment of L-CAM that contains the homophilic binding regions appears as a bent rod–shaped molecule, strikingly similar to the ssd form of N-CAM (Becker et al., 1989). This overall shape may thus be important for the function of cell adhesion molecules.

Ng-CAM

Ng-CAM is less well defined than N- or L-CAM; like N-CAM, it is a member of the immunoglobulin superfamily (M. Grumet, M. P. Burgoon, V. Mauro, G. M. Edelman, and B. A. Cunningham, unpublished data). In mice, it appears as a glycoprotein of ~200 kD (Rathjen and Schachner, 1984; Grumet and Edelman, 1988), but in chickens its expression is more complex. Generally in chickens, a species of 135 kD predominates, with minor components of 200 kD and 80 kD detected in variable amounts (Grumet et al., 1984a). Immunological data, peptide maps, and amino-terminal sequence data indicate that the 135-kD and and 80-kD species are derived from the 200-kd component by proteolysis (Faissner et al., 1985; Sorkin et al., 1985). Ng-CAM is structurally related to N-CAM and L1 (Moos et al., 1988) and certain subfractions of Ng-CAM and N-CAM display the carbohydrate epitope HNK-1 (Hoffman et al., 1984; Kruse et al., 1984; Tucker et al., 1984; Chou et al., 1985; Noronha et al., 1986). In glycolipids this epitope includes sulfated glycuronyl residues (Chou et al., 1985); it was first seen in lymphocytes (Abo and Balch, 1981) and has been found on certain SAMs such as cytotactin (Grumet et al., 1985) and cytotactin-binding proteoglycan (Hoffman and Edelman, 1987). Its functional significance is unknown.

CAM BINDING: TRANSFECTION WITH CAM cDNAs

The complexity of the CAMs and their dependence on membrane attachment for full function have made it difficult to obtain thermodynamic data, but CAM binding has been analyzed by kinetic assays and by cellular transfection experiments. These studies have shown that binding events mediated by N-CAM, L-CAM, and Ng-CAM are specific and that the binding specificities are characteristic and independent of each other. N-CAM and Ng-CAM bind by calcium-independent mechanisms, whereas binding of the L-CAM family is calcium dependent. In all three cases, binding is homophilic (a CAM on one cell binds to the same CAM on the apposing cell). This conclusion was suggested from assays using N-CAM and Ng-CAM linked to Covaspheres or in artificial lipid vesicles (Hoffman and Edelman, 1983; Grumet and Edelman, 1988) and has been verified for N-CAM and L-CAM by transfecting mouse L-cells, which lack these CAMs (Edelman et al., 1987; Nagafuchi et al., 1987), with appropriate cDNAs (see below). In addition to mediating homophilic binding between neurons, Ng-CAM binds neurons to astrocytes; this binding is probably heterophilic because astrocytes lack Ng-CAM (Grumet et al., 1984a,b; Grumet and Edelman, 1988).

So far *in vitro* assays (Hoffman and Edelman, 1983; Grumet and Edelman, 1988) have measured only the *rates* of aggregation for N-CAM and Ng-CAM. They show, however, that a 2-fold increase in surface density can result in a greater than 30-fold increase in binding rates, indicating that binding is dependent on CAM concentration in a highly nonlinear fashion. Such nonlinear changes could lead to rapid changes in cell interactions *in vivo*; for example, in altering the linkage of cells bound in a collective by a given CAM or in forming borders between two collectives that utilize CAMs of different specificity. A similar though less dramatic influence is found for E-to-A conversion of N-CAM: although the polysialic acid is not directly involved in binding, decreases in the amount of this negatively charged sugar can result in a 3- to 4-fold increase of the binding rate (Hoffman and Edelman, 1983).

In N-CAM, alterations in binding may also result from local changes in the conformation of the molecule. Binding probably involves *trans* pairing between two or more Ig-like domains of N-CAM molecules on opposing cells. The detailed mechanism is not known, but could include any of the first four domains. The fifth includes the polysialic acid that occurs at or near the hinge (Becker et al., 1989). Regardless of the detailed binding mechanism, such hinged structures could facilitate *trans*-homophilic binding during changes of cell shape that would otherwise sterically hinder homophilic binding. The polysialic acid may have multiple functions such as influencing *cis* spacing of molecules on the same cell, altering the freedom of the hinge, thus affecting the apposition of terminal binding domains, and modulating cell-to-cell spacing by excluded volume and charge effects. As mentioned before, E forms with their apparently weaker binding predominate when adhesion is being established (such as in the formation of new neural connections), whereas A forms seem to predominate when stabilization is required (as in the adult brain).

The factors that influence CAM binding indicate that the binding properties of homophilic CAMs are under the control of the cells that they ligate. This is in contrast to certain SAMs, which are extracellular molecules, and are hence more indirect in their effects. CAMs, however, may have additional subtleties in their binding: for example, Ng-CAM, which is synthesized by neurons and not by glia (Grumet and Edelman, 1988), is homophilic in neuron–neuron binding and heterophilic in neuron–glia binding; moreover, N-CAM can bind heparan sulfate (apparently within its two amino-terminal Ig-like domains), and it has been suggested that this influences N-CAM activity (Cole and Glaser, 1986; Cole et al., 1986).

To demonstrate more directly that CAMs ligate the cells that synthesize them, mouse L-cells, which ordinarily do not make N-CAM or L-CAM, were transfected with a cDNA for each CAM under control of the SV40 early promoter. Transfected cells that expressed a CAM on their surfaces aggregated specifically. Such experiments have been done (Figure 7) for each of the chicken N-CAM chains (Edelman et al., 1987) and for chicken and mouse L-CAM (Edelman et al., 1987; Nagafuchi et al., 1987). Although transfected with a chimeric cDNA, cells transfected with chicken L-CAM cDNA expressed properly processed, mature L-CAM (Figure 7A) on their cell surfaces (Figure 7D). Other transfectants expressed the sd and ld chains of N-CAM (Figure 7B) in the appropriate form at the cell surface (Figure 7F), but the ssd chain (Figure 7B) was detected only in the cytoplasm or was secreted. This result suggests that the mechanism for the addition of the phosphatidylinositol linkage to the ssd chain was not present or was not activated in these cells; control experiments using monkey kidney (COS; Gluzman, 1981) cells transiently transfected with cDNA for ssd chains indicated that such a mechanism could be activated, yielding ssd chains linked by phosphatidylinositol at the cell surface (B. Murray, J. Hemperly, B. A. Cunningham, and G. M. Edelman, unpublished data).

Transfected cells expressing L-CAM or the sd or ld chains of N-CAM at their respective surfaces were capable of linking cells expressing the homologous CAM in aggregates. L-CAM-mediated aggregation was particularly striking and provided strong evidence that it occurs by a homophilic mechanism inasmuch as untransfected L-cells did not bind with transfected cells expressing L-CAM. As expected, cells expressing either form of N-CAM did not bind to cells expressing L-CAM, but could bind to each other. Cells expressing either form of N-CAM also bound to brain vesicles containing N-CAM. In all cases binding was inhibited by appropriately specific antibody fragments.

Phenotypic changes in shape were seen in cells that possessed the N-CAM ld or sd chains (Figure 7F) on their surfaces, but not in those synthesizing the ssd chain and releasing them into the medium (Edelman et al., 1987). Whether the phenotypic changes are due to an association of the carboxy-terminal domains of the ld and sd chains with the cell cortex or to cis-binding effects resulting from the lack of the polysialic acid remains to be determined. (L-cells probably do not have a mechanism for generating polysialic acid; the forms of N-CAM detected in the transfected cells resembled chicken N-CAM that had been treated with neuraminidase.)

Recent studies (Mege et al., 1988) indicate that mouse sarcoma (S180) cells can be transfected with L-CAM cDNA and express L-CAM, followed by linkage into epithelioid sheets. Both gap and adherens junctions were found in such sheets; junctions diminished when the sheets were dissociated with anti-L-CAM antibody fragments. This is consistent with the precedence hypothesis, which proposes that L-CAM linkage is a prerequisite for junction formation.

CELL SURFACE MODULATION

Local cell surface modulation is the alteration over time of the amount, distribution, or chemical properties of a particular kind of molecule at the cell

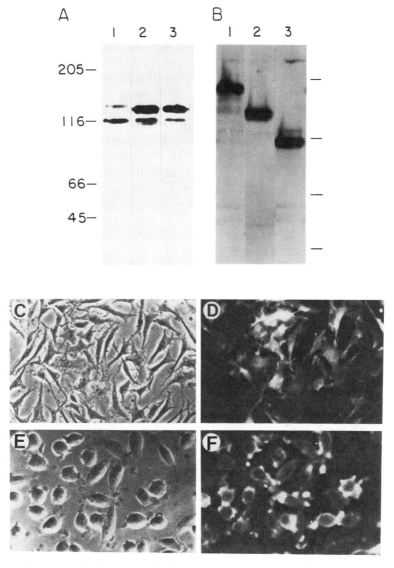

Figure 7. *Expression of L-CAM and N-CAM by mouse L-cells transfected with chicken cDNAs under control of the SV40 early promoter. A:* Immunoblots of extracts from transfected cell lines expressing the precursor and processed form of L-CAM; photomicrographs of one cell line stained with chicken anti-L-CAM are shown in C (phase contrast) and D (fluorescence). *B:* 21 immunoblots of extracts from transfected cell lines expressing the ld (*lane 1*), sd (*lane 2*), and ssd (*lane 3*) forms of N-CAM; both the ld and sd forms appeared on the cell surface as detected by antibody staining, and the line expressing the ld form had an altered morphology as shown in *E* (phase contrast) and *F* (fluorescence). (Adapted from Edelman et al., 1987.)

surface (Edelman, 1976). CAMs can change in amount (prevalence modulation), in position or distribution (polarity modulation), or in molecular structure (chemical modulation) during development. As indicated above, nonlinear binding properties of CAMs at the cell surface result from cell surface modulation, consistent with the cell-regulatory nature of cell–cell adhesion. Examples of prevalence and polarity modulation are given below in our description of the expression sequences of CAMs. An example of chemical modulation is the E-to-A conversion of N-CAM discussed above. Another form of modulation is the differential cellular expression (Pollerberg et al., 1985) of the ld polypeptide (Murray et al., 1986b) of N-CAM, which may be related in some cases to its possible pairing with Ng-CAM on the same cell surface (*cis* interaction). While evidence for this is circumstantial, it raises the possibility that differential cytoskeletal binding or *cis* interactions of the same or different cell surface CAMs could alter their *trans* binding to CAMs on apposed cells and their polarity modulation in the same cell.

Cell surface modulation appears to be a key mechanism in CAM action at the cellular level, and a large variety of such mechanisms may operate. The existence of modulation mechanisms shifts attention from the binding specificities of different CAMs to those parts of each molecule related to secretion, membrane insertion, cytoskeletal interaction, *cis* interaction, and alternative RNA splicing, all of which can yield different forms of the same CAM and each of which could have a significant role in signaling events.

The potential influence of cell surface modulation in various forms also indicates that cell shape and specialization can constrain and alter the effects of cell adhesion in different contexts. In accord with this notion, the ld chain of N-CAM appears to be enriched in neural growth cones (Chuong et al., 1982; Ellis et al., 1985; Wallis et al., 1985) and changes in the relative expression of Ng-CAM and N-CAM in chick dorsal root ganglion cells and retinal explants in culture alter the relative contributions of each CAM to cell binding (Hoffman et al., 1986). For example, in the ganglia, where the Ng-CAM/N-CAM ratio at the cell surface is higher than in the retinas and increases with time, fasciculation of neurites is perturbed mainly by anti-Ng-CAM antibodies; in the retina, where the Ng-CAM/N-CAM ratio is low, retinal layering is perturbed mainly by anti-N-CAM antibodies. Anti-NILE (i.e., anti-Ng-CAM) antibodies and anti-N-CAM antibodies also differentially inhibit fasciculation (Stallcup and Beasley, 1985; Stallcup et al., 1985).

These considerations indicate that where, when, and on what cells the CAMs are expressed are factors as important as their binding specificities, and therefore any definition of the role of CAM-mediated adhesion must take into account the structure of actual tissues and cell morphology. Moreover because the combinations of mechanisms that can influence morphogenesis are virtually limitless, there is no need for a very large number of molecular specificities or definite molecular cell addresses (Sperry, 1963; Edelman, 1984b).

REGULATION OF CAM EXPRESSION *IN VIVO*

CAM expression is controlled epigenetically, and it changes with cellular context, reflecting different tissue interactions and developmental stages. This conclusion has been substantiated by examining CAM expression throughout development, using immunohistochemical methods (Chuong and Edelman, 1984; Crossin et al., 1985; Daniloff et al., 1986a,b; Levi et al., 1987; Richardson et al., 1987). In addition, it has been shown that blocking CAM binding with antibodies can alter morphology. For example side-to-side interactions of axonal fibers of neurons can be perturbed by Fab' fragments to Ng-CAM (Hoffman et al., 1986), layer formation in the retina in organ culture can be disrupted by Fab' fragments of antibodies to N-CAM (Buskirk et al., 1980), and the orderly mapping of the retina to the optic tectum *in vivo* can be altered by anti-N-CAM fragments (Fraser et al., 1984). Moreover, mechanical disruption of morphology can in some cases lead to changes in CAM expression. From these studies it has become apparent that, during development, primary CAMs (and possibly secondary CAMs) follow rules according to which they are expressed or down-regulated at the cell surface during border formation between tissues (Crossin et al., 1985; Edelman, 1986a,b,c).

The significance of these rules is that they are correlated with cell division, movement, and death, which are the major driving forces in the formation of cell patterns. The regulation of binding via CAMs (and SAMs) can alter the sequence and influences of such forces by aggregating cells into collectives or releasing groups of cells to migrate. The most specific such transformation is from epithelia (sheets of cells linked by CAMs, SAMs, and CJMs) to mesenchyme (collections of loosely associated cells that interact with SAMs while the CAMs are generally down-regulated); in mesenchyme, condensation of cells usually coincides with expression of a specific CAM (see Mege et al., 1988). Separate epithelia and mesenchymes linked by CAMs interact with each other, exchanging signals that lead to embryonic induction, a complex process of milieu-dependent differentiation (Gurdon, 1987).

Both N-CAM and L-CAM are expressed on all cells at early stages (e.g., blastoderm or blastula) of chick (Figure 8) and frog embryos (Crossin et al., 1985; Edelman, 1986a,b,c; Levi et al., 1987). As cells migrate at the onset of gastrulation both CAMs are generally down-regulated (Figure 8A,B). After gastrulation, CAM expression is segregated: N-CAM is expressed on cells of the neural plate (Figure 8C) and reappears on condensing mesoderm, whereas L-CAM is seen alone or together with N-CAM in epithelia (Figure 8D). Upon formation of mesenchyme, N-CAM is down-regulated (often during cell movement), only to be reexpressed when mesenchymal cells condense to form various structures (N → O → N, rule I). On early epithelia, as in the blastoderm, one often observes both N-CAM and L-CAM; subsequently, one or the other is lost (NL → L or NL → N, rule II). At many sites where mesenchyme acts to induce epithelia, a collective of cells obeying rule I adjoins

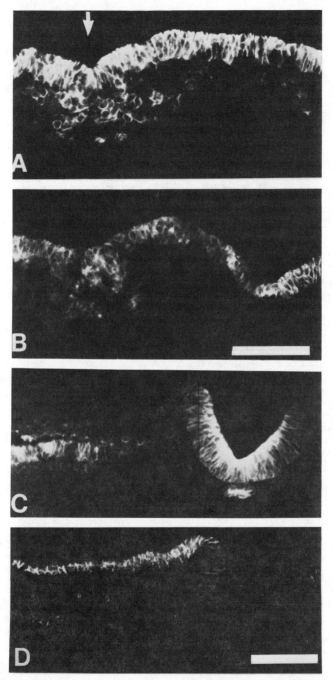

Figure 8. *Immunofluorescent staining of chicken embryo at the head-fold stage in the region of the primitive streak using anti-N-CAM (A) and anti-L-CAM (B) antibodies.* The cells in the epiblast stain with both, whereas cells through the streak stain with neither. Later, during the formation of the neural groove, N-CAM staining is enhanced in the groove (C), whereas L-CAM staining is lost from these cells (D). (Adapted from Crossin et al., 1985.)

a collective of cells obeying rule II (Crossin et al., 1985; Edelman, 1986b). Often in the same developing tissue (e.g., the kidney or the feather) there can be multiple cycles in which the primary CAMs are differentially up-regulated and down-regulated at each stage of histogenesis.

CAMs IN BORDER AND PATTERN FORMATION

After neural induction, L-CAM is excluded from neural derivatives (rule II), and the secondary Ng-CAM is expressed on postmitotic neurons. In the central nervous system (CNS), Ng-CAM appears on extending neurites and in very slight amounts on cell somata. Where neurons migrate on guide glia, however, Ng-CAM is strongly expressed on somata and leading processes as well as on neurites (Thiery et al., 1985a; Daniloff et al., 1986a). Ng-CAM is not seen on glia in the CNS but is found in the peripheral nervous system (PNS), where Ng-CAM and N-CAM are both present on Schwann cells and neurons. In the PNS, however, Ng-CAM is more uniformly expressed on the surface of neurons than in the CNS.

During development there is a site-specific microsequence (Daniloff et al., 1986a) of CAM expression that results in altered distributions of N-CAM and Ng-CAM with time. This sequence includes the down-regulation of Ng-CAM in presumptive myelinating regions (prevalence modulation) in the CNS, and the perinatal E-to-A conversion of N-CAM in tracts. In addition, neural crest cells that form the PNS (among other structures) show a remarkable expression of rule I as applied to ectomesenchyme: N-CAM disappears from the surface of migrating crest cells and reappears at sites where ganglion formation occurs (Thiery et al., 1982, 1985b). All of these sequences reveal coordinated cell surface modulation events during the formation of particular neural structures, including prevalence modulation, polarity modulation, and chemical modulation (E-to-A conversion).

The differential splicing of the N-CAM gene to produce the various polypeptides is tissue specific and could be a critical influence on development. The ld polypeptide is nervous tissue specific; it is expressed on selected cell types (Williams et al., 1985), and it appears differentially in certain layers during the development of the cerebellum and retina (Pollerberg et al., 1985; Murray et al., 1986b). The ssd chain of N-CAM is also differentially expressed; it appears late and is detected primarily in glia. Its phosphatidylinositol anchor (Nybroe et al., 1985; He et al., 1986; Hemperly et al., 1986b) seems to be a feature specialized for nonneural cells consistent with the fact that other forms of N-CAM with a similar mode of attachment have been detected in human muscle (Dickson et al., 1987) and chick cardiac and skeletal muscle (Prediger et al., 1988).

In general, the sd form of N-CAM seems to be expressed in nearly all tissues in which the molecule is seen, the ld chain is most predominant in neurons, and the ssd chain and its variants are expressed in selected nonneural tissues. The utilization of alternative RNA splicing (Hemperly et al., 1986b; Murray et al.,

1986a), the different mode of attachment of the ssd chain to the membrane, and the possibility that the large cytoplasmic domain of the ld polypeptide may interact differentially with the cytoskeleton suggest that differential cell surface modulation of the different chains could lead to altered patterns of cell interaction, migration, and layering. The adequate signal may be one that controls alternative splicing in a local tissue region, another mechanism for cell surface modulation.

A striking example of the recursive use of the rules for primary CAM expression is seen in the feather (Figure 9). Feathers are induced through the formation of dermal condensations of mesodermally derived mesenchyme, which act upon ectodermal cells to form placodes (Sengel, 1976). As feather induction proceeds, the placodes and condensations become hexagonally close-packed in rows from medial to lateral aspects of the skin. Within each induced placode a dermal papilla subsequently forms by repeated inductive interactions between mesoderm and ectoderm. The cellular proliferation of barb ridges, followed by barbule plate formation, then yields the basis for three levels of branching: rachis, ramus, and barbule.

At each of these levels collectives of cells linked by L-CAM are coupled with collectives of cells linked by N-CAM in a series of events involving either cell movement and adhesion or division and adhesion (Chuong and Edelman, 1985a,b). Initially (Figure 9A,B), L-CAM-linked ectodermal cells are approached by CAM-negative mesenchyme cells, which become N-CAM-positive in the ectodermal vicinity (rule I). As the N-CAM-positive cells accumulate to form condensations, placodes are induced in the L-CAM-linked cells. In the papilla which subsequently forms, a similar coupling of L-CAM-linked ectodermal cells and N-CAM-linked mesenchymal collectives is seen. N-CAM-positive mesenchyme cells are then excluded by a basement membrane; the collar cells, derived from the L-CAM-positive papillar ectoderm, express both N-CAM and L-CAM (rule II).

These events are followed by a series of site-restricted applications of rule II: cells derived from papillar ectoderm form barb ridges and express L-CAM (Figure 9C), whereas the basilar cells lose L-CAM and express N-CAM (Figure 9D) to form the marginal plate. The ridge cells also organize into L-CAM-positive barbule plates, and N-CAM is expressed in cells between the barbule plate cells to form the axial plate. The end result is that cell collectives linked by L-CAM alternate with those expressing N-CAM at both the secondary barb level and the tertiary barbule level (Figure 9C,D). After further extension of the barb ridges into rami, the L-CAM-positive cells keratinize, and the N-CAM-positive cells die, leaving spaces between barbules and yielding the characteristic feather morphology. In this way, borders become edges by coordination of CAM expression, cytodifferentiation, and cell death.

Overall, there is periodic CAM modulation, successive formation of L-CAM-linked and N-CAM-linked cell collectives, and the definite association of gene expression events during cytodifferentiation with particular kinds of CAMs (e.g., the expression of keratins only in L-CAM-containing cells).

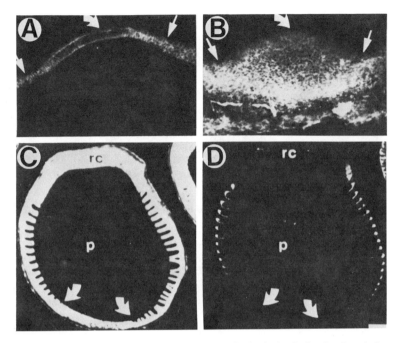

Figure 9. *Examples of differential expressions of CAMs in the developing feather.* In placode formation the placode cells stain with L-CAM (*A*), predominantly on the upper and lower surfaces, whereas the underlying dermal condensation expresses N-CAM (*B*). During barb ridge formation the stratified feather epithelium is mainly L-CAM positive (*C*), and N-CAM appears in the valleys between pairs of barb ridges (*D*). (Adapted from Chuong and Edelman, 1985a,b.)

Throughout this histogenetic process there is an intimate connection between the regulatory process of adhesion and the epigenetic sequences of those primary processes that act as driving forces: morphogenetic movement in the original mesenchymal induction, mitosis in the formation of papillar ectoderm and barb ridges, and death of the N-CAM-linked collectives at the end of feather formation.

The feather induction system also has provided the most striking of the perturbation experiments (Gallin et al., 1986). When anti-L-CAM antibodies were added to chick skin explants, the pattern of N-CAM-linked dermal condensations was altered from a sixfold symmetry pattern to one of fourfold symmetry, and the feather placodes fused mediolaterally into stripes (Figure 10A,B). In long-term cultures with anti-L-CAM present, scalelike plates formed rather than the featherlike patterns seen in unperturbed controls (Figure 10C,D). The N-CAM-linked cells of the inducing mesoderm lack L-CAM; apparently perturbation of the linkages among the overlying epidermal cells by anti-L-CAM Fab' fragments altered the pattern of these mesodermal condensations. This suggests that the antibody effects were indirect,

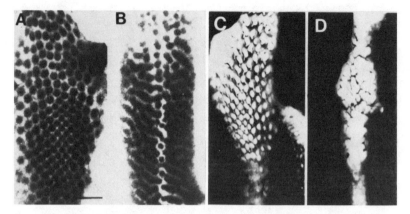

Figure 10. *Perturbation of feather placodes in explants of chicken skin by antibodies to L-CAM.* Cultures of skin from seven-day-old embryos after three days in culture are shown in the absence (*A*) and presence (*B*) of immense rabbit Fab'; similar explants after 10 days in culture are shown with no antibodies (*C*) or with anti-L-CAM Fab' present during the first two days of culture. (Adapted from Gallin et al., 1986.)

altering the response to signals from the dermis as well as the production of ectodermal signals, to influence both the timing and control of mesodermal growth patterns. Thus, it appears that the linkage of cells by a specific CAM can regulate their ability to produce and to respond to other signals. A computer model of feather development based on this idea has been constructed and generates altered patterns that resemble those seen experimentally (Gallin et al., 1986).

These observations suggest that CAMs have a primary role in linking cells in the proper sequence during morphogenesis and in stabilizing formed structures. Moreover, mechanical disruption of signaling between two different tissues, as in the feather, can also affect CAM expression; in another example, breaking nerve–muscle contact causes a marked increase in N-CAM in muscle which down-regulates when contact is reestablished (Daniloff et al., 1986b). The combined observations that perturbation of binding leads to morphological change and that perturbation of morphology alters CAM expression suggest that there are complex interactions across several levels of cellular organization that affect the primary processes leading to form. This level of complexity is not limited to the action of CAMs, but also can involve cellular binding to various SAMs. The intricate play among the primary processes of development, cellular responses to signals, and the modulatory mechanochemical effects of CAMs and SAMs are thus sufficient to yield tissue pattern, and it is not necessary to have large ensembles of molecular binding specificities that generate patterns through "built-in" (genetically preprogrammed) mechanisms.

THE MORPHOREGULATOR HYPOTHESIS

The idea that cell surface modulation with transmembrane control underlies pattern formation has been incorporated into a morphoregulator hypothesis (Edelman, 1984a, 1986a, 1988). This hypothesis states that the essential genetic component controlling pattern formation is the response of genes determining the appearance and function of morphoregulatory molecules (CAMs, SAMs, and CJMs). These sets of genes are assumed to be under separate control from those specifying specific tissue products, and they respond to signals across borders between collectives of cells linked by CAMs of different specificities. The appearance of the CAMs and the action of networks of SAMs alter the growth and movement of these cells. Thus, the successive synthesis, turnover, or degradation of CAMs and SAMs provides new contexts for inductive signaling which in turn results in further gene expression altering borders and subsequent signaling events. The number of such signals need not be large because these events could combine with the general irreversibility of prior gene expression to yield a large number of potential patterns. Of course, such morphoregulatory sequences, while essential, could not in themselves lead to detailed tissue patterns. Therefore, at the same time, batteries of develop-mentally important *tissue-specific* genes, probably controlled by selector genes resembling homeotic genes (see Scott and O'Farrell, 1986), must be expressed to yield sequences of *tissue differentiation* within the patterns established by the morphoregulatory genes.

The morphoregulator hypothesis couples developmental genetics to the mechanochemistry of pattern by connecting CAM and SAM modulation in cell collectives to gene expression, linking the ensuing signals and responses of the cell collectives that emerge to particular states of adhesion. This general picture can also apply to the functions of CJMs (for example, in tight junctions, adherens junctions, and gap junctions). Their expression and regulation, however, govern cell communication and epithelial stabilization and may therefore depend upon the prior expression and action of CAMs even though some such junctions (for example, adherens junctions) may include some CAMs (for example, A-CAM, L-CAM). It has been proposed that the order necessary to form fully functioning epithelia is (1) CAMs, (2) SAMs, and (3) CJMs, a precedence hypothesis subsidiary to the morphoregulator hypothesis and more immediately open to testing. Recent work (Mege et al., 1988) has lent support to this notion.

EVOLUTIONARY CONSIDERATIONS

The idea of a dynamic connection between cell adhesion, morphogenesis, and gene signaling is beginning to provide some new insights into the evolution of form and the emergence of specialized tissue systems. SAMs are found in both vertebrates and invertebrates (Fessler et al., 1984), and it is now known that invertebrate CAMs exist as well (Harrelson and Goodman, 1988; Seeger et al., 1988). In the vertebrates, N-CAM-binding functions are highly conserved (Hoffman et al., 1984), and the CAM expression rules are generally conserved. There are, however, large differences between species in the timing and extent of gene expression for a given CAM at comparable stages of development and morphogenesis (Levi et al., 1987). Morphological changes seen in evolutionarily related forms as well as in metamorphosis may thus occur as a result of molecular heterochrony, that is, differences in the time at which homologous CAMs and SAMs are expressed during the evolution of developmental morphogenetic sequences. Consistent with the morphoregulator hypothesis, changes in the expression of relatively few genes could thus yield large morphogenetic changes even at early developmental stages.

The analyses of N-CAM structure have also provided an example of the emergence in evolution of one tissue function from another. As indicated above, the structural gene specifying N-CAM is closely related to the precursor for the highly versatile group of proteins now collectively called the immunoglobulin superfamily (Hemperly et al., 1986a,b; Cunningham et al., 1987). N-CAM contains segments that are closely related in sequence and in the organization of their disulfide bonds to immunoglobulin domains, although in N-CAM each segment is specified by two exons whereas those in immunoglobulins are specified by one. This observation suggests that the adaptive aspects of immunity which involve distinction between self and nonself may have emerged from the early origins of cell adhesion which evolved primarily for self recognition (Edelman, 1987). In support of this notion, several other types of molecules, including certain growth factor receptors, proteoglycan link proteins, and nervous system–specific molecules such as the myelin-associated glycoprotein (MAG; Arquint et al., 1987; Salzer et al., 1987), and the P_0 myelin-associated glycoprotein (Lemke and Axel, 1985) share this evolutionary origin (Lai et al., 1987). Moreover, molecules with sequences similar to the immunoglobulinlike domains of N-CAM have recently been detected in *Drosophila* (Harrelson and Goodman, 1988; Seeger et al., 1988).

The evolutionary relationship of molecules involved in cell adhesion to those that mediate adaptive immune responses is not surprising in that both morphoregulation and immune regulation require cell–cell interactions at the cell surface. This particular example also exemplifies the opportunistic character of evolution: Genes originally established for metazoan transcendence over the life of the solitary cell are used subsequently to provide the genes which, by the evolution of additional special mechanisms, generate some of the most sophisticated molecular recognition phenomena yet analyzed.

Further study of morphoregulatory molecules should provide other examples and fill important gaps in our knowledge of gene formation. It will certainly give additional insights into the evolution of highly specialized tissue functions. The study of CAMs, SAMs and CJMs could thus dramatically transform our view of anatomy and development and their relation to evolution. Such analyses together with the elucidation of the signals and signal pathways involved in regulating CAMs and SAMs should have a major impact on most areas of metazoan biology and could revolutionize specific areas such as developmental neurobiology, histology, and embryology. Indeed, as the studies outlined in this book indicate, a robust field of molecular histology is well on its way.

ACKNOWLEDGMENTS

Our work was supported by National Institutes of Health grants HD-16550, AM-04256, and HD-09635, and Senator Jacob Javits Center of Excellence in Neuroscience grant NS-22789.

REFERENCES

Abo, T., and C. M. Balch (1981) A differentiation antigen of human NK and K cells identified by a monoclonal antibody (HNK-1). *J. Immunol.* **127**:1024.

Arquint, M., S. Roder, L. S. Chia, J. Down, D. Wilkinson, H. Bayley, P. Braun, and R. Dunn (1987) Molecular cloning and primary structure of myelin-associated glycoprotein. *Proc. Natl. Acad. Sci. USA* **84**:600–604.

Balsamo, J., and J. Lilien (1974) Functional identification of three components which mediate tissue-type specific embryonic cell adhesion. *Nature* **251**:522–524.

Barthels, D., M.-J. Santoni, W. Wille, C. Ruppert, J.-C. Chaix, R. Hirsch, J. C. Fontecilla-Camps, and C. Goridis (1987) Isolation and nucleotide sequence of mouse N-CAM cDNA that codes for a Mr 79 000 polypeptide without a membrane-spanning region. *EMBO J.* **6**:907–914.

Becker, J. W., H. P. Erickson, S. Hoffman, B. A. Cunningham, and G. M. Edelman (1989) Topobiology of cell adhesion molecules. *Proc. Natl. Acad. Sci. USA* **86**:1088–1092.

Bertolotti, R., U. Rutishauser, and G. M. Edelman (1980) A cell surface molecule involved in aggregation of embryonic liver cells. *Proc. Natl. Acad. Sci. USA* **77**:4831–4835.

Bock, E., C. Richter-Landsberg, A. Faissner, and M. Schachner (1985) Demonstration of immunochemical identity between the nerve growth factor–inducible large external (NILE) glycoprotein and the cell adhesion molecule-L1. *EMBO J.* **4**:2765–2768.

Brackenbury, R., J. P. Thiery, U. Rutishauser, and G. M. Edelman (1977) Adhesion among neural cells of the chick embryo. I. An immunological assay for molecules involved in cell–cell binding. *J. Biol. Chem.* **252**:6835–6840.

Buskirk, D. R., J. P. Thiery, U. Rutishauser, and G. M. Edelman (1980) Antibodies to a neural cell adhesion molecule disrupt histogenesis in cultured chick retinae. *Nature* **285**:488–489.

Chalkey, H. W. (1945) Quantitative relation between the number of organized centers and tissue volume in regenerating masses of minced body sections of *Hydra*. *J. Natl. Cancer Inst.* **6**:191–195.

Chou, K. H., A. A. Ilyas, J. E. Evans, R. H. Quarles, and F. B. Jungalwala (1985) Structure of a glycolipid reacting with monoclonal IgM in neuropathy and with HNK-1. *Biochem. Biophys. Res. Commun.* **128**:383–388.

Chuong, C.-M., and G. M. Edelman (1984) Alterations in neural cell adhesion molecules during development of different regions of the nervous system. *J. Neurosci.* **4**:2354–2368.

Chuong, C.-M., and G. M. Edelman (1985a) Expression of cell adhesion molecules in embryonic induction. I. Morphogenesis of nestling feathers. *J. Cell Biol.* **101**:1009–1026.

Chuong, C.-M., and G. M. Edelman (1985b) Expression of cell adhesion molecules in embryonic induction. II. Morphogenesis of adult feathers. *J. Cell Biol.* **101**:1027–1043.

Chuong, C.-M., D. A. McClain, P. Streit, and G. M. Edelman (1982) Neural cell adhesion molecules in rodent brains isolated by monoclonal antibodies with cross-species reactivity. *Proc. Natl. Acad. Sci. USA* **79**:4234–4238.

Cole, G. J., and L. Glaser (1986) A heparin-binding domain from N-CAM is involved in neural cell–substratum adhesion. *J. Cell Biol.* **102**:403–412.

Cole, G. J., A. Loewy, N. V. Cross, R. Akeson, and L. Glaser (1986) Topographic localization of the heparin-binding domain of the neural cell adhesion molecule N-CAM. *J. Cell Biol.* **103**:1739–1744.

Crossin, K. L., G. M. Edelman, and B. A. Cunningham (1984) Mapping of three carbohydrate attachment sites in embryonic and adult forms of the neural cell adhesion molecule. *J. Cell Biol.* **99**:1848–1855.

Crossin, K. L., C.-M. Chuong, and G. M. Edelman (1985) Expression sequences of cell adhesion molecules. *Proc. Natl. Acad. Sci. USA* **82**:6942–6946.

Cunningham, B. A., S. Hoffman, U. Rutishauser, J. J. Hemperly, and G. M. Edelman (1983) Molecular topography of the neural cell adhesion molecule N-CAM: Surface orientation and location of sialic acid–rich and binding regions. *Proc. Natl. Acad. Sci. USA* **80**:3116–3120.

Cunningham, B. A., Y. Leutzinger, W. J. Gallin, B. C. Sorkin, and G. M. Edelman (1984) Linear organization of the liver cell adhesion molecule L-CAM. *Proc. Natl. Acad. Sci. USA* **81**:5787–5791.

Cunningham, B. A., J. J. Hemperly, B. A. Murray, E. A. Prediger, R. Brackenbury, and G. M. Edelman (1987) Neural cell adhesion molecule: Structure, immunoglobulin-like domains, cell surface modulation, and alternative RNA splicing. *Science* **236**:799–806.

D'Eustachio, P., G. Owens, G. M. Edelman, and B. A. Cunningham (1985) Chromosomal location of the gene encoding to the neural cell adhesion molecule (N-CAM) in the mouse. *Proc. Natl. Acad. Sci. USA* **82**:7631–7635.

Damsky, C. H., J. Richa, D. Solter, K. Knudsen, and C. A. Buck (1983) Identification and purification of a cell surface glycoprotein involved in cell–cell interactions. *Cell* **34**:455–466.

Damsky, C. H., K. A. Knudsen, and C. A. Buck (1984) Integral membrane proteins in cell–cell and cell–substratum adhesion. In *The Biology of Glycoproteins*, R. J. Ivatt, ed., pp. 1–64, Plenum, New York.

Daniloff, J. K., C.-M. Chuong, G. Levi, and G. M. Edelman (1986a) Differential distribution of cell adhesion molecules during histogenesis of the chick nervous system. *J. Neurosci.* **6**:739–758.

Daniloff, J. K., G. Levi, M. Grumet, F. Rieger, and G. M. Edelman (1986b) Altered expression of neuronal cell adhesion molecules induced by nerve injury and repair. *J. Cell Biol.* **103**:929–945.

DeMorgan, W., and H. Drew (1914) A study of the restitution masses formed by the dissociated cells of the hybrids *Antennularia ramosa* and *A. antennine*. *J. Mar. Biol. Assoc. UK* **10**:440–463.

Dickson, G., H. J. Gower, C. H. Barton, H. M. Prentice, V. L. Elsom, S. E. Moore, R. D. Cox, C. Quinn, W. Putt, and F. S. Walsh (1987) Human muscle neural cell adhesion molecule (N-CAM): Identification of a muscle specific sequence in the extracellular domain. *Cell* **50**:1119–1130.

Edelman, G. M. (1976) Surface modulation in cell recognition and cell growth. *Science* **192**:218–226.

Edelman, G. M. (1983) Cell adhesion molecules. *Science* **219**:450–457.

Edelman, G. M. (1984a) Cell adhesion and morphogenesis: The regulator hypothesis. *Proc. Natl. Acad. Sci. USA* **81**:1460–1464.

Edelman, G. M. (1984b) Cell surface modulation and marker multiplicity in neural patterning. *Trends Neurosci.* **7**:78–84.

Edelman, G. M. (1985) Cell adhesion and the molecular processes of morphogenesis. *Annu. Rev. Biochem.* **54**:135–169.

Edelman, G. M. (1986a) Molecular mechanisms of morphogenetic evolution. In *Chemica Scripta of the Royal Academy* **36b**:363, Cambridge Univ. Press, Cambridge, England.

Edelman, G. M. (1986b) Epigenetic rules for expression of cell adhesion molecules during morphogenesis. *CIBA Found. Symp.* **125**:192–216.

Edelman, G. M. (1986c) Cell adhesion molecules in the regulation of animal form and tissue pattern. *Annu. Rev. Cell Biol.* **2**:81–116.

Edelman, G. M. (1987) CAMs and Igs: Cell adhesion and the evolutionary origins of immunity. *Immunol. Rev.* **100**:11–45.

Edelman, G. M. (1988) *Topobiology: An Introduction to Molecular Embryology,* Basic Books, New York.

Edelman, G. M., and C.-M. Chuong (1982) Embryonic to adult conversion of neural cell-adhesion molecules in normal and *staggerer* mice. *Proc. Natl. Acad. Sci. USA* **79**:7036–7040.

Edelman, G. M., and J. P. Thiery, eds. (1985) *The Cell in Contact: Adhesions and Junctions as Morphogenetic Determinants,* Wiley, New York.

Edelman, G. M., W. J. Gallin, A. Delouvée, B. A. Cunningham, and J. P. Thiery (1983a) Early epochal maps of two different cell adhesion molecules. *Proc. Natl. Acad. Sci. USA* **80**:4384–4388.

Edelman, G. M., S. Hoffman, C.-M. Chuong, J. P. Thiery, R. Brackenbury, W. J. Gallin, M. Grumet, M. E. Greenberg, J. J. Hemperly, C. Cohen, and B. A. Cunningham (1983b) Structure and modulation of neural cell adhesion molecules in early and late embryogenesis. *Cold Spring Harbor Symp. Quant. Biol.* **68**:515–526.

Edelman, G. M., B. A. Murray, R.-M. Mege, B. A. Cunningham, and W. J. Gallin (1987) Cellular expression of liver and neural cell adhesion molecules after transfection with their cDNAs results in specific cell–cell binding. *Proc. Natl. Acad. Sci. USA* **84**:8502–8506.

Ellis, L., I. Wallis, E. Abreu, and K. H. Pfenninger (1985) Nerve growth cones isolated from fetal rat brain. IV. Preparation of a membrane subfraction and identification of a membrane glycoprotein expressed on sprouting neurons. *J. Cell Biol.* **101**:1977–1989.

Faissner, A., D. B. Teplow, D. Kubler, G. Keilhauer, V. Kinzel, and M. Schachner (1985) Biosynthesis and membrane topography of the neural cell adhesion molecule L1. *EMBO J.* **4**:3105–3113.

Fessler, J. H., G. Lunstrum, K. G. Duncan, A. G. Campbell, R. Sterne, H. P. Bächinger, and L. I. Fessler (1984) Evolutionary constancy of basement membrane components. In *The Role of Extracellular Matrix in Development,* R. L. Trelstad, ed., pp. 207–219, Alan R. Liss, New York.

Finne, J., U. Finne, H. Deagostini-Bazin, and C. Goridis (1983) Occurrence of $\alpha 2$-8-linked polysialosyl units in a neural cell adhesion molecule. *Biochem. Biophys. Res. Commun.* **112**:482–487.

Fischer, G., J. Künemund, and M. Schachner (1986) Neurite outgrowth patterns in cerebellar microexplant cultures are affected by antibodies to the cell surface glycoprotein L1. *J. Neurosci.* **6**:605–612.

Fraser, S. E., B. A. Murray, C.-M. Chuong, and G. M. Edelman (1984) Alteration of the retinotectal map in *Xenopus* by antibodies to neural cell adhesion molecules. *Proc. Natl. Acad. Sci. USA* **81**:4222–4226.

Frazier, W., and L. Glaser (1979) Surface components and cell recognition. *Annu. Rev. Biochem.* **48**:491–523.

Friedlander, D. R., R. Brackenbury, and G. M. Edelman (1985) Conversion of embryonic forms of N-CAM *in vitro* results from *de novo* synthesis of adult forms. *J. Cell Biol.* **101**:412–419.

Friedlander, D. R., M. Grumet, and G. M. Edelman (1986) Nerve growth factor enhances expression of neuron–glia cell adhesion molecule in PC12 cells. *J. Cell Biol.* **102**:413–419.

Gallin, W. J., G. M. Edelman, and B. A. Cunningham (1983) Characterization of L-CAM, a major cell adhesion molecule from embryonic liver cells. *Proc. Natl. Acad. Sci. USA* **80**:1038–1042.

Gallin, W. J., C.-M. Chuong, L. H. Finkel, and G. M. Edelman (1986) Antibodies to L-CAM perturb inductive interactions and alter feather pattern and structure. *Proc. Natl. Acad. Sci. USA* **83**:8235–8239.

Gallin, W. J., B. C. Sorkin, G. M. Edelman, and B. A. Cunningham (1987) Sequence analysis of a cDNA clone encoding the liver cell adhesion molecule, L-CAM. *Proc. Natl. Acad. Sci. USA* **84**:2808–2812.

Gerisch, G., and D. Malchow (1976) Cyclic AMP receptors and the control of cell aggregation in *Dictyostelium. Adv. Cyclic Nucleotide Res.* **7**:49–68.

Gluzman, Y. (1981) SV40-transformed simian cells support the replication of early SV40 mutants. *Cell* **23**:175–182.

Grumet, M., and G. M. Edelman (1984) Heterotypic binding between neuronal membrane vesicles and glial cells is mediated by a specific neuron–glia cell adhesion molecule. *J. Cell Biol.* **98**:1746–1756.

Grumet, M., and G. M. Edelman (1988) Neuron–glia cell adhesion molecule interacts with neurons and astroglia via different binding mechanisms. *J. Cell Biol.* **106**:487–503.

Grumet, M., S. Hoffman, and G. M. Edelman (1984a) Two antigenically related neuronal CAMs of different specificities mediate neuron–neuron and neuron–glia adhesion. *Proc. Natl. Acad. Sci. USA* **81**:267–271.

Grumet, M., S. Hoffman, C.-M. Chuong, and G. M. Edelman (1984b) Polypeptide components and binding functions of neuron–glia cell adhesion molecules. *Proc. Natl. Acad. Sci. USA* **81**:7989–7993.

Grumet, M., S. Hoffman, K. L. Crossin, and G. M. Edelman (1985) Cytotactin, an extracellular matrix protein of neural and non-neural tissues that mediates glia–neuron interaction. *Proc. Natl. Acad. Sci. USA* **82**:8075–8079.

Gurdon, J. B. (1987) Embryonic induction—molecular prospects. *Development* **99**:285–306.

Hall, A. K., and U. Rutishauser (1987) Visualization of neural cell adhesion molecule by electron microscopy. *J. Cell Biol.* **104**:1579–1586.

Harrelson, A. L., and C. S. Goodman (1988) Growth cone guidance in insects: Fasciclin II is a member of the immunoglobulin superfamily. *Science* **242**:700–708.

Hatta, K., and M. Takeichi (1986) Expression of N-cadherin adhesion molecules associated with early morphogenetic events in chick development. *Nature* **320**:447–449.

Hatta, K., T. S. Okada, and M. Takeichi (1985) A monoclonal antibody disrupting calcium-dependent cell–cell adhesion of brain tissues: Possible role of its target antigen in animal pattern formation. *Proc. Natl. Acad. Sci. USA* **82**:2789–2793.

Hatta, K., A. Nose, M. Nagafuchi, and M. Takeichi (1988) Cloning and expression of cDNA encoding a neural calcium-dependent cell adhesion molecule: Its identity in the cadherin gene family. *J. Cell Biol.* **106**:873–881.

Hausman, R. E., and A. A. Moscona (1976) Isolation of a retina-specific cell-aggregating factor from membranes of embryonic neural retina tissue. *Proc. Natl. Acad. Sci. USA* **73**:3594–3598.

He, H. T., J. Barbet, J. C. Chaix, and C. Goridis (1986) Phosphatidylinositol is involved in the membrane attachment of NCA-120, the smallest component of the neural cell adhesion molecule. *EMBO J.* **5**:2489–2494.

Hemperly, J. J., B. A. Murray, G. M. Edelman, and B. A. Cunningham (1986a) Sequence of a cDNA clone encoding the polysialic acid–rich and cytoplasmic domains of the neural cell adhesion molecule N-CAM. *Proc. Natl. Acad. Sci. USA* **83**:3037–3041.

Hemperly, J. J., G. M. Edelman, and B. A. Cunningham (1986b) cDNA clones of the neural cell adhesion molecule (N-CAM) lacking a membrane-spanning region consistent with evidence for membrane attachment via a phosphatidylinositol intermediate. *Proc. Natl. Acad. Sci. USA* **83**:9822–9826.

Hemperly, J. J., B. A. Murray, G. M. Edelman, and B. A. Cunningham (1988) Sequence of a cDNA clone encoding the polysialic acid–rich and cytoplasmic domains of the neural cell adhesion molecule N-CAM (Correction). *Proc. Natl. Acad. Sci. USA* **85**:2008.

Hirn, M., M. S. Ghandour, H. Deagostini-Bazin, and C. Goridis (1983) Molecular heterogeneity and structural evolution during cerebellar ontogeny detected by monoclonal antibody of the mouse cell surface antigen BSP-2. *Brain Res.* **265**:87–100.

Hoffman, S., and G. M. Edelman (1983) Kinetics of homophilic binding by E and A forms of the neural cell adhesion molecule. *Proc. Natl. Acad. Sci. USA* **80**:5762–5766.

Hoffman, S., and G. M. Edelman (1987) A proteoglycan with HNK-1 antigenic determinants is a neuron-associated ligand for cytotactin. *Proc. Natl. Acad. Sci. USA* **84**:2523–2527.

Hoffman, S., B. C. Sorkin, P. C. White, R. Brackenbury, R. Mailhammer, U. Rutishauser, B. A. Cunningham, and G. M. Edelman (1982) Chemical characterization of a neural cell adhesion molecule purified from embryonic brain membranes. *J. Biol. Chem.* **257**:7720–7729.

Hoffman, S., C.-M. Chuong, and G. M. Edelman (1984) Evolutionary conservation of key structures and binding functions of neural cell adhesion molecules. *Proc. Natl. Acad. Sci. USA* **81**:6881–6885.

Hoffman, S., D. R. Friedlander, C.-M. Chuong, M. Grumet, and G. M. Edelman (1986) Differential contributions of Ng-CAM and N-CAM to cell adhesion in different neural regions. *J. Cell Biol.* **103**:145.

Holtfreter, J. (1939) Gewebeaffinität, ein Mittel der embryonalen Formbildung. *Arch. Exp. Zellforsch. Besonders Gewebezuecht.* **23**:169–209.

Holtfreter, J. (1948a) The mechanism of embryonic induction and its relation to parthenogenesis and malignancy. *Symp. Soc. Exp. Biol.* **11**:17.

Holtfreter, J. (1948b) Significance of the cell membrane in embryonic processes. *Ann. N.Y. Acad. Sci.* **49**:709–760.

Hyafil, F., C. Babinet, and F. Jacob (1981) Cell–cell interactions in early embryogenesis: A molecular approach to the role of calcium. *Cell* **26**:447–454.

Imhof, B. A., H. P. Vollmers, S. L. Goodman, and W. Birchmeier (1983) Cell–cell interaction and polarity of epithelial cells: Specific perturbation using a monoclonal antibody. *Cell* **35**:667–675.

Jorgensen, O. S., A. Delouvée, J. P. Thiery, and G. M. Edelman (1980) The nervous system specific protein D2 is involved in adhesion among neurites from cultured rat ganglia. *FEBS Lett.* **111**:39–42.

Kruse, J., R. Mailhammer, H. Wernecke, A. Faissner, I. Sommer, C. Goridis, and M. Schachner (1984) Neural cell adhesion molecules and myelin-associated glycoprotein share a common carbohydrate moiety recognized by monoclonal antibodies L2 and HNK-1. *Nature* **311**:153–155.

Lai, C., M. A. Brow, K.-A. Nave, A. B. Noronha, R. H. Quarles, F. E. Bloom, R. J. Milner, and J. G. Sutcliffe (1987) Two forms of 1B236/myelin-associated glycoprotein, a cell adhesion molecule for postnatal neural development, are produced by alternative splicing. *Proc. Natl. Acad. Sci. USA* **84**:4337–4341.

Lemke, G., and R. Axel (1985) Isolation and sequence of a cDNA encoding the major structural protein of peripheral myelin. *Cell* **40**:501–508.

Levi, G., K. L. Crossin, and G. M. Edelman (1987) Expression sequences and distribution of two primary cell adhesion molecules during embryonic development of *Xenopus laevis. J. Cell Biol.* **105**:2359–2372.

Mege, R.-M., F. Matsuzaki, W. J. Gallin, J. I. Goldberg, B. A. Cunningham, and G. M. Edelman (1988) Construction of epithelioid sheets by transfection of mouse sarcoma cells with cDNAs for chicken cell adhesion molecules. *Proc. Natl. Acad. Sci. USA* **85**:7274–7278.

Merrell, R., D. I. Gottlieb, and L. Glaser (1975) Embryonal cell surface recognition. Extraction of an active plasma membrane component. *J. Biol. Chem.* **250**:5655–5659.

Moos, M., R. Tacke, H. Scherer, D. B. Teplow, K. Früh, and M. Schachner (1988) Neural adhesion molecule L1 as a member of the immunoglobulin superfamily with binding domains similar to fibronectin. *Nature* **334**:701–703.

Moscona, A. A. (1952) Cell suspensions from organ rudiments of chick embryos. *Exp. Cell Res.* **3**:536–539.

Moscona, A. A. (1962) Analysis of cell recombinations in experimental synthesis of tissues *in vitro.* *J. Cell Comp. Physiol. (Suppl. 1)* **60**:65–80.

Murray, B. A., J. J. Hemperly, E. A. Prediger, G. M. Edelman, and B. A. Cunningham (1986a) Alternatively spliced mRNAs code for different polypeptide chains of the chicken neural cell adhesion molecule (N-CAM). *J. Cell Biol.* **102**:189–193.

Murray, B. A., G. C. Owens, E. A. Prediger, K. L. Crossin, B. A. Cunningham, and G. M. Edelman (1986b) Cell surface modulation of the neuronal cell adhesion molecule resulting from alternative mRNA splicing in a tissue-specific developmental sequence. *J. Cell Biol.* **103**:1431–1439.

Nagafuchi, A., Y. Shirayoshi, K. Okazaki, K. Yasuda, and M. Takeichi (1987) Transformation of cell adhesion properties by exogenously introduced E-cadherin cDNA. *Nature* **329**:341–343.

Nguyen, C., M. G. Mattei, C. Goridis, J. F. Mattei, and B. R. Jordan (1985) Localization of the human N-CAM gene to chromosome 11 by *in situ* hybridization with a murine N-CAM cDNA probe. *Cytogenet. Cell Genet.* **40**:713.

Noronha, A. B., A. Ilyas, H. Antonicek, M. Schachner, and R. H. Quarles (1986) Molecular specificity of L2 monoclonal antibodies that bind to carbohydrate determinants of neural cell adhesion molecules and their resemblance to other monoclonal antibodies recognizing the myelin-associated glycoprotein. *Brain Res.* **385**:237–244.

Nose, A., and M. Takeichi (1986) A novel cadherin cell adhesion molecule: Its expression patterns associated with implantation and organogenesis of mouse embryos. *J. Cell Biol.* **103**:2649–2658.

Nose, A., A. Nagafuchi, and M. Takeichi (1987) Isolation of placental cadherin cDNA: Identification of a novel gene family of cell–cell adhesion molecules. *EMBO J.* **6**:3655–3661.

Nybroe, O., M. Albrechtsen, J. Dahlin, D. Linneman, J. M. Lyles, C. J. Møller, and E. Bock (1985) Biosynthesis of the neural cell adhesion molecule: Characterization of polypeptide C. *J. Cell Biol.* **101**:2310–2315.

Oppenheimer, S. B. (1975) Functional involvement of specific carbohydrate in teratoma cell adhesion factor. *Exp. Cell Res.* **92**:122–126.

Owens, G. C., G. M. Edelman, and B. A. Cunningham (1987) Organization of the neural cell adhesion molecule (N-CAM) gene: Alternative exon usage as the basis for different membrane-associated domains. *Proc. Natl. Acad. Sci. USA* **84**:294–298.

Petersen, T. E., H. C. Thøgersen, K. Skorstengaard, K. Vibe-Pedersen, P. Sahl, L. Sottrup-Jensen, and S. Magnusson (1983) Partial primary structure of bovine plasma fibronectin, three types of internal homology. *Proc. Natl. Acad. Sci. USA* **80**:137–141.

Peyrieras, N., F. Hyafil, D. Louvard, H. L. Ploegh, and F. Jacob (1983) Uvomorulin: A nonintegral membrane protein of early mouse embryo. *Proc. Natl. Acad. Sci. USA* **80**:6274–6277.

Pierschbacher, M. D., and E. Ruoslahti (1984) Cell attachment activity of fibronectin can be duplicated by small synthetic fragments of the molecule. *Nature* **309**:30–33.

Pollerberg, E. G., R. Sadoul, C. Goridis, and M. Schachner (1985) Selective expression of the 180-kD component of the neural cell-adhesion molecule N-CAM during development. *J. Cell Biol.* **101**:1921–1929.

Pollerberg, G. E., M. Schachner, and S. Davoust (1986) Differentiation state-dependent surface mobilities of two forms of the neural cell adhesion molecule. *Nature* **324**:462–465.

Prediger, E. A., S. Hoffman, G. M. Edelman, and B. A. Cunningham (1988) Four exons encode a 93bp insert in three N-CAM mRNAs specific for chick heart and skeletal muscle. *Proc. Natl. Acad. Sci. USA* **85**:9616–9620.

Rathjen, F. G., and M. Schachner (1984) Immunocytological and biochemical characterization of a new neuronal cell surface component (L1 antigen) which is involved in cell adhesion. *EMBO J.* **3**:1–10.

Richardson, G., K. L. Crossin, C.-M. Chuong, and G. M. Edelman (1987) Expression of cell adhesion molecules during embryonic induction. III. Development of the otic placode. *Dev. Biol.* **119**:217–230.

Ringwald, M., R. Schuh, D. Vestweber, H. Eistetter, F. Lottspeich, J. Engel, R. Dölz, F. Jähnig, J. Eppler, S. Mayer, C. Müller, and R. Kemler (1987) The structure of cell adhesion molecule uvomorulin. Insights into the molecular mechanism of Ca^{2+}-dependent cell adhesion. *EMBO J.* **6**:3647–3653.

Rothbard, J. B., R. Brackenbury, B. A. Cunningham, and G. M. Edelman (1982) Differences in the carbohydrate structures of neural cell adhesion molecules from adult and embryonic chicken brains. *J. Biol. Chem.* **257**:11064–11069.

Rougon, G., H. Deagostini-Bazin, M. Hirn, C. Goridis (1982) Tissue and developmental stage-specific forms of neural cell surface antigen linked to differences in glycosylation of a common polypeptide. *EMBO J.* **1**:1239–1244.

Salton, S. R. J., C. Richter-Landsberg, L. A. Greene, and M. L. Shelanski (1983a) Nerve growth factor–inducible large external (NILE) glycoprotein: Studies of a central and peripheral neuronal marker. *J. Neurosci.* **3**:441–454.

Salton, S. R. J., M. L. Shelanski, and L. A. Greene (1983b) Biochemical properties of the nerve growth factor–inducible large external (NILE) glycoprotein. *J. Neurosci.* **3**:2420–2430.

Salzer, J. G., W. P. Holmes, and D. R. Colman (1987) The amino acid sequences of the myelin-associated glycoproteins: Homology to the immunoglobulin gene superfamily. *J. Cell Biol.* **104**:957–965.

Santoni, M.-J., D. Barthels, J. A. Barbas, M.-R. Hirsh, M. Steinmetz, C. Goridis, and W. Wille (1987) Analysis of cDNA clones that code for the transmembrane forms of the mouse neural cell adhesion molecule(NCAM) and are generated by alternative RNA splicing. *Nucleic Acids Res.* **15**:8621–8641.

Scott, M. P., and P. H. O'Farrell (1986) Spatial programming of gene expression in early *Drosophila* embryogenesis. *Annu. Rev. Cell Biol.* **2**:49–80.

Seeger, M. A., L. Haffley, and T. C. Kaufman (1988) Characterization of amalgam: A member of the immunoglobulin superfamily from *Drosophila*. *Cell* **55**:589–600.

Sengel, P. (1976) *Morphogenesis of Skin*, Cambridge Univ. Press, Cambridge, England.

Shirayoshi, Y., K. Hatta, M. Hosoda, S. Tsunasawa, F. Sakiyama, and M. Takeichi (1986) Cadherin cell adhesion molecules with distinct binding specificities share a common structure. *EMBO J.* **5**:2485–2488.

Shur, B. D., and S. Roth (1975) Cell surface glycosyltransferases. *Biochim. Biophys. Acta* **415**:473–512.

Small, S. J., G. E. Shull, M.-J. Santoni, and R. Akeson (1987) Identification of a cDNA clone that contains the complete coding sequence for a 140-kD rat NCAM polypeptide. *J. Cell Biol.* **105**:2335–2345.

Sorkin, B. C., S. Hoffman, G. M. Edelman, and B. A. Cunningham (1984) Sulfation and phosphorylation of the neural cell adhesion molecule N-CAM. *Science* **225**:1476–1478.

Sorkin, B. C., M. Grumet, B. A. Cunningham, and G. M. Edelman (1985) Structures of two neuronal cell adhesion molecules. *Soc. Neurosci. Abstr.* **11**:1138.

Sorkin, B. C., J. J. Hemperly, G. M. Edelman, and B. A. Cunningham (1988) Structure of the gene for the liver cell adhesion molecule, L-CAM. *Proc. Natl. Acad. Sci. USA* **85**:7617–7621.

Sperry, R. W. (1963) Chemoaffinity in the orderly growth of nerve fiber patterns and connections. *Proc. Natl. Acad. Sci. USA* **50**:703–710.

Stallcup, W. B., and L. L. Beasley (1985) Involvement of the nerve growth factor–inducible large external glycoprotein (NILE) in neurite fasciculation in primary cultures of rat brain. *Proc. Natl. Acad. Sci. USA* **82**:1276–1280.

Stallcup, W. B., L. L. Beasley, and J. M. Levine (1985) Antibody against nerve growth factor–

inducible external (NILE) glycoprotein labels nerve fiber tracts in the developing rat nervous system. *J. Neurosci.* **5**:1090–1101.

Tacke, R., M. Moos, D. B. Teplow, K. Früh, H. Scherer, A. Bach, and M. Schachner (1987) Identification of cDNA clones of the mouse neural cell adhesion molecule L1. *Neurosci. Lett.* **82**:89–94.

Thiery, J. P., R. Brackenbury, U. Rutishauser, and G. M. Edelman (1977) Adhesion among neural cells of the chick embryo. II. Purification and characterization of a cell adhesion molecule from neural retina. *J. Biol. Chem.* **252**:6841–6845.

Thiery, J. P., J. L. Duband, U. Rutishauser, and G. M. Edelman (1982) Cell adhesion molecules in early chicken embryogenesis. *Proc. Natl. Acad. Sci. USA* **79**:6737–6741.

Thiery, J. P., A. Delouvée, W. J. Gallin, B. A. Cunningham, and G. M. Edelman (1984) Ontogenetic expression of cell adhesion molecules: L-CAM is found in epithelia derived from the three primary germ layers. *Dev. Biol.* **102**:61–78.

Thiery, J. P., A. Delouvée, M. Grumet, and G. M. Edelman (1985a) Initial appearance and regional distribution of the neuron–glia cell adhesion molecule in the chick embryo. *J. Cell Biol.* **100**:442–456.

Thiery, J. P., J.-L. Duband, and A. Delouvée (1985b) The role of cell adhesion in morphogenetic movements during early embryogenesis. In *The Cell in Contact: Adhesions and Junctions as Morphogenetic Determinants,* G. M. Edelman and J. P. Thiery, eds., pp. 169–196, Wiley, New York.

Tucker, G. C., H. Aoyama, M. Lipinski, T. Tursz, and J. P. Thiery (1984) Identical reactivity of monoclonal antibodies HNK-1 and NC-1: Conservation in vertebrates on cells derived from neural primordium and on some leukocytes. *Cell Differ.* **14**:223–230.

Volk, T., and B. Geiger (1984) A 135-kD membrane protein of intercellular adherens junctions. *EMBO J.* **3**:2249–2260.

Volk, T., and B. Geiger (1986) A-CAM: A 135-kD receptor of intercellular adherens junctions. I. Immuno-electron microscopic localization and biochemical studies. *J. Cell Biol.* **103**:1441–1450.

Wallis, I., L. Ellis, K. Suh, and K. H. Pfenninger (1985) Immunolocalization of a neuronal growth–dependent membrane glycoprotein. *J. Cell Biol.* **101**:1990–1998.

Williams, A. F. (1987) A year in the life of the immunoglobulin superfamily. *Immunol. Today* **8**:298–303.

Williams, A. F., A. N. Barday, M. J. Clark, and J. Gagnon (1985) *Gene Expression During Normal and Malignant Differentiation,* L. C. Anderson, C. G. Gathimberg, and P. Ekblom, eds., p. 125, Academic, New York.

Wilson, H. V. (1907) On some phenomena of coalescence and regeneration in sponges. *J. Exp. Zool.* **5**:245–258.

Yoshida, C., and M. Takeichi (1982) Teratocarcinoma cell adhesion: Identification of a cell-surface protein involved in calcium-dependent cell aggregation. *Cell* **28**:217–224.

Chapter 2

The Cell Adhesion Molecule Uvomorulin

ROLF KEMLER
ACHIM GOSSLER
AHMED MANSOURI
DIETMAR VESTWEBER

ABSTRACT

The cell adhesion molecule uvomorulin is involved in cell–cell interaction in the early mouse embryo and in cells of the epithelial cell lineage. During preimplantation development, uvomorulin is redistributed only on the surface of cells destined for epithelial cell differentiation. It is suggested that the polarization and clustering of uvomorulin during early embryogenesis is the initial event in the expression of its adhesive function. In order to learn more about the molecular mechanisms of uvomorulin-mediated adhesion, we have focused on a detailed structural analysis of both uvomorulin protein and its DNA coding sequences and have established the complete amino acid sequence of the molecule. Uvomorulin has three internally repeated domains, a single membrane-spanning region, and a cytoplasmic domain. The membrane integration provides a molecular basis for a link between uvomorulin and cytoplasmic components.

The role of cell adhesion molecules (CAMs) in cell–cell interaction during development and in the maintenance of the histoarchitecture of multicellular organisms has been the subject of several recent reviews (Edelman, 1983a; Edelman and Thiery, 1985; Öbrink, 1986). Most CAMs have been defined using an immunological approach in which antibodies that perturb cell–cell contacts are generated. This approach was first successfully used by Gerisch and his coworkers and led to the characterization of contact sites A and B in *Dictyostelium* (Gerisch, 1977). Uvomorulin was identified by a similar strategy. Since the number of CAMs identified in this way is rather small, it was proposed that a relatively limited number of these molecules, which are temporally and structurally modulated during development, are sufficient to regulate all necessary cell interaction phenomena (Edelman, 1983b). This hypothesis is attractive, but it should be kept in mind that most CAMs have been detected using a similar cell aggregation assay in combination with antibodies. It might well be that this experimental approach limits the detection of additional CAMs, and it will be of interest to develop new functional assays through which other cell–cell interaction mechanisms can be studied.

The presently known CAMs seem to be good immunogens, and their precise role in the adhesion process is generally not well understood. Antibodies may induce changes of protein conformation or block cell aggregation by steric hindrance. At this point one would like to know what the possible molecular counterparts for a certain CAM on adjacent cells are and how many components interact with a particular CAM during the adhesion process. In addition, one would like to know what, if any, kind of link exists between a particular CAM and cytoplasmic components. Answers to these questions should be facilitated by more precise structural analyses. In a first attempt toward better understanding the molecular mechanism of uvomorulin-mediated adhesion, we have focused on a detailed analysis of both uvomorulin protein and its DNA. In this chapter we summarize our results and discuss how uvomorulin-mediated cell adhesion might be regulated.

UVOMORULIN AND PREIMPLANTATION DEVELOPMENT

Uvomorulin expresses its adhesive function very early in mouse development during the compaction of preimplantation embryos. Compaction is a pre-requisite for the generation of cell diversity in the preimplantation embryo and leads to the generation of trophectodermal and inner cell mass cell lineages (Johnson and Ziomek, 1981). During compaction, loosely associated spherical blastomeres flatten at their contact sites and establish junctional complexes composed of tight and gap junctions and desmosomes (Figure 1). It is this process that is blocked by anti-uvomorulin antibodies (Kemler et al., 1977). When two-cell embryos are cultured in the presence of anti-uvomorulin antibodies, cell division is not affected, but the antibodies inhibit compaction and lead to grapelike embryos (Figure 2A,B). This effect is specific for anti-uvomorulin antibodies; removing the antibodies allows subsequent com-paction and the formation of blastocysts (Figure 2B,C), which develop into newborn mice after reimplantation into pseudopregnant females. Thus anti-uvomorulin antibodies exhibit a specific, reversible, and nontoxic effect on mouse preimplantation development.

Studies on uvomorulin synthesis and its cell surface localization in early embryos demonstrate that the protein is already present on the cell surface of unfertilized eggs (Vestweber et al., submitted). However, metabolic labeling experiments indicate that uvomorulin is not synthesized at this stage, suggesting that it might be maternally derived and be a remnant from oogenesis. During egg growth and maturation, granulosa cells contact the oocytes by cytoplasmic processes through the intervening zona pellucida.

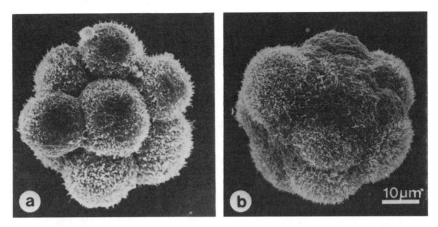

Figure 1. *Mouse preimplantation embryos before (a) and during (b) the compaction process.* During compaction, loosely associated blastomeres maximize their cell–cell contacts, and the outer cells establish intercellular junctions such as tight and intermediate junctions and desmosomes.

Specialized membrane areas such as gap junctions and desmosomes have been reported at these sites of oophorus oocyte contacts (Gilula et al., 1978). It is tempting to think of a possible involvement of uvomorulin in the establishment of these contacts. Since uvomorulin was not found in the developing follicle and stroma of ovaries of 16-day-old embryos (Damjanov et al., 1986), this would suggest a role for uvomorulin in the development of primordial follicles and primordial oocytes.

Uvomorulin synthesis starts in late two-cell-stage embryos and can be inhibited by the drug α-amanitin. The two-cell stage is characterized by a general breakdown of stored maternal mRNA and the activation of the embryonal genome (Clegg and Piko, 1983). Therefore the start of uvomorulin synthesis seems to follow the general activation of zygotic transcription and seems not to be directly correlated with the onset of compaction. Although uvomorulin is present on the cell surface of all preimplantation stages, it does not express its adhesive properties until the late eight-cell stage. There are several possible explanations for this apparent discrepancy. It may well be that adhesion is triggered by the developmentally regulated expression of a hypothetical uvomorulin receptor, although there is at present no experimental evidence for such a molecular counterpart of uvomorulin on the cell surface, and posttranslational modifications of uvomorulin do not seem to play a major role in adhesion (see below). We propose instead that it is the redistribution of uvomorulin on the cell surface that is the critical step in bringing the molecule into its functional state.

The subcellular localization of uvomorulin in preimplantation-stage embryos has been examined using immunogold labeling technique followed by electron microscopy. We have found that uvomorulin is characteristically redistributed, which is a developmentally regulated event since it occurs in

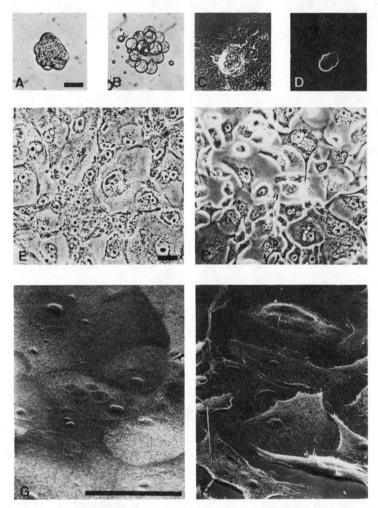

Figure 2. *Uvomorulin is involved in cell–cell interaction of the early mouse embryo and of cells of the epithelial cell lineage.* Compact morulae (*A*) become decompacted (*B*) in the presence of the anti-uvomorulin monoclonal antibody DECMA-1. Recompacted embryos form blastocysts with trophectoderm and inner cell mass cells. *C:* Blastocyst outgrowth. *D:* Immunofluorescence staining for inner cell mass cells. MDCK cells (*E–H*) lose their cell–cell contacts in the presence of DECMA-1 (*F,H*), phase contrast (*E,F*), and scanning electron microscopy (*G,H*). *Calibration bar* = 50 μm. (From Vestweber and Kemler, 1985.)

compact morulae only in cells committed to trophectoderm formation. Figure 3 schematically summarizes these results. Uvomorulin is uniformly distributed over the cell surfaces of early-stage embryos. The redistribution of uvomorulin is confined to the outer cells of compact morulae, where it progressively vanishes from the apical membrane and becomes restricted to the basolateral membrane domain. With the appearance of intercellular junctions, uvomorulin becomes enriched in the intermediate junctions. In contrast, the inner cells of compact morulae retain uniform uvomorulin distribution on their surfaces. This distribution, once established during compaction, is maintained and also present in the blastocyst. On trophectodermal cells, uvomorulin is restricted to basolateral membrane areas, whereas inner cell mass cells show a uniform distribution. The described redistribution does not occur as a sudden event with the onset of compaction. A concentration of uvomorulin can be detected as early as the four-cell stage in membrane areas that are in close contact with adjacent cells. These structures resemble the focal contacts described by Magnuson et al. (1977) and represent so far the earliest signs of uvomorulin redistribution.

With the onset of compaction, a transition from a nonpolar to a polarized cell type takes place. After compaction, the cells of the outer layer exhibit distinct basolateral and apical membrane domains. This layer of epitheliumlike cells gives rise to the trophectoderm and surrounds a core of unpolarized cells that form the inner cell mass. The localization of uvomorulin on preimplantation embryos before, during, and after compaction clearly shows that redistribution occurs only in cells destined for epithelial differentiation. During polarization a redistribution of cellular material takes place both on the cell surface and in the cytoplasm. Surface microvilli and Con A receptors are concentrated at the apical pole, and microfilaments, microtubules, endosomes, and coated vesicles are deposited in the apical region above the nucleus (Johnson et al., 1986a). It may well be that uvomorulin redistribution is influenced by cytoplasmic structures. This would imply some kind of link between cytoplasmic components and the adhesion protein on the cell surface. Since cell adhesion seems to play a role in the synchronization and orientation of polarization (Johnson et al., 1986b), it is conceivable that the mechanism(s) underlying this polarization also controls uvomorulin redistribution on the cell surface. Taken together, all the data available at this writing on uvomorulin during preimplantation development suggest that it is most likely the redistribution of this cell adhesion molecule on the cell surface that triggers compaction.

UVOMORULIN AND EPITHELIA

Uvomorulin is present in the early embryo and is also expressed during later stages of development. The first cell type from which uvomorulin disappears during development is mesodermal cells at the egg cylinder stage of eight-day-old embryos (Vestweber and Kemler, 1984a). Later in develop-

SYNTHESIS AND DISTRIBUTION OF UVOMORULIN DURING MOUSE PREIMPLANTATION DEVELOPMENT

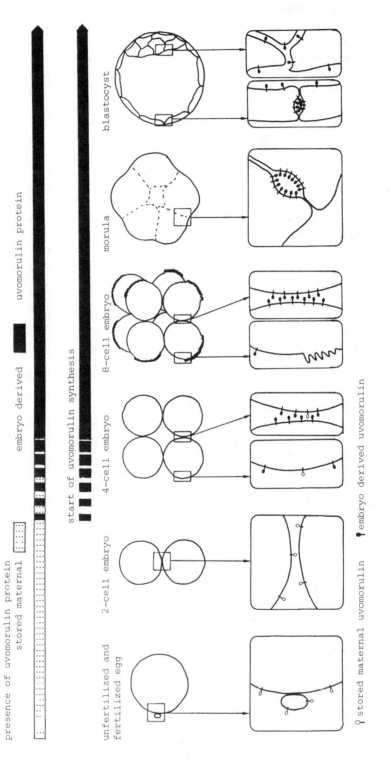

ment and in adult tissues uvomorulin is found only in epithelial cells, independent of their germ layer origin. Anti-uvomorulin antibodies interfere with cell–cell contact of 14-day-old embryonic hepatocytes and MDCK (canine kidney epithelial) cells in tissue culture (Vestweber and Kemler, 1984b, 1985), demonstrating that uvomorulin acts as an adhesive molecule between epithelial cells. MDCK cells polarize in culture and form a dense epithelial layer in which cell boundaries can hardly be distinguished (Figure 2C,G). The addition of anti-uvomorulin antibodies disrupts this epithelial layer, and the individual cells become detached from each other (Figure 2F,H). These experiments demonstrate the importance of uvomorulin for the maintenance of an epithelial layer and moreover suggest that uvomorulin-analogue molecules exist in other species.

Epithelia are composed of highly specialized cells in which distinct membrane domains and intercellular junctions can be defined. The plasma membrane is divided into an apical and a basolateral domain, each of which is functionally and structurally specialized (Simons and Fuller, 1985). Intercellular junctions of the zonula adherens and the zonula occludens are largely responsible for sealing the epithelium and maintaining the polarized barrier (Palade, 1983). Since anti-uvomorulin antibodies disrupt this epithelial layer, the subcellular localization of uvomorulin on the plasma membrane of epithelial cells was investigated (Boller et al., 1985). Uvomorulin was found exclusively on the basolateral membrane domains and was remarkably concentrated in the intermediate junctions of the junctional complex (Figure 4). The predominant localization of uvomorulin in the intermediate junctions demonstrates its biological importance for preserving the integrity of the epithelial sheet. Intestinal epithelial cells exhibit a unique form of motility that involves the terminal web and the corresponding junctional cell contact sides (Burgess, 1982). Ultrastructural analysis revealed a contractile ring of microfilaments at the level of the intermediate junctions (Hull and Staehelin, 1979). The predominant localization of uvomorulin at this site suggests that it may play a role in this motile system, for example, by acting as a counterforce

---

Figure 3. *Schematic representation of uvomorulin synthesis and cell surface localization during mouse preimplantation development.* Uvomorulin is present on the cell surface throughout all stages of preimplantation development. Up to the late two-cell stage there exists only maternally derived uvomorulin protein. Uvomorulin synthesis starts in the late two-cell-stage embryo. On the unfertilized and fertilized egg, as well as on the two-cell-stage embryo, uvomorulin is uniformly distributed on the cell surface. On four-cell-stage embryos, concentration of uvomorulin first occurs in membrane areas, which are in close contact with adjacent blastomeres. During compaction, uvomorulin vanishes from the outer surface of the embryo and is found predominantly in membrane domains that are involved in cell–cell contact. In compact morulae uvomorulin is no longer present on the outer surface of the embryo but is located exclusively on the basolateral membranes of the outer cells, concentrated in their intermediate junctions. Uvomorulin remains evenly distributed on inner cells. This distribution, once established during compaction, is maintained and also found in the blastocyst.

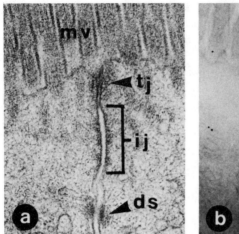
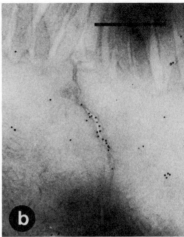

Figure 4. *Subcellular localization of uvomorulin on the cell surface of intestinal epithelial cells.* The zonula adherens of intestinal epithelial cells (*a*) is composed of tight junctions (tj), intermediate junctions (ij), and desmosomes (ds). Uvomorulin is highly concentrated in the intermediate junctions of the junctional complex (*b*). *Calibration bar* = 1 μm.

against the tension of the contractile ring and thus preserving the integrity of the epithelial sheet. If so, one would predict a kind of link between uvomorulin and cytoplasmic structures, as we have suggested occurs during uvomorulin redistribution in preimplantation embryos.

The formation of epithelia is a major morphogenetic event throughout embryogenesis. Morphogenesis of the embryonic kidney represents a very attractive model for the formation of epithelia, since it can be studied in an *in vitro* culture system (Grobstein, 1954). During development of the kidney, the metanephrogenic mesenchyme is converted into an epithelium. This conversion is initiated by an inductive interaction between the epithelial ureter bud and the mesenchyme. Since this conversion can accurately be followed in the *in vitro* model (Saxén et al., 1968), the time course of uvomorulin appearance was examined (Vestweber et al., 1985). Uvomorulin was absent in uninduced mesenchyme, and the first uvomorulin-positive cells were observed 12 hours after induction, when cells started to aggregate. Distinct differences in uvomorulin expression were seen in different parts of the developing nephron. In mesenchymally derived epithelia, uvomorulin could be detected only in the tubules, whereas the epithelium of the glomeruli remained negative at all stages of development. Therefore uvomorulin distribution allowed us to distinguish different epithelial cell subpopulations very early in the developing nephron. In addition, this *in vitro* culture system gave us the opportunity to study the effect of anti-uvomorulin antibodies on the developing kidney. However, no interference with histogenesis was observed. Since the same antibodies disrupt the epithelial layer of cultured MDCK cells, this may indicate that in the organ culture system other adhesive proteins may override uvomorulin function (Ekblom et al., 1986).

UVOMORULIN PROTEIN

Uvomorulin is a 120-kD cell surface glycoprotein (GP120). Trypsin digestion in the presence of calcium releases a discrete fragment of 84 kD, which carries the epitopes recognized by the antibodies that disturb cell–cell contact. *In vitro* translation or pulse-labeling experiments demonstrate the existence of a 135-kD precursor molecule (Peyrieras et al., 1983; Vestweber and Kemler, 1984b) that is already glycosylated. In pulse-chase experiments, the 135-kD precursor can be detected after a 3-min pulse and is subsequently converted into mature GP120, which arrives at the cell surface after 20–30 min of pulse chase. Studies on posttranslational modifications reveal that uvomorulin is glycosylated, phosphorylated, and sulfated at sugar residues. It can be labeled with the most common sugars, and in the presence of tunicamycin, which blocks N-linked glycosylation, it shifts to a molecular weight of 112 kD. Phosphorylation occurs only on serine and threonine residues; the 84-kD uvomorulin fragment is not phosphorylated. Uvomorulin is sulfated at sugar residues. However, these posttranslational modifications do not seem to be of importance to the adhesive properties of uvomorulin. For instance, cells grown in the presence of tunicamycin aggregate similarly to control cells. Furthermore, antibodies block the cell adhesion of tunicamycin-treated cells. Since the 84-kD uvomorulin fragment inhibits the antibody effect on cell adhesion, it is likely that parts of the protein backbone of the 84-kD fragment are responsible for the adhesive function of the protein. At present, no obvious, distinct differences between embryonic and adult molecular forms of uvomorulin have been distinguished.

A putative cell adhesion domain of uvomorulin was identified with the help of monoclonal antibodies (Vestweber and Kemler, 1985). The rationale behind these experiments was that an anti-uvomorulin monoclonal antibody (DECMA-1, which blocks cell aggregation) recognized an epitope very close or even identical to a putative adhesive domain. When different protease-resistant domains of uvomorulin were generated, the DECMA-1 target was localized to a 26-kD fragment of uvomorulin obtained after chymotrypsin digestion. Interestingly, the epitopes of two other monoclonal antibodies, anti-Arc-1 (Behrens et al., 1985) and rr-1 (Gumbiner and Simons, 1986), are located on the same 26-kD fragment. These latter antibodies were selected not for uvomorulin binding but for disassociating MDCK cells, and in subsequent comparison with anti-uvomorulin antibodies it turned out that both recognize uvomorulin in MDCK cells. Thus the antigenic targets of three independently selected anti-adhesive monoclonal antibodies map rather closely together. This suggests that the 26-kD uvomorulin domain bears at least one adhesive site, which is conserved in mouse and dog uvomorulin.

One of the key questions about the functional role of uvomorulin is how it might be associated with the plasma membrane. Experiments with the detergent Triton X-114 revealed the overall hydrophilic nature of uvomorulin (Peyrieras et al., 1983; Vestweber and Kemler, 1984b), which suggests that it might be a nonintegral membrane protein. The strong negative charges resulting from sulfation and phosphorylation and the acidic isoelectric point of pH 4.6 underline its hydrophilic nature. However, recent experimental evidence unambiguously proves that uvomorulin is in fact an integral membrane protein. Most of this evidence comes from the sequencing of uvomorulin cDNA and is discussed below. The transmembrane character of uvomorulin is also evident from pulse-chase experiments with microsomal vesicles prepared from embryonal carcinoma cells. When microsomal vesicles were treated with protease a molecular weight shift from 120 kD to around 103 kD was observed, indicating that uvomorulin has a 17-kD cytoplasmic tail.

UVOMORULIN GENE

Coding sequences for uvomorulin have been identified with affinity-purified anti-uvomorulin antibodies from a cDNA library in the expression vector λgt11 (Schuh et al., 1986). These sequences were used to isolate larger cDNA inserts from a size-selected cDNA library (Ringwald et al., in preparation). The largest cDNA clone of 2.5 kb (F5) hybridized to a single 4.3-kb poly(A)$^+$RNA in Northern blot experiments. The 4.3-kb poly(A)$^+$RNA was detected only in cells that did in fact express the uvomorulin protein. The determination of the clone F5 nucleotide sequence yielded important insights into the uvomorulin protein. The codon usage showed an unusually high percentage (287 of 712 codons = 40%) of the RNY codon (R = purine, N = pyrimidine or purine, and Y = pyrimidine). Such usage has been found in rather conserved genes with low mutation frequencies such as histone genes, genes for ribosomal proteins, and actin genes (Shepherd, 1984).

Clone F5 has an open reading frame of 2134 base pairs. This codes for 711 amino acids, which represent 78,381 D of the carboxy-terminal part of uvomorulin. To prove that the open reading frame is actually used for the generation of uvomorulin, about 400 base pairs of clone F5 3' sequences were expressed as fusion proteins in two different vectors. Antibodies were produced against one fusion protein, and anti-uvomorulin antibodies were affinity purified with the second fusion protein, which was coupled to Sepharose 4B. These antibodies recognized uvomorulin in immunoprecipitation and immunoblot experiments. The most important finding from the clone F5 sequence was the identification of 23 hydrophobic amino acids that may represent a transmembrane segment of the uvomorulin protein. This hydrophobic region is followed by three positively charged residues, which probably represent a stop signal for protein translocation into the endoplasmic reticulum (Sabatini et al., 1982). This hydrophobic domain shows homologies to the transmembrane region of several integral membrane proteins. Hydrophobicity plots indicated that uvomorulin has an α-helix transmembrane domain homologous to the transmembrane region of glycophorin (Figure 5). More direct evidence for the transmembrane character of uvomorulin came from studies with the anti-fusion protein antibodies directed against the carboxy-terminal domain of uvomorulin. These antibodies stained the cell membrane only if cells had been permeabilized. The deduced amino acid sequence of clone F5 shows five consensus sequences for N-linked glycosylation, one of which is most likely not used since it is located in the cytoplasmic domain of uvomorulin. The cytoplasmic domain contains a relatively large number of serine residues that may be used for phosphorylation.

Protein sequence analysis of intact uvomorulin defined the first 26 amino acids from the amino terminus. Alignment of this sequence with the deduced amino acid sequence of clone F5 cDNA made the determination of the complete amino acid sequence of the protein possible (Ringwald et al., in preparation). Uvomorulin contains three internal repeated domains that are most likely generated by gene duplication. The structure of uvomorulin and its plasma membrane association are schematically represented in Figure 6.

Uvomorulin 3' cDNA sequences were used to study the organization of the uvomorulin gene. The characterization of genomic clones from a λ genomic library and cosmid clones from a cosmid library allowed us to determine the exon–intron organization of the uvomorulin gene. Genomic restriction mapping experiments revealed a unique genetic locus for uvomorulin. We have localized the uvomorulin gene to mouse chromosome 8; this mapping was confirmed by two independent experimental approaches. The progeny of an interspecies backcross ([*Mus spretus* × C57BL/6] × C57BL/6) was analyzed for the presence of *Mus spretus*–specific uvomorulin DNA fragments by Southern blotting (Eistetter et al., in preparation). The segregation of the uvomorulin gene in 75 offspring was compared to those previously determined for other polymorphic loci (Guenet, 1986). In determining the recombination frequency between the uvomorulin gene and the cosegregating loci serum

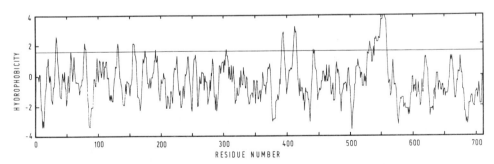

Figure 5. *Hydrophobicity plot of the deduced amino acid sequence from uvomorulin cDNA clone F5.* The sequence shows a single hydrophobic region, which is homologous to the membrane-spanning domain of several integrated membrane proteins.

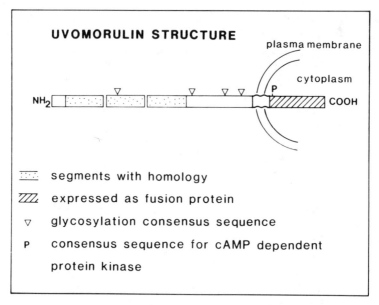

Figure 6. *Schematic representation of the uvomorulin structure and its association with the plasma membrane.* The uvomorulin amino acid sequence was determined by protein sequence analysis in combination with the deduced amino acid sequence from uvomorulin cDNA. Uvomorulin and the 84-kD uvomorulin fragment are identical at their N-terminal sequences. The uvomorulin protein is composed of three domains with internal homology, a membrane-spanning region, and a cytoplasmic domain.

esterase-1 (Es-1) and tyrosine aminotransferase (Tat) on chromosome 8, uvomorulin was placed between Es-1 and Tat 14 cM distal to Es-1. These results are in agreement with *in situ* hybridization to mouse metaphase chromosomes using ^{3}H-labeled uvomorulin cDNA as a probe.

UVOMORULIN-RELATED CAMs

Uvomorulin belongs to the class of CAMs that have in common the fact that they exhibit their cell adhesive function only in the presence of calcium. In addition, many members of this calcium-dependent class show similarities with respect to structure and tissue distribution. They all have apparent molecular weights of 120 kD, from which a roughly 80-kD fragment can be generated upon trypsin digestion in the presence of calcium. They are expressed rather early during development, and expression in later stages is confined exclusively to epithelial cells. Several of these CAMs have been shown to be identical to uvomorulin by comparing their respective antibodies. Takeichi and his coworkers have identified an adhesive glycoprotein from mouse teratocarcinoma cells called E-cadherin that is identical to uvomorulin

(Yoshida-Noro et al., 1984). Two other adhesive glycoproteins, Arc-1 antigen (Behrens et al., 1985) and rr-1 antigen (Gumbiner and Simons, 1986), have been described which when compared to anti-uvomorulin antibodies were identified as the canine homologue of uvomorulin. Two other well-known epithelial CAMs, human cell-CAM 120/80 (Damsky et al., 1983) and chicken L-CAM (Gallin et al., 1983), have not yet been compared to uvomorulin on the protein level. However, the uvomorulin nucleotide sequence shows a strong homology to the recently determined nucleotide sequence of L-CAM (Gallin et al., 1987). Mouse uvomorulin cDNA was used to isolate a 2.1-kb human cDNA clone from a human liver cDNA library (Mansouri et al., submitted). Sequencing data and the alignment of mouse and human uvomorulin sequences revealed 82% of the nucleotide sequence and 83% of the deduced amino acid composition to be identical. Thus the examples cited here seem to have a common origin or are even directly homologous to one another.

SUMMARY AND PERSPECTIVES

In the past, most efforts to achieve a better understanding of the role of CAMs during development and in adult tissues were directed toward defining their adhesive function in different cell types and studying their distribution in the organism. Studies on preimplantation embryos have revealed the key role of uvomorulin in the developing embryo. Redistribution of uvomorulin is a developmentally regulated event, since it occurs only in cells committed for epithelial cell differentiation. The early embryo therefore represents a unique model in which to study the transition from nonpolarized stem cells to epithelial cells and to study the involvement of uvomorulin in this differentiation process. Furthermore, it has been suggested that uvomorulin redistribution and the assembly of the zonula adherens via uvomorulin-mediated interactions play a key role in the establishment of the occluding barrier during the formation of an epithelial cell layer (Gumbiner and Simons, 1987).

The redistribution of uvomorulin may be influenced by the polarization of underlying cytoplasmic structures. This implies some kind of link between cytoplasmic components and uvomorulin located on the cell surface. Our structural analysis provides a molecular basis for such a link. Uvomorulin is an integral membrane protein with a roughly 17-kD cytoplasmic domain. It will be interesting to characterize the cytoplasmic components that can interact with the carboxy-terminal domain of uvomorulin. Uvomorulin has a single membrane-spanning region and therefore represents a classic integral membrane protein with respect to its orientation to the plasma membrane. The amino acid sequence indicates a multidomain structure that may have arisen from gene duplication. The isolation of cDNA probes has been an important step in the analysis of the regulation of uvomorulin expression and uvomorulin-mediated adhesion. Uvomorulin is expressed very early during development, and expression ceases in particular cell lineages. The availability of complete cDNA clones for uvomorulin will allow us to study the regulation of uvomorulin expression. Using the techniques of molecular biology and DNA transfection experiments, questions concerning the effect of uvomorulin expression on the behavior of individual cells can be asked. For example, it will be interesting to see which morphological changes will occur when the gene is expressed in an uvomorulin-negative cell. Similar questions can be asked by introducing antisense constructs into epithelial cells. By placing the uvomorulin gene under the control of inducible regulatory sequences, *in vivo* expression studies can be performed by using genetically transformed embryonic stem cells that can produce transgenic mice (Gossler et al., 1986). The possibility of a uvomorulin cytoplasmic domain for uvomorulin redistribution on the cell surface can be examined by constructing chimeric proteins. For example, uvomorulin 3′ sequences can be fused with 5′ sequences coding for the extracellular domains of some proteins for which antibodies are available. Will such chimeric proteins assemble in the intermediate junctions of epithelial cells? Site-directed mutagenesis of uvomorulin 3′ sequences could subsequently make it possible to determine the recognition site for cytoplasmic components. In conclusion, tools are now available for detailed molecular analyses of the role uvomorulin plays during development and in the maintenance of the tissue histoarchitecture to be fully undertaken.

ACKNOWLEDGMENTS

We thank Steven Cohen and Walter Birchmeier for critically reading and thereby improving the manuscript and Renate Brodbeck and Carla Fickenscher for typing it. This work has been supported by the Deutsche Forschungsgemeinschaft and the Deutsche Krebsforschung/Dr.-Mildred-Scheel-Stiftung.

REFERENCES

Behrens, J., W. Birchmeier, S. L. Goodman, and B. A. Imhof (1985) Dissociation of MDCK epithelial cells by the monoclonal antibody anti-Arc-1: Mechanistic aspects and identification of the antigen as a component related to uvomorulin. *J. Cell Biol.* **101**:1307–1315.

Boller, K., D. Vestweber, and R. Kemler (1985) Cell-adhesion molecule uvomorulin is localized in the intermediate junctions of adult intestinal epithelial cells. *J. Cell Biol.* **100**:327–332.

Burgess, D. R. (1982) Reactivation of intestinal epithelial cell brush border motility: ATP-dependent contraction via a terminal web contractile ring. *J. Cell Biol.* **95**:853–863.

Clegg, K. B., and L. Piko (1983) Quantitative aspects of RNA synthesis and polyadenylation in 1-cell and 2-cell mouse embryos. *J. Embryol. Exp. Morphol.* **74**:169–182.

Damjanov, I., A. Damjanov, and C. H. Damsky (1986) Developmentally regulated expression of the cell–cell adhesion glycoprotein cell CAM 120/80 in peri-implantation mouse embryos and extraembryonic membranes. *Dev. Biol.* **116**:194–202.

Damsky, C. H., J. Richa, D. Solter, K. A. Knudsen, and C. A. Buck (1983) Identification and purification of a cell surface glycoprotein mediating intercellular adhesion in embryonic and adult tissue. *Cell* **34**:455–466.

Edelman, G. M. (1983a) Cell adhesion molecules. *Science* **219**:450–457.

Edelman, G. M. (1983b) Cell adhesion and morphogenesis: The regulatory hypothesis. *Proc. Natl. Acad. Sci. USA* **81**:1460–1464.

Edelman, G. M., and J.-P. Thiery, eds. (1985) *The Cell in Contact: Adhesions and Junctions as Morphogenetic Determinants,* Wiley, New York.

Ekblom, P., D. Vestweber, and R. Kemler (1986) Cell–matrix interactions and cell adhesion during development. *Annu. Rev. Cell Biol.* **2**:27–47.

Gallin, W. J., G. M. Edelman, and B. A. Cunningham (1983) Characterization of L-CAM, a major cell adhesion molecule from embryonic liver cells. *Proc. Natl. Acad. Sci. USA* **80**:1038–1042.

Gallin, W. J., B. C. Sorkin, G. M. Edelman, and B. A. Cunningham (1987) Sequence analysis of a cDNA clone encoding the liver cell adhesion molecule. *Proc. Natl. Acad. Sci. USA* **84**:2808–2812.

Gerisch, G. (1977) Univalent antibody fragments as tools for analysis of cell–cell interactions in *Dictyostelium. Curr. Top. Dev. Biol.* **14**:243–270.

Gilula, N. B., M. L. Epstein, and W. H. Beers (1978) Cell to cell communication and ovulation: A study of the cumulus cell–oocyte complex. *J. Cell Biol.* **78**:58–75.

Gossler, A., T. Doetschman, R. Korn, E. Serfling, and R. Kemler (1986) Transgenesis by means of blastocyst-derived embryonic stem cell lines. *Proc. Natl. Acad. Sci. USA* **83**:9065–9069.

Grobstein, C. (1954) Tissue interaction in the morphogenesis of mouse embryonic rudiments *in vitro.* In *Aspects of Synthesis and Order of Growth,* G. Rudnick, ed., pp. 233–256, Princeton Univ. Press, Princeton, New Jersey.

Guenet, J.-L. (1986) The contribution of wild derived mouse inbred strains to gene mapping methodology. *Curr. Top. Microbiol. Immunol.* **127**:109–113.

Gumbiner, B., and K. Simons (1986) A functional assay for proteins involved in establishing an epithelial occluding barrier: Identification of a uvomorulin-like polypeptide. *J. Cell Biol.* **102**:457–468.

Gumbiner, B., and K. Simons (1987) The role of uvomorulin in the formation of epithelial occluding junctions. *Ciba Found. Symp.* **125**:168–186.

Hull, B. E., and A. Staehelin (1979) The terminal web: A reevaluation of its structure and function. *J. Cell Biol.* **81**:67–82.

Johnson, M. H., and C. A. Ziomek (1981) The foundation of two distinct cell lineages within the mouse morula. *Cell* **24**:71–80.

Johnson, M. H., B. Maro, and M. Takeichi (1986a) The role of cell adhesion in the synchronization and orientation of polarization in 8-cell mouse blastomeres. *J. Embryol. Exp. Morphol.* **93**:239–255.

Johnson, M. H., J. C. Chisholm, T. P. Fleming, and E. Houliston (1986b) A role for cytoplasmic determinants in the development of the mouse early embryo? *J. Embryol. Exp. Morphol. (Suppl.)* **97**:97–121.

Kemler, R., C. Babinet, H. Eisen, and F. Jacob (1977) Surface antigen in early differentiation. *Proc. Natl. Acad. Sci. USA* **74**:4449–4452.

Magnuson, T., A. Demsey, and C. W. Stackpole (1977) Characterization of intercellular junctions in the preimplantation mouse embryo by freeze fracture and thin section electron microscopy. *Dev. Biol.* **61**:252–261.

Öbrink, B. (1986) Epithelial cell adhesion molecules. *Exp. Cell Res.* **163**:1–21.

Palade, G. E. (1983) Membrane biogenesis: An overview. *Methods Enzymol.* **96**:xxix–lv.

Peyrieras, N., F. Hyafil, D. Louvard, H. Ploegh, and F. Jacob (1983) Uvomorulin: A nonintegral membrane protein of early mouse embryo. *Proc. Natl. Acad. Sci. USA* **80**:6274–6277.

Sabatini, D. D., G. Kreibich, T. Morimoto, and M. Adesnik (1982) Mechanisms for the incorporation of proteins in membranes and organelles. *J. Cell Biol.* **92**:1–22.

Saxén, L., O. Koskimies, R. Lahti, H. Miettinen, J. Rapola, and J. Wartiovaara (1968) Differentiation of kidney mesenchyme in an experimental model system. *Adv. Morphol.* **7**:251–293.

Schuh, R., D. Vestweber, I. Riede, M. Ringwald, U. B. Rosenberg, H. Jäckle, and R. Kemler (1986) Molecular cloning of the mouse cell adhesion molecule uvomorulin: cDNA contains a B1-related sequence. *Proc. Natl. Acad. Sci. USA* **83**:1364–1368.

Shepherd, J. C. W. (1984) Fossil remnants of a primeval genetic code in all forms of life? *Trends Biochem. Sci.* **1**:8–10.

Simons, K., and S. Fuller (1985) Cell surface polarity in epithelia. *Annu. Rev. Cell Biol.* **1**:243–288.

Vestweber, D., and R. Kemler (1984a) Rabbit antiserum against a purified surface glycoprotein decompacts mouse preimplantation embryos and reacts with specific adult tissues. *Exp. Cell Res.* **152**:169–178.

Vestweber, D., and R. Kemler (1984b) Some structural and functional aspects of cell adhesion molecule uvomorulin. *Cell Differ.* **15**:269–273.

Vestweber, D., and R. Kemler (1985) Identification of a putative cell adhesion domain of uvomorulin. *EMBO J.* **4**:3393–3398.

Vestweber, D., R. Kemler, and P. Ekblom (1985) Cell-adhesion molecule uvomorulin during kidney development. *Dev. Biol.* **112**:213–221.

Yoshida-Noro, C., N. Suzuki, and M. Takeichi (1984) Molecular nature of the calcium-dependent cell–cell adhesion system in mouse teratocarcinoma and embryonic cells studied with a monoclonal antibody. *Dev. Biol.* **101**:19–27.

Chapter 3

A-CAM: An Adherens Junction–Specific Cell Adhesion Molecule

BENJAMIN GEIGER
TOVA VOLBERG
ILANA SABANAY
TALILA VOLK

ABSTRACT

We have previously presented studies on the molecular and cytoarchitectural properties of adherens-type junctions (Geiger et al., 1985a). This unique group of cell contacts shares a considerable degree of molecular homology, manifested by the junctions' ubiquitous association with actin filaments and by the presence of vinculin along their cytoplasmic surfaces (Geiger, 1983; Geiger et al., 1983). The interaction of specialized junctional components within the plasma membrane and the cytoplasmic plaque domain and the membrane-bound microfilaments during the assembly of adherens junctions represents a cascade of cellular processes that accompany cell contact formation and are probably triggered and regulated by it (Geiger, 1982; Geiger et al., 1984). Attempts to dissect adherens junctions into their elementary structural units have revealed the presence of at least three distinct junctional subdomains. These include a cytoskeletal domain composed of actin and a variety of actin-associated proteins; a membrane-bound plaque with vinculin and probably additional peripheral membrane constituents, such as talin and plakoglobin (Burridge and Connell, 1983; Geiger et al., 1985b; Cowin et al., 1986); and intrinsic membrane-bound "contact receptor(s)" (Geiger, 1982; Avnur et al., 1983; Geiger et al., 1985a). Moreover, most of the evidence thus far accumulated suggests that junction assembly is a polar process triggered by cell contact with extracellular surfaces, which leads to the assembly of the vinculin-rich plaque and consequently to the assembly of the microfilament system. These cellular interactions and contact formation thus may lead not only to the establishment of long-range intercellular and cell–matrix adhesions but also to the spatially regulated assembly of the cytoskeleton (Geiger, 1982; Avnur et al., 1983; Geiger et al., 1984).

Studies on the molecular properties of a variety of adherens-type junctions have revealed two major subclasses that differ with respect to the nature of the surface molecule(s) with which they interact and with respect to the composition of the junctional plaque (Geiger et al., 1985b). We have shown that adherens junctions formed with noncellular surfaces (i.e., focal contacts, dense plaques of smooth muscle, attachments to the basement membrane, etc.) differ in their composition from the corresponding intercellular junctions; the former contain vinculin and talin in their membrane-bound plaques (Geiger et al., 1985b), while the latter contain vinculin and plakoglobin but not talin (Geiger et al., 1985b; Cowin et al., 1986). In addition, the two junctional sub-families differ markedly in their integral membrane constituents. Recent studies on the mechanism of adhesion to extracellular matrices indicate that these interactions are mediated through specific surface receptors now commonly known as integrins (Hynes, 1987; Horwitz et al., this volume; Hynes et al., this volume). Immunocytochemical localization of these molecules points to their abundance in and around focal contacts and related attachment sites to basement membranes (Chen et al., 1985; Damsky et al., 1985; Horwitz et al., 1985) and to their apparent absence from cell–cell contacts. Moreover, *in vitro* binding studies indicated that integrins are capable of forming ternary complexes both with extracellular matrix (ECM) components (such as fibronectin) and with talin (Horwitz et al., 1985, 1986). These findings raised the possibility that integrins indeed provide a physical link between the ECM at the cell exterior and elements of the junctional plaque.

As pointed out above, intercellular junctions of the adherens type are molecularly distinguishable from their cell-matrix counterparts. Previously we described a new membrane glycoprotein (Volk and Geiger, 1984; Geiger et al., 1985a) with an apparent molecular weight of 135 kD that was specifically associated with intercellular adherens-type junctions. Recent studies indicate that this 135-kD protein has properties of a cell adhesion molecule, and because of its specific association with adherens junctions it has been termed A-CAM (Volk and Geiger, 1986a,b). We have investigated a variety of aspects related to the structure and function of A-CAM. Among these are the general structure of the molecule, its surface topology, its cell-type-restricted expression in adult and developing tissues, and its involvement in morphogenesis and epithelial remodeling. It appears that A-CAM shares some similar properties with the liver cell adhesion molecule (L-CAM or uvomorulin), including its cellular mode of action, calcium dependence, specificity, and so on (see below). In this chapter we summarize presently available information on the functional and structural features of A-CAM and the mode of its involvement in cell contact formation.

A-CAM AS A JUNCTIONAL CELL ADHESION MOLECULE

The experimental approach that led to the discovery of A-CAM included the production of monoclonal antibodies against membrane constituents of chicken cardiac muscle intercalated discs. These antibodies were selected by radioimmunoassay and immunofluorescence and were used, in turn, for the identification of specific junctional molecule(s). One antibody in that series, denoted ID 7.2.3, strongly reacted with a single molecule with an apparent molecular weight of 135 kD that was specifically localized in intercellular contact areas in cultured cells (Figure 1) or in a variety of tissues (Volk and Geiger, 1984). In the past, nearly all the data we obtained on this 135-kD molecule were based on the use of this monoclonal antibody. A broad library of antibodies reactive with this component has since been prepared for experimental use.

The association of A-CAM with cell contact areas was first detected by immunofluorescence microscopy of different tissues and cultured cells and was later followed by immunoelectron microscope labeling (Volk and Geiger, 1986a). The latter examination, namely immunogold staining of ultrathin frozen sections of chicken cardiac muscle, provided an unequivocal indication of the selective association of A-CAM with authentic adherens junctions. Further immunocytochemical support of its junctional specificity was obtained from double fluorescent labeling experiments in which A-CAM was colocalized with actin (Figure 2d,e) and vinculin (Figure 2a,b; see also Geiger et al., 1985a) in the same cells. A comparison of the fluorescent images obtained by this analysis confirmed that A-CAM is localized at the termini of actin cables,

Figure 1. *Immunofluorescent labeling of a confluent monolayer of cultured chicken lens cells with anti-A-CAM (monoclonal antibody ID 7.2.3).* Notice the extensive staining of cell–cell contact regions. *Calibration bar* = 10 μm.

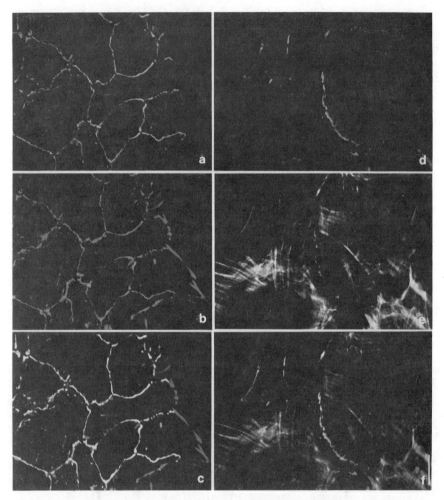

Figure 2.   *Double labeling of the same cultured chicken lens cells for A-CAM (d) and actin (e) or for A-CAM (a) and vinculin (b). c and f* show labeling superpositions. Notice the association of A-CAM with actin- and vinculin-rich sites of intercellular contacts and its apparent absence from vinculin-rich focal contacts *(arrow)* and actin-containing stress fibers. *Calibration bar = 10 μm.*

essentially coinciding with the cell–cell junction-associated vinculin. It should be reemphasized that the immunoelectron microscope labeling and the double immunofluorescence labeling were absolutely essential for the assignment of A-CAM to adherens-type junctions. The abundance of other intercellular contacts that are mediated through a variety of other direct and indirect mechanisms and that probably employ diverse molecular constituents in the junctional region emphasized the need for the higher-resolution immunolocalization and the direct demonstration of close interrelationships with vinculin and actin. Moreover, careful examination of cells labeled for A-CAM revealed that besides the predominant distribution of the molecule in the junctional areas, low yet significant levels of specific labeling were also occasionally detected along the plasma membrane in extrajunctional sites. We have concluded from these observations that the partitioning or recruitment of A-CAM into the junctional domain depends on additional factors (besides the contact itself) that may coordinately regulate the assembly or disassembly of the various junctional subdomains. The identity of these putative factors and their mode of operation are not known at present. Yet we anticipate that future research addressing these aspects will reveal such elements and point to their roles in the cellular regulation of morphogenetic events.

A characteristic property of intercellular adherens-type junctions that reflects on the functional properties of A-CAM is their strict dependence on the presence of extracellular calcium ions. It has been previously shown by us and by others (Cereijido et al., 1978; Kartenbeck et al., 1982; Pitelka et al., 1983; Volberg et al., 1986) that reduction of the calcium concentration of the medium leads to splitting of these junctions, followed by detachment of the plaque and microfilament bundle from the endofacial surfaces of the membrane (Kartenbeck et al., 1982; Volberg et al., 1986).

The modulation of calcium concentration has also served as a useful tool for the study of the fine topology and function of A-CAM. For example, we have shown that reduction of free calcium concentration in the medium to 0.5 mM or lower results in rapid splitting of the junctions into halves with A-CAM exposed on their surfaces (Figure 3), as can be judged from positive staining for this protein in both permeabilized and nonpermeabilized cells (Volk and Geiger, 1984, 1986a). It is also noteworthy that the stress fibers anchored to cell–substrate focal contacts were hardly affected by this treatment within a comparable time frame (Figure 3b).

The ability to modulate junction formation by altering extracellular calcium concentrations has yielded some important insights into the cellular functions of A-CAM. Epithelial sheets of cultured lens cells were treated with EGTA until essentially all the intercellular junctions dissociated. Normal calcium-containing medium was then added to the cells, and incubation proceeded for five hours. Labeling of the cells for A-CAM at that stage indicated that intercellular adherens junctions were effectively reestablished, and that the cells reformed an epithelial monolayer essentially indistinguishable from that of a normal, untreated culture (Figure 4a,b). However, the recovery of junctions could be completely inhibited by the introduction of monovalent anti-A-CAM to the culture medium, as shown in Figure 4c,d. It is also notable that in the absence of new junctions the entire microfilament system essentially collapsed, showing only a few fragmented bundles and essentially no stress fibers (Figure 4c; see also Volk and Geiger, 1986b). This inhibition was specific and could not be obtained with the divalent antibody. In fact, we found that the treatment of cells during the reformation of adherens junctions with an intact antibody (ID 7.2.3) rendered the junctions largely calcium independent,

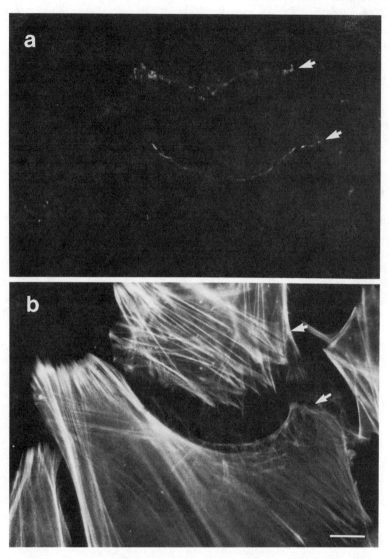

Figure 3. *Double immunofluorescent labeling of cultured chick lens cells for A-CAM (a) and actin (b), following 5-min treatment with 5 mM EGTA.* Notice that the A-CAM-rich junction has dissociated and the actin bundles associated with it are distorted. The stress fibers attached to cell–substrate focal contacts were considerably less affected by the removal of calcium ions. *Calibration bar = 10 μm.*

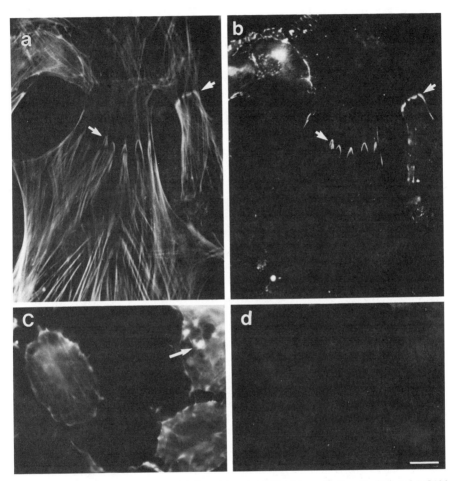

Figure 4. *Double immunofluorescent labeling of the same cultured lens cells for actin (a,c) and A-CAM (b,d).* The labeled cells were either untreated (*a,b*) or briefly treated with EGTA and then incubated for five hours with calcium-containing medium in the presence of anti-A-CAM (Fab fragment). In the untreated culture, as well as in cultures incubated with an irrelevant antibody (Volk and Geiger, 1986b), apical stress fibers are abundant, apparently terminating in the reformed A-CAM-containing contact regions (see, for example, *matched arrows* in *a* and *b*). In the anti-A-CAM-treated cells actin cables are apparently absent, presenting only a few distorted bundles or aggregates (*arrow* in *c*). A-CAM in the same cells is essentially undetectable (*d*).

suggesting that the divalent antibodies either provided direct links between A-CAM molecules on the two adjacent cells or altered the sensitivity of A-CAM to calcium. The former possibility, which we find more likely, implies that indirect (antibody-mediated) A-CAM–A-CAM interactions may be sufficient for the transmembrane induction of junction formation (for further discussion, see Volk and Geiger, 1986b).

In conclusion, we have shown that A-CAM is not merely an inert component of adherens-type junctions but actually functions as a calcium-dependent cell adhesion molecule. Some of the more recently prepared monoclonal antibodies were, however, disruptive in their intact, divalent forms.

MOLECULAR PROPERTIES OF A-CAM

Until the present time, A-CAM had not been isolated in sufficient quantities and purity to enable a direct, extensive biochemical analysis. Therefore the approaches used to identify A-CAM or its fragments were largely based on immunoblotting analyses with the A-CAM-specific monoclonal antibody. Immunoblotting analysis of cultured chick lens cells or of cultured cells treated with trypsin in the presence of 1.8 mM calcium revealed pronounced reactivity of the antibody with the intact 135-kD band (Figure 5). However, brief trypsinization of the same lens cells in the presence of EGTA resulted in a rapid cleavage of A-CAM into several major species of cell-attached proteolytic

---

Figure 5. *a: Immunoblotting analysis of A-CAM and its fragments in cultured chick lens cells.* The samples tested included a control of untreated cells containing the 135-kD band (C); extracts of cultured lens cells pretreated for 5 min with 0.5 mg/ml trypsin (TR) in the presence of 1.8 mM calcium (+) or in the presence of 5 mM EGTA (−) 1.8 mM. A-CAM in the former sample remained largely intact, while in the cells treated with trypsin and EGTA, three major immunoreactive cell-bound fragments of A-CAM were detected with apparent molecular weights of 78 kD, 60 kD, and 46 kD. Analysis of the medium in which the lens cells were cultured for two days revealed a 100-kD immunoreactive fragment that was released from the cells (R). In order to identify carbohydrate-rich regions in A-CAM, lens cell extracts were applied to Sepharose-Con A columns and chromatographed in the absence (−αMM) or presence (+αMM) of 0.2 M α-methylmannoside. The bands eluted from the Con A column were then identified by immunoblotting analysis, revealing specific binding of intact A-CAM (C) as well as of all the three major tryptic EGTA cell-bound fragments (TR/E) and the 100-kD released polypeptide (R). The *arrowheads* mark the positions of the polypeptides, with apparent molecular weights of 135 kD, 100 kD, 78 kD, 60 kD, and 46 kD. *b:* A putative linear model of A-CAM showing the sites cleaved by trypsin in the presence of EGTA (78T, 60T, 46T) as well as the cleavage site produced by the endogenous protease, which cleaves off the 100-kD peptide released into the medium (100R). The expected region in which the antigenic epitope of ID 7.2.3 is present is marked by the *asterisk.* Definitive carbohydrate-containing regions (CHO) as well as less definitive sites (CHO?) are shown. The location of the membrane-spanning domain (m) and the size of the cytoplasmic tail are rather speculative and are based, to some extent, on the overall similarity between A-CAM and L-CAM (see text).

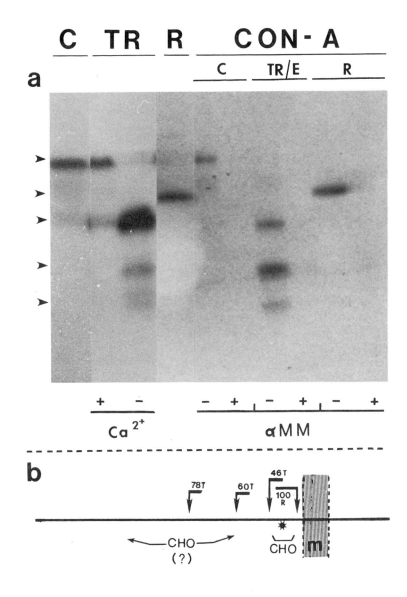

fragments. The most prominent immunoreactive tryptic polypeptides that were retained on the cell surface had apparent molecular weights of 78 kD, 60 kD, and 46 kD (Figure 5). Interestingly, prolonged treatment with EGTA without exogenous proteases or even prolonged culturing of cells in normal medium resulted in the release into the medium of an approximately 100-kD fragment (Figure 5). These findings suggest that a portion of the A-CAM molecule associated with the cell surface is relatively insensitive to trypsin, yet the greater part of the molecule (>100 kD) is exposed on the outer cell surface and may be cleaved off by endogenous or exogenous proteolytic enzymes. Based on this analysis, it appears that the antigenic epitope recognized by the anti-A-CAM monoclonal antibody ID 7.2.3 is probably present in a region of the molecule located between 90 and 100 kD away from the extracellular terminus of the molecule (see putative model in Figure 5b).*

Further information was obtained with a Triton X-114 partitioning assay. This biphasic partitioning system has been a useful tool for identifying and isolating intrinsic membrane proteins (Bordier, 1981; Coudrier et al., 1983) that tend to accumulate in the detergent phase. This analysis indicated that the intact A-CAM molecule was inclined to partition fully into the buffer phase (Volk and Geiger, 1986a), yet a similar examination of the partitioning profile of EGTA-trypsinized cells indicated that the 78-kD and 60-kD fragments showed a tendency to partition into the detergent phase and to be largely excluded from the lighter buffer phase. It was thus concluded that the extended extracellular moieties of A-CAM confer hydrophilic properties on the intact molecule that favor its partitioning into the buffer fraction. The proteolytic removal of these regions allows for the putative membrane-associated regions to affect the partitioning of the cell-bound fragments into the detergent phase. It is noteworthy that another calcium-dependent cell adhesion molecule, namely L-CAM (or uvomorulin), exhibits a similar tendency to partition into the buffer phase (Peyrieras et al., 1983), and recent data concerning the primary structure of this protein point to the presence of putative trans-membrane sequences at a position located at three-quarters of the length of the molecule toward the carboxyl end (Gallin et al., 1987).

The reaction of electrophoretic gels with iodinated concanavalin A (Con A) and the fractionation of extracted proteins on Sepharose-Con A column indicated that A-CAM is a glycoprotein. Furthermore, chromatography of trypsinized A-CAM on Sepharose-Con A revealed that all three major cell-bound proteolytic fragments, as well as the released 100-kD fragment, contain carbohydrate moieties (Figure 5). Based on the limited information presented here and on the idea that A-CAM might share some homology with L-CAM (see below), it was possible to draft a hypothetical linear model for the A-CAM molecule, as shown in Figure 5b.*

---

*The cloning and sequencing of N-cadherin cDNA, which occurred after this chapter was written (Hatta et al., 1988), opens new possibilities for analyzing the proteolytic cleavage of A-CAM and A-CAM's relationship to other CAMs.

A-CAM DISTRIBUTION IN THE DEVELOPING CHICKEN EMBRYO

Immunofluorescent labeling of a large variety of adult chicken tissues indicates that A-CAM is present in some but not all adherens junction–containing cells. Thus A-CAM was abundant in heart, lens, and several other tissues but was apparently absent from a large variety of polar epithelia, such as those of the intestine, kidney tubules, pancreas, and liver. The latter tissues contain prominent adherens junctions and are known to express another calcium-dependent CAM, L-CAM (Hyafil et al., 1980; Gallin et al., 1983); in at least some of these tissues L-CAM was reported to be specifically associated with the zonula adherens of the cells (Boller et al., 1985).

Immunolocalization of A-CAM in early chick embryos revealed extensive labeling of ectodermal placodes as well as a prominent but transient expression in several mesoderm-derived epithelia. A comprehensive survey of the embryonic expression of A-CAM was carried out in collaboration with J.-L. Duband and J. P. Thiery (Duband et al., 1988); only two examples are discussed here. It is also noteworthy that the distribution of A-CAM in chick embryos is similar to that of N-cadherin (Hatta and Takeichi, 1986; Duband et al., 1987), suggesting that the two molecules may be closely related or possibly even identical; a direct comparison of A-CAM and N-cadherin is now in progress (T. Volk, M. Takeichi, and B. Geiger, unpublished data). Here we discuss the involvement of A-CAM in two processes: epithelial remodeling during neurulation and mesenchymal–epithelial transition during somite formation. For further details and examples from other systems the reader is referred to the detailed study by Duband et al. (1988).

The ectoderm of the early chick embryo is largely negative for A-CAM, and its cells express mainly L-CAM on their surface (Duband et al., 1987, 1988). However, careful examination indicates that areas of epithelial infoldings such as the primitive streak and the neural groove (and later the neural tube) as well as the different ectodermal placodes express elevated levels of A-CAM. Figure 6 demonstrates that the level of labeling for A-CAM is limited during the initial stages of neurulation (Figure 6a) and becomes quite extensive as the process proceeds (Figure 6b–d). The extensive expression of A-CAM in the neural tube is apparently accompanied by the assembly of a conspicuous zonula adherens along the lumenal (apical) borders of the cells. This is clearly visible in low-power (Figure 6e) and high-power (Figure 6f) electron micrographs. Adherens-type junctions were also present along the L-CAM-rich ectoderm of the same embryos. These junctions are readily detected by electron microscopy (Figure 7). However, the density of microfilaments associated with them was markedly lower than that found in the A-CAM-rich junctions of the adjacent neural tube (compare, e.g., Figure 7b with Figure 6f). Furthermore, labeling of similar frozen sections with fluoresceinated phalloidin revealed elevated concentrations of actin along the A-CAM-rich regions compared to the ectoderm (not shown).

The close correlation between the spatially and temporally regulated expression of A-CAM and the concomitant assembly of a massive, actin-rich

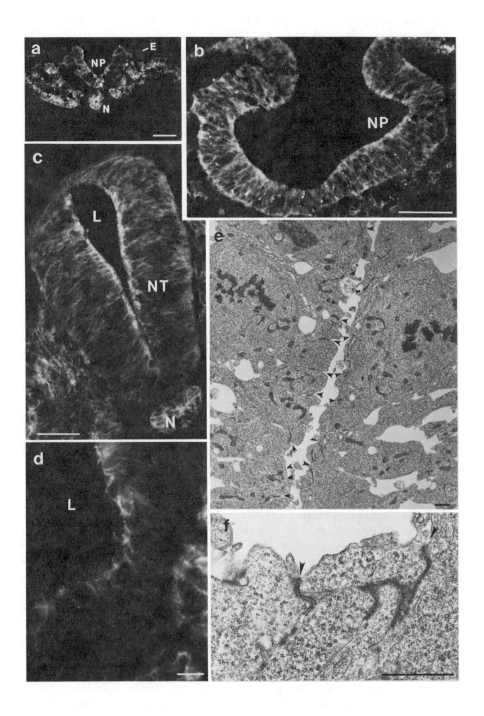

68

junction raises some interesting possibilities concerning the mechanism of epithelial reorganization such as neural tube formation; it is widely accepted that the topological modulation of the neural plate and its infolding may be driven by localized contraction forces that operate at the apical aspects of the neuroepithelial cells (i.e., Baker and Schroeder, 1967; Jacobson, 1985). We would like to suggest that this apical contractility is largely generated by the adherens junction–associated actin bundles and that the local expression of A-CAM plays an important role in initiating the formation of these micro-filament-bound junctions.

In the case of the neural tube, the appearance of A-CAM is apparently associated with the reorganization of an already assembled epithelium. In other processes such as somite formation, loosely packed mesodermal cells assemble into apparently homogenous mesenchymal rods, namely the seg-mental plates. These structures subsequently metamerize, forming epithelial somites. Examination of A-CAM levels throughout this process indicates that while positive surface labeling is noted in the cells of the segmental plate, the labeling markedly increased in the forming somites and is predominantly concentrated close to the lumenal aspect of the somite cells (Figure 8a). Electron microscopy confirmed the presence of conspicuous adherens junc-tions in these regions, as shown in Figure 8c,d. Thus the mesenchymal–epithelial transformation was accompanied by an elevated expression of A-CAM and the formation of an elaborate network of adherens junctions. Further examination of the somites at later developmental stages enabled us to detect an apparent reversal of this process: Following the assembly of the somite epithelia, the medioventral region of the somite (i.e., the sclerotome) loses its epithelial organization, while the dorsolateral dermomyotome retains its epithelial features. Labeling for A-CAM indicates that, prior to their dissoci-ation, the cells of the sclerotome markedly lose A-CAM expression, in contrast to the epithelial cells of the dermomyotome, which remain strongly positive (Figure 8b).

Figure 6.   *Immunofluorescent labeling of A-CAM in frozen sections (a–d), and corresponding transmission electron microscopy (e,f), of the developing chick neural tube. a:* Early stages in the invagination of the neural plate (NP) in a 30-hour-old chick embryo showing strong labeling of the neural plate cells and of the notochord (N), in contrast to the largely negative ectoderm (E). The labeling is notably enriched in the apical region of the cells. *b:* Deep invagination of the neural plate in the head region of an embryo of the same stage disclosing intense labeling of the apical portion of the neural plate cells (NP). *c:* Mature neural tube (NT) of a two-day-old embryo showing strong labeling close to the lumen. The apical labeling close to the lumen (L) is shown at high magnification in *d. e:* Low-power and *(f)* high-power examination of neural tube of a three-day-old chick embryo by transmission electron microscopy. Examination reveals large adherens junctions located at the subapical region of the lumenal cells (*arrowheads* in *e* and *f*). These junctions contain conspicuous plaques and associated actin bundles. *Calibration bars* = 50 μm (*a–d*) and 1 μm (*e–f*).

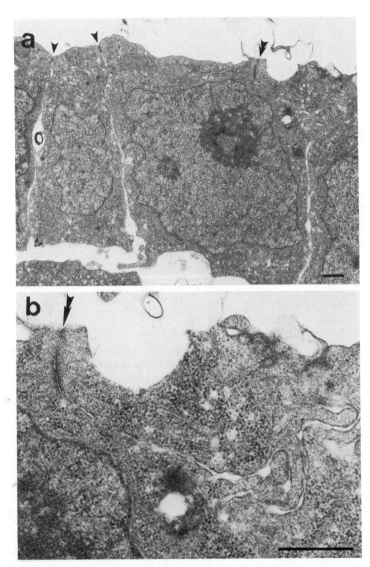

Figure 7. *Electron micrographs of the ectoderm of a three-day-old embryo.* The *arrowheads* mark the location of adherens intercellular junctions of ectodermal cells, and the *double arrowheads* in *a* and *b* point to the same junction shown at low- and high-power magnification, respectively. Notice that the electron-dense junctional plaques in the ectodermal epithelium and the associated microfilaments are considerably less elaborate than those found in the same type of junction located along the neural plate or neural tube, as shown in Figure 6. *Calibration bars* = 0.2 μm.

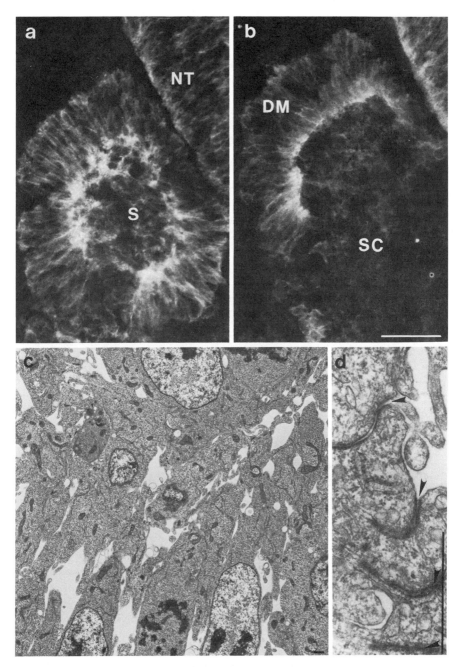

Figure 8. *Immunofluorescent labeling with anti-A-CAM (a,b) and the corresponding transmission electron microscopy (c,d) of somites in two-day-old chick embryos.* In the mature somite (S) intense labeling for A-CAM is detected, especially along the subapical borders of the cells. At later stages of development the sclerotome (SC) apparently loses A-CAM, in contrast to the dermomyotome (DM), which remains intensely positive. Electron microscopy of somites reveals numerous well-developed adherens-type junctions *(arrowheads* in *d). Calibration bars =* 50 μm *(b)* and 1 μm *(c,d).*

In conclusion, the results given here as well as additional detailed information presented elsewhere (Duband et al., 1988) support the view that the expression of A-CAM and its spatiotemporal modulation may play cardinal roles in at least two types of morphogenetic events: the remodeling of epithelial layers and mesenchymal–epithelial transformations. In both morphogenetic processes A-CAM apparently contributes to intercellular adhesion, to the assembly of a complex, cytoskeleton-bound junction, and to the generation and coordination of mechanical, transcellular forces.

FORMATION OF CHIMERIC JUNCTIONS BETWEEN A-CAM-CONTAINING CELLS AND L-CAM-CONTAINING CELLS

One of the functions often attributed to CAMs is the selective sorting of cells. According to this view, cells bearing specific CAMs on their surface may interact selectively with certain cells but not with others, based on the binding specificity of the particular CAM or CAMs present. Such selective recognition may be the basis for the homotypic interactions of dissociated embryonic cells as well as normal histogenesis *in vivo* (Steinberg, 1970; Moscona, 1974; Edelman, 1983; Takeichi et al., 1985).

In order to study the fine binding specificity of A-CAM-mediated interactions, we cocultured chick lens cells (with A-CAM on their surface) together with L-CAM-containing cells derived from chicken liver. Examination of the mixed cultures by light and electron microscopy revealed that besides homotypic contacts between cells of the same origin there were numerous heterotypic associations between liver and lens cells: The identity of the individual cells in such complexes was determined by a variety of assays and by the distinct morphologies of the two cell types (Figure 9; for further details and controls, see Volk et al., 1987). In addition, we covalently modified one of the cell populations with a fluorescent probe prior to coculturing and used this labeling as a marker for the origins of individual cells within the mixed monolayer.

The electron micrographs such as those in Figure 9 indicate that the heterotypic contacts contain adherens-type junctions with typical plaques and membrane-bound microfilaments. These indications were further substantiated by fluorescence labeling for vinculin and actin, the characteristic components of the junctional plaque and the cytoskeletal subdomains, respectively (see Volk et al., 1987).

The most interesting observation made with the mixed cultures was obtained with double immunolabeling of the cells for A-CAM and L-CAM: Lens cells express A-CAM and nearly no L-CAM, while the reverse is true for the cultured liver cells. It should be mentioned that in the liver cell cultures there was a minority of cells that did express A-CAM in addition to L-CAM. The source of these cells is not clear to us, yet they have contributed an additional insight into the molecular parameters of the heterotypic interaction, as is discussed below.

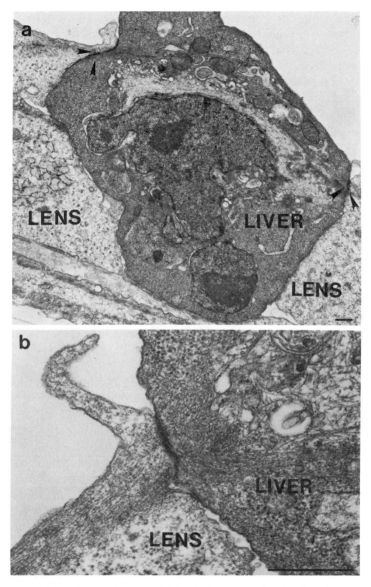

Figure 9. *Transmission electron microscopy of chimeric adherens-type junctions formed between cultured lens and liver cells at low-power (a) and high-power (b) magnification.* The locations of the junctions are indicated by *arrowheads*. The high-power photomicrograph shows the junctional area and the microfilament bundles associated with it. *Calibration bars* = 0.2 μm.

The double immunofluorescent labeling of the mixed (lens–liver) cultures mentioned above revealed three major molecular forms of adherens junctions: liver–liver contacts characterized by the presence of L-CAM only; lens–lens junctions displaying A-CAM; and chimeric lens–liver junctions, positively labeled for both A-CAM and L-CAM (Figure 10a–c). To determine the exact cellular origins of each of the CAMs present in the mixed junctions, the cultures were briefly treated with EGTA prior to their fixation and labeling. This treatment resulted in a limited splitting of the intercellular junctions. Examination of the EGTA-treated cells indicated that each of the two halves of the split mixed junction contained a different CAM; the liver cell partner contained L-CAM, while the lens cell had A-CAM on its junctional surface (Figure 10d–f). The fluorescence thus obtained showed mirror-image patterns, indicating that A-CAM and L-CAM in the mixed junctions have close spatial relationships.

Inhibition experiments further suggested that the two CAMs were both obligatorily involved in the intercellular interaction. Thus the addition of either anti-L-CAM or anti-A-CAM (Fab fragments) effectively inhibited the formation of chimeric junctions (as well as the relevant homotypic junctions; see Volk et al., 1987). On the basis of these observations, we have concluded that A-CAM and L-CAM are both functionally involved in the formation of mixed junctions. Obviously this is not sufficient to prove that the two CAMs interact directly with each other in these junctions, but such a possibility appears to be quite likely.

Some fortuitous side observations made with the mixed cultures merit additional discussion, since they shed some light on the putative A-CAM- and L-CAM-mediated contacts. First, we noticed that all three adherens-type junctions formed in such cultures (lens–lens, liver–liver, and lens–liver) were comparably sensitive to calcium withdrawal. Thus these junctions showed similar calcium concentration dependence, and all three dissociated at a comparable rate upon its removal. A second aspect is related to the possible involvement of A-CAM and L-CAM in selective intercellular interactions (i.e., cell sorting). Examination of mixed cultures indicated that homotypic adhesions were more frequent than the heterotypic ones. However, these differences could be attributed to a large number of factors besides the particular CAMs discussed here, for example, the proliferation of cells after plating as well as the differential rates of attachment of the two cell populations to the substrate. Both processes could, in principle, lead to the formation of homogeneous islands.

A more definitive analysis of the selectivity of apparently homophilic versus heterophilic interactions was obtained in another series of experiments in which either lens or liver cells were plated in culture together with kidney cells. Unlike the former two cell types, each of which expresses only one type of calcium-dependent CAM, kidney cells express both A-CAM and L-CAM. Thus chimeric junctions between kidney cells and either liver or lens cells presented a unique situation in which the two associated cells shared one CAM, yet the kidney cell partner contained, in addition, the heterologous

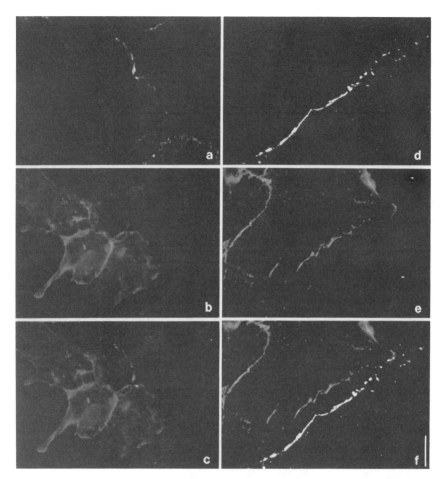

Figure 10. *Double labeling for A-CAM and L-CAM in mixed cultures of lens and liver cells.* The cells were fixed and labeled either without any pretreatment (*a–c*) or following a brief (30 sec) treatment with 5 mM EGTA in order to dissociate the junctions (*d–f*). Superpositions of the two labels are shown in *c* and *f,* allowing for a close comparison of the relative distributions of the two antigens. It is apparent that besides homotypic contacts (lens–lens and liver–liver), there are many cases of heterotypic junctions containing A-CAM on the lens cell partner and L-CAM on the neighboring liver cell. *Calibration bar* = 10 μm.

protein. Keeping in mind the intrinsic symmetry of junctional interactions, this system enabled us to examine whether the relative presentations of A-CAM and L-CAM on the junctional surfaces of the kidney cells were affected by the nature of the junctional partner cell and the particular CAM present on its surface. Careful examination of such cultures following a brief EGTA treatment indicated that the relative labeling intensities for the two CAMs on the kidney cell junctions were essentially the same, irrespective of the junctional partner. This feature, schematically summarized in Figure 11, suggests that A-CAM and

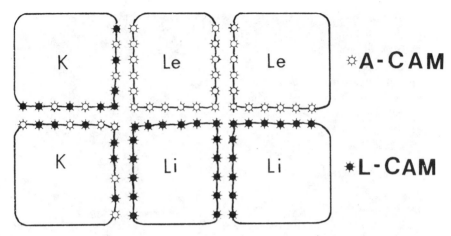

Figure 11. *A scheme showing the various junction-associated CAMs in homotypic and heterotypic cell junctions formed between kidney cells (K), lens cells (Le), and liver cells (Li).* Evidently each of the three cell populations can form adherens-type intercellular junctions with the other two cell types, and the relative association of the CAM(s) expressed by it within the junction is not affected by the nature of the junctional partner.

L-CAM can similarly bind (directly or indirectly) either to the homologous or to the heterologous protein under these culture conditions.

The results presented above provide some insight into the gross cellular specificities of A-CAM- and L-CAM-mediated interactions. Intensive biochemical, molecular genetic, and ultrastructural analyses will undoubtedly be required to determine the fine specificity of these interactions at the molecular level. It will also be important to determine whether such heterotypic and heterophilic interactions bear any significance for native cellular interactions in adult or embryonic tissues.

CONCLUSIONS AND FUTURE PROSPECTS*

In this chapter we have outlined part of the currently available information concerning A-CAM, its general structure, cellular distribution, physiological functions, and molecular specificity. The preliminary studies presented here on its distribution in early chick embryos clearly suggests that the tight regulation of A-CAM expression is correlated with major morphogenetic events related to the formation and modulation of epithelia. The second phase of these processes, namely the disappearance of A-CAM from various epithelia

---

*Since these lines were written, major experimental progress has been made in cloning and sequencing various calcium-dependent CAMs (cadherins) and analyzing their expression. These experiments address many of the issues presented here.

at later stages of maturation, has not yet been studied. It will be extremely important to examine the control mechanism of A-CAM expression vis-à-vis the expression of other CAMs and to characterize it at the cellular and molecular–genetic levels.

Another topic, only briefly discussed here, concerns the molecular relationships between various calcium-dependent CAMs. The formation of mixed junctions in which A-CAM is present on one cell and L-CAM on the other suggests that the two CAMs may show a significant degree of molecular and functional homology. This suggestion is indirectly corroborated by the recent findings of Shirayoshi et al. (1986), who showed that the N-terminal sequences of N-cadherin, a molecule most likely related to A-CAM (Duband et al., 1987, 1988), and L-CAM (E-cadherin) are identical. It therefore seems that there is a family of calcium-dependent CAMs that exhibit a certain degree of functional and possibly structural homology. Obviously the direct evidence for such relationships is still poor, and the extent of homology between the CAMs unknown. Further characterization of this aspect will involve studies at both cellullar and molecular–genetic levels. Such studies are now in progress in several laboratories, and their results will, no doubt, shed much light on the molecular basis for cell–cell interaction and on the mechanism of transmembrane signal transduction in areas of cell contact.

ACKNOWLEDGMENTS

This study was supported by grants from the Muscular Dystrophy Association and from the Weizmann-Rockefeller Foundation. B. G. holds the E. Neter Professorial Chair in Cell and Tumor Biology at the Weizmann Institute of Science.

REFERENCES

Avnur, Z., J. V. Small, and B. Geiger (1983) Actin-independent association of vinculin with the cytoplasmic aspects of the plasma membrane in cell–substrate contacts. *J. Cell Biol.* **96**:1622–1630.

Baker, P. C., and T. E. Schroeder (1967) Cytoplasmic filaments and morphogenetic movement in the amphibian neural tube. *Dev. Biol.* **15**:432–450.

Boller, K., D. Vestweber, and R. Kemler (1985) Cell adhesion molecule uvomorulin is localized in the intermediate junctions of adult intestinal epithelial cells. *J. Cell Biol.* **100**:327–332.

Bordier, C. (1981) Phase separation of integral membrane proteins in Triton X-114 solution. *J. Biol. Chem.* **256**:1604–1607.

Burridge, K., and L. Connell (1983) Talin: A cytoskeletal component in adhesion plaques and other sites of actin membrane interaction. *Cell Motil.* **3**:405–417.

Cereijido, M., E. S. Robbins, W. J. Dolan, C. A. Rotinno, and D. D. Sabatini (1978) Polarized monolayers formed by epithelial cells on a permeable and translucent support. *J. Cell Biol.* **77**:853–880.

Chen, W.-T., T. Hasegawa, C. Hasegawa, C. Weinstock, and K. M. Yamada (1985) Development of cell surface linkage complexes in cultivated fibroblasts. *J. Cell Biol.* **100**:1103–1114.

Coudrier, E., H. Reggio, and D. Louvard (1983) Characterization of membrane glycoproteins involved in attachment of microfilaments to the microvillar membrane. *Ciba Found. Symp.* **95**:216–232.

Cowin, P., H. P. Kapprell, W. W. Franke, J. Tamkun, and R. O. Hynes (1986) Plakoglobin: A protein common to different kinds of intercellular adhering junctions. *Cell* **46**:1063–1073.

Damsky, C. M., K. A. Knudsen, D. Bradley, C. A. Buck, and A. F. Horwitz (1985) Distribution of the CSAT cell–matrix antigen on myogenic and fibroblastic cells in culture. *J. Cell Biol.* **100**:1528–1539.

Duband, J. L., S. Dufour, K. Hatta, M. Takeichi, G. M. Edelman, and J. P. Thiery (1987) Distribution and role of adhesion molecules in the development of the metameric pattern of the avian embryo. *J. Cell Biol.* **104**:1361–1374.

Duband, J. L., T. Volberg, J. P. Thiery, and B. Geiger (1988) Spatial and temporal distribution of the adherens junction-associated adhesion molecule A-CAM during avian embryogenesis. *Development* **103**:325–344.

Edelman, G. M. (1983) Cell adhesion molecules. *Science* **219**:450–457.

Gallin, W. J., G. M. Edelman, and B. A. Cunningham (1983) Characterization of L-CAM, a major cell adhesion molecule from embryonic liver cells. *Proc. Natl. Acad. Sci. USA* **80**:1038–1042.

Gallin, W. J., B. C. Sorkin, G. M. Edelman, and B. A. Cunningham (1987) Sequence analysis of a cDNA clone encoding the liver cell adhesion molecule. *Proc. Natl. Acad. Sci. USA* **84**:2808–2812.

Geiger, B. (1982) Involvement of vinculin in contact-induced cytoskeletal interactions. *Cold Spring Harbor Symp. Quant. Biol.* **46**:671–682.

Geiger, B. (1983) Membrane–cytoskeleton interaction. *Biochim. Biophys. Acta.* **737**:305–341.

Geiger, B., E. Schmidt, and W. W. Franke (1983) Spatial distribution of proteins specific for desmosomes and adherens junctions in epithelial cells demonstrated by double immunofluorescence microscopy. *Differentiation* **23**:189–205.

Geiger, B., Z. Avnur, G. Rinnerthaler, H. Hinssen, and V. J. Small (1984) Microfilament-organizing centers in areas of cell contact: Cytoskeletal interactions during cell attachment and locomotion. *J. Cell Biol.* **99**:835–915.

Geiger, B., Z. Avnur, T. Volberg, and T. Volk (1985a) Molecular domains of adherens junctions. In *The Cell in Contact: Adhesions and Junctions as Morphogenetic Determinants*, G. M. Edelman and J. P. Thiery, eds., pp. 461–489, Wiley, New York.

Geiger, B., T. Volk, and T. Volberg (1985b) Molecular heterogeneity of adherens junctions. *J. Cell Biol.* **101**:1523–1531.

Hatta, K., and M. Takeichi (1986) Expression of N-cadherin adhesion molecules associated with early morphogenetic events in chick development. *Nature* **320**:447–449.

Hatta, K., A. Nose, A. Nagafuchi, and M. Takeichi (1988) Cloning and expression of cDNA encoding a neural calcium-dependent cell adhesion molecule: Its identity in the cadherin gene family. *J. Cell Biol.* **106**:873–881.

Horwitz, A. F., K. Duggan, R. Greggs, C. Decker, and C. A. Buck (1985) The cell substrate attachment (CSAT) antigen has properties of a receptor for laminin and fibronectin. *J. Cell Biol.* **101**:2134–2144.

Horwitz, A. F., K. Duggan, C. A. Buck, M. C. Beckerle, and K. Burridge (1986) Interaction of plasma membrane fibronectin receptor with talin—a transmembrane linkage. *Nature* **320**:531–533.

Hyafil, F., P. Morello, C. Babinet, and F. Jacob (1980) A cell surface glycoprotein involved in the compaction of embryonal carcinoma cells and cleavage stage embryos. *Cell* **21**:927–934.

Hynes, R. O. (1987) Integrins: A family of cell surface receptors. *Cell* **48**:549–554.

Jacobson, A. G. (1985) Adhesion and movement of cells may be coupled to produce neurulation. In *The Cell in Contact: Adhesions and Junctions as Morphogenetic Determinants,* G. M. Edelman and J. P. Thiery, eds., pp. 49–65, Wiley, New York.

Kartenbeck, J., E. Schmidt, W. W. Franke, and B. Geiger (1982) Different modes of internalization of proteins associated with adherens junctions and desmosomes: Experimental separation of lateral contacts induces endocytosis of desmosome plaque material. *EMBO J.* **1**:725–732.

Moscona, A. A. (1974) Surface specifications of embryonic cells: Lectin receptors, cell recognition, and specific cell ligands. In *The Cell Surface in Development,* A. A. Moscona, ed., pp. 67–99, Wiley, New York.

Pitelka, D. R., B. N. Taggart, and S. T. Karnamoto (1983) Effects of extracellular calcium depletion on membrane topography and occluding junctions of mammary epithelial cells in culture. *J. Cell Biol.* **96**:613–624.

Peyrieras, N., F. Hyafil, D. Louvard, H. L. Ploegh, and F. Jacob (1983) Uvomorulin: A nonintegral membrane protein of early mouse embryo. *Proc. Natl. Acad. Sci. USA* **80**:6274–6277.

Shirayoshi, Y., K. Hatta, M. Hosoda, S. Tsunawasa, F. Sahniyama, and M. Takeichi (1986) Cadherin cell adhesion molecules with distinct binding specificities share a common structure. *EMBO J.* **5**:2485–2488.

Steinberg, M. S. (1970) Does differential adhesion govern self-assembly processes in histogenesis? Equilibrium configuration and the emergence of a hierarchy among population of embryonic cells. *J. Exp. Zool.* **173**:395–433.

Takeichi, M., K. Hatta, and A. Nagafuchi (1985) Selective cell adhesion mechanism: Role of the calcium-dependent cell adhesion system. In *Molecular Determinants of Animal Form,* G. M. Edelman, ed., pp. 223–233, Alan R. Liss, New York.

Volberg, T., B. Geiger, J. Kartenbeck, and W. W. Franke (1986) Changes of membrane–microfilament interaction in intercellular adherens junctions upon removal of extracellular Ca^{2+} ions. *J. Cell Biol.* **102**:1832–1842.

Volk, T., and B. Geiger (1984) A 135-kD membrane protein of intercellular adherens junctions. *EMBO J.* **3**:2249–2260.

Volk, T., and B. Geiger (1986a) A-CAM: A 135-kD receptor of intercellular adherens junctions. I. Immunoelectron microscopic localization and biochemical studies. *J. Cell Biol.* **103**:1441–1450.

Volk, T., and B. Geiger (1986b) A-CAM: A 135-kD receptor of intercellular adherens junctions. II. Antibody-mediated modulation of junction formation. *J. Cell Biol.* **103**:1451–1464.

Volk, T., O. Cohen, and B. Geiger (1987) Formation of heterotypic adherens-type junctions between L-CAM-containing liver cells and A-CAM-containing lens cells. *Cell* **50**:987–994.

Chapter 4

The Carbohydrate Units of Nervous Tissue Glycoproteins: Structural Properties and Role in Cell Interactions*

JUKKA FINNE

ABSTRACT

Studies on the protein-bound carbohydrate units of nervous tissue have revealed several structures that are novel for glycoproteins. These include a terminal sequence containing fucose and sialic acid, polysialic acid chains composed of $\alpha 2$–8-linked N-acetylneuraminic acid residues, an α-galactose-containing O-linked disaccharide, and a series of mannose-containing O-linked oligosaccharides. Comparison with glycans from other tissues indicates that some of the structures described are highly enriched in nervous tissue.

The occurrence of the glycans in molecules involved in cell adhesion, and their developmental regulation, suggests that carbohydrates may be involved in cellular interactions of the nervous system. The glycan of most significant developmental regulation is polysialic acid. It occurs in the neural cell adhesion molecule N-CAM and appears to modulate its adhesivity. In neuronally differentiating cell lines, poly-N-acetyllactosamine glycans, as revealed by a novel cell surface labeling method, occur in an inducible glycoprotein closely similar to the neuron–glia cell adhesion molecule Ng-CAM. The mannose-linked glycans, on the other hand, occur in a chondroitin sulfate proteoglycan involved in laminin and cytotactin interaction.

The glycoproteins are also involved in interactions of bacterial and host cells in the pathogenesis of meningitis. Polysialic acid chains share their structure with the capsular polysaccharide of certain important pathogens that may thus escape host immune defense. Sialic acid–containing O-linked oligosaccharides, on the other hand, may be involved in meningitis by serving as adhesion receptors for a novel adhesion specificity characterized in certain bacterial strains causing meningitis.

The surface of animal cells is covered with a layer of complex carbohydrates consisting of the carbohydrate units of glycoproteins and glycolipids. Whereas the structures of the major glycolipids are well known, information on the properties of the protein-bound glycans, which account for most of the

---

*Dedicated to Professor Johan Järnefelt on the occasion of his 60th birthday.

carbohydrates of cell membranes, has accumulated only during the last few years and is still far from complete. Although a considerable amount of information is available on the carbohydrate units of soluble, mainly plasma glycoproteins (for reviews, see Montreuil, 1980; Vliegenthart et al., 1983), the data cannot be directly applied to membrane-bound molecules. Moreover, examples of the biological significance of the carbohydrate units are scarce, and the physiological role of this major class of biopolymers is still for the most part unresolved.

The highly organized patterns of multiple cell contacts in the nervous system are thought to be based on specific molecular interactions at the cell surfaces. In addition to their participation in the proper organization of the neuronal, glial, and other cells within the nervous system, the surface molecules have to be considered in interactions of the nervous system with other parts of the body, and also with the environment, as in the case of invasion of the tissue by microorganisms. Because their highly complex structures can carry more information than polypeptides or nucleic acids (Sharon, 1984), and because of their abundance in the outer layer of the cell surfaces, carbohydrates are thought to be essential components of cellular recognition phenomena in the nervous system (Hughes, 1976).

A prerequisite for the elucidation of the role of different molecular components in biological interactions is knowledge of the structures and properties of the compounds involved. Several glycoproteins of the nervous system have been identified (see Brunngraber, 1987), and some of them have been implicated in important functions such as cell adhesion (for reviews, see Edelman, 1983; Goridis et al., 1983; Rutishauser, 1984). However, with few exceptions, very little information on the structures of the carbohydrate units of these glycoproteins is available, one reason being the usually low amounts of material that can be isolated. More important, it is difficult, if not impossible, to identify those carbohydrate structures that are characteristic of the nervous system by studying a single glycoprotein species. This information, on the other hand, is essential in efforts directed to the elucidation of the role of carbohydrates in nervous tissue.

In order to compare the carbohydrate components of different tissues, representative samples have to be used. The approach of studying the glycans of whole tissues and cells, rather than single, randomly isolated glycoproteins, has therefore been used for this purpose. By protease treatment of tissue samples, the carbohydrate side chains of glycoproteins can be converted nearly quantitatively to soluble glycopeptides consisting of the intact carbohydrate units and one or a few amino acids. Although the fractionation of such glycopeptides was initially found difficult because of the multitude of structures present, the problem was to a great extent overcome by developing a general fractionation method (Krusius and Finne, 1977; Finne and Krusius, 1982), which is now widely used for the fractionation of protein-bound glycans from many other sources as well.

Most of the structural studies on nervous tissue glycoproteins have been carried out with material derived from rat brain, but many features characterized have been found to apply to other animal species as well. With additional data on adrenal medulla and neural cell lines, as well as comparative studies on other tissue and cell types, it is now possible to delineate the main characteristics of the carbohydrate units of nervous tissue glycoproteins. Several previously unknown carbohydrate structures of glycoproteins have been found, and some of them are highly enriched in the nervous system. By studying brain samples at different phases of development, as well as *in vitro* neuronal differentiation models, some glycans have also been found to be developmentally regulated. The identification of known cell adhesion molecules as the carriers of some of these glycans has given support to the idea that carbohydrate units of glycoproteins participate in cellular interactions of the nervous system. In addition, the carbohydrate units have been shown to play an important role in other types of biological interactions as well, such as the pathogenesis of bacterial meningitis.

GENERAL PROPERTIES OF BRAIN GLYCOPROTEIN CARBOHYDRATES

Main Classes of Glycans

The N- *and* O-*Linked Glycans.*   Most of the glycoprotein carbohydrate in brain, as in many other tissues, is bound by an N-glycosidic linkage between N-acetylglucosamine and asparagine (Margolis and Margolis, 1979). About 10% of the total protein-bound carbohydrate is in the form of O-glycosidic oligosaccharides, but since they are significantly smaller than the N-linked glycan, they account in number for about 30% of the oligosaccharide units. Most of the O-linked glycans are linked to the proteins by the common linkage between N-acetylgalactosamine and serine or threonine, but a smaller fraction is involved in a novel type of linkage involving mannose (see below). Although the brain is known to contain a high concentration of gangliosides, the amount of total protein-bound carbohydrate in the brain is somewhat higher than that of glycolipid-bound carbohydrate (Brunngraber, 1972).

Group Fractionation of the Glycans.   The protein-bound glycans of tissues can be converted almost quantitatively to glycopeptides containing the intact glycans and one or a few amino acid residues by the extensive proteolytic treatment of tissue samples. The fractionation of the glycopeptides is very difficult with classical methods based on size or charge separation, owing to the large number of different carbohydrate structures present and to the additional heterogeneity of the variable peptide portions (Brunngraber, 1972). This difficulty has been partially overcome by the development of a fractionation procedure that separates the glycans into well-defined groups according to their core structures (see Finne et al., 1980a). The method is based on the separation of the N-glycosidic glycopeptides by affinity chromatography with concanavalin A into three groups: multi-antennary, bi-antennary, and high-mannose glycopeptides (Krusius and Finne, 1977). In addition, after the affinity chromatography step, the O-glycosidically linked glycans can be separated as reduced oligosaccharides from the N-linked glycans, because of their generally smaller size (Finne, 1975; Finne and Krusius, 1982). Although originally developed for the fractionation of glycopeptides from brain and other tissues, the method has become a widely used routine procedure in the study of glycopeptides of many different types of samples (see Finne and Krusius, 1982). The structures of the core portions of the main glycopeptide types are shown in Figure 1.

The relative proportions of the four main types of glycans isolated by the fractionation procedure from different sources are shown in Figure 2. In the brain, as in most tissues, the main fraction is represented by the tri- and tetra-atennary N-linked glycans. This is in contrast to the previously held view that the bi-antennary glycans were representative of this class of glycans, which was apparently due to the fact that most glycans structurally characterized were derived from the soluble glycoproteins of plasma. It is clear that the proportions of the glycans in this source are different from those in the tissues (Figure 2) (Finne and Krusius, 1979). In all subcellular fractions studied, the ratios of the main glycan classes are on the whole rather similar, although there is a relative enrichment of the O-glycosidic oligosaccharide units and the bi-antennary N-linked glycans in the soluble fractions (Figure 2).

Terminal Sugar Sequences of the N-Linked Glycans

Structural Characterization by Methylation Analysis. The usually low amounts of material available in the study of glycans isolated from samples of biological origin have made it necessary to apply improved techniques for the characterization of the structural properties of nervous tissue glycopeptides. By applying potassium butoxide as a novel reagent for the methylation of carbohydrates (Finne et al., 1980b) the procedure has been made simpler to perform and more reliable, and the amount of interfering impurities has been reduced. In addition, the use of multiple ion detection in the gas-liquid chromatography–mass spectromeric analysis of the methylated derivatives of neutral and amino sugars (Krusius and Finne, 1977), reduced amino sugars (Finne and Rauvala, 1977), sialic acids (Finne et al., 1977a), and di- and oligosaccharides (Finne et al., 1977c; Krusius and Finne, 1978; Mononen et al., 1978) has increased the sensitivity and specificity of detection considerably. The combination of this technique with the specific degradation of the glycans with glycosidases (Finne et al., 1977a; Krusius and Finne, 1978), chromium trioxide oxidation (Finne, 1975), Smith degradation (Krusius and Finne, 1981), partial acid degradation (Finne et al., 1977c), deamination (Krusius and Finne, 1978), and uronic acid degradation (Krusius and Finne, 1978) has made it

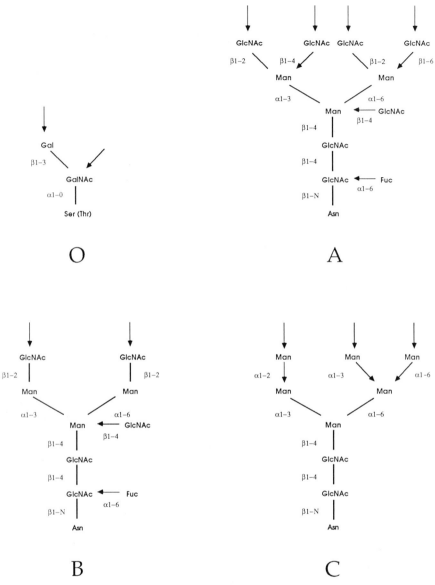

Figure 1. *Structure of the core regions of the main types of nervous tissue glycoprotein glycans.* The structures are based in part on the assumption of structural similarity to glycans from other sources and are supported by data from structural analyses (see text). The main positions of variable or incomplete glycosylation are indicated by *arrows*. O: O-glycosidic glycans. A: Tri- and tetra-antennary N-glycosidic glycans. B: Bi-antennary N-glycosidic glycans. C: High-mannose N-glycosidic glycans. Notice that the bisecting GlcNAc residue affects the interaction of the bi-antennary glycans with concanavalin A (see Finne and Krusius, 1982).

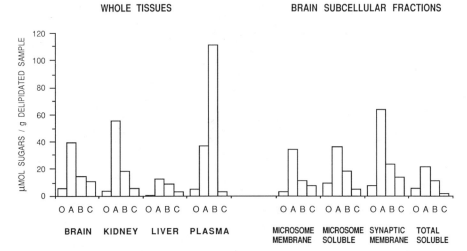

Figure 2. *Amounts of the main glycan fractions in different tissues and brain subcellular fractions in the rat.* The microsomal samples are represented by the light microsomal subfraction. The data are collected from Krusius and Finne (1977), Krusius et al. (1978a), and Finne and Krusius (1979).

possible to determine several of the terminal sugar sequences present in the glycans (Figure 3).

Structural Properties. Most of the terminal sequences of the N-linked glycans are derivatives of the disaccharide N-acetyllactosamine, Gal(β1–4)GlcNAc. However, about one-fifth of the disaccharides released from rat brain glycopeptides by acid degradation have been identified as Gal(β1–3)GlcNAc (Krusius and Finne, 1978), suggesting that this disaccharide also occurs in N-linked glycans, as found later to be the case for some glycoproteins of other sources (Takasaki and Kobata, 1986). The possible sugar substituents of this disaccharide in brain glycoproteins are not known. The N-acetyllactosamine branches occur mainly in their sialylated forms, but may also occur unsubstituted or even in a form lacking the galactose residue (Figure 3). In addition, sulfate may occur bound to galactose or N-acetylglucosamine residues (Margolis and Margolis, 1979).

In rat brain, two additional structural units were described for the first time as constituents of N-linked glycans. One of them consists of fucose residues bound by an α1–3 linkage to N-acetylglucosamine (Krusius and Finne, 1978). The terminals containing this fucose substituent seem to be mainly in a sialic acid–containing form (Figure 3). Another structure discovered in brain glycoproteins consists of N-acetylneuraminic acid residues bound to each other by an α2–8 linkage (Finne et al., 1977a). This linkage was thought before to occur only in gangliosides, but in developing brain its concentration is in fact even higher in glycoproteins than in glycolipids (Finne et al., 1977b; Finne, 1982). The sugar sequence was originally discovered in glycopeptide fractions

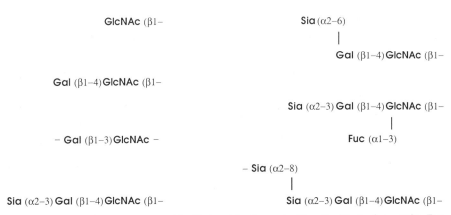

Figure 3. *Terminal sugar sequences of the N-glycosidic oligosaccharide units of brain glycoproteins.* For references, see Finne et al. (1980a).

containing relatively normal amounts of sialic acid, but later it was found as a constituent of the polysialic acid–containing glycopeptides (see below).

As suggested by recent immunological data, a sulfate-containing glucuronic acid residue also occurs in the brain, and constitutes the HNK-1 or L2 antigen determinant present in a subfraction of glycoproteins involved in cell adhesion (Keilhauer et al., 1985). Its structure in glycoproteins has not been determined, but from the structure of glycolipids containing the same antigen determinant (Chou et al., 1986), the presence of the SO_4^--3GlcUA$(\beta 1$–3)Gal$(\beta 1$–4)GlcNAc $(\beta 1$–) sequence or part of it can also be predicted in glycoproteins.

Similar termini are present in both the multi-antennary and the bi-antennary N-linked glycans, but their relative proportions vary. The fucose-containing sequence and the α2–8-linked sialic acid residues are clearly more enriched in fraction A. Also, the α2–3-bound sialic acid residues, as compared with the α2–6-bound residues, are enriched in the multi-antennary fraction, which according to recent observations may be explained by the substrate specificities of the corresponding sialyl transferases (Joziasse et al., 1987). The α2–3-linked residues, which are especially abundant in brain tissue, represent in fact the most common sialyl linkage in tissue glycoproteins. Only in the bi-antennary glycans of plasma glycoproteins does the α2–6 linkage predominate (Finne and Krusius, 1979). Since most structurally characterized glycoproteins originate from plasma and are enriched in bi-antennary glycans (Figure 2), this sialyl linkage was erroneously regarded before as most typical for glycoproteins in general.

All available evidence suggests that the sugar termini of the high-mannose glycans are similar to those described for other sources (Krusius and Finne, 1977). In addition to terminal mannose residues, phosphate groups may also be present (see Margolis and Margolis, 1979), as well as small amounts of terminal N-acetylglucosamine residues, suggesting the presence of some hybrid-type glycans as well.

Tissue-Dependent Differences. Comparison of the sugar termini of N-linked glycans in different tissues of the rat reveals differences in their estimated relative proportions (Figure 4) (Finne et al., 1977b; Krusius and Finne, 1977, 1978). Of the tissues studied, the brain seems to contain glycans with the most structural variation. The terminal sequences are mainly sialylated with α2–3-linked N-acetylneuraminic acid residues. In some animal species, N-glycolyl-neuraminic acid may replace N-acetylneuraminic acid in varying proportions, as exemplified by the glycoproteins of bovine adrenal chromaffin granules (Kiang et al., 1982; Margolis et al., 1984). As judged from the methylation analysis data, the α1–3-linked fucose residues bound to N-acetylglucosamine

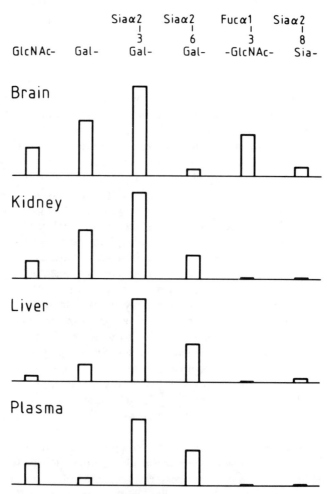

Figure 4. *Relative proportions of some terminal sugar sequences in the tri- and tetra-antennary glycan fraction (fraction A) in different tissues of the adult rat.* The values are recalculated from the published data discussed in the text.

are 20–30 times as abundant in the brain as in the kidney and liver of the rat (Krusius and Finne, 1978). Although this sequence (see Montreuil, 1980; Vliegenthart, 1983) and its sialylated form (Lamblin et al., 1984) have later been found in other sources as well, this sugar substitution seems to remain one of the characteristic features of brain glycoproteins. The same applies to the α2–8-bound sialic acid residues, which, in addition to constituting a characteristic feature of the normal-size glycopeptides (Figure 4), also occur as extended polymers in the polysialic acid chains (see below).

The characteristic patterns of sugar termini in the N-linked glycans, as well as many features of the O-linked glycans described below, are expressed in many subcellular fractions, and in both membrane-bound and soluble glycoproteins (Finne et al., 1977b; Krusius and Finne, 1977; Krusius et al., 1978a). Although some differences in the relative proportions of the sugar substituents exist, it appears that the relatively consistent glycosylation patterns of each tissue type are due to basic differences in the glycosylation machineries that may affect a number of different glycoproteins in a similar way. Thus, predictions based on the analysis of whole tissue protein-bound carbohydrates (Krusius and Finne, 1977), such as the presence or absence of "bisecting" N-acetylglucosamine residues, may be seen at the level of single glycoproteins isolated from different tissues (Yamashita et al., 1983). On the other hand, many of the structures identified as characteristic of the brain in the rat have also been found in similar amounts in the brains of other animal species. However, it is likely that differences in the brains of different species will be found as well.

Developmental Changes.   The relative proportions of total monosaccharides in brain glycoproteins, representing mainly those of N-linked glycans, stay relatively constant during postnatal brain development (Krusius et al., 1974). N-acetylgalactosamine, an integral component of the O-linked chains, shows a slightly different developmental pattern, as does sialic acid, owing to the presence of the α2–8-linked sialic acid units. Also, the proportions of the main types of glycans display relatively small changes, the main change being the slight increase in the multi-antennary fraction during development (T. Krusius and J. Finne, unpublished data). The relatively constant patterns of carbohydrates during brain development should be contrasted with the major change in the amount of polysialic acid that occurs during brain development (see below).

O-Linked Oligosaccharides

Fractionation.   The O-glycosidically linked carbohydrate units are nearly quantitatively liberated as reduced oligosaccharides by the alkaline borohydride treatment of glycoproteins or glycopeptides. During this treatment the N-acetylgalactosamine residues originally involved in the carbohydrate–peptide linkage are converted to their reduced form, N-acetylgalactosaminitol.

A portion of the isolated oligosaccharides contain not N-acetylgalactosaminitol but mannitol, and represent a novel class of O-glycosidic oligosaccharides. Their properties are described below in the section dealing with brain chondroitin sulfate proteoglycan.

The loss of the heterogeneic peptide portions and the relatively small size of the oligosaccharides have made the isolation and fractionation of the O-linked glycans simpler than that of the N-linked glycans. The oligosaccharides are conveniently fractionated on the basis of their sialic acid content by ion-exchange chromatography using a volatile buffer (Finne, 1975; Finne and Krusius, 1982), a procedure adapted for the fractionation of the O-glycosidic oligosaccharides of many other sources as well (Saito et al., 1981; Hull et al., 1984). Final purification of the individual oligosaccharides is achieved by conventional methods such as thin-layer chromatography and high-pressure liquid chromatography.

Structure.   The structures of the main N-acetylgalactosaminitol-containing O-glycosidic oligosaccharides of rat brain and bovine adrenal medulla are shown in Figure 5. The characterization of the structures has been performed with the aid of combined gas–liquid chromatography and mass spectrometry in the analysis of the core disaccharide units (Finne et al., 1977c) and the methylated monosaccharide derivates (Finne and Rauvala, 1977). Most oligosaccharides are derivatives of the disaccharide Gal(β1–3)GalNAcol, which occurs mainly in its sialylated forms and contains one or two sialic acid units (Finne, 1975). A small amount of an oligosaccharide containing three sialic acid units is also present (J. Finne and T. Krusius, unpublished data). In bovine adrenal medulla, N-glycolylneuraminic acid is present in addition to N-acetylneuraminic acid in the oligosaccharides, thus adding to the number of individual oligosaccharides (Kiang et al., 1982).

The ratios of the free disaccharide and the forms containing one and two sialic acid units in glycoproteins of different tissues are shown in Figure 6. Each tissue seems to contain its own pattern of oligosaccharides, whereas the ratio of the oligosaccharides in the brain of the three different animal species is very similar (Finne and Krusius, 1976). This suggests that the set of oligosaccharides

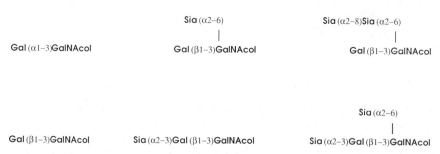

Figure 5. *Structure of O-glycosidic oligosaccharides isolated from nervous tissue glycoproteins.* For references, see Finne et al. (1980a) and Kiang et al. (1982).

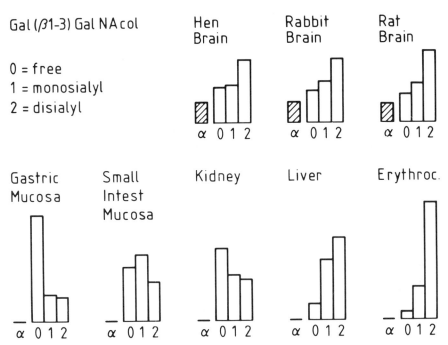

Figure 6. *Relative proportions of O-glycosidic oligosaccharides obtained from whole tissue glycoproteins of the rat. 0, 1,* and *2 refer to the free, monosialyl, and disialyl forms of Gal(β1–3)GalNAcol; α refers to Gal(α1–3)GalNAcol (Finne and Krusius, 1976).*

present in the glycoproteins is not randomly controlled but rather reflects the physiological properties of the tissues.

α-Galactose-Containing Disaccharide. The disaccharide Gal(α1–3)GalNAcol has been identified as a novel glycoprotein oligosaccharide in brain glycoproteins (Finne, 1975; Finne et al., 1977c). The α-galactose moiety seems to prevent substitution by sialic acid units, since this disaccharide, in contrast to its β-linked isomer, occurs only in its nonsialylated form. The presence of α-linked galactose, a relatively rare form of galactose in glycoproteins, was confirmed by chromium trioxide oxidation (Finne, 1975) and affinity chromatography using an α-galactose-specific lectin (Finne, 1977). A protein conjugate containing the chemically synthesized disaccharide (H. Paulsen, University of Hamburg) has recently been shown to elicit the production of specific antibodies to the α-galactose-containing disaccharide (J. Finne and M. Nikulin, unpublished data).

Analysis of glycopeptides from several tissues of the rat revealed the presence of the α-galactose-containing disaccharide only in the brain (Finne and Krusius, 1976), but small amounts were later found in the muscle and the adrenal medulla (Kiang et al., 1982). The disaccharide is present in relatively similar amounts in the brains of rats, rabbits, and hens (Figure 6). It is also present, although in a slightly lower ratio, in the frog and fish brain (J. Finne, unpublished data). The disaccharide has so far been reported only from one cultured cell line, a human teratocarcinoma line (Leppänen et al., 1986).

In the brain the α-galactose-containing disaccharide and the differentially sialylated forms of the β-linked form are found in all subcellular fractions and in relatively similar ratios, although the proportion of the α-linked isomer is low in the soluble fraction, suggesting a primarily membranous location for its carrier glycoproteins (Krusius et al., 1978a). The ratios of the O-linked oligosaccharides remain relatively constant during postnatal brain development.

POLYSIALIC ACID CHAINS OF THE NEURAL CELL ADHESION MOLECULE

Structural Properties

About one-tenth of the total protein-bound sialic acid in developing rat brain has been found to occur in a novel type of glycans containing polymers of sialic acid (Finne, 1982). For a long time the occurrence of these glycans in the brain was overlooked, since they were lost, because of their polyanionic nature, in the glycosaminoglycan fraction during the purification of glycopeptides. Structural studies have indicated that polysialic acid units are composed of polymers of $\alpha2$–8-linked N-acetylneuraminic acid units. This structure was concluded on the basis of gas–liquid chromatography and mass spectrometry of the methylation analysis products of native and neuraminidase-degraded glycans (Finne, 1982; Finne et al., 1983a).

The structure of the polysialic acid chains, a polymer of $\alpha2$–8-linked N-acetylneuraminic acid units, is the same as that of the capsular polysaccharides of group B meningococci and Escherichia coli K1 (Jennings, 1983). In accordance with this observation, antibodies to group B meningococci were found to specifically recognize the polysialic acid units of brain (Finne et al., 1983b). In addition, an endosialidase associated with a bacteriophage specific for the K1 capsule specifically cleaves the polysialic acid units (Finne and Mäkelä, 1985). The specificity of these two prokaryote-derived probes for brain polysialic acid was confirmed by Vimr et al. (1984), who also reported the use of a polysialic acid synthesizing system of E. coli in the elongation of brain polysialic acid chains.

The polysialic acid chains occur as polymers of variable length, the longest containing at least 12 sialic acid units (Finne and Mäkelä, 1985). Probably much longer chains also exist, judging from immunological data and from preliminary results from the NMR analysis of the glycans (H. van Halbeek and J. Finne, unpublished data). The polysialic acid unit seems to be bound by an $\alpha2$–3 linkage to a galactose residue in N-linked glycans of the tetra- and triantennary type (Figure 7) (Finne, 1982).

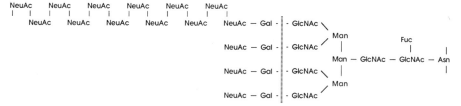

Figure 7. *Structural features of the polysialic acid–containing glycans of N-CAM.* The polysialic acid is represented by one chain of α2–8-linked N-acetylneuraminic acid units, although the presence of some additional short oligosialosyl units cannot be ruled out (Finne and Mäkelä, 1985). Sugar and methylation analysis data are in accordance with a tri- and tetra-antennary core glycan structure, but other substituents may also occur (Finne, 1982). The presence of a core GlcNAc–GlcNAc sequence is also supported by the susceptibility of the glycan to cleavage by peptide N-glycosidase F (McCoy and Troy, 1987).

Occurrence in N-CAM

A major glycoprotein containing polysialylated glycans has been identified as the neural cell adhesion molecule N-CAM, as shown by methylation analysis of the purified glycans (Finne et al., 1983a). The presence of polysialic acid in N-CAM is in accordance with the reported slow release of sialic acid from N-CAM by mild acid or neuraminidase (Rothbard et al., 1982), but because of several different factors that may influence the release, the form of the sialic acid units cannot be concluded on the basis of such experiments. The polysialic acid has been located in the middle domain of the extracellular part of N-CAM and occurs in one or more of its three asparagine-linked glycans (Edelman, 1985).

Comparisons of the amounts of polysialic acid found in whole brain with the amounts present in N-CAM (Finne et al., 1983a), as well as experiments with crossed immunoelectrophoresis (Lyles et al., 1984), suggest that N-CAM is the main carrier of polysialic acid in the brain. Minor components containing polysialic acid have been observed in immunoelectrophoresis and immuno-blotting (Finne et al., 1987), but the possibility that they represent degradation products of N-CAM cannot yet be excluded.

As judged from the mobility of N-CAM in gel electrophoresis as well as immunological reactivity, polysialic acid chains are present in a wide variety of species down to elasmobranchs (Edelman, 1985).

Developmental Regulation and Tissue Distribution

As discussed above, only moderate quantitative differences are observed in the relative amounts of different monosaccharide components or glycan types of glycoproteins during the postnatal development of rat brain. In contrast, there is a dramatic decrease in polysialylated glycans during postnatal development (Finne, 1982). A similar developmental decrease also occurs in mouse and human brain (Finne et al., 1983a,b) and, as suggested by more indirect evidence, in other animal species as well (Edelman, 1985). The change of the polysialylated form of N-CAM to the adult form containing significantly less sialic acid is correlated with the development of nervous tissue, and occurs at different times in different parts of the nervous system (Rougon et al., 1982; Edelman, 1985). An interesting observation is that in the neurologically defective *staggerer* genetic mutant of mice, the conversion to the adult form is delayed in the cerebellum (Edelman, 1985).

By the analysis of polysialic acid in the form of glycopeptides isolated from whole tissue, the polymer was initially thought to be specific to developing brain or neural cells in general (Finne, 1982; Margolis and Margolis, 1983). By using the more sensitive immunological approach, the enrichment of polysialic acid in developing brain was confirmed, but a small amount of polysialic acid was also found in adult brain, as well as in some developing extraneural tissues (Figure 8) (Finne et al., 1987). In the muscle the polysialic acid–containing component has an electrophoretic mobility similar to that of N-CAM, and probably corresponds to the sialic acid–rich form of N-CAM reported to occur

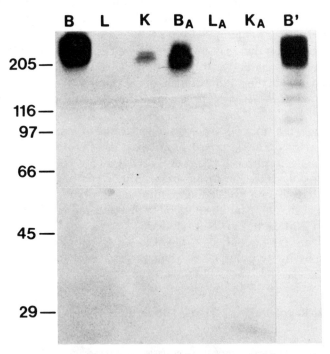

Figure 8. *Polysialic acid–containing components in newborn and adult mouse tissues.* Tissue homogenates of newborn mouse brain (B), liver (L), and kidney (K) and the corresponding tissue of adult mouse (B_A, L_A, and K_A) were subjected to SDS–gel electrophoresis, and the polysialic acid–containing components were revealed by immunoblotting, using a monoclonal antibody to polysialic acid (Finne et al., 1987). B', newborn mouse brain components detected with antibody to N-CAM.

in skeletal muscle during development (Covault and Sanes, 1985; Moore and Walsh, 1985; Rieger et al., 1985).

In the kidney the polysialic acid also corresponds by mobility to a component reacting with poly- and monoclonal antibodies to N-CAM (Finne et al., 1987). During postnatal developmental it is enriched in the medulla and in the part of the kidney derived from the ureteric bud, and occurs specifically at the basal surface of the developing collecting ducts, that is, at the surface facing the surrounding mesenchyme-derived tissue (Roth et al., 1987). In cortical regions polysialic acid is present transiently in the cells developing to proximal tubular cells. The polysialic acid disappears from the kidney concomitantly with tissue maturation. The specific cellular location and developmental regulation suggest that polysialic acid is involved in the development of some extraneural tissues, such as kidney, in addition to its previously suggested role in neural development.

In salmonid fishes, polymers of sialic acid are present, interestingly, in an egg sialoglycoprotein (Inoue and Iwasaki, 1980). However, they differ from the polysialic acid chains discussed in two respects: the sialic acid is in the form of N-glycolylneuraminic acid, and the chains occur in O-glycosidically linked glycans. The physiological function of the glycans and their possible relationship to brain polysialic acid are not known.

Mechanism of Molecular Interactions

Minimum Chain Length.   The digestion of polysialic acid with the bacteriophage PK1A endosialidase results in the generation of a fragment mixture containing oligomers three to seven sialic acid residues long. The failure of further degradation is explained by the unusual substrate specificity of the endosialidase (Finne and Mäkelä, 1985). The minimum requirement for cleavage is the presence of eight sialic acid residues, whereas oligomers containing seven or fewer residues are not cleaved during prolonged incubation at "physiological" bacteriophage concentrations (Figure 9). Experiments with individual oligomers of different lengths modified in their reducing or nonreducing ends suggest that the eight sialyl residues are cleaved so that three residues are left in the nonreducing end and five in the reducing end of the molecule (Figure 9). Another endosialidase recently characterized, PK1F, differs from PK1A in requiring only five sialic acid residues for cleavage, and in displaying differential requirements for the reducing end–modified molecule (Hallenbeck et al., 1987).

A long segment of polysialic acid is also required for the binding to anti-polysialic acid antibodies (Figure 9). A minimum of about 10 residues is required for the binding of labeled sialyl oligomers to horse polyclonal IgM antibodies, and the binding is increased by increasing chain length to far above 10 residues (Finne and Mäkelä, 1985). A similar conclusion was drawn by Jennings et al. (1985) for the same antibodies, using binding inhibition with unlabeled oligomers. Mouse monoclonal IgG antibodies similarly require a long fragment, about 8 sialic acid residues, for binding (Figure 9) (Finne et al.,

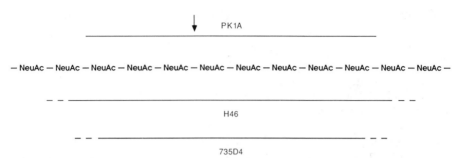

Figure 9. *Requirement of a long oligosaccharide segment in some molecular interactions of polysialic acid.* A minimum of 8 sialic acid residues is needed for cleavage by the PK1A endosialidase (site of cleavage is indicated by the *arrow*) (Finne and Mäkelä, 1985). Approximately 8–10 residues are required for binding to mouse monoclonal IgG antibody (735D4) and horse IgM antibody (H46) (Finne and Mäkelä, 1985; Finne et al., 1987).

1987). Preliminary results with Fab' fragments of the antibodies indicate binding similar to that of the whole antibody, excluding the possibility that the requirement for a long polysialic acid chain is due to the simultaneous binding of both antibody-binding sites to the same oligosaccharide chain (J. Häyrinen, D. Bitter-Suermann, and J. Finne, unpublished data). Not all antibodies, however, are similar to these two antibodies. A human myeloma IgM, although apparently specific for α2–8-linked polysialic acid, also reacts strongly with certain polynucleotides (Kabat et al., 1986). A mouse monoclonal IgM, on the other hand, seems to require a shorter segment of polysialic acid for binding, as indicated by its already weak reactivity with a ganglioside containing three sialic acid residues (Rougon et al., 1986).

To the list of molecular interactions dependent on long polysialic acid chains should be added an *O*-acetylating enzyme recently characterized in *E. coli* (Higa and Varki, 1987). This enzyme reportedly transfers acetyl groups to the hydroxyl at C-9 in polysialic acid chains of a minimum length of approximately 15 sialic acid units.

The requirement of a long oligosaccharide for cleavage by endosialidase is unusual for glycosidases, which often recognize a much shorter oligosaccharide segment of the substrate. Even more striking is the need for a long oligosaccharide for binding to anti-polysialic acid antibodies. It is generally believed that the antibody binding sites can maximally adapt oligosaccharides 6–7 sugar units long (Sharon et al., 1982). For such antibodies, submaximal binding is already reached with fragments containing 5–6 sugar residues, which is drastically different from the properties of the anti-polysialic acid antibodies, where binding is still increased with fragments more than 10 residues long. A reasonable explanation for the discrepancy could be that long polysialic acid chains form a conformational epitope, which is recognized by the antibodies and possibly also other molecules interacting with the polysaccharide. However, the conformation of the polysialic acid chains and their role in different molecular interactions have so far not been characterized.

Lactonization.   At acid pH, internal esters are spontaneously formed in oligo-
and polysaccharides containing α2–8-linked sialic acid residues (Lifely et al.,
1981). Esterification starts below pH 6 and increases progressively as the pH is
lowered. At pH 2.9 about 90% of the carboxyl groups are involved in lactones.
The ease of lactonization seems to be a characteristic feature of structures
containing sialic acid residues bound to one another, since oligosaccharide
structures containing single sialic acid residues bound to other sugars require
more drastic chemical treatments for the formation of internal esters. It has
been suggested that internal esters are relatively abundant in brain gangliosides
(Acquotti et al., 1987), but the degree of lactonization of polysialic acid *in vivo* is
not known.

The formation of lactones affects the structure of polysialic acid in several
ways. New bonds are formed by the esterification of the carboxyl groups to the
hydroxyl groups at C-9, the negative charges of the carboxyl groups involved
are lost, and the conformation of the polysaccharide is consequently also
altered. These changes critically affect the interaction of the molecule with its
specific antibodies, as shown by the loss of the immunoreactivity of molecules
with as little as 9% internal esters (Lifely et al., 1981), a finding correlated with
the requirement of long polysialic acid segments for antibody binding, as
discussed above.

Role in N-CAM

Inhibitory Modulation.   Model experiments *in vitro* suggest that the polysialic
acid chains of N-CAM may function as negative modulators of the adhesive
property of the molecule. By using liposomes containing purified N-CAM,
Sadoul et al. (1983) have shown that the molecule binds specifically to cells of
neural origin but not to fibroblasts. Binding is obtained with liposomes
containing the adult, less sialylated form of N-CAM, whereas liposomes
containing the polysialylated form show little binding. The binding of the latter
is, however, induced if the sialic acid residues are first removed with
neuraminidase. Similar results were reported by Hoffman and Edelman (1983)
in a study of the kinetics of the reaggregation of vesicles containing the
different forms of N-CAM.

Spinal ganglion neurons kept in cell culture in the presence of endosialidase
exhibit increased neuronal fasciculation, suggesting increased adhesiveness
between the extending neurites (Rutishauser et al., 1985). The injection of
endosialidase into the developing eye was also reported to induce major
disturbances of the cellular organization in the retina, but the mechanism of
these changes is not known.

The idea of the polysialic acid chains as negative modulators of N-CAM-
mediated adhesion is in accord with the gradual disappearance of the
polysialylated form of N-CAM concomitantly with development, not only in
neural tissue but also in extraneural tissue such as the kidney. In the latter the
specific location of the polysialic acid chains in the collecting ducts at the
border of cells derived from the two anlagen (Roth et al., 1987) is also in line

with such an inhibitory role. In this regard, the observation that growing axons extending from the retina toward the tectum contain the polysialylated form of N-CAM, while the perikarya and neurite shafts of the same cells in the retina contain a less sialylated form of the molecule, is highly interesting (Rutishauser, 1984).

Molecular Mechanism. In view of the suggested modulatory role of the polysialic acid chains, a central question is why polysialic acid chains are needed for inhibition. According to the "charge perturbation" hypothesis of Edelman (1983) the polysialic acid units provide the molecule with negative charges that cause repulsion of the molecules. Microheterogeneity of the polysialic acid chains would thus provide the molecule with a diversity of binding strengths necessary during development. The "critical size" model, on the other hand, suggests that the important factor is the size of the polysialic acid chain that changes reactivity at a certain length, possibly as a result of a conformational change (Finne, 1985). The specific conformation of the polysialic acid would then be needed in interactions of the chains with other molecules or other parts of the same N-CAM molecule. The generation of a series of affinities could be generated by several mechanisms, such as the number of molecules containing polysialic acid, the total number and type of N-CAM molecules expressed, their state of aggregation, and modulation by other molecules like heparan sulfate (Cole et al., 1986).

Although much speaks for the inhibitory role of polysialic acid in N-CAM-mediated cell adhesion, the question still remains whether this is the only reason for the occurrence of this unusual type of carbohydrate. If the only function of polysialic acid were to inhibit adhesion, it would seem more economical for the cell to achieve this in ways other than by synthesizing this rather complex structure, for which a specific sialyltransferase so far not known to act on any other glycoprotein is needed as well. Also, in view of the suggested general role of sialic acids as inhibitors of many molecular interactions (Schauer, 1985), it is striking that these effects are as a rule caused by single sialic acid residues bound to the sugar termini of the molecules in question. Other roles for polysialic acid could be to serve as a specific recognition ligand or to undergo local modulation by, for example, an endosialidase. Such endogenous binding factors or enzymes have, however, so far not been described.

POLY-*N*-ACETYLLACTOSAMINE GLYCANS OF NEURAL CELL LINES

Linear and Branched Poly-*N*-Acetyllactosamine Chains

Before the discovery of the polysialic acid–containing carbohydrate units, the only class of large protein-bound glycans known to exist was the poly-*N*-acetyllactosamine glycans. These were originally discovered in the human erythrocyte membrane (Finne et al., 1978) and mouse embryonal cells (Muramatsu et al., 1978), which are rich sources of these carbohydrate units, but are now known to occur in varying amounts in a number of different cell types (see Fukuda, 1985). The chains are built of the repeating disaccharide N-acetyllactosamine, Gal(β1–4)GlcNAc, and occur in either a linear or branched form (Figure 10). These two forms are distinguished by antibodies from patients with the cold agglutinin syndrome, the branched form having I-antigenic reactivity and the linear form i-antigenic reactivity (Feizi, 1985). In most reported cases, the chains are bound to N-linked glycans of glycoproteins (Järnefelt et al., 1978; Krusius et al., 1978b), but O-linked forms may also exist. Glycolipids also contain poly-N-acetyllactosamine chains (Koscielak et al., 1976). Sialic acid, fucose, ABH blood-group determinants, and other sugars have been found in terminal positions of the glycans (see Finne et al., 1980c).

In contrast to the normal glycans of glycoproteins, the poly-N-acetyl-lactosamine chains undergo major changes during development. In the early mouse embryo, a major part of all protein-bound glycans is in the form of poly-N-acetyllactosamine glycans, but with development the proportion dramatically decreases. A similar decrease is seen in embryonal carcinoma cells after induced differentiation *in vitro* (Muramatsu et al., 1978). At the same time, the branched glycans present in the undifferentiated cells are partially replaced by unbranched chains. An opposite developmental pattern is seen in human erythrocytes, where the linear chains present during the fetal period are replaced by the branched glycans present in adult erythrocytes (Fukuda, 1985).

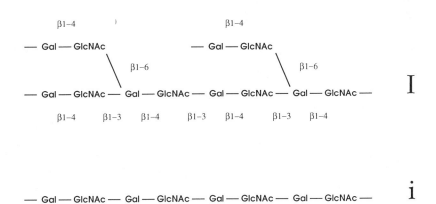

Figure 10. *Branched and linear forms of poly-N-acetyllactosamine glycans.* The chain length and degree of branching are variable, as well as the substitution by different sugar termini. The branched forms are reactive with anti-I, and the linear chains with anti-i, antibodies (see Feizi, 1985).

Occurrence in Neural Cell Lines

Part of the protein-bound glycans of mouse neuroblastoma lines N-18 and N1E-115 and rat PC12 pheochromocytoma cells are present as poly-N-acetyllactosamine glycans (Spillmann and Finne, 1987). In contrast to other cell lines studied, where these glycans are present mainly in the branched form, the poly-N-acetyllactosamine glycans of the cells of neural origin were found to be mainly in the linear form. Furthermore, the proportion of the linear glycans increased during neuronal differentiation induced by serum deprivation of the neuroblastoma cells or by nerve growth factor in PC12 cells.

The expression of the linear poly-N-acetyllactosamine glycans by the neural cell lines is correlated with the strong reactivity with anti-i antibodies specific to this form of the glycans (Figure 10) (Spillmann and Finne, 1987). In immunofluorescence microscopy the i-reactive glycans are detected all over the cells, including the cell bodies and neurites. In contrast to other cell lines, anti-I antibodies specific for the branched glycans show little binding to these neural cells. Interestingly, mouse teratocarcinoma cells induced to differentiate into cells expressing some neuronal properties (Kuff and Fewell, 1980) concomitantly also change from the i-negative to the i-positive form.

In contrast to the neural cell lines, few poly-N-acetyllactosamine chains are found in adult rat brain. On the other hand, poly-N-acetyllactosamine glycans accumulate in the brain in GM1-gangliosidosis, indicating that such chains are synthesized in brain tissue (Berra et al., 1986). Also, poly-N-acetyllactosamine glycans are present in sympathetic neurons (Margolis et al., 1986), and sulfated, mannose-linked poly-N-acetyllactosamine chains are present in a soluble brain proteoglycan (see below). It is possible that the higher amounts of the glycans in the neural cell lines reflect their normal presence on developing neural cells, and that the low amounts in the adult brain are due to a developmentally determined decrease.

Characterization by a Novel Cell Surface Labeling Method

In order to identify the cell surface glycoproteins carrying poly-N-acetyllactosamine chains, a labeling method specific for this class of glycans has been developed (Viitala and Finne, 1984; Spillmann and Finne, 1987). The method is based on the specificity of the enzyme endo-β-galactosidase for poly-N-acetyllactosamine chains. The enzyme only cleaves chains containing repeating N-acetyllactosamine units, and as a result of the enzyme action, the remaining parts of the digested glycans have N-acetylglucosamine as their terminal sugar residue (Figure 10). In a second step, the terminal N-acetylglucosamine residues are labeled by transfer of radioactively labeled galactose from UDP-galactose in a galactosyltransferase-catalyzed reaction. Since the amount of terminal N-acetylglucosamine in cells not treated with endo-β-galactosidase is usually very low, essentially all labeling detected is due to poly-N-acetyllactosamine chains. A further advantage of the labeling method is that it employs readily available commercial reagents.

As revealed by the labeling method, only a few glycoproteins of PC12 cells appear to be carriers of the poly-N-acetyllactosamine glycans (Figure 11) (Spillmann and Finne, 1987). During induced neuronal differentiation, the components are differentially modulated, as exemplified by the large decrease in a broad component in the molecular weight range of 40–60 kD, or the appearance of a 230-kD molecular weight component in the differentiated cells. The latter appears to be the nerve growth factor–inducible glycoprotein NILE, recently identified as closely related to or the same as the L1 adhesion glycoprotein (Bock et al., 1985) and the neuron–glia cell adhesion molecule Ng-CAM (Friedlander et al., 1986). The occurrence of poly-N-acetyllactosamine glycans in this cell adhesion molecule is of particular interest in view of the postulated regulatory role of polysialic acid in N-CAM-mediated cell adhesion and of poly-N-acetyllactosamine chains in fibronectin–gelatin interaction (Zhu and Laine, 1985).

MANNOSE-LINKED GLYCANS OF BRAIN CHONDROITIN SULFATE PROTEOGLYCAN

Structural Properties of the Proteoglycan

The chondroitin sulfate proteoglycan of brain is a soluble molecule with a molecular weight of about 300 kD (Kiang et al., 1981). It consists of 56% protein, 24% glycosaminoglycans, and 20% glycoprotein-type oligosaccharides. Of the glycosaminoglycans, more than 93% is chondroitin 4-sulfate. An average of four chondroitin sulfate chains are bound to a single core protein.

There are on average approximately 13 N-glycosidic, mainly tri- and tetra-antennary, complex oligosaccharides, 15 O-glycosidic N-acetylgalactosamine-linked oligosaccharides, and 40 O-glycosidic mannose-linked units in the proteoglycan (Table 1) (Finne et al., 1979; Krusius et al., 1986). The N-acetyl-galactosamine-linked oligosaccharides are mainly differentially sialylated derivatives of the disaccharide Gal(β1–3)GalNAc, common to brain and other tissues. In contrast, the mannose-linked oligosaccharides represent a novel type of oligosaccharides of vertebrate glycoproteins.

In alkaline borohydride–treated glycopeptides isolated from the proteoglycan, there is a decrease in mannose, which is compensated for by an equivalent amount of mannitol found as the proximal sugar of oligosaccharides released by this treatment (Krusius et al., 1986). The decrease in mannose is accompanied by a corresponding decrease in serine and threonine, which are converted to their β-elimination products. These findings strongly suggest that the mannitol-containing oligosaccharides are derived from glycans bound by mannosyl-O-serine/threonine linkages to the proteoglycan, although an indirect linkage via some other residue (such as phosphate or sulfate) cannot be ruled out.

Structures of the mannitol-containing oligosaccharides and their estimated numbers in the chondroitin sulfate proteoglycan are shown in Table 1. The

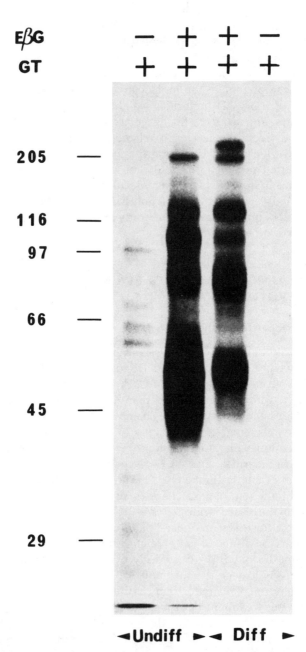

Figure 11. *Poly-N-acetyllactosamine-containing cell surface components of undifferentiated and differentiated PC12 pheochromocytoma cells.* The components were revealed by endo-β-galactosidase followed by labeling with galactosyltransferase and radioactively labeled UDP-galactose (Spillmann and Finne, 1987).

Table 1.   Carbohydrate Units of Rat Brain Chondroitin Sulfate
Proteoglycan and Their Estimated Average Number per Molecule of
Proteoglycan (300 kD)[a]

Component	Mol/Mol
Chondroitin sulfate (18 kD)	4
N-linked tri- and tetra-antennary chains	13
O-linked (Ser/Thr) oligosaccharides	
Gal(β1–3)GalNAcol	3
NeuAc$_m$/SO$_{4n}^-$ ..... Gal(β1–3)GalNAcol[b]	12
Manol	4
GlcNAc(β1–3)Manol	4
Gal(β1–4)[Fucα1–3]GlcNAc(β1–3)Manol	4
NeuAc(α2–3)Gal(β1–4)GlcNAc(β1–3)Manol	7
NeuAc$_x$/SO$_{4y}^-$ ..... Manol[b]	20
Keratan sulfate (3–10 kD) ..... Manol	1

[a]Based on previously published data (Finne et al., 1979; Krusius et al., 1986).
[b]Oligosaccharides containing sialic acid and/or sulfate.

structures of the major small oligosaccharides have been determined, whereas
the somewhat larger oligosaccharides containing varying amounts of sialic acid
and/or sulfate have not yet been characterized. In addition, the proteoglycan
contains an average of one mannose-linked keratan sulfate chain with a
molecular weight of 3–10 kD. Brain proteoglycan has also been found to carry
the sulfate- and glucuronic acid–containing HNK-1 epitope (Hoffman and
Edelman, 1987).

Biological Properties

Over half of the proteoglycan occurs in the soluble fraction of rat brain and
accounts for less than 1% of the protein in this fraction (Kiang et al., 1981).
After subcellular fractionation, most of the remainder is loosely associated with
the low-density microsomal membrane fraction. It is not known whether other
proteoglycans or glycoproteins in brain contain mannose-linked oligosaccharides similar to those of the chondroitin sulfate proteoglycan. However, it
seems that the oligosaccharides are clearly enriched in the proteoglycan and
that the subcellular distribution of the oligosaccharides follows that of the
proteoglycan (Finne et al., 1979). Mannitol-containing oligosaccharides are
present not only in rat brain but also in rabbit and chicken brain.

As revealed by immunoelectron microscopy, the proteoglycan is exclusively
extracellular in the newborn rat brain (Aquino et al., 1984). During early
postnatal development it gradually assumes an intracellular localization. It is
then found in the axoplasm and cytoplasm of certain neurons and in astrocytes.

Model experiments with microspheres containing covalently bound molecules indicate that brain chondroitin sulfate proteoglycan binds specifically to

the cell adhesion molecule cytotactin (Hoffman and Edelman, 1987). Further-more, similar experiments also suggest that the proteoglycan binds laminin, whereas cytotactin does not. On the other hand, cytotactin but not the proteoglycan binds specifically to fibronectin. Thus, macromolecular com-plexes may be formed by the specific interaction of these molecules with each other. The findings are also of special interest because of the suggested involvement of laminin in neurite outgrowth (see Roberts et al., 1986; Laitinen et al., 1987).

The role of the oligosaccharide units of the proteoglycan in its molecular interactions is not yet known. However, it is of interest that laminin is reported to bind to sulfate- and sialic acid–containing glycolipids (Roberts et al., 1986; Laitinen et al., 1987), and that both sulfate and sialic acid are present in the O- and N-linked glycans of the proteoglycan. On the other hand, antibodies to the L2/HNK-1 antigen have been found to inhibit neuron–astrocyte and astrocyte–astrocyte adhesion (Keilhauer et al., 1985), suggesting the in-volvement of carbohydrates in these interactions.

ROLES OF CARBOHYDRATES IN BACTERIAL MENINGITIS

Role of Bacterial Polysaccharide

Among the numerous bacterial polysaccharide structures that have been characterized, only very few show a resemblance to the carbohydrate units of vertebrate glycoconjugates. With the discovery of the polysialic acid–containing brain glycopeptides, however (Finne, 1982), a case of striking structural similarity was revealed. The polysialic acid structure of these glycans, a polymer of α2–8-linked N-acetylneuraminic acid residues, is exactly the same as that previously reported for the capsular polysaccharides of E. coli K1 and group B meningococci (Jennings, 1983). Both bacteria are important causes of human meningitis. Two more bacteria have also recently been found to have a similar polysaccharide haemolytica serotype A2, Moraxella nonliquefaciens and Pasteurella (Adlam et al., 1987). The clinical significance of the former is not known, whereas the latter is an important ovine pathogen and is responsible for the majority of outbreaks of pasteurellosis in young, pre-weaned lambs.

As suggested by its chemical structure, the capsular polysaccharides of E. coli K1 and group B meningococci also immunologically cross-react with brain polysialic acid (Finne et al., 1983b, 1987). In contrast to many other bacterial polysaccharides including α2–9-linked polysialic acid, the polysialic acid capsules of these bacteria are known to be very poor immunogens (Table 2). An explanation for the poor immunogenicity could be immunological tolerance owing to the presence of similar polysialic acid chains in the brain and other tissues (Finne et al., 1983b, 1987). Although it seems that the presence of a high concentration of sialic acid on the bacterial surface is itself an advantage for the bacteria in inhibiting the activation of the alternative pathway of complement

Table 2.   Structure and Immunogenicity of Some Bacterial Polysaccharides

Polysaccharide	Bacterium	Clinical Significance	Effective Vaccine
$(ManNAc\alpha1\text{-}P\text{-}6)_n$	*Neisseria meningitidis* group A	Human meningitis	+
$(NeuAc\alpha2\text{-}8)_n$	*Neisseria meningitidis* group B	Human meningitis	−
	E. coli K1	Human newborn meningitis	−
	Pasteurella haemolytica	Pasteurellosis in young lambs	−
$(NeuAc\alpha2\text{-}9)_n$	*Neisseria meningitidis* group C	Human meningitis	+
$(Gal\alpha1\text{-}4NeuAc\alpha2\text{-}6)_n$	*Neisseria meningitidis* group W_{135}	Human meningitis	+
$(Glc\alpha1\text{-}4NeuAc\alpha2\text{-}6)_n$	*Neisseria meningitidis* group Y	Human meningitis	+

Source: Jennings (1983); Adlam et al. (1987).

activation (Jennings, 1983), it is possible that in the case of the bacteria containing an $\alpha2$–8-linked polysialic acid capsule, an additional advantage is achieved by mimicking a structure present in tissue glycoproteins of the host. The question of immunological cross-reactivity and tolerance is of major practical importance in attempts to develop chemically modified vaccines against group B meningococci (Jennings, 1983), or in suggested immuno-therapy with the corresponding antibodies (Cross et al., 1983), since harmful effects might be produced as a result of the presence of cross-reacting tissue components.

An interesting question concerning meningitis caused by *E. coli* K1 is its age dependence. *E. coli* K1 is a major cause of meningitis and septicemia in the newborn infant, but almost never causes meningitis later (Robbins et al., 1974). In experimental meningitis of the rat, the production of meningitis is also dependent on the K1 polysialic acid capsule as well as the age of the rats (Glode et al., 1977). Meningitis is produced in the rats only during the first two postnatal weeks, which is also the time period of the presence of a high concentration of polysialic acid in the brain. The age correlation suggests that the production of meningitis by the bacteria containing polysialic acid on their surface might be linked to the presence of polysialic acid in the host. The possible role of polysialic acid in directing the bacteria to the meninges is supported by the fact that there is a much higher proportion of K1 polysialic acid–containing *E. coli* in isolates from meningitis (80%) as compared with septicemia (30–40%) or gut flora (15%) (Robbins et al., 1974; McCabe et al., 1978). Elucidation of the molecular mechanisms responsible for the possible specificity of the polysialic acid–containing bacteria for the meninges may not only reveal pathogenic mechanisms of meningitis, but might also give indications of the physiological roles of the endogenous tissue polysialic acid units.

Tissue Oligosaccharide Structures as Bacterial Binding Sites

A prerequisite for the establishment of bacterial infections is that the bacteria are able to bind to host tissues (Beachey, 1981). The ability of the bacteria to adhere to host cells may not only be decisive at the initial stage of the infection at epithelial surfaces, but might also become important at later stages in the tissue localization of the infection, such as in the establishment of meningitis. In most cases of the few known bacteria-binding adhesion specificities characterized at the molecular level, the adhesion receptors have been shown to be carbohydrates (see Korhonen and Finne, 1985). By studying the adhesion properties of E. coli strains isolated from newborn meningitis and septicemia, a novel adhesion specificity recognizing sialic acid has been identified (Parkkinen et al., 1983). This adhesion specificity is associated with a specific type of fimbriae (S fimbriae), and is found in much higher proportion in these strains than in strains isolated from other sources (Korhonen et al., 1984). Furthermore, because of phase variation, the bacteria express S fimbriae only in part of the bacterial cells in vitro, whereas 60–70% of originally nonfimbriated and all S-fimbriated bacterial cells express S fimbriae, as studied in the mouse peritonitis model (Nowicki et al., 1986). This suggests that the presence of S fimbriae is induced in the bacteria during systemic infection and is thus linked to the pathogenic mechanism.

Studies using free oligosaccharides isolated from milk and urine (Parkkinen and Finne, 1987) and human erythrocytes (Korhonen and Finne, 1985) as a model have indicated that the S-fimbrial adhesin best recognizes oligosaccharides containing the NeuAc(α2–3)Gal(β1–3)GalNAc sequence (Table 3), that is, structures present in the O-glycosidically linked carbohydrate units of

Table 3. Inhibitory Activity of Sialyloligosaccharides on the Binding of S-Fimbriated E. coli to Glycophorin A

Oligosaccharide	Minimum Inhibitory Concentration (mM)[a]
NeuAc(α2–3)Gal(β1–3)GalNAc	2
NeuAc(α2–3)Gal(β1–3)[NeuAc(α2–6)]GalNAc	2
NeuAc(α2–3)Gal(β1–4)GlcNAc	5
NeuAc(α2–3)Gal(β1–4)Glc	5
NeuAc(α2–3)Gal(β1–4)Glcol	5
NeuAc(α2–6)Gal(β1–4)GlcNAc	NI[b]
NeuAc(α2–6)Gal(β1–4)Glc	NI[b]
NeuAc(α2–8)NeuAc(α2–3)Gal(β1–4)Glc	NI[b]

[a]Concentration giving 50% inhibition of the binding of bacteria to microtiter plates coated with glycophorin A (Parkkinen et al., 1986).
[b]No inhibition at the highest concentration (10 mM) tested.

glycoproteins in the brain and other tissues (Figures 5 and 6) (Parkkinen et al., 1986). In accordance with this finding, erythrocytes desialylated with sialidase lose their activity, and regain it after resialylation with the specific $\alpha2$–3 sialyltransferase. By a novel application of the blot overlay method, the receptor glycoprotein on erythrocytes has been identified as glycophorin A, the major sialoglycoprotein rich in O-glycosidic carbohydrate units (Figure 12) (Parkkinen et al., 1986). The molecular nature of the adhesion receptor in meningeal tissue is so far not known. Regarding the binding specificity, it is of

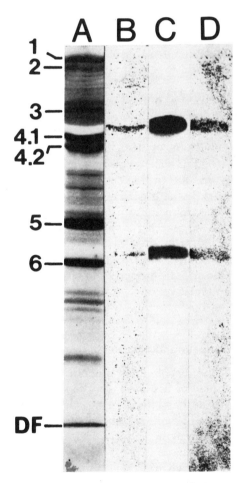

Figure 12. *Binding of radiolabeled S-fimbriated E. coli to erythrocyte membrane proteins and purified glycophorin A. Lane A:* Erythrocyte membrane protein blot stained by amido black. *Lane B:* Erythrocyte membrane protein blot incubated with radiolabeled bacteria. *Lane C:* Purified glycophorin A blot incubated with radioactive wheat germ agglutinin. *Lane D:* Purified glycophorin A blot incubated with radiolabeled bacteria (Parkkinen et al., 1986). The numbering of the bands is according to the common nomenclature of erythrocyte membrane proteins. DF, dye front.

interest that polysialic acid is a very poor inhibitor of the adhesin, which indicates that the possible association of tissue polysialic acid with the production of meningitis is not associated with the sialic acid–binding property of S fimbriae.

DISCUSSION AND CONCLUSIONS

Structure of the Carbohydrate Units of Nervous Tissue Glycoproteins

Focusing on the protein-bound glycans of whole tissue samples, cells, and subcellular fractions rather than single purified glycoproteins has proved to be very fruitful in the study of nervous tissue glycoproteins. Using this strategy, it has been possible to obtain a general picture of the properties of nervous tissue glycoprotein carbohydrates, and to identify structures that are characteristic of this tissue type.

The type and general properties of the protein-bound glycans of nervous tissue are for the most part similar to those present in many other tissues. However, several novel structures have been characterized for the first time in nervous tissue glycoproteins. Some of them, like the Fuc(α1–3) substituent in complex N-linked glycans and its sialic acid–containing sequence, have been found later in comparable amounts in glycoproteins of some other sources. In contrast, others, represented by the α-galactose-containing disaccharide and polysialic acid, have so far been found only in low amounts outside the nervous tissue. The mannose-linked oligosaccharide units, on the other hand, have so far been found only in the brain, but their occurrence in other tissues has not been thoroughly studied. In conclusion, the contribution of the carbohydrate units of glycoproteins to the specific properties of nervous tissue seems to be related to their absolute and relative concentrations as well as their spatial and temporal expression rather than strict tissue specificity.

Physiological Role of Protein-Bound Glycans in Nervous Tissue

The physiological role of the carbohydrate units of glycoproteins is for the most part still obscure. It may be assumed that the same general properties ascribed to the glycans, such as their effects on the solubility and conformation of glycoproteins and the role of mannose phosphate in lysosomal targeting, also operate in the nervous system. The significance of the glycans to the specific properties of nervous tissue is, however, not known.

One possibility is that the carbohydrate units function as specific recognition ligands in molecular interactions, as in the binding of different types of glycoproteins to their hepatic carbohydrate-specific receptors (see Ashwell and Harford, 1982). Although specific cellular interactions could in principle also be mediated through carbohydrate–carbohydrate interactions (Misevic et al., 1987), it is generally assumed that such interactions would require the presence of carbohydrate-binding proteins like lectins or glycosyltransferases. Lectin activities have indeed been found in nervous tissue (Joubert et al., 1985), but their role in cellular interactions remains to be established. Glycosyltransferase, on the other hand, has also been shown to be capable of supporting cell adhesion (Rauvala et al., 1983; Bayna et al., 1986), but it is not known whether this mechanism functions in nervous tissue *in vivo*. It is also of interest that antibodies to the L2/HNK-1 epitope have been reported to inhibit the cell adhesion of neural cells, suggesting that this carbohydrate structure could be directly involved in the interaction. With regard to the possible role of carbohydrates as specific recognition ligands, it is important to notice that even seemingly small differences in the concentration of the structure involved may, because of threshold effects, have a dramatic influence on cell adhesion (Weigel et al., 1979).

Besides functioning as specific recognition ligands, carbohydrates may also play another type of role by serving as modulators of molecular interactions. An example of this is the suggested role of polysialic acid in N-CAM. The adhesive property in this model is mediated by the polypeptide part of the molecule, but the strength of the interaction would be regulated by the carbohydrate. Other carbohydrates, such as heparan sulfate, also seem to be required (Cole et al., 1986). Another example of a suggested modulatory role of carbohydrates is the reported effect of poly-*N*-acetyllactosamine chains of fibronectin on gelatin binding (Zhu and Laine, 1985). This phenomenon is of special interest in view of the presence of poly-*N*-acetyllactosamine glycans in neural cells, and in particular in a known cell adhesion molecule. It is possible that the proposed modulatory role of carbohydrates in molecular interactions may turn out to represent an important part of the physiological roles of carbohydrates.

The Role of Carbohydrates in Bacterial Meningitis

Considerations of the biological roles of carbohydrates cannot be restricted to functions endogenous to the organism, since each living organism is a part of the external biological system it is living in. One important question is the role of carbohydrates in cellular interactions involving infectious processes. The structure of the carbohydrates present in host tissues, and their possible similarity to carbohydrates covering the bacterial surface, may have an important role in the pathogenesis of bacterial infections like meningitis, as indicated by the immunochemical mimicry of polysialic acid. Furthermore, additional molecular mechanisms may be involved in determining the tissue specificity of the infections.

Another important role of carbohydrates in the pathogenesis of infections is that of adhesion receptors. The sialic acid–containing glycans recognized by the adhesin of *E. coli* strains that cause meningitis is a good example of this. In addition, this adhesion system serves as an example of a biologically important interaction where the carbohydrate units serve as specific recognition ligands.

Finally, the fact that carbohydrates are involved in the pathogenesis of microbial infections may be regarded as additional evidence of the physiological importance of carbohydrates, since unless they were needed for the normal functioning of the organism, they should have been eliminated during evolution owing to the selection caused by pathogenic microbes.

ACKNOWLEDGMENTS

I wish to thank J. Järnefelt, T. Krusius, J. Kärkkäinen, S. Pelkonen, J. Parkkinen, H. Rauvala, and D. Spillmann, as well as D. Bitter-Suermann (University of Mainz), W. Dahr (University Hospital Cologne), C. Goridis (Centre d'Immunologie INSERM-CNRS de Marseille-Luminy), T. K. Korhonen (University of Helsinki), P. H. Mäkelä (University of Helsinki), R. K. and R. U. Margolis (New York University Medical Center), G. N. Rogers (University of California, Los Angeles), J. Roth (Biocenter of the University of Basel), and their colleagues for fruitful collaboration. Financial support has been received from the Sigrid Jusélius Foundation, Finland, the Swiss National Foundation, and the Academy of Finland.

REFERENCES

Acquotti, D., G. Fronza, L. Riboni, S. Sonnino, and G. Tettamanti (1987) Ganglioside lactones: ^{1}H-NMR determination of the inner ester position of G_{D1b}-ganglioside lactone naturally occurring in human brain or produced by chemical synthesis. *Glycoconjugate J.* **4**:119–127.

Adlam, C., J. M. Knights, A. Mugridge, J. M. Williams, and J. C. Lindon (1987) Production of colominic acid by *Pasteurella haemolytica* serotype A2 organisms. *FEBS Microbiol. Lett.* **42**:23–25.

Aquino, D. A., R. U. Margolis, and R. K. Margolis (1984) Immunocytochemical localization of a chondroitin sulfate proteoglycan of nervous tissue. II. Studies in developing brain. *J. Cell Biol.* **99**:1130–1139.

Ashwell, G., and J. Harford (1982) Carbohydrate-specific receptors of the liver. *Annu. Rev. Biochem.* **51**:531–554.

Bayna, E. M., R. B. Runyan, N. F. Scully, J. Reichner, L. C. Lopez, and B. D. Shur (1986) Cell surface galactosyltransferase is a recognition molecule during development. *Mol. Cell Biochem.* **72**:141–151.

Beachey, E. H. (1981) Bacterial adherence: Adhesin-receptor interactions mediating the attachment of bacteria to mucosal surfaces. *J. Infect. Dis.* **143**:325–345.

Berra, B., R. DeGasperi, S. Rapelli, S. Okada, S. C. Li, and Y. T. Li (1986) Presence of glycoproteins containing polylactosamine structure in brain and liver of G_{M1} gangliosidosis patients: Comparative study between clinical types I and II, using endo-β-galactosidase enzyme. *Neurochem. Pathol.* **4**:107–118.

Bock, E., C. Richter-Landsberg, A. Faissner, and M. Schachner (1985) Demonstration of immunochemical identity between the nerve growth factor–inducible large external (NILE) glycoprotein and the cell adhesion molecule L1. *EMBO J.* **4**:2765–2768.

Brunngraber, E. G. (1972) Biochemistry, function, and neuropathology of the glycoproteins in brain tissue. In *Functional and Structural Proteins of the Nervous System,* A. N. Davison, P. Mandel, and I. G. Morgan, eds., pp. 109–133, Plenum, New York.

Brunngraber, E. G. (1987) Function of glycoprotein carbohydrate. *Neurotoxicology* 8:181–198.

Chou, D. K. H., A. A. Ilyas, J. E. Evans, C. Costello, R. H. Quarles, and F. B. Jungalwala (1986) Structure of sulfated glucuronyl glycolipids in the nervous system reacting with HNK-1 antibody and some IgM paraproteins in neuropathy. *J. Biol. Chem.* 261:11717–11725.

Cole, G. J., A. Loewy, and L. Glaser (1986) Neuronal cell–cell adhesion depends on interactions of N-CAM with heparin-like molecules. *Nature* 320:445–447.

Covault, J., and J. R. Sanes (1985) Neural cell adhesion molecule (N-CAM) accumulates in denervated and paralyzed skeletal muscles. *Proc. Natl. Acad. Sci. USA* 82:4544–4548.

Cross, A. S., W. Zollinger, R. Mandrell, P. Gemski, and J. Sadoff (1983) Evaluation of immunotherapeutic approaches for the potential treatment of infections caused by K1-positive *Escherichia coli. J. Infect. Dis.* 147:68–76.

Edelman, G. M. (1983) Cell adhesion molecules. *Science* 219:450–457.

Edelman, G. M. (1985) Cell adhesion and the molecular processes of morphogenesis. *Annu. Rev. Biochem.* 54:135–169.

Feizi, T. (1985) Demonstration by monoclonal antibodies that carbohydrate structures of glycoproteins and glycolipids are oncodevelopmental antigens. *Nature* 314:53–57.

Finne, J. (1975) Structure of the O-glycosidically linked carbohydrate units of rat brain glycoproteins. *Biochim. Biophys. Acta* 412:317–325.

Finne, J. (1977) Studies on the alkali-labile O-glycosidically linked carbohydrate units of brain glycoproteins. Thesis, University of Helsinki.

Finne, J. (1982) Occurrence of unique polysialosyl carbohydrate units in glycoproteins of developing brain. *J. Biol. Chem.* 257:11966–11970.

Finne, J. (1985) Polysialic acid—A glycoprotein carbohydrate involved in neural adhesion and bacterial meningitis. *Trends Biochem. Sci.* 10:129–132.

Finne, J., and T. Krusius (1976) O-glycosidic carbohydrate units from glycoproteins of different tissues: Demonstration of a brain-specific disaccharide, α-galactosyl-$(1 \rightarrow 3)$-N-acetylgalactosamine. *FEBS Lett.* 66:94–97.

Finne, J., and T. Krusius (1979) Structural features of the carbohydrate units of plasma glycoproteins. *Eur. J. Biochem.* 102:583–588.

Finne, J., and T. Krusius (1982) Preparation and fractionation of glycopeptides. *Methods Enzymol.* 83:269–277.

Finne, J., and H. Mäkelä (1985) Cleavage of the polysialosyl units of brain glycoproteins by a bacteriophage endosialidase: Involvement of a long oligosaccharide segment in molecular interactions of polysialic acid. *J. Biol. Chem.* 260:1265–1270.

Finne, J., and H. Rauvala (1977) Determination (by methylation analysis) of the substitution pattern of 2-amino-2-deoxyhexitols obtained from O-glycosidic carbohydrate units of glycoproteins. *Carbohydr. Res.* 58:57–64.

Finne, J., T. Krusius, and H. Rauvala (1977a) Occurrence of disialosyl groups in glycoproteins. *Biochem. Biophys. Res. Commun.* 74:405–410.

Finne, J., T. Krusius, H. Rauvala, and K. Hemminki (1977b) The disialosyl group of glycoproteins: Occurrence in different tissues and cellular membranes. *Eur. J. Biochem.* 77:319–323.

Finne, J., I. Mononen, and J. Kärkkäinen (1977c) Analysis of hexosaminitol-containing disaccharide alditols from rat brain glycoproteins and gangliosides as O-trimethylsialyl derivates by gas chromatography mass spectrometry. *Biomed. Mass Spectrom.* 4:281–283.

Finne, J., T. Krusius, H. Rauvala, R. Kekomäki, and G. Myllylä (1978) Alkali-stable blood group A- and B-active poly(glycosyl)-peptides from human erythrocyte membrane. *FEBS Lett.* 89:111–115.

Finne, J., T. Krusius, R. K. Margolis, and R. U. Margolis (1979) Novel mannitol-containing oligosaccharides obtained by alkaline borohydride treatment of a chondroitin sulfate proteoglycan from brain. *J. Biol. Chem.* 254:10295–10300.

Finne, J., T. Krusius, and J. Järnefelt (1980a) Fractionation of glycopeptides. In *IUPAC 27th International Congress of Pure and Applied Chemistry*, A. Varmavuori, ed., pp. 147–159, Pergamon, Oxford, England.

Finne, J., T. Krusius and H. Rauvala (1980b) Use of potassium *tert*-butoxide in the methylation of carbohydrates. *Carbohydr. Res.* **80**:336–339.

Finne, J., T. Krusius, H. Rauvala, and J. Järnefelt (1980c) Molecular nature of the blood-group ABH antigens of the human erythrocyte membrane. *Blood Transfus. Immunohaematol.* **23**:545–552.

Finne, J., U. Finne, H. Deagostini-Bazin, and C. Goridis (1983a) Occurrence of α2–8-linked polysialosyl units in a neural cell adhesion molecule. *Biochem. Biophys. Res. Commun.* **112**:482–487.

Finne, J., M. Leinonen, and P. H. Mäkelä (1983b) Antigenic similarities between brain components and bacteria causing meningitis: Implications for vaccine development and pathogenesis. *Lancet* **2**:355–357.

Finne, J., D. Bitter-Suermann, C. Goridis, and U. Finne (1987) An IgG monoclonal antibody to group B meningococci cross-reacts with developmentally regulated polysialic acid units of glycoproteins in neural and extraneural tissues. *J. Immunol.* **138**:4402–4407.

Friedlander, D. R., M. Grumet, and G. M. Edelman (1986) Nerve growth factor enhances expression of neuron–glia cell adhesion molecule in PC12 cells. *J. Cell Biol.* **102**:413–419.

Fukuda, M. (1985) Cell surface glycoconjugates as oncodifferentiation markers in hematopoietic cells. *Biochim. Biophys. Acta* **780**:119–150.

Glode, M. P., A. Sutton, E. R. Moxon, and J. B. Robbins (1977) Pathogenesis of neonatal *Escherichia coli* meningitis: Induction of bacteremia and meningitis in infant rats fed *E. coli* K1. *Infect. Immun.* **16**:75–80.

Goridis, C., H. Deagostini-Bazin, M. Hirn, M.-R. Hirsch, G. Rougon, R. Sadoul, O. K. Langley, G. Gombos, and J. Finne (1983) Neural surface antigens during nervous system development. *Cold Spring Harbor Symp. Quant. Biol.* **48**:527–537.

Hallenbeck, P. C., E. C. Vimr, F. Yu, B. Bassler, and F. A. Troy (1987) Purification and properties of a bacteriophage-induced endo-N-acetylneuraminidase specific for poly-α-2,8-sialosyl carbohydrate units. *J. Biol. Chem.* **262**:3553–3561.

Higa, H., and A. Varki (1987) An *E. coli* polysialosyl O-acetyl-transferase that is responsible for the O-acetyl form variation selectively recognizes large polysialosyl units. *Fed. Proc.* **46**:2203.

Hoffman, S., and G. M. Edelman (1983) Kinetics of homophilic binding by embryonic and adult forms of the neural cell adhesion molecule. *Proc. Natl. Acad. Sci. USA* **80**:5762–5766.

Hoffman, S., and G. M. Edelman (1987) A proteoglycan with HNK-1 antigenic determinants is a neuron-associated ligand for cytotactin. *Proc. Natl. Acad. Sci. USA* **84**:2523–2527.

Hughes, R. C. (1976) *Membrane Glycoproteins*, Butterworths, London.

Hull, S. R., R. A. Laine, T. Kaizu, I. Rodriguez, and K. L. Carraway (1984) Structures of the O-linked oligosaccharides of the major cell surface sialoglycoprotein of MAT-B1 and MAT-C1 ascites sublines of the 13762 rat mammary adenocarcinoma. *J. Biol. Chem.* **259**:4866–4877.

Inoue, S., and M. Iwasaki (1980) Characterization of a new type of glycoprotein saccharides containing polysialosyl sequence. *Biochem. Biophys. Res. Commun.* **93**:162–165.

Järnefelt, J., J. Rush, Y.-T. Li, and R. A. Laine (1978) Erythroglycan, a high molecular weight glycopeptide with the repeating structure [galactosyl-$(1 \rightarrow 4)$-2-deoxy-2-acetamido-glucosyl-$(1 \rightarrow 3)$] comprising more than one-third of the protein-bound carbohydrate of human erythrocyte stroma. *J. Biol. Chem.* **253**:8006–8009.

Jennings, H. J. (1983) Capsular polysaccharides as human vaccines. *Adv. Carbohydr. Chem. Biochem.* **41**:155–208.

Jennings, H. J., R. Roy, and F. Michon (1985) Determinant specificities of the groups B and C polysaccharides of *Neisseria* meningitidis. *J. Immunol.* **134**:2651–2657.

Joubert, R., M. Caron, M. A. Deugnier, F. Rioux, M. Sensenbrenner, and J. C. Bisconte (1985) Effects of adult rat brain extracts on cultures of mouse brain: Consequence of the depletion in a carbohydrate-binding fraction. *Cell Mol. Biol.* **31**:131–138.

Joziasse, D. H., W. E. C. M. Schiphorst, D. H. van den Eijnden, J. A. van Kuik, H. van Halbeek, and J. F. G. Vliegenthart (1987) Branch specificity of bovine colostrum CMP-sialic acid: Galβ1 → 4GlcNAc-R α2→6-sialyltransferase. Sialylation of bi-, tri-, and tetraantennary oligosaccharides and glycopeptides of the N-acetyllactosamine type. *J. Biol. Chem.* **262**:2025–2033.

Kabat, E. A., K. G. Nickerson, J. Liao, L. Grossbard, E. F. Osserman, E. Glickman, L. Chess, J. B. Robbins, R. Schneerson, and Y. Yang (1986) A human monoclonal macroglobulin with specificity for α(2→8)-linked poly-N-acetyl neuraminic acid, the capsular polysaccharide of group B meningococci and *Escherichia coli* K1, which cross-reacts with polynucleotides and with denatured DNA. *J. Exp. Med.* **164**:642–654.

Keilhauer, G., A. Faissner, and M. Schachner (1985) Differential inhibition of neurone–neurone, neurone–astrocyte, and astrocyte–astrocyte adhesion by L1, L2, and N-CAM antibodies. *Nature* **316**:728–730.

Kiang, W.-L., R. U. Margolis, and R. K. Margolis (1981) Fractionation and properties of a chondroitin sulfate proteoglycan and the soluble glycoproteins of brain. *J. Biol. Chem.* **256**:10529–10537.

Kiang, W.-L., T. Krusius, J. Finne, R. U. Margolis, and R. K. Margolis (1982) Glycoproteins and proteoglycans of the chromaffin granule matrix. *J. Biol. Chem.* **257**:1651–1659.

Korhonen, T. K., and J. Finne (1985) Agglutination assays for detecting bacterial binding specificities. In *Enterobacterial Surface Antigens: Methods for Molecular Characterization*, T. K. Korhonen, E. A. Dawes, and P. H. Mäkelä, eds., pp. 301–313, Elsevier, Amsterdam.

Korhonen, T. K., V. Väisänen-Rhen, M. Rhen, A. Pere, J. Parkkinen, and J. Finne (1984) *Escherichia coli* fimbriae recognizing sialyl galactosides. *J. Bacteriol.* **159**:762–766.

Koscielak, J., H. Miller-Podraza, R. Krauze, and A. Piasek (1976) Isolation and characterization of poly(glycosyl)-ceramides (megaloglycolipids) with A, H, and I blood-group activities. *Eur. J. Biochem.* **71**:9–18.

Krusius, T., and J. Finne (1977) Structural features of tissue glycoproteins: Fractionation and methylation analysis of glycopeptides derived from rat brain, kidney, and liver. *Eur. J. Biochem.* **78**:369–379.

Krusius, T., and J. Finne (1978) Characterization of a novel sugar sequence from rat-brain glycoproteins containing fucose and sialic acid. *Eur. J. Biochem.* **84**:395–403.

Krusius, T., and J. Finne (1981) Use of the Smith degradation in the study of the branching pattern in the complex-type carbohydrate units of glycoproteins. *Carbohydr. Res.* **90**:203–214.

Krusius, T., J. Finne, J. Kärkkäinen, and J. Järnefelt (1974) Neutral and acidic glycopeptides in adult and developing rat brain. *Biochim. Biophys. Acta* **365**:80–92.

Krusius, T., J. Finne, R. U. Margolis, and R. K. Margolis (1978a) Structural features of microsomal, synaptosomal, mitochondrial, and soluble glycoproteins of brain. *Biochemistry* **17**:3849–3854.

Krusius, T., J. Finne, and H. Rauvala (1978b) The poly(glycosyl) chains of glycoproteins: Characterization of a novel type of glycoprotein saccharides from human erythrocyte membrane. *Eur. J. Biochem.* **92**:289–300.

Krusius, T., J. Finne, R. K. Margolis, and R. U. Margolis (1986) Identification of an O-glycosidic mannose-linked sialylated tetrasaccharide and keratan sulfate oligosaccharides in the chondroitin sulfate proteoglycan of brain. *J. Biol. Chem.* **261**:8237–8242.

Kuff, E. L., and J. W. Fewell (1980) Induction of neural-like cells and acetylcholinesterase activity in cultures of F9 teratocarcinoma treated with retinoic acid and dibutyryl cyclic adenosine monophosphate. *Dev. Biol.* **77**:103–115.

Laitinen, J., R. Löppönen, J. Merenmies, and H. Rauvala (1987) Binding of laminin to brain gangliosides and inhibition of laminin–neuron interaction by the gangliosides. *FEBS Lett.* **217**:94–100.

Lamblin, G., A. Boersma, A. Klein, P. Roussel, H. van Halbeek, and J. F. G. Vliegenthart (1984) Primary structure determination of the five sialylated oligosaccharides derived from bronchial mucus glycoproteins of patients suffering from cystic fibrosis: The occurrence of the NeuAcα(2→3)Galβ(1→4)[Fucα(1→3)]GlcNAcβ(1→·) structural element revealed by 500-MHz ^{1}H NMR spectroscopy. *J. Biol. Chem.* **259**:9051–9058.

Leppänen, A., A. Korvuo, K. Puro, and O. Renkonen (1986) Glycoproteins of human teratocarcinoma cells (PA1) carry both anomers of O-glycosyl-linked D-galactopyranosyl-(1→3)-2-acetamido-2-deoxy-α-D-galactopyranosyl group. *Carbohydr. Res.* **153**:87–95.

Lifely, M. R., A. S. Gilbert, and C. Moreno (1981) Sialic acid polysaccharide antigens of *Neisseria meningitidis* and *Escherichia coli*: Esterification between adjacent residues. *Carbohydr. Res.* **94**:193–203.

Lyles, J. M., D. Linnemann, and E. Bock (1984) Biosynthesis of the D2-cell adhesion molecule: Post-translational modifications, intracellular transport, and developmental changes. *J. Cell Biol.* **99**:2082–2091.

Margolis, R. K., and R. U. Margolis (1979) Structure and distribution of glycoproteins and glycosaminoglycans. In *Complex Carbohydrates of Nervous Tissue*, R. U. Margolis and R. K. Margolis, eds., pp. 45–73, Plenum, New York.

Margolis, R. K., and R. U. Margolis (1983) Distribution and characteristics of polysialosyl oligosaccharides in nervous tissue glycoproteins. *Biochem. Biophys. Res. Commun.* **116**:889–894.

Margolis, R. K., J. Finne, T. Krusius, and R. U. Margolis (1984) Structural studies on glycoprotein oligosaccharides of chromaffin granule membranes and dopamine β-hydroxylase. *Arch. Biochem. Biophys.* **228**:443–449.

Margolis, R. K., L. A. Greene, and R. U. Margolis (1986) Poly(N-acetyllactosaminyl) oligosaccharides in glycoproteins of PC12 pheochromocytoma cells and sympathetic neurons. *Biochemistry* **25**:3463–3468.

McCabe, W. R., B. Kaijser, S. Olling, M. Uwaydah, and L. A. Hanson (1978) *Escherichia coli* in bacteremia: K and O antigens and serum sensitivity of strains from adults and neonates. *J. Infect. Dis.* **138**:33–41.

McCoy, R. D., and F. A. Troy (1987) CMP-NeuNAc: Poly-α-2,8-sialosyl sialytransferase in neural cell membranes. *Methods Enzymol.* **138**:627–637.

Misevic, G. N., J. Finne, and M. M. Burger (1987) Involvement of carbohydrates as multiple low affinity interaction sites in the self-association of the aggregation factor from the marine sponge *Microciona prolifera*. *J. Biol. Chem.* **262**:5870–5877.

Mononen, I., J. Finne, and J. Kärkkäinen (1978) Analysis of permethylated hexopyranosyl-2-acetamido-2-deoxyhexitols by g.l.c.-m.s. *Carbohydr. Res.* **60**:371–375.

Montreuil, J. (1980) Primary structure of glycoprotein glycans. *Adv. Carbohydr. Chem. Biochem.* **37**:157–223.

Moore, S. E., and F. S. Walsh (1985) Specific regulation of N-CAM/D2-CAM cell adhesion molecule during skeletal muscle development. *EMBO J.* **4**:623–630.

Muramatsu, T., G. Gachelin, J. F. Nicolas, H. Condamine, H. Jakob, and F. Jakob (1978) Carbohydrate structure and cell differentiation: Unique properties of fucosyl-glycopeptides isolated from embryonal carcinoma cells. *Proc. Natl. Acad. Sci. USA* **75**:2315–2319.

Nowicki, B., J. Vuopio-Varkila, P. Viljanen, T. K. Korhonen, and P. H. Mäkelä (1986) Fimbrial phase variation and systemic E. coli infection in the mouse peritonitis model. *Microb. Pathogen.* **1**:335–347.

Parkkinen, J., and J. Finne (1987) Isolation of sialyl oligosaccharides and sialyl oligosaccharide phosphates from bovine colostrum and human urine. *Methods Enzymol.* **138**:289–300.

Parkkinen, J., J. Finne, M. Achtman, M. Väisänen, and T. K. Korhonen (1983) *Escherichia coli* strains recognizing neuraminyl α2–3 galactosides. *Biochem. Biophys. Res. Commun.* **111**:456–461.

Parkkinen, J., G. N. Rogers, T. Korhonen, W. Dahr, and J. Finne (1986) Identification of the

O-linked sialyloligosaccharides of glycophorin A as the erythrocyte receptors for S-fimbriated *Escherichia coli. Infect. Immun.* **54**:37–42.

Rauvala, H., J.-P. Prieels, and J. Finne (1983) Cell adhesion mediated by a purified fucosyltransferase. *Proc. Natl. Acad. Sci. USA* **80**:3991–3995.

Rieger, F., M. Grumet, and G. M. Edelman (1985) N-CAM at the vertebrate neuromuscular junction. *J. Cell Biol.* **101**:285–293.

Robbins, J. B., G. H. McCracken, E. C. Gotschlich, F. Ørskov, I. Ørskov, and L. A. Hanson (1974) *Escherichia coli* capsular polysaccharide associated with neonatal meningitis. *N. Engl. J. Med.* **290**:1216–1220.

Roberts, D. D., L. A. Liotta, and V. Ginsburg (1986) Gangliosides indirectly inhibit the binding of laminin to sulfatides. *Arch. Biochem. Biophys.* **250**:498–504.

Roth, J., D. Taatjes, D. Bitter-Suermann, and J. Finne (1987) Polysialic acid units are spatially and temporally expressed in developing postnatal rat kidney. *Proc. Natl. Acad. Sci. USA* **84**:1969–1973.

Rothbard, J. B., R. Brackenbury, B. A. Cunningham, and G. M. Edelman (1982) Differences in the carbohydrate structures of neural cell-adhesion molecules from adult and embryonic chicken brains. *J. Biol. Chem.* **257**:11064–11069.

Rougon, G., H. Deagostini-Bazin, M. Hirn, and C. Goridis (1982) Tissue- and developmental stage-specific forms of a neural cell surface antigen linked to differences in glycosylation of a common polypeptide. *EMBO J.* **1**:1239–1244.

Rougon, G., C. Dubois, N. Buckley, J. L. Magnani, and W. Zollinger (1986) A monoclonal antibody against meningococcus group B polysaccharide distinguishes embryonic from adult N-CAM. *J. Cell Biol.* **103**:2429–2437.

Rutishauser, U. (1984) Developmental biology of a neural cell adhesion molecule. *Nature* **310**:549–554.

Rutishauser, U., M. Watanabe, J. Silver, F. A. Troy, and E. R. Vimr (1985) Specific alteration of NCAM-mediated cell adhesion by an endoneuraminidase. *J. Cell Biol.* **101**:1842–1849.

Sadoul, R., M. Hirn, H. Deagostini-Bazin, G. Rougon, and C. Goridis (1983) Adult and embryonic mouse neural cell adhesion molecules have different binding properties. *Nature* **304**:347–349.

Saito, T., T. Itoh, S. Adachi, T. Suzuki, and T. Usui (1981) The chemical structure of neutral and acidic sugar chains obtained from bovine colostrum κ-casein. *Biochim. Biophys. Acta* **678**:257–267.

Schauer, R. (1985) Sialic acids and their role as biological masks. *Trends Biochem. Sci.* **10**:357–360.

Sharon, N. (1984) Glycoproteins. *Trends Biochem. Sci.* **9**:198–202.

Sharon, S., E. Kabat, and S. L. Morrison (1982) Immunochemical characterization of binding sites of hybridoma antibodies specific for $\alpha(1 \to 6)$-linked dextran. *Mol. Immunol.* **19**:375–388.

Spillmann, D., and J. Finne (1987) Poly-N-acetyllactosamine glycans of cellular glycoproteins: Predominance of linear chains in mouse neuroblastoma and rat pheochromocytoma cell lines. *J. Neurochem.* **49**:874–883.

Takasaki, S., and A. Kobata (1986) Asparagine-linked sugar chains of fetuin: Occurrence of tetrasialyl triantennary sugar chains containing the Galβ1 → 3GlcNAc sequence. *Biochemistry* **25**:5709–5715.

Viitala, J., and J. Finne (1984) Specific cell surface labeling of polyglycosyl chains in human erythrocytes and HL-60 cells using endo-β-galactosidase and galactosyltransferase. *Eur. J. Biochem.* **138**:393–397.

Vimr, E. R., R. D. McCoy, H. F. Vollger, N. C. Wilkison, and F. A. Troy (1984) Use of prokaryotic-derived probes to identify poly(sialic acid) in neonatal neuronal membranes. *Proc. Natl. Acad. Sci. USA* **81**:1971–1975.

Vliegenthart, J. F. G., L. Dorland, and H. van Halbeek (1983) High-resolution, ^{1}H-nuclear magnetic resonance spectroscopy as a tool in the structural analysis of carbohydrates related to glycoproteins. *Adv. Carbohydr. Chem. Biochem.* **41**:209–374.

Weigel, P. H., R. L. Schnaar, M. S. Kuhlenschmidt, E. Schmell, R. T. Lee, Y. C. Lee, and S. Roseman (1979) Adhesion of hepatocytes to immobilized sugars: A threshold phenomenon. *J. Biol. Chem.* **254**:10830–10838.

Yamashita, K., A. Hitoi, N. Taniguchi, N. Yokosawa, Y. Tsukada, and A. Kobata (1983) Comparative study of the sugar chains of γ-glutamyltranspeptidase from rat liver and rat AH-66 hepatoma cells. *Cancer Res.* **43**:5059–5063.

Zhu, B. L. R., and R. A. Laine (1985) Polylactosamine glycosylation on human placental fibronectin weakens binding affinity of fibronectin to gelatin. *J. Biol. Chem.* **260**:4041–4045.

Chapter 5

Linkage Genetics of Cell Adhesion Molecules

PETER D'EUSTACHIO

ABSTRACT

Among mammals, mice are uniquely well suited to genetic analysis. The analysis of somatic cell hybrids allows the assignment of genes to chromosomes, providing a straightforward means for assessing the extent to which potentially related genes have been dispersed over the genome. Linkage analysis, facilitated in laboratory mice by an abundance of restriction fragment length polymorphism, allows a more precise localization of genes. Applied to three genes involved in neural development—Ncam (neural cell adhesion molecule), Mag (myelin-associated glycoprotein), and Hox-1 (homeobox complex-1)—these techniques have allowed each to be localized. Furthermore, each localization has raised the intriguing possibility of a functional link between the locus and one or more nearby loci defined by phenotypically visible mutations.

A chromosome is as much an anatomical structure in the cell as are, for example, gap junctions or the elements of the Golgi complex, and despite the obvious differences in the technologies employed, determining the linkage relationships among the genes on a chromosome is the same sort of exercise as anatomizing one of the latter structures. Given a linkage map, its interpretation is likewise analogous. Much as one needs criteria to determine, for example, whether the association of cytoskeletal elements with a purified cell junction complex *in vitro* occurs as well in the native complex *in vivo*, and if so, whether the association is functionally important, one needs criteria for assessing the possible biological importance of observed physical relationships among genes in linkage maps.

Given the limited information now available, defining general criteria is impossible, but data from several lines of work can be amalgamated to suggest the form criteria might take. Molecular experiments have yielded evidence of biologically important interactions in *cis* among DNA sequence elements separated by distances up to a few thousand kilobases. An entire gene, including its introns and flanking promoter and enhancer regulatory elements, can easily span tens of kilobases, and gene families all of whose members might be tightly and coordinately regulated in a tissue- or developmental stage-specific manner often span hundreds of kilobases. Large families of homologous genes, such as those encoding ribosomal RNAs, immunoglobulin heavy chains, and the class I antigens of the major histocompatibility complex in the mouse, are genetically unstable, showing high frequencies of mutations and gene rearrangements. While the frequencies of these events are too low to affect gene expression appreciably within an individual, the effect on a population over a span of even a few generations is probably substantial. Indirect evidence suggests that the large numbers of homologous genes within each complex lead to instability by facilitating events such as unequal crossing over and gene conversion. These complexes span thousands of kilobases of DNA (Geliebter and Nathenson, 1987).

What fraction of a total mammalian genome do these distances represent? The physical size of the haploid mouse genome is approximately 2.7×10^6 kb (Lewin, 1980), and its genetic size, estimated from counting chiasmata in meiotic figures, is 1600 cM (Henderson and Edwards, 1968; Nesbitt and Francke, 1973), an average of 1700 kb of DNA per centimorgan. Thus if the assignment of a gene to a position in a linkage map is to be interpretable in functional terms, given the kinds of interactions among genes that are experimentally accessible now, the localization needs to be precise on a scale of centimorgans. Consistent with this assessment, while the conservation of genetic linkage relationships is extensive among mammals, the conserved segments (except for the X chromosome) rarely exceed 10–15 cM in length and are often much smaller (Nadeau and Taylor, 1984).

In this chapter I discuss a variety of approaches to the problem of constructing precise and detailed genetic linkage maps in mammals. The discussion is focused almost entirely on the laboratory mouse. This is partly due to the particular interests of this laboratory, but also reflects the unique properties of the mouse as an object of genetic analysis. The mouse has a short generation time and large numbers of offspring. Mice tolerate inbreeding well, and large numbers of highly inbred, well-characterized strains are available. Large numbers of mutations with visible effects, including more than 70 with neurological or neuromuscular effects, have been described; many of these have been placed in a linkage map that in total incorporates about 650 loci defined by mutation or biochemical or serological polymorphism, distributed over all of the autosomes and the X chromosome (Green, 1981b; Davisson and Roderick, 1986).

Analysis of interspecies somatic cell hybrids provides a direct approach to mapping genes and gene families. Mouse- and Chinese hamster-derived cells can be physically fused in culture, forming hybrids carrying complete chromosome sets derived from both species. As these hybrids are propagated, chromosomes derived from the mouse parent are progressively lost, while a complete chromosome set from the Chinese hamster parent is retained. After an initial period of rapid chromosome loss, relatively stable hybrid lines often arise. Panels of clonal cell lines can thus be derived that contain various combinations of mouse chromosomes on the background of a complete Chinese hamster genome. Correlating the presence or absence of the mouse form of a gene with the presence or absence of the various mouse chromosomes allows the gene to be mapped to a chromosome.

To apply the mapping assay to genes expressed in differentiated cell types, two strategies are possible. If both parental cells are of the appropriate differentiated type, cell type–specific genes may be expressed in the hybrids, allowing mouse-derived genes to be detected phenotypically. In several cases in which established cell lines that maintain a differentiated phenotype are available, this approach has yielded useful data. Indeed, in the case of liver parenchymal cells, the approach has been exploited further to identify *trans*-acting regulatory genes that repress particular liver-specific genes in nonhepatic tissues (Killary and Fournier, 1984). In general, however, mapping genes expressed in differentiated tissues at the phenotype level in somatic cell hybrids is impractical.

The alternative strategy is to score the presence or absence of mouse genes at the level of the hybrid cell genotype. A cloned DNA fragment corresponding to the gene (or, equally useful, to part of it or to a segment of unique-sequence DNA known to flank the gene) is used to probe a Southern blot of genomic DNA extracted from the members of a hybrid cell panel. This assay is unaffected by the presence or absence of gene expression in the hybrid cell lines. Equally important, even where coding sequences for a gene are very closely conserved between species (e.g., histone genes), sufficient point mutations have accumulated in flanking and intronic DNA sequences so that the mouse and Chinese hamster homologs of the gene are likely to fall on restriction enzyme fragments of different sizes. An example of such an analysis, using a DNA fragment derived from the mouse *Ncam* (neural cell adhesion molecule) gene, is shown in Figure 1. Data from analyses using DNA probes for *Ncam*, *Mag* (myelin-associated glycoprotein), *Mbp* (myelin basic protein) (P. D'Eustachio, J. Salzer, and D. Colman, unpublished observations), *Hox-1* (homeobox complex-1) (Rubin et al., 1986), and the L3T4 protein (Field et al., 1987) are summarized in Table 1.

The particular strength of this method is its ability to determine whether genes localize together to a single chromosome or are dispersed to several in one experiment. Thus the three mouse genomic DNA fragments reactive with our *Ncam* probe were either present as a group or absent as a group from all hybrid lines examined, strongly suggesting that they are located on one mouse chromosome (Figure 1). Similarly, the concordant segregation of *Hox-1* and L3T4 DNA fragments in the panel suggests that these two genes are located on the same chromosome. Conversely, the existence of hybrids containing *Mag*-derived fragments and lacking *Ncam*-derived fragments (Table 1) is compelling evidence that *Mag* and *Ncam* are not linked in the genome. The assay is especially useful when applied to a multigene family to determine whether family members remain clustered at a single site or have become dispersed over multiple sites.

Table 1. Gene Mapping in Interspecies Somatic Cell Hybrids[a]

Hybrid	Reaction with Probes				
	Ncam	*Mag*	*Hox-1*	*L3T4*	*Mbp*
BEM1-4	−	−	+	+	+
MACH4B31Az3	−	+	−	−	−
MACH2A2B1	+	+	+	+	−
MACH2A2C2	+	+	−	−	−
MACH2A2H3	+	+	+	+	+
MAE28	−	−	−	−	−
R44-1	−	−	−	−	+
ECm4e	−	−	−	−	−

[a]Chromosomal contents of somatic cell hybrids are known from karyotypic and isoenzyme analyses (D'Eustachio et al., 1981; Goff et al., 1982) but are omitted here to emphasize the fact that conclusions as to the colocalization of markers can be drawn independently of this information. DNA from hybrid cell lines was scored for the presence or absence of mouse-specific DNA fragments by Southern blotting as shown in Figure 1.

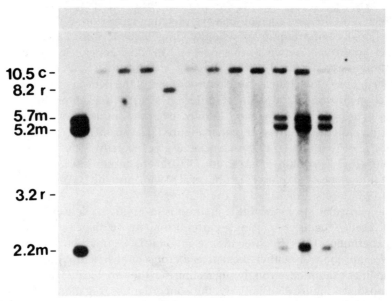

Figure 1. *Detection of mouse N-CAM-specific DNA fragments in interspecies somatic hybrid cell lines.* DNA from C57BL/6J mouse liver (A), the Chinese hamster cell line E36 (B), and the hybrid cell lines ABm11 (C), ABm14 (D), F(11)U (E), BEM1-4 (F), MAE28 (G), MACH4A63 (H), MACH4B31Az3 (I), MACH2A2B1 (J), MACH2A2C2 (K), MACH2A2H3 (L), ECm4e (M), and R44-1 (N) was digested with Hind III restriction endonuclease and analyzed by Southern blotting. Somatic hybrid cell lines contained various numbers of mouse chromosomes on the background of a complete rat genome [F(11)U, *lane E*] or Chinese hamster genome (all other lanes). Sizes of fragments are shown in kilobases, and the species of origin of each is indicated: *m*, mouse; *c*, Chinese hamster; *r*, rat. (From D'Eustachio et al., 1985.)

While the analysis of somatic cell hybrids can suggest biologically significant linkage relationships, the technique in general provides no information as to the organization of genes along a chromosome. Mendelian genetic analysis is needed to obtain this information. Two factors—a high frequency of restriction fragment length polymorphism and the availability of sets of recombinant inbred strains—facilitate this analysis in laboratory mice.

Between individuals within a species (as between species), accumulated mutations should alter restriction enzyme cleavage sites in the genomic DNA, giving rise to alternate DNA fragment patterns when DNA samples from several individuals of the same species are analyzed by Southern blotting. If these mutations cause silent changes in the third positions of codons, or fall in intronic or flanking DNA sequences, they may not affect gene function and could thus persist in a population as neutral polymorphisms (Botstein et al., 1980). This model of restriction fragment length polymorphism leads to the prediction that the frequency of such polymorphism should be low, and in most instances should be due to point substitutions. While changes in the copy number of tandem arrays of short reiterated sequence elements have also been shown to contribute to restriction fragment polymorphism among humans (Jeffreys et al., 1985; Nakamura et al., 1987), a large proportion of observed human polymorphism appears to fit the model (Barker et al., 1984).

Both the frequency of polymorphism and the nature of the DNA sequence variation associated with a typical polymorphism are strikingly different in the laboratory mouse. Most of the loci we have examined (e.g., D'Eustachio, 1984; D'Eustachio et al., 1984, 1985) are associated with polymorphisms when DNA from inbred strains of mice was analyzed by Southern blotting. These loci include structural genes as well as dispersed pseudogenes and loci defined by cloned arbitrary DNA fragments. The only group of loci that has consistently proved monomorphic in our hands is defined by a series of arbitrary DNA fragments from the X chromosome. This excess polymorphism is not confined to loci characterized at the DNA level. Surveys of protein and serological polymorphisms have likewise suggested that the genetic heterogeneity of a "population" composed of the standard laboratory strains of mice significantly exceeds that found in wild mouse populations (Rice and O'Brien, 1980; Fitch and Atchley, 1985).

Furthermore, the polymorphisms are complex. That is, the alternate patterns of restriction fragments cannot be explained by single-point mutations but require the occurrence of multiple-point mutations, often affecting several different restriction enzyme cleavage sites, or insertions, deletions, or rearrangements of DNA at the locus. *Hba-4ps*, an α-globin-like pseudogene located on chromosome 17 proximal to the major histocompatibility complex, is typical (Leder et al., 1981; Fox et al., 1983; D'Eustachio et al., 1984; Mann et al., 1986). As shown in Figure 2, polymorphisms revealed by digestion with Taq I endonuclease allow four alleles of the locus to be defined in a survey of standard inbred strains of mice as well as inbred strains derived from wild *Mus molossinus* (MOL/Ei) and *Mus castaneus* (CAST/Ei). Two variable Taq I cleavage sites are required to explain the observed fragment patterns. In addition, the DNA

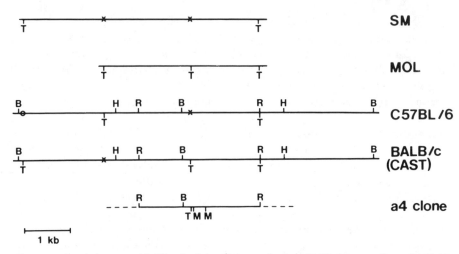

Figure 2.   *Restriction maps of the* Hba-4ps *locus.* The map of a cloned DNA fragment from a BALB/cJ
genomic library (Leder et al., 1981) is shown at the *bottom. Above* are restriction maps deduced
from genomic Southern blotting experiments for four alleles of the locus designated here by
the name of the type inbred strain in which each is found. Cleavage sites are shown for the
restriction enzymes Bam HI (*B*), Eco RI (*R*), Hind III (*H*) Msp I (*M*), and Taq I (*T*). An *x*
indicates the loss of a cleavage site; the *circle* indicates a site whose presence or absence could
not be determined.

sequence variation at the rightward Taq I site affects not only the cleavage site
for Taq I but an adjoining pair of Msp I cleavage sites as well. The trio of sites
appears to be present or absent as a group, consistent with the deletion of a small
DNA segment in negative strains such as C57BL/6J and SM/J (Mann et al., 1986;
P. D'Eustachio, unpublished observations).

These data suggest possible explanations for both the excess polymorphism
and its complexity. Note that one of the alleles of *Hba-4ps* is shared between an
inbred strain (BALB/c) and another mouse species (*Mus castaneus*). A survey
of catalogs of polymorphisms indicates that at about one-fifth of all loci
examined, one or more inbred strains of mice have fixed alleles derived not
from *Mus domesticus*, but from another species of the genus *Mus*. This
observation is consistent with known histories of inbred mouse strains and
suggests the hypothesis that these strains are in fact interspecies hybrids,
deriving portions of their genomes from at least the species *Mus domesticus*,
Mus musculus, *Mus castaneus*, and *Mus molossinus* (Blank et al., 1986).

Whatever its source, the extensive polymorphism greatly facilitates linkage
analysis in mice. This analysis is helped further by the existence of sets of
recombinant inbred (RI) strains of mice. RI strains are derived by crossing mice
of two existing inbred strains and using pairs of the F_1 offspring of this cross to
derive new inbred strains by multiple generations of brother–sister mating.
Each new strain produced by this breeding scheme is homozygous for a
patchwork of chromosomal segments derived from the two progenitor strains.

Patch boundaries, determined by randomly occurring crossover events early in the inbreeding process, vary from strain to strain. As a result, the closer two loci are to each other on a chromosome, the more frequently they will be inherited concordantly (i.e., an RI strain will carry alleles at both loci derived from the same progenitor strain). The fraction of RI strains concordant for a pair of loci can be used to calculate the distance between the loci in centimorgans. Because each strain is a homozygous inbred one, typing data obtained for a novel locus can be compared directly to data for the same strains obtained previously by other workers, so that linkage maps can be built progressively. The availability of multiple genetically identical individuals from each recombinant inbred strain also makes it possible to analyze genetic traits such as the induction of audiogenic seizures or tumor susceptibility that are variably expressed (Taylor, 1978; Bailey, 1981).

Three examples will illustrate the use of RI strains for linkage mapping. These are the localization of *Ncam* (D'Eustachio et al., 1985), of *Mag* (D'Eustachio et al., 1988), and of *Hox-1* (Bucan et al., 1986; Rubin et al., 1987) (Table 2). Of 64 mice typed for inheritance of alleles of *Ncam* and *Alp-1* (Apolipoprotein-1, a marker of chromosome 9; Eicher et al., 1980; Nadeau et al., 1981), 57 inherited alleles derived from the same progenitor strain at both loci. The odds of observing this degree of concordance or a greater one, were the two loci not linked, are approximately 8×10^{-7} (Silver and Buckler, 1986; Blank et al., 1988). The observed recombination fraction (7/64) yields an estimate of the distance between the two loci of 3.3 cM, with 99% confidence limits of 0.9 cM and 9.6 cM (Taylor, 1978; Silver, 1985). Similarly, the observed recombination fraction between *Mag* and *Abpa* (androgen-binding protein α, a marker of chromosome 7; Dlouhy et al., 1987), 1/49, is unlikely to be a chance event ($p = 2 \times 10^{-7}$) and yields an estimated linkage distance of 0.5 cM with 99% confidence limits of 0 cM and 4.5 cM. The recombination fraction between *Hox-1* and *Ggc* (gamma glutamyl cyclotransferase, a marker of chromosome 6; Tulchin and Taylor, 1981), 33/44, is unlikely to be a chance event ($p = 11 \times 10^{-7}$) and yields an estimated linkage distance of 1.9 cM with 99% confidence limits of 0.2 cM and 8.7 cM.

Conversely, a comparison of the strain distribution patterns observed for each of these loci with ones known for several hundred unlinked loci yielded no patterns of concordant inheritance better than would be expected by chance alone. For example, 28 of 49 mice informative for *Ncam* and *Mag* showed discordant inheritance of alleles at the two loci (Table 2). The odds of this outcome, were the two loci linked (i.e., less than 50 cM apart on one chromosome), are 0.017 (Silver and Buckler, 1986; Blank et al., 1988).

Each of the three linkage results is intriguing. The localization of *Ncam* near *Alp-1* on chromosome 9 places it near *Thy-1*, another member of the immunoglobulin gene superfamily (Seki et al., 1985; Hunkapiller and Hood, 1986; Cunningham et al., 1987) also expressed on neural cell surfaces, and near the locus defined by the *staggerer* (*sg*) mutation. Homozygous *sg/sg* mice fail to produce appropriate amounts of N-CAM protein showing the adult glycosyla-

Table 2. Inheritance of *Hox-1*, *Mag*, *Ncam*, and Linked Loci in Recombinant Inbred Strain Sets[a]

AKXD Strain

Locus	1	2	3	6	7	8	9	10	11	12	13	14	15	16	17	18	20	21	22	23	24	25	26	27	28
Mag	A	D	A	A	D	A	D	A	A	A	A	A	A	A	D	A	A	A	A	A	A	D	D	D	D
Abpa	A	D	A	A	D	A	D	A	A	A	A	A	A	A	D	A	A	A	A	A	A	A	A	D	D
Alp-1	D	D	A	A	D	—	D	—	A	A	D	D	D	—	D	A	D	D	D	A	A	D	A	D	A
Ncam	D	D	A	D	D	D	A	A	A	A	D	D	D	D	D	A	D	D	D	A	A	D	A	D	A

BXD Strain

Locus	1	2	5	6	8	9	11	12	13	14	15	16	18	19	22	23	24	25	27	28	29	30	31	32	33
Ggc	B	B	B	D	D	D	B	B	D	B	B	B	B	D	B	D	B	D	D	D	B	D	B	B	B
Hox-1	B	B	B	D	D	D	B	B	D	B	B	B	B	D	B	D	B	D	D	D	B	D	B	B	B
Mag	B	D	B	D	B	B	D	D	B	B	D	B	B	B	D	D	B	B	B	D	B	B	D	B	D
Abpa	B	D	B	D	B	B	B	D	B	B	D	B	B	B	B	D	B	B	B	D	B	B	D	B	D
Alp-1	B	D	B	D	D	B	D	D	B	B	B	D	D	D	D	B	D	D	B	B	D	D	B	B	D
Ncam	B	D	B	B	B	B	B	D	B	B	B	D	D	D	B	B	D	B	B	B	D	B	B	B	D

BXH Strain

CXB Strain

	2	3	4	6	7	8	9	10	11	12	14	19	D	E	G	H	I	J	K
Ggc	H	B	B	B	B	B	—	B	B	B	B	B	B	B	C	B	C	C	C
Hox-1	H	B	B	B	B	B	B	H	B	B	B	B	B	B	C	B	C	C	C
Alp-1	B	B	H	H	B	B	H	H	H	H	B	H	C	C	C	B	B	B	B
Ncam	B	B	H	B	H	B	H	H	H	H	H	H	C	B	C	B	B	B	C

[a]Recombinant inbred strains of mice were derived (Taylor, 1978, 1981; Bailey, 1981) by inbreeding pairs of F_1 hybrid mice from a cross between two inbred progenitor strains: AKXD, AKR/J × DBA/2J; BXD, C57BL/6J × DBA/2J; BXH, C57BL/6J × C3H/HeJ; CXB, BALB/cBy × C57BL/6By. To type mice for restriction fragment length polymorphisms associated with the *Mag*, *Hox-1*, and *Ncam* loci, liver or spleen DNA was digested with restriction endonucleases, fractionated by electrophoresis on 0.8% agarose gels, blotted onto nitrocellulose paper, and probed with plasmids radiolabeled by nick translation. The final wash of filters before autoradiography was in 0.075 M NaCl, 0.0075 M NaCitrate, and 0.1% sodium dodecyl sulfate, at 65°C. All strains tested were homozygous for one of the progenitor strain forms of each locus, as indicated by the letters. A, AKR/J-like; B, C57BL/6J-like; C, BALB/cJ-like; D, DBA/2J-like; H, C3H/HeJ-like; —, not tested. Typing data for other loci are from published sources. *Ggc*: Dlouhy et al., 1987. *Alp-1*: Eicher et al., 1980; Nadeau et al., 1981. *Abpa*: Tulchin and Taylor, 1981.

125

tion patterns, although the N-CAM polypeptide itself is not detectably altered, and apparently normal amounts of N-CAM showing the embryonic glycosylation patterns are produced (Edelman and Chuong, 1982; Chuong and Edelman, 1984). The localization of *Mag* places it near the mutant locus *quivering* (*qv*). Mice homozygous for this mutation show a variety of neurological defects, although myelin appears morphologically normal and the *Mag* gene, MAG mRNA, and MAG protein all appear normal as assessed by Southern, Northern, and Western blotting (Yoon and Les, 1957; McNutt, 1962; P. D'Eustachio, D. Colman, and J. Salzer, unpublished observations). Finally, the localization of the *Hox-1* complex suggests a close linkage between it and one or both of the pleiotropic developmental mutants *hypodactyly* (*Hd*) and *postaxial hemimelia* (*px*) (Searle, 1964; Hummel, 1970).

Interpretation of these linkages is difficult, however. In all three cases the nominal linkage distances obtained by interpolating the newly mapped gene into the linkage map of the chromosome (Davisson and Roderick, 1986) suggests a distance between genes (*Ncam*, *Thy-1*, and *sg*; *Mag* and *qv*; *Hox-1* and *Hd* and/or *px*) small enough to be biologically meaningful, as discussed at the beginning of this chapter. At the same time, the uncertainties associated with the estimates of linkage distances are so great in every case that no firm conclusions can be drawn.

These uncertainties are intrinsic to linkage mapping. An experimentally observed recombination fraction is a binomial variable, and its confidence limit is therefore large for small sample sizes (Green, 1981a). The option available to *Drosophila* geneticists, that of increasing sample size in breeding experiments by two or more orders of magnitude in order to construct precise linkage maps, is rarely possible in mouse genetics, and essentially never possible in human genetics.

Analysis of linkage data for pairs of loci, however, does not fully use the available information. If recombination fractions have been measured between a test locus and each of several marker loci, then the best placement for the test locus is the one that simultaneously maximizes the probabilities of observing all of the recombination fractions. Several algorithms for such multilocus analysis of linkage data have been devised for use with human pedigrees (e.g., Lathrop et al., 1984; Bishop, 1985). We have recently devised an algorithm for use with data from backcrosses or RI strain sets (Blank et al., 1988).

To use the algorithm, each marker locus is provisionally fixed in place in a linkage map, and the test locus is successively placed in each of its 1600 possible positions (moving in 1-cM steps through the entire genome). At each position, the probability of observing all of the recombination fractions, assuming that position for the test locus, is calculated. The result is a plot of probability as a function of position. The test locus is then provisionally fixed at the peak position of the probability distribution, and each marker locus is treated as a test locus in turn. This process is continued until the test locus and all of the markers come to rest.

The algorithm allows a novel locus to be added to an existing map. Equally important, it allows the construction of linkage maps from scratch. Two loci are fixed provisionally at the distance from each other corresponding to the recombination fraction between them. Additional loci are added one at a time, and the map is cycled to stability after each addition. Using the recombination data from RI mice summarized in Table 3, a map of chromosome 9 was constructed in this fashion (Figure 3).

The area under the probability distribution curve for each locus treated as a test locus, with all others treated as markers, can be summed to obtain a confidence interval for the placement of the locus. The overlines in Figure 3 indicate 99% confidence limits about each locus. Because these limits are made with respect to the entire map, rather than with respect to a single flanking locus, they can be used to evaluate alternative gene orders. Some pairs of closely spaced loci (e.g., *Env-2* and *Ncam*, *Pgm-3* and *Mod-1*) cannot be unambiguously ordered. Overall, however, with no more than 75 informative RI strains typed in any case, loci spaced 2–3 cM apart can be ordered with 99% confidence (e.g., *Lap-1–Ncam–Alp-1*, *Ltw-3–Fv-2–Bgl*). The gaps in several of the confidence interval overlines indicate positions within the interval that can be excluded. In the case of *Thy-1*, for example, the four recombinants found among 25 RI strains informative for that locus and *Ncam* allow the conclusion, at 99% confidence, that the two loci cannot reside within 0.3 cM of each other.

A comparison of Figure 3 with Table 3 illustrates the power of multilocus mapping strategies. Pairwise analysis of the data (Table 3) provides convincing evidence for linkage of the loci and allows relatively proximal and distal clusters of loci to be distinguished. It provides no basis for unambiguously ordering loci within clusters, however. At the same time, the pairwise and multilocus maps are consistent: Each interval between loci fixed in the multilocus map agrees, within sampling fluctuation, with the interval estimated by pairwise analysis.

A map of this region of chromosome 9 has also been constructed by pairwise analysis of data from two backcross experiments (Antonucci et al., 1984; Nadeau, 1986). Approximately 300 informative backcross mice were typed in each case so that loci could be ordered unambiguously. The backcross

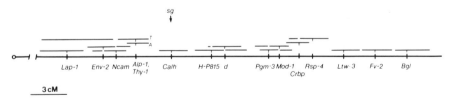

Figure 3. *Linkage map of mouse chromosome 9, compiled from RI typing data (Table 3) using a multilocus algorithm.* Each gene is placed at its most probable location; the 99% confidence interval for this localization is shown by the *associated overline*. The centromere is at the *extreme left*. Nothing in this analysis allows its distance from the most proximal marker, *Lap-1*, to be estimated, but data from other experiments (Davisson and Roderick, 1986) suggest that the distance is approximately 20 cM. The mutant locus *sg* is interpolated into the map on the basis of data placing it 4.5 cM proximal to *d* (Green and Lane, 1967).

Table 3. Recombination Fractions Among Loci on Chromosome 9 Measured in RI Strains of Mice[a]

	Lap-1	Env-2	Ncam	Alp-1	Thy-1	Calh	HP815	d	Pgm-3	Mod-1	Crbp	Rsp-4	Ltw-3	Fv-2	Bgl
Lap-1	—	7.6 1.3 44.6	8.2 2.3 29.2	5.1 1.4 16.1	ni	11.7 3.5 45.6	ns	ns	ns	ns	ns	ns	ns	ns	ns
Env-2	5/24	—	2.4 0.1 16.6	2.3 0.1 15.5	ni	5.9 0.8 35.4	ns	ns	ns	ns	ns	ns	ns	ns	ns
Ncam	9/41	2/24	—	3.3 0.9 9.8	5.3 0.7 29.0	6.7 1.8 23.3	ns	ns	ns	ns	ns	ns	ns	ns	ns
Alp-1	8/51	2/25	7/64	—	0.0 0.0 7.9	5.7 1.4 20.6	ns	12.5 4.3 41.9	ns	ns	ns	ns	ns	ns	ns
Thy-1	0/0	0/0	4/25	0/22	—	ni	ni	ns	ns	ns	ns	ns	ns	ns	ns
Calh	11/40	4/23	8/42	7/41	0/0	—	ns	ns	ns	9.6 2.8 35.0	ns	ns	ns	ns	ns
HP815	6/23	9/23	9/21	7/23	0/0	5/20	—	1.2 0.0 12.2	ns	ns	ns	ns	ns	ns	ns
d	6/24	10/26	18/49	14/49	8/25	5/23	1/23	—	4.3 1.1 13.9	5.3 1.5 16.7	7.1 1.2 39.8	ns	ns	ns	ns

Table: Recombination data for pairs of loci. Below the diagonal, each entry is the recombination fraction (discordant/total informative strains); above the diagonal each cell lists the estimated distance (cM) followed by its lower and upper 99% confidence limits; the diagonal is shown as "—". The first eight columns correspond to loci whose labels are not shown on this portion of the table.

									Pgm-3	Mod-1	Crbp	Rsp-4	Ltw-3	Fv-2	Bgl
Pgm-3	8/24	12/26	20/47	19/49	18/41	8/23	5/23	7/51	—	1.6, 0.2, 7.3	2.0, 0.0, 8.9	4.1, 0.4, 25.2	6.5, 1.8, 22.2	ns	ns
Mod-1	19/50	13/25	23/65	26/75	15/34	10/41	7/23	8/50	3/50	—	5.6, 0.8, 31.9	6.3, 0.9, 39.8	3.9, 0.6, 17.8	10.4, 3.2, 36.9	12.8, 4.2, 46.1
Crbp	9/23	13/25	12/23	10/24	10/18	8/22	6/22	5/25	3/43	4/24	—	2.6, 0.1, 19.3	5.6, 0.0, 19.7	12.2, 3.7, 49.4	12.2, 3.7, 49.4
Rsp-4	7/21	10/23	9/21	8/22	0/0	7/20	5/20	6/23	3/23	4/22	2/22	—	6.3, 0.9, 39.8	ns	6.7, 0.9, 45.4
Ltw-3	14/31	16/25	14/30	15/32	10/18	11/29	10/23	9/25	8/43	4/32	7/42	4/22	—	3.2, 0.6, 11.6	6.7, 2.0, 21.7
Fv-2	22/44	15/24	21/41	22/44	7/16	15/40	12/24	12/24	13/40	11/43	11/39	7/21	5/47	—	5.7, 1.6, 18.7
Bgl	12/31	12/24	11/29	13/31	6/16	10/28	11/23	11/24	13/40	13/45	13/40	11/39	9/47	8/47	—

[a]Below the diagonal, each entry shows the recombination fraction (discordant strains found/total informative strains examined) for the two loci. The corresponding cell above the diagonal shows the estimated distance between the two loci in centimorgans, calculated according to Taylor (1978), and the lower and upper 99% confidence limits of that estimate in centimorgans, calculated according to Silver (1985). Strain distribution patterns for Ncam, Thy-1 (part), and Calh were determined by Southern blotting analysis (D'Eustachio et al., 1985; P. D'Eustachio, T. Hunter, and B. Tack, unpublished observations). Those for other loci are from Taylor's (1981) compilation and from other publications. d: Jenkins et al., 1981. Alp-1, d, Pgm-3, Mod-1, Ltw-3: Nadeau et al., 1981. HP815: Russell et al., 1981. Env-2: Blatt et al., 1983. Lap-1, Alp-1: Lusis et al., 1983. Thy-1: Hayes et al., 1984. Rsp-4: Elliott, 1986. Alp-1: Goldman and Pikus, 1986. Crbp: Demmer et al., 1987. ni, not informative (no RI mice were informative for this pair of loci). ns, not significant (the 99% upper confidence limit for this recombination fraction is greater than or equal to 50 cM).

map and the multilocus RI map yield the same order for all loci examined, and the same estimates of distances in the proximal region of the chromosome between *Lap-1* and *Alp-1*. The distances measured among more distal markers were consistently larger in the backcross map than in the RI map. The cause of this difference is unknown. It is unlikely to reflect selection events occurring during the construction of the RI strains (Blank et al., 1986; Taylor, 1978). It could reflect real differences in chromosome length among different inbred strains of mice. This latter possibility seems plausible in light of the hetero-geneous origins of these strains discussed above.

The *sg* mutant locus can be incorporated into the multilocus map by interpolation. Analysis of animals from an intercross informative for *sg* and *d* indicated a distance between these two loci of 4.5 ± 0.8 cM, with *sg* proximal (Green and Lane, 1967). The position thus derived for *sg*, approximately 4 cM distal to *Ncam*, is consistent with the data discussed above, suggesting that the *sg* function might affect the processing of the N-CAM polypeptide rather than the polypeptide itself.

This chapter has discussed various approaches to the problem of relating a gene's physical localization to its function in a mammalian system. Each of the approaches—analysis of interspecies somatic cell hybrids, the use of restriction fragment length polymorphism to search for linkage relationships among genes, and the use of multilocus statistical analyses to construct and refine linkage maps—provides distinct information. Hybrid cell panels provide a one-step assay for distinguishing linked from unlinked genes and thus also for distinguishing clustered from dispersed multigene families. Linkage analysis, facilitated by the abundant polymorphism found among inbred strains of mice, allows genes to be localized more precisely and tested for relationships to known mutations. Finally, the ability to construct detailed and accurate maps raises the possibility, by no means fully realized at this point, of beginning to attack fundamental questions concerning the ordering of genes in the genome: how the order arose over evolutionary time, what sorts of forces act to maintain or disrupt it, and how it affects present-day gene function.

ACKNOWLEDGMENTS

Work in the author's laboratory was supported by grant GM-32105 from the National Institutes of Health and research grant 1-1018 from the National Science Foundation/March of Dimes.

REFERENCES

Antonucci, T. K., O. H. von Deimling, B. B. Rosenblum, L. C. Skow, and M. H. Meisler (1984) Conserved linkage within a 4-cM region of mouse chromosome 9 and human chromosome 11. *Genetics* **107**:463–475.

Bailey, D. W. (1981) Recombinant inbred strains and bilineal congenic strains. In *The Mouse in Biomedical Research*, Vol. I, H. L. Foster, J. D. Small, and J. G. Fox, eds., pp. 223–239, Academic, New York.

Barker, D., M. Schafer, and R. White (1984) Restriction sites containing CpG show a higher frequency of polymorphism in human DNA. *Cell* **36**:131–138.

Bishop, D. T. (1985) The information content of phase-known matings for ordering genetic loci. *Genet. Epidemiol.* **2**:349–361.

Blank, R. D., G. R. Campbell, and P. D'Eustachio (1986) Possible derivation of the laboratory mouse genome from multiple wild *Mus* species. *Genetics* **114**:1257–1269.

Blank, R. D., G. R. Campbell, A. Calabro, and P. D'Eustachio (1988) A linkage map of mouse chromosome *12*: Localization of *Igh* and effects of sex and interference on recombination. *Genetics* **120**:1073–1083.

Blatt, C., K. Mileham, M. Haas, M. N. Nesbitt, M. E. Harper, and M. I. Simon (1983) Chromosomal mapping of the mink cell focus-inducing and xenotropic *env* gene family in the mouse. *Proc. Natl. Acad. Sci. USA* **80**:6298–6302.

Botstein, D., R. White, M. Skolnick, and R. W. Davis (1980) Construction of a genetic map in man using restriction fragment length polymorphisms. *Am. J. Hum. Genet.* **32**:314–331.

Bucan, M., T. Yang-Feng, A. M. Colberg-Poley, D. J. Wolgemuth, J.-L. Guenet, U. Francke, and H. Lehrach (1986) Genetic and cytogenetic localisation of the homeo box containing genes on mouse chromosome 6 and human chromosome 7. *EMBO J.* **5**:2899–2905.

Chuong, C.-M., and G. M. Edelman (1984) Alterations in neural cell adhesion molecules during development of different regions of the nervous system. *J. Neurosci.* **4**:2354–2368.

Cunningham, B. A., J. J. Hemperly, B. A. Murray, E. A. Prediger, R. Brackenbury, and G. M. Edelman (1987) Neural cell adhesion molecule: Structure, immunoglobulin-like domains, cell surface modulation, and alternative RNA splicing. *Science* **236**:799–806.

Davisson, M. T., and T. H. Roderick (1986) Mouse linkage map. *Mouse News Lett.* **77**:11–15.

Demmer, L. A., E. H. Birkenmeier, D. A. Sweetser, M. S. Levin, S. Zollman, R. S. Sparkes, T. Mohandas, A. J. Lusis, and J. I. Gordon (1987) The cellular retinol binding protein II gene. Sequence analysis of the rat gene, chromosomal localization in mice and humans, and documentation of its close linkage to the cellular retinol binding protein gene. *J. Biol. Chem.* **262**:2458–2467.

D'Eustachio, P. (1984) A genetic map of mouse chromosome 12 composed of polymorphic DNA fragments. *J. Exp. Med.* **160**:827–838.

D'Eustachio, P., A. L. M. Bothwell, T. K. Takaro, D. Baltimore, and F. H. Ruddle (1981) Chromosomal location of structural genes encoding murine immunoglobulin λ light chains. *J. Exp. Med.* **153**:795–800.

D'Eustachio, P., B. Fein, J. Michaelson, and B. A. Taylor (1984) The α-globin pseudogene on mouse chromosome 17 is closely linked to *H-2*. *J. Exp. Med.* **159**:958–963.

D'Eustachio, P., G. C. Owens, G. M. Edelman, and B. A. Cunningham (1985) Chromosomal location of the gene encoding the neural cell adhesion molecule (N-CAM) in the mouse. *Proc. Natl. Acad. Sci. USA* **82**:7631–7635.

D'Eustachio, P., D. R. Colman, and J. L. Salzer (1988) Chromosomal location of the mouse gene that encodes the myelin-associated glycoproteins. *J. Neurochem.* **50**:589–593.

Dlouhy, S. R., B. A. Taylor, and R. C. Karn (1987) The genes for mouse salivary androgen-binding protein (ABP) subunits alpha and gamma are located on chromosome 7. *Genetics* **115**:535–543.

Edelman, G. M., and C.-M. Chuong (1982) Embryonic to adult conversion of neural cell adhesion molecules in normal and staggerer mice. *Proc. Natl. Acad. Sci. USA* **79**:7036–7040.

Eicher, E. M., B. A. Taylor, S. C. Leighton, and J. E. Womack (1980) A serum protein polymorphism determinant on chromosome 9 of *Mus musculus*. *Mol. Gen. Genet.* **177**:571–576.

Elliott, R. W. (1986) A mouse "minisatellite." *Mouse News Lett.* **74**:115–117.

Field, E. H., B. Tourvieille, P. D'Eustachio, and J. R. Parnes (1987) The gene encoding the mouse T cell differentiation antigen L3T4 is located on chromosome 6. *J. Immunol.* **138**:1968–1970.

Fitch, W. M., and W. R. Atchley (1985) Evolution in inbred strains of mice appears rapid. *Science* **228**:1169–1175.

Fox, H. S., L. M. Silver, and G. R. Martin (1983) An alpha-globin pseudogene is located within the mouse *t* complex. *Immunogenetics* **19**:129–135.

Geliebter, J., and S. G. Nathenson (1987) Recombination and the concerted evolution of the murine MHC. *Trends Genet.* **3**:107–112.

Goff, S. P., P. D'Eustachio, F. H. Ruddle, and D. Baltimore (1982) Chromosomal assignment of the endogenous proto-oncogene c-*Abl*. *Science* **218**:1317–1319.

Goldman, D., and H. J. Pikus (1986) Fourteen genetically variant proteins of mouse brain: Discovery of two new variants and chromosomal mapping of four loci. *Biochem. Genet.* **24**:183–194.

Green, E. L. (1981a) *Genetics and Probability in Animal Breeding Experiments,* Oxford Univ. Press, New York.

Green, M. C., ed. (1981b) *Genetic Variants and Strains of the Laboratory Mouse,* Gustav Fischer, Stuttgart.

Green, M. C., and P. W. Lane (1967) Linkage group II of the house mouse. *J. Hered.* **58**:225–228.

Hayes, C. E., K. K. Klyczek, D. P. Krum, R. M. Whitcomb, D. A. Hullett, and H. Cantor (1984) Chromosome 4 *Jt* gene controls murine T cell surface I-J expression. *Science* **223**:550–563.

Henderson, S. A., and R. G. Edwards (1968) Chiasma frequency and maternal age in mammals. *Nature* **218**:22–28.

Hummel, K. P. (1970) *Hypodactyly,* a semidominant lethal mutation in mice. *J. Hered.* **61**:219–220.

Hunkapiller, T., and L. Hood (1986) The growing immunoglobulin gene superfamily. *Nature* **323**:15–16.

Jeffreys, A. J., V. Wilson, and S. L. Thein (1985) Hypervariable "minisatellite" regions in human DNA. *Nature* **314**:67–73.

Jenkins, N. A., N. G. Copeland, B. A. Taylor, and B. K. Lee (1981) Dilute (*d*) coat colour mutation of DBA/2J mice is associated with the site of integration of an ecotropic MuLV genome. *Nature* **293**:370–374.

Killary, A. M., and R. E. K. Fournier (1984) A genetic analysis of extinction: *Trans*-dominant loci regulate expression of liver-specific traits in hepatoma hybrid cells. *Cell* **38**:523–534.

Lathrop, G. M., J. M. Lalouel, C. Julier, and J. Ott (1984) Strategies for multilocus linkage analysis in humans. *Proc. Natl. Acad. Sci. USA* **81**:3443–3446.

Leder, A., D. Swan, F. Ruddle, P. D'Eustachio, and P. Leder (1981) Dispersion of α-like globin genes of the mouse to three different chromosomes. *Nature* **293**:196–200.

Lewin, B. (1980) *Gene Expression,* 2nd Ed., pp. 962–964, Wiley, New York.

Lusis, A. J., B. A. Taylor, R. W. Wangenstein, and R. C. LeBoeuf (1983) Genetic control of lipid transport in mice. II. Genes controlling structure of high density lipoproteins. *J. Biol. Chem.* **258**:5071–5078.

Mann, E. A., L. M. Silver, and R. W. Elliott (1986) Genetic analysis of a mouse *t* complex locus that is homologous to a kidney cDNA clone. *Genetics* **114**:993–1006.

McNutt, W. (1962) Urinary amino acid excretion in quivering mice (*qvqv*). *Anat. Rec.* **142**:257.

Nadeau, J. H. (1986) A chromosomal segment conserved since divergence of lineages leading to man and mouse: The gene order or aminoacylase-1, transferrin, and beta-galactosidase on mouse chromosome 9. *Genet. Res.* **48**:175–178.

Nadeau, J. H., and B. A. Taylor (1984) Lengths of chromosomal segments conserved since divergence of man and mouse. *Proc. Natl. Acad. Sci. USA* **81**:814–818.

Nadeau, J. H., J. Kömpf, G. Siebert, and B. A. Taylor (1981) Linkage of *Pgm-3* in the house mouse and homologies of three phosphoglucomutase loci in mouse and man. *Biochem. Genet.* **19**:465–474.

Nakamura, Y., M. Leppert, P. O'Connell, R. Wolff, T. Holm, M. Culver, C. Martin, E. Fujimoto, M. Hoff, E. Kumlin, and R. White (1987) Variable number of tandem repeat (VNTR) markers for human gene mapping. *Science* **235**:1616–1622.

Nesbitt, M. N., and U. Francke (1973) A system of nomenclature for band patterns of mouse chromosomes. *Chromosoma* **41**:145–158.

Rice, M. C., and S. J. O'Brien (1980) Genetic variance of laboratory outbred Swiss mice. *Nature* **283**:157–161.

Rubin, M. R., L. E. Toth, M. D. Patel, P. D'Eustachio, and M. C. Nguyen-Huu (1986) A mouse homeo box gene is expressed in spermatocytes and embryos. *Science* **233**:663–667.

Rubin, M. R., W. King, L. E. Toth, I. S. Sawczuk, M. S. Levine, P. D'Eustachio, and M. C. Nguyen-Huu (1987) Murine *Hox-1.7* homeo box gene: Cloning, chromosomal location, and expression. *Mol. Cell Biol.* **7**:3836–3841.

Russell, J. H., C. B. Dobos, R. J. Graff, and B. A. Taylor (1981) Genetic control of cross-reactive cytotoxic T-lymphocyte responses to a BALB/c tumor. *Immunogenetics* **14**:263–272.

Searle, A. G. (1964) The genetics and morphology of two "luxoid" mutants in the house mouse. *Genet. Res.* **5**:171–197.

Seki, T., H.-C. Chang, T. Moriuchi, R. Denome, H. Ploegh, and J. Silver (1985) A hydrophobic transmembrane segment at the carboxyl terminus of Thy-1. *Science* **227**:649–651.

Silver, J. (1985) Confidence limits for estimates of gene linkage based on analysis of recombinant inbred strains. *J. Hered.* **72**:436–440.

Silver, J., and C. E. Buckler (1986) Statistical considerations for linkage analysis using recombinant inbred strains and backcrosses. *Proc. Natl. Acad. Sci. USA* **83**:1423–1427.

Taylor, B. A. (1978) Recombinant inbred strains: Use in gene mapping. In *Origins of Inbred Mice*, H. C. Morse, III, ed., pp. 423–438, Academic, New York.

Taylor, B. A. (1981) Recombinant inbred strains. In *Genetic Variants and Strains of the Laboratory Mouse*, M. C. Green, ed., pp. 397–407, Gustav Fischer, Stuttgart.

Tulchin, N., and B. A. Taylor (1981) Gamma-glutamyl cyclotransferase: A new genetic polymorphism in the mouse (*Mus musculus*) linked to *Lyt-2*. *Genetics* **99**:109–116.

Yoon, C. H., and E. P. Les (1957) Quivering, a new first chromosome mutation in mice. *J. Hered.* **48**:176–180.

Section 2

Structure and Interactions of Substrate Adhesion Molecules

Chapter 6

Cell Interaction Sites of Fibronectin in Adhesion and Metastasis

MARTIN J. HUMPHRIES
KENNETH M. YAMADA

ABSTRACT

The interaction of cells with extracellular matrices is important for the regulation of cell morphology, migration, and morphogenesis. A number of glycoprotein factors that mediate such events have now been identified, and structure–function analyses of their adhesive activity are progressing rapidly. In this chapter we discuss one of the principal adhesion proteins, fibronectin. Immunological studies have previously implicated fibronectin in a plethora of developmental events, and in order to furnish specific probes for obtaining more information about the adhesive functions of fibronectin in vivo, there has been considerable interest in identifying and characterizing those regions of the fibronectin molecule that interact with cell surfaces. The current state of our knowledge with regard to cell-binding sequences in fibronectin is presented, together with a comprehensive summary of biological processes thought to employ each of these sites.

The adhesion of cells to extracellular matrices is now acknowledged to be a critical component of many biological processes. This specialized class of cellular interactions is particularly interesting, not only because it plays a role in migratory events during embryonic development and in tissue organization, but also because it affects cellular differentiation, growth promotion, and tumor metastasis (Hay, 1981; Liotta et al., 1983; Edelman and Thiery, 1985; McCarthy et al., 1985; Reddi, 1985; Trelstad, 1985). It is apparent that a detailed understanding of these processes at the molecular level will yield important information regarding the functional role of the extracellular matrix *in vivo*.

Fibronectin is the most intensely studied noncollagenous adhesion factor, and has served as a prototype for studies analyzing the biochemistry and molecular biology of cell adhesion (for recent reviews, see Furcht, 1983; Yamada, 1983; Mosher, 1984, 1988; Hynes, 1985, 1987; Yamada et al., 1985; Ruoslahti and Pierschbacher, 1986; Akiyama and Yamada, 1987; Buck and Horwitz, 1988). In this chapter we describe in detail those studies that have led to the identification of peptide sequences within fibronectin that mediate its interactions with cells. A current summary of the biological processes thought to employ each of these cell-binding sites is also presented, together with our perspective of the areas in which future investigations are likely to be concentrated.

STRUCTURE OF FIBRONECTIN

Fibronectin is a large, dimeric glycoprotein with an approximate subunit molecular weight of 250–280 kD. Two interchain disulfide bonds link the subunits together at sites close to their C-termini (Figure 1). Fibronectin is widely distributed in the interstitial matrices of normal tissues and is also present at high concentration in plasma and other body fluids. Historically, the two major forms of fibronectin have been termed *plasma* and *cellular* in deference to their sites of origin. However, these and other forms of fibronectin, for example, amniotic fluid fibronectin (Balian et al., 1979b; Ruoslahti et al., 1981b) and astrocyte fibronectin (Price and Hynes, 1985), share similar biological, biochemical, and immunochemical properties; these findings suggest the existence of only slight structural differences between isotypes. Indeed, it now appears that there is only a single fibronectin gene (Kornblihtt et al., 1983; Tamkun et al., 1984; Akiyama and Yamada, 1985c) and that each variant form of the protein arises as a result of the complex alternative splicing of mRNA transcripts (see below).

Fibronectin is both N- and O-glycosylated (Mosesson et al., 1975; Yamada et al., 1977; Fukuda and Hakomori, 1979; Takasaki et al., 1979; Zhu et al., 1984; Skorstengaard et al., 1986b), sulfated (Dunham and Hynes, 1978; Wilson et al., 1981; Paul and Hynes, 1984; Liu and Lipmann, 1985), and phosphorylated (Teng and Rifkin, 1979; Ali and Hunter, 1981; Ledger and Tanzer, 1982). While the oligosaccharide chains of fibronectin have been implicated in protection against proteolysis (Olden et al., 1978, 1979; Bernard et al., 1982), enhancement of cell adhesive function (Jones et al., 1986), and modulation of gelatin-binding activity (Zhu and Laine, 1985), the function of the other posttranslational

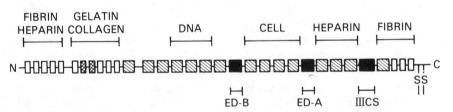

Figure 1. *Modular structure of fibronectin.* Each box represents one primary structure homology unit: Type I units are *open,* type II are *double-hatched,* and type III are *single-hatched.* The locations of interchain disulfide bonds and ligand-binding domains are shown, together with sites of alternative splicing (*solid boxes*).

modifications is currently unknown. With the exception of the spliced segments and other posttranscriptional modifications, the structure of each fibronectin molecule appears to be identical.

Physicochemical analysis using fluorescence and circular dichroism techniques originally suggested a structural model for fibronectin in which each subunit chain was composed of a series of tightly folded globular domains interconnected by short, flexible polypeptide segments (Alexander et al., 1978, 1979; Colonna et al., 1978). It is in the globular domains that many of the functional ligand-binding activities of fibronectin now appear to be located. Since extended polypeptide segments are usually more accessible to proteolytic attack than tightly folded region are, limited proteolysis has been the method of choice for isolation of each of the structural domains of fibronectin. The ability of many laboratories to identify and purify discrete fragments of fibronectin that retained many of the functional activities of the intact molecule has been instrumental not only in mapping active domains along the subunit, but also in furthering our understanding of how fibronectin might perform many of its biological functions (for reviews, see Mosesson and Amrani, 1980; Mosher, 1980; Pearlstein et al., 1980; Ruoslahti et al., 1981a; Hynes and Yamada, 1982; Vartio et al., 1983b; Yamada, 1988).

Primary Structure of Fibronectin

Extensive primary structure data, obtained by a combination of protein and DNA sequencing, is currently available from four species (human, rat, bovine, and chicken). Protein sequencing of bovine plasma fibronectin was the first approach to yield structural information (Skorstengaard et al., 1982, 1984, 1986a,b; Petersen et al., 1983), and subsequent limited protein sequencing of human fibronectin (Pande and Shively, 1982; Pierschbacher et al., 1982; Garcia-Pardo et al., 1983, 1984, 1985, 1987; Gold et al., 1983; Calaycay et al., 1985; Pande et al., 1985, 1987) and cDNA and genomic cloning data for human (Kornblihtt et al., 1983, 1984a,b, 1985; Oldberg et al., 1983; Vibe-Pedersen et al., 1984; Bernard et al., 1985; Umezawa et al., 1985; Sekiguchi et al., 1986), rat (Schwarzbauer et al., 1983; Tamkun et al., 1984; Odermatt et al., 1985), and chicken fibronectins (Hirano et al., 1983, 1986) have provided additional insight into the structure of the fibronectin subunit.

An intriguing finding arising from sequencing of bovine plasma fibronectin was that the vast majority of the molecule is composed of a series of homologous polypeptide repeats. These units fall into three basic forms, termed types I, II, and III (Petersen et al., 1983). Both type I and type II repeats are double-disulfide loop structures of approximately 40–50 amino acids in length, and are now thought to be encoded by single exons (Hirano et al., 1983; Owens and Baralle, 1986b). Type III repeats are longer (approximately 90 residues), lack disulfide bonds, and are usually encoded by two exons (Vibe-Pedersen et al., 1984; Odermatt et al., 1985; Oldberg and Ruoslahti, 1986). The repeat structure of the fibronectin subunit is consistent with the results of

R-loop analysis of the exon structure of the chicken fibronectin gene (Hirano et al., 1983). This study identified at least 48 exons, the central 46 of which possessed very similar sizes (147 ± 37 base pairs). Overall, there are 12 type I, 2 type II, and at least 15 type III homology units in fibronectin (Figure 1). In addition, there are three sites in the molecule where alternative splicing takes place. Two of the spliced segments comprise entire type III units; the third is nonhomologous.

It is interesting that the different homology units that make up fibronectin are concentrated in discrete domains of the molecule. The N-terminal domain, which possesses binding sites for fibrin (Hormann and Seidl, 1980; Sekiguchi et al., 1981), heparin (Hormann and Seidl, 1980; Sekiguchi and Hakomori, 1980), actin (Keski-Oja and Yamada, 1981), gangliosides (Thompson et al., 1986), and some bacterial cell surfaces (Mosher and Proctor, 1980), is composed of a tandem array of five type I repeats (Figure 1). Four more type I repeats are located in the adjacent collagen-binding domain (Balian et al., 1979a; Gold et al., 1979; Hahn and Yamada, 1979b; Ruoslahti et al., 1979). This region also contains the only two type II repeats in fibronectin (Figure 1). The unique presence of type II homologies is suggestive evidence of their involvement in collagen binding, and recent studies with fusion proteins containing short segments of the collagen-binding domain seem to confirm this hypothesis (Owens and Baralle, 1986a). The double disulfide loop motif of the type II homologies of fibronectin closely resembles that of a "kringle" structure (Patthy et al., 1984), a primordial protein-binding unit that appears to have been incorporated into a number of proteins through gene duplication and has subsequently diversified with time to yield specialized ligand-binding domains.

After a 50-amino-acid stretch with no apparent homology to the rest of the fibronectin molecule, there follows a continuous array of all 15 type III homology units, interrupted only by alternatively spliced segments (Figure 1). Within this central region are situated consecutive binding sites for DNA (Pande and Shively, 1982), eukaryotic cell surfaces (Hahn and Yamada, 1979a; Ruoslahti and Hayman, 1979), and heparin (Hayashi et al., 1980; Sekiguchi and Hakomori, 1980; Yamada et al., 1980). Each of these domains has been separated by limited proteolysis and affinity chromatography, and each region has been sequenced at the protein level from both bovine and human plasma fibronectins (Pierschbacher et al., 1982; Pande et al., 1985, 1987; Skorstengaard et al., 1986a). The resistance of type III units to proteolysis suggests that they can form tightly folded structures even in the absence of disulfide bonds. Furthermore, the finding that those regions of type III homology that possess biological activity can be separated from each other following digestion suggests that the junctions between certain repeats are unfolded and therefore more accessible to proteolytic attack.

The final three type I repeats follow the heparin-binding domain and form a region specialized for the binding of fibrin (Figure 1) (Sekiguchi et al., 1981; Seidl and Hormann, 1983). It is notable that both fibrin-binding sites are located in regions containing exclusively type I repeats, while the two sites interacting with heparin are found in regions of type I and type III homology. A structure homologous to the type I repeat has been identified in tissue plasminogen activator (t-PA), but not in the related enzyme, urokinase (Banyai et al., 1983; Ny et al., 1984). Interestingly, t-PA binds to fibrin but urokinase does not. One common denominator between the two heparin-binding regions is their net positive charge ($+2$ for the N-terminal domain and $+9.5$ for the higher affinity C-terminal domain). This cationic character would be expected for a region supporting a predominantly hydrophilic interaction with a negatively charged biopolymer such as a glycosaminoglycan. The extreme C-terminus of fibronectin is composed of 26 amino acids containing the half-cystines involved in interchain bonding (Figure 1) (Petersen et al., 1983; Garcia-Pardo et al., 1984).

Subunit Variation

While it is now well established that all fibronectins contain the same arrangement of ligand-binding domains, electrophoretic analysis of fibro-nectins isolated from different tissues has revealed variations in subunit molecular weight (Balian et al., 1979b; Yamada and Kennedy, 1979; Hayashi and Yamada, 1981; Ruoslahti et al., 1981b; Price and Hynes, 1985), and in certain cases such as plasma fibronectin, a doublet band profile is obtained that cannot be accounted for by the most common posttranslational modifications (Yamada and Kennedy, 1979; Paul and Hynes, 1984). Furthermore, some internal proteolytic fragments display size differences when isolated either from different subunits or from different fibronectin isotypes (Hayashi and Yamada, 1981, 1983; Sekiguchi et al., 1981, 1985; Sekiguchi and Hakomori, 1983a,b), and there is also some immunological evidence for differences between cellular and plasma fibronectins (Atherton and Hynes, 1981; Vartio et al., 1983a). The structural basis for subunit differences has now been shown conclusively, through the analysis of cDNA clones, to result from alternative mRNA splicing of internal segments of the fibronectin molecule (Schwarzbauer et al., 1983, 1985; Kornblihtt et al., 1984a,b, 1985; Tamkun et al., 1984; Vibe-Pedersen et al., 1984; Umezawa et al., 1985). It should be noted that in addition to these modifications, there is also evidence that fibronectin is initially synthesized as a prepropolypeptide with a 26-amino-acid signal sequence and a 5-amino-acid prosequence (Gutman et al., 1986; Dean et al., 1987).

The alternative splicing pattern of fibronectin is complex. In rat fibronectin, one region of difference (termed V) has been pinpointed to a site between the final two type III repeats at the junction of the C-terminal heparin- and fibrin-binding domains. Inserts of 0, 285, or 360 base pairs have been detected that encode 0, 95, or 120 amino acids, respectively (Figure 2) (Schwarzbauer et al., 1983). The 95-residue insert lacks the N-terminal 25 amino acids of the 120-residue insert. Nucleotide sequencing revealed consensus splice junctions consistent with the processing of fibronectin mRNA to these products, and S1

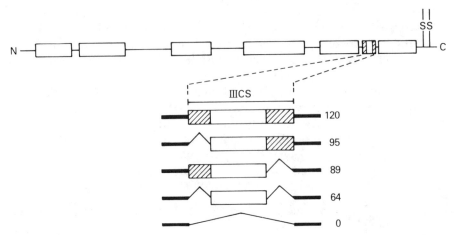

Figure 2. *Potential spliced products from the IIICS/V region of fibronectin.* In human fibronectin all five products containing the indicated number of amino acid residues can be made; in rat fibronectin only the 120, 95, and 0 variants are possible. The *hatched* segments contain cell adhesion sites for melanoma cells (see text).

nuclease mapping demonstrated that all forms are synthesized *in vivo* (Schwarzbauer et al., 1983; Tamkun et al., 1984). In human fibronectin, this region of the molecule has been termed the type III connecting segment or IIICS. An additional splice site is present in the human mRNA molecule (Kornblihtt et al., 1984b; Bernard et al., 1985; Vibe-Pedersen et al., 1986), and clones containing nucleotide sequences encoding 0, 64, 89, or 120 amino acids have been detected (Kornblihtt et al., 1984b; Bernard et al., 1985; Umezawa et al., 1985; Sekiguchi et al., 1986). The 64-amino-acid insert is homologous to the first 64 residues of the rat 95-residue segment, and the 89-amino-acid insert contains both the 64-amino-acid and the N-terminal 25-residue segments (Figure 2). A fifth variant in this region is probably made but has not been detected as yet. This segment would be equivalent to the rat 95-residue insert. In both rat and human fibronectins, the V/IIICS region is encoded in one exon (Tamkun et al., 1984; Vibe-Pedersen et al., 1986). Hence, variants arise by alternative splicing of a single complex coding region rather than by simple exon skipping.

Two other regions of variation have been detected in human and rat cDNAs. These are termed extra domains or EDs (Kornblihtt et al., 1984a,b, 1985; Odermatt et al., 1985; Umezawa et al., 1985; R. O. Hynes, A. R. Kornblihtt, and F. E. Baralle, personal communications). In contrast to nonspliced type III units, both comprise entire type III homology repeats present as single exons in the genome (Vibe-Pedersen et al., 1984; Odermatt et al., 1985) and are spliced in or out of mRNA in toto. Both ED regions are located adjacent to the central cell-binding domain, with ED-B at the N-terminus and ED-A at the C-terminus. If all possible combinations of spliced products are made, there are potentially 20 different human fibronectins and 12 different rat fibronectins. When this complexity is combined with potential differences in carbohydrate, phosphate, and sulfate, there is ample scope for the subtle regulation of function through minor changes in subunit structure.

Currently, there are only limited data concerning the relative distribution of the different forms of fibronectin within tissues. S1 nuclease mapping of cDNA/RNA hybrids showed that liver cells produced only mRNA lacking ED-A, whereas a number of other cell types, including fibroblasts, synthesized some fibronectin molecules that contained the spliced region (Kornblihtt et al., 1984b; Colombi et al., 1986; Sekiguchi et al., 1986). By raising antibodies to fusion proteins containing the ED regions of fibronectin, it has proved possible, in a limited number of cases, to determine which molecules contain which spliced segments. Antibodies to either ED-A or ED-B do not recognize plasma fibronectin but both cross-react with cellular fibronectin (Paul et al., 1986; R. O. Hynes, A. R. Kornblihtt, and F. E. Baralle, personal communications). Since plasma fibronectin appears to originate in the liver (Tamkun and Hynes, 1983), these findings are consistent with the S1 nuclease mapping studies discussed above, and suggest that the ED regions may be exclusively included in the tissue form of the fibronectin molecule. In addition, it is notable that the quantity of ED-A incorporated into fibronectin is elevated after the transformation of human cells (Castellani et al., 1986). Currently, the function of the ED regions is unknown.

A similar approach to the V/IIICS region suggests that this sequence constitutes the difference between the subunits of the fibronectin dimer. Two-dimensional electrophoresis has permitted the separation of fibronectin subunits that differ in both molecular weight and isoelectric point (Tamkun and Hynes, 1983). Antibodies raised against the rat 95-amino-acid segment stain only the large subunits of rat plasma and cellular fibronectins in immunoblotting experiments (Schwarzbauer et al., 1985; Paul et al., 1986). Quantitation of the relative abundance of each possible variant of rat fibronectin revealed that 64% of the plasma molecules contain at least part of the V region, while this figure is 91% in cellular fibronectin (Paul et al., 1986). Furthermore, based on an approach quantitating different proteolytic products arising from variations in the content of spliced polypeptides, the subunits of human cellular fibronectin have been shown to contain more of the IIICS than the plasma fibronectin subunits (Castellani et al., 1986). The distribution of the various forms of the IIICS between fibronectin isotypes is currently unclear. However, liver mRNA variants containing 0, 64, or 95 amino acids of the IIICS have been identified (Umezawa et al., 1985; Sekiguchi et al., 1986), and there are data from protein sequencing of proteolytic fragments to suggest that at least one variant exists that contains the first 25 amino acids of the IIICS (Garcia-Pardo et al., 1987; Pande et al., 1987). As discussed in detail below, the identification of cell interaction sites in this region of fibronectin suggests that regulation of the processing of mRNA transcripts may have implications for certain cell adhesive processes. It is also likely that in the future, important functions will be identified for the ED regions, perhaps in fibronectin fibrillogenesis or matrix assembly.

Even from this brief review it is clear that the modular structure of the fibronectin molecule is ideally suited to performing biological functions through cooperative ligand-binding activities. Among the many questions that can now be posed are: Which amino acid residues make up active sites for the binding of different ligands? To what extent do the separate binding domains interact to perform biological functions? Owing to the fact that fibronectin is primarily a cell adhesion molecule, it is the eukaryotic cell-binding activity of the molecule that has received the most attention in terms of answering these questions. The remainder of this review centers on those studies that have (1) identified cell-binding sequences in fibronectin, and (2) examined which biological processes are mediated through each of these active sites.

THE CENTRAL CELL-BINDING DOMAIN

Proteolytic Fragment and Synthetic Peptide Studies

By employing the technique of limited proteolysis, several laboratories have isolated large polypeptides that do not bind to gelatin affinity columns but that continue to bind to cells (Hahn and Yamada, 1979a; Ruoslahti and Hayman, 1979; McDonald and Kelley, 1980; Sekiguchi and Hakomori, 1980; Ehrismann et al., 1981; Hayashi and Yamada, 1983). Assignment of cell-binding activity was based on assays measuring either the attachment or the spreading of fibroblastic cells on the substrate-adsorbed fragments. These adhesion-promoting polypeptides ranged in molecular weight from 75 kD to 160 kD. Where tested, cell-binding fragments were generally found to possess specific activity similar to that of intact fibronectin, suggesting that they contained the principal cell-binding site(s) in the parent molecule. Analysis of partial digests containing overlapping fragments subsequently mapped the active region to the center of the fibronectin subunit (Figure 3) (Hahn and Yamada, 1979a; McDonald and Kelley, 1980; Ehrismann et al., 1981).

In a complementary approach to adhesion studies, the direct binding of tritiated fibronectin, or its cell-binding fragments, to fibroblastic cells in suspension has been used to examine the nature of the interaction of fibronectin with the cell surface. Under these conditions, there is no influence of either the extracellular matrix or substrate immobilization of the ligand on the interaction, as occurs in other assays. Standard ligand-binding experiments were performed, and the resulting Scatchard analysis predicted a single class of fibronectin receptor with a moderate affinity (0.8 μM; Figure 3) (Akiyama and Yamada, 1985a). Similar studies with a purified 75-kD cell-binding fragment revealed a slightly higher affinity (0.4 μM; Figure 3) (Akiyama et al., 1985). This may be due to increased exposure of cell-binding sequences after proteolysis. The active region in the central cell-binding domain has been narrowed down further by two approaches:

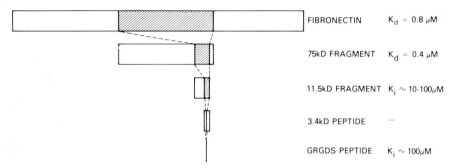

Figure 3. *The central cell-binding domain of human plasma fibronectin.* One site mediating the adhesion of fibroblastic cells to this region of fibronectin has been narrowed down progressively from intact fibronectin, through proteolytic fragments of molecular weight 75 kD and 11.5 kD, to the synthetic peptide GRGDS. Estimates of the affinities of each of these agents for the fibroblast cell surface are shown at the *right.*

1. Exhaustive peptic cleavage of a 120-kD cell adhesion–promoting chymotryptic polypeptide produced a series of fragments, one of which was specifically retained by an affinity matrix composed of a monoclonal antibody with adhesion-blocking activity (Pierschbacher et al., 1981). This fragment was sequenced and found to contain 108 amino acids with a calculated molecular weight of 11.5 kD (Figure 3) (Pierschbacher et al., 1982). Under suitable conditions, the 11.5-kD fragment supported cell adhesion, but it was less active on a molar basis than either intact fibronectin or larger cell-binding fragments (Pierschbacher et al., 1982), and has been shown to bind with lower affinity to the fibroblast cell surface (Figure 3) (Akiyama et al., 1985). From analysis of a series of four synthetic peptides that together spanned the entire 11.5-kD fragment, one active site was localized to a 3.4-kD polypeptide comprising its C-terminal sequence (Figure 3) (Pierschbacher et al., 1983). Systematic testing of progressively smaller synthetic peptides subsequently identified a short determinant that now appears to represent the minimal active sequence within this region of fibronectin. The active site contains the tetrapeptide Arg-Gly-Asp-Ser (RGDS; Figure 3) (Pierschbacher and Ruoslahti, 1984a; Yamada and Kennedy, 1984). RGDS or longer peptides bearing short extensions at either terminus were found to block both cell adhesion to fibronectin (Pierschbacher and Ruoslahti, 1984a,b; Yamada and Kennedy, 1984, 1985; Akiyama and Yamada, 1985b; Hayman et al., 1985; Horwitz et al., 1985; Silnutzer and Barnes, 1985) and the direct binding of tritiated fibronectin to fibroblastic cells in suspension (Akiyama and Yamada, 1985b), indicating that this sequence is indeed critical for the cell adhesive function of the parent molecule. The presence of an N-terminal glycine residue before RGDS (e.g., in GRGDS or GRGDSP) generally yields significantly higher activities (Pierschbacher and Ruoslahti, 1984a; Yamada and Kennedy, 1985).

2. In a biological approach based on competitive inhibition of cell adhesion, fibronectin, the 75-kD fragment representing the central cell-

binding domain, and hydrophilic synthetic peptides containing the RGDS sequence were all found to autoinhibit fibroblastic cell spreading on fibronectin (Yamada and Kennedy, 1984). This approach has provided evidence for the biological interaction of the RGDS sequence with surface fibronectin receptors; furthermore, the finding that the inhibitory activity of each of these cell-binding polypeptides could be overcome by increasing the level of fibronectin adsorbed onto the substrate suggested that this interaction was a direct one, and was not mediated through surface molecules unrelated to the fibronectin receptor (Yamada and Kennedy, 1984).

Subsequent to the identification of the RGDS recognition site in fibronectin, the sequences of other adhesion factors have been examined for related signals. A number of these proteins contain RGD sequences, including fibrinogen, collagens (Pierschbacher and Ruoslahti, 1984a), serum-spreading factor/vitronectin (Hayman et al., 1985; Jenne and Stanley, 1985; Suzuki et al., 1985), von Willebrand factor (Sadler et al., 1985), and osteopontin (Oldberg et al., 1986). The somewhat surprising finding that such a small peptide segment accounts for a key aspect of the biological activity of certain adhesion factors has nonetheless been supported by a series of careful structure–function studies that have examined the specificity of cell interactions with the RGD sequence (reviewed in detail by Ruoslahti and Pierschbacher, 1986; Akiyama and Yamada, 1987).

A general consensus that has emerged from studies of the inhibition of fibroblast adhesion to fibronectin by peptide variants of RGDS is that no alterations in the first three residues of the tetrapeptide can be tolerated, but that the fourth residue, while it must be present, can be replaced by a variety of alternative amino acids (Pierschbacher and Ruoslahti, 1984a,b; Yamada and Kennedy, 1984, 1985; Hayman et al., 1985; Pierschbacher et al., 1985; Silnutzer and Barnes, 1985). The absolute requirement for RGD is apparent from analysis of the activities of peptide homologues bearing very conservative substitutions in each position. For example, KGDS (lysine for arginine), RADS (alanine for glycine), and RGES (glutamic acid for aspartic acid) possess negligible inhibitory activity in adhesion assays. Similarly, both the order and the spacing of the charged residues within RGDS are critical, since the synthetic peptides GDGRS, GRDGS, and GRGGDS are inactive. The RGD sequence alone, with or without a blocked C-terminus, displays either good (M. D. Pierschbacher and E. Ruoslahti, personal communication) or minimal (D. Kennedy, K. M. Yamada, and M. J. Humphries, unpublished data) activity. It may also be relevant that over 100 proteins (3% of those sequenced) contain the tripeptide RGD, many of which are not adhesion proteins (for discussion, see Yamada, 1988).

Thus, despite the obvious role of the RGDS signal in fibronectin-mediated adhesion, this sequence is unlikely to be the sole polypeptide region contributing to the activity of the intact molecule. As discussed above, the 11.5-kD fragment containing RGDS is less active in supporting adhesion than whole fibronectin is, and the direct binding of either tritiated 11.5-kD fragment or tritiated GRGDS to fibroblastic cells could not be measured because their affinity is substantially lower than the parent molecule (Akiyama and Yamada, 1985b; Akiyama et al., 1985). Nevertheless, both the 11.5-kD fragment and GRGDS competitively inhibit the direct binding of labeled fibronectin, and from double reciprocal analyses, estimated K_i values of 10–100 μM have been deduced (Figure 3) (Akiyama et al., 1985). These findings imply that in the transition from intact fibronectin/75-kD fragment to 11.5-kD fragment/RGDS, polypeptide information required for the maintenance of optimal activity is lost. The losses are large, ranging from 50- to >100-fold differences in estimated affinities (Akiyama et al., 1985). Two equally plausible explanations to account for this discrepancy are: (1) that the 75-kD fragment contains a second site, distinct from RGDS, that contributes significantly to the receptor-binding activity of intact fibronectin, and (2) that the 75-kD fragment contains polypeptide regions required for the correct folding of the RGDS determinant into a fully active conformation.

DNA Cloning and Expression Studies

The second site apparently required for enhancing the binding of the GRGDS recognition site to the cell surface has been mapped more precisely by recombinant DNA methods (Obara et al., 1987). This approach potentially permits extremely accurate mapping of key sites and the ability to mutate at will any polypeptide region in order to determine its function. A cDNA clone spanning 60% of the protein sequence of human fibronectin was constructed by ligating preexisting clones. As a first test of the system, cDNA segments corresponding to the 75-kD and 11.5-kD proteolytic fragments described above were expressed in the λgt11 bacteriophage expression system in E. coli (Obara et al., 1987). The protein products consisted of fibronectin fragments fused to the "carrier" protein β-galactosidase. These fusion proteins were purified by antibody affinity chromatography using a monoclonal antibody directed against the 11.5-kD region, and their biological activities were quantitated in a standard cell-spreading assay.

The large cDNA segment containing the full 75-kD cell-binding domain was nearly as active as an equimolar amount of intact fibronectin (Table 1) (Obara et al., 1987). This result is important in demonstrating that the various eukaryotic posttranscriptional modifications of fibronectin, for example, glycosylation, phosphorylation, and sulfation, have little role in the final biological activity of fibronectin as measured by this in vitro adhesion assay; the small decrease in activity may reflect a minor contribution from N-linked glycosylation to adhesive function (Jones et al., 1986). Fusion proteins produced from a cDNA segment corresponding to the 11.5-kD fragment displayed approximately 50-fold less activity (Table 1) (Obara et al., 1987), consistent with the results above using proteolytic fragments; the presence of the large β-galactosidase carrier rules out any artifacts that are due to adsorption of

Table 1. Activities of Adhesion Proteins[a]

Protein	Relative Specific Activity
Fibronectin	74
λCBD1	54
λCBD20	1
β-Galactosidase	0

[a]The relative adhesion-promoting activities of fibronectin and two fusion proteins, λCBD1 and λCBD20, the sequences of which cover approximately the 75-kD (Hayashi and Yamada, 1983) and 11.5-kD (Pierschbacher et al., 1981, 1982) cell-binding polypeptides, respectively, were estimated using a standard BHK cell-spreading assay (Obara et al., 1987). The results are consistent with the direct binding data presented in Figure 3.

only a short peptide to the substrate, and ELISA assays showed that equal amounts of antigen actually coated the substrates (Obara et al., 1987).

This analysis was extended by constructing a nested set of deletion mutants spanning the region between the largest and the smallest cell-binding domain fragments using Ba1-31 nuclease digestion of a cDNA construct, followed by expression of the truncated protein in the λgt11 system (Obara et al., in preparation). Adhesion-promoting activity remained high as the protein was shortened, until a point equivalent to approximately 25–30 kD of polypeptide chain N-terminal to the RGDS site. Activity dropped precipitously as this region was deleted, and it remained constant past the junction of the 11.5-kD fragment. These results indicate that the crucial second site is located a substantial distance along the polypeptide backbone away from the RGDS adhesive recognition site. This second site will need to be mapped further by local deletions and mutagenesis to establish whether it is indeed an adhesive recognition sequence, or whether it functions as a conformational determinant. It may be pertinent that even if fibronectin is denatured with 8 M urea, then immobilized by adsorption onto the substrate in a denatured form, then renatured by washing away the urea, it displays substantial activity (M. Kang, unpublished data); this result suggests that conformational effects might not be involved, or that they can be rapidly reconstituted after denaturation.

An alternative approach to this question is to examine the function of the cell-binding domain after inactivation of the GRGDS site by mutation, then to do complementation analysis with the putative first and second adhesive recognition sequences located in separate polypeptides. In preliminary experiments, the Asp residue in RGDS was mutated to a Glu residue (to match the synthetic peptide RGES) by a single base substitution using site-directed mutagenesis. The fibronectin fusion protein containing this mutation showed greatly reduced activity, confirming the critical importance of this primary adhesive recognition site (M. Obara and M. Kang, unpublished data). However, this fusion protein displayed some residual biological activity, consistent with either the presence of a second adhesion site or incomplete inactivation of RGDS by the mutation.

Two complementary mutated polypeptides of very low biological activity were thus available, one mutated in the first site (RGES fusion protein) and one missing the putative second site (11.5-kD fragment). In very preliminary experiments, assaying a mixture of these mutated proteins revealed a dramatic synergism of activity (M. Obara and M. Kang, preliminary data). Since such "trans" complementation of activity does not involve any covalent association of one site with the other, it is highly unlikely that one site is a conformational determinant of the other. Instead, there now appear to be two distinct adhesive recognition sites in this major cell-binding domain. This unexpected complexity may provide an explanation for the fact that many proteins contain RGD sequences, yet only fibronectin is recognized by the fibronectin receptor—specificity and sufficient affinity for binding would be provided by this crucial second site. This hypothesis can be tested rigorously in the future by site-directed mutagenesis of this interesting new sequence.

BIOLOGICAL FUNCTIONS OF THE CENTRAL CELL-BINDING DOMAIN

A ubiquitous tissue distribution has been frequently cited as suggestive evidence for the involvement of fibronectin in biological processes *in vivo*. Correlative studies linking the localization of fibronectin-containing matrices with the presence of migratory cell populations have implicated fibronectin in the migration of amphibian primordial germ cells (Heasman et al., 1981), avian neural crest cells (Newgreen and Thiery, 1980; Mayer et al., 1981; Duband and Thiery, 1982a; Thiery and Duband, 1982), and avian precardiac mesoderm cells (Linask and Lash, 1986). Fibronectin also appears to play an early role in amphibian and avian gastrulation (Critchley et al., 1979; Duband and Thiery, 1982b; Boucaut and Darribere, 1983; Boucaut et al., 1984a), avian somitogenesis (Lash et al., 1984), and sea urchin ingression (Katow et al., 1982); a later temporal accumulation of fibronectin occurs at sites of tissue remodeling during formation of feather buds (Mauger et al., 1982).

During development, fibronectin is lost from differentiating cartilage and muscle cells (Linder et al., 1975; Stenman and Vaheri, 1978). The re-addition of exogenous cellular fibronectin blocks chondrogenesis and myoblast fusion, demonstrating that the loss of adhesion factors may also have profound effects on organogenesis (Pennypacker et al., 1979; Podleski et al., 1979; West et al., 1979). In the adult organism, fibronectin appears soon after vascular wounding, and, in combination with fibrin, provides a substrate for fibroblast adhesion (Grinnell, 1984). Epithelial cells also appear to utilize fibronectin during the healing of corneal wounds (Nishida et al., 1983).

Historically, treatment with purified antibodies has been employed for specific abrogation of the function of molecules *in vivo* and *in vitro*. Thus, treatment with anti-fibronectin antibodies has been shown to inhibit a variety of morphogenetic events, including avian neural crest cell migration (Rovasio et al., 1983) and amphibian gastrulation (Boucaut et al., 1984a). Nevertheless, despite these clear effects, the conclusions that can be drawn from an immunological approach are necessarily limited, because it can never be proved absolutely that the observed effects of the antibody are due to specific inhibition of antigen-related functions, rather than resulting from secondary effects such as cross-linking, cell surface modulation, or mild cytotoxicity.

The generation of synthetic peptide inhibitors therefore represents a novel, complementary approach for examining the role of fibronectin and other molecules *in vivo*. Peptides would be expected to function differently from antibodies in that their inhibitory activities should be competitive and readily reversible. Critical to this approach is that the peptides be specific. Fortunately, specificity can be tested rather convincingly by synthesis of homologues differing only in the substitution or deletion of critical amino acids.

The adhesion of many different cell types has now been examined for sensitivity to RGDS-containing peptides. These include fibroblasts (Piersch-bacher and Ruoslahti, 1984a; Yamada and Kennedy, 1984; Hayman et al., 1985; Horwitz et al., 1985; Silnutzer and Barnes, 1985), endothelial cells (Hayman et al., 1985), platelets (Gartner and Bennett, 1985; Ginsberg et al., 1985; Haverstick et al., 1985), pigmented epithelial cells (Avery and Glaser, 1986), thymocytes (Cardarelli and Pierschbacher, 1986), macrophages and other hemopoietic cells (Wright and Meyer, 1985; Giancotti et al., 1986; Patel and Lodish, 1986), *Dictyostelium discoideum* (Springer et al., 1984), and a variety of tumor cell lines (Hayman et al., 1985; Silnutzer and Barnes, 1985; Akeson and Warren, 1986; McCarthy et al., 1986). In each case adhesion to fibronectin was inhibited by peptides containing the authentic sequence, but when tested was unaffected by control peptides based on an RGES sequence. These results therefore provide presumptive evidence for recognition of the RGDS sequence in the central cell-binding domain by each of these cell types. In addition, synthetic peptides modeled on the RGDS sequence have received considerable attention for their use in examining the role of this adhesive signal *in vivo*.

RGDS and Development

In studies using whole embryos, microinjection of RGDS-containing peptides was found to prevent gastrulation in both amphibian and insect embryos (Boucaut et al., 1984b, 1985; Naidet et al., 1987), as well as neural crest cell migration in avian embryos (Boucaut et al., 1984b). In the amphibian a layer of fibronectin-containing extracellular matrix is deposited on the roof of the blastocoel just prior to invagination. It is located along the path of active cell migration during formation of the mesodermal cell layer. Using *Pleurodeles waltlii* as a model, early studies that examined the effects of microinjection of the peptides RGDSPASSKP and GRGDSPC into blastula or early gastrula stage embryos revealed dramatic morphological effects (Boucaut et al., 1984b). In scanning electron micrographs, ectodermal cells were seen to form a convoluted cap with deep furrows consistent with the piling up of non-invaginating mesodermal cells. The usual circular blastopore and yolk plug were replaced by a wide slit that separated ectodermal and endodermal cells.

Sectioning of embryos confirmed the accumulation of rounded, noninvading mesodermal cells. Injection of an equal concentration of a control peptide unrelated to the RGDS peptides did not affect gastrulation, demonstrating the specificity of RGDS peptide activity (Boucaut et al., 1984b).

A distortion of morphogenetic events similar to those seen in amphibia was also observed in recent studies in which three different peptides containing the RGDS sequence were injected into *Drosophila* embryos (Naidet et al., 1987). No ventral furrow was formed, development of the cephalic furrow was delayed, and the dorsoventral polarity that usually develops at this time was lost. Deep folding of ectodermal cell layers took place over the entire surface of the embryo.

The neural crest is an embryonic tissue that lies on the dorsal edge of the neural tube. Neural crest cells are highly motile and undergo temporal detachment and migration away from the neural tube to form a variety of cell types, including peripheral neurons, pigment cells, and several mesenchymal tissues in the head and neck region (Le Douarin, 1982). Injection of the decapeptide RGDSPASSKP into chick embryos was found to dramatically inhibit neural crest cell migration (Boucaut et al., 1984b). Rather than passing ventrally through the cell-free spaces that lie laterally under the ectoderm, neural crest cells failed to leave the dorsal side of the neural tube, instead forming a compact colony at their site of origin.

The chick system has also provided evidence to suggest that fibronectin, and in particular the RGDS signal, may be involved in the compaction of presomitic cells, a process that takes place prior to the actual separation of individual somites from the segmental plate. Culture of isolated segmental plates in the presence of low concentrations of GRGDS resulted in accelerated cell condensation and morphogenetic segmentation, and when the culture was repeated with individual segmental plate cells, dramatic, reversible increases in cell–cell adhesion were observed that mirror the changes ordinarily observed *in vivo* (Lash and Yamada, 1986; Lash et al., 1987).

In an assay system designed to mimic the process of implantation, trophoblast cells from the mammalian blastocyst have been found to attach to and grow out onto fibronectin-coated substrata (Armant et al., 1986a). This adhesion is blocked by the hexapeptide GRGDSP, but not by GRGESP (Armant et al., 1986b), suggesting that the acquisition of adhesiveness by the blastocyst may be related to induction of surface molecules that specifically recognize this sequence.

Yet another biological process that appears to use the RGDS signal *in vivo* is that of hemopoietic precursor cell homing to the embryonic thymus gland. During normal development, in a process resembling that of tumor metastasis, T-cell precursors migrate to the thymus via either the vasculature or lymphatic system, extravasate through a basement membrane, and intercalate into the thymic epithelium. A model system based on the human amnion basement membrane was used to show that the invasion of thymic cells was inhibited by a number of anti-adhesive agents, including anti-fibronectin and anti-laminin antibodies, anti-fibronectin receptor antibodies, and the synthetic peptide GRGDS (but not GRGES; Savagner et al., 1986).

RGDS and Cell Interactions in Diseases

In addition to developmental studies, RGDS-containing peptides have also been used *in vivo* for studies investigating other biological events thought to involve cell adhesion. One such process is the colonization of target organs during tumor metastasis. The pentapeptide GRGDS was found to be an active inhibitor of this phase of the metastatic cascade, using a model system in which cultured B16–F10 melanoma cells are injected into the tail vein of syngeneic mice, and pulmonary metastases are scored 14–21 days later (Humphries et al., 1986a). The inhibitory effect of GRGDS was dependent on the number of melanoma cells injected, and if the inoculum size was restricted to $1-3 \times 10^4$, essentially complete blockage of colonization was observed (Humphries et al., submitted). This degree of inhibition translates into a substantial prolongation of life for mice receiving peptide (Figure 4) (Humphries et al., submitted), and indicates that agents based on the RGDS sequence may have future applications as prophylactics or adjuvants for the treatment of metastasis. Potential uses of such an agent would be to prevent seeding of malignant tumor cells either during surgical removal of a primary tumor or during infusion of blood cells as part of an autologous bone marrow transplantation protocol.

Kinetic analysis of the number of melanoma cells bound to the target organ revealed a rapid effect of the peptide in lowering retention, a finding that would be consistent with a mechanism of action involving interference in cell adhesion or cell invasion (Humphries et al., 1986a). RGDS has been shown previously to disrupt the haptotactic migration of B16–F10 melanoma cells in an assay that may be relevant for the extravasation phase of metastatic colonization (McCarthy et al., 1986). An early effect of GRGDS is also

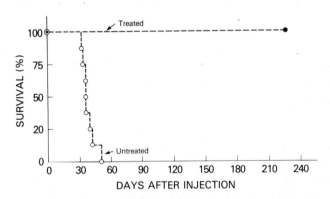

Figure 4. *Effect of GRGDS on survival of mice challenged intravenously with B16–F10 melanoma cells.* 1×10^4 melanoma cells were injected with (treated) or without (untreated) 3 mg GRGDS into the lateral tail vein of groups of eight mice, and survival was examined as a function of time.

predicted from its rapid vascular clearance rate; the half-life of the labeled molecule in the circulation was only eight minutes (Humphries et al., submitted). The inhibition of experimental metastasis by GRGDS was highly specific, since several peptides unrelated to the RGDS sequence and two homologs bearing minor sequence changes (GRGES and GRDGS) were almost completely inactive (Humphries et al., 1986a). Recently, this specificity was tested further by examining the effects of a family of homologs bearing substitutions, deletions, or inversions within the GRGDS sequence (Humphries et al., submitted). The overall profile of inhibition by these molecules closely matched their established capacity to inhibit cell adhesion *in vitro*, further supporting a role for the RGDS recognition signal in metastatic colonization; even the reverse tetrapeptide SDGR (RGDS inverted) possessed moderate activity (Humphries et al., submitted).

Although it is most likely that RGDS-containing peptides abrogate metastasis by interfering with tumor cell adhesive or migratory processes, there are several other potential sites of action. One of these is the inhibition of tumor cell–platelet binding, an event known to be important in mediating the lodgment of tumor cell emboli (Gasic, 1984; Mehta, 1984). There are now many studies documenting the ability of platelets to utilize the RGD sequences found in a variety of extracellular ligands, including fibronectin, vitronectin, fibrinogen, von Willebrand factor, and thrombospondin (Gartner and Bennett, 1985; Ginsberg et al., 1985; Haverstick et al., 1985; Plow et al., 1985a,b; Lam et al., 1987). However, when the effect of GRGDS on experimental metastasis was investigated in animals with compromised platelet function (either thrombocytopenic mice treated with an anti-platelet antiserum or mice fed acetylsalicylic acid), there was no loss of activity, indicating that platelets were not required for the biological effect (Humphries et al., submitted). This conclusion is also supported indirectly by the finding that SDGR possesses antimetastatic activity; this peptide has already been shown to have no effect on platelet adhesion (Haverstick et al., 1985). RGDS does, however, block platelet aggregation, and related peptides may have potential applications as antithrombotic agents; initial attempts have been made to increase the activity of the RGDS molecule specifically for this purpose (Ruggeri et al., 1986).

Based on a number of recent studies, it now appears that certain bacteria and parasitic organisms have evolved to employ the cell-binding activity of fibronectin as a means of infecting their host. The protozoan parasite that causes Chagas' disease, *Trypanasoma cruzi*, has been shown to utilize fibronectin as a bridge for mediating its interaction with mammalian cells; this binding was blocked by a variety of synthetic peptides based on the RGDS sequence (Ouaissi et al., 1986). Furthermore, the immunization of mice with the decapeptide AVTGRGDSPC conjugated to tetanus toxin induced a significant protection against *T. cruzi* infection, suggesting that the RGDS sequence is recognized *in vivo* by the parasitic cells (Ouaissi et al., 1986). In similar studies, the syphilis bacterium, *Treponema pallidum*, has been shown to bind specifically to the isolated cell-binding domain of fibronectin (Thomas et al., 1985a). This recognition is thought to be the mechanism by which the bacterium parasitizes host cells. The heptapeptide GRGDSPC specifically inhibited the binding of isolated cell-binding domain to *T. pallidum,* and reduced bacterial attachment to monolayers of cultured epithelial or fibrosarcoma cells (Thomas et al., 1985b).

THE TYPE III CONNECTING SEGMENT ADHESION SITE

Following the identification of RGDS as a crucial adhesive determinant in the central cell-binding domain of fibronectin, extensive analyses of the sequence specificity of the activity of this peptide have been performed. Based on the consensus sequence required for activity, it has proved possible to design a select library of peptide molecules containing various deletions, substitutions, or inversions that can theoretically be used to test the amino acid specificity of cell adhesive events on ligands containing RGDS-related sequences. One such library contains the peptides GRGDS, RGDS, SDGR, GRGES, GRGD, and GRDGS (for sequence definitions, see Table 2). GRGDS and RGDS contain the authentic recognition determinant, SDGR is an inversion of the RGDS sequence, GRGES contains a conservative substitution of glutamic acid for aspartic acid, GRGD lacks the C-terminal serine residue, and GRDGS contains a transposition of the central glycine and aspartic acid residues.

Differences were observed in the inhibitory activities of the components of this library when either BHK cells or chick embryo fibroblasts were examined for spreading on substrates coated with fibronectin, laminin, vitronectin, or native collagen gels. For fibronectin and vitronectin, (G) RGDS was the most active peptide, whereas for laminin and native collagen gels the reverse peptide SDGR was most effective (Yamada and Kennedy, 1987). These results demonstrated that adhesion to different ligands depends on distinct peptide recognition events, and that information can be obtained concerning the nature of an unknown substrate from the peptide sensitivity of cell adhesion to that substrate.

This approach can be used in a similar way to compare the fibronectin recognition properties of different cell types. When BHK fibroblasts were compared with B16–F10 melanoma cells for the relative inhibitory activity of

Table 2. Activities of Synthetic Peptides for Fibroblastic Cell Spreading on Fibronectin

Peptide	Relative Activity
Gly-Arg-Gly-Asp-Ser (GRGDS)	+++
Arg-Gly-Asp-Ser (RGDS)	++
Ser-Asp-Gly-Arg (SDGR)	+
Gly-Arg-Gly-Glu-Ser (GRGES)	−
Gly-Arg-Gly-Asp (GRGD)	−
Gly-Arg-Asp-Gly-Ser (GRDGS)	−

the synthetic peptide library on fibronectin-mediated cell spreading, a number of differences were noted (Humphries et al., 1986b). Although GRGDS, RGDS, and SDGR were all inhibitory for both cell types, the relative activities of each were different. For BHK cells the order of activity was GRGDS > RGDS > SDGR. For B16–F10 cells it was RGDS > SDGR > GRGDS. Furthermore, while GRDGS, GRGD, and GRGES were almost completely inactive for BHK cells, two peptides, GRGD and GRGES, possessed substantial activity for B16–F10 cells (Humphries et al., 1986b). In particular, the inhibitory activity of GRGES contrasts markedly with all previous studies of fibroblastic cell adhesion (e.g., Pierschbacher and Ruoslahti, 1984a; Yamada and Kennedy, 1985). Two of the most likely explanations for these results were: (1) that the two cell types might be recognizing different sites within the fibronectin molecule, and (2) that the surface fibronectin receptors on B16–F10 cells might be modified structurally, resulting in a less specific recognition of the RGDS signal. To test the first hypothesis directly, different proteolytic fragments of fibronectin were tested for their ability to support adhesion of the two cell types.

The purified fragments [numbered according to the nomenclature of Hayashi and Yamada (1983) and shown diagrammatically in Figure 5] were

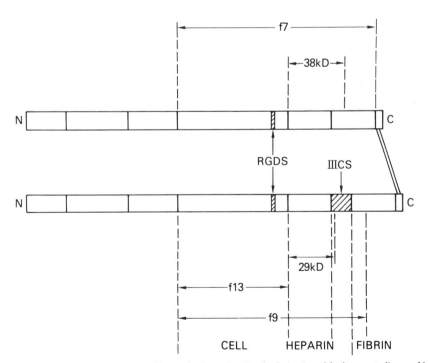

Figure 5.   *Map of the human plasma fibronectin dimer showing the derivation of the fragments discussed in the text. The sites of the RGDS tetrapeptide and the IIICS region are* hatched.

f13 (a 75-kD polypeptide containing RGDS but devoid of heparin-binding activity), f9 (a 113-kD fragment from the large fibronectin subunit incorporating f13 together with the adjacent heparin-binding domain, the alternatively spliced IIICS region, and part of the C-terminal fibrin-binding domain), and f7 (a 146-kD fragment from the small fibronectin subunit containing f13 and the heparin- and fibrin-binding domains). Since the small fibronectin subunit appears to lack the alternatively spliced V/IIICS (Schwarzbauer et al., 1985; Paul et al., 1986), this region represents the only polypeptide segment present in f9 but absent in f7 (Figure 5).

From analysis of the adhesion-promoting activity of these fragments for BHK fibroblasts and B16–F10 melanoma cells, three general findings were apparent (Humphries et al., 1986b): (1) BHK cells spread as well on all three fragments as on intact fibronectin. This result was expected, since all four molecules contain the RGDS recognition signal. (2) B16–F10 cells spread extremely poorly on f13: intact fibronectin was approximately 75-fold better on a molar basis. This finding suggested that GRGDS was not the primary recognition site in fibronectin for B16–F10 cells, a result confirmed by monoclonal antibody and f13 autoinhibition studies (Humphries et al., 1986b). (3) B16–F10 cells spread to an equivalent extent on fibronectin and f9, while spreading on f7 was again poor (10-fold worse than f9). The differential activity of f7 and f9 implies that any sequence common to both fragments would be unlikely to act as the melanoma cell recognition site. Thus, by deduction, the principal region of human plasma fibronectin recognized by B16–F10 cells was determined to be the IIICS. However, since previous studies had implicated the C-terminal heparin-binding domain in cell adhesion and cell contact formation (Laterra et al., 1983a,b; Beyth and Culp, 1984; Lark et al., 1985; Izzard et al., 1986; Woods et al., 1986; Waite et al., 1987), proteolytic fragments of 29 kD and 38 kD representing this region from the large and small fibronectin subunits, respectively, were isolated and tested for promotion of B16–F10 cell adhesion (Figure 5). Neither fragment displayed any activity, apparently discounting the possibility that melanoma cells were recognizing the heparin-binding domain (Humphries et al., 1986b).

The active site(s) mediating melanoma cell adhesion were narrowed down further by synthesizing six peptides that spanned the entire 120 amino acids of the IIICS and testing them using complementary approaches for quantitating adhesion-promoting and -inhibiting activity (Figure 6). Only two of the six peptides were able to specifically and reversibly block fibronectin-mediated spreading of B16–F10 melanoma cells, CS1 (residues 1–25 of the IIICS), and CS5 (residues 90–109). CS1 was approximately 70-fold more active than CS5 on a molar basis, the former exhibiting half-maximal inhibition at only 5.5 μM (Humphries et al., 1987). None of the six CS peptides had any significant inhibitory activity for BHK fibroblast spreading on fibronectin.

As a more direct test of the cell recognition properties of these peptides, all six molecules were synthesized with an N-terminal cysteine residue, and were covalently coupled to IgG via the heterobifunctional cross-linker N-succinimidyl-

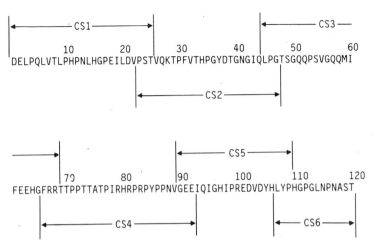

Figure 6. *Scheme for synthesis of peptides spanning the IIICS.* Overlaps of three amino acids were introduced to ensure that every possible tetrapeptide was contained within at least one peptide. The two internal splice sites are between residues 25/26 and 89/90.

3(2-pyridyl-dithio) propionate. CS-IgG conjugates were then tested for adhesion-promoting activity. Again, CS1-IgG and CS5-IgG were active, with a similar difference in relative activity being observed (CS1-IgG was approximately two orders of magnitude more active than CS5-IgG; Humphries et al., 1987). Consistent with the results of inhibition studies, none of the CS-IgG conjugates was able to support BHK fibroblast spreading. Quantitation of the amount of CS1-IgG required to support optimal spreading revealed that the conjugate was only 2.5-fold less active than the parent fibronectin molecule (Humphries et al., 1987). This unexpected finding of unusually high activity implicates the first 25 residues of the IIICS as being quantitatively the principal melanoma cell recognition site in human plasma fibronectin.

The adhesion-promoting activities of CS1-IgG and CS5-IgG were additive, as were the inhibitory activities of the free peptides, suggesting that both sites can function in tandem, and that either site alone or both in combination could comprise a cell recognition sequence. As would be predicted from this scenario, the inhibitory activity of the CS1 and CS5 peptides could be overcome by increasing the amount of fibronectin adsorbed onto the tissue culture dish surface prior to adhesion assays (Humphries et al., 1987).

The importance of the identification of the IIICS as a cell interaction site is heightened by the potential for alternative splicing of this region in precursor mRNA. Of the five potential spliced variants of the IIICS in human fibronectin mRNA, four have been identified to date (Kornblihtt et al., 1984b; Bernard et al., 1985; Umezawa et al., 1985; Sekiguchi et al., 1986). Although the relative abundance of each derivative in different tissues has not yet been determined in detail, it is notable that potential products exist that contain both the CS1 and

CS5 sequences, each sequence alone, or neither sequence (Figure 2). It will be instructive in the future to determine whether separable effects on cell function can be ascribed to the spliced segments containing CS1 and CS5, and whether the presence of either region in the intact fibronectin molecule can modulate the activity of the other.

Analysis of the hydrophilicity of the IIICS revealed three prominent stretches containing charged amino acids (Humphries et al., 1986b). These regions might be expected to be exposed on the surface of the fibronectin molecule and to be available for binding to a cell surface receptor. One of these hydrophilic peaks was located within the CS5 peptide and was represented by the tetrapeptide Arg-Glu-Asp-Val (REDV), a sequence somewhat related to RGDS. To test the idea that REDV was the sequence in CS5 recognized by B16–F10 cells, synthetic REDV was tested for its ability to inhibit cell spreading on fibronectin. The REDV peptide was indeed active, exhibiting half-maximal activity at approximately 350 μg/ml (Humphries et al., 1986b). However, as would be expected for a peptide in which the central glycine residue is replaced by a glutamic acid residue, REDV was without activity in BHK cell-spreading assays. The molar activities of CS5 and REDV were similar for melanoma cells, suggesting that REDV was the functionally important site within CS5 (Humphries et al., 1986b, 1987).

The structural similarity between the REDV and RGDS recognition signals appears to explain the profile of inhibition of melanoma cell spreading by the RGDS-related peptide library, and provides information on the degree of specificity required for normal functioning of each of these sequences. As discussed above, the overall pattern of inhibition of BHK and B16–F10 cells is similar, but with some notable exceptions. Unlike BHK, melanoma cells tolerate some substitutions in the fourth position (S or V) or the second position (G or E) of the tetrapeptide, and indeed the fourth position may not in fact be required at all (GRGD is active; Humphries et al., 1986b). Thus, it may be that the REDV sequence in the human IIICS is held in a less restricted conformation than RGDS; Chou-Fassman analysis of the polypeptide region around RGDS has predicted the tetrapeptide to be located at a β-turn (Pierschbacher and Ruoslahti, 1984a).

A further test of the sequence specificities of the REDV signal is provided by rat fibronectin, in which this sequence in the V region is present as RGDV. Schwarzbauer et al. (1983) originally called attention to this sequence, suggesting that it might be a second cell adhesion site, since synthetic RGDV was inhibitory for fibroblast adhesion to fibronectin (Pierschbacher and Ruoslahti, 1984b). Such a site would be expected to have the same cell-type specificity as RGDS. However, melanoma cell spreading on rat fibronectin appears identical to that on human fibronectin; in particular, both REDV and RGDV cross-inhibit B16–F10 cell adhesion to human and rat plasma fibronectins (Humphries et al., 1986b; Humphries et al., in preparation). It may be, therefore, that the REDV and RGDV sequences in human and rat fibronectins are functionally equivalent. If true, this finding suggests that the stereochemical orientation of the arginine and aspartic acid residues is critical, and that the second residue of the tetrapeptide is of lesser importance.

An interesting analogy is therefore apparent between the central cell-binding domain, active for fibroblasts, and the IIICS. Both sites contain an RXDX recognition sequence that appears to have much lower affinity for cell surfaces compared to intact fibronectin, and both appear to contain an additional high-affinity interaction site (CS1 in the IIICS, and the putative high-affinity site N-terminal to RGDS in the central cell-binding domain) (Figure 7).

BIOLOGICAL FUNCTIONS OF THE IIICS ADHESION SITE

The ability of B16–F10 melanoma cells, but not BHK fibroblasts, to adhere specifically to the IIICS region of human plasma fibronectin suggests that usage of this region may be cell-type specific. Melanoma cells are of neural crest origin, and it is conceivable that other, if not all, neural crest derivatives are able to use this region of fibronectin at some point in development. As a further test of the importance of the IIICS to neural crest–derived cells, we have recently examined which regions of fibronectin are required for promotion of neurite outgrowth from peripheral nervous system ganglia.

As reported previously (Rogers et al., 1983, 1985), both dorsal root ganglia and sympathetic ganglia were found to extend stable neurites on a plastic surface coated with fibronectin. Similarly, isolated neuronal cells derived from ganglia by trypsinization retained this activity. The ability of proteolytic

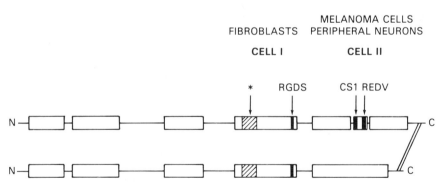

Figure 7. *Summary of the cell interaction sites in human fibronectin.* The central cell-binding domain is composed of the RGDS tetrapeptide together with a second region as far as 25–30 kD from its N-terminal site (*) that appears to contribute greatly to the affinity of this domain for the fibroblast cell surface. The less well-defined nature of the * site is indicated by *hatching*. A second cell-binding domain is located in the IIICS region. This cell type–specific domain for melanoma cells and peripheral neurons is again composed of two interaction sites, represented by the synthetic peptide sequences in CS1 and CS5. The active site in CS5 is the tetrapeptide REDV.

fragments and synthetic peptide–IgG conjugates to support neurite outgrowth was investigated to determine the active regions within fibronectin. In quantitative assays comparing the radius of neurite outgrowth on fibronectin and f13 (representing the central cell-binding domain), fibronectin was approximately twice as active on a molar basis (Figure 8) (Humphries et al., submitted). In addition, CS1-IgG, but no other CS-IgG conjugate, was able to mediate neurite extension (Figure 8). Recently, the V region of chicken fibronectin has been sequenced at the nucleotide level. Whereas the sequence of the region corresponding to CS1 is similar in human and chicken fibro-

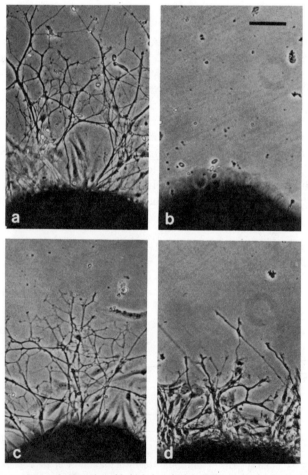

Figure 8.   *Neurite outgrowth of 11-day-old chick dorsal root ganglia incubated for 24 hours on a control substrate (b), or on substrates coated with either 20 µg/ml human plasma fibronectin (a), a molar equivalent of 75-kD cell-binding fragment (c), or 500 µg/ml CS1-IgG conjugate (d). These concentrations provided near-maximal outgrowth for each ligand. Calibration bar = 100 µm.*

nectins, chicken fibronectin appears to lack the REDV/RGDV motif found in CS5 from human and rat fibronectins (P. Norton and R. O. Hynes, personal communication). This finding appears to explain the inactivity of CS5-IgG in neurite outgrowth assays. The activity of CS1-IgG reached a maximal level equivalent to approximately 25–30% of the degree of outgrowth obtained with fibronectin (Figure 8). Co-coating of the substrate with maximally active concentrations of f13 and CS1-IgG produced an additive response, indicating the likelihood that both regions of fibronectin were functioning in the parent molecule (Humphries et al., submitted).

These studies demonstrate a number of important points. First, usage of the IIICS adhesion site is not restricted to melanoma cells, and neuronal cells are also able to recognize this region of fibronectin. Second, cells explanted directly from the embryo can employ this adhesion site and that its usage is, therefore, not an artifact of cells maintained in tissue culture. Third, additional neural crest–derived cell populations can recognize the IIICS. Currently, there is no further evidence for the usage of the IIICS by other cell types, but this question will be an important one to study in the future.

FUTURE DIRECTIONS

Although the extent of our knowledge of cell interactions with fibronectin has grown dramatically over the last several years, there are still many areas requiring additional investigation. These topics include further studies of the sites in fibronectin recognized by cell surface receptors, identification and characterization of these fibronectin-binding molecules, and investigation of the biochemical mechanisms by which cell–fibronectin interactions are modulated.

In this review we have discussed in detail the current state of our knowledge with regard to cell-binding sequences in fibronectin. In the central cell-binding domain the RGDS tetrapeptide is now well established as a crucial recognition determinant for fibroblastic and other cell types, while in the IIICS/V domain the RGDV/REDV tetrapeptide appears to account for the adhesion-promoting activity of melanoma cells in the C-terminal spliced segment of this region. Nevertheless, the critical residues that appear to be responsible for maintaining much of the affinity, and possibly specificity, of the two adhesion sites are still unknown. In the IIICS the spliced polypeptide corresponding to synthetic peptide CS1 possesses surprisingly high activity in adhesion assays, and it is likely that a number of amino acids within this peptide will be found to be crucial for cell binding.

At this writing, the least well defined site is that localized by cDNA cloning studies to the N-terminal region of the central cell-binding domain. This site appears to act in conjunction with RGDS, but the particular peptide sequences mediating its activity await identification. It is likely that by a combination of molecular biology and synthetic peptide approaches, the cell-binding sites in fibronectin will be more concisely delineated in the near future.

As a result of studies examining the role of cell surface proteoglycans and gangliosides in cell adhesion, there is now substantial evidence supporting a significant role for these molecules in determining the architecture of contact sites, as well as in supporting cell adhesion directly (Kleinman et al., 1979; Yamada et al., 1981; Laterra et al., 1983a,b; Beyth and Culp, 1984; Lark et al., 1985; Spiegel et al., 1985, 1986; Cheresh and Klier, 1986; Cheresh et al., 1986; Izzard et al., 1986; Woods et al., 1986; Waite et al., 1987). It is therefore conceivable that the heparin- and ganglioside-binding domains in fibronectin might act cooperatively with cell-binding sites to influence cell behavior. Since cell adhesion is such a complex, multifaceted phenomenon, such cooperativity could provide an additional mechanism for the generation of biological specificity. In the future, identification of these binding sites, followed by deletion or site-directed mutagenesis of critical residues and regions and testing of the mutated molecules in relevant biological assays, should permit a clearer understanding of the roles of these regions in cell adhesive events.

Owing to the relatively low affinity of fibronectin for the cell surface, it is only recently that glycoproteins with the properties expected of a fibronectin receptor have been identified. Dimeric complexes of molecular weight of approximately 140 kD appear to be involved in recognition of the RGDS-containing cell-binding site (for recent reviews, see Ruoslahti and Piersch-bacher, 1986; Akiyama and Yamada, 1987; Hynes, 1987; Horwitz and Buck, 1988; Yamada, 1988), but there is also evidence that other molecules may function as fibronectin receptors (Aplin et al., 1981; Hughes et al., 1981; Oppenheimer-Marks and Grinnell, 1981, 1982; McKeown-Longo and Mosher, 1985; Urushihara and Yamada, 1986). Structural analysis of these molecules, together with the identification of fibronectin-binding sites, should follow in the next few years. At present, the identity of the receptor for the IIICS adhesion site is unknown, but its identification will be an obvious area to explore further.

As discussed previously, the apparent cell-type specificity of the IIICS adhesion site indicates that it will be worthwhile to examine which cell types can use this region. At present only melanoma cells and peripheral neurons are known to recognize this region, and so it will be instructive to test which other neural crest–derived cell types are able to respond to the CS1 signal, and to examine whether other nonneural crest lineages are also active. Depending on the outcome of these studies, there may be many new avenues opened up for exploring the regulation of morphogenesis by alternative splicing.

Finally, the mechanisms by which adhesion to fibronectin is regulated need to be investigated. Potential sites at which modulation could occur include regulation of alternative splicing, transcriptional control of fibronectin gene expression, and fluctuations in receptor number or state of aggregation. Of particular interest will be studies examining whether changes in expression of variant forms of fibronectin occur *in vivo* during such processes as embryonic development and wound healing.

REFERENCES

Akeson, R., and S. L. Warren (1986) PC12 adhesion and neurite formation on selected substrates are inhibited by some glycosaminoglycans and a fibronectin-derived tetrapeptide. *Exp. Cell Res.* **162**:347–362.

Akiyama, S. K., and K. M. Yamada (1985a) The interaction of plasma fibronectin with fibroblastic cells in suspension. *J. Biol. Chem.* **260**:4492–4500.

Akiyama, S. K., and K. M. Yamada (1985b) Synthetic peptides competitively inhibit both direct binding to fibroblasts and functional biological assay for the purified cell-binding domain of fibronectin. *J. Biol. Chem.* **260**:10402–10405.

Akiyama, S. K., and K. M. Yamada (1985c) Comparisons of evolutionarily distinct fibronectins: Evidence for the origin of plasma and fibroblast cellular fibronectins from a single gene. *J. Cell. Biochem.* **27**:97–107.

Akiyama, S. K., and K. M. Yamada (1987) Fibronectin. In *Advances in Enzymology and Related Areas of Molecular Biology*, A. Meister, ed., pp. 1–57, Wiley, New York.

Akiyama, S. K., E. Hasegawa, T. Hasegawa, and K. M. Yamada (1985) The interaction of fibronectin fragments with fibroblastic cells. *J. Biol. Chem.* **260**:13256–13260.

Alexander, S. S., Jr., G. Colonna, K. M. Yamada, I. Pastan, and H. Edelhoch (1978) Molecular properties of a major cell surface protein from chick embryo fibroblasts. *J. Biol. Chem.* **253**:5820–5824.

Alexander, S. S., Jr., G. Colonna, and H. Edelhoch (1979) The structure and stability of human plasma cold-insoluble globulin. *J. Biol. Chem.* **254**:1501–1505.

Ali, I. U., and T. Hunter (1981) Structural comparison of fibronectins from normal and transformed cells. *J. Biol. Chem.* **256**:7671–7677.

Aplin, J. D., R. C. Hughes, C. L. Jaffe, and N. Sharon (1981) Reversible cross-linking of cellular components of adherent fibroblasts to fibronectin and lectin-coated substrata. *Exp. Cell Res.* **134**:488–494.

Armant, D. R., H. A. Kaplan, and W. J. Lennarz (1986a) Fibronectin and laminin promote *in vitro* attachment and outgrowth of mouse blastocysts. *Dev. Biol.* **116**:519–523.

Armant, D. R., H. A. Kaplan, H. Mover, and W. J. Lennarz (1986b) The effect of hexapeptides on attachment and outgrowth of mouse blastocysts cultured *in vitro:* Evidence for the involvement of the cell recognition tripeptide Arg-Gly-Asp. *Proc. Natl. Acad. Sci. USA* **83**:6751–6755.

Atherton, B. T., and R. O. Hynes (1981) A difference between plasma and cellular fibronectins located with monoclonal antibodies. *Cell* **25**:133–141.

Avery, R. L., and B. M. Glaser (1986) Inhibition of retinal pigment epithelial cell attachment by a synthetic peptide derived from the cell-binding domain of fibronectin. *Arch. Ophthalmol.* **104**:1220–1222.

Balian, G., E. M. Click, E. Crouch, J. M. Davidson, and P. Bornstein (1979a) Isolation of a collagen binding fragment from fibronectin and cold-insoluble globulin. *J. Biol. Chem.* **254**:1429–1432.

Balian, G., E. Crouch, E. M. Click, W. G. Carter, and P. Bornstein (1979b) Comparison of the structures of human fibronectin and plasma cold-insoluble globulin. *J. Supramol. Struct.* **12**:505–516.

Banyai, L., A. Varadi, and L. Patthy (1983) Common evolutionary origin of the fibrin-binding structures of fibronectin and tissue-type plasminogen activator. *FEBS Lett.* **163**:37–41.

Bernard, B. A., K. M. Yamada, and K. Olden (1982) Carbohydrates selectively protect a specific domain of fibronectin against proteases. *J. Biol. Chem.* **257**:8549–8554.

Bernard, M. P., M. Kolbe, D. Weil, and M.-L. Chu (1985) Human cellular fibronectin: Comparison of the carboxyl-terminal portion with rat identifies primary structural domains separated by hypervariable regions. *Biochemistry* **24**:2698–2704.

Beyth, R. J., and L. A. Culp (1984) Complementary adhesive responses of human skin fibroblasts to the cell-binding domain of fibronectin and the heparan sulfate-binding protein, platelet factor-4. *Exp. Cell Res.* **155**:537–548.

Boucaut, J. C., and T. Darribere (1983) Fibronectin in early amphibian embryos: Migrating mesodermal cells contact fibronectin established prior to gastrulation. *Cell Tissue Res.* **234**:135–145.

Boucaut, J. C., T. Darribere, H. Boulekbache, and J. P. Thiery (1984a) Prevention of gastrulation but not neurulation by antibodies to fibronectin in amphibian embryos. *Nature* **307**:364–367.

Boucaut, J. C., T. Darribere, T. J. Poole, M. Aoyama, K. M. Yamada, and J. P. Thiery (1984b) Biologically active synthetic peptides as probes of embryonic development: A competitive peptide inhibitor of fibronectin function inhibits gastrulation in amphibian embryos and neural crest cell migration in avian embryos. *J. Cell Biol.* **99**:1822–1830.

Boucaut, J. C., T. Darribere, S. D. Li, H. Boulekbache, K. M. Yamada, and J. P. Thiery (1985) Evidence for the role of fibronectin in amphibian gastrulation. *J. Embryol. Exp. Morphol.* **89**:211–227.

Buck, C. A., and A. F. Horwitz (1987) Cell surface receptors for extracellular matrix molecules. *Annu. Rev. Cell Biol.* **3**:179–205.

Calaycay, J., H. Pande, T. Lee, L. Borsi, A. Siri, J. E. Shively, and L. Zardi (1985) Primary structure of a DNA- and heparin-binding domain (Domain III) in human plasma fibronectin. *J. Biol. Chem.* **260**:12136–12141.

Cardarelli, P. M., and M. D. Pierschbacher (1986) T-lymphocyte differentiation and the extracellular matrix: Identification of a thymocyte subset that attaches specifically to fibronectin. *Proc. Natl. Acad. Sci. USA* **83**:2647–2651.

Castellani, P., A. Siri, C. Rosellini, E. Infusini, L. Borsi, and L. Zardi (1986) Transformed human cells release different fibronectin variants than do normal cells. *J. Cell Biol.* **103**:1671–1677.

Cheresh, D. A., and F. G. Klier (1986) Disialoganglioside GD_2 distributes preferentially into substrate-associated microprocesses on human melanoma cells during their attachment to fibronectin. *J. Cell Biol.* **102**:1887–1897.

Cheresh, D. A., M. D. Pierschbacher, M. A. Herzig, and K. Mujoo (1986) Disialogangliosides GD2 and GD3 are involved in the attachment of human melanoma and neuroblastoma cells to extracellular matrix proteins. *J. Cell Biol.* **102**:688–696.

Colombi, M., S. Barlati, A. R. Kornblihtt, F. E. Baralle, and A. Vaheri (1986) A family of fibronectin mRNAs in human normal and transformed cells. *Biochim. Biophys. Acta* **868**:207–214.

Colonna, G., S. S. Alexander, K. M. Yamada, I. Pastan, and H. Edelhoch (1978) The stability of cell surface protein to surfactants and denaturants. *J. Biol. Chem.* **253**:7787–7790.

Critchley, D. R., M. A. England, J. L. Wakely, and R. O. Hynes (1979) Distribution of fibronectin in the ectoderm of gastrulating chick embryos. *Nature* **280**:498–500.

Dean, D. C., C. L. Bowlus, and S. Bourgeois (1987) Cloning and analysis of the promoter region of the human fibronectin gene. *Proc. Natl. Acad. Sci. USA* **84**:1876–1880.

Duband, J.-L., and J. P. Thiery (1982a) Distribution of fibronectin in the early phase of avian cephalic neural crest cell migration. *Dev. Biol.* **93**:308–323.

Duband, J.-L., and J. P. Thiery (1982b) Appearance and distribution of fibronectin during chick embryo gastrulation and neurulation. *Dev. Biol.* **94**:337–350.

Dunham, J. S., and R. O. Hynes (1978) Differences in the sulfated macromolecules synthesized by normal and transformed hamster fibroblasts. *Biochim. Biophys. Acta* **506**:242–255.

Edelman, G. M., and J. P. Thiery, eds. (1985) *The Cell in Contact: Adhesions and Junctions as Morphogenetic Determinants,* Wiley, New York.

Ehrismann, R., M. Chiquet, and D. C. Turner (1981) Mode of action of fibronectin in promoting chicken myoblast attachment: $M_r = 60,000$ gelatin-binding fragment binds native fibronectin. *J. Biol. Chem.* **256**:4056–4062.

Fukuda, M., and S. Hakomori (1979) Carbohydrate structure of galactoprotein a, a major transformation-sensitive glycoprotein released from hamster embryo fibroblasts. *J. Biol. Chem.* **254**:5451–5457.

Furcht, L. T. (1983) Structure and function of the adhesive glycoprotein fibronectin. In *Modern Cell Biology*, Vol. 1, B. Satir, ed., pp. 53–117, Alan R. Liss, New York.

Garcia-Pardo, A., E. Pearlstein, and B. Frangione (1983) Primary structure of human plasma fibronectin: The 29,000-dalton NH_2-terminal domain. *J. Biol. Chem.* **258**:12670–12674.

Garcia-Pardo, A., E. Pearlstein, and B. Frangione (1984) Primary structure of human plasma fibronectin—Characterization of the 6,000 dalton C-terminal fragment containing the interchain disulfide bridges. *Biochem. Biophys. Res. Commun.* **120**:1015–1021.

Garcia-Pardo, A., E. Pearlstein, and B. Frangione (1985) Primary structure of human plasma fibronectin: Characterization of a 31,000-dalton fragment from the COOH-terminal region containing a free sulfhydryl group and a fibrin-binding site. *J. Biol. Chem.* **260**:10320–10325.

Garcia-Pardo, A., A. Rostagno, and B. Frangione (1987) Primary structure of human plasma fibronectin: Characterization of a 38 kDa domain containing the C-terminal heparin-binding site (Hep III site) and a region of molecular heterogeneity. *Biochem. J.* **241**:923–928.

Gartner, T. K., and J. S. Bennett (1985) The tetrapeptide analogue of the cell attachment site of fibronectin inhibits platelet aggregation and fibrinogen binding to activated platelets. *J. Biol. Chem.* **260**:11891–11894.

Gasic, G. J. (1984) Role of plasma, platelets, and endothelial cells in tumor metastasis. *Cancer Metastasis Rev.* **3**:99–116.

Giancotti, F. G., P. M. Comoglio, and G. Tarone (1986) Fibronectin–plasma membrane interaction in the adhesion of hemopoietic cells. *J. Cell Biol.* **103**:429–437.

Ginsberg, M. H., M. D. Pierschbacher, E. Ruoslahti, G. Marguerie, and E. F. Plow (1985) Inhibition of fibronectin binding to platelets by proteolytic fragments and synthetic peptides which support fibroblast adhesion. *J. Biol. Chem.* **260**:3931–3936.

Gold, L. I., A. Garcia-Pardo, B. Frangione, E. C. Franklin, and E. Pearlstein (1979) Subtilisin and cyanogen bromide cleavage products of fibronectin that retain gelatin-binding activity. *Proc. Natl. Acad. Sci. USA* **76**:4803–4807.

Gold, L. I., B. Frangione, and E. Pearlstein (1983) Biochemical and immunological characterization of three binding sites on human plasma fibronectin with different affinities for heparin. *Biochemistry* **22**:113–119.

Grinnell, F. (1984) Fibronectin and wound healing. *J. Cell. Biochem.* **26**:107–116.

Gutman, A., K. M. Yamada, and A. Kornblihtt (1986) Human fibronectin is synthesized as a pre-propolypeptide. *FEBS Lett.* **207**:145–148.

Hahn, L.-H. E., and K. M. Yamada (1979a) Isolation and biological characterization of active fragments of the adhesive glycoprotein fibronectin. *Cell* **18**:1043–1051.

Hahn, L.-H. E., and K. M. Yamada (1979b) Identification and isolation of a collagen-binding fragment of the adhesive glycoprotein fibronectin. *Proc. Natl. Acad. Sci. USA* **76**:1160–1163.

Haverstick, D. M., J. F. Cowan, K. M. Yamada, and S. A. Santoro (1985) Inhibition of platelet adhesion to fibronectin, fibrinogen, and von Willebrand factor substrates by a synthetic tetrapeptide derived from the cell-binding domain of fibronectin. *Blood* **66**:946–952.

Hay, E. D., ed. (1981) *Cell Biology of Extracellular Matrix*, Plenum, New York.

Hayashi, M., and K. M. Yamada (1981) Differences in domain structures between plasma and cellular fibronectin. *J. Biol. Chem.* **256**:11292–11300.

Hayashi, M., and K. M. Yamada (1983) Domain structure of the carboxyl-terminal half of human plasma fibronectin. *J. Biol. Chem.* **258**:3332–3340.

Hayashi, M., D. H. Schlesinger, D. W. Kennedy, and K. M. Yamada (1980) Isolation and characterization of a heparin-binding domain of cellular fibronectin. *J. Biol. Chem.* **255**:10017–10020.

Hayman, E. G., M. D. Pierschbacher, and E. Ruoslahti (1985) Detachment of cells from culture substrate by soluble fibronectin peptides. *J. Cell Biol.* **100**:1948–1954.

Heasman, J., R. O. Hynes, A. P. Swan, V. Thomas, and C. C. Wylie (1981) Primordial germ cells of *Xenopus* embryos: The role of fibronectin in their adhesion duriing migration. *Cell* **27**:437–447.

Hirano, H., Y. Yamada, M. Sullivan, B. de Crombrugghe, I. Pastan, and K. M. Yamada (1983) Isolation of genomic DNA clones spanning the entire fibronectin gene. *Proc. Natl. Acad. Sci. USA* **80**:46–50.

Hirano, H., S. Kobomura, M. Obara, S. Gotoh, K. Higashi, and K. M. Yamada (1986) Structure of the chicken fibronectin gene. In *International Symposium on Biology and Chemistry of Basement Membranes*, Elsevier, Amsterdam.

Hormann, H., and M. Seidl (1980) Affinity chromatography of immobilized fibrin monomer. III. The fibrin affinity center of fibronectin. *Hoppe-Seyler's Z. Physiol. Chem.* **361**:1449–1452.

Horwitz, A. F., K. Duggan, R. Greggs, C. Decker, and C. A. Buck (1985) The cell substrate attachment (CSAT) antigen has properties of a receptor for laminin and fibronectin. *J. Cell Biol.* **101**:2134–2144.

Hughes, R. C., T. D. Butters, and J. D. Aplin (1981) Cell surface molecules involved in fibronectin-mediated adhesion. A study using specific antisera. *Eur. J. Cell Biol.* **26**:198–207.

Humphries, M. J., K. Olden, and K. M. Yamada (1986a) A synthetic peptide from fibronectin inhibits experimental metastasis of murine melanoma cells. *Science* **233**:467–470.

Humphries, M. J., S. K. Akiyama, A. Komoriya, K. Olden, and K. M. Yamada (1986b) Identification of an alternatively spliced site in human plasma fibronectin that mediates cell type–specific adhesion. *J. Cell Biol.* **103**:2637–2647.

Humphries, M. J., A. Komoriya, S. K. Akiyama, K. Olden, and K. M. Yamada (1987) Identification of two distinct regions of the type III connecting segment of human plasma fibronectin that promote cell type–specific adhesion. *J. Biol. Chem.* **262**:6886–6892.

Hynes, R. O. (1985) Molecular biology of fibronectin. *Annu. Rev. Cell Biol.* **1**:67–90.

Hynes, R. O. (1987) Integrins: A family of cell surface receptors. *Cell* **48**:549–554.

Hynes, R. O., and K. M. Yamada (1982) Fibronectins: Multifunctional modular glycoproteins. *J. Cell Biol.* **95**:369–377.

Izzard, C. S., R. Radinsky, and L. A. Culp (1986) Substratum contacts and cytoskeletal reorganization of BALB/c 3T3 cells on a cell-binding fragment and heparin-binding fragments of plasma fibronectin. *Exp. Cell Res.* **165**:320–336.

Jenne, D., and K. K. Stanley (1985) Molecular cloning of S-protein, a link between complement, coagulation, and cell–substrate adhesion. *EMBO J.* **4**:3153–3157.

Jones, G. E., R. G. Arumugham, and M. L. Tanzer (1986) Fibronectin glycosylation modulates fibroblast adhesion and spreading. *J. Cell Biol.* **103**:1663–1670.

Katow, H., K. M. Yamada, and M. Solursh (1982) Occurrence of fibronectin on the primary mesen-chyme cell surface during migration in the sea urchin embryo. *Differentiation* **28**:120–124.

Keski-Oja, J., and K. M. Yamada (1981) Isolation of an actin-binding fragment of fibronectin. *Biochem. J.* **193**:615–620.

Kleinman, H. K., G. R. Martin, and P. H. Fishman (1979) Ganglioside inhibition of fibronectin-mediated cell adhesion to collagen. *Proc. Natl. Acad. Sci. USA* **76**:3367–3371.

Kornblihtt, A. R., K. Vibe-Pedersen, and F. E. Baralle (1983) Isolation and characterization of cDNA clones for human and bovine fibronectins. *Proc. Natl. Acad. Sci. USA* **80**:3218–3222.

Kornblihtt, A. R., K. Vibe-Pedersen, and F. E. Baralle (1984a) Human fibronectin: Molecular cloning evidence for two mRNA species differing by an internal segment coding for a structural domain. *EMBO J.* **3**:221–226.

Kornblihtt, A. R., K. Vibe-Pedersen, and F. E. Baralle (1984b) Human fibronectin: Cell specific alternative mRNA splicing generates polypeptide chains differing in the number of internal repeats. *Nucleic Acids Res.* **12**:5853–5868.

Kornblihtt, A. R., K. Umezawa, K. Vibe-Pedersen, and F. E. Baralle (1985) Primary structure of human fibronectin: Differential splicing may generate at least 10 polypeptides from a single gene. *EMBO J.* **4**:1755–1759.

Lam, S. C., E. F. Plow, M. A. Smith, A. Andrieux, J. J. Ryckwaert, G. A. Marguerie, and M. H. Ginsberg (1987) Evidence that arginyl-glycyl-aspartate peptides and fibrinogen gamma chain peptides share a common binding site on platelets. *J. Biol. Chem.* **262**:947–950.

Lark, M. W., J. Laterra, and L. A. Culp (1985) Close and focal contact adhesions of fibroblasts to a fibronectin-containing matrix. *Fed. Proc.* **44**:394–403.

Lash, J. W., and K. M. Yamada (1986) The adhesion recognition signal of fibronectin: A possible trigger mechanism for compaction during somitogenesis. In *Somites in Developing Embryos*, R. Bellairs, D. A. Ede, and J. W. Lash, eds., pp. 201–208, Plenum, New York.

Lash, J. W., A. W. Seitz, C. M. Cheney, and D. Ostrovsky (1984) On the role of fibronectin during the compaction stage of somitogenesis in the chick embryo. *J. Exp. Zool.* **232**:197–206.

Lash, J. W., K. K. Linask, and K. M. Yamada (1987) Synthetic peptides that mimic the adhesive recognition signal of fibronectin: Differential effects on cell–cell and cell–substratum adhesion in embryonic chick cells. *Dev. Biol.* **123**:411–420.

Laterra, J., E. K. Norton, C. S. Izzard, and L. A. Culp (1983a) Contact formation by fibroblasts adhering to heparan sulfate-binding substrata (fibronectin or platelet factor 4). *Exp. Cell Res.* **146**:15–27.

Laterra, J., J. E. Silbert, and L. A. Culp (1983b) Cell surface heparan sulfate mediates some adhesive responses of glycosaminoglycan-binding matrices, including fibronectin. *J. Cell Biol.* **96**:112–123.

Ledger, P. W., and M. L. Tanzer (1982) The phosphate content of human fibronectin. *J. Biol. Chem.* **257**:3890–3895.

Le Douarin, N. (1982) *The Neural Crest,* Cambridge Univ. Press, Cambridge, England.

Linask, K. K., and J. W. Lash (1986) Precardiac cell migration: Fibronectin localization at mesoderm-endoderm interface during directional movement. *Dev. Biol.* **114**:87–101.

Linder, E., A. Vaheri, E. Ruoslahti, and J. Wartiovaara (1975) Distribution of fibroblast surface antigen in the developing chick embryo. *J. Exp. Med.* **142**:41–49.

Liotta, L. A., C. N. Rao, and S. H. Barsky (1983) Tumor invasion and the extracellular matrix. *Lab. Invest.* **49**:636–649.

Liu, M. C., and F. Lipmann (1985) Isolation of tyrosine-O-sulfate by pronase hydrolysis from fibronectin secreted by Fujinami sarcoma virus-infected rat fibroblasts. *Proc. Natl. Acad. Sci. USA* **82**:34–37.

Mauger, A., M. Demarchez, D. Herbage, J.-A. Grimaud, M. Druguet, D. Hartmann, and P. Sengel (1982) Immunofluorescent localization of collagen types I and III, and of fibronectin during feather morphogenesis in the chick embryo. *Dev. Biol.* **94**:93–105.

Mayer, B. W., E. D. Hay, and R. O. Hynes (1981) Immunocytochemical localization of fibronectin in embryonic chick trunk and area vasculosa. *Dev. Biol.* **82**:267–286.

McCarthy, J. B., M. L. Basara, S. L. Palm, D. F. Sas, and L. T. Furcht (1985) The role of cell adhesion proteins—laminin and fibronectin—in the movement of malignant and metastatic cells. *Cancer Metastasis Rev.* **4**:125–152.

McCarthy, J. B., S. T. Hagen, and L. T. Furcht (1986) Human fibronectin contains distinct adhesion- and motility-promoting domains for metastatic melanoma cells. *J. Cell Biol.* **102**:179–188.

McDonald, J. A., and D. G. Kelley (1980) Degradation of fibronectin by human leukocyte elastase. Release of biologically active fragments. *J. Biol. Chem.* **255**:8848–8858.

McKeown-Longo, P. J., and D. F. Mosher (1985) Interaction of the 70 kilodalton amino-terminal fragment of fibronectin with the matrix-assembly receptor of fibroblasts. *J. Cell Biol.* **100**:364–374.

Mehta, P. (1984) Potential role of platelets in the pathogenesis of tumor metastasis. *Blood* **63**:55–63.

Mosesson, M. W., and D. L. Amrani (1980) The structure and biological activities of plasma fibronectin. *Blood* **56**:145–158.

Mosesson, M. W., A. B. Chen, and R. M. Huseby (1975) The cold-insoluble globulin of human plasma: Studies of its essential structural features. *Biochim. Biophys. Acta* **386**:509–524.

Mosher, D. F. (1980) Fibronectin. *Prog. Hemost. Thromb.* **5**:111–151.

Mosher, D. F. (1984) Physiology of fibronectin. *Annu. Rev. Med.* **35**:561–575.

Mosher, D. F., ed. (1988) *Fibronectin,* Academic, New York.

Mosher, D. F., and R. A. Proctor (1980) Binding and factor XIIIa–mediated cross-linking of a 27-kilodalton fragment of fibronectin to *Staphylococcus aureus. Science* **209**:927–929.

Naidet, C., M. Semeriva, K. M. Yamada, and J. P. Thiery (1987) Peptides containing the cell-attachment recognition signal Arg-Gly-Asp prevent gastrulation in *Drosophila* embryos. *Nature* **325**:348–350.

Newgreen, D., and J. P. Thiery (1980) Fibronectin in early avian embryos: Synthesis and distribution along the migration pathways of neural crest cells. *Cell Tissue Res.* **211**:269–291.

Nishida, T., S. Nakagawa, T. Awata, Y. Ohashi, K. Watanabe, and R. Manabe (1983) Fibronectin promotes epithelial migration of cultured rabbit cornea *in situ. J. Cell Biol.* **97**:1653–1663.

Ny, T., F. Elgh, and B. Lund (1984) The structure of the human tissue–type plasminogen activator gene: Correlation of intron and exon structures of functional and structural domains. *Proc. Natl. Acad. Sci. USA* **81**:5355–5359.

Obara, M., M. S. Kang, S. Rocher-Dufour, A. Kornblihtt, J. P. Thiery, and K. M. Yamada (1987) Expression of the cell-binding domain of human fibronectin in *E. coli. FEBS Lett.* **213**:261–264.

Odermatt, E., J. W. Tamkun, and R. O. Hynes (1985) The repeating modular structure of the fibronectin gene: Relationship to protein structure and subunit variation. *Proc. Natl. Acad. Sci. USA* **82**:6571–6575.

Oldberg, A., and E. Ruoslahti (1986) Evolution of the fibronectin gene: Exon structure of cell attachment domain. *J. Biol. Chem.* **261**:2113–2116.

Oldberg, A., E. Linney, and E. Ruoslahti (1983) Molecular cloning and nucleotide sequence of a cDNA clone coding for the cell attachment domain in human fibronectin. *J. Biol. Chem.* **258**:10193–10196.

Oldberg, A., A. Franzen, and D. Heinegard (1986) Cloning and sequence analysis of rat bone sialoprotein (osteopontin) cDNA reveals an Arg-Gly-Asp cell-binding sequence. *Proc. Natl. Acad. Sci. USA* **83**:8819–8823.

Olden, K., R. M. Pratt, and K. M. Yamada (1978) Role of carbohydrates in protein secretion and turnover. Effects of tunicamycin on the major cell surface glycoprotein of chick embryo fibroblasts. *Cell* **13**:461–473.

Olden, K., R. M. Pratt, and K. M. Yamada (1979) Role of carbohydrate in biological function of the adhesive glycoprotein fibronectin. *Proc. Natl. Acad. Sci. USA* **76**:3343–3347.

Oppenheimer-Marks, N., and F. Grinnell (1981) Effects of plant lectins on the adhesive properties of baby hamster kidney cells. *Eur. J. Cell Biol.* **23**:286–294.

Oppenheimer-Marks, N., and F. Grinnell (1982) Inhibition of fibronectin receptor function by antibodies against baby hamster kidney cell wheat germ agglutinin receptors. *J. Cell Biol.* **95**:876–884.

Ouaissi, M. A., J. Cornette, D. Afchain, A. Capron, H. Gras-Masse, and A. Tartar (1986) *Trypanosoma cruzi* infection inhibited by peptides modeled from a fibronectin cell attachment domain. *Science* **234**:603–607.

Owens, R. J., and F. E. Baralle (1986a) Mapping the collagen-binding site of human fibronectin by expression in *Escherichia coli. EMBO J.* **5**:2825–2830.

Owens, R. J., and F. E. Baralle (1986b) Exon structure of the collagen-binding domain of human fibronectin. *FEBS Lett.* **204**:318–322.

Pande, H., and J. E. Shively (1982) NH$_2$-terminal sequences of DNA-, heparin-, and gelatin-binding tryptic fragments from human plasma fibronectin. *Arch. Biochem. Biophys.* **213**:258–265.

Pande, H., J. Calaycay, D. Hawke, C. M. Ben-Avram, and J. E. Shively (1985) Primary structure of a glycosylated DNA-binding domain in human fibronectin. *J. Biol. Chem.* **260**:2301–2306.

Pande, H., J. Calaycay, T. D. Lee, K. Legesse, J. E. Shively, A. Siri, L. Borsi, and L. Zardi (1987) Demonstration of structural differences between the two subunits of human plasma fibronectin in the carboxyl-terminal heparin-binding domain. *Eur. J. Biochem.* **162**:403–411.

Patel, V. P., and H. F. Lodish (1986) The fibronectin receptor on mammalian erythroid precursor cells: Characterization and developmental regulation. *J. Cell Biol.* **102**:449–456.

Patthy, L., M. Trexler, Z. Vali, L. Banyai, and L. Varadi (1984) Kringles: Modules specialized for protein binding. Homology of the gelatin-binding region of fibronectin with the kringle structure of proteases. *FEBS Lett.* **171**:131–136.

Paul, J. I., and R. O. Hynes (1984) Multiple fibronectin subunits and their post-translational modifications. *J. Biol. Chem.* **259**:13477–13488.

Paul, J. I., J. E. Schwarzbauer, J. W. Tamkun, and R. O. Hynes (1986) Cell type–specific fibronectin subunits generated by alternative splicing. *J. Biol. Chem.* **261**:12258–12265.

Pearlstein, E., L. I. Gold, and A. Garcia-Pardo (1980) Fibronectin: A review of its structure and biological activity. *Mol. Cell. Biochem.* **29**:103–128.

Pennypacker, J. P., J. R. Hassell, K. M. Yamada, and R. M. Pratt (1979) The influence of an adhesive cell surface protein on chondrogenic expression *in vitro. Exp. Cell Res.* **121**:411–415.

Petersen, T. E., H. C. Thogersen, K. Skorstengaard, K. Vibe-Pedersen, L. Sottrup-Jensen, and S. Magnusson (1983) Partial primary structure of bovine plasma fibronectin, three types of internal homology. *Proc. Natl. Acad. Sci. USA* **80**:137–141.

Pierschbacher, M. D., and E. Ruoslahti (1984a) Cell attachment activity of fibronectin can be duplicated by small synthetic fragments of the molecule. *Nature* **309**:30–33.

Pierschbacher, M. D., and E. Ruoslahti (1984b) Variants of the cell recognition site of fibronectin that retain attachment-promoting activity. *Proc. Natl. Acad. Sci. USA* **81**:5985–5988.

Pierschbacher, M. D., E. G. Hayman, and E. Ruoslahti (1981) Location of the cell-attachment site in fibronectin with monoclonal antibodies and proteolytic fragments of the molecule. *Cell* **26**:259–267.

Pierschbacher, M. D., E. Ruoslahti, J. Sundelin, P. Lind, and P. A. Peterson (1982) The cell attachment domain of fibronectin. Determination of the primary structure. *J. Biol. Chem.* **257**:9593–9597.

Pierschbacher, M. D., E. G. Hayman, and E. Ruoslahti (1983) Synthetic peptide with cell attachment activity of fibronectin. *Proc. Natl. Acad. Sci. USA* **80**:1224–1227.

Pierschbacher, M. D., E. G. Hayman, and E. Ruoslahti (1985) The cell attachment determinant in fibronectin. *J. Cell. Biochem.* **28**:115–126.

Plow, E. F., R. P. McEver, B. S. Coller, V. L. Woods, Jr., G. A. Marguerie, and M. H. Ginsberg (1985a) Related binding mechanisms for fibrinogen, fibronectin, von Willebrand factor, and thrombospondin on thrombin-stimulated human platelets. *Blood* **66**:724–727.

Plow, E. F., M. D. Pierschbacher, E. Ruoslahti, G. A. Marguerie, and M. H. Ginsberg (1985b) The effect of Arg-Gly-Asp-containing peptides on fibrinogen and von Willebrand factor binding to platelets. *Proc. Natl. Acad. Sci. USA* **82**:8057–8061.

Podleski, T. R., I. Greenberg, J. Schlessinger, and K. M. Yamada (1979) Fibronectin delays the fusion of L6 myoblasts. *Exp. Cell Res.* **123**:104–126.

Price, J., and R. O. Hynes (1985) Astrocytes in culture synthesize and secrete a variant form of fibronectin. *J. Neurosci.* **5**:2205–2211.

Reddi, A. H., ed. (1985) *Extracellular Matrix: Structure and Function,* Alan R. Liss, New York.

Rogers, S. L., P. C. Letourneau, S. L. Palm, J. McCarthy, and L. Furcht (1983) Neurite extension by peripheral and central nervous system neurons in response to substratum-bound fibronectin and laminin. *Dev. Biol.* **98**:212–220.

Rogers, S. L., J. B. McCarthy, S. L. Palm, L. T. Furcht, and P. C. Letourneau (1985) Neuron-specific interactions with two neurite-promoting fragments of fibronectin. *J. Neurosci.* **5**:369–378.

Rovasio, R. A., A. Delouvee, K. M. Yamada, R. Timpl, and J. P. Thiery (1983) Neural crest cell migration: Requirement for exogenous fibronectin and high cell density. *J. Cell Biol.* **96**:462–473.

Ruggeri, Z. M., R. A. Houghton, S. R. Russell, and T. S. Zimmerman (1986) Inhibition of platelet function with synthetic peptides designed to be high-affinity antagonists of fibrinogen binding to platelets. *Proc. Natl. Acad. Sci. USA* **83**:5708–5712.

Ruoslahti, E., and E. G. Hayman (1979) Two active sites with different characteristics in fibronectin. *FEBS Lett.* **97**:221–224.

Ruoslahti, E., and M. D. Pierschbacher (1986) Arg-Gly-Asp: A versatile cell recognition signal. *Cell* **44**:517–518.

Ruoslahti, E., E. G. Hayman, P. Kuusela, J. E. Shively, and E. Engvall (1979) Isolation of a tryptic fragment containing the collagen-binding site of plasma fibronectin. *J. Biol. Chem.* **254**:6054–6059.

Ruoslahti, E., E. Engvall, and E. G. Hayman (1981a) Fibronectin: Current concepts of its structure and functions. *Coll. Relat. Res.* **1**:95–128.

Ruoslahti, E., E. Engvall, E. G. Hayman, and R. G. Spiro (1981b) Comparative studies on amniotic fluid and plasma fibronectins. *Biochem. J.* **193**:295–299.

Sadler, J. E., B. B. Shelton-Inloes, J. M. Sorace, J. M. Harlan, K. Titani, and E. W. Davie (1985) Cloning and characterization of two cDNAs coding for human von Willebrand factor. *Proc. Natl. Acad. Sci. USA* **82**:6394–6398.

Savagner, P., B. A. Imhof, K. M. Yamada, and J. P. Thiery (1986) Homing of hemopoietic precursor cells to the embryonic thymus: Characterization of an invasive mechanism induced by chemotactic peptides. *J. Cell Biol.* **103**:2715–2727.

Schwarzbauer, J. E., J. W. Tamkun, I. R. Lemischka, and R. O. Hynes (1983) Three different fibronectin mRNAs arise by alternative splicing within the coding region. *Cell* **35**:421–431.

Schwarzbauer, J. E., J. I. Paul, and R. O. Hynes (1985) On the origin of species of fibronectin. *Proc. Natl. Acad. Sci. USA* **82**:1424–1428.

Seidl, M., and H. Hörmann (1983) Affinity chromatography on immobilized fibrin monomer. IV. Two fibrin-binding peptides of a chymotryptic digest of human plasma fibronectin. *Hoppe-Seyler's Z. Physiol. Chem.* **364**:83–92.

Sekiguchi, K., and S. Hakomori (1980) Functional domain structure of fibronectin. *Proc. Natl. Acad. Sci. USA* **77**:2661–2665.

Sekiguchi, K., and S. Hakomori (1983a) Topological arrangements of four functionally distinct domains in hamster plasma fibronectin: A study with combination of S-cyanylation and limited proteolysis. *Biochemistry* **22**:1415–1422.

Sekiguchi, K., and S. Hakomori (1983b) Domain structure of human plasma fibronectin. Differences and similarities between human and hamster fibronectins. *J. Biol. Chem.* **258**:3967–3973.

Sekiguchi, K., M. Fukuda, and S. Hakomori (1981) Domain structure of hamster plasma fibronectin. Isolation and characterization of four functionally distinct domains and their unequal distribution between two subunit polypeptides. *J. Biol. Chem.* **256**:6452–6462.

Sekiguchi, K., A. Siri, L. Zardi, and S. Hakomori (1985) Differences in domain structure between human fibronectins isolated from plasma and from culture supernatants of normal and transformed fibroblasts. *J. Biol. Chem.* **260**:5105–5114.

Sekiguchi, K., A. M. Alos, K. Kurachi, S. Yoshitake, and S. Hakomori (1986) Human liver fibronectin complementary DNAs: Identification of two different messenger RNAs possibly encoding the alpha and beta subunits of plasma fibronectin. *Biochemistry* **25**:4936–4941.

Silnutzer, J. E., and D. W. Barnes (1985) Effects of fibronectin-related peptides on cell spreading. *In Vitro* **21**:73–78.

Skorstengaard, K., H. C. Thogersen, K. Vibe-Pedersen, T. E. Petersen, and S. Magnusson (1982) Purification of twelve cyanogen bromide fragments from bovine plasma fibronectin and the amino acid sequence of eight of them. *Eur. J. Biochem.* **128**:605–623.

Skorstengaard, K., H. C. Thogersen, and T. E. Petersen (1984) Complete primary structure of the collagen-binding domain of bovine fibronectin. *Eur. J. Biochem.* **140**:235–243.

Skorstengaard, K., M. S. Jensen, T. E. Petersen, and S. Magnusson (1986a) Purification and complete primary structures of the heparin-, cell-, and DNA-binding domains of bovine plasma fibronectin. *Eur. J. Biochem.* **154**:15–29.

Skorstengaard, K., M. S. Jensen, P. Sahl, T. E. Petersen, and S. Magnusson (1986b) Complete primary structure of bovine plasma fibronectin. *Eur. J. Biochem.* **161**:441–453.

Spiegel, S., K. M. Yamada, B. E. Hom, J. Moss, and P. H. Fishman (1985) Fluorescent gangliosides as probes for the retention and organization of fibronectin by ganglioside-deficient mouse cells. *J. Cell Biol.* **100**:721–726.

Spiegel, S., K. M. Yamada, B. E. Hom, J. Moss, and P. H. Fishman (1986) Fibrillar organization of fibronectin is expressed coordinately with cell surface gangliosides in a variant murine fibroblast. *J. Cell Biol.* **102**:1898–1906.

Springer, W. R., D. N. W. Cooper, and S. H. Barondes (1984) Discoidin I is implicated in cell–substratum attachment and ordered cell migration of *Dictyostelium discoideum* and resembles fibronectin. *Cell* **39**:557–564.

Stenman, S., and A. Vaheri (1978) Distribution of a major connective tissue protein, fibronectin, in normal human tissue. *J. Exp. Med.* **147**:1054–1064.

Suzuki, S., A. Oldberg, E. G. Hayman, M. D. Pierschbacher, and E. Ruoslahti (1985) Complete amino acid sequence of human vitronectin deduced from cDNA. Similarity of cell attachment sites in vitronectin and fibronectin. *EMBO J.* **4**:2519–2524.

Takasaki, S., K. Yamashita, K. Suzuki, S. Iwanaga, and A. Kobata (1979) The sugar chains of cold-insoluble globulin. A protein related to fibronectin. *J. Biol. Chem.* **254**:8548–8553.

Tamkun, J. W., and R. O. Hynes (1983) Plasma fibronectin is synthesized and secreted by hepatocytes. *J. Biol. Chem.* **258**:4641–4647.

Tamkun, J. W., J. E. Schwarzbauer, and R. O. Hynes (1984) A single rat fibronectin gene generates three different mRNAs by alternative splicing of a complex exon. *Proc. Natl. Acad. Sci. USA* **81**:5140–5144.

Teng, M.-H., and D. B. Rifkin (1979) Fibronectin from chicken embryo fibroblasts contains covalently bound phosphate. *J. Cell Biol.* **80**:784–791.

Thiery, J. P., and J.-L. Duband (1982) Pathways and mechanisms of avian trunk neural crest cell migration and localization. *Dev. Biol.* **93**:324–343.

Thomas, D. D., J. B. Baseman, and J. F. Alderete (1985a) Fibronectin mediates *Treponema pallidum* cytadherence through recognition of fibronectin cell-binding domain. *J. Exp. Med.* **161**:514–528.

Thomas, D. D., J. B. Baseman, and J. F. Alderete (1985b) Fibronectin tetrapeptide is target for syphilis spirochete cytadherence. *J. Exp. Med.* **162**:1715–1719.

Thompson, L. K., P. M. Horowitz, K. L. Bentley, D. D. Thomas, J. F. Alderete, and R. J. Klebe (1986) Localization of the ganglioside-binding site of fibronectin. *J. Biol. Chem.* **261**:5209–5214.

Trelstad, R. L., ed. (1985) *The Role of Extracellular Matrix in Development,* Alan R. Liss, New York.

Umezawa, K., A. R. Kornblihtt, and F. E. Baralle (1985) Isolation and characterization of cDNA clones for human liver fibronectin. *FEBS Lett.* **186**:31–34.

Urushihara, H., and K. M. Yamada (1986) Evidence for involvement of more than one class of glycoprotein in cell interactions with fibronectin. *J. Cell. Physiol.* **126**:323–332.

Vartio, T., S. Barlati, G. DePetro, V. Miggiano, C. Stahli, B. Takacs, and A. Vaheri (1983a) Evidence for preferential proteolytic cleavage of one of the two fibronectin subunits and for immunological localization of a site distinguishing them. *Eur. J. Biochem.* **135**:203–207.

Vartio, T., A. Vaheri, G. DePetro, and S. Barlati (1983b) Fibronectin and its proteolytic fragments. *Invasion Metastasis* **3**:125–138.

Vibe-Pedersen, K., A. R. Kornblihtt, and F. E. Baralle (1984) Expression of a human α-globin/fibronectin gene hybrid generates two mRNAs by alternative splicing. *EMBO J.* **3**:2511–2516.

Vibe-Pedersen, K., S. Magnusson, and F. E. Baralle (1986) Donor and acceptor splice signals within an exon of the human fibronectin gene: A new type of differential splicing. *FEBS Lett.* **207**:287–291.

Waite, K. A., G. Mugnai, and L. A. Culp (1987) A second cell-binding domain on fibronectin (RGDS-independent) for neurite extension of human neuroblastoma cells. *Exp. Cell Res.* **169**:311–327.

West, C. M., R. Lanza, J. Rosenbloom, M. Lowe, H. Holtzer, and N. Avdalovic (1979) Fibronectin alters the phenotypic properties of cultured chick embryo chondroblasts. *Cell* **17**:491–501.

Wilson, B. S., G. Ruberto, and S. Ferrone (1981) Sulfation and molecular weight of fibronectin shed by human melanoma cells. *Biochem. Biophys. Res. Commun.* **101**:1047–1051.

Woods, A., J. R. Couchman, S. Johansson, and M. Hook (1986) Adhesion and cytoskeletal organisation of fibroblasts in response to fibronectin fragments. *EMBO J.* **5**:665–670.

Wright, S. D., and B. C. Meyer (1985) Fibronectin receptor of human macrophages recognizes the sequence Arg-Gly-Asp-Ser. *J. Exp. Med.* **162**:762–767.

Yamada, K. M. (1983) Cell surface interactions with extracellular materials. *Annu. Rev. Biochem.* **52**:761–799.

Yamada, K. M. (1988) Fibronectin domains and receptors. In *Fibronectin,* D. F. Mosher, ed., pp. 47–121, Academic, New York.

Yamada, K. M., and D. W. Kennedy (1979) Fibroblast cellular and plasma fibronectins are similar but not identical. *J. Cell Biol.* **80**:492–498.

Yamada, K. M., and D. W. Kennedy (1984) Dualistic nature of adhesive protein function: Fibronectin and its biologically active peptide fragments can autoinhibit fibronectin function. *J. Cell Biol.* **99**:29–36.

Yamada, K. M., and D. W. Kennedy (1985) Amino acid sequence specificities of an adhesive recognition signal. *J. Cell. Biochem.* **28**:99–104.

Yamada, K. M., and D. W. Kennedy (1987) Peptide inhibitors of fibronectin, laminin, and other adhesion molecules: Unique and shared features. *J. Cell. Physiol.* **130**:21–28.

Yamada, K. M., D. H. Schlesinger, D. W. Kennedy, and I. Pastan (1977) Characterization of a major fibroblast cell surface glycoprotein. *Biochemistry* **16**:5552–5559.

Yamada, K. M., D. W. Kennedy, K. Kimata, and R. M. Pratt (1980) Characterization of fibronectin interactions with glycosaminoglycans and identification of active proteolytic fragments. *J. Biol. Chem.* **255**:6055–6063.

Yamada, K. M., D. W. Kennedy, G. R. Grotendorst, and T. Momoi (1981) Glycolipids: Receptors for fibronectin? *J. Cell. Physiol.* **109**:343–351.

Yamada, K. M., S. K. Akiyama, T. Hasegawa, E. Hasegawa, M. J. Humphries, D. W. Kennedy, K. Nagata, H. Urushihara, K. Olden, and W.-T. Chen (1985) Recent advances in research on fibronectin and other cell attachment proteins. *J. Cell. Biochem.* **28**:79–97.

Zhu, B. C.-R., and R. A. Laine (1985) Polylactosamine glycosylation on human fetal placental fibronectin weakens the binding affinity of fibronectin to gelatin. *J. Biol. Chem.* **260**:4041–4045.

Zhu, B. C.-R., S. F. Fisher, H. Pande, J. Calaycay, J. E. Shively, and R. A. Laine (1984) Human placental (fetal) fibronectin: Increased glycosylation and higher protease resistance than plasma fibronectin. Presence of polylactosamine glycopeptides and properties of a 44-kilodalton chymotryptic collagen-binding domain: Difference from human plasma fibronectin. *J. Biol. Chem.* **259**:3962–3970.

Chapter 7

Nectins and Integrins: Versatility in Cell Adhesion

RICHARD O. HYNES
DOUGLAS W. DeSIMONE
JACK J. LAWLER
EUGENE E. MARCANTONIO
PAMELA A. NORTON
ERICH ODERMATT
RAMILA S. PATEL
JEREMY I. PAUL
JEAN E. SCHWARZBAUER
MARY ANN STEPP
JOHN W. TAMKUN

ABSTRACT

A single fibronectin gene encodes up to 20 different protein subunits by a process of alternative splicing. Different cell types splice differently, producing variations in the subunit composition of fibronectins from different sources. All fibronectins contain a conserved cell adhesion site; some forms contain one or more additional cell adhesion sites in the alternatively spliced V region. The structure of fibronectin and its gene is modular with structural units encoded as exonic units, which have undergone exon shuffling during evolution. Another adhesive protein, thrombospondin, has an analogous modular structure but is largely nonhomologous with fibronectin except in its cell adhesion site. The same appears to be true for other adhesive proteins (nectins). These proteins are recognized by a family of cell surface receptors known as integrins. These are transmembrane glycoproteins composed of two subunits, α and β. Both α- and β-subunits occur as homologous gene families. The ligand specificity of different integrin receptors is a function of their αβ-subunit composition; different cell types express different receptors, contributing further versatility in cell adhesion.

Most cells adhere to extracellular matrices at all times. This cell–matrix adhesion plays a central role in cell structure, growth, differentiation, migration, and other behavior. It is now clear that a variety of extracellular proteins function in these adhesive processes. The best understood are fibronectins, and they will be a major concern of this chapter. However, interesting parallels have been discovered between fibronectins and other adhesive proteins (nectins), and these will also be briefly discussed as a prelude to considering recent advances in understanding of a group of cell surface receptors (integrins) that mediate the interactions of cells with fibronectins and with other nectins.

Fibronectins are large glycoproteins that perform many functions of physiological importance, both during development and in adult organisms (Hynes and Yamada, 1982; Yamada, 1983; Hynes, 1986). Their primary role is to promote the adhesion of cells to extracellular matrices, of which fibronectins are major constituents. Adhesion is a complex phenomenon, consisting not only of the attachment of a cell to a surface but also of cell spreading, the development of cellular asymmetry, and the organization of the cytoskeleton. Fibronectins affect all of these processes. In turn, these basic cellular phenomena are important in higher-order functions such as cell migration and cellular differentiation, and, not surprisingly, fibronectins affect these as well. It is abundantly clear that fibronectins promote cell migration both during embryogenesis and in wound healing; they also affect terminal differentiation of several cell types and probably play important roles in other aspects of morphogenesis.

In addition to these roles in development, fibronectins are important in hemostasis and thrombosis, promoting platelet interactions with surfaces, and participating in the formation of blood clots (later used as substrates for cell migration during wound healing). A major impetus for work on fibronectins comes from the observation that tumor cells frequently have much less fibronectin, or none at all, associated with them (Hynes, 1973, 1976). This loss of fibronectin correlates with the reduced adhesion, altered morphology, and disordered cytoskeleton of tumor cells. It has been shown in some cases that fibronectin can be restored to the surfaces of tumor cells and that this produces a temporary reversion of these other phenotypic properties toward normal (Yamada et al., 1976; Ali et al., 1977). Closer examination of the effects of fibronectin on cytoskeletal organization has shown that the extracellular fibronectin-rich matrix fibrils are somehow connected across the membrane to actin-based microfilaments (Hynes and Destree, 1978; Singer, 1979, 1982; Hynes, 1981; Hynes et al., 1982).

The participation of fibronectins in this wide variety of physiologically important cellular processes obviously raises questions about the structure and origin of these proteins and about the ways in which cells interact with them. In this chapter we review work done over the past four to five years that has led to a detailed understanding of the structure of fibronectins, providing important insights into their functions. We also discuss recent work on the structures of another adhesive protein, thrombospondin, and of the family of cell surface receptors known as integrins, which interact with fibronectins and with other analogous proteins. We concentrate on work from our own laboratory but relate it to the work of others. We then discuss briefly how the molecular complexity revealed by this work is likely to lead to further insights into the variety of effects of cell–matrix adhesions.

STRUCTURE OF FIBRONECTINS

The first level of complexity arises from the fact that there exist several distinct types of fibronectin (FN). While all forms of FN consist of dimers of subunits of around 250 kD in mass, there exist subtle differences between the FNs made by different cells and found in different places. A major form of FN is plasma fibronectin (pFN), which is synthesized by hepatocytes (Tamkun and Hynes, 1983) and is found in plasma at 300 μg/ml (0.6 μM). Slightly different forms of FN are synthesized by many other cells (cellular fibronectin, or cFN). pFN has subunits of two different sizes, which can be resolved on two-dimensional gels into at least four distinct forms. cFN is slightly larger than pFN and displays a more complex profile of subunits on high-resolution gels (Tamkun and Hynes, 1983; Paul and Hynes, 1984). We now know that much of this complexity and the differences between pFN and cFN arise by alternative splicing of the RNA transcript of a single FN gene, as will be discussed later.

We have cloned and sequenced cDNAs encoding all known forms of rat FN (Schwarzbauer et al., 1983, 1987b; Patel et al., 1987) and partial clones of chicken FN (Norton and Hynes, 1987). Others have sequenced human FN cDNA clones (Bernard et al., 1985; Kornblihtt et al., 1985) and bovine pFN at the protein level (Skorstengaard et al., 1986). The results are all in agreement and are diagramed in Figure 1.

FNs are composed of a series of three different kinds of repeating modules. Figure 2 shows these sequence homologies. Repeat types I and II are disulfide-bonded loops 45–50 amino acids long that make up the N-terminal fibrin- and collagen-binding domains and the C-terminal fibrin-binding domain. Type III repeats (~90 amino acids long) make up the central 150–180 kD of each FN subunit. There is only one significant stretch of sequence that does not fit one of these three homologies. This segment, which we term the V region (it has also been called the III-cs segment), lies between the last two type III repeats and represents one of the regions of variation among FNs (see below). Several rather short regions of nonhomologous sequence fall between domains of FN and are sites of ready proteolytic cleavage. They appear to be extended, hingelike segments between groups of repeating modules. The modules themselves each appear to be tightly folded into 2–3-nm globules, each one largely independent of its neighbors. FNs can be viewed as a tightly strung string of beads. The two subunits of a dimer are held together by disulfide bonds in the flexible C-terminal segment. Biophysical and electron microscope data indicate that the subunits are elongated and flexible (Odermatt et al., 1982).

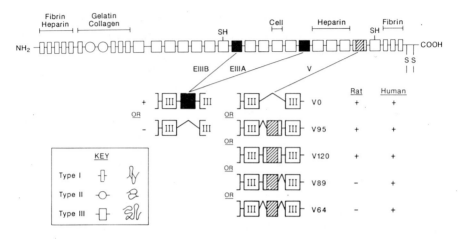

Fibronectin Structure & Variants

Figure 1. *Structure of fibronectin and its variants.* Diagram of the modular structure of fibronectin subunits. Each box represents a repeating unit that is one of three types (see text). The known binding sites of fibronectin are marked, as are the two free sulfhydryl groups and the two C-terminal interchain disulfide bonds. Each module is encoded in the gene by one or two exons, and introns fall precisely between modules. Three modules marked by *shading* are sometimes present and sometimes absent (see text, Table 1, and Figure 4). These segments (EIIIB, EIIIA, and V) are alternatively spliced in the primary transcript. EIIIA and EIIIB are encoded precisely by single exons that can be spliced in or out. Splicing of the V region is complex, with multiple variants possible (see also Figure 3). (After Hynes, 1985; Kornblihtt et al., 1985; Schwarzbauer et al., 1987b.)

Figure 2. *Sequence of rat fibronectin.* The sequence is arranged to display the homologies among the different repeats; each *line* represents a repeat. *Vertical lines* mark conserved cystine residues (type I and II repeats) and aromatic residues (type III repeats). Some of the other conserved residues are *underlined*. The repeats are also arranged to display the exon structure of the gene; positions of known introns are marked by *arrowheads*. It is readily seen that most repeats are separated by introns (the same is probably true for I-8 and I-9 and III-4–6, which have not been analyzed directly). The three alternatively spliced exons are *underlined*. The N-terminus of the mature protein is marked by a *star*. It is preceded by a prepro sequence, with the likely site of signal peptidase cleavage marked by an *arrow*. Potential carbohydrate addition sites are marked by *triangles* (*N*-linked) or a *diamond* (*O*-linked); the free sulfhydryl residues are marked by *filled circles*, and the cell adhesion site in repeat III-10 is *boxed*. The figure is based on data from Schwarzbauer et al. (1983, 1987b), Tamkun et al. (1984), Odermatt et al. (1985), and Patel et al. (1987). Virtually complete sequences (minus prepro peptide and EIIIB) are available for human (Kornblihtt et al., 1985) and bovine (Skorstengaard et al., 1986) FN, and there is a partial sequence of chicken FN (Norton and Hynes, 1987). All show strong homology (see text).

176

```
N-TERM                    MLRGPGPGRLLLLAVLCLGTSVRCTETGKSKRQAQQIVQPPSPVAVSQSK

I-1               PGCFDN--GKHYQINQQWERTYL--GNALVCTCYGGSRG-FNCESKPE

I-2               PEETCFDKYTGNTYKVGDTYERPKDS--MIWDCTCIGAGRGRISCTIA

I-3     FIBRIN    NRCHEG--GQSYKIGDKWRRPHETGGYMLECLCLGNGKGEWTCKPI

I-4               AEKCFDHAAGTSYVVGETWEKPYQ-GWMMVDCTCLGEGNGRITCTSR

I-5               NRCNDQDTRTSYRIGDTWSKKDNR-GNLLQCVCTGNGRGEWKCERHVLQSASA

I-6          GSGSFTDVRTAIYQPQTHPQPAPYGHCVTDS-GVVYSVGMQWLKSQ--GDKQMLCTCLGNG---VSCQET

II-1              AVTQTYGGNSNGEPCVLPFHYNGRTFYSCTTEGRNDGHLWCSTTSNYENDQKYSFCTDHA

II-2    COLLAGEN  VLVQTRGGNSNGALCHFPFLYSNRSYSDCTSEGRRDNMKWCGTTQNYDADQKFCFCPMA

I-7               AHEEICTTNE-GVMYRIGDQWDKQHDL-GHMMRCTCVGNGRGQWACIPYSQLR

I-8               DQCIVD--DITYNVNDTFHKRHEE-GHMLNCTCFGQGRGRWKCD

I-9               PIDRCQDSETRTFYQIGDSWEKFVH--GVRYQCYCYGRGIGEWHCQPLQTYPGT

III-1        TGPVQVIITETPSQPNSHPIQWNAPEPSHITKYILRWRPKTSTGRWKEATIPGHLNSYTIK-GLTºGVIYEGQLISIQQYGHQEVTRFDFTTSASTPVT

III-2   SNTVTGETAPFSPVVATSESVTEITASSFVVSWVSASDT-VSGFRVEYELSEEGDEPQYLDLPSTATSVNIP-DLLPGRKYIVNVYQISEEGKQSLILSTSQTT

III-3        APDAPPDPTVDQVDDTSIVVRWSRPQAP-ITGYRIVYSPSVEGSSTELNLPETANSVTLS--DLQPGVQYNITIYAVEENQESTPVFIQQETTGVPRS

III-4        DDVPAPKDLQFVEVTDVKVTIHWTPPNSA-VTGYRVDVLPVNLPGEHGQRLPVNRNTFAEVT-GLSPGVTYLFKVFAVHQGRESKPLTAQQTT

III-5   "DNA"    KLDAPTNLQFVNETDRTVLVTWTPPRAR-IAGYRLTVGLTRGGQPKQYNVGPMASKYPLR--NLQPGSEYTVTLHAVKGNQQSPKATGVFTTL

III-6        QPLRSIPPYNTEVTETTIVITWTPAPR---IGPKLGVRPSQGGEAPREVTSDSGSIVVS---GLTPGVEYTYTIQVLRDGQERDAPIVNRVVTP

III-7        LSPPTNLHLEANPDTGVLTVSWERSTTPDITGYRITITPTNGQQGTALEEVVHADQSSCTFENRNPGLEYNVSVYTVKDDKESAPISDTVIP

EIIIB        EVPQLTDLSFVDITDSSIGLRWTPLNSSTIIGYRITVVAAGEGIPIFEDFVDSSVGYYTVT-GLFPGIDYDISVITLINGGESAPTTLTQQT

III-8        AVPPPTDLRFTNIGPDTMRVTWAPPPSIELTNLLVRYSPVKNEEDVAELSISPSDNAVVLT-NLLPGTEYLVSVSSVYEQHESIPLRGRQKT

III-9   CELL     GLDSPTGFDSSDVTANSFTVHWVAPRAP-ITGYIIRHHAEHSAGRPRQDRVPPSRNSITLT-NLNPGTEYIVTIIAVNGREESPPLIGQQST

III-10       VSDVPRDLEVIASTPTSLLISWEPPAVS-VRYYRITYGETGGNSPVQEFTVPGSKSTATIN-NIKPGADYTITLYAVT[RGDS]PASSKPVSINYQT

III-11       EIDKPSQHQVTDVQDNSISVRWLPSTSP-VTGYRVTTAPKNGLGPTKSQTVSPDQTEHTIE-GLQPTVEYVVSVYAQNRRNGESQPLVQTAVT

EIIIA        NIDRPKGLAFTDVDVDSIKIAWESPQGQ-VSRYRVTYSSPEDGIHELFPAPDGEDEDTAELH-GLRPGSEYTVSVVALHGGCHESQPLIGVQST

III-12       TIPAPTNLKFTQVSPTTLTAQWTAPSVK-LTGYRVRVTPKEKTGPMKEINLSPDSTSVIVS-GLHVATKYEVSVYALKDTLTSRPAQGVVTTLE

III-13  HEPARIN   NVSPPRRARVTDATETTITISWRTKTET-ITGFQVDAIPANGQTPVQRTISPD-VRSYTIT-GLQPGTDYKIHLYTLNDNARSSPVVIDAST

III-14       AIDAPSNLRFLTTTPNSLLVSWQAPRAR-ITGYIIKYEKPGSPPREVVPRPRPGCVTEATIT-GLFPGTEYTIYVIALKNNQKSEPLIGRKKT

                  DELPQLVTLPHPNLHGPEILDVPSTVQKTPFVTNPGYDTENGIQLPGTSHQQPSVGQQMI
V SEGMENT
                  FEEHGFRRTTPPTAATPVRLRPRPYLPNVDEEVQIGHVPRGDVDYHLYPHVPGLNPNAST

III-15       GQEALSQTTISWTPFQES--SEYIISCQPVGTDEEPLQFQVPGTSTSATLT-GLTRGVTYNIIVEALHNQRRHKVREEVVTVGNT

I-10              VNEGLNQPTDDSCFDPYTVSHYAVGEEWERLSDS-GFKLTCQCLGFGSGHFRCDSS

I-11    FIBRIN    KWCHDN--GVNYKIGEKWDRQGEN-GQRMSCTCLGNCKGEFKCDP

I-12 + HINGE      HEATCYDD--GKTYHVGEQWQKEYL--GAICSCTCFGQGR-GWRCDNCRRPGAAEPSPDGTTGHTYNQYTQRYHQRTNT

C-TERM            NVNCPIECFMPLDVQADRDDSRE
```

177

While the repeating homologous modules within the FN of a given species are typically 20–50% identical with one another, the homology between species is much higher. Over most of the sequence the amino acids are >90% identical in different mammals, and the chicken sequence is around 80% identical with those of mammals (a few exceptional segments will be mentioned later). This means that a given repeat within the sequence is clearly recognizable irrespective of the species. In other words, the endoduplication of primordial modules and their divergence, which must have given rise to the present structure of FN, took place long before vertebrates diverged from one another.

VARIANT FIBRONECTIN SUBUNITS

As diagramed in Figure 1, the sequences of different cDNA clones reveal that not all FN subunits are the same. This had been suspected from earlier protein chemical studies but was extremely difficult or impossible to sort out without the sequence information. It is now clear that there are three positions in FN where sequence variations can occur. An extensive survey failed to reveal any others (Schwarzbauer et al., 1987b). At two positions (EIIIA and EIIIB), entire type III repeats are either included or omitted, while at the third position, the V segment can be omitted or included, wholly or in part. Whereas the inclusion or omission of the EIII segments occurs in a similar fashion in all species examined to date, the pattern of V region variation is complex and varies among species (Figure 3).

The EIII segments have sequences typical of other type III repeats. The EIIIB segment, which was initially discovered during studies of the rat FN gene (Schwarzbauer et al., 1987b), has now been identified in chicken FN (Norton and Hynes, 1987) and human FN (F. E. Baralle, personal communication; A. R. Kornblihtt, personal communication). It is striking that the amino acid sequences of the EIIIB segment are 100% identical between rat and human and 96% identical between these two species and chicken. This high degree of conservation suggests a particularly important function for this segment. The EIIIA segment is not more conserved than the rest of FN. In contrast, the V region is less well conserved in sequence than the rest (90–92% identity among mammals and 50% identity between mammals and chickens).

These variant forms of FN cDNA have been shown to reflect variant mRNAs by nuclease protection experiments (Schwarzbauer et al., 1983, 1987b; Kornblihtt et al., 1984a,b). Furthermore, antibodies raised to fusion proteins encompassing the EIIIA and V regions recognize particular forms of FN (Schwarzbauer et al., 1985; Paul et al., 1986). This immunological approach has yet to succeed for EIIIB because of its high degree of conservation. cDNA probes for the three variable segments and antibodies against EIIIA and V segments reveal that the FN mRNAs and proteins expressed by different cells differ markedly. Thus, the mRNA from liver cells does not contain the EIIIA and EIIIB

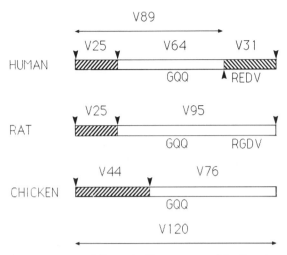

Figure 3. *Variation in the V region of fibronectin.* The structure of the V region of three species is shown. In all cases, this region can be alternatively spliced, but the allowed patterns of splicing are different. Splice sites are marked by *arrowheads*. The rat and human segments are each encoded in exons that also contain part of the following type III repeat. In this case deletion of the entire V segment arises by splicing to a 3' splice site within the exon, at the end of the V segment. Deletion of the V25 segment similarly arises by use of a 3' splice site at the end of this segment. Neither of these splice options occurs in chicken FN, which instead has only a single 3' splice site 132 bases into the V segment. Use of this splice site deletes the V44 segment, leaving the V76 segment; splicing at the end of V120 to give a V0 form does not occur in chickens. The human FN gene (and not the others) contains an internal 5' splice site at the beginning of the V31 segment, allowing an internal splice to delete this segment; this can occur in concert with, or separately from, the V25 deletion to give, respectively, a V64 or V89 form. Possible functional sites in this region of FN include apparent cell adhesion sites in the V25 segment (Yamada et al., this volume), an RGD-related site in the V31 or V95 segment of human and rat FNs (see text), and a potential transglutaminase cross-linking site (GQQ). See text for further details and references.

segments (Kornblihtt et al., 1984a,b; Norton and Hynes, 1987; Schwarzbauer et al., 1987b), and pFN, which is made by liver cells, does not react with antibodies to the EIIIA segment (Schwarzbauer et al., 1985; Paul et al., 1986).

The situation in the V region is more complex (Figure 3). The entire segment is 360 nucleotides, or 120 amino acids, long. Forms of FN containing the entire segment (V120) are found in all species and cell types tested to date. The V0 form, in which the whole segment is omitted, is rare in most cell types but prevalent in mammalian liver mRNAs and in pFN (Schwarzbauer et al., 1983, 1987b; Sekiguchi et al., 1986). However, no V0 form occurs in chickens (Norton and Hynes, 1987). Various partially deleted forms of V also occur, and these differ between species. The first 75 nucleotides/25 amino acids (V25) can be omitted in rats and humans, while the last 93 nucleotides/31 amino acids (V31) can be omitted by humans but not by rats or chickens. Thus, three V region variants (V120, V95, V0) are found in rat FN, and these plus two additional forms (V64, V89), arising by omission of the V31 segment, are found in human FN. Chickens display only two V region variants, V120 and a form (V76) in which the first 44 amino acids are omitted. This complexity, and the inclusion/exclusion of the EIII segments, arises via alternative splicing (see next section).

By a combination of nuclease protection experiments and antibody blotting analyses, we have screened the expression of all three variable segments in rat FNs from various sources. The major results are summarized in Table 1 and Figure 4. Less extensive analyses in other species are consistent with this picture. In summary, the form of FN that lacks all inserts is prevalent only in liver mRNA and in pFN, and it represents the smallest class of pFN subunits. The partial or total inclusion of the V region produces the larger subunits of pFN (Schwarzbauer et al., 1985; Paul et al., 1986). Thus, V region variation accounts for the well-known doublet pattern of pFN subunits. As mentioned, the inclusion of the EIII segments never occurs in liver mRNA or pFN. When these segments are included in other cell types, the resulting cFN subunits are larger and more acidic. All combinations of EIIIA, EIIIB, and V appear to be allowed. This combinatorial variation can account for 20 different subunits in humans, 12 in rats, and 8 in chickens. The largest and smallest forms of FN differ in size by 360 amino acids, and the larger ones have significantly more acidic pIs. This variation accounts for much of the complexity of the two-dimensional gel profiles of FNs from different sources. Carbohydrate differences between pFN and cFN contribute further heterogeneity (Paul and Hynes, 1984). We are still unable to explain the pairs of identically sized subunits that differ in pI (see Figure 4). Since we cannot detect a fourth region of alternative splicing, this difference may be due to some yet to be determined posttranslational modification.

Table 1. Distribution of Three Alternatively Spliced Segments of Rat Fibronectin[a]

Type	pFN		cFN	
Source	Hepatocytes	Fibroblassts	Astrocytes	Platelets
EIIIB	−	+/−	+/−	ND
EIIIA	−	+/−	+/−	+/−
V	+/−	+	+	+/−

[a]The data come from nuclease protection experiments, which assay the presence of the segments in FN mRNA (Schwarzbauer et al., 1983, 1987b), and from the use of segment-specific antibodies prepared against fusion proteins, which assay the presence of the segments in FN subunits separated on gels (Schwarzbauer et al., 1985; Paul et al., 1986); + indicates that the segment is present in most mRNAs and FN subunits, − that it is virtually always absent; +/− indicates that both + and − forms are present in significant quantities; ND indicates that the experiment has not been done because platelets lack mRNA, and no antibody specific for EIIIB is available. Note that the different segments are differentially expressed in different cells and that alternative splicing of a given segment can be found within one cell type.

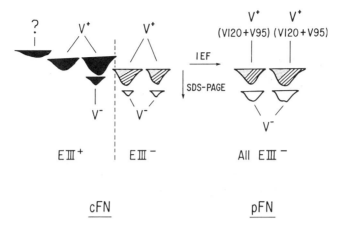

cFN   pFN

Figure 4. *Schematic of rat FN variants.* The figure diagrams the FN subunits identifiable on two-dimensional gels and using antibodies specific for alternatively spliced segments. pFN includes no subunits containing EIII segments. The smaller subunits (*unshaded*) contain no V region insert, while the larger subunits (*shaded*) contain either V95 or V120 segments (these alternative forms comigrate). The basis for the doubling of each set of spots (V⁻ and V⁺) is not known. cFN includes the same set of subunits as pFN, although V⁻ subunits are rare. In addition, cFN includes larger and more acidic subunits, which contain EIIIA (*black*). Since cells synthesizing cFN also contain EIIIB⁺ mRNA, EIIIB⁺ subunits are also likely to be present and probably comigrate with the EIIIA⁺ subunits. Since EIIIA⁺/EIIIB⁺ mRNA species also occur (Schwarzbauer et al., 1987b), it is likely that the largest, most acidic subunits (marked ?) are EIIIA⁺/EIIIB⁺. The presence of the EIIIB segment has not been tested at the protein level. cFN also differs from pFN in its carbohydrates, so that the EIII⁻ subunits of cFN do not comigrate with pFN unless synthesized in the presence of tunicamycin, which blocks glycosylation. (Based on the data of Tamkun and Hynes, 1983; Paul and Hynes, 1984; Paul et al., 1986, Schwarzbauer et al., 1987b.)

The existence of these FN variants raises two questions: (1) How do they arise? (2) What are the functional differences between them? Analyses of the structure of the FN gene provide information relevant to the first question. We will return later to questions of function.

THE FIBRONECTIN GENE: ALTERNATIVE SPLICING AND EXON SHUFFLING

We have cloned and analyzed in some detail the entire rat FN gene (Tamkun et al., 1984; Odermatt et al., 1985; Patel et al., 1987; Schwarzbauer et al., 1987b). The chicken FN gene has also been cloned and analyzed by R looping (Hirano et al., 1983), and parts of the human gene have been studied (Vibe-Pedersen et al., 1984, 1987; Oldberg and Ruoslahti, 1986; Owens and Baralle, 1986).

In all species examined, there appears to be a single FN gene. Therefore, one must explain the FN variants in terms of alternative RNA processing. The gene has almost 50 exons. In rats it is 71 kb in size; in chickens it is 50 kb. Extensive analysis of the rat gene shows that most (probably all) type I and II repeats are encoded by single exons, while most, but not all, type III repeats are encoded by *pairs* of exons (Figures 1 and 2). Each repeat is separated from adjacent repeats by introns, and all these introns fall after the first base of a codon. Those introns that have been located in the human gene conform to this pattern and, indeed, are in positions identical to those in the rat gene.

The two EIII repeats are exceptional in being encoded entirely by one exon each (Vibe-Pedersen et al., 1984; Odermatt et al., 1985; Schwarzbauer et al., 1987b). This presumably facilitates the alternative splicing that generates variations at these positions. The V region is encoded by a large exon that also encodes part of the last type III repeat (Tamkun et al., 1984; Vibe-Pedersen et al., 1987). Splice acceptor sequences *within* this exon allow splicing out of the V25 or V120 segments in rat and human FN and the V44 segment in chickens (Norton and Hynes, 1987). A splice donor sequence present only in human FN allows splicing out of the V31 segment (Vibe-Pedersen et al., 1987).

These patterns of alternative splicing are cell type–specific as mentioned and must, therefore, be regulated. This regulation appears to involve segments of the gene close to the alternatively spliced segments. The gene is transcribed from a single initiation site to a single polyadenylation site (Patel et al., 1987). Transfections of genomic segments encompassing EIIIA and EIIIB into cultured cells give rise to mRNAs both including and lacking these exons (Vibe-Pedersen et al., 1984; Schwarzbauer et al., 1987a,b). These results focus attention on the sequence of the gene surrounding the alternatively spliced exons. The nature of the regulatory elements is currently under study.

A second interesting feature of the FN gene arises from the fact that each structural module of the protein is encoded by a separate exonic unit (1 or 2 exons). Homologous type I and II repeats occur also in other proteins (Banyai et al., 1983; Esch et al., 1983; McMullen and Fujikawa, 1985). In one of these, tissue plasminogen activator, a single type I repeat is also encoded by a single exon (Ny et al., 1984), entirely consistent with the idea of exons shuffling between genes. It seems highly probable that the type I and II repeats in other proteins are also encoded by single exons. Type III repeats have not yet been detected in proteins other than FN.

The promoter of the FN gene is relatively complex (Dean et al., 1987; Patel et al., 1987), consistent with the known regulation of FN synthesis by a variety of mediators including cell differentiation, cell density, oncogenic trans-formation, glucocorticoids, TGF-β, interferon, and others. The rat FN promoter region contains elements matching the consensus sequences characteristic of SP1 binding sites, and of genes regulated by cAMP, glucocorticoids, inter-feron, acute phase response, and heat shock (Patel et al., 1987). Some but not all of these elements are also seen in the human FN promoter (Dean et al., 1987). It will be of interest to analyze the role of these potential regulatory elements.

In summary, the FN gene is large and complex. It clearly arose by endoduplication and divergence of small primordial units representative of the current exonic units encoding the three repeating modules. The gene represents a good example of the idea suggested by Gilbert (1978) that exons encode structural units in proteins that can reassort during evolution and, in this case, can vary in a cell type–specific fashion by alternative splicing.

POTENTIAL FUNCTIONS OF FIBRONECTIN VARIANTS

The three regions of variation in FN presumably have specific functions. What are they? In truth, we have information concerning only the V region, which has the advantage of being the first one discovered and of being present in pFN. Because the EIIIA and EIIIB segments are absent from pFN, by far the most accessible and most extensively studied form of FN, these regions have not been analyzed in the vast majority of studies of the structure–function relationships of FN. The EIIIB region is of particular interest given its strong conservation, but both segments seem likely to confer on cFN functions that are not needed or might actually be detrimental in pFN. One such is fibrillogenesis. It may be of some interest that the 4000 molecules of FN in platelet α-granules include EIIIA$^+$ forms (Paul et al., 1986) that may confer some property absent in pFN. An appealing idea is that these pFN molecules might nucleate fibrillogenesis. Another possibility is that EIIIA$^+$ and EIIIB$^+$ forms of FN may be incorporated into specific extracellular matrices and have specific effects on the differentiation, migration, or adhesion of certain cell types. These hypotheses and others are now testable using recombinant DNA methods, but at present they remain speculative.

Functions of the V region have been analyzed to some extent. Earlier data suggest that this region of FN may contain sites for FN–FN interaction (Ehrismann et al., 1982) and for transglutaminase cross-linking (Richter et al., 1981). The V region does contain a conserved GQQ sequence homologous with known transglutaminase sites (Figure 3). It is possible, therefore, that the presence of the V region in two copies per dimer of cFN and one per dimer of pFN might render cFN more competent to polymerize into matrix fibrils.

Another likely function for the V region is in cell adhesion. The V regions of rat and bovine FNs contain a sequence, RGDV, homologous with the known cell-binding site (RGDS) in repeat III-10 of FN. A fusion protein containing the rat V95 region promotes cell adhesion (J. W. Tamkun, unpublished data). However, the RGDV sequence is absent in human FN and is replaced by REDV, which was thought to be inactive in cell adhesion based on peptide analogue studies (Pierschbacher and Ruoslahti, 1984a,b; Yamada and Kennedy, 1984, 1985). Humphries et al. (1986) have recently shown that this segment of human pFN will promote the adhesion of melanoma cells. They have also shown that the V25 region may contain a third cell-binding site (Yamada et al., this volume). The V25 site cannot be RGD-related, since no sequence of this type occurs here. Thus, the human V region contains two potential cell-binding sites, the V25 segment and the RGDV, which is in the V31 segment (cf. Figure

3). Both can be alternatively spliced selectively. Rat and bovine FN probably contain the same or closely related sites. Chicken FN contains no sequence related to RGDV in its V region, and the V25 segment is significantly divergent. It is currently unclear whether chicken FN has a cell adhesion site in the V region. Thus, there is possibly interspecies variation in the functions of this region, just as there is variation in the splicing pattern. It is clear that the functions of these variable segments offer a fruitful area for investigation. We have begun a series of studies on the expression of the variant forms of FN using retroviral vectors (Schwarzbauer et al., 1987a). This approach offers the potential for detailed analyses of the functions both of these variable segments and of other regions of FN.

ANALOGOUS PROTEINS

The diversity of forms of FN discussed above suggests that the extracellular ligands that mediate cell adhesion are complex and variable. In fact, this is only the beginning of the complexity. There are numerous other adhesive glyco-proteins. Some of the best studied are listed in Table 2. Many of these proteins also contain RGD sites that appear to be involved in their cell adhesion function. We will briefly discuss one further example on which we have worked.

The glycoprotein thrombospondin (TSP) was originally described in platelets but is synthesized by a variety of cell types and is present in extracellular matrices (Lawler, 1986). It is clearly involved in platelet aggrega-tion. We have isolated and sequenced cDNA clones for human TSP (Lawler

Table 2.   Adhesive Glycoproteins (Nectins)[a]

Fibronectin (FN)	RGDS (RGDV, REDV)
Vitronectin (VN)	RGDV
Fibrinogen (FB)	RGDS, RGDF
von Willebrand factor (VWF)	RGDS (RGDC?)
Thrombospondin (TSP)	RGDA
Laminin (LM)	
Tenascin (TN)	
Entactin (EN)	
Collagen (CO)	

[a]The table lists the best-studied extracellular matrix proteins that promote cell adhesion. Proteins in the first group have all been sequenced and contain the indicated sites homologous with the RGDS site in fibronectin. In each case it has been shown that the RGD site plays a role in cell adhesion. Proteins in the second group have not been completely sequenced, and/or it has not been proved that they contain an RGD-related cell adhesion site. See text and the review by Ruoslahti and Pierschbacher (1986) for references.

and Hynes, 1986). The structure of TSP, like that of FN, is made up of a series of repeating modules, although thrombospondin does contain two domains, one at each end, which are not repeated (Figure 5). Although the structure of TSP is analogous with that of FN, it is striking that it is not homologous: the repeating modules are different. In the case of TSP, the modules are homologous with malaria sporozoite surface glycoproteins, EGF, and calmodulin. It is not known whether the TSP modules are encoded by separate exons, although this seems a reasonable prediction. Despite the overall lack of homology between TSP and FN, TSP does contain an RGDA sequence, and we have shown that this site functions in cell adhesion (Lawler et al., 1988). The RGD of TSP is found in a very different environment from that in FN; it is located in a calcium-binding repeat, probably in an ω loop conformation. The function of the RGD site in TSP is dependent on the calcium-dependent conformation of TSP. In contrast, the primary RGD site in FN appears to be located in a β-hairpin

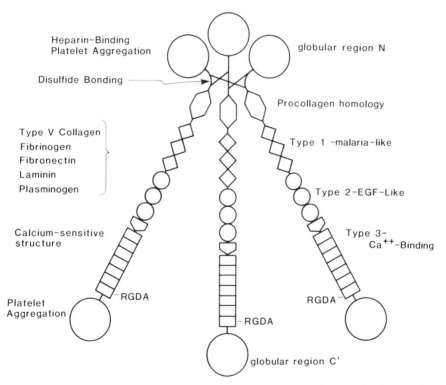

Figure 5. *Structure of thrombospondin.* Thrombospondin is a trimer of identical subunits. The subunits are made up of a series of repeating modules of three types with unique globular regions at each end and an interchain disulfide-bonding segment toward the N-terminus. The repeating modules are homologous with other proteins as marked but are not homologous with fibronectin. Locations of known binding sites are marked. (Reproduced from Lawler and Hynes, 1987.)

(Pierschbacher et al., 1982; Odermatt et al., 1985), while the RGD in the V region is in a proline-rich segment with no predicted regular structure (Odermatt et al., 1985). Similarly, the RGD sites in the other proteins listed in Table 2 are located in different nonhomologous segments in each case. Like FN and TSP, these other proteins are analogues, not homologues, and have little or no sequence homology with each other. An exception to this generalization is that a short, cysteine-rich segment of thrombospondin appears homologous with a repeat unit of von Willebrand factor.

The most likely explanation for the presence of RGD-containing cell adhesion sites in these basically nonhomologous proteins is convergent evolution. Presumably one protein at an early stage in evolution contained an RGD site recognized by a cell surface receptor that mediated cell adhesion to the extracellular protein. As other extracellular proteins acquired RGD sites by mutation, perhaps these were also recognized by the cell surface receptor, and the RGD sites were then maintained by natural selection. This hypothesis immediately raises the question of the identity of the cell surface receptor(s) for FN and analogous proteins.

INTEGRINS: CELL SURFACE ADHESION RECEPTORS

In the past two or three years there has been a great deal of progress in studies of the surface receptors. Prior to 1985, evidence existed that implicated several glycoproteins of around 140 kD in adhesion to FN. This evidence came from antibodies to cell surfaces that interfered with adhesion to FN (Knudsen et al., 1981, 1985; Greve and Gottlieb, 1982; Neff et al., 1982). These antibodies selected a set of proteins from cell extracts. Subsequent immunofluorescence studies showed that these glycoproteins codistributed both with FN outside cells and with cytoskeletal proteins inside cells (Chen et al., 1985; Damsky et al., 1985). This strongly suggested that they might mediate the connection between FN and the cytoskeleton, which had previously been demonstrated by several immunolocalization methods (see Figure 6) (Hynes and Destree, 1978; Singer, 1979, 1982; Burridge and Feramisco, 1980; Singer and Paradiso, 1981; Burridge and Connell, 1983).

Corroborative evidence came from affinity chromatography experiments in which cell extracts were passed through columns of FN cell-binding fragments and bound material was specifically eluted with RGD-containing peptides (Pytela et al., 1985a,b, 1986). Different sets of proteins were selected (1) from different cell types and (2) from the same cell type using FN or vitronectin. For example, FN fragments selectively bind the glycoprotein complex IIb/IIIa from platelet extracts (Gardner and Hynes, 1985; Pytela et al., 1986), while the same FN fragments select a different protein complex from other cell types (Pytela et al., 1985a, 1986; Ruoslahti et al., this volume).

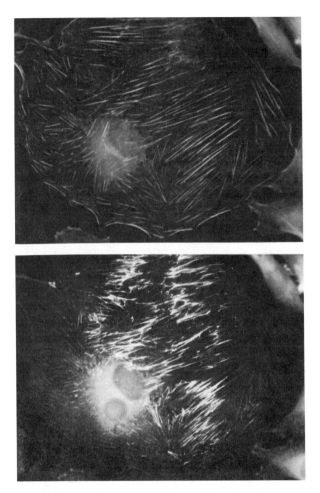

Figure 6. *Codistribution of the extracellular matrix and the cytoskeleton.* The figure shows double-label immunofluorescence staining for fibronectin and actin in the NIL8 hamster cell line. The two sets of fibrils show extensive codistribution, indicating some form of physical connection between them. Similar studies with antibodies to various cytoskeletal proteins (talin, vinculin, α-actinin) and to integrins show that they also codistribute.

Analysis *in vitro* of the protein complexes isolated by adhesion-blocking antibodies and by affinity chromatography showed them to be very similar. Both would bind to FN in various *in vitro* binding assays (Horwitz et al., 1985, 1986; Pytela et al., 1985a, 1986; Akiyama et al., 1986). It is now clear that all these receptors and a variety of others are members of a large family of cell surface proteins, which we term *integrins* because they are integral membrane proteins that appear to integrate the extracellular organization of the extracellular matrix and the cytoskeleton (Tamkun et al., 1986; Hynes, 1987).

The indications from the biochemical studies that the various protein complexes were related have been extensively confirmed by cDNA sequencing studies. We have cloned and sequenced cDNAs encoding integrins from chickens (Tamkun et al., 1986, unpublished data), *Xenopus* (DeSimone and Hynes, 1988), and mice (D. W. DeSimone, V. Patel, H. F. Lodish, and R. O. Hynes, unpublished data), and a number of laboratories have isolated and sequenced cDNA clones for various human integrins (Argraves et al., 1986, 1987; Suzuki et al., 1986; Fitzgerald et al., 1987; Kishimoto et al., 1987; Law et al., 1987; Poncz et al., 1987; Ruoslahti et al., this volume). All these proteins consist of heterodimers of two subunits, α and β, both of which are transmembrane glycoproteins (Figure 7). The β-subunits contain four highly characteristic cystine-rich repeats (Figure 7), and it has become clear that there are three different β-subunits (Table 3). The β_1-subunits of various species are greater than 80% identical in amino acid sequence, and all 56 cytines in the external domain, including the four cystine-rich repeats, are conserved. The other β-subunits also contain all 56 cystines and are around 45% identical with the β_1-subunit (Table 3). The high cystine content is presumably responsible for the characteristic decrease in SDS-gel mobility that occurs on reduction of these subunits.

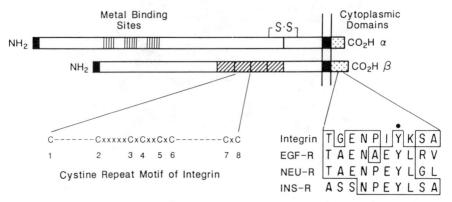

Figure 7. *Structure of integrin receptors.* These receptors are heterodimers of two subunits (α and β), each of which is a transmembrane protein with a short cytoplasmic domain. The large extracellular domains of each subunit contain repeating motifs. The β-subunits have four contiguous repeats of a characteristic cystine-rich motif around 40 amino acids long. The α-subunits contain three repeats around 60 amino acids long that contain apparent divalent cation-binding sites. Some α-subunits, but not all, are cleaved posttranslationally into an extracellular heavy chain and a transmembrane light chain, which remain attached via disulfide bonds. The cystine-rich nature of the β-subunits causes them to migrate faster when nonreduced, while reduction of the interchain disulfide bonds in the α-subunits causes them to migrate faster. These characteristic mobility shifts have been helpful in initial identifications of integrins. A tyrosine residue (Y) in the cytoplasmic domain of some integrin β-subunits can be phosphorylated (see text).

Table 3. Homologies Among Integrin β-Subunits[a]

	Integrin Type	Cysteines	Conserved CYS Repeats	Tyrosine Consensus Sequence	% Homology
β_1	Chicken				100
	Xenopus	56	4	yes	83
	Mouse				~85
	Human				85
β_2	Human LFA/Mac1/p150,95	56	4	no	45
β_3	Human IIIA	56	4	yes	47

[a]Key features of the sequences are indicated. Note the complete conservation of half-cystine residues. Overall the homology between species within the β_1-class is greater than 80%, while conservation between different β-chains within one species (human) is around 45%, suggesting divergence of the β-chains before divergence of vertebrates. Sources of the sequence information are as follows: Chicken β_1: Tamkun et al. (1986); Xenopus β_1: DeSimone and Hynes (1988); Mouse β_1: V. Patel, D. W. DeSimone, H. F. Lodish, and R. O. Hynes, unpublished data; Human β_1: Argraves et al. (1987); Human β_2: Kishimoto et al. (1987); Human β_3: Fitzgerald et al. (1987).

The integrin α-subunits are even more numerous; there are at least 10 (Table 4). It appears that integrins can be classified into three subfamilies, each characterized by one of the three β-subunits. Each β-subunit can occur in association with one of a subset of the α-subunits. The α-subunits are frequently cleaved posttranslationally to produce a heavy and light chain joined by disulfide bonds. The light chain contains the transmembrane domain; the sequences of several are compared in Figure 8. Although the conservation of sequence among α-subunits appears to be somewhat less than that among β-subunits, around 30%, certain characteristic features occur in all of them. The light chains contain, rather surprisingly, conserved transmembrane domains and a set of three cystine residues, at least one of which must be involved in the attachment to the heavy chain (Figures 7 and 8). The heavy chains contain a series of metal-binding sites in their extracellular domains (Ruoslahti et al., this volume).

A comparison of sequences of the various β-chains has revealed that they are particularly well conserved in their C-terminal segments, comprising the putative transmembrane and cytoplasmic domains (Figure 9). We have prepared antibodies against a synthetic peptide from the cytoplasmic domain sequence and have used it to demonstrate two things (E. E. Marcantonio, D. W. DeSimone, and R. O. Hynes, unpublished data). First, this antibody does not stain intact cells but does stain patterns of striae characteristic of integrins when the cells are permeabilized. This result confirms the cytoplasmic location of this segment and should be contrasted with staining results with other anti-integrin antibodies that readily stain and even detach live cells. In particular, we have tested antibodies against fusion proteins from the N-terminal third of

Table 4. Integrins: A Cell Surface Receptor Family[a]

	Subunits	Distribution	Function	Inhibited by RGDS
Chicken integrin	α_1, α_2	Widespread	Adhesion to FN, LM, VN	+
"FN-R"	$\alpha_F,$ β_1	Widespread	Adhesion to FN	+
VLA	α_{1-5}	Widespread	Adhesion to FN, LM	?
Ic/IIa	α_{Ic}	Platelets	Adhesion to FN	+
Ia/IIa	α_{Ia}	Platelets	Adhesion to collagen	
LFA-1	α_L	Lymphoid, myeloid	Leukocyte adhesion	
Mac-1	$\alpha_M,$ β_2	Lymphoid, myeloid	Leukocyte adhesion, C3bi receptor	
p150,95	α_X	Lymphoid, myeloid	?	
"VN-R"	α_V $\Big\}\,\beta_3$	Fibroblasts, etc.	Adhesion to VN	+
IIb/IIIa	α_{IIb}	Platelets (etc.?)	Adhesion to FN, FB, VWF, VN	+

[a]The different receptors in this homologous family are classified into three groups or subfamilies by their common β-subunits (cf. Table 3). Each individual receptor is an $\alpha\beta$-complex (see Figure 7) (Hynes, 1987). The β_1-subfamily is widespread and contains at least five α-subunits. It is likely that several of the listed subunits are in fact, identical with each other. For example, α_F is probably the same as VLA α_5, and the chicken α_2-subunit may be the equivalent of the human VLA α_3-subunit (Takada et al., 1987a). α_{Ic} and α_{IIa} are probably ALSO VLA α-subunits (Pischel et al., 1988). The VLA set of proteins has been studied particularly by Hemler et al. (1987; Takada et al., 1987a,b). The β_2-subfamily is apparently restricted to leukocytes and is involved in various aspects of leukocyte adhesion, both cell–cell and cell–substratum (Anderson and Springer, 1987). LFA-1 participates in lymphocyte interactions during T-cell help, cytotoxicity, and extravasation. Mac-1 is involved in macrophage and neutrophil adhesion and can also function as a C3bi receptor. p150,95 may have functions similar to those of Mac-1 but has a somewhat different cell-type distribution. The β_3-subfamily comprises the vitronectin receptor isolated by affinity chromatography on VN (Pytela et al., 1985a,b; Argraves et al., 1986) and the GPIIb/IIIa protein isolated from platelets (Fitzgerald et al., 1985, 1987). These two receptors appear to share a common β-subunit (Ginsberg et al., 1987). For other references, consult the text, reviews (Ruoslahti and Pierschbacher, 1986; Hynes, 1987), and chapters by Horwitz et al. and Ruoslahti et al. in this volume.

Figure 8. *Homologies among integrin α-subunit light chains.* Note the conserved group of three cysteine residues near the N-terminus of the light chains, the conserved transmembrane domains (*under-* and *overlined*), and homologies near the beginning of the cytoplasmic domains (− and +). Other homologies exist, and some are marked. The chicken (CHK) sequence (M. A. Stepp, R. S. Patel, and R. O. Hynes, unpublished data) and the FN-R sequence (Suzuki et al., 1986) are from the β₁-subfamily. The VN-R (Argraves et al., 1986) and IIB (Poncz et al., 1987) sequences are from the β₃-subfamily. Known cleavage sites (VN-R and IIB) are marked by *solid arrowheads,* and potential cleavages (chicken and FN-R) are marked by *open arrowheads.*

the chicken integrin β-subunit and have shown that they bind to the surfaces of live cells (Tamkun et al., 1986). Thus, results with the segment-specific antibodies confirm the transmembrane organization of the β-subunit shown in Figure 7.

A second series of results obtained with the anti-cytoplasmic domain antibody concerns the phylogenetic distribution of integrins. This antibody immunoprecipitates integrin αβ-complexes from cells of avian, rodent, human, and *Drosophila* origins (Marcantonio et al., unpublished data). In parallel, Southern blotting data have confirmed the presence of sequences cross-reactive with chicken integrin β₁ in all of these species plus amphibians, echinoderms, and nematodes (DeSimone et al., unpublished data).

It is clear, therefore, that at least the integrin β₁-subfamily is widely distributed and strongly conserved in different phyla. Our results confirm and extend earlier indications that the so-called position-specific antigens in *Drosophila* are related to the vertebrate integrins (Wilcox et al., 1984; Leptin et al., 1987). It remains to be seen whether all three integrin subfamilies are as generally distributed as the β₁-family.

INTEGRIN BETA CHAIN C—TERMINAL HOMOLOGIES

```
                        1                                                    40
XENOPUS  BETA-1         ePECPsGPDI  IPIVAGVVAG  IVLIGLALLL  IWKLLMIIHD
CHICKEN  BETA-1         tPECPsGPDI  IPIVAGVVAG  IVLIGLALLL  IWKLLMIIHD
MOUSE    BETA-1         tPdCPtGPDI  IPIVAGVVAG  IVLIGLALLL  IWKLLMIIHD
HUMAN    BETA-1         nPECPtGPDI  IPIVAGVVAG  IVLIGLALLL  IWKLLMIIHD

HUMAN    BETA-2         srECvaGPnI  aaIVgGtVAG  IVLIGIILLv  IWKaLIhIsD
HUMAN    BETA-3         ePECPkGPDI  IvvIIsVmga  IILIGLAaLL  IWKLLItIHD

                                                         ●
                        RREFAKFEKE  KMNAKWDTGE  NPIYKSAVTT  VVNPKYEGK
                        RREFAKFEKE  KMNAKWDTGE  NPIYKSAVTT  VVNPKYEGK
                        RREFAKFEKE  KMNAKWDTGE  NPIYKSAVTT  VVNPKYEGa
                        RREFAKFEKE  KMNAKWDTGE  NPIYKSAVTT  VVNPKYEGK

                        IREyrrFEKE  KIksqWn-nd  NPIfKSAtTT  VmNPKfaes
                        RkEFAKFEeE  rarAKWDTan  NPIYKeAtsT  ftNItYrGt
                                                         ●
```

Figure 9.   *Homologies among integrin β-subunits.* The figure shows the sequences of the C-termini of several integrin β-subunits, including the transmembrane domains (*over-* and *underlined*) and the cytoplasmic domain. *Uppercase letters* denote residues conserved in at least three sequences. Note the very high degree of conservation among the β_1-subunits of various species and the divergence of the β_2- and β_3-subunits. The *black dots* mark the tyrosine residues present in β_1- and β_3-subunits that are potential sites for phosphorylation (see text). An antibody prepared against a synthetic peptide based on the β_1 cytoplasmic domain sequence is widely species cross-reactive as expected but does not recognize β_2- or β_3-subunits. For references to sequence data, see Table 3.

FUTURE PROSPECTS

We have reviewed here two complementary lines of research: one on the extracellular matrix glycoproteins (nectins) to which cells attach, and a second concerning the cell surface glycoproteins (integrins) that mediate the attachment. These two sets of proteins clearly interact in a ligand–receptor fashion during many biological processes. The ligands represent a set of analogues, while many of the receptors comprise a homologous family. We have not discussed evidence of other groups for several other cell surface receptors for matrix proteins that appear to be unrelated to integrins.

The work on the ligands has reached a point where complete sequences are available for many of the proteins, and we now understand the structural basis of the diversity of fibronectin isoforms. Two clear lines of research derive from these results. In the first it will be necessary to determine the basis for the cell type–specific control of the alternative splicing of fibronectins. This is now an accessible project. It remains to be seen whether other ligands are alternatively spliced, but it is clear that this mechanism of variation is widely used in other systems, including N-CAM (Cunningham et al. and Walsh et al., this volume). Thus, an understanding of the regulation of this process in the case of fibronectins should have broad implications. The second line of investigation of ligands such as fibronectin and thrombospondin is the detailed analysis of their structure–function relationships. This will now be possible at a much more sophisticated level, given the primary sequences and the application of recombinant DNA methods such as expression vectors (e.g., Schwarzbauer et al., 1987a).

Research on the integrin receptors is at a less well developed stage, although extremely rapid progress is being made. The sequences of several of these receptors are complete, and many others should be completed soon. The methods alluded to above will be applicable here also. Areas of considerable interest will be the nature of the ligand–receptor interaction and the basis of the specificities of different integrins. Present data suggest that much of this specificity lies in the α-chains, since integrins with common β-chains but differing α-chains have different ligand specificities (see Table 4). However, the β-chains also interact with ligands, as shown by cross-linking studies (Bennett et al., 1982; Marguerie et al., 1984; Gardner and Hynes, 1985; Lam et al., 1987; Santoro and Lawing, 1987).

The integrins also probably interact specifically with cytoskeletal proteins. This has been shown directly for the interaction of chicken integrin with the cytoskeletal protein talin (Horwitz et al., 1986). The high conservation of the cytoplasmic domains (Figure 9) indicates that the same will be true in other species, and one suspects it will also be true for the β_2- and β_3-subfamilies. It is of some interest that two of these subunits (β_1 and β_3) contain tyrosine residues homologous with the autophosphorylation site of the EGF receptor (see Figure 7) (Tamkun et al., 1986; Fitzgerald et al., 1987). It has been shown that this site is phosphorylated in the chicken integrin β_1-chain in transformed cells (Hirst et al., 1986). This phosphorylation appears to affect the ability of integrin to bind to talin (Horwitz et al., this volume). It will be of interest to determine whether the homologous tyrosine residue in the β_3-subunit plays a role in regulating the function of the vitronectin receptor and the IIb/IIIa receptor of platelets. It seems clear that the functions of integrin receptors need to be regulated in various situations, such as cell adhesion and migration. The roles of phosphorylation by tyrosine kinases, and perhaps other kinases, and of proteolysis by extracellular proteases need investigation.

The wide array of ligands and receptors, all of which appear to play roles in cell–substratum adhesion, in principle allows a great deal of combinatorial diversity in the interactions of cells with matrices during adhesion and migration. This could clearly have important consequences in morphogenesis and differentiation. It will be important to investigate the regulation of expression (coordinate or otherwise) of the various ligands and receptors during developmental processes. One can anticipate that subtle variations in the patterns of ligands to which cells are exposed and the patterns of receptors by which they interact will have crucial effects on their behavior. The availability of cDNA and antibody probes for the various molecules involved will make possible detailed analyses of these patterns of expression. Furthermore, the possibility of applying genetic approaches, offered by the insect and nematode systems, is an exciting new avenue to be pursued.

In summary, the molecular biological results of the last five or six years have had a major impact on our understanding of the molecular basis of cell adhesion. While many questions have been answered by the work to date, many more have been raised or made newly accessible by our current molecular understanding and it is clear that the next five years will provide many new insights.

ACKNOWLEDGMENTS

The research described in this article was supported largely by grants from the United States Public Health Service (P01CA26712 and R01CA17001 to R. O. H. and R01HL28749 to J. J. L.); by grants from the Whitaker Foundation and postdoctoral fellowships from the National Institutes of Health (D. W. D., E. E. M., P. A. N.), the American Cancer Society (M. A. S.), and the Charles A. King Trust (J. E. S.); and by grants to the MIT Cancer Center from the National Cancer Institute, Ajinomoto, and Bristol-Meyers.

REFERENCES

Akiyama, S. K., S. S. Yamada, and K. M. Yamada (1986) Characterization of a 140-kD avian cell surface antigen as a fibronectin-binding molecule. *J. Cell Biol.* **102**:442–448.

Ali, I. U., V. M. Mautner, R. P. Lanza, and R. O. Hynes (1977) Restoration of normal morphology, adhesion, and cytoskeleton in transformed cells by addition of a transformation-sensitive surface protein. *Cell* **11**:115–126.

Anderson, D. C., and T. A. Springer (1987) Leukocyte adhesion deficiency: An inherited defect in the Mac-1, LFA-1, and p150,95 glycoproteins. *Annu. Rev. Med.* **38**:175–194.

Argraves, W. S., R. Pytela, S. Suzuki, J. L. Millan, M. D. Pierschbacher, and E. Ruoslahti (1986) cDNA sequences from the β subunit of the fibronectin receptor predict a transmembrane domain and a short cytoplasmic peptide. *J. Biol. Chem.* **261**:12922–12924.

Argraves, W. S., S. Suzuki, H. Arai, K. Thompson, M. D. Pierschbacher, and E. Ruoslahti (1987) Amino acid sequence of the human fibronectin receptor. *J. Cell Biol.* **105**:1183–1190.

Banyai, L., A. Varadi, and L. Patthy (1983) Common evolutionary origin of the fibrin-binding structures of fibronectin and tissue-type plasminogen activator. *FEBS Lett.* **163**:37–41.

Bennett, J. S., G. Vilaire, and D. B. Cines (1982) Identification of the fibrinogen receptor on human platelets by photoaffinity labeling. *J. Biol. Chem.* **257**:8049–8054.

Bernard, M. P., M. Kolbe, D. Weil, and M. L. Chu (1985) Human cellular fibronectin: Comparison of the C-terminal portion with rat identifies primary structural domains separated by hypervariable regions. *Biochemistry* **24**:2698–2704.

Burridge, K., and L. Connell (1983) A new protein of adhesion plaques and ruffling membranes. *J. Cell Biol.* **97**:359–367.

Burridge, K., and J. R. Feramisco (1980) Microinjection and localization of a 130K protein in living fibroblasts: A relationship to actin and fibronectin. *Cell* **19**:587–595.

Chen, W.-T., T. Hasegawa, C. Hasegawa, C. Weinstock, and K. M. Yamada (1985) Development of cell surface linkage complexes in cultivated fibroblasts. *J. Cell Biol.* **100**:1103–1114.

Damsky, C. M., K. A. Knudsen, D. Bradley, C. A. Buck, and A. F. Horwitz (1985) Distribution of the CSAT cell–matrix antigen on myogenic and fibroblastic cells in culture. *J. Cell Biol.* **100**:1528–1539.

Dean, D. C., C. L. Bowlus, and S. Bourgeois (1987) Cloning and analysis of the promoter region of the human fibronectin gene. *Proc. Natl. Acad. Sci. USA* **84**:1876–1880.

DeSimone, D. W., and R. O. Hynes (1988) Structural conservation and evolutionary divergence of integrin β subunits. *J. Biol. Chem.* **263**:5333–5340.

Ehrismann, R., D. E. Roth, H. M. Eppenberger, and D. C. Turner (1982) Arrangement of attachment-promoting, self-association, and heparin-binding sites in horse serum fibronectin. *J. Biol. Chem.* **257**:7381–7387.

Esch, F. S., N. C. Ling, P. Bohlen, S. Y. Ying, and R. Guillemin (1983) Primary structure of PDC-109, a major protein constituent of bovine seminal plasma. *Biochem. Biophys. Res. Commun.* **113**:861–867.

Fitzgerald, L. A., I. F. Charo, and D. R. Phillips (1985) Human and bovine endothelial cells synthesize membrane proteins similar to human platelet glycoproteins IIb and IIIa. *J. Biol. Chem.* **260**:10893–10896.

Fitzgerald, L. A., B. Steiner, S. C. Rall, S.-S. Lo, and D. R. Phillips (1987) Protein sequence of endothelial glycoprotein IIIa derived from a cDNA clone: Identity with platelet glycoprotein IIIa and similarity to integrin. *J. Biol. Chem.* **262**:3936–3939.

Gardner, J. M., and R. O. Hynes (1985) Interaction of fibronectin with its receptor on platelets. *Cell* **42**:439–448.

Gilbert, W. (1978) Why genes in pieces? *Nature* **271**:501.

Ginsberg, M. H., J. Loftus, M. D. Pierschbacher, E. Ruoslahti, and E. F. Plow (1987) Immuno-chemical and N-terminal sequence comparison of two cytoadhesins indicates they contain similar or identical beta subunits and distinct alpha subunits. *J. Biol. Chem.* **262**:5437–5440.

Greve, J. M., and D. I. Gottlieb (1982) Monoclonal antibodies which alter the morphology of cultured chick myogenic cells. *J. Cell. Biochem.* **18**:221–230.

Hemler, M. E., C. Huang, and L. Schwarz (1987) The VLA protein family: Characterization of five distinct cell surface heterodimers each with a common 130,000 M_r subunit. *J. Biol. Chem.* **262**:3300–3309.

Hirano, H., Y. Yamada, M. Sullivan, B. deCrombugghe, I. Pastan, and K. M. Yamada (1983) Isolation of genomic DNA clones spanning the entire fibronectin gene. *Proc. Natl. Acad. Sci. USA* **80**:46–50.

Hirst, R., A. F. Horwitz, C. A. Buck, and L. Rohrschneider (1986) Phosphorylation of the fibronectin receptor complex in cells transformed by oncogenes that encode tyrosine kinases. *Proc. Natl. Acad. Sci. USA* **83**:6470–6474.

Horwitz, A. F., K. Duggan, R. Greggs, C. Decker, and C. A. Buck (1985) The cell substrate attachment (CSAT) antigen has properties of a receptor for laminin and fibronectin. *J. Cell Biol.* **101**:2134–2144.

Horwitz, A. F., E. Duggan, C. A. Buck, M. C. Beckerle, and K. Burridge (1986) Interaction of plasma membrane fibronectin receptor with talin—A transmembrane linkage. *Nature* **320**:531–533.

Humphries, M. J., S. K. Akiyama, A. Komoriya, K. Olden, and K. M. Yamada (1986) Identification of an alternatively spliced site in human plasma fibronectin that mediates cell type–specific adhesion. *J. Cell Biol.* **103**:2637–2647.

Hynes, R. O. (1973) Alteration of cell-surface proteins by viral transformation and by proteolysis. *Proc. Natl. Acad. Sci. USA* **70**:3170–3174.

Hynes, R. O. (1976) Cell surface proteins and malignant transformation. *Biochim. Biophys. Acta* **458**:73–107.

Hynes, R. O. (1981) Relationships between fibronectin and the cytoskeleton. *Cell Surf. Rev.* **7**:97–139.

Hynes, R. O. (1985) Molecular biology of fibronectin. *Annu. Rev. Cell Biol.* **1**:67–90.

Hynes, R. O. (1986) Fibronectins. *Sci. Am.* **254**:42–51.

Hynes, R. O. (1987) Integrins: A family of cell surface receptors. *Cell* **48**:549–554.

Hynes, R. O., and A. T. Destree (1978) Relationships between fibronectin (LETS protein) and actin. *Cell* **15**:875–886.

Hynes, R. O., and K. M. Yamada (1982) Fibronectins: Multifunctional modular glycoproteins. *J. Cell Biol.* **95**:369–377.

Hynes, R. O., A. T. Destree, and D. D. Wagner (1982) Relationships between fibronectin, actin, and cell–substratum adhesion. *Cold Spring Harbor Symp. Quant. Biol.* **46**:659–670.

Kishimoto, T. K., K. O'Connor, A. Lee, T. M. Roberts, and T. A. Springer (1987) Cloning the β subunit of the leukocyte adhesion proteins: Homology to an extracellular matrix receptor defines a novel supergene family. *Cell* **48**:681–690.

Knudsen, K. A., P. E. Rao, C. H. Damsky, and C. A. Buck (1981) Membrane glycoproteins involved in cell–substratum adhesion. *Proc. Natl. Acad. Sci. USA* **78**:6071–6075.

Knudsen, K. A., A. F. Horwitz, and C. A. Buck (1985) A monoclonal antibody identifies a glycoprotein complex involved in cell–substratum adhesion. *Exp. Cell Res.* **157**:218–228.

Kornblihtt, A. R., K. Vibe-Pedersen, and F. E. Baralle (1984a) Human fibronectin: Molecular cloning evidence for two mRNA species differing by an internal segment coding for a structural domain. *EMBO J.* **3**:221–226.

Kornblihtt, A. R., K. Vibe-Pedersen, and F. E. Baralle (1984b) Human fibronectin: Cell specific alternative mRNA splicing generates polypeptide chains differing in the number of internal repeats. *Nucleic Acids Res.* **12**:5853–5868.

Kornblihtt, A. R., K. Umezawa, K. Vibe-Pedersen, and F. E. Baralle (1985) Primary structure of human fibronectin: Differential splicing may generate at least 10 polypeptides from a single gene. *EMBO J.* **4**:1755–1759.

Lam, S. C.-T., E. F. Plow, M. A. Smith, A. Andrieux, J. J. Ryckwaert, G. Marguerie, and M. H. Ginsberg (1987) Evidence that arginyl-glycyl-aspartate peptides and fibrinogen γ chain peptides share a common binding site on platelets. *J. Biol. Chem.* **262**:947–950.

Law, S. K. A., J. Gagnon, J. E. K. Hildreth, C. E. Wells, A. C. Willis, and A. J. Wong (1987) The primary structure of the beta-subunit of the cell surface adhesion glycoproteins LFA-1, CR3 and p150,95 and its relationship to the fibronectin receptor. *EMBO J.* **6**:915–919.

Lawler, J. (1986) Review: The structural and functional properties of thrombospondin. *Blood* **67**:1197–1209.

Lawler, J., and R. O. Hynes (1986) The structure of human thrombospondin, an adhesive glycoprotein with multiple calcium-binding sites and homologies with several different proteins. *J. Cell Biol.* **103**:1635–1648.

Lawler, J., and R. O. Hynes (1987) Structure-function relationships of thrombospondin. *Semin. Thromb. Hemost.* **13**:245–254.

Lawler, J., R. Weinstein, and R. O. Hynes (1988) Arg-Gly-Asp and calcium-dependent mechanisms for cell attachment to thrombospondin. *J. Cell Biol.* **107**:2351–2361.

Leptin, M., R. Aebersold, and M. Wilcox (1987) *Drosophila* position-specific antigens resemble the vertebrate fibronectin-receptor family. *EMBO J.* **6**:1037–1043.

Marguerie, G. A., N. Thomas-Maison, M. H. Ginsberg, and E. F. Plow (1984) The platelet-fibrinogen interaction: Evidence for proximity of the A chain of fibrinogen to platelet membrane glycoproteins IIb/IIIa. *Eur. J. Biochem.* **139**:5–11.

McMullen, B. A., and K. Fujikawa (1985) Amino acid sequence of the heavy chain of human β-factor XIIIa (activated Hageman factor). *J. Biol. Chem.* **260**:5328–5341.

Neff, N. T., C. Lowrey, C. Decker, A. Tovar, C. Damsky, C. Buck, and A. F. Horwitz (1982) A monoclonal antibody detaches embryonic skeletal muscle from extracellular matrices. *J. Cell Biol.* **95**:654–666.

Norton, P. A., and R. O. Hynes (1987) Alternative splicing of chicken fibronectin in embryos and in normal and transformed cells. *Mol. Cell. Biol.* **7**:4297–4307.

Ny, T., F. Elgh, and B. Lund (1984) The structure of the human tissue-type plasminogen activator gene: Correlation of intron and exon structures to functional and structural domains. *Proc. Natl. Acad. Sci. USA* **81**:5355–5359.

Odermatt, E., J. Engel, H. Richter, and H. Hormann (1982) Shape, conformation, and stability of fibronectin fragments determined by electron microscopy, circular dichroism, and ultra-centrifugation. *J. Mol. Biol.* **159**:109–123.

Odermatt, E., J. W. Tamkun, and R. O. Hynes (1985) The repeating modular structure of the fibronectin gene: Relationship to protein structure and subunit variation. *Proc. Natl. Acad. Sci. USA* **82**:6571–6575.

Oldberg, A., and E. Ruoslahti (1986) Evolution of the fibronectin gene: Exon structure of cell attachment domain. *J. Biol. Chem.* **261**:2113–2116.

Owens, R. J., and F. E. Baralle (1986) Exon structure of the collagen-binding domain of human fibronectin. *FEBS Lett.* **204**:318–322.

Patel, R. S., E. Odermatt, J. E. Schwarzbauer, and R. O. Hynes (1987) Organization of the fibronectin gene provides evidence for exon shuffling during evolution. *EMBO J.* **6**:2565–2572.

Paul, J. I., and R. O. Hynes (1984) Multiple fibronectin subunits and their posttranslational modifications. *J. Biol. Chem.* **259**:13477–13487.

Paul, J. I., J. E. Schwarzbauer, J. W. Tamkun, and R. O. Hynes (1986) Cell-type-specific fibronectin subunits generated by alternative splicing. *J. Biol. Chem.* **261**:12258–12265.

Pierschbacher, M. D., and E. Ruoslahti (1984a) The cell attachment activity of fibronectin can be duplicated by small synthetic fragments of the molecule. *Nature* **309**:30–33.

Pierschbacher, M. D., and E. Ruoslahti (1984b) Variants of the cell recognition site of fibronectin that retain attachment-promoting activity. *Proc. Natl. Acad. Sci. USA* **81**:5985–5988.

Pierschbacher, M. D., E. Ruoslahti, J. Sundelin, P. Lind, and P. A. Peterson (1982) The cell attachment domain of fibronectin: Determination of the primary structure. *J. Biol. Chem.* **257**:9593–9597.

Pischel, K. D., H. G. Bluestein, and V. L. Woods, Jr. (1988) Platelet glycoproteins Ia, Ic, and IIa are physicochemically indistinguishable from the very late activation adhesion-related proteins of lymphocytes and other cell types. *J. Clin. Invest.* **81**:505–513.

Poncz, M., R. Eisman, R. Heidenreich, S. M. Silver, G. Vilaire, S. Surrey, E. Schwartz, and J. S. Bennett (1987) Structure of the platelet membrane glycoprotein IIb: Homology to the alpha subunits of the vitronectin and fibronectin membrane receptors. *J. Biol. Chem.* **262**:8476–8482.

Pytela, R., M. D. Pierschbacher, and E. Ruoslahti (1985a) Identification and isolation of a 140 kd cell surface glycoprotein with properties expected of a fibronectin receptor. *Cell* **40**:191–198.

Pytela, R., M. D. Pierschbacher, and E. Ruoslahti (1985b) A 125/115-kDa cell surface receptor specific for vitronectin interacts with the arginine-glycine-aspartic acid adhesion sequence derived from fibronectin. *Proc. Natl. Acad. Sci. USA* **82**:5766–5770.

Pytela, R., M. D. Pierschbacher, M. H. Ginsberg, E. F. Plow, and E. Ruoslahti (1986) Platelet membrane glycoprotein IIb/IIIa: Member of a family of Arg-Gly-Asp-specific adhesion receptors. *Science* **231**:1559–1562.

Richter, H., M. Seidl, and H. Hormann (1981) Location of heparin-binding sites of fibronectin: Detection of a hitherto unrecognized transamidase-sensitive site. *Hoppe-Seyler's Z. Physiol. Chem.* **362**:399–408.

Ruoslahti, E., and M. D. Pierschbacher (1986) Arg-Gly-Asp: A versatile cell recognition signal. *Cell* **44**:517–518.

Santoro, S. A., and W. J. Lawing, Jr. (1987) Competition for related but nonidentical binding sites on the glycoprotein IIb–IIIa complex by peptides derived from platelet adhesive proteins. *Cell* **48**:867–873.

Schwarzbauer, J. E., J. W. Tamkun, I. R. Lemischka, and R. O. Hynes (1983) Three different fibronectin mRNAs arise by alternative splicing within the coding region. *Cell* **35**:421–431.

Schwarzbauer, J. E., J. I. Paul, and R. O. Hynes (1985) On the origin of species of fibronectin. *Proc. Natl. Acad. Sci. USA* **82**:1424–1428.

Schwarzbauer, J. E., R. C. Mulligan, and R. O. Hynes (1987a) Efficient and stable expression of recombinant fibronectin polypeptides. *Proc. Natl. Acad. Sci. USA* **84**:754–758.

Schwarzbauer, J. E., R. S. Patel, D. Fonda, and R. O. Hynes (1987b) Multiple sites of alternative splicing of the rat fibronectin gene transcript. *EMBO J.* **6**:2573–2580.

Sekiguchi, K., A. M. Klos, K. Kurachi, S. Yoshitake, and S. Hakomori (1986) Human liver fibronectin complementary DNAs: Identification of two different messenger RNAs possibly encoding the α and β subunits of plasma fibronectin. *Biochemistry* **25**:4936–4941.

Singer, I. I. (1979) The fibronexus: A transmembrane association of fibronectin-containing fibers and bundles of 5nm microfilaments in hamster and human fibroblasts. *Cell* **16**:675–685.

Singer, I. I. (1982) Association of fibronectin and vinculin with focal contacts and stress fibers in stationary hamster fibroblasts. *J. Cell Biol.* **92**:398–408.

Singer, I. I., and P. R. Paradiso (1981) A transmembrane relationship between fibronectin and vinculin (130 kd protein): Serum modulation in normal and transformed hamster fibroblasts. *Cell* **24**:481–492.

Skorstengaard, K., M. S. Jensen, P. Sahl, T. E. Petersen, and S. Magnusson (1986) Complete primary structure of bovine plasma fibronectin. *Eur. J. Biochem.* **161**:441–453.

Suzuki, S., W. S. Argraves, R. Pytela, H. Arai, T. Krusius, M. D. Pierschbacher, and E. Ruoslahti (1986) cDNA and amino acid sequences of the cell adhesion protein receptor recognizing vitronectin reveal a transmembrane domain and homologies with other adhesion protein receptors. *Proc. Natl. Acad. Sci. USA* **83**:8614–8618.

Takada, Y., C. Huang, and M. E. Hemler (1987a) Fibronectin receptor structures are included within the VLA family of heterodimers. *Nature* **326**:607–609.

Takada, Y., J. L. Strominger, and M. E. Hemler (1987b) The very late antigen family of heterodimers is part of a superfamily of molecules involved in adhesion and embryogenesis. *Proc. Natl. Acad. Sci. USA* **84**:3239–3243.

Tamkun, J. W., and R. O. Hynes (1983) Plasma fibronectin is synthesized and secreted by hepatocytes. *J. Biol. Chem.* **258**:4641–4647.

Tamkun, J. W., J. E. Schwarzbauer, and R. O. Hynes (1984) A single rat fibronectin gene generates three different mRNAs by alternative splicing of a complex exon. *Proc. Natl. Acad. Sci. USA* **81**:5140–5144.

Tamkun, J. W., D. W. DeSimone, D. Fonda, R. S. Patel, C. A. Buck, A. F. Horwitz, and R. O. Hynes (1986) Structure of integrin, a glycoprotein involved in the transmembrane linkage between fibronectin and actin. *Cell* **42**:271–282.

Vibe-Pedersen, K., A. R. Kornblihtt, and T. E. Petersen (1984) Expression of a human β-globin/fibronectin gene hybrid generates two mRNAs by alternative splicing. *EMBO J.* **3**:2511–2516.

Vibe-Pedersen, K., A. R. Kornblihtt, and T. E. Petersen (1984) Expression of a human β-globin/fibronectin gene hybrid generates two mRNAs by alternative splicing. *EMBO J.* **3**:2511–2516.

Wilcox, M., N. Brown, M. Piovant, R. J. Smith, and R. A. H. White (1984) The *Drosophila* position-specific antigens are a family of cell surface glycoprotein complexes. *EMBO J.* **3**:2307–2313.

Yamada, K. M. (1983) Cell surface interactions with extracellular materials. *Annu. Rev. Biochem.* **52**:761–799.

Yamada, K. M., and D. W. Kennedy (1984) Dualistic nature of adhesive protein function: Fibronectin and its biologically active peptide fragments can autoinhibit fibronectin function. *J. Cell Biol.* **99**:29–36.

Yamada, K. M., and D. W. Kennedy (1985) Amino acid sequence specificities of an adhesive recognition signal. *J. Cell. Biochem.* **28**:99–104.

Yamada, K. M., S. S. Yamada, and I. Pastan (1976) Cell surface protein partially restores morphology, adhesiveness, and contact inhibition of movement to transformed fibroblasts. *Proc. Natl. Acad. Sci. USA* **73**:1217–1221.

Chapter 8

The Arg-Gly-Asp Sequence and Its Receptors: A Versatile Recognition System

ERKKI RUOSLAHTI
W. SCOTT ARGRAVES
KURT R. GEHLSEN
JAMES GAILIT
MICHAEL D. PIERSCHBACHER

ABSTRACT

Originally identified as the cellular recognition site in fibronectin, the tripeptide sequence Arg-Gly-Asp has now been found to serve the same purpose in many other extracellular matrix and platelet adhesion proteins. The list of such proteins currently includes vitronectin, collagens, osteopontin, thrombospondin, fibrinogen, and von Willebrand factor. Despite the similarity of their cell attachment sites, these proteins can be recognized individually by cell surface receptors. The receptors are heterodimeric proteins that belong to a family of related proteins, integrins. At least 10 different members of this receptor family have already been identified, and several of them have been cloned and sequenced. The individual recognition specificity of the receptors may depend on conformational differences among the RGD sequences of the various adhesion proteins. The receptors are differentially susceptible to inhibition by a series of peptides containing the RGD sequence, and their functions can be probed individually with such peptides. Thus the integrin-type receptors and their adhesion protein ligands constitute a versatile recognition system. The cell surface interactions mediated by this recognition system are likely to play an important role in the anchorage, polarity, migration, and possibly differentiation and growth of cells.

Extracellular matrices are insoluble structures composed of collagens, elastin, various glycoproteins, proteoglycans, and hyaluronic acid. Collagens and many, perhaps all, of the extracellular glycoproteins interact with cells, and this interaction can promote cell adhesion. The glycoproteins include fibronectin, laminin, vitronectin, thrombospondin, von Willebrand factor, osteopontin, and tenascin.

Our studies on the interaction of fibronectin and vitronectin with cells have led to the discovery that these two proteins as well as many other adhesive

201

proteins share a tripeptide Arg-Gly-Asp (RGD) as their cell attachment site (Pierschbacher and Ruoslahti, 1984a,b; Ruoslahti and Pierschbacher, 1986) and that these RGD sites are recognized by a family of related cell surface adhesion receptors. The purpose of this chapter is to summarize some of our recent work on the recognition system constituted by the RGD adhesion proteins and their receptors.

THE FIBRONECTIN CELL ATTACHMENT SITE

The structure of the cell attachment site in fibronectin was elucidated in a series of experiments that included the isolation and sequencing of an 11.5-kD fragment of fibronectin containing the cell attachment site, and the duplication of its cell attachment–promoting activity with peptides containing the amino acid sequence RGD. The peptides can not only duplicate the attachment-promoting function of fibronectin, they can also inhibit it. Which result is obtained depends on the presentation of the peptide; when coated on a surface the peptides promote cell attachment, but in solution they can inhibit the attachment of cells to a surface coated with fibronectin or with the peptides themselves. Changes as small as the exchange of alanine for glycine or glutamic acid for aspartic acid, which constitute the addition of a single methyl or methylene group to the RGD tripeptide, can eliminate the activity of the peptide (Pierschbacher et al., 1981; Pierschbacher and Ruoslahti, 1984a,b; Ruoslahti and Pierschbacher, 1986).

THE RGD SEQUENCE IN OTHER PROTEINS

The RGD tripeptide occurs in proteins other than fibronectin. Recent structural studies on adhesive proteins have identified this sequence in a number of extracellular matrix and platelet adhesion proteins, where it probably plays a functional role (Table 1).

The proteins known to carry functional RGD sequences include the platelet adhesion proteins fibrinogen and von Willebrand factor (Gartner and Bennett, 1985; Ginsberg et al., 1985; Haverstick et al., 1985; Plow et al., 1985; Ruggeri et al., 1986), as well as type I collagen (Dedhar et al., 1987), vitronectin (Hayman et al., 1985; Suzuki et al., 1985), and osteopontin (Oldberg et al., 1986). In addition, thrombospondin (Lawler and Hynes, 1986) and other types of collagens contain RGD sequences that are likely to be the cell attachment sites of these proteins. Finally, there is evidence from receptor work indicating that one of the cell attachment sites of laminin is also an RGD site (see below). Thus the RGD sequence is the cell attachment site of many different adhesive proteins in the vertebrate species. Among lower life forms, an active RGD sequence is present in the slime mold aggregation protein, discoidin I

Table 1.   RGD Adhesion Proteins

Fibronectin
Vitronectin
Fibrinogen
von Willebrand factor
Collagens
Osteopontin
Thrombospondin
Laminin

(Springer et al., 1984). This shows that the function of the RGD sequence as a cell attachment signal became established during early stages of evolution.

In addition to being the cell attachment site of various proteins, the cell attachment–promoting activity of which appears to be physiologically signi-ficant, the RGD sequence imparts cell attachment–promoting activity to some proteins in which such activity has no apparent significance. One such protein is the γII crystallin of the eye lens (Pierschbacher et al., unpublished data). Several other proteins show no demonstrable activity, including epidermal growth factor receptor, human α-fetoprotein, and *E. coli* β-galactosidase (Ruoslahti et al., unpublished data). We discuss later the possible reasons for the varied activities of RGD sequences. Finally, there are a number of RGD-containing proteins in which the RGD sequence may be important but that have not yet been tested. These include various complement and viral envelope proteins.

Despite the similarity of the cell attachment sequence in the various adhesive proteins, cells can recognize these proteins individually. This specificity is provided by the existence of a number of receptors, each of which is capable of recognizing only a single RGD-containing protein ligand, or in some cases a limited number of ligands. The RGD-containing peptides have been instrumental in the identification of these receptors.

PURIFICATION OF RGD-DIRECTED RECEPTORS

We have isolated RGD-directed cell surface receptors from various types of cells by using affinity chromatography on Sepharose carrying the appropriate, covalently bound adhesion protein. Specific elution of the materials bound to the affinity matrix was accomplished with a hexapeptide containing the RGD sequence. The use of fibronectin as the affinity ligand yielded a receptor that is a heterodimer of a 160-kD α-subunit and a 140-kD β-subunit (Pytela et al., 1985a, 1986). Similar affinity chromatography experiments in which vitronectin was used as the ligand yielded a different heterodimeric receptor protein, a vitronectin receptor (Pytela et al., 1985b). Electrophoretically homogeneous receptors can be prepared by adding a lectin chromatography step (Pytela et al., 1987) after the affinity purification procedure (Figure 1). These receptors are immunologically distinct proteins (Figure 2). Yet another receptor binds to type I collagen and to an RGD-containing peptide that assumes a collagenlike triple helical structure (Dedhar et al., 1987). All these receptors can be obtained from the same cloned osteosarcoma cells. Thus these cells possess at least three different adhesion receptors recognizing the RGD sequence in their individual ligands.

Fractionation of platelet extracts by affinity chromatography on insolubilized fibrinogen and fibronectin yields a fourth RGD-directed receptor (Gardner and Hynes, 1985; Pytela et al., 1986). This receptor is indistinguishable from the previously characterized platelet protein GP IIb/IIIa (Parise and Phillips, 1985; Nurden et al., 1986). GP IIb/IIIa and the vitronectin receptor can also be isolated by employing Sepharose with an RGD-containing heptapeptide substituted as the affinity matrix (Pytela et al., 1986). The fibronectin and collagen receptors, however, do not have sufficient affinities for the short peptides to allow them to bind to the peptide matrix. The advantage of these affinity procedures is that they allow one to isolate individual adhesion receptors. Adhesion receptor complexes that have been independently identified by using cell attachment–inhibiting antibodies and that display multiple receptor specificities may represent mixtures of receptors (Horwitz et al., 1985; Akiyama et al., 1986; Burridge, 1986; Hall et al., 1987).

POLYPEPTIDE COMPOSITION OF RGD-DIRECTED RECEPTORS

The mammalian RGD-directed receptors are typically heterodimers of two subunits, α and β. Sodium dodecyl sulfate–polyacrylamide gel electrophoresis (SDS–PAGE) analysis of the receptors with and without prior treatment with a reducing agent shows that the α-subunits of the fibronectin and vitronectin receptors and GP IIb/IIIa consist of two polypeptides, heavy and light chain, disulfide bonded to one another. These chains of the α-subunit arise from proteolytic cleavage of a precursor polypeptide (see below). The sizes of the α-subunit polypeptides range between 120 and 140 kD for the heavy chain and are about 20 kD for the light chain. The β-subunits, on the other hand, are single polypeptides that range in molecular weight from 90 to 140 kD. The β-subunit migrates much faster in SDS–PAGE when it is not reduced, suggesting a compact structure resulting from extensive disulfide bonding. As it was realized that this heterodimeric structure was characteristic of the RGD-directed receptors, it became apparent that three protein families—a group of adhesive leukocyte surface proteins (Springer et al., 1987), a group of proteins called very late antigens (VLA) of activation (Takada et al., 1987), and the so-called position-specific antigens of *Drosophila* (Leptin et al., 1987)—were very

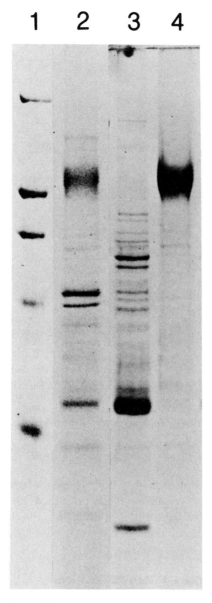

Figure 1. *SDS–PAGE analysis of human placental fibronectin receptor.* Separation of a placental octylglucoside extract by affinity chromatography on a 120-kD cell attachment–promoting fragment followed by elution with an RGD-containing hexapeptide gives a protein fraction with a prominent band at 140 kD after reduction (*lane 2*). This band contains the two receptor subunits that migrate together at 140 kD after reduction. Further fractionation of this eluted fraction on wheat germ agglutinin–Sepharose yields a nonbound fraction devoid of the receptor (*lane 3*) and a fraction containing the receptor. *Lane 1* shows molecular weight standards of 200 kD, 116 kD, 94 kD, 68 kD, and 44 kD.

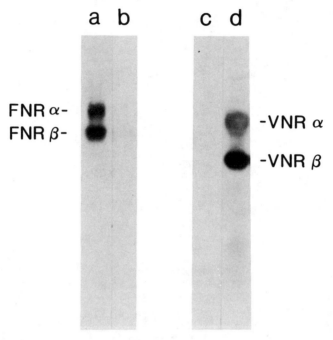

Figure 2. *Immunological comparison of two adhesion receptors.* The fibronectin receptor (*lanes a, c*) and the vitronectin receptor (*lanes b, d*) were isolated from human placental tissue (Pytela et al., 1987), separated by SDS–PAGE without reduction, blotted on nitrocellulose filters, and immunoblotted with rabbit antibodies prepared against the fibronectin receptor (*lanes a, b*) or the vitronectin receptor (*lanes c, d*). The positions of the α- and β-subunits of the two receptors are indicated.

similar to the RGD receptors in structure. These molecules, together with the RGD-directed receptors we have already discussed, form the integrin super-family (Hynes, 1987; Ruoslahti and Pierschbacher, 1987). The known integrins are listed in Table 2.

PRIMARY STRUCTURE OF THE INTEGRINS

Structural analysis of the integrins by cDNA and amino acid sequencing has progressed rapidly. We have recently completed the amino acid sequence of the human fibronectin receptor (Argraves et al., 1986, 1987). Work done in our laboratory and others makes it possible to deduce the sequences of vitronectin and platelet receptors because the α-chains as well as the β-chain they appear to share (Ginsberg et al., 1987) have each been cloned and sequenced (Suzuki et al., 1986, 1987; Fitzgerald et al., 1987; Poncz et al., 1987). In addition, the human LFA family-shared β-subunit (Kishimoto et al., 1987) and one

Table 2. Integrin Receptor Superfamily

Protein[a]	Ligands	Function
Fibronectin receptor[b]	Fibronectin	Cell attachment, phagocytosis
VLA-1	?	
VLA-2	?	
VLA-3	?	
VLA-4	?	
Vitronectin receptor	Vitronectin	Cell attachment, phagocytosis?
GP IIb/IIIa	Fibrinogen, fibronectin, von Willebrand factor, vitronectin	Platelet aggregation
LFA-1	ICAM-1	Cell–cell adhesion
Mac-1	C3bi	Complement binding
p150,95	?	Cell–cell adhesion

[a]The integrins have been grouped in families according to the β-subunit they contain. Three β-subunits are known: The fibronectin receptor family shares one (Takada et al., 1987), the vitronectin receptor and GP IIb/IIIa have very similar β-subunits (Ginsberg et al., 1987), and the LFA group of receptors shares a third β-subunit (Springer et al., 1987).
[b]The integrin complex in chicken, also known as the CSAT complex (Horwitz et al., 1985; Hynes, 1987), is likely to be equivalent to the fibronectin receptor family, but individual receptors have not yet been isolated from it. The fibronectin receptor is identical to VLA-5 (Takada et al., 1987).

polypeptide of the chicken integrin complex (Tamkun et al., 1986) have also been sequenced. Although much more sequence information on integrins will undoubtedly be published in the near future, the existing sequences allow a number of general conclusions to be made about the structure and relationships of integrins. There is no sequence homology between the α- and β-subunits of any one of these individual integrins, but each α-subunit is homologous to the other α-subunits and each β-subunit to the other β-subunits. The extent of this homology is 40–50% at the amino acid level. An exception is the high degree of homology (85%) between the chicken integrin polypeptide and the human fibronectin receptor β-subunit. Because the extent of the homology in this case is far greater than that between homologous subunits of distinct integrins in the superfamily, the chicken polypeptide probably represents the β-subunit of the fibronectin receptor family. Amino-terminal amino acid sequencing and immunological comparisons have shown that the VLA antigens are related to the fibronectin receptor in that they share the same β-subunit, and each has its own α-subunit. Since five VLA proteins have been identified (Takada et al., 1987), the fibronectin receptor family appears to include at least four other members.

Based on amino acid sequencing, each subunit of each integrin appears to contain a large extracellular domain, a membrane-spanning segment, and a short cytoplasmic domain (Figure 3). The location within the extracellular domain of the binding site for the adhesion protein ligand is not known, but both subunits appear to contribute to ligand binding. The sequence of the extracellular domain of the α-subunit of each receptor contains several sites that are homologous to calcium-binding sites in other proteins such as calmodulin, and these sequences are therefore likely to represent calcium-binding sites. The presence of such sites is in agreement with the known divalent cation requirement for the binding of integrins to their ligand.

The amino acid sequences of the α- and β-subunits strongly suggest that both subunits span the cell membrane because each polypeptide has near its carboxyl terminus a segment with the characteristics of a transmembrane domain. It is very likely that these segments are indeed embedded in the cell membrane because the isolated receptors can be readily incorporated into liposome membranes (Parise and Phillips, 1985; Pytela et al., 1985a,b, 1987).

The portions of the integrin polypeptides extending from the carboxy-terminal ends of the transmembrane domains are probably cytoplasmic. These cytoplasmic tails range between 28 and 47 amino acids in length. Again, there is a complete conservation of sequence in the cytoplasmic domains of the human fibronectin receptor β-subunit and its chicken homolog, suggesting a function dependent on that structure. The cytoplasmic tails of this subunit contain a short amino acid sequence that is homologous to a tyrosine

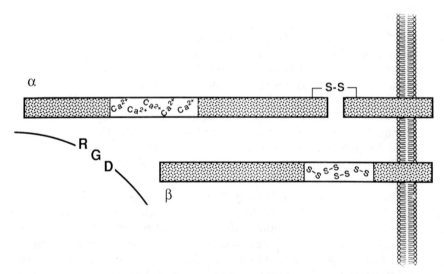

Figure 3. *Schematic representation of the general structure of RGD receptors. The α- and β-subunits are* associated with each other through divalent cation-dependent and other interactions. The α-subunits of many of the receptors consist of two polypeptides that are disulfide bonded to each other. Each subunit contains a typical transmembrane domain that traverses the cell membrane and brings the carboxyl termini of the subunits into the cytoplasmic side of the membrane. The bulk of the receptor is extracellular and contains the binding site for RGD.

phosphorylation site in the epidermal growth factor receptor and the insulin receptor (Tamkun et al., 1986). Phosphorylation of the fibronectin receptor has been observed in virally transformed chicken fibroblasts (Hirst et al., 1986). This receptor has also been shown to have an affinity for talin, a cytoskeletal protein associated with the actin filament network (Horwitz et al., 1986). Thus adhesion receptors may provide a link between the extracellular matrix and the cytoskeleton, and phosphorylation of the cytoplasmic domain may regulate the binding functions. Such regulatory mechanisms could explain the apparent misregulation of adhesion receptor function in malignantly transformed cells that lack both an extracellular matrix and an organized cytoskeleton (Vaheri and Mosher, 1978; Ruoslahti, 1984; Burridge, 1986). However, the β-subunit of the leukocyte receptor family lacks the critical tyrosine residue in its cytoplasmic tail (Kishimoto et al., 1987), indicating that phosphorylation at this site cannot regulate the interactions of these receptors.

LIGAND SPECIFICITY OF ADHESION RECEPTORS

Affinity chromatography on each of the various insolubilized adhesive proteins yields a different adhesion receptor from the same cell extract, indicating that although they have similar target sequences, each receptor has a mutually exclusive specificity at the protein level (Ruoslahti and Pierschbacher, 1987). Moreover, by using an assay in which one of the receptors is incorporated into liposome membranes and the binding of the liposomes to various surfaces is examined, we have shown that the fibronectin receptor-containing liposomes bind only to a surface coated with fibronectin and not to a surface coated with other adhesive proteins (Pytela et al., 1985a, 1986). The vitronectin (Pytela et al., 1985b) and collagen (Dedhar et al., 1987) receptors in liposomes are similarly specific for their own ligands. Despite this specificity, liposome binding can in each case be inhibited with RGD-containing synthetic peptides, revealing common underlying mechanisms for these interactions.

GP IIb/IIIA from platelets, on the other hand, has a different pattern of reactivity. Liposomes containing this receptor can bind to several RGD-containing proteins. These include fibrinogen, fibronectin, vitronectin, von Willebrand factor, and possibly also thrombospondin (Pytela et al., 1986). While GP IIb/IIIa appears to be exceptional in its broad specificity, it remains possible that other receptors also have additional, as yet unrecognized, ligands or that other broadly specific, RGD-directed receptors exist. A possible example is the chicken integrin complex that can be isolated with monoclonal antibodies CSAT and JG22. These antibodies inhibit the attachment of cells to fibronectin, laminin, and type IV collagen (Horwitz et al., 1985; Akiyama et al., 1986; Hall et al., 1987), and the complex that can be isolated by affinity chromatography on these antibodies binds to at least fibronectin and laminin. However, the complex may be a mixture of several receptors and other integrins because it appears to have a more complex subunit composition than mammalian RGD-directed receptors.

A major question concerns the ability of the adhesion receptors to distinguish among the various adhesive ligands despite the fact that many, perhaps all, of them have the same RGD cell attachment signal. One explanation for the ligand specificity displayed by RGD-directed receptors could be that the RGD sequence serves as a shared binding site, while the specificity is generated by a second binding site unique to each protein ligand. The main piece of evidence arguing against this possibility is that the smallest fibronectin fragments, including the synthetic RGD-containing peptides, interact better with the vitronectin receptor than with the fibronectin receptor, although fibronectin itself is not recognized by the vitronectin receptor (Table 3) (Pytela et al., 1985b). It is difficult to see how the vitronectin-specific second site necessitated by the two-site hypothesis would have been acquired by the fragments. Some similarities have been detected in the region of the RGD sequences in fibronectin and vitronectin (Wright et al., 1987). However, if the RGD sequence is omitted as it should be, since this homology was used to select the alignment, the flanking sequences have no significant homology. We favor the alternative hypothesis that the RGD tripeptide would provide most of the information needed and that the role of the surrounding sequences is to force the RGD determinant into an appropriate conformation for the receptor to recognize.

We have been able to show that at the level of short RGD-containing peptides, conformation is important for activity and specificity (Pierschbacher and Ruoslahti, 1987), but it is also clear that important contributions come from the amino acids immediately adjacent to the RGD sequence, especially from the residue following this sequence (Pierschbacher and Ruoslahti, 1984a,b; Ruoslahti and Pierschbacher, 1987; Pierschbacher and Ruoslahti,

Table 3. Binding of Adhesion Receptor Liposomes to Various Substrata

Microtiter Well Coating	Vitronectin Receptor	Fibronectin Receptor
Fibronectin	0	100
Vitronectin	100	0
120-kD fibronectin fragment	0	65
11.5-kD fibronectin fragment	0[a]	0
30-residue synthetic cell attachment peptide	24	0
GRGDSP (C)	80	0

Source: Modified from Pytela et al., 1985a,b.

[a]When coupled to Sepharose, the 11.5-kD fragment binds the vitronectin receptor but not the fibronectin receptor from a detergent extract of human MG-63 osteosarcoma cells (Ruoslahti et al., unpublished data).

1987). An example of the importance of conformation is that the type I collagen receptor we have identified (Dedhar et al., 1987) binds to triple helical collagen but not to collagen in which the triple helix has been unwound by heating. The same receptor also illustrates the importance of the amino acid that follows the RGD sequence in the short peptides; the fibronectin-derived peptide GRGDSP only slightly inhibits the attachment of osteosarcoma cells to type I collagen, whereas the same peptide with a threonine substitution at the serine position is quite active (Dedhar et al., 1987). Since one of the RGD sequences in type I collagen is followed by a threonine residue, this could be a sequence-specific effect, but it is also possible that the substitution modifies the conformations that the peptide can assume in solution. If the conformation of the RGD sequence is the main factor in determining ligand binding, this would readily explain why some RGD proteins promote cell attachment for no apparent physiological reason.

BIOLOGICAL ROLES OF ADHESION RECEPTORS

The adhesive forces generated by RGD interactions may either immobilize a cell or provide traction for cell migration. A delicate balance probably exists between the attachment and detachment of cells, determining whether a cell will remain stationary, migrate through tissues, or be a circulating cell. Accumulating evidence (Ruoslahti, 1984) suggests that a cell that lays down its own extracellular matrix and interacts with it through the various adhesion receptors is likely to be a stationary cell, whereas a cell that lacks both of these functions is likely to circulate. Tumor cells and migratory embryonal cells, as a rule, express an intermediate phenotype; they lack their own extracellular matrix, but are capable of interactions with matrix components. These characteristics are likely to make it easier for the cell to leave its position and migrate through tissues. That receptor–RGD interactions are important in the migration of cells through tissues is suggested by the recent observation that the invasion of tumor cells (Gehlsen et al., 1988) and lymphocytes (Savagner et al., 1986) through amniotic membrane in an invasion assay (Gehlsen and Hendrix, 1986) can be inhibited with RGD-containing peptides (Figure 4).

It is clear that cell–extracellular matrix (and cell–cell) interactions exert profound effects on cellular behavior (Grobstein, 1975; Hay, 1981). Signals from the extracellular matrix may affect cells just as much as hormones and other soluble mediators do; the main difference is that the extracellular matrix (and the surface of an adjacent cell) is insoluble, and it therefore exerts its effects at a short range and in a geometrically exact manner. The versatility of the RGD–adhesion receptor system suggests that it could be responsible for many of the signals from extracellular matrices to cells. These receptors might be particularly well suited to deliver positional signals that determine the location and polarity of cells in the body.

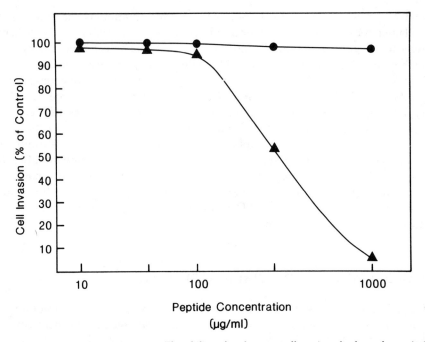

Figure 4. *Tumor cell invasion assay.* The ability of melanoma cells to invade through amniotic membrane tissue was examined in an *in vitro* invasion assay (Gehlsen and Hendrix, 1986). About 8% of the tumor cells passed through the membrane in 78 hours under standard conditions and in the presence of the control peptide GRGESP, whereas the invasion was almost completely inhibited when the cell attachment peptide GRGDSP was included in the assay.

ACKNOWLEDGMENTS

The preparation of this chapter and our original work was supported by grants CA-42507, CA-28896, and Cancer Center Support Grant CA-30199 from the National Cancer Institute, Department of Health and Human Services.

REFERENCES

Akiyama, S. K., S. S. Yamada, and K. M. Yamada (1986) Characterization of a 140-kD avian cell surface antigen as a fibronectin-binding molecule. *J. Cell Biol.* **102**:442–448.

Argraves, W. S., R. Pytela, S. Suzuki, J. L. Millan, M. D. Pierschbacher, and E. Ruoslahti (1986) cDNA sequences from the α subunit of the fibronectin receptor predict a transmembrane domain and a short cytoplasmic peptide. *J. Biol. Chem.* **261**:12922–12924.

Argraves, W. S., S. Suzuki, H. Arai, K. Thompson, M. D. Pierschbacher, and E. Ruoslahti (1987) Amino acid sequence of the human fibronectin receptor. *J. Cell Biol.* **105**:1183–1190.

Burridge, K. (1986) Substrate adhesion in normal and transformed fibroblasts: Organization and regulation of cytoskeletal membrane and extracellular matrix components at focal contacts. *Cancer Rev.* **4**:18–78.

Dedhar, S., E. Ruoslahti, and M. D. Pierschbacher (1987) A cell surface receptor for collagen type I recognizes the Arg-Gly-Asp sequence. *J. Cell Biol.* **104**:585–593.

Fitzgerald, L. A., B. Steiner, S. C. Rall, Jr., S.-S. Lo, and D. R. Phillips (1987) Protein sequence of endothelial glycoprotein IIIa derived from a cDNA clone. *J. Biol. Chem.* **262**:3936–3939.

Gardner, J. M., and R. O. Hynes (1985) Interaction of fibronectin with its receptor on platelets. *Cell* **42**:439–448.

Gartner, T. K., and J. S. Bennett (1985) The tetrapeptide analogue of the cell attachment site of fibronectin inhibits platelet aggregation and fibrinogen binding to activated platelets. *J. Biol. Chem.* **260**:11891–11894.

Gehlsen, K. R., and M. J. C. Hendrix (1986) *In vitro* assay demonstrates similar invasion profiles for B16F1 and B16F10 murine melanoma cells. *Cancer Lett.* **30**:207–212.

Gehlsen, K. R., W. S. Argraves, M. D. Pierschbacher, and E. Ruoslahti (1988) Inhibition of *in vitro* tumor cell invasion by Arg-Gly-Asp-containing synthetic peptides. *J. Cell Biol.* **106**:925–930.

Ginsberg, M. H., M. D. Pierschbacher, E. Ruoslahti, G. Marguerie, and E. Plow (1985) Inhibition of fibronectin binding to platelets by proteolytic fragments and synthetic peptides which support fibroblastic adhesion. *J. Biol. Chem.* **260**:3931–3936.

Ginsberg, M. H., J. Loftus, J.-J. Ryckwaert, M. Pierschbacher, R. Pytela, E. Ruoslahti, and E. F. Plow (1987) Immunochemical and N-terminal sequence comparison of two cytoadhesins indicates they contain similar or identical beta subunits and distinct alpha subunits. *J. Biol. Chem.* **262**:5437–5440.

Grobstein, C. (1975) Developmental role of intercellular matrix: Retrospection and prospection. In *Extracellular Matrix Influences on Gene Expression*, H. C. Slavkin and R. C. Greulich, eds., pp. 9–16, Academic, New York.

Hall, D. E., K. M. Neugebauer, and L. F. Reichardt (1987) Embryonic neural retinal cell response to extracellular matrix proteins: Developmental changes and effects of the cell substratum attachment antibody (CSAT). *J. Cell Biol.* **104**:623–634.

Haverstick, D. M., J. F. Cowan, K. M. Yamada, and S. A. Santoro (1985) Inhibition of platelet adhesion to fibronectin, fibrinogen, and von Willebrand factor substrates by a synthetic tetrapeptide derived from the cell-binding domain of fibronectin. *Blood* **66**:946–952.

Hay, E. D. (1981) Collagen and embryonic development. In *Cell Biology of Extracellular Matrix*, E. D. Hay, ed., pp. 379–405, Plenum, New York.

Hayman, E. G., M. D. Pierschbacher, and E. Ruoslahti (1985) Detachment of cells from culture substrate by soluble fibronectin peptides. *J. Cell Biol.* **100**:1948–1954.

Hirst, R., A. Horwitz, C. Buck, and L. Rohrschneider (1986) Phosphorylation of the fibronectin receptor complex in cells transformed by oncogenes that encode tyrosine kinases. *Proc. Natl. Acad. Sci. USA* **83**:6470–6474.

Horwitz, A., K. Duggan, R. Greggs, C. Decker, and C. Buck (1985) The cell substrate attachment (CSAT) antigen has properties of a receptor for laminin and fibronectin. *J. Cell Biol.* **101**:2134–2144.

Horwitz, A., K. Duggan, C. Buck, M. C. Beckerle, and K. Burridge (1986) Interaction of plasma membrane fibronectin receptor with talin—A transmembrane linkage. *Nature* **320**:531–533.

Hynes, R. O. (1987) Integrins: A family of cell surface receptors. *Cell* **48**:549–554.

Kishimoto, T. K., K. O'Connor, A. Lee, T. M. Roberts, and T. A. Springer (1987) Cloning of the beta subunit of the leukocyte adhesion proteins: Homology to an extracellular matrix receptor defines a novel supergene family. *Cell* **48**:681–690.

Lawler, J., and R. O. Hynes (1986) The structure of human thrombospondin, an adhesive glycoprotein with multiple calcium-binding sites and homologies with several different proteins. *J. Cell Biol.* **103**:1635–1648.

Leptin, M., R. Aebersold, and M. Wilcox (1987) *Drosophila* position-specific antigens resemble the vertebrate fibronectin-receptor family. *EMBO J.* **6**:1037–1043.

Nurden, A. T., J. N. George, and D. R. Phillips (1986) In *Biochemistry of Platelets*, D. R. Phillips and M. A. Shuman, eds., pp. 111–151, Academic, New York.

Oldberg, Å., A. Franzen, and D. Heinegård (1986) Cloning and sequence analysis of rat bone sialoprotein (osteopontin) cDNA reveals an Arg-Gly-Asp cell-binding sequence. *Proc. Natl. Acad. Sci. USA* **83**:8819–8823.

Parise, L. V., and D. R. Phillips (1985) Reconstitution of the purified platelet fibronectin receptor: Fibrinogen binding properties of the glycoprotein IIb–IIIa complex. *J. Biol. Chem.* **260**:10698–10707.

Pierschbacher, M. D., and E. Ruoslahti (1984a) The cell attachment of fibronectin can be duplicated by small synthetic fragments of the molecule. *Nature* **309**:30–33.

Pierschbacher, M. D., and E. Ruoslahti (1984b) Variants of the cell recognition site of fibronectin that retain attachment-promoting activity. *Proc. Natl. Acad. Sci. USA* **81**: 5985–5988.

Pierschbacher, M. D. and E. Ruoslahti (1987) Influence of stereochemistry of the sequence Arg-Gly-Asp-Xxx on binding specificity in cell adhesion. *J. Biol. Chem.* **262**:17294–17298.

Pierschbacher, M. D., E. G. Hayman, and E. Ruoslahti (1981) Location of the cell attachment site in fibronectin using monoclonal antibodies and proteolytic fragments of the molecules. *Cell* **26**:259–261.

Plow, E. F., M. D. Pierschbacher, E. Ruoslahti, G. A. Marguerie, and M. H. Ginsberg (1985) The effect of Arg-Gly-Asp containing peptides on fibrinogen and von Willebrand factor binding to platelets. *Proc. Natl. Acad. Sci. USA* **82**:8057–8061.

Poncz, M., R. Eisman, R. Heidenreich, S. M. Silver, G. Vilaire, S. Surrey, E. Schwartz, and J. S. Bennett (1987) Structure of the platelet membrane glycoprotein IIb: Homology of the alpha subunits of the vitronectin and fibronectin membrane receptors. *J. Biol. Chem.* **262**:8476–8482.

Pytela, R., M. D. Pierschbacher, and E. Ruoslahti (1985a) Identification and isolation of a 140 kd cell surface glycoprotein with properties expected of a fibronectin receptor. *Cell* **40**:191–198.

Pytela, R., M. D. Pierschbacher, and E. Ruoslahti (1985b) A 125/115 kD cell surface receptor specific for vitronectin interacts with the Arg-Gly-Asp adhesion sequence derived from fibronectin. *Proc. Natl. Acad. Sci. USA* **82**:5766–5770.

Pytela, R., M. D. Pierschbacher, M. H. Ginsberg, E. F. Plow, and E. Ruoslahti (1986) Platelet membrane glycoprotein IIb/IIIa is a member of a family of Arg-Gly-Asp-specific adhesion receptors. *Science* **231**:1559–1562.

Pytela, R., M. D. Pierschbacher, W. S. Argraves, S. Suzuki, and E. Ruoslahti (1987) Arg-Gly-Asp adhesion receptors. *Methods Enzymol.* **144**:475–489.

Ruggeri, Z. M., R. A. Houghten, S. R. Russell, and T. S. Zimmerman (1986) Inhibition of platelet function with synthetic peptides designed to be high-affinity antagonists of fibrinogen binding to platelets. *Proc. Natl. Acad. Sci. USA* **83**:5708–5712.

Ruoslahti, E. (1984) Fibronectin in cell adhesion and invasion. *Cancer Metastasis Rev.* **3**:43–51.

Ruoslahti, E., and M. D. Pierschbacher (1986) Arg-Gly-Asp: A versatile cell recognition signal. *Cell* **44**:517–518.

Ruoslahti, E., and M. D. Pierschbacher (1987) New perspectives in cell adhesion: RGD and integrins. *Science* **238**:491–497.

Savagner, P., B. A. Imhof, K. M. Yamada, and J.-P. Thiery (1986) Homing of hemopoietic precursor cells to the embryonic thymus: Characterization of an invasive mechanism induced by chemotactic peptides. *J. Cell Biol.* **103**:2715–2727.

Springer, W. R., D. N. W. Cooper, and S. H. Barondes (1984) Discoidin I is implicated in cell–substratum attachment and ordered cell migration of *Dictyostelium discoidium* and resembles fibronectin. *Cell* **39**:557–564.

Springer, T. A., M. L. Dustin, T. K. Kishimoto, and S. D. Marlin (1987) The lymphocyte function-associated LFA-1, CD2, and LFA-3 molecules: Cell adhesion receptors of the immune system. *Annu. Rev. Immunol.* **5**:223–252.

Suzuki, S., Å. Oldberg, E. G. Hayman, M. D. Pierschbacher, and E. Ruoslahti (1985) Complete amino acid sequence of human vitronectin deduced from cDNA: Similarity of cell attachment sites in vitronectin and fibronectin. *EMBO J.* **4**:2519–2524.

Suzuki, S., W. S. Argraves, R. Pytela, H. Arai, T. Krusius, M. D. Pierschbacher, and E. Ruoslahti (1986) cDNA and amino acid sequences of the cell adhesion receptor recognizing vitronectin reveal a transmembrane domain and homologies with other adhesion receptors. *Proc. Natl. Acad. Sci. USA* **83**:8614–8618.

Suzuki, S., W. S. Argraves, H. Arai, L. R. Languino, M. D. Pierschbacher, and E. Ruoslahti (1987) Amino acid sequence of the vitronectin receptor α-subunit and comparative expression of adhesion receptor mRNAs. *J. Biol. Chem.* **262**:14080–14085.

Takada, Y., J. Strominger, and M. E. Hemler (1987) The VLA family of heterodimers are members of a superfamily of molecules involved in adhesion and embryogenesis. *Proc. Natl. Acad. Sci. USA* **84**:3239–3243.

Tamkun, J. W., D. W. DeSimone, D. Fonda, R. S. Patel, C. Buck, A. F. Horwitz, and R. O. Hynes (1986) Structure of integrin, a glycoprotein involved in the transmembrane linkage between fibronectin and actin. *Cell* **46**:271–282.

Vaheri, A., and D. F. Mosher (1978) High molecular weight, cell surface-associated glycoprotein (fibronectin) lost in malignant transformation. *Biochim. Biophys. Acta* **516**:1–25.

Wright, S. D., P. A. Reddy, M. T. C. Jong, and B. W. Erickson (1987) C3bi receptor (complement receptor type 3) recognizes a region of complement protein C3 containing the sequence Arg-Gly-Asp. *Proc. Natl. Acad. Sci. USA* **84**:1965–1968.

Chapter 9

The Integrin Family and Neighbors

ALAN F. HORWITZ
DONNA BOZYCZKO
CLAYTON A. BUCK

ABSTRACT

Avian integrin, known also as the CSAT antigen, 140-kD complex, and fibronectin receptor, is a heteromeric complex found in adherens-type junctions and at the site of focal contacts in cultured cells. Functionally, it appears to be involved in cell–substratum adhesion, serving as a transmembrane bridge between the extracellular matrix and elements of the cytoskeleton. As such, it possesses an extracellular domain that binds to fibronectin, laminin, and vitronectin, and a cytoplasmic domain that binds to talin. The binding to matrix molecules is sensitive to the fibronectin cell-binding peptide RGD. The binding to talin is competed by a synthetic decapeptide corresponding to the tyrosine kinase phosphorylation site on the lower-molecular-weight β-subunit of integrin. Transformation of chick fibroblasts with Rous sarcoma virus results in the phosphorylation of the β-subunit and reduces the ability of integrin to bind talin and fibronectin. At this time, it is not clear whether integrin is a single promiscuous heteromeric receptor capable of binding several matrix molecules, or a mixture of structurally similar receptors each specific for a single matrix molecule. Recent competitive binding experiments favor the former hypothesis.

Cell surface receptors that interact with the extracellular matrix are receiving increasing attention as their structure, function, and relationships with other cell surface molecules continue to be elucidated (Buck and Horwitz, 1987). These receptors serve as adhesive molecules, function in cell motility over matrices, and may also transduce inductive signals from the extracellular matrix. The focus of our recent work is an avian cell surface receptor that goes by many names: the CSAT antigen, 140-kD complex, fibronectin receptor, and most recently, integrin. The latter is emerging as a consensus name and will be used here. This molecular complex was identified simultaneously in two different laboratories by screening for monoclonal antibodies (mabs) that inhibit the adhesion of skeletal myoblasts to tissue culture dishes (Greve and Gottlieb, 1982; Horwitz et al., 1982; Neff et al., 1982). Three such mabs were reported: CSAT, JG-22, and JG-9. These antibodies inhibit adhesion of fibroblasts and other cell types to a number of different substrates, including laminin, fibronectin, collagen (types I and IV), and vitronectin (Neff et al., 1982; Decker et al., 1984; Bronner-Fraser, 1985; Chen et al., 1985a; Horwitz et al., 1985; Bozyczko and Horwitz, 1986; Duband et al., 1986). They have little detectable effect on the adhesion of these cells on poly-L-lysine.

INTEGRIN LOCALIZATION ON MUSCLE AND FIBROBLASTS

Immunofluorescence localization using monoclonal antibodies as well as polyclonal antisera shows that integrin tends to concentrate in regions where junctions of the adherens type occur (Chen et al., 1985a,b; Damsky et al., 1985). On fibroblasts, integrin localizes in the vicinity of focal contacts and close contacts, along portions of stress fibers, and at the edge of leading lamellae. In general, this localization is very similar to that of vinculin and nearly identical to that of talin (Burridge and Feramisco, 1980; Geiger et al., 1980; Burridge and Connell, 1983). There is some ambiguity concerning the localization of integrin at the focal contact. In older cultures of well-spread cells, integrin is often seen to surround the focal contact in a needle's-eye kind of arrangement. Apparently, the needle's-eye configuration is present only in a mature or fully organized contact. This could arise either from an intrinsic structural feature of the contact or from poor penetration of the antibodies into this contact region. Other cells, especially those with less well developed focal contacts, appear to have integrin coinciding with the focal contact regions. Integrin distribution does seem to differ between highly motile and more stationary cells. On motile neural crest cells, myoblasts, and short-term cultures of somitic fibroblasts, it is diffusely distributed, reflecting the absence of highly organized vinculin-rich adhesion plaques (Damsky et al., 1985; Duband et al., 1986).

We have also studied integrin distribution on adult skeletal muscle (Bozyczko et al., 1989). On slow-twitch, anterior latissimus dorsi, integrin is found over most of the cell surface, including the myotendinous and neuromuscular junctions. On fast-twitch, posterior latissimus dorsi and thigh muscles, integrin is regionalized and appears prominently in the vicinity of the neuromuscular and myotendinous junctions. Occasional oblique sections of the sarcolemma reveal a banding pattern that colocalizes with vinculin. This pattern is similar to that of costameres, which are thought to represent sites of z-band linkage to the cell surface (Pardo et al., 1983).

This striking regionalization in adult muscle contrasts with the distribution on embryonic muscle. In the embryo, integrin is seen in a punctate distribution around the myotubes and young muscle fibers. The distribution changes on older embryonic muscle, where it becomes more uniform with local densities that correspond to known junctional areas. On slow-twitch muscles in the adult, integrin distribution is increasingly uniform, obscuring any local prominent densities. On fast-twitch muscles, the distribution becomes more regionalized, as discussed above.

Aspects of these developmental alterations are mirrored in muscle cultures. From an initially punctate distribution along the cell surface and at the myotube ends, the distribution on older cultures becomes more uniform along the surface, and then in much older cultures it becomes patchy. Acetylcholine receptor clusters generally colocalize with the local densities (Figure 1). These observations are especially striking when the cells are extracted with Triton X-100, which reveals stabilized integrin molecules. Occasionally, in very mature

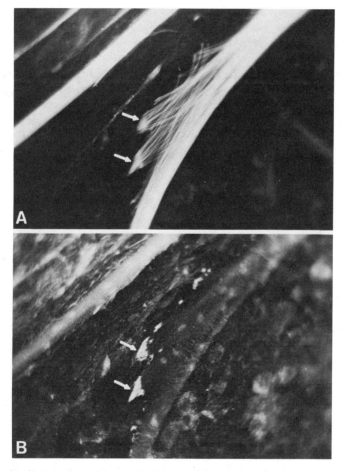

Figure 1. *Localization of integrin at myofibrillar termini.* Myotubes were grown on gelatin-coated tissue culture dishes in culture medium containing eight parts DMEM, one part embryo extract, and one part horse serum. After 8–12 days in culture, the myotubes were first fixed in 1% paraformaldehyde for 10 min and then treated with 0.5% Triton X-100 in phosphate-buffered saline for 10 min. The cells were first stained with a rabbit polyclonal antibody raised against integrin followed by a secondary, rhodamine-labeled goat anti-rabbit antibody (*B*). They were then stained with fluorescein phalloidin to identify actin-containing myofibrils (*A*). *Arrows* designate myofibrillar termini.

cultures, a weak, banded staining pattern is seen along portions of the myotube surface.

INTEGRIN STRUCTURE

Integrin has been purified by immunoaffinity chromatography and shown to resolve into three bands when electrophoresed on nonreduced SDS gels (Greve and Gottlieb, 1982; Neff et al., 1982; Chapman, 1984; Hasegawa et al., 1985; Knudsen et al., 1985; Akiyama et al., 1986). These constituent proteins have been characterized only partially, but have already revealed some interesting properties. All three bands are glycosylated and behave as integral membrane proteins. The lower-molecular-weight band (band 3, or the β-band) has been cloned and sequenced using recombinant DNA technologies (Tamkun et al., 1986). It has the following properties. The polypeptide consists of 803 amino acids with a single 20-amino-acid stretch that is the putative membrane-spanning region. The cytoplasmic domain contains only about 47 amino acids, including a tyrosine kinase phosphorylation site similar to that found on the EGF and insulin receptors. *In vitro* and *in vivo* experiments, to be discussed below, have demonstrated that this site is indeed phosphorylated by oncogenes encoding tyrosine kinase. Other features include a large extracellular domain with four cysteine-rich repeats. Similar cysteine-rich repeats have been observed in the β-subunits of other integrinlike receptors (Hynes, 1987).

ECM-BINDING INTERACTIONS OF PURIFIED INTEGRINS

The binding of purified integrin to a number of potential ligands has been studied. The extracellular ligands that have been investigated are fibronectin, laminin, and vitronectin (Horwitz et al., 1985). These are well-characterized molecules that have been purified and are available in large quantities. Fibronectin is a 220-kD dimer that contains a modular domain structure. A major cell-binding domain has been identified, and through a series of proteolytic and synthetic peptide studies, the cell attachment site has been shown to reside in a short sequence of amino acids consisting of Arg-Gly-Asp (RGD) (Pierschbacher and Ruoslahti, 1984a,b). However, it now appears that other regions of the fibronectin molecule can also participate in cell adhesion (Humphries et al., 1986; Obara et al., 1986). Laminin is less well characterized, but it appears to have two cell-binding sites, one in the "cross" region and the other near the heparin-binding domain (Edgar et al., 1984; Graf et al., 1987). The former appears to be where the 68-kD, high-affinity laminin receptor binds, and the latter is required for neurite extension. Vitronectin is a less well characterized molecule. It is smaller than the others, 60 kD, and also contains an RGD sequence (Suzuki et al., 1986).

The binding data of Akiyama et al. (1985) suggested that the affinity of some of these ligands may be in the micromolar range. With large interacting molecules, this implies potentially rapid equilibriums. We have used the technique of equilibrium molecular sieving to circumvent potential problems arising from the rapid dissociation of the complex (Horwitz et al., 1985). In this method a column containing Ultrogel AcA22 is preequilibrated with the desired concentration of ligand to be studied. Thus the receptor is always surrounded by a high concentration of ligand and will rapidly re-ligate after dissociation. This ensures virtually continual receptor occupancy.

All three of the extracellular matrix receptors bind to the integrin complex using this assay. The binding of laminin is inhibited by the CSAT monoclonal antibody, which also inhibits the adhesion of several cell types to laminin-coated substrates. The binding of fibronectin and vitronectin is inhibited specifically by RGD. It should be pointed out that the binding of vitronectin differs from that of the other ligands (Figure 2). Its binding is apparent without preequilibration of the gel filtration column with the ligand; thus its complex with integrin does not dissociate as rapidly.

CYTOSKELETAL-ASSOCIATED PROTEIN-BINDING INTERACTIONS WITH PURIFIED INTEGRIN

The location of the integrin complex in junctions of the adherens type suggests that it may also interact with cytoskeletal components. Three cytoskeletal-associated molecules, hypothesized to participate in the connection between the actin filaments and the cell surface (Figure 3), have been purified and characterized (Burridge, 1987): talin (230 kD), α-actinin (110-kD dimer), and vinculin (130 kD). Only talin shows a detectable interaction with purified

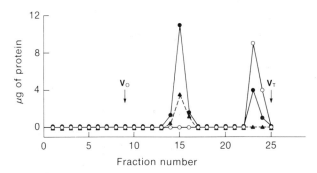

Figure 2. *Binding of vitronectin and integrin.* 5 μg of integrin and 10 μg of vitronectin were mixed and incubated at room temperature for 30 min in a total volume of 30 μl. The mixture was then passed over Ultrogel AcA22, and 60 μl fractions were collected and assayed for protein. The gel filtration elution profile of integrin alone *(filled triangles)* and vitronectin alone *(open circles)* and of a mixture of integrin and vitronectin *(filled circles)* is shown.

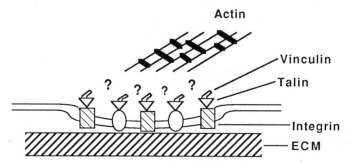

Figure 3. *Model of the possible arrangement of molecules participating in transmembrane linkage, which is thought to connect actin filaments to the extracellular matrix.*

integrin, using the equilibrium gel filtration assay (Figure 4) (Horwitz et al., 1986). To demonstrate specificity, we used two approaches. In one we showed that both fibronectin and talin can bind simultaneously to purified integrin and thus occupy spatially distinct sites (Horwitz et al., 1986). An analogous experiment with talin demonstrated that integrin and vinculin bind simultaneously to talin, showing that talin also has two distinct binding sites (Horwitz et al., 1986).

INTEGRIN AS A TARGET FOR pp^{60src}

The location of the integrin complex in adhesion plaques and its receptor function suggested that it, like other adhesion plaque molecules, may be a target of the pp^{60src} kinase oncogene product (Sefton et al., 1981; Pasquale et al., 1986). This kinase was also found in adhesion plaques (Rohrschneider, 1980). Prominent phenotypic alterations noted upon transformation included alterations in adhesion, morphology, cytoskeletal organization, and fibronectin binding (Burridge, 1987). Integrin played a role in all of these phenomena. Its organization changed to a less organized, diffuse distribution upon transformation (Chen et al., 1986; Hirst et al., 1986).

Following transformation with viruses encoding tyrosine kinase, the integrin complex was phosphorylated primarily on band 3 with some phosphorylation on band 2 (Hirst et al., 1986). Analysis of the phosphoamino acids showed that the phosphorylation was largely on serine of band 2 and on tyrosine of band 3. In the presence of vanadate, roughly half of the integrin band-3 molecules were phosphorylated, with 80% of the phosphate on tyrosine. In the absence of vanadate, the level of labeling was reduced to about 3–6%, with 80–90% of this labeling on serine residues. This vanadate effect suggests that the phosphotyrosine appears to be quite labile to phosphatases and is readily turned over *in vivo*.

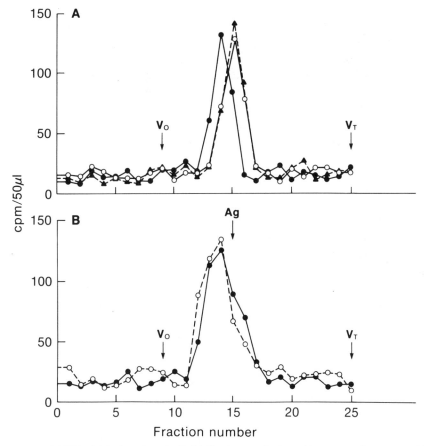

Figure 4. *Inhibition of the integrin and talin binding by a synthetic peptide corresponding to a tyrosine-containing sequence from near the carboxyl terminus of integrin.* A mixture of ligand and metabolically labeled [^{35}S] integrin was passed through a column of Ultrogel AcA22 that had been preloaded with ligand to be assayed (see Horwitz et al., 1985). *A:* Equilibrium gel filtration elution profile of integrin alone (*filled triangles*); integrin plus 0.4 mg/ml talin (*filled circles*); integrin plus talin (0.4 mg/ml) plus peptide (1 mg/ml) (*open circles*). *B:* Equilibrium gel filtration elution profile of integrin plus fibronectin (0.8 mg/ml) (*filled circles*); integrin plus fibronectin (0.8 mg/ml) plus peptide (1.0 mg/ml) (*open circles*). The sequence of the peptide is Trp-Asp-Thr-Gly-Glu-Asn-Pro-Ile-Tyr-Lys.

Tryptic peptide analysis of the phosphorylated integrin complex showed a single major phosphorylated peptide. When the synthetic peptide corresponding to the portion of the cytoplasmic domain containing the conserved tyrosine kinase phosphorylation site was phosphorylated *in vitro* using purified kinases, it comigrated with the tryptic peptide from the phosphorylated integrin band 3 (Tapley et al., 1989).

The binding properties of the phosphorylated integrin were explored using the equilibrium gel filtration assay described above. For these experiments, integrin was isolated from transformed cells labeled with ^{32}P just prior to isolation. The ^{32}P labeling ensured that only phosphorylated molecules were being assayed for binding. In contrast to antigen isolated from untransformed cells, the transformed ^{32}P-labeled antigen did not show a detectable interaction with talin (Figure 5). Binding of "transformed" integrin to fibronectin, though detectable, was reduced about fourfold compared to that of the untransformed integrin. The alterations in binding seemed to correlate best with phos-

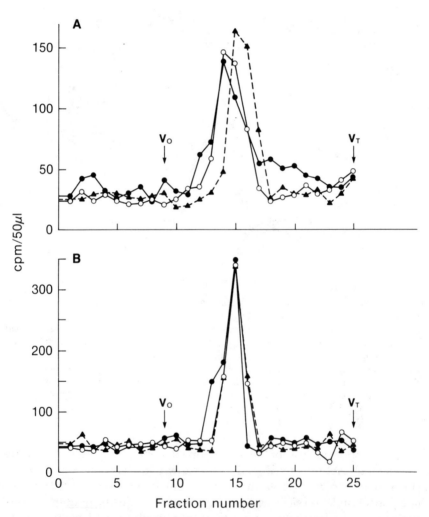

Figure 5. *Binding of integrin purified from transformed cells to talin and fibronectin. A:* Equilibrium gel filtration profiles of metabolically labeled [^{35}S] nontransformed integrin alone (*filled triangles*); plus 0.4 mg/ml talin (*open circles*); 0.8 mg/ml fibronectin (*filled circles*). *B:* Equilibrium gel filtration profiles of [^{32}P] transformed integrin alone (*filled triangles*); plus 0.4 mg/ml talin (*open circles*); 0.8 mg/ml fibronectin (*filled circles*).

phorylation on tyrosine. Integrin isolated from RSV transformed cells labeled with $^{32}PO_4$ in the absence of vanadate were phosphorylated primarily on serine and showed nearly normal binding to talin and fibronectin. It seems likely that the phosphorylation of integrin contributes to the altered morphology and adhesion accompanying viral transformation. However, other changes very likely also contribute to the transformed phenotype (Tapley et al., 1989).

THE REQUIREMENT OF OLIGOMERIC STRUCTURE FOR INTEGRIN FUNCTION

Gel filtration and sedimentation studies have pointed to the oligomeric nature of the integrin complex (Horwitz et al., 1985). A monoclonal antibody called "G" has provided further evidence for its oligomeric nature and, most significantly, a route to assaying the function of the individual subunits (Buck et al., 1986). This mab immunoblots band 3 specifically. When the purified integrin complex is passed over an affinity column to which the G mab is attached, a mixture of bands 1 and 2 passes through the column, and band 3, which remains on the column, can then be eluted. When run on gel filtration columns, the bands 1 plus 2 fraction and band 3 fraction each elute at a smaller Stokes radius than does the native antigen. Both fractions have been assayed for binding to laminin, fibronectin, and talin, the three ligands that interact with the integrin complex. None of them binds with affinities sufficient to be detected in our assays. However, on combining the band 1 plus 2 and band 3 fractions, the mixture elutes with the same Stokes radius as the native integrin complex, and all three ligands bind. The binding of the CSAT and JG-22 antibodies has also been assayed. Both antibodies bind only to band 3, the low-molecular-weight (β) band.

THE INTEGRIN FAMILY—RELATIONSHIPS

Recently, it has become apparent that the avian integrin complex is a member of a family of cell surface receptors which in turn is a member of a superfamily of cell surface receptors (Charo et al., 1986; Leptin, 1986; Plow et al., 1986; Ruoslahti and Pierschbacher, 1986; Ginsberg et al., 1987; Takada et al., 1987).

The members of the families have several features in common. They include similar molecular weights and oligomeric structure, an anomalous migration of the lower-molecular-weight band on nonreduced SDS–PAGE, sequence homology, and, in general, recognition of the RGD sequence in their ligands. The avian integrin has some features that appear to distinguish it from most other members of the family: All of the other receptors appear to be dimers, whereas the integrin complex is comprised of three bands. Furthermore, the mammalian receptors appear to be quite specific in their ligand interactions, and the avian integrin complex interacts with several different extracellular matrix constituents.

   Two hypotheses represent opposing models for the structure of the avian integrin (Figure 6). In one, the complex is a heteromeric trimer that functions as a promiscuous receptor for a number of different extracellular molecules. In the other, the complex consists of two or more heterodimers, each specific for a particular ligand: fibronectin, laminin, vitronectin, and so on. We have tested the hypothesis that the receptor is promiscuous, using inhibition experiments with the various ligands. Vitronectin appears to compete with fibronectin and laminin binding to integrin *in vivo* (Figure 7). Likewise, fibronectin competes with laminin binding. From these competition experiments, it seems likely that the ligands bind to the same molecule. This argues for a promiscuous receptor.

   The next issue is whether the antigen is dimeric or trimeric. Analyses of the molecular weight of the complex using gel filtration and sucrose density gradient sedimentation suggest that the complex is in the molecular weight range of 200–250 kD (Table 1). This value is most compatible with a dimeric molecule. Assuming this to be true, it appears then that the integrin complex is likely comprised of heterodimers that share a common lower-molecule-weight β-subunit in a manner similar to that of the VLA antigens described by Takada et al. (1987). One of these oligomers would be promiscuous and bind to many different ligands—that is, all of the ligands we have assayed to date. The other oligomer(s) would then have functions and binding specificities that remain to be identified.

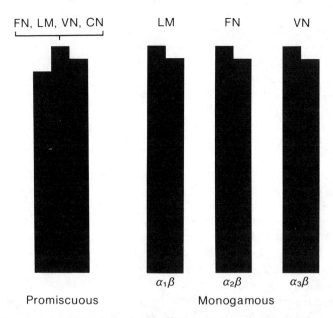

Figure 6.   *Two extreme hypotheses for the functional structure of integrin.* In the promiscuous model, integrin is a dimer or trimer (as shown) that interacts with several different ligands: fibronectin, laminin, vitronectin, and so on. The monogamous hypothesis envisions a number of different heterodimeric receptors, each with restricted ligand specificity and possessing different α-chains and a common β-chain.

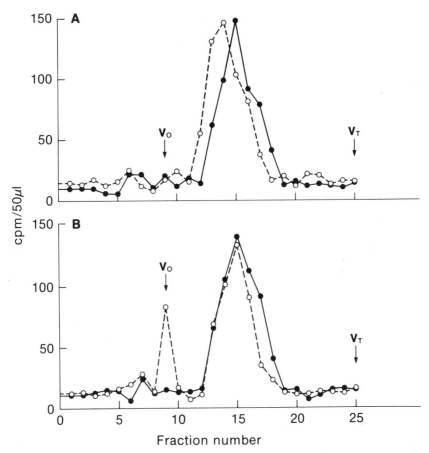

Figure 7. *Competition among ligands for binding to integrin. A:* Equilibrium gel filtration profile of integrin plus 0.8 mg/ml fibronectin (*open circles*); integrin–vitronectin mixture plus 0.8 mg/ml fibronectin (*filled circles*). *B:* Equilibrium gel filtration profile of integrin plus 0.4 mg/ml laminin (*open circles*); integrin–vitronectin mixture plus 0.4 mg/ml laminin (*filled circles*). For experiments involving vitronectin, integrin was first incubated with 0.5 mg/ml vitronectin for 30 min at room temperature prior to gel filtration.

Table 1.   Hydrodynamic Properties of Integrin[a]

Stokes radius (nm)	6.0	±0.2
Sedimentation coefficient (S)	8.6	±0.4
Partial molar volume (ml/g)	0.75	±0.03
Molecular weight (detergent + integrin)	236,000	±31,000
Molecular weight (integrin)	212,000	±10,000
Frictional ratio	1.51	

[a]The sedimentation and gel filtration measurements and analyses were performed as described in Horwitz et al. (1985). The frictional ratio and molecular weights were calculated as described by Clark (1975).

227

ACKNOWLEDGMENTS

This work was supported by National Institutes of Health grant GM-23244, the H. M. Watts Neuromuscular Disease Research Center (A. F. H.), and National Cancer Institute grants CA-19144 and CA-10818 (C. A. B.). We also thank our colleagues L. Rohrschneider and K. Burridge for their continued collaboration and for allowing us to cite unpublished data, and Ms. Marie Lennon for preparing the manuscript.

REFERENCES

Akiyama, S. K., E. Hasegawa, T. Hasegawa, and K. M. Yamada (1985) The interaction of fibronectin fragments with fibroblastic cells. *J. Biol. Chem.* **260**:13256–13260.

Akiyama, S. K., S. S. Yamada, and K. M. Yamada (1986) Characterization of a 140 kD avian cell surface antigen as a fibronectin-binding molecule. *J. Cell Biol.* **102**:442–448.

Bozyczko, D., and A. F. Horwitz (1986) The participation of a putative cell surface receptor for laminin and fibronectin in peripheral neurite extension. *J. Neurosci.* **6**:1241–1251.

Bozyczko, D., C. Decker, J. Meschler, and A. F. Horwitz (1989) Integrin on developing adult skeletal muscle. *Exp. Cell Res.* (in press).

Bronner-Fraser, M. (1985) Alterations in neural crest migration by a monoclonal antibody that affects cell adhesion. *J. Cell Biol.* **101**:610–617.

Buck, C., and A. F. Horwitz (1987) Receptors for extracellular matrix molecules. *Annu. Rev. Cell Biol.* **3**:179–205.

Buck, C. A., E. Shea, K. Duggan, and A. F. Horwitz (1986) Integrin (the CSAT antigen): Functionality requires oligomeric integrity. *J. Cell Biol.* **103**:2421–2428.

Burridge, K. (1987) Substrate adhesions in normal and transformed fibroblasts: Organization and regulation of cytoskeletal, membrane, and extracellular matrix components at focal contacts. *Cancer Rev.* **4**:18–78.

Burridge, K., and J. Feramisco (1980) Microinjection and localization of a 130 kD protein in living fibroblasts: A relationship to actin and fibronectin. *Cell* **19**:587–595.

Burridge, K., and L. Connell (1983) A new protein of adhesion plaques and ruffling membranes. *J. Cell Biol.* **94**:359–367.

Chapman, A. E. (1984) Characterization of a 140 kD cell surface glycoprotein involved in myoblast adhesion. *J. Cell. Biochem.* **25**:109–121.

Charo, I. F., L. A. Fitzgerald, B. Steiner, S. C. Rall, Jr., L. S. Bekaert, and D. R. Phillips (1986) Platelet glycoproteins IIb and IIIa: Evidence for a family of immunologically and structurally related glycoproteins in mammalian cells. *Proc. Natl. Acad. Sci. USA* **83**:8351–8355.

Chen, W.-T., E. Hasegawa, T. Hasegawa, C. Weinstock, and K. M. Yamada (1985a) Development of cell surface linkage complexes in culture fibroblasts. *J. Cell Biol.* **100**:1103–1114.

Chen, W.-T., J. M. Greve, D. I. Gottlieb, and S. J. Singer (1985b) Immunocytochemical localization of 140 kD cell adhesion molecules in cultured chicken fibroblasts and in chicken smooth muscle and intestinal epithelial tissues. *J. Histochem. Cytochem.* **33**:576–586.

Chen, W.-T., J. Wang, T. Hasegawa, S. S. Yamada, and K. M. Yamada (1986) Regulation of fibronectin receptor distribution by transformation, exogenous fibronectin, and synthetic peptides. *J. Cell Biol.* **103**:1649–1661.

Clark, S. (1975) The size and detergent binding of membrane proteins. *J. Biol. Chem.* **250**:5459–5469.

Damsky, C. H., K. A. Knudsen, D. Bradley, C. A. Buck, and A. F. Horwitz (1985) Distribution of the cell–substratum attachment (CSAT) antigen on myogenic and fibroblastic cells in culture. *J. Cell Biol.* **100**:1528–1539.

Decker, C., R. Greggs, K. Duggan, J. Stubbs, and A. F. Horwitz (1984) Adhesive multiplicity in the interaction of embryonic fibroblasts and myoblasts with extracellular matrices. *J. Cell Biol.* **99**:1398–1404.

Duband, J. L., S. Rocker, W.-T., Chen, K. M. Yamada, and J. P. Thiery (1986) Cell adhesion and migration in the early vertebrate embryo: Location and possible role of the putative fibronectin receptor complex. *J. Cell Biol.* **102**:160–178.

Edgar, D., R. Timpl, and H. Thoenen (1984) The heparin binding domain of laminin is responsible for its effects on neurite outgrowth and neuronal survival. *EMBO J.* **3**:1463–1468.

Geiger, B., K. T. Tokuyasu, A. H. Dutton, and S. J. Singer (1980) Vinculin, an intracellular protein localized at specialized sites where microfilament bundles terminate at cell membranes. *Proc. Natl. Acad. Sci. USA* **77**:4127–4131.

Ginsberg, M. H., J. Loftus, J.-J. Ryckwaert, M. D. Pierschbacher, R. Pytela, E. Ruoslahti, and E. F. Plow (1987) Immunochemical and N-terminal sequence comparison of two cytoadhesins indicates they contain similar or identical beta subunits and distinct alpha subunits. *J. Biol. Chem.* **262**:5437–5440.

Graf, J., Y. Iwamoto, M. Sasaki, G. R. Martin, H. Kleinman, F. Robey, and Y. Yamada (1987) Identification of an amino acid sequence in laminin mediating cell attachment, chemotaxis, and receptor binding. *Cell* **48**:989–996.

Greve, J. M., and D. I. Gottlieb (1982) Monoclonal antibodies which alter the morphology of culture chick myogenic cells. *J. Cell. Biochem.* **18**:221–230.

Hasegawa, T., E. Hasegawa, W.-T. Chen, and K. M. Yamada (1985) Characterization of a membrane-associated glycoprotein complex implicated in cell adhesion to fibronectin. *J. Cell. Biochem.* **28**:307–318.

Hirst, R., A. F. Horwitz, C. A. Buck, and L. Rohrschneider (1986) Phosphorylation of the fibronectin receptor complex in cells transformed by oncogenes that encode tyrosine kinases. *Proc. Natl. Acad. Sci. USA* **83**:6470–6474.

Horwitz, A. F., N. Neff, A. Sessions, and C. Decker (1982) Cellular interactions in myogenesis. In *Muscle Development: Molecular and Cellular Control,* M. L. Pearson and H. F. Epstein, eds., pp. 291–299, Cold Spring Harbor Press, Cold Spring Harbor, New York.

Horwitz, A. F., K. Duggan, R. Greggs, C. Decker, and C. A. Buck (1985) Cell substrate attachment (CSAT) antigen has properties of a receptor for laminin and fibronectin. *J. Cell Biol.* **103**:2134–2144.

Horwitz, A. F., K. Duggan, C. A. Buck, K. Beckerle, and K. Burridge (1986) Interaction of plasma membrane fibronectin receptor with talin—A transmembrane linkage. *Nature* **320**:531–533.

Humphries, M. H., S. K. Akiyama, A. Komoriya, K. Olden, and K. M. Yamada (1986) Identification of an alternatively spliced site in human plasma fibronectin that mediates cell type–specific adhesion. *J. Cell Biol.* **103**:2637–2648.

Hynes, R. O. (1987) Integrins: A family of cell surface receptors. *Cell* **48**:549–554.

Knudsen, K. A., A. F. Horwitz, and C. A. Buck (1985) A monoclonal antibody identifies a glycoprotein complex involved in cell–substratum adhesion. *Exp. Cell Res.* **157**:218–226.

Leptin, M. (1986) The fibronectin receptor family. *Nature* **321**:728–729.

Neff, N. T., C. Lowrey, C. Decker, A. Tovar, C. Damsky, C. A. Buck, and A. F. Horwitz (1982) A monoclonal antibody detaches skeletal muscle from extracellular matrices. *J. Cell Biol.* **95**:654–666.

Obara, M., M. S. Kang, S. K. Akiyama, A. Komoriya, K. Olden, and K. M. Yamada (1986) Expression of the cell binding domain of human fibronectin in *E. coli. FEBS Lett.* **213**:261–264.

Pardo, J. V., D. Siliciano, and S. W. Craig (1983) A vinculin-containing cortical lattice in skeletal muscle: Transverse lattice elements ("costameres") mark sites of attachment between myofibrils and the sarcolemma. *Proc. Natl. Acad. Sci. USA* **80**:1008–1012.

Pasquale, E. B., P. A. Maher, and S. J. Singer (1986) Talin is phosphorylated on tyrosine in chicken embryo fibroblasts transformed by Rous sarcoma virus. *Proc. Natl. Acad. Sci. USA* **83**:5507–5511.

Pierschbacher, M. D., and E. Ruoslahti (1984a) Cell attachment activity of fibronectin can be duplicated by small synthetic fragments of the molecule. *Nature* **309**:30–33.

Pierschbacher, M. D., and E. Ruoslahti (1984b) Variants of the cell recognition site of fibronectin that retain attachment-promoting activity. *Proc. Natl. Acad. Sci. USA* **81**:5985–5988.

Plow, E. F., J. C. Loftus, E. G. Levin, D. S. Fair, D. Dixon, J. Forsyth, and M. H. Ginsberg (1986) Immunologic relationship between platelet membrane glycoprotein gpIIb/IIIa and cell surface molecules expressed by a variety of cells. *Proc. Natl. Acad. Sci. USA* **83**:6002–6006.

Rohrschneider, L. (1980) Adhesion plaques of Rous sarcoma virus-transformed cells contain the *src* gene product. *Proc. Natl. Acad. Sci. USA* **77**:3514–3518.

Ruoslahti, E., and M. D. Pierschbacher (1986) Arg-Gly-Asp: A versatile cell recognition signal. *Cell* **44**:517–518.

Sefton, B., T. Hunter, E. Ball, and S. J. Singer (1981) Vinculin: A cytoskeletal target of the transforming protein of Rous sarcoma virus. *Cell* **24**:165–174.

Suzuki, S., A. Oldberg, E. Hayman, M. D. Pierschbacher, and E. Ruoslahti (1986) Complete sequence of human vitronectin deduced from cDNA: Similarity of attachment sites in vitronectin and fibronectin. *EMBO J.* **4**:2519–2524.

Takada, Y., J. L. Strominger, and M. E. Hemler (1987) The VLA family of heterodimers are members of a superfamily of molecules involved in adhesion and embryogenesis. *Proc. Natl. Acad. Sci. USA* **84**:3239–3243.

Tamkun, J. W., D. W. DeSimone, D. Fonda, R. S. Patel, C. A. Buck, A. F. Horwitz, and R. O. Hynes (1986) Structure of integrin, a glycoprotein involved in the transmembrane linkage between fibronectin and actin. *Cell* **46**:271–282.

Tapley, R., A. F. Horwitz, C. A. Buck, K. Burridge, K. Duggan, R. Hirst, and L. Rohrschneider (1989) Integrins isolated from rous sarcoma virus transform chicken embryo fibroblasts. *Oncogene* (in press).

Chapter 10

Laminin: Structure, Expression, and Cell-Binding Sequence

YOSHIHIKO YAMADA
JEANNETTE GRAF
YUKIHIDE IWAMOTO
SEISHI KATO
HYNDA K. KLEINMAN
KIMITOSHI KOHNO
GEORGE R. MARTIN
KOHEI OGAWA
MAKOTO SASAKI

ABSTRACT

Laminin is a large glycoprotein specific to basement membranes. Recently, its chains have been cloned, and the complete primary sequence of the B1 and B2 chains and most of the sequence of the A chain of mouse laminin has been completed. Computer analysis of the sequence data revealed that each chain of laminin has a multidomain structure with α-helical segments, cysteine-rich homologous repeats, and globules. Sequence homology suggests that the genes for the B1, B2, and A chains arose from a common ancestor. The B1 chain gene is about 63 kb and contains at least 36 exons. The promoter regions of the B1 and B2 chain genes contain several unique features with little homology between the two, consistent with their independent regulation as suggested also by variations in mRNA levels in tissues. A B1 chain pentapeptide Y1GSR (Tyr-Ile-Gly-Ser-Arg) was found to be active in cell binding, to be chemotactic, and to displace the laminin receptor (67 kD) from a laminin affinity column. This peptide also inhibits the formation of metastases in mice injected with malignant tumor cells, presumably by preventing their binding to basement membranes.

Laminin is the major glycoprotein of basement membranes. The most extensively studied laminin molecule is from a mouse tumor, the Engelbreth-Holm-Swarm (EHS) tumor (Timpl et al., 1979). Laminin from this tumor, as well as from human placenta and mouse teratocarcinoma cells, contains three chains designated A (400 kD), B1 (230 kD), and B2 (220 kD), which are joined

231

by disulfide bonds (Timpl et al., 1983a). Examination of rotary-shadowed laminin showed a cross-shape molecule with globules at the end of each arm (Engel et al., 1981).

Laminin binds to various other proteins, including collagen IV, heparan sulfate proteoglycan, entactin (nidogen), and itself, creating an integrated structure within the basement membrane. Rotary-shadowing studies show that collagen IV binds to the ends of the short arms of laminin, while entactin (nidogen) binds at the intersections of the three chains. Heparin binds to the end of the long arm, and heparan proteoglycans bind to the same region of the molecule. Laminin also binds to certain cell surface components, including a laminin receptor (Lesot et al., 1983; Malinoff and Wicha, 1983; Rao et al., 1983), heparan sulfate, sulfatides, and gangliosides.

Laminin has a variety of biological activities. It promotes cell adhesion, migration, growth, and differentiation (for a review, see Kleinman et al., 1985). Laminin also induces morphological changes in a variety of cells. For example, when cultured in the presence of laminin, neural cells produce long axonlike processes, whereas Sertoli cells become twofold more columnar. Laminin promotes metastatic activity when tumor cells are cultured with this glyco-protein or when it is injected together with the cells (Barsky et al., 1984; Terranova et al., 1984). On the other hand, a proteolytic fragment of laminin (Barsky et al., 1984) and antibodies to laminin (Terranova et al., 1982) reduce the metastatic activity of tumor cells by inhibiting the binding of the cells.

These biological effects of laminin occur most likely through its binding to cell surface receptor(s). A 67-kD laminin cell surface receptor has been isolated from muscle cells (Lesot et al., 1983) and from tumor cells (Malinoff and Wicha, 1983; Rao et al., 1983), cloned, and partially sequenced (Wewer et al., 1986). A major cell-binding site was also identified on the central part of laminin (Rao et al., 1982; Timpl et al., 1983b) and shown to interact with the 67-kD receptor (Terranova et al., 1983; Graf et al., 1987). As described below, a pentapeptide sequence, YIGSR (Tyr-Ile-Gly-Ser-Arg), from the B1 chain of laminin has been identified as the active site for cell binding. More recently, several laboratories have shown that other regions of laminin are active in cell attachment (Goodman et al., 1987; M. Aumailley, personal communication). Similarly, neural laminin receptors other than 67-kD receptors have been observed (H. K. Kleinman, unpublished data). Thus, it is likely that multiple interactions determine the variety of responses of cells to the laminin.

STRUCTURE OF LAMININ

Progress on the structure of laminin has been made recently by cDNA cloning. cDNA clones encoding the entire B1 and B2 chain of murine laminin have been isolated and sequenced (Sasaki et al., 1987; Yamada et al., 1987). We also have characterized several overlapping cDNA clones that encode about 80% of the A chain (M. Sasaki, unpublished data). Sequence analysis revealed that B1 and

B2 chains consist of 1786 amino acids ($M_r = 196{,}903$) and 1607 amino acids ($M_r = 177{,}541$), respectively. The B1 and B2 chains have a similar structure, with each containing six distinctive domains designated I to VI (Figure 1). The sequences of the C-terminal domains I and II in the B1 and B2 chains reveal α-helical structures. These domains consist of a heptad repeat of hydrophobic and polar amino acids characteristic for proteins with coiled-coil configurations, such as tropomyosin and myosin. The heptad repeat is more perfect for domain I than for domain II, but both are likely to form a coiled-coil structure. The coiled-coil helical structure of the B1 and B2 chains was previously predicted by partial sequence data (Barlow et al., 1984; Paulsson, 1985). The B1 chain has a small segment (α-domain) of about 30 amino acids containing 6 cysteines that interrupts the α-helical coiled-coil structure. Domains III and V contain homologous cysteine-rich repeats, each of which has about 50 amino acids with 8 cysteines at regular positions. Domains III and V of the B1 chain are

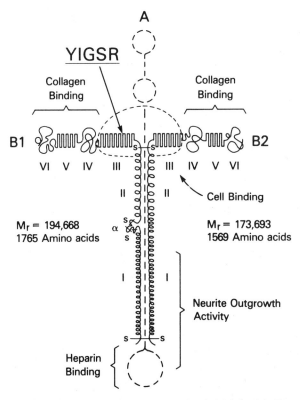

Figure 1.   *Schematic model for the structure of the B1, B2, and A chains in laminin. Roman numerals and α* designate the various domains. The location of the A chain is represented by a *broken line.* Domain III of the B1 chain contains a sequence, YIGSR, that is active in cell attachment, receptor binding, chemotaxis, and inhibiting the metastasis of tumor cells.

composed of 8 and 5 repeats, respectively, while domains III and V of the B2 chain consist of 6 and 4.5 repeats, respectively. It is interesting to note that a part of the repeat in domain III of the B1 chain is homologous to EGF and to TGFα. The N-terminal domains IV and VI are likely to form the globular structures observed by electron microscopy.

The B1 and B2 chains have 13 and 14 Asn-X-Ser or Asn-X-Thr sequences, respectively, which are potential sites for N-linked oligosaccharide attachment. This is in good agreement with predictions from biosynthetic studies (Howe, 1984) that estimate this level of carbohydrate in the protein.

Preliminary analysis of A chain sequences also predicts a coiled-coil helix, cysteine-rich homologous repeats, and globular structures similar to those of the B1 and B2 chains. In addition to these homologous structures, the A chain has a unique large globular domain with a molecular weight of about 100 kD at its C-terminus. In this domain a stretch of hydrophobic amino acids creates a potential transmembrane segment whose function is not clear. Considerable homology exists between the three chains, suggesting that these genes arose from a common ancestor.

A model of laminin, based on analyses of its cDNA sequence, electron microscopy, and protein fragment data, is shown in Figure 1. In this model the N-terminal domains III to VI of each chain are present in the short arms, where they form a rodlike structure between globules. Domains I and II of the B chains form part of the long arm. A cysteine at the C-terminus of the B1 and B2 chains links the two chains by a single disulfide bond (Paulsson et al., 1985). In addition, there are two cysteines in domain II of the B1 and B2 chains that could also form an interchain disulfide bond. Thus the interaction of the three chains is stabilized not only by disulfide bonds but also by their forming a double- or triple-helix coiled-coil structure.

EXPRESSION OF LAMININ GENES

Laminin is the first extracellular matrix component to appear during mouse embryogenesis. The B1 and B2 chains appear at the 2-cell stage, and the synthesis of the A chain occurs at the 16-cell stage (Cooper and MacQueen, 1983). A number of cultured cells produce laminin. Some cells (i.e., PYS2, differentiating F9) synthesize all three chains, whereas other cells (i.e., melanoma, 3T3, HT-1080, Schwannoma) synthesize only B chains. Several cell lines have been used to study laminin gene expression. For example, when mouse F9 teratocarcinoma stem cells are induced to differentiate by treatment with both retinoic acid and dibutyryl cAMP, the synthesis of all three chains increases at least 20-fold (Strickland et al., 1980). Similarly, the synthesis of the B chains is increased in human breast carcinoma cells (MCF7) when treated with estrogen (A. Albini, personal communication).

During the differentiation of F9 cells, the steady-state levels of mRNA for the B1, B2, and A chains increase in parallel and reach a maximum at almost the same time. These results suggest that genes for these chains are coordinately regulated, at least in the F9 cells (Kleinman et al., 1987). In contrast, tissue levels of mRNA for the laminin chains have been found to vary considerably. In kidney the level of B1 chain mRNA was higher than that of the B2 chain, whereas in heart the level of the B2 chain mRNA was higher. The levels of A chain mRNA were very low in all tissues examined, including tissues such as kidney, heart, and lung, with a prominent content of basement membranes. On the other hand, similar levels of mRNA for the $\alpha 1$ (IV) collagen chain, a basement membrane–specific gene, were found in these same tissues. The results suggest that the expression of genes for laminin chains is not coordinately regulated at all times. It is also possible that some tissues produce laminins with differing chain composition. It is of interest that laminin secreted from Schwann cells consists only of B chains and is active in promoting neurite outgrowth (D. Edgar and R. Timpl, personal communication). Thus, heterodimeric forms of laminin may exist and function in tissues.

LAMININ GENES

The entire gene for the B1 chain and the 5' portion of the gene for the B2 chain have been isolated. The overall structure of the B1 chain gene was determined by electron microscope analysis of RNA–DNA hybrids. The gene is about 63 kb long and contains at least 36 exons (Figure 2). Since the B1 chain is a multidomain protein, it was of interest to compare the exon and domain organization. Some of the exons coding for domains III and V of the B1 chain, which contains the cysteine-rich homologous repeats, have been determined by DNA sequencing. Three out of seven repeats are encoded by a single exon, suggesting that these domains arose by a multiple duplication of a single primordial unit. Further sequence analysis will clarify whether the B1 chain gene is another example of a gene derived by a process of exon shuffling.

The transcription initiation sites of the B1 and B2 chain genes were determined by both S1 mapping and primer extension and showed some microheterogeneity of the transcription start site in both genes. The biological significance of this is not clear at present, although it is possible that certain of the transcription initiation sites predominate in different tissues.

The promoters of the B1 and B2 chain genes were characterized by DNA sequence analysis, and certain of their features are schematically diagramed in

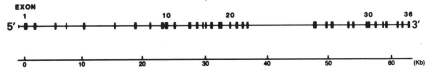

Figure 2. *Schematic representation of the structure of the B1 chain gene. Vertical bars* represent exons.

Figure 3. While both promoters had high GC content, no strong sequence homology was found between them except for a few segments that included the "GC box" motif and the cAMP consensus sequence. A striking feature in the B2 chain promoter was the presence of an 11-nucleotide sequence that was repeated nine times between nucleotides −200 to −450, with a consensus sequence of 5'-CCCNCCCNCCT-3'. These repeats did not occur in the B1 promoter region. The importance of these repeats to the promoter activity is not clear, although the promoter activity of the B2 chain gene was strong in transfection experiments in comparison to the RSV-LTR promoter, which is a strong promoter. The repeats and the number of GC boxes may account for the high transcription of the B2 promoter CAT constructs.

A LAMININ CELL-BINDING SEQUENCE

Cell Adhesion

In the past, laminin was proteolytically cleaved in order to localize sites on the molecule responsible for its various biological activities. Using this approach, the cell attachment and cell growth activities of laminin were localized to the P1

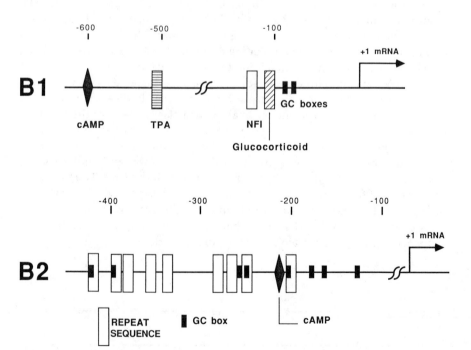

Figure 3. *Schematic diagrams of the promoter regions of the B1 and B2 chain genes.* +1 represents the transcription initiation site.

fragment, which comprises the intersection of the three chains (Rao et al., 1982; Timpl et al., 1983b). With the advance of protein cloning, these activities were further localized to a sequence of five amino acids.

The identification of the active site was achieved by using synthetic peptides and peptide-specific antibodies in various biological assays. Seven peptides of approximately 20 amino acids in size were synthesized corresponding to the various structural domains in the B1 chain (Figure 4). These peptides were not active when tested in cell attachment and cell migration assays. Antibodies that cross-reacted with native laminin were raised against their BSA conjugates, and the antibody corresponding to the peptide from domain III inhibited cell adhesion to a laminin substrate. This prompted a closer analysis of the amino acid sequence making up domain III and resulted in the synthesis of two more peptides from this region. One of these peptides, a nonapeptide, CDPGYIGSR (Cys-Asp-Pro-Gly-Tyr-Ile-Gly-Ser-Arg), was found to have biological activity (Graf et al., 1987). Subsequently, smaller peptides showed YIGSR to be the minimally fully active sequence (J. Graf, unpublished data).

Most of the initial experiments performed to assess the adhesion activity of the peptide were done using HT-1080 cells, a human fibrosarcoma cell line (Figure 5). Adhesion activity was performed in a variety of ways. Direct adhesion was assessed by either coating dishes with peptide or adding the peptide directly to serum-free media. The peptide was also tested to see if it could block adhesion to laminin. After a short preincubation of the peptide with the cells, this suspension was then added to a laminin-coated dish. In many adhesion experiments, CDPGYIGSR conjugated to BSA was used, a form consisting of approximately 20% peptide by weight and more potent than the peptide itself.

Once the active nonapeptide in domain III was identified, a series of peptides was produced to determine the minimal sequence with biological activity. A variety of modifications, substitutions, and deletions established that YIGSR is the minimal sequence with biological activity. This sequence was assayed for direct adhesion and was found to mediate the adhesion of a variety

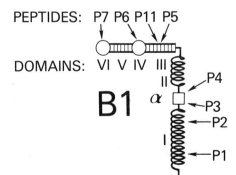

Figure 4. *Domain model of the B1 chain and locations of the synthetic peptides. Roman numerals* designate the domains, and *arabic numerals* represent the sequences chosen for the synthesis of peptide.

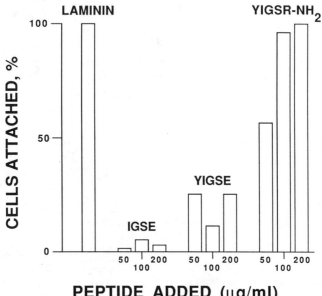

Figure 5. *Diagram of the ability of synthetic peptides from the laminin B1 chain to promote HT-1080 cell adhesion.* Laminin (10 μg/ml) and various concentrations of synthetic peptides were first incubated in Eagle's minimal essential medium containing 0.02% bovine serum albumin for 60 min on tissue culture dishes. Then the cells (1 $\times$ 10^5) in the same medium were added to dishes and incubated for 60 min. The unattached cells were removed, and the attached cells were released with 0.1% trypsin–EDTA and counted in an electronic cell counter.

of cell types, but not all cell types; nor did it stimulate neurite formation, indicating the presence of other attachment sites on laminin. YIGSR is unique in that only one other protein, a plant protein present in the data bank, contains the exact same sequence. In contrast, the fibronectin cell-binding sequence RGDS (Arg-Gly-Asp-Ser) is found in many proteins (for a review, see Ruoslahti and Pierschbacher, 1986). Therefore, the mechanism of cell binding of laminin is likely to be quite different from that of fibronectin.

Receptor Binding

A laminin receptor (67 kD) on the surface of cells has been described (for a review, see von der Mark and Kühl, 1985), but its direct role in cell adhesion was not demonstrated. Studies were carried out to determine if YIGSR could bind to this receptor. Membrane extracts were applied to a laminin affinity column, and the 67-kD laminin receptor was eluted from the column with the active peptide YIGSR, whereas other peptides or modifications of the YIGSR motif were inactive (Figure 6). These data suggested that YIGSR not only recognizes the 67-kD laminin receptor but is quite specific in its interaction.

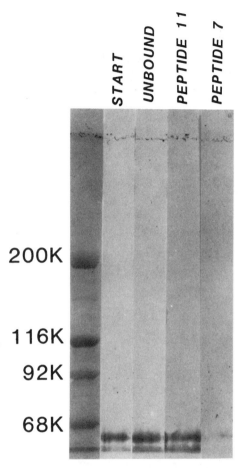

Figure 6. *Western immunoblot of laminin receptor eluted from a laminin affinity column by synthetic peptides from the B1 chain of laminin.* Cell membranes were prepared by sucrose density centrifugation from cells isolated from the EHS tumor. After extraction with detergent, the solubilized membrane proteins were circulated through laminin–Sepharose columns for 18 hours. The unbound material was washed from the columns with low salt, and then peptides at 1 mg/ml were passed through the column. The unbound and peptide-eluted material was collected, dialyzed, electrophoresed on a 5% polyacrylamide gel, and transferred by Western blot to nitrocellulose. The nitrocellulose was reacted with an antibody to the 67-kD laminin receptor according to established procedures. Shown are the immunoblot of the membrane extract before the column (*start*), the unbound material from the column, and the material eluted by peptide 11 (CDPGYIGSR) and by peptide 7 (PERDIRDNPLCEPCTCDPAGSE). All peptides designated in Figure 4 were tested, but only peptide 11 eluted the 67-kD laminin receptor.

Cell Migration

Laminin stimulates the haptotactic migration of a variety of cells, including B16–F10 melanoma cells (McCarthy and Furcht, 1984). Using a modified Boyden chamber, we assessed the ability of the synthetic peptides to affect laminin sequences promoting cell migration (Figure 7). CDPGYIGSR-NH$_2$ stimulated the migration of B16–F10 melanoma cells, showing about 35% of the response observed with laminin. Other peptides tested in this manner had no effect on cell migration. A variation of this assay was performed by placing laminin or fibronectin in the lower compartment of the Boyden chamber and adding the peptide with the cells to the upper compartment. This peptide inhibited migration to laminin by 87%, while a control peptide did not alter the cells' movement (Figure 7). The specificity of the YIGSR peptide was also shown by its failure to inhibit the migration of the cells to fibronectin as expected, since the YIGSR sequence is not found in fibronectin.

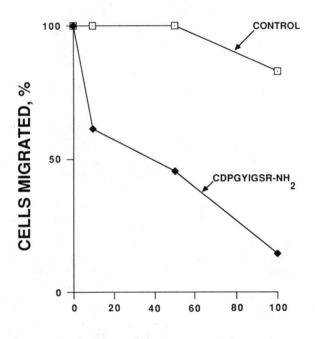

PEPTIDE ADDED (μg/ml)

Figure 7. *Diagram of the inhibition of migration of B16–F10 melanoma cells to laminin by synthetic peptides from the sequence of the B1 chain of laminin.* Cell migration was measured by using a Boyden chamber with the lower well containing 20 μg of laminin. Peptides of various concentrations were added to the upper well along with the cells. After 5 hours the cells that had migrated to the lower surface of the filter were counted. All the peptides shown in Figure 4 were tested, but only peptide 11, CDPGYIGSR-NH$_2$, blocked cell migration. Shown are the data for peptide 11 and peptide 3.

Cell Invasion

A distinct series of sequential events leads to the metastasis of tumor cells (Liotta, 1984; Terranova et al., 1986). In this metastatic cascade, tumor cells first detach from the primary site, survive in the circulation system, penetrate through vessel walls, and migrate and proliferate in distant tissues. Basement membranes are major barriers to the invasion of tumor cells, and laminin plays a major role in this invasion. Current concepts suggest that tumor cells bind to the basement membrane through the interaction of the 67-kD laminin receptor. This interaction induces the secretion of collagenase and proteases, which degrade the basement membrane and enable the cells to pass through it. We have tested the ability of the synthetic peptides to inhibit B16–F10 melanoma tumor cell invasion *in vitro* and metastasis *in vivo*.

An *in vitro* invasion assay using a layer of reconstituted basement membrane (matrigel) in a modified Boyden chamber was developed (Albini et al., 1987). This assay measures the ability of the B16–F10 mouse melanoma cells to attach, degrade, and migrate through the matrigel. By adding YIGSR to the upper compartment of the Boyden chamber, *in vitro* invasion through the reconstituted basement membrane was inhibited by about 85%, whereas the control peptide did not inhibit invasion.

When B16–F10 melanoma cells were injected into the tail veins of syngeneic mice, pulmonary metastases were formed. However, when the nonapeptide CDPGYIGSR or YIGSR was injected with the cells or soon after, these peptides inhibited the formation of pulmonary metastases by 80–90% (Figure 8). Injection with a control peptide resulted in no inhibition of metastases. The inhibitory effects were dose dependent, with more than 70% of the tumors blocked by 100 μg of peptides. The inhibition of metastasis by the peptides was due to neither cytotoxicity nor blocking of the tumorigenicity of the cells.

The mechanism of inhibition of metastases by YIGSR is probably via binding to the laminin receptor. By recognizing the laminin receptor, the adhesion of the cell to the basement membrane *in vivo* is inhibited, and thus the subsequent invasion of the cells is blocked. The fibronectin cell-binding sequence GRGDS has been found to inhibit the metastasis of B16–F10 cells (Humphries et al., 1986). Although the mechanism of this inhibition is not clear, these results suggest that metastasis involves multiple steps and interactions with multiple matrix components.

CONCLUSIONS

Laminin is a multidomain protein with diverse biological functions. The primary structure obtained by cloning the chains of laminin has been useful in understanding the structure of the protein and in identifying the sequences involved in its biological functions. The use of synthetic peptides and of cDNA expression systems in prokaryotes and eukaryotes opens up new directions in

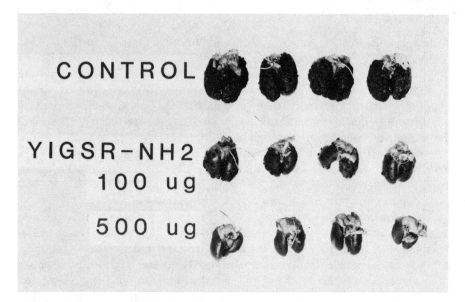

Figure 8. *Scheme of the inhibition of the formation of pulmonary metastases by YIGSR.* B16–F10 cells
(5×10^5) were mixed with YIGSR and then injected into the tail veins of C57BL/6 female mice.
Two weeks after injection, the lungs were fixed and photographed.

elucidating the functions of laminin. While very little is known about the
regulation of laminin gene expression, the isolation of the promoters for the B1
and B2 chains is the initial step in identifying *cis*-acting elements that regulate
expression of the genes.

REFERENCES

Albini, A., Y. Iwamoto, H. K. Kleinman, G. R. Martin, S. A. Aaronson, J. M. Kozlowski, and R. N.
    McEwan (1987) A rapid *in vitro* assay for quantitating the invasive potential of tumor cells.
    Cancer Res. **47**:3239–3245.

Barlow, D. P., N. M. Green, M. Kurkinen, and B. M. Hogan (1984) Sequencing of laminin B chain
    cDNA reveals C-terminal regions of coiled-coil alpha-helix. *EMBO J.* **3**:2355–2362.

Barsky, S. H., C. N. Rao, J. E. Williams, and L. A. Liotta (1984) Laminin molecular domains which
    alter metastasis in a murine model. *J. Clin. Invest.* **74**:843–848.

Cooper, A. R., and H. A. MacQueen (1983) Subunits of laminin are differentially synthesized in
    mouse eggs and early embryos. *Dev. Biol.* **96**:467–471.

Engel, J., E. Odermatt, A. Engel, J. A. Madri, H. Furthmayr, H. Rohde, and R. Timpl (1981) Shapes,
    domain organizations, and flexibility of laminin and fibronectin, two multifunctional proteins
    of the extracellular matrix. *J. Mol. Biol.* **150**:97–120.

Goodman, S. L., R. Deutzmann, and K. von der Mark (1987) Two distinct cell-binding domains in
    laminin can independently promote nonneuronal cell adhesion and spreading. *J. Cell Biol.*
    105:589–598.

Graf, J., Y. Iwamoto, M. Sasaki, G. R. Martin, H. K. Kleinman, F. A. Robey, and Y. Yamada (1987) Identification of an amino acid sequence in laminin mediating cell attachment, chemotaxis, and receptor binding. *Cell* 48:989–996.

Howe, C. C. (1984) Functional role of laminin carbohydrate. *Mol. Cell. Biol.* 4:1–7.

Humphries, M. J., K. Olden, and K. M. Yamada (1986) A synthetic peptide from fibronectin inhibits experimental metastasis of murine melanoma cells. *Science* 25:467–470.

Kleinman, H. K., F. B. Cannon, G. W. Laurie, J. R. Hassell, M. Aumailley, V. P. Terranova, and G. R. Martin (1985) Biological activities of laminin. *J. Cell. Biochem.* 27:317–325.

Kleinman, H. K., I. Ebihara, P. D. Killen, M. Sasaki, F. B. Cannon, Y. Yamada, and G. R. Martin (1987) Genes for basement membrane proteins are coordinately expressed in differentiating F9 cells but not in adult murine tissues. *Dev. Biol.* 122:373–378.

Lesot, H., U. Kühl, and K. von der Mark (1983) Isolation of a laminin binding protein from muscle cell membranes. *EMBO J.* 2:861–865.

Liotta, L. A. (1984) Tumor invasion and metastases: Role of the basement membrane. *Am. J. Pathol.* 117:339–348.

Malinoff, H. L., and M. S. Wicha (1983) Isolation of a cell surface receptor protein for laminin from murine fibrosarcoma cells. *J. Cell Biol.* 96:1475–1479.

McCarthy, J. B., and L. T. Furcht (1984) Laminin and fibronectin promote the haptotactic migration of B16 mouse melanoma cells *in vitro*. *J. Cell Biol.* 98:1474–1480.

Paulsson, M., R. Deutzmann, R. Timpl, D. Dalzoppo, E. Odermatt, and J. Engel (1985) Evidence for coiled-coil α-helical regions in the long arm of laminin. *EMBO J.* 4:309–316.

Rao, C. N., I. M. K. Margulies, T. S. Tralka, V. P. Terranova, J. A. Madri, and L. A. Liotta (1982) Isolation of a subunit of laminin and its role in molecular structure and tumor cell attachment. *J. Biol. Chem.* 257:9740–9744.

Rao, C. N., S. H. Barsky, V. P. Terranova, and L. A. Liotta (1983) Isolation of a tumor cell laminin receptor. *Biochem. Biophys. Res. Commun.* 111:804–808.

Ruoslahti, E., and M. D. Pierschbacher (1986) Arg-Gly-Asp: A versatile cell recognition signal. *Cell* 44:517–518.

Sasaki, M., K. Kohno, S. Kato, G. R. Martin, and Y. Yamada (1987) Sequence of the cDNA encoding the laminin B1 chain reveals a multidomain protein containing cysteine-rich repeats. *Proc. Natl. Acad. Sci. USA* 84:935–939.

Strickland, S., K. K. Smith, and K. R. Marotti (1980) Hormonal induction of differentiation in teratocarcinoma stem cells: Generation of parietal endoderm by retinoic acid and dibutyryl cAMP. *Cell* 21:347–355.

Terranova, V. P., L. A. Liotta, R. G. Russo, and G. R. Martin (1982) Role of laminin in the attachment and metastasis of murine tumor cells. *Cancer Res.* 42:2265–2269.

Terranova, V. P., C. N. Rao, T. Kalebic, M. K. Margulies, and L. A. Liotta (1983) Laminin receptor on human breast carcinoma cells. *Proc. Natl. Acad. Sci. USA* 80:444–448.

Terranova, V. P., J. E. Williams, L. A. Liotta, and G. R. Martin (1984) Modulation of the metastatic activity of melanoma cells by laminin and fibronectin. *Science* 226:982–985.

Terranova, V. P., E. S. Hujanen, and G. R. Martin (1986) Basement membranes and the invasive activity of metastatic tumor cells. *J. Nat. Cancer Inst.* 77:311–316.

Timpl, R., H. Rohde, P. Gehron Robey, S. I. Rennard, J.-M. Foidart, and G. R. Martin (1979) Laminin–A glycoprotein from basement membranes. *J. Biol. Chem.* 254:9933–9937.

Timpl, R., J. Engel, and G. R. Martin (1983a) Laminin, a multifunctional protein of basement membranes. *Trends Biochem. Sci.* 8:207–209.

Timpl, R., S. Johansson, V. van Delden, I. Oberbäumer, and M. Hook (1983b) Characterization of protease resistant fragments of laminin mediating attachment and spreading of rat hepatocytes. *J. Biol. Chem.* 258:8922–8927.

von der Mark, K., and U. Kühl (1985) Laminin and its receptor. *Biochim. Biophys. Acta* **823**:147–160.

Wewer, U. M., L. A. Liotta, M. Jaye, G. A. Ricca, W. N. Drohan, A. P. Claysmith, C. N. Rao, P. Wirth, J. E. Coligan, R. Albrechtsen, M. Mudrji, and M. E. Sobel (1986) Altered levels of laminin receptor mRNA in various human carcinoma cells that have different abilities to bind laminin. *Proc. Natl. Acad. Sci. USA* **83**:7137–7141.

Yamada, Y., A. Albini, I. Ebihara, J. Graf, S. Kato, P. Killen, H. K. Kleinman, K. Kohno, G. R. Martin, C. Rhodes, F. A. Robey, and M. Sasaki (1987) Structure and expression of mouse laminin. In *Mesenchymal–Epithelial Interactions in Neural Development*, J. R. Wolff, J. Sievers, and M. Berry, eds., pp. 31–43, Springer-Verlag, Berlin.

Chapter 11

Structure and Expression of SPARC (Osteonectin, BM-40): A Secreted Calcium-Binding Glycoprotein Associated with Extracellular Matrix Production

BRIGID L. M. HOGAN
PETER W. H. HOLLAND
JÜRGEN ENGEL

ABSTRACT

The parietal endoderm of the mouse embryo is highly specialized for the synthesis and remodeling of a basement membrane. Differential screening of a cDNA library led to the identification of SPARC, a secreted, acidic, cysteine-rich glycoprotein of 43 kD, which constitutes about 25% of the total secreted protein of parietal endoderm cells. SPARC is identical to the calcium-binding protein osteonectin, previously thought to be bone specific, and to BM-40, a major product of the mouse EHS tumor. Two putative calcium-binding domains have been identified in SPARC. One is an N-terminal region containing clusters of glutamic acid residues, and the other is an EF-hand domain near the C-terminus. In situ hybridization with single-stranded RNA probes has been used to follow the temporal and tissue-specific pattern of SPARC expression during embryonic development. These results suggest that SPARC may play a rather general role in calcium-dependent processes in the extracellular matrix.

The parietal endoderm of the mouse embryo is a useful model system for studying the synthesis, assembly, and remodeling of the extracellular matrix (for reviews, see Hogan et al., 1984, 1986). Parietal endoderm cells, which first differentiate at around 4.5 days after fertilization, are highly specialized for the production of a thick basement membrane known as Reichert's membrane. This surrounds the rapidly growing fetus and for most of gestation forms a barrier to the passage of cells and maternal blood, allowing the filtration of nutrients and some large molecules. The structural glycoproteins produced by the parietal endoderm, such as type IV collagen, laminin, entactin, and heparan sulfate proteolgycan, are not unique to Reichert's membrane, but are present in all basement membranes throughout the embryo and adult. They are also found in a variety of other locations where the extracellular matrix is thought to play an important role in morphogenetic processes, including cell migration and epithelial folding and branching (Chen and Little, 1987; Loring and Erickson, 1987).

The differentiation of parietal endoderm *in vivo* can be mimicked *in vitro* by treating monolayers of F9 teratocarcinoma stem cells with retinoic acid and cyclic AMP. Over a period of about five days the cells differentiate into derivatives with many of the morphological characteristics of parietal endoderm, and there is a concomitant increase in the synthesis of basement membrane glycoproteins and the accumulation of their messenger mRNAs (for a review, see Hogan et al., 1983). This culture system therefore provides a convenient source of mRNAs for matrix proteins and a means of assaying for both *cis*-acting DNA sequences and *trans*-acting factors that regulate their expression during cell differentiation.

In order to isolate cDNAs for proteins produced in large amounts by parietal endoderm, we and others have constructed cDNA libraries from both differentiated F9 cells and parietal endoderm dissected from the 13.5-day mouse embryo. Screening of the libraries, either differentially with cDNA probes or with antibodies to matrix proteins, has led to the isolation of cDNAs for the well-characterized matrix components laminin A and B chains, $\alpha 1$ and $\alpha 2$ type IV collagen, and entactin (for reviews, see Kurkinen et al., 1983, 1985; Wang and Gudas, 1983; Barlow et al., 1984; Chung et al., 1985; Barlow et al., 1987; Sasaki et al., 1987).

Recently, we have also isolated a cDNA for a smaller protein differentially expressed by parietal endoderm. Analysis of the predicted amino acid sequence showed that it coded for a secreted protein that is acidic and rich in cysteine, and it was therefore called SPARC. SPARC is a major product of parietal endoderm and constitutes about 0.5% of the total poly A^+ RNA and about 25% of the total ^{35}S-methionine-labeled protein secreted by the cells over a 14-hour culture period (Mason et al., 1986a,b). In this chapter we describe the structure of SPARC and its identity with both the calcium-binding protein osteonectin (Termine et al., 1981; Romberg et al., 1985) and BM-40, produced by the EHS mouse tumor, which synthesizes a large amount of basement membrane material (Dziadek et al., 1986; Mann et al., 1987). We review the evidence that SPARC is synthesized by a wide variety of embryonic and adult tissues *in vivo*, especially those that produce basement membranes or matrices rich in type I collagen. Possible functions of SPARC in the assembly or modification of the matrix are also discussed.

DOMAIN STRUCTURE OF SPARC

After removal of the 17 amino acids of the putative leader sequence, mature SPARC consists of 285 residues with a predicted molecular weight of 33,062.

The single carbohydrate side chain added in parietal endoderm cells has a molecular weight of about 1300, and consists predominantly of a di-antennary complex type chain containing variable amounts of sialic acid (Hughes et al., 1987). The discrepancy between the predicted relative molecular mass of SPARC and that estimated from its electrophoretic mobility in SDS-polyacrylamide gels under reducing conditions (approximately 43 kD) probably results from the acidic nature of the protein (pI 4.3), owing to both the high content of glutamic and aspartic acid residues and the phosphorylation of serine residues (Engel et al., 1987).

Figure 1 summarizes our current model of SPARC structure. Domain I, covering the first 52 residues after removal of the signal peptide, features two segments of 14–15 residues, each of which contains seven to eight glutamic acids in short clusters. According to secondary structure predictions, these segments are potentially α-helical at physiological pH but only if the high charge density is neutralized.

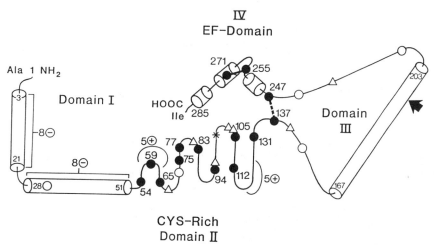

Figure 1. *Schematic representation of features of the mouse SPARC sequence.* Numbering of the residues differs from that in Mason et al. (1986a,b) in that the Ala at position 18 immediately following the signal sequence of the translated protein is now designated residue 1. The N-terminal residue of bovine and porcine osteonectin, bovine 43-kD protein, and mouse BM-40 is Ala. The cysteine residues (*filled circles*) are also numbered in relation to Ala as 1. The only disulfide linkage established is that between Cys 255 and Cys 271. It is assumed that Cys 247 forms a bridge between the C-terminal region and the Cys-rich domain II. The site of cleavage by an endogenous protease in the EHS tumor is shown with an *arrow*. Domain II contains two clusters of five positively charged residues (HHCKHGK and KKGHKLH), the unique glycosylation site at Asn 98 (*asterisk*), and six of the seven serine residues (*triangles*) in SPARC. In domain I there are two regions, each of which contains eight negatively charged residues. Segments for which an α-helical secondary structure is predicted by the methods of Chou and Fasman (1978) and Garnier et al. (1978) are indicated as *cylinders*, with, for the three longest, the numbers of the first and last residues.

Domain II contains 11 cysteine residues of which 10 are probably involved in disulfide bonds within the domain and one in linkage with the C-terminal region. Secondary structure predictions suggest that domain II consists of short segments of β-structure alternating with β-bends. The Cys-rich domain also contains six of the seven serine residues in SPARC; these are all potential phosphorylation sites, including the sequence DSS that is close to the Asn residue involved in glycosylation.

Domain III is predominantly helical, but has no other outstanding features. It is connected to region IV by a nonhelical stretch of 32 residues containing five prolines.

REGION IV CONTAINS THE EF-HAND CALCIUM-BINDING DOMAIN

In Table 1, residues 251–272 of mature SPARC are compared with the consensus sequence of an EF-hand calcium-binding domain, as first described for cytoplasmic calcium-binding proteins such as calmodulin, parvalbumin, and intestinal calcium-binding protein (for reviews, see Kretsinger, 1979, 1980; Van Eldik et al., 1982). The only deviation in the SPARC sequence from the general motif is the exchange of a normally conserved glycine by lysine. This occurs in the loop region in which the calcium ion is held in coordination with oxygens from the carboxyl side chains of the aspartic and glutamic acid residues and a carbonyl oxygen from the peptide backbone (Kretsinger, 1980). However, the presence of lysine is not expected to interfere with calcium binding. This can be shown by substituting in a computer model the SPARC EF-hand sequence for residues in the EF-hand domain II of the intestinal calcium-binding protein, the three-dimensional structure of which is known from X-ray crystallography (Szebenyi and Moffat, 1986). From such modeling it can be seen that the lysine side chain points away from the center of the EF-hand loop. It is also clear that the cysteines in the SPARC sequence are brought close enough to allow a disulfide bond to form without a need for changing the three-dimensional EF-hand structure (Engel et al., 1987). This might help stabilize the hand in the extracellular environment. The only other extracellular protein in which a putative EF-hand calcium-binding domain has been identified is fibrinogen (Table 2) (Dang et al., 1985), and it is interesting to note that here a glycine → lysine replacement is observed in the same position as in SPARC. The putative calcium-binding domains of thrombospondin are thought to be in the configuration of ω loops (Leszczynski and Rose, 1986; Hynes et al., this volume).

The EF-hand protein domain is contained within the penultimate of the 10 exons of the mouse SPARC gene (J. McVey, personal communication). This exon extends from amino acids 228 to 275, and therefore extends 19 residues N-terminal from the EF-hand. When the nucleotide sequence of this exon was

Table 1. Comparison of the amino acid sequence of mouse SPARC with the EF-hand domain of various calcium-binding proteins[a]

	E-helix	Calcium-Binding Loop +X +Y +Z −Y −X −Z	F-helix
		* * * * * *	
EF-hand motif	h h X X h	D X D G X I D X — X E	h X X h h
Mouse SPARC	F F E T C	D L D N D K Y I A L — E E	W A G C F
Rat calmodulin	A F R V F	D K D G N G Y I S A — E L	R H V M
Human skeletal muscle troponin C	I I E E V	D E D G S G T I D F — E E	F L V M M
Human fibrinogen α-chain	Q F S T W	D N D N D K F E G N C — A E	Q D G S G
Bovine intestinal calcium-binding protein	L F E E L	D K N G D G E V S F — E E	F Q V L V

[a]Asterisks indicate proposed calcium-liganding residues according to the model of Kretsinger (1980). In the case of a Gly at position −X, the calcium ligand is water (Kretsinger, 1980). By inference, this would also apply to the Ala at −X in SPARC.

used to search the EMBL data base (December 1986), no significant homology was found with other EF-hand-containing proteins.

A SECOND CALCIUM-BINDING DOMAIN IN SPARC PREDICTED FROM STUDIES ON CIRCULAR DICHROISM

Recently, a second potential calcium-binding domain has been identified in SPARC from studies on calcium-dependent conformational change in the structure of BM-40 (Engel et al., 1987). This protein can be extracted from the transplantable EHS mouse tumor with physiological salt containing 10 mM EDTA and then purified in milligram amounts (Dziadek et al., 1986; Mann et al., 1987). Sequence analysis of proteolytic peptides of BM-40 strongly suggests that it is identical to mouse SPARC, and therefore provides a convenient source of material for structure and functional studies (Mann et al., 1987). When BM-40 (SPARC) is purified without prolonged exposure to denaturing agents and without "nicking" by endogenous proteases, it shows a distinctive circular dichroism spectrum in the presence of calcium, indicating 25–30% α-helical content. This is close to the value predicted for SPARC if domain I is in the α-helical configuration. Upon removal of calcium from BM-40, the mean residue ellipticity at 220 nM decreases to a value corresponding to an α-helix content of 16–19%. Such a decrease would be expected if the glutamic acid–rich domain I is now in a random coil configuration. The change in configuration is completely reversible upon the re-addition of calcium, and appears to be a cooperative process, with the binding of one calcium increasing the probability of the region binding a second. Assuming that at least two glutamic acid residues are required for binding one calcium, then it is possible that up to eight calcium atoms could be held in domain I of SPARC.

No homology has been found between domain I and other proteins. N-terminal clusters of glutamic acids flanked by hydrophobic residues are, however, a characteristic feature of the so-called Gla-domains in factors VII, IX, and X, prothrombin, protein C, and protein S, all vitamin K–dependent proteins of the blood-clotting system. In these proteins a fraction of the Glu residues are γ-carboxylated, a modification that appears to enhance binding of the molecule via calcium bridges to phospholipid on the cell surface. No γ-carboxylation has been found in BM-40, however, and SPARC lacks the approximately 20-amino-acid basic propeptide sequence immediately following the leader sequence in vitamin K–dependent proteins that appears to signal modification of subsequent Glu residues by γ-carboxylation (Pan and Price, 1985).

The binding of about six calcium ions to the Gla domain of prothrombin is cooperative and is accompanied by a large increase in α-helicity of the molecule (Nelsestuen et al., 1981; Deerfield et al., 1986). Likewise, a calcium-dependent α-helical transition has been observed in the Gla domain of osteocalcin (Hauschka and Carr, 1982).

COMPARISON OF SPARC WITH OSTEONECTIN

Comparison of the cDNA sequences of mouse SPARC and bovine osteonectin shows a high level of homology over the coding sequence (M. Bolander and J. D. Termine, personal communication). At the amino acid level this homology reaches 92%, with most differences being accommodated by small deletions and insertions near the C-terminal end which do not, however, change the overall acidic nature of the domain. The bovine sequence also has an additional cysteine residue. Since it is known that SPARC and osteonectin are coded for by single genes in the mouse and cow, respectively (Mason et al., 1986b; Young et al., 1986), it is very likely that the proteins are identical. Analysis of the sequence of human SPARC (predicted from a placental cDNA) also shows a very high degree of homology with respect to the mouse protein (S. Anand and U. Francke, personal communication). This evolutionary conservation suggests that there must be strong selection pressure to maintain the structure of the protein.

Romberg et al. have used changes in the fluorescence of osteonectin purified from fetal bovine bone to show that the protein has a single calcium-binding site of about $3 \times 10^{-7}\ M$ (Romberg et al., 1985). This may be tentatively attributed to the EF-hand domain, with tryptophan as the source of the fluorescence signal. Although only $0.5\ M$ EDTA and no denaturing agents were used to isolate the osteonectin, the prolonged (five to seven days) extraction method appears to have resulted in irreversible denaturation of the N-terminal Glu-rich domain I, which, as described in the previous section, has been tentatively identified as a second calcium-binding domain in SPARC. Romberg et al. also show that bovine osteonectin is able to inhibit the growth of hydroxylapatite crystals "seeded" into a saturated solution of calcium phosphate. Fifty percent inhibition was observed at a concentration of about $0.3\ \mu M$ osteonectin (about $5\ \mu g/ml$). This is presumably achieved by osteonectin binding tightly to the surface of the crystals and preventing their enlargement.

TEMPORAL AND TISSUE-SPECIFIC EXPRESSION
OF THE SPARC GENE

SPARC was first identified as a major product of parietal endoderm, a cell type highly specialized for the synthesis of a thick basement membrane (Mason et al., 1986a,b). As shown in Figure 2, parietal endoderm cells contain high levels of SPARC mRNA, as revealed by in situ hybridization with ^{35}S-labeled anti-sense RNA. Control experiments with sense RNA show only background levels of autoradiographic grains over the cytoplasm of the cells.

Studies on the distribution of SPARC, osteonectin, and BM-40 mRNA and/or protein have also provided evidence for expression in a wide variety of other cells and tissues (Termine et al., 1981; Sage et al., 1984; Wasi et al., 1984; Tung et al., 1985; Dziadek et al., 1986; Mason et al., 1986a,b; Holland et al.,

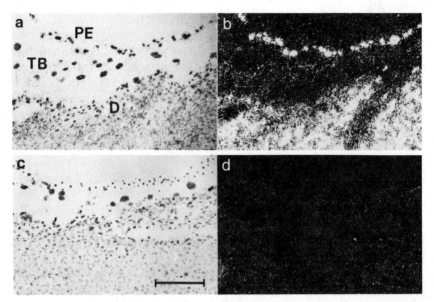

Figure 2. *SPARC RNA in parietal endoderm and maternal decidual cells visualized by* in situ *hybridization. a:* Detail of autoradiograph of a 7-μm cryostat section of an 8.5-day p. c. mouse conceptus showing the maternofetal junction. PE, parietal endoderm; TB, trophoblast giant cells; D, maternal decidual cells. Processed for *in situ* hybridization with −strand probe and photographed under bright-field illumination after staining with toluidine blue. *b:* Same field as *a* photographed under dark-ground illumination. *c, d:* Corresponding sections processed with +strand probe. Slides were exposed to emulsion for two days. Details can be found in Holland et al. (1987). Calibration bar = 200 μm.

1987; Mann et al., 1987). A number of common features emerge from these studies. First, expression of the gene is always high in cells producing basement membrane components and synthesizing laminin, type IV collagen, entactin, and heparan sulfate proteoglycan. These include cultured aortic endothelial cells, astrocytes, and Schwann cells, and, *in vivo,* kidney glomeruli (both endothelial cells adjacent to the glomerular basement membrane and cells on the wall of Bowman's capsule) (Figure 3 and unpublished data), decidual cells in the uterus (Figure 3), and dermal fibroblasts in the newborn mouse (Holland et al., 1987). High levels of expression in Schwann cells *in vivo* may account for the strong hybridization of antisense SPARC RNA to the Vth cranial nerves and other large peripheral nerve bundles in the newborn mouse (Figure 4). SPARC RNA levels are very low in most cells of the central nervous system, but high expression is seen locally in some regions, for example in the pia mater and choroid plexus (Figure 5). It is not known if these cells also produce matrix glycoproteins such as laminin and type IV collagen.

High levels of SPARC expression are also seen in tissues producing a matrix rich in type I collagen. Predominant among these is fetal bone. During so-called endochondral bone formation the calcified type II collagen matrix of

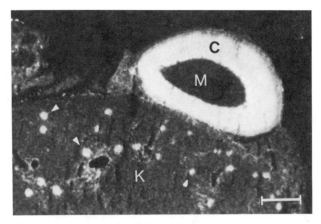

Figure 3. *SPARC RNA distribution in the adult adrenal gland and kidney.* Autoradiograph of section processed for *in situ* hybridization with –strand probe photographed with dark-ground illumination. In the adrenal gland, hybridization is much stronger in the cortex (C) than in the medulla (M) (Holland et al., 1987), and in the kidney (K) a strong signal is associated with cells adjacent to the glomerular basement membrane and wall of Bowman's capsule. Exposure time was two days. Calibration bar = 500 μm.

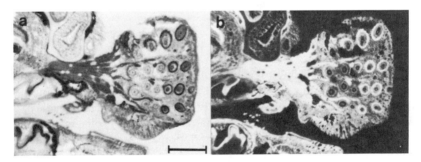

Figure 4. *SPARC RNA distribution in the snout region of the newborn mouse.* Bright-field (*a*) and dark-ground (*b*) illumination. Note strong hybridization in the Vth cranial nerve bundles running from the whiskers, and in the connective tissue sheath around the whisker follicles themselves. Exposure time was two days. Calibration bar = 1 mm.

cartilage is removed by the action of osteoclasts and replaced by the type I collagen matrix fibrils characteristic of bone that are synthesized by the osteoblasts. Some areas of bone such as the skull and jaw are laid down without a cartilage model, in a process known as membrane bone formation. High levels of SPARC expression are seen in the osteoblasts of both endochrondal and membrane bone. In the mature osteocytes, which are surrounded by mineralizing collagen I matrix, SPARC expression declines, as judged by a decrease in the number of autoradiographic grains associated with the cells after *in situ* hybridization (Holland et al., 1987).

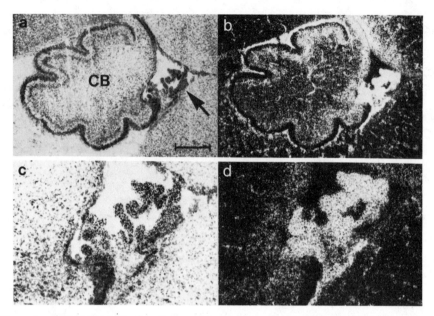

Figure 5. *SPARC RNA in the choroid plexus of the newborn mouse.* Section through the cerebellum (CB) and choroid plexus (*arrow*) of the newborn mouse. Autoradiograph of section hybridized with –strand probe and photographed under bright-field (*a,c*) or dark-ground (*b,d*) illumination (*c* and *d* show higher magnification details of *a* and *b*). Exposure time was two days. Calibration bar = 300 μm.

In spite of the very high levels of SPARC expression seen in developing bone, it is important to emphasize that SPARC expression is not confined to tissues that produce a matrix destined for mineralization. High levels of RNA are also seen in tissues that produce a matrix rich in type I collagen that is never normally mineralized. These include periodontal ligament (Wasi et al., 1984; Tung et al., 1985), ligaments of the developing mouse limb (S. Nomura and B. L. M. Hogan, unpublished data), and cultured fibroblasts derived from bovine ligamentum nuchae (Sage et al., 1984).

Finally, SPARC expression has also been observed in a number of cell types specialized for the synthesis and secretion of steroids. *In situ* hybridization and *in vitro* labeling studies suggest that SPARC RNA levels are high in cells of the zona fasciculata of the adrenal (Figure 3), Leydig cells of the testis, and granulosa cells of the ovary (Holland et al., 1987; B. L. M. Hogan, unpublished data).

IS SPARC (BM-40, OSTEONECTIN) A COMPONENT OF THE EXTRACELLULAR MATRIX?

In the previous section we presented evidence that expression of the SPARC gene is particularly high in tissues and cells synthesizing and modeling basement membranes or matrices rich in type I collagen. However, it is at present unclear whether the protein is itself a component of the extracellular matrix or is only transiently associated with it during some critical phase in assembly or modification. Specific polyclonal antibodies to three proteins, osteonectin, BM-40, and a 43-kD glycoprotein produced by endothelial cells (Sage et al., 1984), all of which are identical to SPARC, have been used to analyze protein distribution by immunohistochemistry or radioimmunoassay. In summary, these antibodies show strong cytoplasmic staining in a number of tissues (e.g., periodontal ligament and odontoblasts) (Tung et al., 1985) but stain the extracellular matrix only in a few, for example, fetal bone (Termine et al., 1981), Reichert's basement membrane, and the matrix of the EHS tumor (Dziadek et al., 1986). Antibodies to BM-40 do not stain other basement membranes such as the lens capsule or kidney basement membrane, nor the extracellular matrix produced by the PYS parietal endoderm cell line and a variety of other cells in culture (Sage et al., 1984; Dziadek et al., 1986).

There are several possible explanations for the lack of immunological reaction in the extracellular matrix around cells expressing SPARC. First, antigenic sites may be masked due to tight complexing with other matrix components, particularly if these have undergone cross-linking or other secondary modifications after being laid down. Alternatively, SPARC may not be a matrix component at all but may normally have only a short half-life in the extracellular environment, and after playing a critical role in some assembly or processing step, be removed by diffusion or proteolytic degradation. Slow diffusion from thick avascular basement membranes or matrices such as Reichert's membrane, bone, and the EHS tumor matrix *in vivo* might account for the strong immunofluorescence reaction in these areas. Studies with parietal endoderm cultured *in vitro* suggest that a small proportion of SPARC is incorporated into the matrix compared with the total amount secreted into the medium, at least under the conditions used (Mason et al., 1986a).

In conclusion, there is at present little information about the local concentration of either SPARC (BM-40, osteonectin) or, indeed, calcium, in the extracellular environment around cells. The EHS tumor contains about 500 μg BM-40 per gram wet weight, and most of this can be extracted under mild conditions with a physiological salt solution containing 10 mM EDTA, which also extracts laminin and entactin as a complex (Mann et al., 1986). However, as discussed above, the unusual thickness of the tumor matrix may prevent rapid diffusion of BM-40, and much lower levels may normally be present in extracellular matrices *in vivo*. Serum levels of BM-40 and osteonectin are low, at around 33 ng/ml, but increase in mice bearing the EHS tumor (Termine et al., 1981; Dziadek et al., 1986).

SPECULATION CONCERNING THE FUNCTION OF SPARC

At present, the precise function of SPARC is not known. Originally, it was suggested that osteonectin plays a role in initiating matrix mineralization in bone and teeth, by binding to both type I collagen and hydroxylapatite (Termine et al., 1981). However, the finding that SPARC is also expressed at high levels by a variety of nonmineralized tissues clearly demands a reassessment of this proposal, and suggests that the protein plays a wider role in some calcium-dependent process in the extracellular matrix. Alternative possibilities for the function of SPARC have been discussed by Engel et al. (1987) and will only be summarized briefly here. These are (1) the inhibition of ectopic or premature mineralization by binding to hydroxylapatite and preventing crystal growth, (2) a role in some calcium-dependent step in the assembly of laminin and/or collagen molecules into three-dimensional arrays or aggregates, and (3) a calcium-dependent protease or protease inhibitor regulating matrix assembly or turnover. In order to test these hypotheses, we are attempting to isolate mouse mutants defective in SPARC expression, to study the developmental effects of specific perturbations in the expression of SPARC mRNA, and to study in more detail the physicochemical properties of the purified protein.

ACKNOWLEDGMENTS

We thank Sarah Harper for assistance with the *in situ* hybridization experiments and William Taylor, Mats Paulsson, Jeffrey Raisman, Helene Sage, and John McVey for helpful discussion.

REFERENCES

Barlow, D. P., N. M. Green, M. Kurkinen, and B. L. M. Hogan (1984) Sequencing of laminin B chain cDNAs reveals C-terminal regions of coiled-coil alpha-helix. *EMBO J.* 3:2355–2362.

Barlow, D. P., J. McVey, and B. L. M. Hogan (1987) Molecular cloning of laminin. *Methods Enzymol.* 144:464–474.

Chen, J.-M., and C. D. Little (1987) Cellular events associated with lung branching morphogenesis including the deposition of collagen type IV. *Dev. Biol.* 120:311–321.

Chou, P. Y., and G. D. Fasman (1978) Prediction of the secondary structure of proteins from their amino acid sequence. *Adv. Enzymol.* 47:45–148.

Chung, A. E., S. L. Phillips, M. E. Durkin, S. M. Gardner, B. B. Bartos, J. S. Vergnes, and P. Labriola (1985) The regulation of expression of laminin: Molecular cloning and characterization of cDNA complementary to laminin B2 mRNA. In *Basement Membranes: Proceedings of the International Symposium on Basement Membranes*, S. Shibata, ed., pp. 155–165, Elsevier, Amsterdam.

Dang, C. V., R. F. Erbert, and W. O. Bell (1985) Localization of a fibrinogen calcium binding site between α-subunit positions 311 and 336 by terbium fluorescence. *J. Biol. Chem.* 260:9713–9719.

Deerfield, D. W., P. Berkowitz, D. L. Olsen, S. Wells, R. A. Hoke, K. A. Koehler, L. G. Pedersen, and R. G. Hiskay (1986) The effect of divalent metal ions on the electrophoretic mobility of bovine prothrombin and bovine prothrombin fragment 1. *J. Biol. Chem.* 261:4833–4839.

Dziadek, M., M. Paulsson, M. Aumailley, and R. Timpl (1986) Purification and tissue distribution of a small protein (BM-40) extracted from a basement membrane tumor. *Eur. J. Biochem.* **161**:455–464.

Engel, J., W. Taylor, M. Paulsson, H. Sage, and B. L. M. Hogan (1987) Calcium binding domains and calcium-induced conformational transition of SPARC (BM-40, osteonectin), an extracellular glycoprotein expressed in mineralized and non-mineralized tissues. *Biochemistry* **26**:6958–6965.

Garnier, J., D. J. Osguthorpe, and B. Robson (1978) Analysis of the accuracy and implications of simple methods for predicting the secondary structure of globular proteins. *J. Mol. Biol.* **120**:97–120.

Hauschka, P. V., and S. A. Carr (1982) Calcium-dependent α-helical structure in osteocalcin. *Biochemistry* **21**:2538–2547.

Hogan, B. L. M., D. P. Barlow, and R. Tilly (1983) Teratocarcinoma cells as a model for the differentiation of parietal and visceral endoderm in the mouse embryo. *Cancer Surv.* **2**:115–140.

Hogan, B. L. M., D. P. Barlow, and M. Kurkinen (1984) Reichert's membrane as a model for studying the biosynthesis and assembly of basement membrane components. *Ciba Found. Symp.* **108**:60–69.

Hogan, B. L. M., F. Costantini, and E. Lacy (1986) *Manipulating the Mouse Embryo: A Laboratory Manual*, Cold Spring Harbor Laboratory, Cold Spring Harbor, New York.

Holland, P. W. H., S. J. Harper, J. H. McVey, and B. L. M. Hogan (1987) *In vivo* expression of mRNA for the Ca^{++}-binding protein SPARC (osteonectin) revealed by *in situ* hybridization. *J. Cell Biol.* **105**:473–482.

Hughes, R. C., A. Taylor, H. Sage, and B. L. M. Hogan (1987) Distinct patterns of glycosylation of colligin, a collagen-binding glycoprotein, and SPARC (osteonectin) a secreted Ca^{++}-binding glycoprotein. *Eur. J. Biochem.* **163**:57–65.

Kretsinger, R. H. (1979) The informational role of calcium in the cytosol. In *Advances in Cyclic Nucleotide Research*, Vol. II, P. Greengard and G. A. Robinson, eds., pp. 1–26, Raven, New York.

Kretsinger, R. H. (1980) Structure and evolution of calcium-modulated proteins. *CRC Crit. Rev. Biochem.* **8**:119–174.

Kurkinen, M., D. P. Barlow, D. M. Helfman, J. G. Williams, and B. L. M. Hogan (1983) Isolation of cDNA clones for basal lamina components: Type IV collagen. *Nucleic Acids Res.* **11**:6199–6209.

Kurkinen, M., M. P. Bernard, D. P. Barlow, and L. T. Chow (1985) *Nature* **317**:177–179.

Leszczynski, J. F., and G. D. Rose (1986) Loops in globular proteins: A novel category of secondary structure. *Science* **234**:849–855.

Loring, J. F., and C. A. Erikson (1987) Neural crest cell migratory pathways in the trunk of the chick embryo. *Dev. Biol.* **121**:220–236.

Mann, K., R. Deutzmann, M. Paulsson, and R. Timpl (1987) Solubilization of protein BM-40 from a basement membrane tumor with chelating agents and evidence for its identity with osteonectin and SPARC. *FEBS Lett.* **218**:167–172.

Mason, I. J., A. Taylor, J. G. Williams, H. Sage, and B. L. M. Hogan (1986a) Evidence from molecular cloning that SPARC, a major product of mouse embryo parietal endoderm, is related to an endothelial cell "culture shock" glycoprotein of $M_r = 43,000$. *EMBO J.* **5**:1465–1472.

Mason, I. J., D. Murphy, M. Münke, U. Francke, R. W. Elliott, and B. L. M. Hogan (1986b) Developmental and transformation-sensitive expression of the SPARC gene on mouse chromosome 11. *EMBO J.* **5**:1831–1837.

Nelsestuen, G. L., R. M. Resnick, G. J. Wei, C. H. Pletcher, and V. A. Bloomfield (1981) Metal ion interactions with bovine prothrombin and prothrombin fragment 1: Stoichiometry of binding, protein self-association, and conformational change induced by a variety of metal ions. *Biochemistry* **20**:351–358.

Pan, L. C., and P. A. Price (1985) The propeptide of rat bone γ-carboxy-glutamic acid protein shares homology with other vitamin K–dependent protein precursors. *Proc. Natl. Acad. Sci. USA* 82:6109–6113.

Romberg, R. W., P. G. Werness, P. Lollar, B. L. Riggs, and K. G. Mann (1985) Isolation and characterization of native adult osteonectin. *J. Biol. Chem.* 260:2728–2736.

Sage, H., C. Johnson, and P. Bornstein (1984) Characterization of a novel serum albumin-binding glycoprotein secreted by endothelial cells in culture. *J. Biol. Chem.* 259:3993–4007.

Sasaki, M., S. Kato, K. Kohno, G. R. Martin, and Y. Yamada (1987) Sequence of the cDNA encoding the laminin B1 chain reveals a multidomain protein containing cysteine-rich repeats. *Proc. Natl. Acad. Sci. USA* 84:935–939.

Szebenyi, D. M. E., and K. Moffat (1986) The refined structure of vitamin D–dependent calcium-binding protein from bovine intestine. *J. Biol. Chem.* 261:8761–8777.

Termine, J. D., H. K. Kleinman, S. W. Whitson, K. M. Conn, M. L. McGarvey, and G. R. Martin (1981) Osteonectin, a bone-specific protein linking mineral to collagen. *Cell* 26:99–105.

Tung, P. S., C. Domenicucci, S. Wasi, and J. Sodek (1985) Specific immunohistochemical localization of osteonectin and collagen types I and III in fetal and adult porcine dental tissues. *J. Histochem. Cytochem.* 33:531–540.

Van Eldik, L. J., J. G. Zendegui, D. R. Marshak, and D. M. Watterson (1982) Calcium-binding proteins and the molecular basis of calcium action. *Int. Rev. Cytol.* 77:1–61.

Wang, S.-Y., and L. J. Gudas (1983) Isolation of cDNA clones specific for collagen IV and laminin from mouse teratocarcinoma cells. *Proc. Natl. Acad. Sci. USA* 80:5880–5884.

Wasi, S., K. Otsuka, K.-L. Yao, P. S. Tung, J. E. Aubin, and J. Sodek (1984) An osteonectin-like protein in porcine periodontal ligament and its synthesis by periodontal ligament fibroblasts. *Can. J. Biochem. Cell Biol.* 62:470–477.

Young, M. F., M. E. Bolander, A. A. Day, C. I. Ramis, P. G. Robey, Y. Yamada, and J. D. Termine (1986) Osteonectin mRNA: Distribution in normal and transformed cells. *Nucleic Acids Res.* 14:4483–4497.

Chapter 12

Coordinate Expression and Function of Cytotactin and Its Proteoglycan Ligand

STANLEY HOFFMAN
KATHRYN L. CROSSIN
GERALD M. EDELMAN

ABSTRACT

Analyses of molecules involved in cell adhesion have suggested that these molecules play a role in the control of the primary cellular process of development that leads to pattern formation. This category includes molecules involved in both cell–cell adhesion (CAMs) and cell–substrate adhesion (SAMs), several of which are present at any given site in a developing embryo. A knowledge of the patterns of coordinate expression of these molecules and the resulting effects on cell behavior is therefore necessary for understanding the role of cell adhesion in morphogenesis.

The binding properties, distribution, and effects on cell behavior of cytotactin, a recently identified SAM, make it particularly useful for evaluating how the coordinate expression of adhesive molecules may affect development. Cytotactin is an extracellular matrix protein that binds to receptors on cell surfaces as well as to other matrix proteins, including fibronectin and cytotactin-binding (CTB) proteoglycan. These molecules appear to be part of an interactive network of extracellular matrix proteins whose abilities to bind to cells are modulated by their interactions with each other. Cytotactin has a relatively restricted distribution during development, a property shared by several CAMs but not by well-characterized SAMs such as collagen and fibronectin. The pattern and levels of adhesive proteins coexpressed with cytotactin vary greatly during development and between organs. In the developing central nervous system, cytotactin is present at high levels and is synthesized by glia, while its ligand, CTB proteoglycan, is synthesized by neurons. Cytotactin is involved in neuron–glia adhesion, although by a molecular mechanism distinct from that involving the neuron–glia cell adhesion molecule Ng-CAM. Nevertheless, both Ng-CAM and cytotactin are involved in the migration of neurons along glia that occurs during the formation of histological layers in the cerebellar cortex.

Cytotactin also affects a very different mode of cell migration: the migration of neural crest cells into the developing sclerotome, a primary step in the segmentation of the peripheral nervous system. Neural crest cells populate only the rostral half of each sclerotome, where cytotactin is localized, and are absent from the caudal halves, where CTB proteoglycan later becomes localized. Cytotactin and CTB proteoglycan are the first specific proteins to be identified with these distributions. In vitro experiments have indicated that cytotactin is a poor substrate for neural crest cell spreading and migration, unlike fibronectin, which is present throughout the sclerotome and promotes cell spreading and migration. Moreover, increasing doses of cytotactin in the presence of a constant amount of fibronectin decrease the extent of neural crest cell migration, suggesting that cytotactin can modulate movement on fibronectin substrates.

These results indicate that the colocalization of cytotactin and neural crest cells in vivo *may occur as a result of the ability of cytotactin to halt the migration of these cells on fibronectin and that this process must depend on the proper coordinate expression of CAMs and SAMs, including N-CAM, cytotactin, fibronectin, and CTB proteoglycan. As other morphogenetic systems are further analyzed, similar events involving the coordinate expression of CAMs and SAMs are likely to be found, supporting the hypothesis that an interactive network of SAMs can differentially affect the behavior of cell collectives maintained by various CAMs.*

Morphogenesis in regulative development results from milieu-dependent cues to cells in collectives that drive subpopulations of cells along distinct developmental pathways by differentially affecting the primary processes of development (cell adhesion, migration, proliferation, death, and differentiation) (Needham, 1933). At the cellular level, these signals are believed to be transduced by an array consisting of the nucleus, the cytoskeleton, and specific cell surface and surface-associated proteins involved in cell adhesion and communication (Edelman, 1976, 1988). These proteins include integral membrane proteins that mediate cell–cell adhesion (CAMs), integral membrane proteins that form specialized junctions between cells (CJMs), and extracellular matrix proteins that mediate cell–substrate adhesion (SAMs) and their receptors.

Several examples have already been identified in which the differential expression of specific CAMs and SAMs at particular times during development is correlated with patterned cell behavior (for a review, see Edelman, 1986). At sites of primary and secondary induction, borders between tissues expressing N-CAM and L-CAM are observed; later, during histogenesis, N-CAM–L-CAM borders mark the boundaries between tissues that will undergo alternate paths of terminal differentiation (Edelman et al., 1983; Chuong and Edelman, 1985a,b; Crossin et al., 1985; Richardson et al., 1987). In another example involving a CAM, N-CAM, and a SAM, fibronectin, neural crest cells express N-CAM when they first differentiate, lose N-CAM while they migrate as single cells on a fibronectin-rich substratum, and finally re-express N-CAM and aggregate, forming ganglionic rudiments (Thiery et al., 1982). Thus, in each of these systems, complex morphogenetic patterns may result from changes in cell behavior governed by the integration of independent signals provided by the binding of CAMs and SAMs at the cell surface.

In this chapter, we focus on the coordinate expression and function of cytotactin and other CAMs and SAMs, in particular a chondroitin sulfate proteoglycan that binds to cytotactin, which we have named CTB (cytotactin-binding) proteoglycan (Hoffman and Edelman, 1987). Two considerations suggest that such studies will be particularly enlightening in understanding how adhesive molecules participate in the control of cell behavior. First, although cytotactin is present in neural tissue and at a variety of nonneural sites, it nevertheless has a relatively restricted distribution compared to other SAMs such as collagen and fibronectin (Crossin et al., 1986). Second, it appears that the ability of cytotactin to interact with cells may be modulated by its intermolecular interactions with other components of the extracellular matrix (Hoffman and Edelman, 1987; Hoffman et al., 1988) and it may, therefore, differentially affect cell behavior.

STRUCTURE OF CYTOTACTIN

Perhaps the most unusual feature of cytotactin is its three-dimensional structure observed in electron microscope images (Figure 1A). Cytotactin appears as a six-armed structure known as a hexabrachion (Erickson and

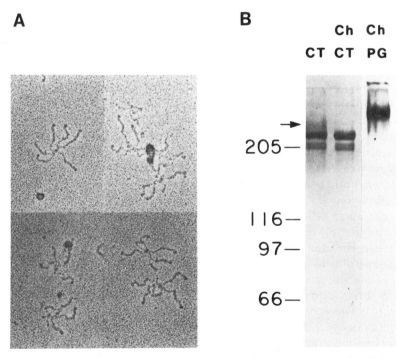

Figure 1. *Electron micrographs of purified cytotactin and electrophoretic comparison of purified cytotactin and CTB proteoglycan. A:* Purified cytotactin was sprayed onto mica, vacuum dried, and rotary-shadowed as described by Fowler and Erickson (1979), and selected hexabrachions were photographed by Dr. Harold P. Erickson, Duke University. *B:* Cytotactin (CT), chondroitinase-treated cytotactin (Ch CT), and chondroitinase-treated CTB proteoglycan (Ch PG) were resolved on SDS-polyacrylamide gels. The *arrow* indicates the migration of a chondroitinase-sensitive component on cytotactin. (Summarized from Hoffman et al., 1988.)

Iglesias, 1984), in which each arm emanates from a central globular region (Hoffman et al., 1988). The appearance of these structures and their molecular weight as estimated from their sedimentation rate (Friedlander et al., 1988) suggest that each structure contains six polypeptide chains. SDS-gel electrophoresis under reducing conditions of cytotactin purified from brain extracts indicates the presence of a major 220-kD component and minor components of 250 kD, 200 kD, and 190 kD (Figure 1B). Unreduced samples of cytotactin fail to enter a 6% polyacrylamide gel (Grumet et al., 1985) indicating that cytotactin polypeptides are disulfide cross-linked and strongly suggesting that such bonds are involved in maintaining the six-armed structures described above. It remains to be determined whether each six-armed structure contains a single polypeptide species or a mixture of polypeptides.

A variety of approaches have provided evidence about the structural relationships between these cytotactin polypeptides. The 250-kD component appears to contain covalently bound chondroitin sulfate because it is altered by chondroitinase treatment (Figure 1B), while the other components of cytotactin are not affected by this treatment. All of the cytotactin polypeptide species purified from brain extracts contain the carbohydrate epitope that is recognized by monoclonal antibody HNK-1 (Hoffman et al., 1988) and that is also present on several molecules involved in cell adhesion, including N-CAM, Ng-CAM, and CTB proteoglycan (Grumet et al., 1984a,b; Hoffman and Edelman, 1987). Recently, cDNA probes specific for cytotactin have been prepared (Jones et al., 1988). Northern blots performed with these probes indicate the presence of at least three cytotactin mRNA species in embryonic brain and therefore suggest that the various cytotactin polypeptides observed may result from the translation of different messages rather than from the differential posttranslational modification or processing of a single polypeptide.

Recent experiments have indicated that cytotactin from other tissues can differ from brain cytotactin in both polypeptide structure (Jones et al., 1988) and carbohydrate modification (Hoffman et al., 1988). For example, gizzard tissue contains a 240-kD form of cytotactin that is larger than any form of brain cytotactin (following chondroitinase treatment). The observation that gizzard tissue also contains a cytotactin mRNA larger than any found in brain is consistent with the possibility that this large mRNA codes for the 240-kD polypeptide and that this polypeptide differs in primary structure from any form of brain cytotactin (Jones et al., 1988). Differences in carbohydrate structure between cytotactin from neural and nonneural sources were detected using monoclonal antibody HNK-1. For example, cytotactin from fibroblast culture medium was readily recognized by monoclonal and polyclonal antibodies specific for the polypeptide portion of cytotactin, but unlike cytotactin from brain, it did not contain the carbohydrate epitope recognized by HNK-1 (Hoffman et al., 1988). The possibility is therefore open that structural variations in cytotactin polypeptides and carbohydrates may contribute to different functions for the molecule in different tissues.

Sequence analysis of clones of cytotactin cDNA has revealed several aspects of the primary structure of the molecule (Jones et al., 1988, 1989). The amino-terminal portion of the molecule contains the cysteines involved in the disulfide bonds that link the six polypeptide chains in the hexabrachion. The remainder of the 220-kD form of the molecule consists of 13 EGF-like cysteine-rich repeats, 11 repeats similar to the type III repeats in fibronectin (Peterson et al., 1983), and a terminal fibrinogenlike segment (Figure 4). Type III repeats appear to be involved both in the structural diversity among cytotactin polypeptides and in its cell-binding function. The 190-kD form of cytotactin differs from the 220-kD form only in that it contains 8 type III repeats instead of 11. The sequence Arg-Gly-Asp, which is known to be involved in cell adhesion in fibronectin (Ruoslahti and Pierschbacher, 1986), is in a similar position within a type III repeat in both molecules. This sequence may also be involved in the binding of cytotactin to cells as indicated by the fact that peptides containing this sequence inhibit the binding of fibroblasts to cytotactin-coated substrates (Friedlander et al., 1988). The extensive similarity between cytotactin and other proteins suggests that cytotactin has an evolutionary and possibly a functional relationship to each of these proteins.

STRUCTURE OF CTB PROTEOGLYCAN

CTB proteoglycan is a large chondroitin sulfate proteoglycan (Hoffman and Edelman, 1987). Like other proteoglycans, it is made up predominantly of carbohydrate ($>60\%$) and, therefore, has a higher buoyant density than most proteins, a property that is important in its purification. Following the enzymatic removal of chondroitin sulfate, the core protein of CTB proteoglycan can be resolved by SDS-PAGE as a single component of 280 kD (Figure 1B); prior to this treatment the molecule is too large to enter a 6% gel. CTB proteoglycan purified from brain is the first proteoglycan to be identified whose core protein bears the carbohydrate epitope recognized by monoclonal antibody HNK-1 (Hoffman and Edelman, 1987). Like cytotactin, however, when CTB proteoglycan is purified from nonneural sources, it lacks the HNK-1 epitope (Hoffman et al., 1988). When anti-CTB proteoglycan antibodies are used to immunoblot chondroitinase-treated extracts of adult brain, several components of lower molecular weight than intact CTB proteoglycan are detected, some of which are similar in molecular weight to breakdown products that appear in chondroitinase-treated CTB proteoglycan preparations during storage. Whether these smaller chondroitin sulfate proteoglycans that cross-react with anti-CTB proteoglycan antibodies are fragments of CTB proteoglycan or arise through other means remains to be determined.

RELATIONSHIPS OF CYTOTACTIN TO OTHER PROTEINS

Two molecules described by other investigators, brachionectin (Erickson and Taylor, 1987) and myotendinous antigen (Chiquet and Fambrough, 1984), also known as tenascin (Chiquet-Ehrismann et al., 1986), appear to be closely related or identical to cytotactin. All three molecules migrate as similar polypeptides on SDS gels, and monoclonal and polyclonal anti-myotendinous antigen antibodies recognize both brachionectin and cytotactin. Brachionectin and tenascin as currently purified probably represent subsets of cytotactin because they are obtained from nonneural sources and therefore lack the HNK-1 carbohydrate epitope. Nevertheless, brachionectin, myotendinous antigen, and brain cytotactin all appear as similar six-armed figures (Erickson and Taylor, 1987; Vaughan et al., 1987; Hoffman et al., 1988) when analyzed by electron microscopy, which suggests that the presence of the HNK-1 carbohydrate epitope on cytotactin does not alter its three-dimensional structure.

It has been suggested that a third molecule, a 160-kD protein purified from adult brain tissue and known as the J1 antigen (Kruse et al., 1985), is related to cytotactin. Recent studies have indicated, however, that the molecules are not structurally related (Hoffman et al., 1988). In early studies antibodies to cytotactin and antibodies to J1 antigen were each reported to recognize 220-kD and 200-kD polypeptides in embryonic brain extracts and 180-kD and 160-kD polypeptides in adult brain extracts (Grumet et al., 1985; Kruse et al., 1985), all of which contained the carbohydrate epitope recognized by monoclonal antibody HNK-1.

Although these polypeptides have distinct one-dimensional peptide maps (Grumet et al., 1985), it was suggested that, like N-CAM, the J1 antigen had structurally related embryonic and adult forms (Kruse et al., 1985). Recent studies using antisera directed against more stringently purified cytotactin have suggested, however, that the 180-kD and 160-kD polypeptides and the 220-kD and 200-kD polypeptides are unrelated. These anti-cytotactin antibodies do not recognize the 180-kD and 160-kD polypeptides (Hoffman et al., 1988). Furthermore, recently prepared specific antibodies to the 180-kD polypeptide do not recognize cytotactin. It will be of interest to determine whether the 180-kD and 160-kD polypeptides interact with cytotactin, providing a basis for their apparent copurification in earlier studies and to determine the function (or lack thereof) of these molecules in cell adhesion.

BINDING PROPERTIES OF CYTOTACTIN

Protein-Coated Beads Facilitate Analyses of Molecular Binding

Many glycoproteins in the extracellular matrix interact both with cell surface receptors and with other components of the extracellular matrix (Yamada, 1983). Cytotactin is no exception to this generalization. Binding studies have demonstrated that it binds to cells, including neurons and fibroblasts, as well as to other extracellular matrix components, including CTB proteoglycan and fibronectin (Hoffman and Edelman, 1987; Hoffman et al., 1988).

Fluorescent beads known as Covaspheres have been used to simplify the analysis of the binding properties of cytotactin. These 0.5-μm-diameter beads have chemically activated surfaces that covalently bind proteins during incubation in phosphate-buffered saline. Their intrinsic red or green fluorescence allows them to be readily identified by fluorescence microscopy. Two basic protocols were used to analyze the cellular and intermolecular binding properties of cytotactin (Figure 2). In cell-binding studies, radioiodinated proteins on Covaspheres were incubated with cells in suspension for short times, and the cells were separated from unbound beads by differential centrifugation. The presence of beads bound to cells was identified by fluorescence microscopy and quantitated by gamma spectroscopy. In intermolecular binding experiments red-fluorescing Covaspheres coated with one molecule were incubated with green-fluorescing Covaspheres coated with a second molecule. Intermolecular interactions between these molecules resulted in the formation of Covasphere aggregates containing both red and green Covaspheres. The extent of aggregation was quantitated using a particle counter set to detect aggregate but not monomer Covaspheres.

The use of Covaspheres to determine the cell-binding properties of a protein has certain obvious advantages over methods in which the protein is adsorbed to a plastic dish in which cells are then cultured. In Covasphere–cell binding experiments in suspension, the short-term initial binding of beads

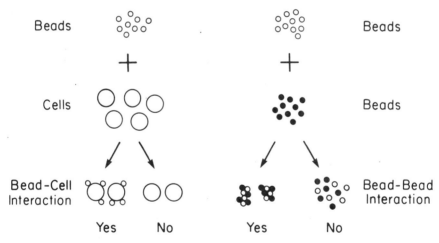

Figure 2. *Schematic description of Covasphere–cell and Covasphere–Covasphere binding assays.* Covaspheres are 0.5-μm diameter, fluorescent beads to which proteins attach covalently when coincubated in phosphate-buffered saline. These protein-coated beads can be used to explore the molecular binding properties of the adsorbed proteins. In cell-binding experiments (*left*), protein-coated beads are coincubated with cells for 20 min, and cells with bound beads are separated from unbound beads by differential centrifugation. The extent of binding characteristic for a particular protein is determined either by fluorescence microscopy or by gamma spectroscopy if radioiodinated proteins are attached to the beads. In intermolecular binding experiments (*right*), green-fluorescing beads (*open circles*) coated with a particular protein are coincubated with red-fluorescing beads (*filled circles*) coated with a second protein. Intermolecular interactions result in the formation of aggregates containing both species of beads. These aggregates can be detected by fluorescence microscopy; the extent of aggregation is quantitated using a particle counter to determine the number of aggregates formed.

bearing a known amount of protein is readily quantitated. In cell–substrate-binding experiments it is more difficult to determine the protein concentration on the substrate: incubation times are significantly longer, increasing the probability of a variety of complications, and most important, effects secondary to initial binding, particularly cell spreading, may selectively influence the quantitation of cell binding to different proteins on the substrate (Friedlander et al., 1988).

Several preliminary experiments were performed to confirm that the binding properties of molecules on Covaspheres were similar to those of the soluble molecules. Molecules known to bind to cells, such as lectins and specific antibodies against cell surface proteins, when coupled to Covaspheres, caused the Covaspheres to bind to cells. Members of known binding couples, such as fibronectin and gelatin or various proteins and specific antibodies to the proteins, were tested for their ability to mediate Covasphere aggregation. When individually coupled to red and green Covaspheres, these pairs of molecules caused the Covaspheres to coaggregate even though each population of Covaspheres bearing a single protein did not self-aggregate.

Cellular Binding Properties

With the validity of this methodology firmly established, the cellular binding properties of cytotactin and other adhesive proteins coupled to Covaspheres were examined (Table 1). Cytotactin-coated Covaspheres bound well to neurons (Hoffman and Edelman, 1987). The specificity of this binding was suggested by the observation that boiling or trypsinization of the Covaspheres or the presence of a lower concentration of cytotactin on the Covaspheres substantially decreased binding. In other controls, Covaspheres coated with

Table 1.   Binding of Protein-Coated Covaspheres to Neurons[a]

Protein on Covaspheres	Relative Binding
Cytotactin	100
Cytotactin (low level)	26
Cytotactin (boiled)	30
Cytotactin (trypsinized)	18
N-CAM	76
Ng-CAM	90
Ovalbumin	10
BSA	8

[a]The relative levels of binding obtained with various protein-coated Covaspheres are expressed as a percentage of the binding obtained with cytotactin-coated Covaspheres. (Summarized from Hoffman and Edelman, 1987; Grumet and Edelman, 1988.)

N-CAM or Ng-CAM, two molecules known to mediate neuron–neuron adhesion, bound well to cells. Covaspheres coated with other proteins such as bovine serum albumin or ovalbumin did not bind. To confirm the specificity of binding, cytotactin-coated Covaspheres and N-CAM-coated Covaspheres were incubated with cells in the presence of Fab' fragments of anti-cytotactin or anti-N-CAM antibodies. As expected, anti-cytotactin antibodies inhibited cytotactin binding but not N-CAM binding, and anti-N-CAM antibodies inhibited N-CAM binding but not cytotactin binding.

To examine the possibility that the presence of other extracellular matrix proteins might alter the ability of cytotactin to bind to cells, cytotactin-coated Covaspheres were incubated with cells in the presence of various proteins (Table 2). Soluble CTB proteoglycan strongly inhibited the binding of cytotactin-coated Covaspheres to cells; a heparan sulfate proteoglycan, chondroitin sulfate (the glycosaminoglycan class on CTB proteoglycan), fibronectin, and cytotactin itself had little or no effect (Hoffman and Edelman, 1987). These results suggested either that CTB proteoglycan binds to cytotactin and thereby decreases its ability to bind to cells or that CTB proteoglycan and cytotactin compete for a common cell surface receptor.

Intermolecular Binding Properties

To evaluate directly the possibility that cytotactin binds to CTB proteoglycan, Covaspheres coated with cytotactin were incubated with Covaspheres coated with CTB proteoglycan (Table 3). Rapid and extensive coaggregation of Covaspheres occurred, even though each population of Covaspheres alone showed little aggregation (Hoffman and Edelman, 1987). The specificity of this Covasphere coaggregation was indicated by the fact that it was inhibited by Fab' fragments of anti-cytotactin antibodies, soluble cytotactin, or soluble CTB proteoglycan, but not by soluble fibronectin. In other control experiments Covaspheres coated with bovine serum albumin (BSA) or laminin did not interact with cytotactin-coated or CTB proteoglycan-coated Covaspheres.

Table 2. Effects of Extracellular Matrix Molecules on the Binding of Cytotactin-Coated Covaspheres to Neurons[a]

Protein on Covaspheres	Soluble Molecule	Relative Binding
Cytotactin	None	100
Cytotactin	CTB proteoglycan	26
Cytotactin	HS proteoglycan	94
Cytotactin	Chondroitin sulfate	91
Cytotactin	Fibronectin	113
Cytotactin	Cytotactin	83

[a]The relative levels of binding of cytotactin-coated Covaspheres to neurons obtained in the presence of various soluble molecules are expressed as a percentage of the control level of binding obtained without such additions. (Summarized from Hoffman and Edelman, 1987.)

Table 3. Coaggregation of Cytotactin-Coated Covaspheres with CTB Proteoglycan-Coated Covaspheres and Fibronectin-Coated Covaspheres[a]

Protein(s) on Covaspheres	Soluble Protein	Covasphere Aggregates
Cytotactin, proteoglycan		49,500
Cytotactin only		1,660
Proteoglycan only		420
Cytotactin, proteoglycan	Anti-cytotactin	2,480
Cytotactin, proteoglycan	Cytotactin	11,800
Cytotactin, proteoglycan	Proteoglycan	14,700
Cytotactin, BSA	Fibronectin	44,800
Cytotactin, BSA		400
Cytotactin, laminin		610
Cytotactin, fibronectin		17,300
Proteoglycan, fibronectin		320

[a]In these experiments Covaspheres coated with a single protein were incubated with other Covaspheres coated with a different protein. In certain combinations aggregation between the two populations of Covaspheres occurred and could be detected using a particle counter. (Summarized from Hoffman and Edelman, 1987.)

Fibronectin-coated Covaspheres, however, did coaggregate with cytotactin-coated Covaspheres but not with CTB proteoglycan–coated Covaspheres, indicating that cytotactin specifically binds to fibronectin. The fact that fibronectin binds to cytotactin but does not inhibit its ability to bind to cells suggests the possibility that the fibronectin-binding site on cytotactin is not sufficiently near to the cell-binding site to alter its function.

Despite the fact that soluble fibronectin did not inhibit the binding of cytotactin-coated Covaspheres to cells (Hoffman and Edelman, 1987), soluble cytotactin did inhibit the binding of fibronectin-coated Covaspheres to cells (Hoffman et al., 1988). Therefore, the results suggest that the binding of cytotactin to cells can be modulated by CTB proteoglycan interacting with the cytotactin and that the binding of fibronectin to cells can be moduled by cytotactin interacting with the fibronectin. In addition, other examples of one extracellular matrix protein altering the cell-binding properties of another were observed. Specifically, soluble laminin inhibited the binding of cytotactin-coated Covaspheres to cells (Hoffman and Edelman, 1987), and soluble CTB proteoglycan inhibited the binding of fibronectin-coated Covaspheres to cells (Hoffman et al., 1988). In all of these cases it remains to be determined whether the effects are mediated solely by interactions among extracellular matrix proteins or whether competition for cell surface receptors is also involved.

The cellular and intermolecular binding properties of cytotactin were also analyzed in terms of their dependence on divalent cations and on covalently attached oligosaccharides (Hoffman et al., 1988). The binding of cytotactin to cells and to CTB proteoglycan and fibronectin was inhibited by EDTA, suggesting that all these mechanisms are divalent cation dependent. Moreover, further experiments strongly suggest that the intermolecular binding mechanisms are specifically calcium dependent. When binding experiments were performed to compare the functions of cytotactin polypeptides containing covalently bound chondroitin sulfate and the HNK-1 epitope with cytotactin polypeptides lacking these carbohydrates, all these forms of cytotactin were similar in their abilities to bind to cells or CTB proteoglycan or fibronectin. These observations indicate that neither of these carbohydrates is directly involved as a ligand in any of these binding mechanisms. The possibility cannot be ruled out, however, that these carbohydrates may indirectly affect binding as observed, for example, for the sialic acid in N-CAM (Hoffman and Edelman, 1983).

The results of these cellular and intermolecular binding studies suggest that interactions among components of the extracellular matrix may modulate their ability to bind to cells. In this manner cell behavior and cellular primary processes such as cell migration, proliferation, and cytodifferentiation may be differentially controlled at various times and sites during development by the proper ratios of a relatively limited set of matrix proteins. This idea and the results leading to it are illustrated diagrammatically in Figure 3.

Fragments of Cytotactin Active in Cell–Substrate Adhesion

The binding experiments described above indicated that cytotactin on beads binds to cells. To confirm that cytotactin can also mediate the binding of cells to substrates, plastic culture dishes were coated with the molecule and the ability of fibroblasts to bind to these dishes was determined. As expected, fibroblasts bound well to cytotactin-coated substrates (Friedlander et al., 1988). Moreover, as in the Covasphere assay, binding was inhibited by CTB proteoglycan and by Fab' fragments of anti-cytotactin antibodies.

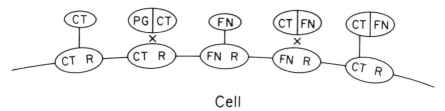

Cell

Figure 3. *Schematic illustration of the effects of the binding of cytotactin to CTB proteoglycan or fibronectin on the binding of cytotactin or fibronectin to their cell surface receptors.* In the absence of other extracellular matrix components, cytotactin (CT) binds to its cell surface receptor (CT R), and fibronectin (FN) binds to its cell surface receptor (FN R). In each case this binding is indicated by a *line* joining the interacting molecules. When cytotactin interacts with CTB proteoglycan (PG), the binding of the complex to the cytotactin receptor is inhibited, as indicated by the X between the CT/PG complex and the CT R. The binding of FN in CT/FN complexes to FN R is also inhibited, while the binding of CT in CT/FN complexes to CT R is not inhibited.

To relate the binding function and structure of cytotactin, proteolytic fragments of the molecule were purified and tested for their ability to mediate cell–substrate adhesion. Chymotryptic digestion produced fragments that were fractionated into two major pools: one (called fraction I) contained disulfide-linked oligomers of a 100-kD fragment, and the second (called fraction II) contained monomeric 90-kD and 65-kD fragments. The 90-kD and 65-kD fragments in fraction II were closely related to each other and were distinct from the 100-kD fragment in fraction I (Friedlander et al., 1988).

In cell-binding experiments, fibroblasts adhered to substrates coated with the components in fraction II, but not to those in fraction I (Friedlander et al., 1988). Furthermore, both the binding to fraction II and to intact cytotactin were inhibited by CTB proteoglycan and by peptides containing the sequence Arg-Gly-Asp. The sites where cytotactin interacts with CTB proteoglycan and fibronectin also appear to be present in or near fraction II.

A structural and functional model of cytotactin is shown in Figure 4. Because fraction I is a disulfide-bonded oligomer, it must include at least a portion of the central core of the molecule where the six polypeptide chains that form the hexabrachion meet. This idea was confirmed by comparison of the amino-terminal sequence of fraction I with the sequence of cytotactin deduced from cDNA clones. Because the polypeptides in fraction II are monomeric, they probably correspond to more distal portions of the free arms of the molecule. Sequence analyses of fraction II suggested that this fragment begins at one of two similar sequences in the region of cytotactin similar to fibronectin type III repeats. Although the binding of cells to fraction II was inhibited by peptides containing the sequence Arg-Gly-Asp, it has not yet been definitively determined whether this sequence is present within fraction II. In summary, the combined structural and functional data suggest that a cell-binding site is present in the distal portion of the arms of the cytotactin hexabrachion.

COORDINATE DISTRIBUTIONS OF CYTOTACTIN AND CTB PROTEOGLYCAN IN NEURAL TISSUE

Cytotactin was originally characterized as a glial molecule involved in neuron–glia adhesion (Grumet et al., 1985). Although cytotactin and CTB proteoglycan appear in a very similar distribution in cerebellar tissue sections, the two molecules are very different in their cellular sites of synthesis (Hoffman et al., 1988). Immunofluorescent staining experiments using brain cell cultures containing both neurons and glia confirmed that cytotactin is specifically synthesized by glia and indicated that CTB proteoglycan is specifically synthesized by neurons (Figure 5). In these experiments cells were identified as neurons and glia both by their morphologies and by their staining with specific marker antibodies: anti-Ng-CAM, which is specific for neurons in the central nervous system (Grumet et al., 1984b; Thiery et al., 1985; Daniloff et al., 1986), and antibodies against a glia-specific cytoskeletal component (Drager et

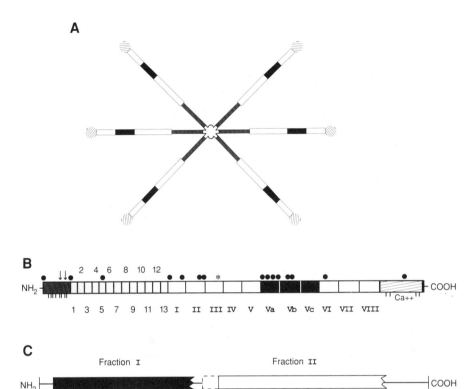

Figure 4. *Functional map of cytotactin. A:* Proposed model of a cytotactin hexamer. The amino-terminal regions of each polypeptide are disulfide-linked and represented as a single structure forming the core of the hexabrachion. The EGF-like repeats are *cross-hatched,* the type III repeats present in both polypeptides are *white* whereas those in the insert are *black,* and the fibrinogenlike nodular region is marked with *slanted lines. B:* Linear model of cytotactin. The epidermal growth factor-like repeats (*lightly stippled boxes*) are numbered 1–13; type III repeats (*open boxes*) are designated by Roman numerals with those in the additional insert (*black boxes*) designated Va, Vb, and Vc. The region similar to fibrinogen (*diagonal lines*) includes a putative calcium-binding domain (Ca++). *Darkly stippled boxes* denote regions that have no extensive homology to any known protein, and include the amino-terminal 142 amino acids and the carboxyl-terminal 13 amino acids. Cysteine residues that are not in epidermal growth factor-like repeats (*short lines*), potential asparagine glycosylation sites (*dark circles*), potential glycosaminoglycan addition sites (*arrows*), and the Arg-Gly-Asp sequence (*asterisk*) are indicated. *C:* Linear map of fragments characterized in binding experiments. Fraction I (*solid box*) includes cysteine residues in the amino-terminal portion of the molecule that form interchain disulfide bonds. Fraction II (*open box*) is located in the distal portion of the arms of the hexabrachion. Direct sequence analyses of fractions I and II were used to align these fragments with the structural model of cytotactin in B. Two possible amino termini are shown for fraction II because its sequence is similar to the two indicated sites in the intact molecule. The carboxyl termini of fractions I and II have been drawn with the *jagged ends* to indicate that their precise locations in the molecule have not yet been determined.

Anti-Cytotactin   Anti-Proteoglycan

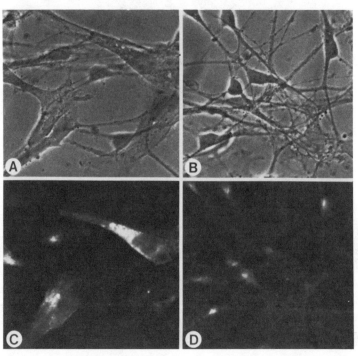

Figure 5. *Immunofluorescent localization of cytotactin and CTB proteoglycan in glia and neurons.* Brain cells from nine-day-old embryonic chickens were cultured and double-labeled with polyclonal antibodies against cytotactin or CTB proteoglycan, monoclonal antibodies specific for glia or neurons, and appropriate second antibodies. The cell type–specific antibodies used were anti-R5, which recognizes a cytoskeletal protein in glia (Drager et al., 1984), and anti-Ng-CAM, which is specific for neurons in the central nervous system (Grumet et al., 1984b; Thiery et al., 1985; Daniloff et al., 1986). The large, flat cells seen in the phase-contrast micrograph in *A* were labeled with anti-R5 and were, therefore, identified as glia; these cells were also heavily labeled with anti-cytotactin antibodies both intracellularly and on their cell surface (*C*). The small, rounded cells in *A* were not labeled with either of these antibodies. In contrast, the small, rounded cells in the phase-contrast micrograph in *B* were identified as neurons by their binding of anti-Ng-CAM antibodies and were also found to bind anti-CTB proteoglycan antibodies intracellularly and on their cell surfaces (*D*). The large, flat cells in the background in *B* were not labeled with anti-Ng-CAM or anti-CTB proteoglycan antibodies.

al., 1984). For both cytotactin and CTB proteoglycan, staining either occurred in a punctate, perinuclear pattern or was associated with the cell surface (Figure 5). Despite their association with the cell surface, the fact that both molecules are readily extracted from brain tissue in neutral, isotonic, detergent-free buffers (Hoffman et al., 1988) indicates that they are extracellular and not integral membrane proteins, and it raises the possibility that each of these molecules binds to cell surface receptors as well as to each other.

In addition to being synthesized by different cell types, cytotactin and CTB proteoglycan are expressed along different time courses and in different molecular forms during neural development (Hoffman et al., 1988). The expression of cytotactin decreases during development; only one-seventh as much cytotactin is present per milliliter of tissue in adult brains as compared to that in six-day-old embryo brains. While the 250-kD, 220-kD, and 200-kD components of cytotactin bearing chondroitin sulfate are present at similar levels up to day 15 of embryonic development, during later development the 220-kD component becomes predominant.

In contrast to cytotactin, the level of CTB proteoglycan in the brain increases about ninefold from six-day-old embryos to 18-day-old embryos and decreases only about 25% in later development (Hoffman et al., 1988). Therefore, the ratio of CTB proteoglycan expression to cytotactin expression increases greater than 40-fold between six-day-old embryos and adults. As described above, several proteoglycan species with lower-molecular-weight core proteins than intact CTB proteoglycan are detectable with anti-CTB proteoglycan antibodies in adult organisms. While these may represent degradation products, it is also possible that they are unique molecular species with specific functions. In any case, in view of the interactive nature of these two molecules as well as their potential to interact with other molecules, modulation in the amounts of either could affect cell interactions mediated by their molecular binding. It seems most likely, given the observation that CTB proteoglycan can block the binding of cytotactin to cells, that these molecules have a dynamic role in neuron–glia adhesion and processes that depend on neuron–glia adhesion, such as cerebellar cell migration (Rakic, 1971), which may be controlled by spatial and developmental variations in their relative expression. These differences are likely to be critical to an understanding of the roles of cytotactin and CTB proteoglycan in neuron–glia adhesion and cerebellar cell migration.

COORDINATE DISTRIBUTIONS AND FUNCTIONS OF CYTOTACTIN, Ng-CAM, AND N-CAM IN NEURAL TISSUES

Cytotactin is not the only molecule to have been implicated in neuron–glia adhesion. The neuronal cell surface protein Ng-CAM is also involved in this process (Grumet and Edelman, 1984; Grumet et al., 1984a). Ng-CAM is also involved in neuron–neuron adhesion, as is N-CAM (Grumet et al., 1984b). Moreover, because Ng-CAM is not present on glia, it has been concluded that the glial ligand for Ng-CAM is a distinct, currently unidentified protein. Despite the fact that cytotactin is a glial protein, it does not appear to be the receptor for Ng-CAM. Rather, the two molecules function in neuron–glia adhesion through independent molecular mechanisms, as evidenced by the observations that the two molecules do not interact with each other and that, in experiments in which neuron–glia adhesion is perturbed by antibodies, the inhibitory effects of a mixture of saturating doses of anti-Ng-CAM and anti-cytotactin antibodies are additive (Grumet et al., 1985). Data obtained from studies on the coordinate distribution and functions of cytotactin, Ng-CAM, and N-CAM in neural tissues are therefore likely to shed light on how these molecules and neuron–neuron and neuron–glia adhesion in general are differentially involved in development.

A particularly revealing example is the development of the cerebellar cortex. The characteristic formation of horizontal cellular and fibrous layers in the developing cerebellar and cerebral cortices depends on the radial migration of neurons along glial fibers (Rakic, 1971). To evaluate the contributions of cell adhesion molecules to this process, the distributions of N-CAM, Ng-CAM, and cytotactin in the cerebellar cortex at the time of cell migration were correlated with the ability of specific antibodies to these molecules to inhibit cell migration in the cerebellum (Hoffman et al., 1986; Chuong et al., 1987). Distribution studies indicated that while N-CAM is present throughout the cerebellum, Ng-CAM is absent from the proliferative zone, the outer half of the external granule cell layer where neurons undergo their final cell division prior to migration. When these neurons leave the cell cycle and enter the inner half of the external granule cell layer or premigratory zone, they begin to express high levels of Ng-CAM. At this time the neurons begin to extend processes that fasciculate, a process known to be mediated by Ng-CAM in other systems (Hoffman et al., 1986). Ng-CAM and cytotactin are both expressed at high levels in the fibrous molecular layer (Chuong et al., 1987).

Cerebellar cell migration can be studied *in vitro* in explant cultures in which neurons are labeled with tritiated thymidine during their final cell division. The progress of external granule cell neurons through the molecular layer and into the internal granule cell layer can be monitored by autoradiography. The function of cell adhesion molecules in this process can be evaluated by incubating the explants with specific antibodies (Hoffman et al., 1986; Chuong et al., 1987). When explants were incubated with anti-N-CAM antibodies, little or no inhibition of cell migration was observed. Anti-Ng-CAM antibodies strongly inhibited cell migration from the premigratory zone; in contrast, anti-cytotactin antibodies had no effect on the early phase of migration but markedly inhibited migration through the molecular layer (Figure 6).

Experiments in which anti-Ng-CAM and anti-cytotactin antibodies were added on different time schedules also supported the idea that the time of sensitivity to anti-Ng-CAM antibodies was earlier than the time of sensitivity to anti-cytotactin antibodies (Chuong et al., 1987). Moreover, in a manner similar to their effects on neuron–glia adhesion *in vitro*, antibodies to Ng-CAM and cytotactin in combination inhibited cell migration through the molecular layer to a greater extent than either antibody alone, further supporting the idea that cytotactin and Ng-CAM function as parts of independent molecular mechanisms. Thus, the main sites of appearance of two molecules involved in neuron–glia adhesion, Ng-CAM and cytotactin, are strongly correlated with the sites at which antibodies to these molecules inhibit cell migration (Chuong

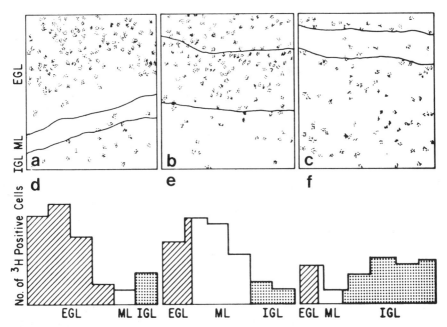

Figure 6. *Effects of anti-Ng-CAM and anti-cytotactin antibodies on the migration of triateted cerebellar granule cells. a–c:* Tracings of autoradiograms of paraffin sections of cerebellar explants that were pulse-labeled with tritiated thymidine for one hour and cultured for three days. Silver grains and the relative thickness of the cortical layers are highlighted. *d–f:* Histograms of the number of tritiated cells versus distance along layers. *a,d:* Anti-Ng-CAM. *b,e:* Anti-cytotactin. *c,f:* Nonimmune control. (Summarized from Chuong et al., 1987.)

et al., 1987). In contrast, although N-CAM, a molecule involved in neuron–neuron adhesion, is present throughout the cerebellum, anti-N-CAM antibodies have little or no effect on neuronal cell migration.

That these molecular mechanisms in cell and neurite migration may be more widespread in the nervous system is suggested by studies of the distribution of cytotactin, Ng-CAM, and N-CAM in the developing acoustic ganglion (Richardson et al., 1987). Specifically, Ng-CAM and cytotactin are expressed at sites where neurons invade the sensory epithelium. Although glial cells are not present in this system, cytotactin is present and probably synthesized by nonneural support cells. It thus appears that cytotactin and Ng-CAM may be functionally correlated with systems that support cell and neurite movement, independent of the cell types involved in this process.

The distributions of cytotactin, N-CAM, and Ng-CAM were also examined in the peripheral nervous system (Rieger et al., 1986). All three molecules were found to be localized in the region of the nodes of Ranvier, the periodic annular narrowing interrupting the myelin sheath in myelinated axons. Moreover, differences in the timing of their sequestration to the node suggested possible functional roles in the development and stabilization of the node. In embryonic sciatic nerve fibers prior to myelination, periodic coincident accumulations of

Ng-CAM and N-CAM, but not cytotactin, occurred. Cytotactin was found on Schwann cells at this time. Interestingly, in mutant mice with dysmyelinating disease, aberrant distributions of Ng-CAM, N-CAM, and cytotactin were observed. These results suggest that the proper structural and functional maturation of the node of Ranvier is dependent on the cell–cell interactions leading to the colocalization of all of these adhesive proteins at the site of this structure.

DISTRIBUTIONS OF CYTOTACTIN AND CTB PROTEOGLYCAN IN NONNEURAL TISSUES

Despite the fact that cytotactin and CTB proteoglycan can bind to each other, in certain tissues the distributions of the two molecules are similar, while in other tissues they are quite different (Hoffman et al., 1988). For example, both molecules are colocalized in the ganglionic, muscular, and tendinous portions of gizzard and in vascular smooth muscle (Figure 7). In contrast, while CTB proteoglycan is prominent throughout developing cartilage, cytotactin is restricted to immature chondrocytes (Figure 7). These data appear to reflect a differential site-related synthesis of these proteins, and they raise the possibility that additional ligands for CTB proteoglycan exist in regions where the proteoglycan is present and cytotactin is absent.

The developing heart is another particularly striking example of the differential expression of cytotactin and CTB proteoglycan. Early in cardiac development, the endothelial cells lining the lumen of the heart and the myocardium both express N-CAM. While the heart muscle continues to express N-CAM throughout development, some endothelial cells lose N-CAM and migrate into the endocardial cushion, originally an acellular region between the endothelium and the myocardium. Within the endocardial cushion is a region rich in CTB proteoglycan that remains relatively free of cells and a region rich in cytotactin in which invading cells accumulate. This endocardial cushion tissue then differentiates into tendonlike connective tissue and participates in the septation of the heart. Later in development CTB proteoglycan is prominent in cardiac muscle, while cytotactin is absent from cardiac muscle, although it is still present in valves that form from endocardial cushion tissue. Thus patterns of expression of N-CAM, cytotactin, and CTB proteoglycan are again correlated with patterns of morphogenesis.

Biochemical differences also exist in cytotactin and CTB proteoglycan from neural tissue or nonneural tissue. While both molecules are readily extracted from neural tissue in neutral, detergent-free buffers containing physiological salt concentrations, only CTB proteoglycan is extracted from nonneural tissues under these conditions (Hoffman et al., 1988). The extraction of cytotactin requires either high pH or strong denaturants such as 4 M guanidine. These results indicate that in nonneural tissues cytotactin is strongly associated with other components of the extracellular matrix that are insoluble under nondenaturing conditions.

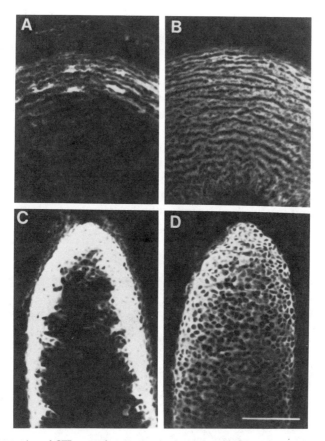

Figure 7. *Cytotactin and CTB proteoglycan expression in 14-day-old embryonic vascular smooth muscle (A,B) and sternal cartilage (C,D).* Sections were processed for immunohistology with chondroitin ABC–lyase treatment and stained with antibodies either to cytotactin *(A,C)* or to CTB proteoglycan *(B,D)*. Cytotactin staining *(A)* is found in moderate amounts in vascular smooth muscle of the great vessels of the heart and is most intense between the outermost layers of smooth muscle fibers. CTB proteoglycan *(B)* is present throughout the layers of muscle, although in a concentric, apparently cell-associated pattern. In cartilage, cytotactin staining is more extensive in extracellular spaces of immature chondrocytes *(C)*, whereas CTB proteoglycan is found surrounding all the cells and filling their extracellular spaces *(D)*. Calibration bar = 100 μm. (Summarized from Hoffman et al., 1988.)

CYTOTACTIN AND CTB PROTEOGLYCAN IN SEGMENTATION AND CELL MIGRATION IN EARLY EMBRYOGENESIS

Segmentation during vertebrate embryogenesis results from the formation of a series of somites in the mesodermal tissue along the neural tube (Bellairs et al., 1986). At first, cells known as the segmental plate condense to form the spherical, epithelial somites; each somite then differentiates into an epithelial portion, the dermomyotome, and a mesenchymal portion, the sclerotome, which gives rise to the axial skeleton. Each sclerotome is invaded by neural crest cells migrating in from the adjacent level of the neural tube. Within each sclerotome the neural crest cells aggregate to form the dorsal root ganglia.

The developmental distributions of cytotactin and CTB proteoglycan suggest that their functions may be intimately involved during certain steps of this segmentation process. In the early formation of the epithelial somites, cytotactin is present in the anteriormost somites, where development is most advanced, but not in the more posterior somites, nor in the more lateral portions of the mesoderm (Crossin et al., 1986). This distribution led us to postulate that cytotactin delineates early neural crest pathways. As somitic differentiation progresses, the distribution of cytotactin and CTB proteoglycan becomes more distinct (Tan et al., 1987). As the mesodermal sclerotome forms from the epithelial somite, the surrounding basement membrane is lost, allowing neural crest cells to migrate into the sclerotome. These neural crest

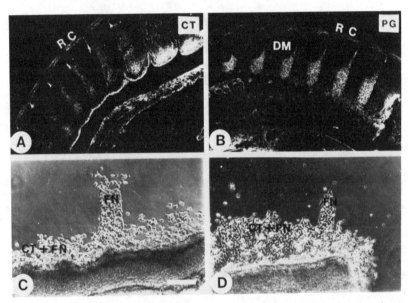

Figure 8. *Distribution* in vivo *and effects on cell migration* in vitro *of cytotactin and other extracellular matrix proteins.* Parasagittal sections through a 42-somite-stage embryo were immunofluorescently labeled with anti-cytotactin (A) or anti-CTB proteoglycan antibodies (B). Cytotactin is expressed in the rostral (R) half of each sclerotome, that is, the region into which neural crest cells migrate. Cytotactin is also seen in a uniform distribution along the dorsal, ventral, and intersomitic borders of the mature somites and surrounding the aorta. CTB proteoglycan is expressed in the caudal (C) half of each sclerotome and in a variety of other sites, but not within the dermomyotome (DM). Neural crest cell migration is seen on alternating tracks of fibronectin (FN) and various mixtures of cytotactin and fibronectin (CT + FN). Neural crest cells migrate well on fibronectin, but barely migrate on a substrate containing a 10:1 cytotactin-to-fibronectin ratio (C). Even on a substrate containing a 1:1 ratio (D), migration is significantly less than on fibronectin alone. (Summarized from Tan et al., 1987.)

cells preferentially enter the rostral half of each sclerotome; interestingly, the distribution of cytotactin is similarly polarized to the rostral half of each sclerotome (Figure 8A). Even in the least mature somites in which neural crest cells can be detected, the distribution of cytotactin is already polarized and is found in the rostral halves of the sclerotomes (Tan et al., 1987).

In contrast, while CTB proteoglycan is relatively uniformly distributed in immature sclerotomes, it later becomes polarized and located in the *caudal* halves after the time of neural crest invasion (Figure 8B). Thus, cytotactin and CTB proteoglycan are the first defined molecules involved in cell adhesion to be identified that have polarized distributions with the sclerotome. Furthermore, the idea that the rostrocaudal localizations of cytotactin and CTB proteoglycan within the sclerotome lead to, rather than result from, the rostral localization of neural crest cells is strongly supported by the observations that this molecular patterning occurs even in regions of embryos in which the neural tube and accompanying neural crest cells have been surgically removed (Tan et al., 1987).

To evaluate the mechanisms through which cytotactin and CTB proteoglycan may affect neural crest cell migration, these molecules were tested for their ability to support cell migration *in vitro* (Tan et al., 1987). When segments of neural tube were cultured on substrates coated with alternate regions of fibronectin and cytotactin or fibronectin and CTB proteoglycan, neural crest cells spread and migrated on fibronectin but avoided cytotactin and CTB proteoglycan. The few cells present on cytotactin or CTB proteoglycan were rounded up. To further investigate the effects of cytotactin and CTB proteoglycan on neural crest cell migration, the ability of these molecules to inhibit cell migration on fibronectin was examined (Figure 8C,D).

The migration on substrates coated with a mixture of fibronectin and cytotactin or fibronectin and CTB proteoglycan was strongly inhibited, compared to the migration on substrates coated only with fibronectin (Tan et al., 1987). This is of interest because fibronectin is present throughout the sclerotome *in vivo* and is normally a highly permissive substrate for neural crest cell migration (Rovasio et al., 1983). These observations raise the possibility that the colocalization of cytotactin and neural crest cells in the rostral half-somite *in vivo* is a consequence of the ability of cytotactin to inhibit the migration of cells on fibronectin.

Several interesting parallels between the migration of cells into the sclerotome and into endocardial cushion tissue in the heart are noteworthy. In both cases cells express N-CAM prior to migration, lose N-CAM, and migrate into structures containing regions rich in cytotactin and adjacent regions rich in CTB proteoglycan. The cells become localized in the regions where cytotactin is expressed and not in the regions rich in CTB proteoglycan. It will be interesting to determine whether these properties are generalized to other sites of cell migration, whether the details of this process differ between tissues, and what the effects of these processes on further cytodifferentiation are in each tissue.

PERSPECTIVES

The studies reviewed in this chapter indicate that cytotactin is present at a variety of sites where cell migration is an important step in pattern formation. Other molecules involved in cell adhesion, such as N-CAM, Ng-CAM, CTB proteoglycan, and fibronectin, are also present in these systems in distinct distributions. Thus, the major issue to be confronted in the future is: How do these coordinate patterns of adhesive proteins affect cell behavior leading to particular patterns of cell migration, cell proliferation, and cytodifferentiation? It has been proposed that the binding of ligands to cells globally alters the status of the cytoskeleton and thereby affects signaling to the nucleus (Edelman, 1976, 1988). This process is known as global cell surface modulation. Various lectins have been shown to affect global cell surface modulation, but these particular molecules are not present *in vivo*. It will be important to determine whether cytotactin and other adhesive proteins are natural ligands that cause global cell surface modulation. Already the evidence suggests an unusual relationship between cell binding to cytotactin and the status of the cytoskeleton. Cells spread on most adhesive substrates such as fibronectin and collagen; cells remain round when they bind to cytotactin (Friedlander et al., 1988). Thus the examination of the ability of cytotactin to act as a natural mediator of global modulation may reveal how it is involved in the transmembrane signaling that controls normal and aberrant development.

ACKNOWLEDGMENTS

This work was supported by United States Public Health Service grants HL-37641, DK-04256, HD-09635, and HD-16550, and Senator Jacob Javits Center for Excellence in Neuroscience grant NS-22789.

REFERENCES

Bellairs, R., D. A. Ede, and J. W. Lash (1986) *Somites in Developing Embryos,* Plenum, New York.

Chiquet, M., and D. M. Fambrough (1984) Chick myotendinous antigen. II. A novel extracellular glycoprotein complex consisting of large disulfide-linked subunits. *J. Cell Biol.* 98:1937–1946.

Chiquet-Ehrismann, R., E. J. Mackie, C. A. Person, and T. Sakakura (1986) Tenascin: An extracellular matrix protein involved in tissue interactions during fetal development and oncogenesis. *Cell* 47:131–139.

Chuong, C.-M., and G. M. Edelman (1985a) Expression of cell adhesion molecules in embryonic induction. I. Morphogenesis of nestling feathers. *J. Cell Biol.* 101:1009–1026.

Chuong, C.-M., and G. M. Edelman (1985b) Expression of cell adhesion molecules in embryonic induction. II. Morphogenesis of adult feathers. *J. Cell Biol.* 101:1027–1043.

Chuong, C.-M., K. L. Crossin, and G. M. Edelman (1987) Sequential expression and differential functions of multiple adhesion molecules during the formation of cerebellar cortical layers. *J. Cell Biol.* 104:331–342.

Crossin, K. L., C.-M. Chuong, and G. M. Edelman (1985) Expression sequences of cell adhesion molecules. *Proc. Natl. Acad. Sci. USA* **82**:6942–6946.

Crossin, K. L., S. Hoffman, M. Grumet, J. P. Thiery, and G. M. Edelman (1986) Site-restricted expression of cytotactin during development of the chicken embryo. *J. Cell Biol.* **102**:1917–1930.

Daniloff, J. K., G. Levi, M. Grumet, F. Rieger, and G. M. Edelman (1986) Differential distribution of cell adhesion molecules during histogenesis of the chicken nervous system. *J. Neurosci.* **6**:739–758.

Drager, U. C., D. L. Edwards, and C. J. Barnstable (1984) Antibodies against filamentous components in discrete cell types of the mouse retina. *J. Neurosci.* **4**:2025–2042.

Edelman, G. M. (1976) Surface modulations in cell recognition and cell growth. *Science* **192**:218–226.

Edelman, G. M. (1986) Cell adhesion molecules in the regulation of animal form and tissue pattern. *Annu. Rev. Cell Biol.* **2**:81–116.

Edelman, G. M. (1988) *Topobiology: An Introduction to Molecular Embryology,* Basic Books, New York.

Edelman, G. M., W. J. Gallin, A. Delouvée, B. A. Cunningham, and J. P. Thiery (1983) Early epochal maps of two different cell adhesion molecules. *Proc. Natl. Acad. Sci. USA* **80**:4384–4388.

Erickson, H. P., and J. L. Iglesias (1984) A six-armed oligomer isolated from cell surface fibronectin preparations. *Nature* **311**:267–269.

Erickson, H. P., and H. C. Taylor (1987) Hexabrachion proteins in embryonic chicken tissues and human tumors. *J. Cell Biol.* **105**:1387–1394.

Fowler, W. E., and H. P. Erickson (1979) Trinodular structure of fibrinogen: Confirmation by both shadowing and negative stain electron microscopy. *J. Mol. Biol.* **134**:241–249.

Friedlander, D. R., S. Hoffman, and G. M. Edelman (1988) Functional mapping of cytotactin: Proteolytic fragments active in cell–substrate adhesion. *J. Cell Biol.* **107**:2329–2340.

Grumet, M., and G. M. Edelman (1984) Heterotypic binding between neuronal membrane vesicles and glial cells is mediated by a specific neuron–glia cell adhesion molecule. *J. Cell Biol.* **98**:1746–1756.

Grumet, M., and G. M. Edelman (1988) Neuron–glia cell adhesion molecule interacts with neurons and astroglia via different binding mechanisms. *J. Cell Biol.* **106**:487–503.

Grumet, M., S. Hoffman, and G. M. Edelman (1984a) Two antigenically related neuronal CAMs of different specificities mediate neuron–neuron and neuron–glia adhesion. *Proc. Natl. Acad. Sci. USA* **81**:267–271.

Grumet, M., S. Hoffman, C.-M. Chuong, and G. M. Edelman (1984b) Polypeptide components and binding functions of neuron–glia cell adhesion molecules. *Proc. Natl. Acad. Sci. USA* **81**:7989–7993.

Grumet, M., S. Hoffman, K. L. Crossin, and G. M. Edelman (1985) Cytotactin, an extracellular matrix protein of neural and nonneural tissues that mediates glia–neuron interaction. *Proc. Natl. Acad. Sci. USA* **82**:8075–8079.

Hoffman, S., and G. M. Edelman (1983) Kinetics of homophilic binding by E and A forms of the neural cell adhesion molecule. *Proc. Natl. Acad. Sci. USA* **80**:5762–5766.

Hoffman, S., and G. M. Edelman (1987) A proteoglycan with HNK-1 antigenic determinants is a neuron-associated ligand for cytotactin. *Proc. Natl. Acad. Sci. USA* **84**:2523–2527.

Hoffman, S., D. R. Friedlander, C.-M. Chuong, M. Grumet, and G. M. Edelman (1986) Differential contributions of Ng-CAM and N-CAM to cell adhesion in different neural regions. *J. Cell Biol.* **103**:145–158.

Hoffman, S., K. L. Crossin, and G. M. Edelman (1988) Molecular forms, binding function, and developmental expression patterns of cytotactin and CTB proteoglycan, an interactive pair of extracellular matrix molecules. *J. Cell Biol.* **106**:519–532.

Jones, F. S., M. P. Burgoon, S. Hoffman, K. L. Crossin, B. A. Cunningham, and G. M. Edelman (1988) A cDNA clone for cytotactin contains sequences apparently homologous to EGF repeats and segments of fibronectin and fibrinogen. *Proc. Natl. Acad. Sci. USA* **85**:2186–2190.

Jones, F. S., S. Hoffman, B. A. Cunningham, and G. M. Edelman (1989) A detailed structural model of cytotactin: Protein homologies, alternative RNA splicing, and binding regions. *Proc. Natl. Acad. Sci. USA* **86**:1905–1909.

Kruse, J., G. Keilhauer, A. Faissner, R. Gimple, and M. Schachner (1985) The J_1 glycoprotein: A novel nervous system cell adhesion molecule of the L2/HNK-1 family. *Nature* **316**:146–148.

Needham, J. (1933) On the dissociability of the fundamental process in ontogenesis. *Biol. Rev. Cambridge Philos. Soc.* **8**:180–223.

Petersen, T. E., H. C. Thøgersen, K. Skorstengaard, K. Vibe-Pedersen, P. Sahl, L. Sottrup-Jensen, and S. Magnusson (1983) Partial primary structure of bovine plasma fibronectin, three types of internal homology. *Proc. Natl. Acad. Sci. USA* **80**:137–141.

Rakic, P. (1971) Neuron–glia relationship during granule cell migration in developing cerebellar cortex. A Golgi and electron microscopic study in *Macacus rhesus. J. Comp. Neurol.* **141**:283–312.

Richardson, G., K. L. Crossin, C.-M. Chuong, and G. M. Edelman (1987) Expression of cell adhesion molecules during embryonic induction. III. Development of the otic placode. *Dev. Biol.* **119**:217–230.

Rieger, F., J. K. Daniloff, M. Pinçon-Raymond, K. L. Crossin, M. Grumet, and G. M. Edelman (1986) Neuronal cell adhesion molecules and cytotactin are colocalized at the node of Ranvier. *J. Cell Biol.* **103**:379–391.

Rovasio, R. A., A. Delouvée, K. M. Yamada, R. Timpl, and J. P. Thiery (1983) Neural crest cell migration: Requirements for exogenous fibronectin and high cell density. *J. Cell Biol.* **96**:462–473.

Ruoslahti, E., and M. D. Pierschbacher (1986) Arg-Gly-Asp: A versatile cell recognition signal. *Cell* **44**:517–518.

Tan, S. S., K. L. Crossin, S. Hoffman, and G. M. Edelman (1987) Asymmetric expression in somites of cytotactin and its proteoglycan ligand is correlated with neural crest cell distribution. *Proc. Natl. Acad. Sci. USA* **84**:7977–7981.

Thiery, J. P., J.-L. Duband, U. Rutishauser, and G. M. Edelman (1982) Cell adhesion molecules in early chick embryogenesis. *Proc. Natl. Acad. Sci. USA* **79**:6737–6741.

Thiery, J. P., A. Delouvée, M. Grumet, and G. M. Edelman (1985) Initial appearance and regional distribution of the neuron–glia cell adhesion molecule (Ng-CAM) in the chick embryo. *J. Cell Biol.* **100**:442–456.

Vaughan, L., S. Huber, M. Chiquet, and K. H. Winerhalter (1987) A major, six-armed glycoprotein from embryonic cartilage. *EMBO J.* **6**:349–353.

Yamada, K. M. (1983) Cell surface interactions with extracellular materials. *Annu. Rev. Biochem.* **52**:761–799.

Section 3

Components of Junctional Complexes

Chapter 13

Subplasmalemmal Plaques of Intercellular Junctions: Common and Distinguishing Proteins

HANS-PETER KAPPRELL
RAINER DUDEN
KATSUSHI OWARIBE
MONIKA SCHMELZ
WERNER W. FRANKE

ABSTRACT

The two major groups of intercellular junctions of the adhering type, the desmosomes and the various forms of intermediate junctions (zonulae adherentes, fasciae adherentes, and puncta adherentia), are characterized by a pair of cytoplasmic plaques of weblike material that provide anchorage plates of cytoskeletal filaments. Compositionally, the desmosomal and nondesmosomal plaques, in which only one common protein (plakoglobin) has been identified, are different. The desmosomal plaque integrates the cytoplasmic portion(s) of at least one transmembrane glycoprotein, desmoglein, and certain entirely cytoplasmic proteins that specifically assemble at this membrane domain, including desmoplakin I. In certain stratified and complex glandular epithelia, desmoplakin II and the basic "band 6 polypeptide" have also been found in the plaque. In the plaques of the intermediate junctions, in which common transmembrane proteins have not yet been identified, plakoglobin colocalizes with vinculin and α-actinin. However, the plaques of different forms of the intermediate junctions are not identical, as demonstrated by an antigen (ZA-1TJ) that is present only in the zonulae adherentes of the polar epithelia. Although plaque proteins are resistant to extractions with various buffers, soluble forms of vinculin, plakoglobin, and desmoplakin have also been identified. It is proposed that a dynamically regulated equilibrium of unassembled and plaque-bound forms of these molecules is involved in the regulation of junction formation, and hence cell architecture. Information on the unassembled state of these components is therefore especially important to our understanding of the regulatory mechanisms for plaque assembly and disassembly and in the topogenic specificity of the plaque proteins.

The interaction and adhesion of cells in tissues involve special, mostly symmetrical domains of the plasma membrane, the intercellular adhering junctions (Farquhar and Palade, 1963; Cowin et al., 1986), which serve dual functions. The extracellular surface of these membrane regions establishes physical contact between neighboring cells, whereas the cytoplasmic side is associated with dense, matlike elements, termed plaques, which provide specific anchorage sites for two kinds of cytoskeletal filaments, the actin microfilaments (MFs) and the intermediate-sized filaments (IFs). In this way the adhering junctions represent major architectural elements of both inter-cellular and intracellular organization, which contribute to the functional coherence and alignment of the cells within given tissues as well as to the positioning and the functional arrangement of the organelles and other structural components within the cells they connect.

In principle, two morphological forms of adhering junction plaques can be distinguished: (1) In tissues in which extended regions of the surface of a cell are closely apposed to the surfaces of adjacent cells, leaving only a narrow interspace, most of the cell surface can be regarded as equivalent to a junctional structure, and hence the entire subplasmalemmal filamentous web ("membrane skeleton"; Marchesi, 1985) appears to be equivalent to a plaque. This view has recently received support by immunolocalization data in ocular lens tissue, the prototype tissue of this kind of tight cell–cell interaction (see Franke et al., 1987b, and below). (2) In most tissues the sites of close attachment of neighboring cells are limited in size and form to individual, distinct junctions that fall into two major groups, the desmosomes (maculae adherentes) associ-ating with IFs and the intermediate junctions (zonulae adherentes, fasciae ad-herentes, puncta adherentia) associating with MFs (Farquhar and Palade, 1963; Staehelin, 1974; Drochmans et al., 1978; Fawcett, 1981; Cowin et al., 1985a,b, 1986; Geiger et al., 1985a,b).

PLAKOGLOBIN—A PLAQUE PROTEIN COMMON TO ADHERING JUNCTIONS OF THE DESMOSOMAL AND INTERMEDIATE TYPE

All the various forms of plaque-bearing intercellular junctions so far tested have at least one major plaque protein in common, plakoglobin (polypeptide ~83 kD; isoelectric pH in native state ~5.3, after denaturation in urea ~6.3). In fractions of isolated bovine snout desmosomes this polypeptide has previously been referred to as "band 5 polypeptide". (For references, see Cowin et al., 1985a,b, 1986; Franke et al., 1987b,c; see also Gorbsky et al., 1985.) Therefore, this protein, which is encoded by a special mRNA of ~3.5 kb (Franke et al., 1983c; Cowin et al., 1986), may serve as the hallmark constituent protein of theadhering junction. While in the desmosomal plaque plakoglobin coexists a number of desmosome-specific proteins (see below), it colocalizes in the plaques of the intermediate junctions with a different set of proteins, including vinculin and α-actinin. Plakoglobin also occurs in the subplasmalemmal coat of the extended junction–equivalent plasma membrane regions of the closely apposed cells of ocular lens tissue, where it is accompanied not only by vinculin, α-actinin, and actin but also by proteins of the spectrin and band 4.1 protein families (Franke et al., 1987b).

COMPOSITIONAL DIFFERENCES OF PLAQUE-BEARING JUNCTIONS

Despite the gross morphological similarities of the adhering junctions and the presence of plakoglobin, the adhering junctions can be readily subdivided into the two types mentioned above.

The intermediate junctions (zonula adherens, fascia adherens, and punctum adherens) usually reveal a relatively electron-transparent layer of about 20 nm in thickness between the membrane bilayers of either cell and a pair of loosely woven plaques associated with vinculin, α-actinin, and, in most cases, bundles of MFs (Geiger et al., 1983, 1985a). Common transmembrane proteins of this type of junction have not yet been identified, but some glycoproteins such as uvomorulin (Damsky et al., 1983; Ogou et al., 1983; Peyriéras et al., 1983; Boller et al., 1985; Gumbiner and Simons, 1986; see these for further references and synonyms for this protein) and a 135-kD glycoprotein (Volk and Geiger, 1984, 1986a,b; Volk et al., 1987) have been found to occur in certain subtypes of intermediate junctions. Other plaque components that allow a further subdivision of these junctions will be discussed below.

The other type of adhering junction, the desmosome, is characterized by an intercellular layer of about 30 nm, which frequently reveals an electron-dense "midline" structure, and a pair of usually well-profiled, electron-dense plaques that anchor the IFs. At present, the composition of the desmosomal plaque is the best studied and, therefore, is discussed in some detail.

TRANSMEMBRANE PROTEINS CONTRIBUTING TO THE DESMOSOMAL PLAQUE

All the diverse morphological variants of desmosomes contain a moderately glycosylated protein that on SDS-polyacrylamide gel electrophoresis (SDS-PAGE) appears as a glycopolypeptide of ~165 kD. The carbohydrate-containing portion of this glycopeptide seems to be oriented to—and to contribute to—the intercellular "cement" of the desmosome, including the midline region; therefore, it has been termed desmoglein (the "band 3 glyco-polypeptide" of desmosomal fractions from bovine nasal epidermis; see Skerrow and Matoltsy, 1974a,b; Franke et al., 1981c; Gorbsky and Steinberg, 1981; Cohen et al., 1983; Cowin and Garrod, 1983; Giudice et al., 1984; Kapprell et al., 1985; Schmelz et al., 1986a,b; Steinberg et al., 1987). The carbo-hydrate composition of this component has been determined for bovine epidermal desmosomes (Kapprell et al., 1985; the significance of the sugar content was overlooked in our earlier analyses; cf. Franke et al., 1981c), and Penn et al. (1987) have recently shown that its glycosylation is sensitive to tunicamycin treatment, suggesting an asparagine linkage.

In previous studies this glycoprotein was reported to resolve into a series of finer bands or to appear as a broad, indistinct band, which suggested molecular heterogeneity, whether natural or a result of a partial breakdown during preparation (Franke et al., 1981c; Cohen et al., 1983; Cowin and Garrod, 1983; Giudice et al., 1984; see also Penn et al., 1987). In addition, certain differences of gel electrophoretic mobility of desmoglein from various tissues and cell types have also been described (Giudice et al., 1984; Suhrbier and Garrod, 1986). With the standard gel electrophoretic systems, desmoglein in desmosomes of different sources showed very similar electrophoretic mobility (Schmelz et al., 1986a,b); however, using a gel system with better resolution (Hubbard and Lazarides, 1979), we have recently detected certain slight but consistent differences in relative molecular weight between desmogleins from different tissues of the same species (Figure 1a). In this gel system, desmoglein from desmosomes of several bovine cultured cell lines such as BMGE and MDBK cells or from tissues containing single-layered epithelia, such as liver, migrated significantly faster (by ~15 kD) than desmoglein from tongue mucosa and muzzle epidermis. Desmoglein from the latter tissues in turn appeared to migrate somewhat faster than esophageal desmoglein (Figure 1a).

These differences in molecular weight, however, were in the opposite direction from those reported by other authors (Giudice et al., 1984; Suhrbier and Garrod, 1986). When desmoglein from different cell lines and species was examined after treatment of the cells with tunicamycin, a molecular weight

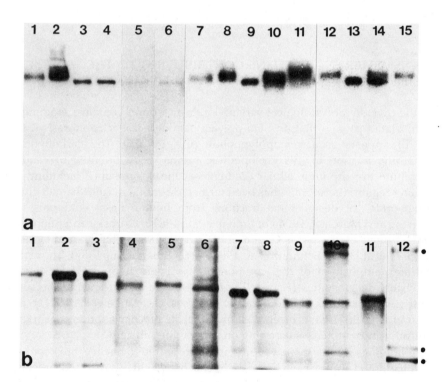

value lowered by approximately 12–14 kD was found in all cases (Figure 1b), which is in reasonable agreement with recent observations of cultured canine (MDCK) and bovine (MDBK) kidney epithelial cells (Penn et al., 1987). These results suggest that the extent of desmoglein glycosylation is comparable in different cell types, and the differences in molecular weight observed probably reflect not different degrees of asparagine glycosylation but rather the synthesis of slightly different desmoglein polypeptides, although O-glycosylations and other modifications cannot be rigorously excluded at present. This interpretation was supported by one- and two-dimensional gel electrophoretic analyses of translation products of epidermal mRNA *in vitro* (Figure 2a,b, lanes 3' and 4'; Figure 3a,b). Moreover, desmoglein immunoprecipitated from tunicamycin-treated cultured cells (Figure 3c,d) showed a reduced molecular

---

Figure 1. *Identification of desmoglein in various tissues and cell lines. a:* Autoradiographs showing the immunoblot reaction of polypeptides of different cytoskeletal fractions after separation by SDS-polyacrylamide gel electrophoresis (SDS-PAGE, using the system of Hubbard and Lazarides, 1979). The transfer to nitrocellulose paper and reaction with monoclonal antibody DG3.10 to desmoglein (band 3 polypeptide) was as described in Schmelz et al. (1986a). Only bands positively reacting with the antibody are seen. Proteins of desmosomal fractions obtained according to Gorbsky and Steinberg (1981) (*lanes 1, 7, 12, 15*) and total snout epidermal tissue proteins directly solubilized in sample buffer (*lanes 2, 8*) are compared with cytoskeletal proteins (for cell lines and methods, see Franke et al., 1981b,c; Schmid et al., 1983b; Cowin et al., 1985b; Achtstaetter et al., 1986) of various bovine cells and tissues such as cultured mammary gland epithelial cells of line BMGE-H (*lanes 3, 4, 9, 13*), kidney epithelial cells (MDBK, *lane 5*), liver tissue (*lane 6*), tongue mucosa (*lane 10*), and esophagus epithelium (*lane 11*). To examine the significance of the slight differences in electrophoretic mobility between desmoglein from different sources, a mixture of proteins from epidermal desmosomes (as in *lanes 1, 7, 12, 15*) and cytoskeletal proteins of BMGE-H cells (as in *lanes 3, 4, 9, 13*) has been loaded in *lane 14*. Note the slightly faster migration of desmoglein from cultured cells (MDBK, BMGE-H) and liver, compared to desmoglein from stratified tissues (snout epidermis, tongue, and esophagus). Note also that in this gel system, tongue desmoglein seems to migrate somewhat faster than esophageal desmoglein. *b:* Autoradiofluorograph showing immuno-precipitates of desmoglein from different ^{35}S-methionine-labeled cultured cells of canine, bovine, and human origin, without (*lanes 1–5*) and with (*lanes 6–10*) treatment of the cells with tunicamycin (0.1 μg/ml) for nine hours. Proteins immunoprecipitated with antibodies to desmoglein of canine kidney epithelial (MDCK) cell line (*lanes 1, 6*), human epidermoid vulvar carcinoma cells of line A-431 (*lanes 2, 7*), human oral squamous cell carcinoma cells of line TR-146 (*lanes 3, 8*), and bovine cells of lines BMGE-H (*lanes 4, 9*) and MDBK (*lanes 5, 10*) are compared. Cell lysates were prepared with detergents and desmoglein immunoprecipitated with murine monoclonal antibody DG3.10 (*lanes 2–5, 7–10*) or guinea pig antibodies (*lanes 1, 6*) to desmoglein, following the principles described by Blose and Meltzer (1981; for details, see Duden, 1987). Immunoprecipitated proteins were separated by SDS-PAGE (10% acrylamide) according to Hubbard and Lazarides (1979), and labeled proteins were visualized by fluorography. For comparison, native, largely unmodified desmoglein obtained after immuno-precipitation of proteins labeled with ^{35}S-methionine during *in vitro* translation from bovine snout epidermal RNA (cf. Kreis et al., 1983; Magin et al., 1983; Cowin et al., 1986) is shown in *lane 11. Lane 12* contains ^{14}C-labeled reference proteins (*dots denote, from top to bottom,* polypeptide markers of ~220 kD, 100 kD, and 92 kD).

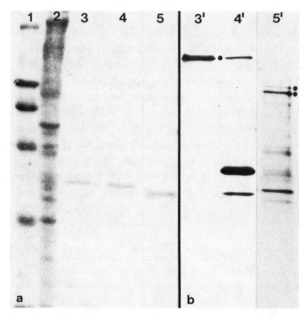

Figure 2. *Identification of* [35]*S-methionine-labeled* in vitro *translation products of mRNAs encoding desmoglein and desmocollins (bands 4a and 4b) by immunoprecipitation, followed by SDS-PAGE and fluorography. a:* Coomassie blue staining of proteins separated by SDS-PAGE. *Lane 1,* reference proteins (from *top* to *bottom,* myosin heavy chain, 220 kD; β-galactosidase, 116 kD; phosphorylase A, 92 kD; bovine serum albumin, 68 kD; α-actin, 43 kD); *lane 2,* crude desmosomal fraction from bovine snout epidermis prepared with the citric acid procedure (Gorbsky and Steinberg, 1981); *lane 3,* immunoprecipitate from the reticulocyte lysate of the *in vitro* translation assay (see Figure 1b) obtained after the addition of monoclonal antibody DG3.10 (for details, see Duden, 1987; only the immunoglobulin heavy chain is seen here); *lane 4,* as in *lane 3,* but here tongue mucosa RNA was used; *lane 5,* as in *lane 3,* but the immunoprecipitation was performed with monoclonal antibody DC4a/b-29.2 against desmocollins (Duden, 1987). *b:* Autoradiofluorograph (*lanes 3'–5'*), corresponding to *lanes 3–5* of *a.* Note that the *in vitro* translation products of desmoglein from bovine snout (*lane 3'*) and tongue (*lane 4'*) are indistinguishable from each other in SDS-PAGE mobility. The *dots* in *lane 5'* denote two *in vitro* translation products of 110 kD and 105 kD, respectively, which probably correspond to the two desmocollins I and II. The lower-molecular-weight bands represent breakdown products or coprecipitated polypeptides.

weight value but a very similar electrical charge (Figure 3a,b) compared with the mature glycosylated polypeptide, indicating that the glycosylation does not involve the addition of considerable numbers of acidic sugar residues (see also Kapprell et al., 1985). Clearly, further comparisons of *in vitro* translation products as well as the identification of mRNAs and amino acid sequence determinations are required to decide definitively whether the desmoglein molecules found in different cell types represent different members of a multigene family or different types of posttranslational modification.

It is reasonable to assume that the asparagine-linked carbohydrate moiety of desmoglein is located on the outer side of the plaque membrane, a view also supported by cytochemical and antibody staining data (Gorbsky and Steinberg, 1981; Shida et al., 1984; Steinberg et al., 1987). However, at least a part of the desmoglein molecule appears to be located on the cytoplasmic side and to contribute to the plaque, as suggested by electron microscope immunolabeling data (Schmelz et al., 1986b; Franke et al., 1987a; Miller et al., 1987; Steinberg et al., 1987). As indicated by the binding of certain desmoglein antibodies microinjected into living cells, a portion of desmoglein extends through the plaque and is exposed on the plaque's cytoplasmic surface (Schmelz et al., 1986b). Therefore, we have concluded that desmoglein is a transmembrane protein with a glycosylated outer domain and an extended, cytoplasmically projecting domain that is an integral part of the plaque structure (Schmelz et al., 1986a,b; see also Miller et al., 1987; Steinberg et al., 1987).

Desmocollins I and II (also described as "bands 4a and 4b" of isolated bovine snout epidermal desmosomes; Franke et al., 1981c) are two similar glycoproteins of ~130 kD and ~115 kD (Skerrow and Matoltsy, 1974b; Gorbsky and Steinberg, 1981; Cowin and Garrod, 1983; Mueller and Franke, 1983; Kapprell et al., 1985; Suhrbier and Garrod, 1986). *In vitro* translation of mRNA from bovine snout epidermis, followed by immunoprecipitation with a monoclonal antibody (DC4a/b-29.2) against desmocollins (Figure 2b, lane 3'; Figure 4a,b), has revealed two distinct polypeptides that probably correspond to the nonglycosylated forms of desmocollins I and II. As expected, these translational products appear to be somewhat smaller than the mature glycosylated molecules, as suggested by their higher electrophoretic mobilities, corresponding to molecular weight values of 100 kD and 110 kD. Remarkably, both translational products are somewhat less acidic than the mature gly-cosylated forms, which suggests the presence of a considerable proportion of negatively charged sugar residues (see also Kapprell et al., 1985).

Although the presence of desmocollins or immunologically related proteins has been reported for a variety of tissues and cell culture lines (e.g., Cowin et al., 1984a,b; Giudice et al., 1984; Suhrbier and Garrod, 1986; Penn et al., 1987), it is not yet clear whether these proteins are general and obligatory constituents of desmosomes, showing the same distribution as desmoglein and desmoplakin (see below). Moreover, the molecular topology of desmocollins is not clear. Most of the antibodies described seem to recognize epitopes located in the "desmoglea" (Gorbsky and Steinberg, 1981), that is, the intercellular cement of the desmosome (Cowin et al., 1984a,b; Miller et al., 1987; Steinberg et al., 1987), but evidence of a cytoplasmic domain of desmocollins and hence a transmembranous character is still lacking.

CYTOPLASMIC PROTEINS SPECIFIC TO THE DESMOSOMAL PLAQUE

Desmoplakin, a nonglycosylated polypeptide of ~250 kD, has been identified as a major component of desmosomal plaques of a variety of tissues, including those of epithelia, myocardium, arachnoidal cells of meninges, and dendritic

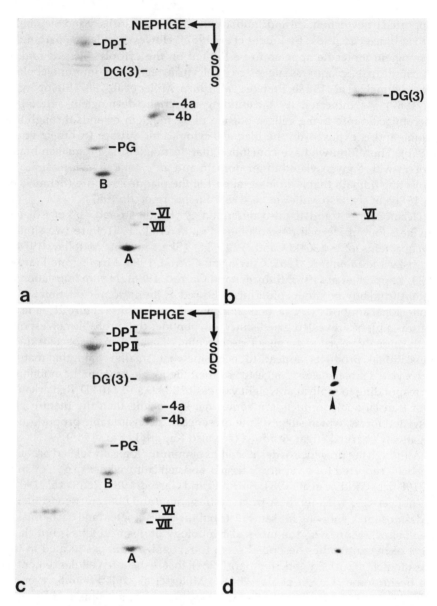

Figure 3. *Two-dimensional gel electrophoresis of* in vitro *translation product of desmoglein from bovine snout epidermal RNA (a,b) and of a mixture containing tunicamycin-arrested and native forms of desmoglein immunoprecipitated from the human A-431 cells.* Nonequilibrium pH gradient electrophoresis (NEPHGE) was used in the first, and SDS-PAGE (7.5% acrylamide) in the second dimension (basic polypeptides to the *left*). Markers used for coelectrophoresis are rabbit α-actin (A) and bovine serum albumin (B). *a*: Coomassie blue staining of polypeptides from a mixture of desmosomal proteins (as in Figure 2) and the immunoprecipitate from the reticulocyte lysate of the *in vitro* translation assay (as in Figures 1, 2) obtained after the addition of monoclonal antibody DG3.10 to desmoglein. Major components are denoted: desmoplakins I and II (DPI, DPII), desmoglein (DG3), desmocollins (4a, 4b), plakoglobin (PG). The positions

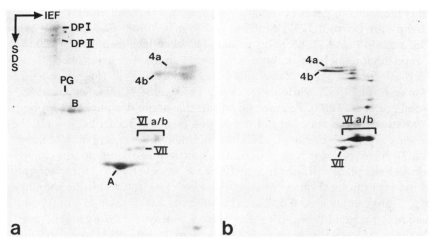

Figure 4. *Analysis of* in vitro *translation products of desmocollins from bovine snout epidermal RNA by two-dimensional gel electrophoresis and immunoprecipitation.* Proteins were analyzed by two-dimensional gel electrophoresis, using isoelectric focusing in the first dimension (basic polypeptides to the *left*) and SDS-PAGE in the second dimension (reference polypeptides as in Figure 3). *a*: Coomassie blue staining of a mixture of desmosomal proteins (as in Figure 3a) and the *in vitro* translation products from total bovine muzzle epidermal RNA (see Figures 2, 3), obtained after immunoprecipitation with monoclonal antibody DC4a/b-29.2 to desmocollins I and II. (Symbols as in Figure 3.) *b*: Autoradiofluorogrpah corresponding to α, showing that the *in vitro* translation products of desmocollin mRNAs appear at a more basic position (∼pH 5.3) than the mature glycosylated desmocollins (pH 4.8). The appearance of two distinct *in vitro* translation products of desmocollins suggests that desmocollins I and II are products of distinct mRNAs.

of some residual acidic (type I) cytokeratins are marked VI and VII (cf. Kreis et al., 1983). *b*: Autoradiofluorograph corresponding to *a*, showing the *in vitro* translation product of desmoglein mRNA. Note its higher electrophoretic mobility in the second-dimension SDS-PAGE, but its similar position in the pH gradient, indicating an electrical charge similar to that of the mature glycoprotein. *c*: Coomassie blue staining of polypeptides of the desmosomal fraction from bovine muzzle epidermis (as in *a*), mixed and coelectrophoresed with two samples of cytoskeletal proteins from ^{35}S-methionine-labeled A-431 cells (nearly equal amounts of protein from cells treated with tunicamycin and untreated cells) obtained after solubilization and immunoprecipitation with monoclonal antibody DG3.10 (symbols as in *a*). *d*: Autoradiofluorograph of *c*, presenting a direct comparison of the normal, that is, glycosylated form (denoted by the *upper arrowhead*) and the tunicamycin-arrested form (lower arrowhead) of desmoglein.

reticulum cells of germ centers of lymph nodes (Franke et al., 1982, 1983b; Cowin and Garrod, 1983; Mueller and Franke, 1983; Cowin et al., 1984a, 1985a,b; Moll et al., 1986). This protein has been localized exclusively to the plaque region in both sections and permeabilized cells as well as upon microinjection of desmoplakin antibodies into living cells (for references, see Franke et al., 1982; Kartenbeck et al., 1983, 1984; Cowin et al., 1985a,b; Steinberg et al., 1987). From observations that some desmoplakin antibodies decorate only the periphery of the plaque (see also Kartenbeck et al., 1984; Cowin et al., 1985a,b; Jones and Goldman, 1985; Miller et al., 1987), Miller et al. (1987) have proposed that desmoplakins are restricted to the interface between the plaque proper and the IFs. This, however, is in contrast to reports of the immunolocalization of desmoplakins close to the plaque membrane (e.g., Franke et al., 1983a) and the retention of desmoplakin in a thin residual plaque at the membrane during progressive removal of IF proteins and most of the plaque material by denaturing agents (Franke et al., 1983a; see also Figure 5 of Steinberg et al., 1987).

While desmoplakin I is clearly a constitutive component of the desmosomal plaque, desmoplakin II ~215 kD), a very closely related polypeptide, has been unequivocally detected only in various stratified epithelia (Cowin et al., 1985a,b). It is still not clear whether this component is a genuine polypeptide or a proteolytically trimmed derivative of desmoplakin I (for discussion, see Cowin et al., 1985a,b). In particular, it remains to be seen whether small amounts of protein of cytoskeletal fractions from various other tissues that migrate, on SDS-PAGE, in a position similar to that of authentic desmoplakin II and show immunoreactivity with antibodies recognizing both desmoplakins really represent distinct desmoplakin II molecules.

Another cytoplasmic desmosomal plaque protein recently localized to the desmosomes of some but not all cells is a polypeptide of ~75 kD that is the only major positively charged component identified on gel electrophoresis under denaturing conditions (isoelectric pH ~8). It has also been identified as a genuine product of translation *in vitro* ("band 6 protein" of SDS-PAGE patterns of desmosomal fractions from bovine snout; Franke et al., 1981c, 1983c; Mueller and Franke, 1983). Antibodies specific for band 6 protein (Figure 5) react with a protein present in all stratified and complex epithelia tested such as epidermis, lingual and esophageal mucosa, cornea, exocervix, vagina, trachea, and bladder urothelium, and in certain cultured cells derived from tumors of stratified squamous epithelia such as human A-431 and TR-146 cells (Figures 6–8). In immunofluorescence microscopy these antibodies localize to punctate arrays, corresponding to individual desmosomes. Furthermore, immuno-electron microscopy has specifically localized band 6 protein to the plaque domain of desmosomes (Figure 9). The protein has not been detected, however, in desmosomes of diverse simple epithelia, myocardium, meninges, and lymph nodes, or in various cell culture lines derived from simple epithelial cells (for details, see Kapprell et al., 1988).

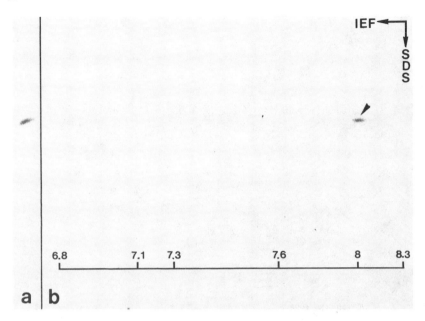

Figure 5. *Autoradiograph of immunoblots showing the specificity of affinity-purified guinea pig antibodies to desmosomal band 6 polypeptide. a:* SDS-PAGE of proteins of desmosomal fraction from bovine snout epidermis (as in Figures 1–4) after immunoblot with antibodies to band 6 protein (for details, see Kapprell, 1987). This sample has been electrophoresed in a special marginal lane in the same gel as the second-dimension electrophoresis shown in *b. b:* Two-dimensional gel electrophoresis of total desmosomal proteins (as in *a*) that had been allowed to renature (IEF system modified for optimal separation in the range pH 6–8.3; for details, see Kapprell, 1987). Determined pH values are given at the *bottom margin;* SDS-PAGE, as in Figure 4. *Arrowhead* denotes antibody binding to band 6 protein.

Other nonmembranous components of the desmosomal plaque are less well characterized, such as the widely distributed antigen D1 (polypeptide of >200 kD; Franke et al., 1981c), or have been shown only in desmosomes of certain tissues such as the calmodulin-binding protein desmocalmin (~240 kD; Tsukita and Tsukita, 1985) in bovine snout epidermis.

CONSTITUENTS OF THE PLAQUES OF NONDESMOSOMAL JUNCTIONS

The composition of the plaques of the adhering junctions of the intermediate type, notably zonulae and fasciae adherentes, is less well studied. It is clear that, besides plakoglobin (see above), vinculin and α-actinin are associated with the plaques of this category and that the desmosome-specific plaque proteins described above are absent (Geiger et al., 1981, 1983, 1985a,b). It is not clear, however, whether and to what extent any of these proteins contributes to the

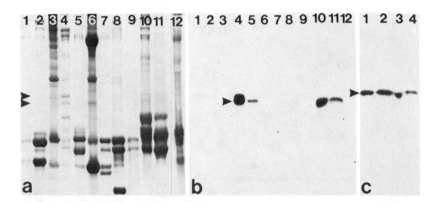

Figure 6. *Identification of band 6 polypeptide in various tissues and cell lines. a*: Coomassie blue staining of SDS-PAGE-separated polypeptides of isolated desmosomes from bovine snout epidermis (*lane 4*) (cf. Figures 1–4) and cytoskeletal fractions (cf. Achtstaetter et al., 1986) from cultured bovine mammary gland epithelium cells of the lines BMGE−H (*lane 1*) and BMGE+H (*lane 2*) (cf. Schmid et al., 1983a), bovine heart tissue (*lane 3*), rat tongue mucosa (*lane 5*), rat heart tissue (*lane 6*), rat intestinal mucosa (*lane 7*), rat liver (*lane 8*), cultured rat hepatoma cells of line MH$_1$C$_1$ (*lane 9*) (cf. Franke et al., 1981a), human epidermal tissue (*lane 10*), human wart tissue (*verruca vulgaris; lane 11*), and human meningeal tissue (*lane 12*). The positions of plakoglobin ("band 5 protein") and band 6 polypeptide are indicated by the *arrowheads* in the *left margin. b*: Autoradiograph corresponding to *a*, showing the immunoblot reaction with guinea pig antibodies to band 6 polypeptide (for details, see Kapprell et al., 1988). Note positive reaction only in stratified epithelia (*lanes 4, 5, 10, 11*), whereas material from simple epithelia, heart tissue, and meninges do not reveal reactivity. *c*: Autoradiograph corresponding to part of *a* (*lanes 1–4*), showing the immunoblot reaction with guinea pig antibodies to plakoglobin. Note the positive reaction of plakoglobin in all samples, including cultured BMGE cells (*lanes 1, 2*) and heart tissue (*lane 3*).

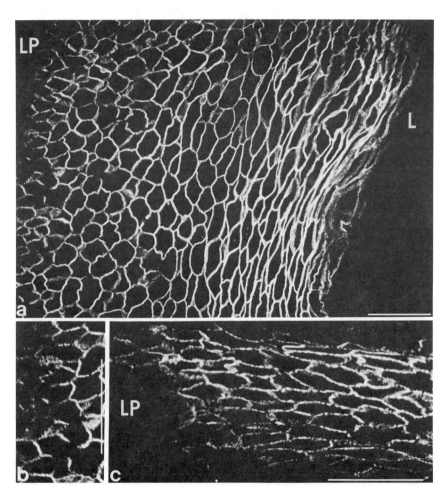

Figure 7. *Immunofluorescence microscopy on frozen sections of human tissues with antibodies to band 6 polypeptide of desmosomes. a,b:* Human esophagus, showing specific staining at cell boundaries (*a*, survey); at higher magnification (*b*), the staining is often resolved into small fluorescent dots representing individual desmosomes. *c:* Human tongue mucosa, showing punctate desmosomal staining along cell borders. L, lumen; LP, lamina propria. Calibration bars = 50 μm.

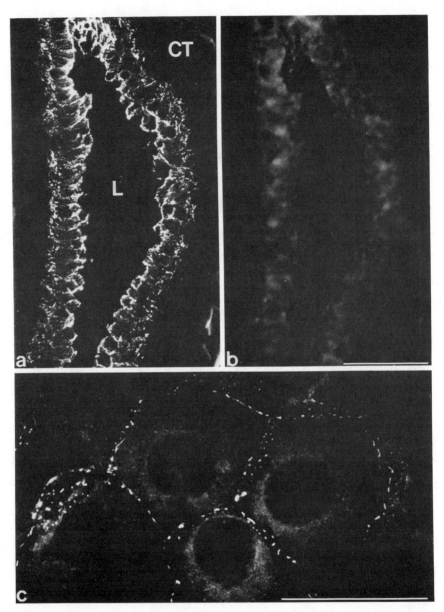

Figure 8. *Immunofluorescence micrographs showing results of reactions with antibodies to desmosomal band 6 protein in simple epithelia and cultured cells. a,b:* Double-label immunofluorescence microscopy of glandular epithelium in human esophagus, using mouse monoclonal antibody dp1 and dp2–2.15 to desmoplakin (*a*) (cf. Cowin et al., 1985b), in combination with affinity-purified guinea pig antibodies to band 6 polypeptide. (*a*) Fluorescein isothiocyanate– and (*b*) Texas Red–labeled secondary antibodies were used. Note that the glandular epithelial cells of this gland are negative for band 6 polypeptide (*b*) and positive for desmoplakin (*a*). L, lumen; CT, connective tissue. *c:* Immunofluorescence microscopy showing the reaction of antibodies to band 6 polypeptide on a whole-mount preparation of a monolayer culture of human A-431 cells, derived from vulvar epidermoid carcinoma. Note the fluorescent reaction in linear punctate arrays along the cell borders in a pattern very similar to those seen with antibodies to other desmosomal proteins such as desmoplakin and desmoglein. Calibration bars = 50 μm.

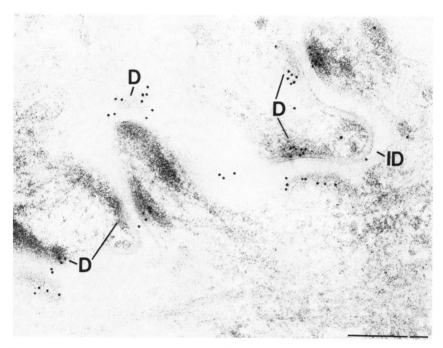

Figure 9. *Immunoelectron micrograph of the reaction of guinea pig antibodies to band 6 polypeptide.* Ultrathin sections of Lowicryl-embedded bovine tongue mucosa (postembedding procedure) and 10-nm gold-particle-coupled goat anti-mouse IgG were used (cf. Schmelz et al., 1986b). Note the specificity of reaction for desmosomes (D). Ideal cross-sections show that the label is restricted to the desmosomal plaques. ID, interdesmosomal region. Calibration bar = 0.5 μm.

buffer extraction–resistant structure. More recently, a protein (ZO-1, poly-peptide ~225 kD) detected by a special monoclonal antibody, R26.4C, has been localized in various species to the cytoplasmic surface of plasma membrane regions corresponding to tight junctions (Stevenson et al., 1986). That this protein is extractable, at least in part, by high-salt buffers suggests that it is a "peripheral" component rather than an integral membrane protein. As tight junctions are usually located in the plaque-covered broader region of a zonula adherens (cf. Gumbiner and Simons, 1986), protein ZO-1 may also represent a component of the plaque.

We have recently selected a monoclonal antibody (ZA-1TJ) that reacts with a junctional protein of zonulae adherentes in a pattern similar to protein ZO-1 (Figures 10–13), although only in certain human cells. For example, in hepatocytes of liver tissue this antibody exclusively reacts with the zonula region of the bile canaliculi (Figure 10), and in the intestine it specifically localizes the entire zonula of the "terminal bar" (Figures 11, 12). However, this antigen is absent from desmosomes as well as from the sparse and small intermediate junctions (puncta adherentia) of the basolateral wall, which can be positively demonstrated by these junctions' reactivity with antibodies to vinculin, α-actinin, and plakoglobin (Figure 11) (see also Cowin et al., 1986; Drenckhahn and Franz, 1986; Franke et al., 1987b,c). In contrast to protein ZO-1 (Stevenson et al., 1986), the zonula component detected by antibody ZA-1TJ has not been seen in vascular endothelium and Sertoli cells of testis (data not shown) that are known to contain zonula adherens–type junctions (Vogl and Soucy, 1985; Franke et al., 1987b,c) and structures similar to tight junctions (for a review, see Fawcett, 1981). Remarkably, the appearance of the ZA-1TJ antigen in the cell periphery is not dependent on the formation of a continuous zonula adherens and/or zonula occludens (tight junctions) encircling the cell, as intense immunostaining is already seen at intercellular boundaries of small cell colonies but not on the "free" margins of these cells (Figure 14a–b').

The ZA-1TJ antigen has been localized by immunoelectron microscopy to the plaque region of the zonula adherens, where it is found along the entire plaque but is restricted to a small plaque segment located close to the tight junction (Figure 15a,b). That the ZA-1TJ immunogold-labeled plaque moiety can detach from the junctional membrane surface under certain situations of artificial shrinkage (Figure 15b) shows that the epitope-bearing protein is part of the plaque and not stably integrated within the membrane.

These findings with antibody ZA-1TJ indicate that zonula plaque structures and intermediate junctions in general are not identical in different cell types. They either contain different proteins or expose their proteins in different arrangements. Moreover, our observations suggest that the zonula adherens plaque, despite its rather homogeneous appearance in the electron microscope, contains immunologically distinguishable subdomains. These preliminary immunological observations emphasize the need for further studies of the

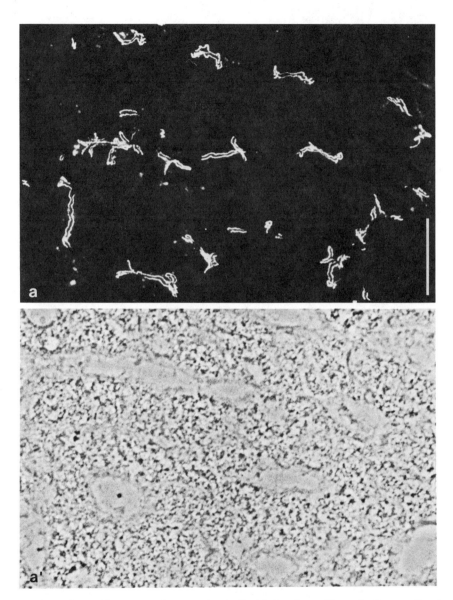

Figure 10. *Immunofluorescence microscopy showing the detection of a specific zonula adherens antigen by monoclonal murine antibody ZA-1TJ on a frozen tissue section through human liver tissue.* Note the specific and intense immunostaining at cell boundaries along the bile canaliculi (same optical field: *a,* epifluorescence optics; *b,* phase contrast). Calibration bar = 25 µm.

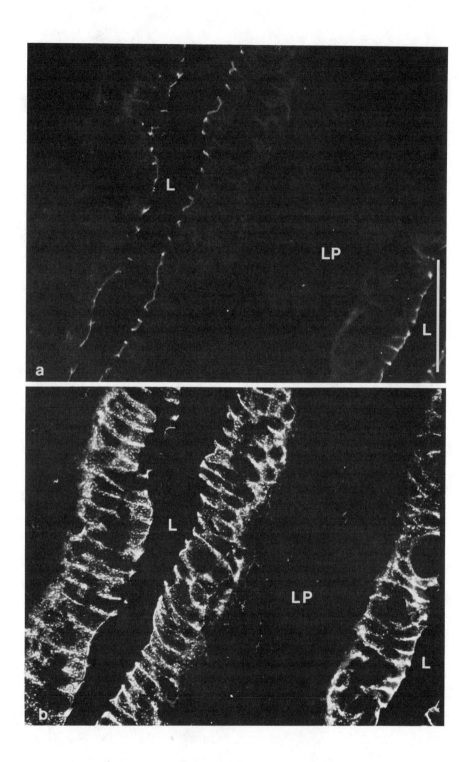

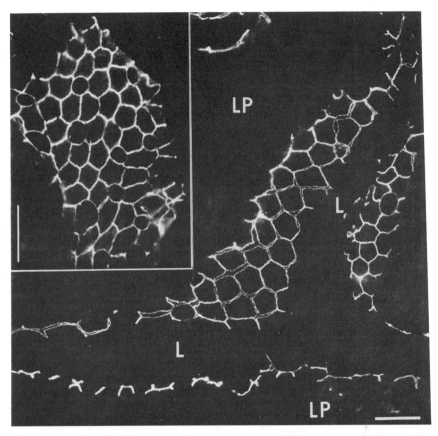

Figure 12. *Immunofluorescence microscopy of frozen section of human colonic tissue, as in Figure 11, after reaction with antibody ZA-1TJ.* The zonula adherens–specific antigen is revealed in longitudinal sections (e.g., *lower left*), in grazing sections (*center*), and in tangential sections including the terminal bar region (*inset*). The separation of the zonula adherens line into two sublines (e.g., at some cell boundaries in the central position) is probably due to artificial symmetrical splitting of the junctional complex (cf. Figure 15). (Symbols as in Figure 11.) Calibration bar = 25 μm.

Figure 11. *Double-label immunofluorescence of frozen tissue section of human colon obtained with murine antibody ZA-1TJ (a) and guinea pig antibodies to plakoglobin (b).* Note specific labeling of subapical zonula adherens by antibody ZA-1TJ (seen mostly in grazing and longitudinal sections in *a*) as compared with the staining of the same zonula plus desmosomes and puncta adherentia on the lateral walls by antibodies to plakoglobin (*b*). L, lumen; LP, lamina propria. Calibration bar = 50 μm.

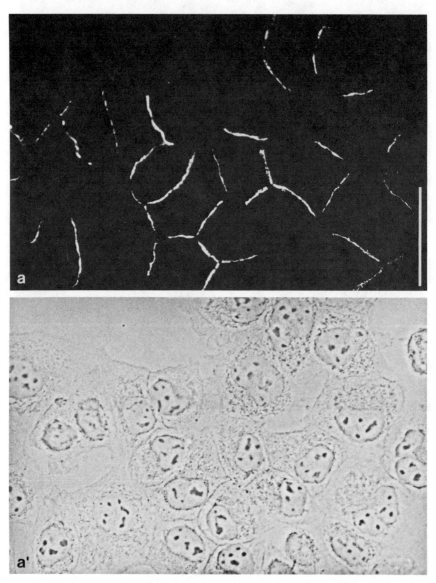

Figure 13. *Immunofluorescence microscopy of monolayer culture of human mammary carcinoma–derived cells of line MCF-7 after reaction with monoclonal antibody ZA-1TJ.* Note the intense and specific labeling of cell–cell boundaries but the absence of staining at free cell edges. Calibration bar = 50 μm.

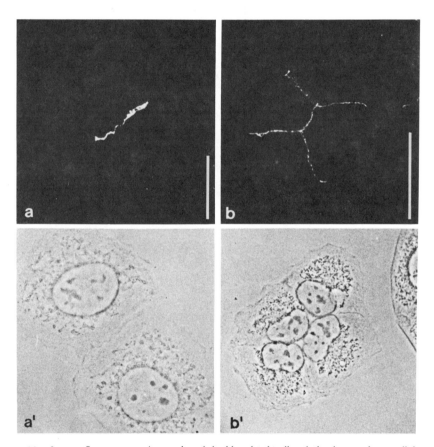

Figure 14. *Immunofluorescence micrographs of freshly plated cells of the human hepatocellular carcinoma–derived line PLC after reaction with antibody ZA-1TJ against a zonula adherens plaque-specific protein.* Small cell colonies show labeling only in regions of close cell–cell apposition but not in regions of the plasma membrane that are free, that is, not highly associated with another cell. *a,b*: Epifluorescence optics. *a',b'*: Same-field contrast optics. Calibration bars = 25 μm (*a*) and 50 μm (*b*).

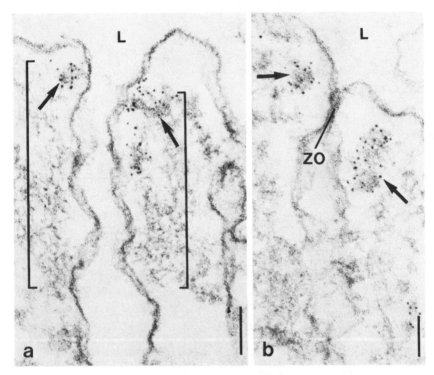

Figure 15. *Immunoelectron micrographs showing the reaction of monoclonal antibody ZA-1TJ with a special portion of the zonula adherens plaque in human colon tissue. a*: Antibody binding, visualized by immunogold particles (pre-embedding technique, using cryostat sections and saponin treatment; for details and some references, see Kartenbeck et al., 1984; Cowin et al., 1986), is seen in an apical portion of the plaque of the zonula adherens, which corresponds to the position of the tight junction (zonula occludens; *arrows* in *a*). The entire plaque is demarcated by brackets; in this region the junctional complex is locally split and separated by an intercellular gap. *b*: In this region the ZA-1TJ-positive portion of the zonula adherens plaque has somewhat detached from the membrane (ZO, zonula occludens), indicating that the antigen is a constituent of the plaque and not of the membrane (*arrows*). L, lumen of colon. Calibration bars = 0.1 μm.

biochemical composition of zonulae adherentes as well as of other intermediate junctions.

UNASSEMBLED PLAQUE PROTEINS

Junctional plaques can be regarded as local assemblies of cytoplasmic proteins that specifically interact with integral membrane proteins clustered to form a topogenic domain. If this is so, one might expect to find a soluble cytoplasmic pool of unassembled plaque protein molecules that exist in a dynamically regulated equilibrium with a plaque-assembled, perhaps polymeric, state. The formation of adhering junctions might require nucleation of plaque proteins at membrane-bound domains, and it has been suggested that calcium ions could be important in such an assembly (e.g., Hennings et al., 1980; Kartenbeck et al., 1982; Hennings and Holbrook, 1983; Watt et al., 1984; Mattey et al., 1986a; Volberg et al., 1986; O'Keefe et al., 1987). In supernatant fractions obtained after high-speed ultracentrifugation of various cell and tissue homogenates, several plaque-forming proteins have been described in distinct soluble forms. These are an ~6.5-S (for similar S values of preparatively isolated vinculin, see Burridge and Mangeat, 1984; Evans et al., 1984), probably monomeric, form of vinculin with a Stokes radius of ~4.4 nm (Figure 16a–c,e,f) (for the existence of soluble vinculin in living cells, see also Rosenfeld et al., 1985) and an ~7-S form of plakoglobin, which appears to represent a homodimeric form, as indicated by cross-linking experiments (Figures 16d, 17a,b) (see also Cowin et al., 1986; Kapprell et al., 1987).

An exchange between the diffusible and the plaque-bound forms of vinculin has been described in microinjection experiments (Kreis et al., 1982; Geiger et al., 1984). The relative amounts of soluble and structure-bound forms of vinculin in different cell types recovered under different environmental conditions can reach remarkably high values (up to 30% soluble plakoglobin in certain cells, and up to 80% soluble vinculin in vascular smooth muscle tissue). A soluble 9-S form of both desmoplakins has recently been identified in lysates of various cultured cells (murine keratinocytes, human A-431 carcinoma cells) kept in medium containing a very low calcium concentration (Duden, 1987; Duden and Franke, 1988).

Further information on the unassembled forms of plaque constituents is needed for a better understanding of the processes involved in junction formations—and hence cell–cell interactions—and their regulation. In addition, progress in the characterization of soluble forms of actin and actin complexes with certain specific actin-binding proteins (for a review, see Pollard and Cooper, 1986) and the demonstrations of distinct soluble forms of IF proteins (Soellner et al., 1985; Franke et al., 1987d) should encourage studies of the dynamics of the interaction of plaques and plaque proteins with MF and IF proteins in the regulation of filament anchorage.

The recruitment of various plaque proteins to their specific plasma membrane domains and the different kinetics of disassembly and re-formation of desmosomal and intermediate junction plaques in the same cell type as demonstrated in calcium depletion experiments (Kartenbeck et al., 1982; Mattey and Garrod, 1986a,b; Volberg et al., 1986) indicate the existence of different assembly mechanisms for these two major groups of adhering junctions. Moreover, although plakoglobin binds to plaque domains of both desmosomal and nondesmosomal junctions, other cytoplasmic plaque components are precisely sorted to only one type of junction. An elucidation of the molecular basis for the specific interaction of cytoplasmic plaque proteins with membrane domains and of plaque-assembled proteins with filament

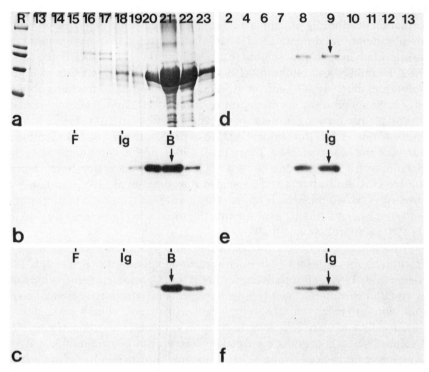

Figure 16. *Identification of soluble states of the plaque proteins vinculin and plakoglobin by gel filtration and sucrose gradient centrifugation analysis of proteins present in supernatant fractions from cell homogenates. a,b:* Gel filtration of proteins present in the 100,000 g × 2 hours supernatant of cultured bovine endothelial cells (line CPAE) obtained after homogenization in "near-physiological buffer" on a Superose 12 column (for conditions, see Kapprell, 1987; Kapprell et al., 1987, 1988). Proteins contained in the individual fractions were precipitated with 10% TCA and analyzed by SDS-PAGE (8% acrylamide). Fraction numbers are given at the *top margin. a:* Coomassie blue staining of proteins of fractions 13–23 (out of a total of 27). R, reference proteins (from *top* to *bottom:* myosin heavy chain, β-galactosidase, phosphorylase A, bovine serum albumin, and actin). *b:* Corresponding autoradiograph of an immunoblot reaction with murine monoclonal antibody to vinculin. Note that the soluble vinculin has eluted at a peak position corresponding to a Stokes radius of 4.4 nm (*arrow*). Markers for calibration are indicated on the *upper margin.* F, bovine spleen ferritin; Ig, rabbit immunoglobulin; B, bovine serum albumin. *c:* Autoradiograph showing the immunoblot reaction of polypeptides present in the 100,000 g × 2 hours supernatants obtained after homogenization of cultured calf lens cells in near-physiological buffer (for details, see Kapprell, 1987), fractionation on a Superose 12 column, and analysis of the fractions by SDS-PAGE (Coomassie blue, not shown). Note elution of vinculin at a position corresponding to a Stokes radius of 4.4 nm (*arrow*), visualized by reaction with vinculin antibodies. *d–f:* Autoradiographs showing the immunoblot reaction of polypeptides present in 100,000 g × 2 hours supernatants of CPAE (*d,e*) and calf lens cells (*f*) obtained after homogenization (see above) and separation on a 5–30% (w/v) sucrose gradient, followed by SDS-PAGE and immunoblotting with guinea pig antibodies to plakoglobin (*d*) or murine monoclonal antibody to vinculin (*e,f*). Note that in both cell types the soluble forms of soluble plakoglobins and vinculin appear at a similar peak position corresponding to 7 S (*arrows*).

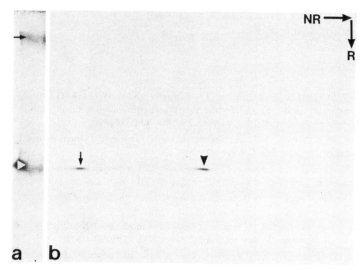

Figure 17.   *Demonstration of homodimers of soluble plakoglobin by chemical cross-linking with cupric 1,10-phenanthroline (Cu-P) of the 6.5–7-S peak fractions of ^{35}S-methionine-labeled BMGE-H cells.* a: Autoradiofluorograph showing the monomeric (*arrowhead*) and the cross-linked (*arrow*) forms of plakoglobin obtained after homogenization of cultured BMGE-H cells in near-physiological buffer, sucrose gradient centrifugation of the 100,000 g $\times$ 2 hours supernatants, cross-linking of the 6.5–7-S peak fraction with Cu-P for four hours; immunoprecipitation with guinea-pig antibodies to plakoglobin and SDS-PAGE without reducing conditions (for chemical cross-linking, see Kapprell et al., 1987). b: Autoradiofluorograph showing cleavage of the cross-linked product of plakoglobin (as described in a) first by SDS-PAGE under nonreducing conditions (NR), and then after cleavage by incubation with 2-mercaptoethanol by SDS-PAGE under reducing conditions (R) (for details, see Quinlan and Franke, 1982, 1983). In this way the dimer is displaced from the expected diagonal. Note that the cleavage of the cross-linked form of plakoglobin results in only one polypeptide spot, which in two-dimensional SDS-PAGE comigrates with monomeric plakoglobin, indicating that the 7-S plakoglobin particle is a homodimer.

proteins will be important for a further understanding of cell and tissue architecture.

ACKNOWLEDGMENTS

We thank Dr. Pamela Cowin (Department of Cell Biology, New York University Medical School) for stimulating discussions, as well as Martina Ittensohn and Signe Mähler for expert technical assistance. We further thank Dr. Frank Longo (Department of Anatomy, University of Iowa) for reading and correcting the manuscript and Friederike Schmitt for careful manuscript preparation.

Note added in proof: Recently, the first plaque protein of desmosomal and intermediate junctions, plakoglobin, has been cDNA-cloned and sequenced

(Franke et al., 1989). This paper also contains partial amino acid sequence information for desmoplakin.

REFERENCES

Achtstaetter, T., M. Hatzfeld, R. A. Quinlan, D. C. Parmelee, and W. W. Franke (1986) Separation of cytokeratin polypeptides by gel electrophoretic and chromatographic techniques and their identification by immunoblotting. *Methods Enzymol.* **134**:355–371.

Blose, S. H., and D. I. Meltzer (1981) Visualization of the 10-nm filament vimentin rings in vascular endothelial cells *in situ*. *Exp. Cell Res.* **135**:299–309.

Boller, K., D. Vestweber, and R. Kemler (1985) Cell-adhesion molecule uvomorulin is localized in the intermediate junctions of adult intestinal epithelial cells. *J. Cell Biol.* **100**:327–332.

Burridge, K., and P. Mangeat (1984) An interaction between vinculin and talin. *Nature* **308**:744–746.

Cohen, S. M., G. Gorbsky, and M. S. Steinberg (1983) Immunochemical characterization of related families of glycoproteins in desmosomes. *J. Biol. Chem.* **258**:2621–2627.

Cowin, P., and D. R. Garrod (1983) Antibodies to epithelial desmosomes show wide tissue and species cross-reactivity. *Nature* **302**:148–150.

Cowin, P., D. Mattey, and D. R. Garrod (1984a) Distribution of desmosomal components in the tissues of vertebrates, studied by fluorescent antibody staining. *J. Cell Sci.* **66**:119–132.

Cowin, P., D. Mattey, and D. R. Garrod (1984b) Identification of desmosomal surface components (desmocollins) and inhibition of desmosome formation by specific Fab'. *J. Cell Sci.* **70**:41–60.

Cowin, P., W. W. Franke, C. Grund, and H.-P. Kapprell (1985a) The desmosome-intermediate filament complex. In *The Cell in Contact: Adhesions and Junctions as Morphogenetic Determinants*, G. M. Edelman and J. P. Thiery, eds., pp. 427–460, Wiley, New York.

Cowin, P., H.-P. Kapprell, and W. W. Franke (1985b) The complement of desmosomal plaque proteins in different cell types. *J. Cell Biol.* **101**:1442–1454.

Cowin, P., H.-P. Kapprell, W. W. Franke, J. Tamkun, and R. O. Hynes (1986) Plakoglobin: A protein common to different kinds of intercellular adhering junctions. *Cell* **46**:1063–1073.

Damsky, C. H., J. Richa, D. Solter, K. Knudsen, and C. A. Buck (1983) Identification and purification of a cell surface glycoprotein mediating intercellular adhesion in embryonic and adult tissue. *Cell* **34**:455–466.

Drenckhahn, D., and H. Franz (1986) Identification of actin-, α-actinin-, and vinculin-containing plaques at the lateral membrane of epithelial cells. *J. Cell Biol.* **102**:1843–1852.

Drochmans, P., C. Freudenstein, J.-C. Wanson, L. Laurent, T. W. Keenan, J. Stadler, R. Leloup, and W. W. Franke (1978) Structure and biochemical composition of desmosomes and tono-filaments isolated from calf muzzle epidermis. *J. Cell Biol.* **79**:427–443.

Duden, R. (1987) Desmosomenproteine in gekoppelten und ungekoppelten Zellkultursituationen. Diploma thesis, Faculty of Biology, University of Heidelberg, pp. 1–30.

Duden, R., and W. W. Franke (1988) Organization of desmosomal plaque proteins in cells growing at low calcium concentrations. *J. Cell Biol.* **107**:1049–1063.

Evans, R. R., R. M. Robson, and M. H. Stromer (1984) Properties of smooth muscle vinculin. *J. Biol. Chem.* **259**:3916–3924.

Farquhar, M. G., and G. E. Palade (1963) Junctional complexes in various epithelia. *J. Cell Biol.* **17**:375–412.

Fawcett, D. W. (1981) *The Cell*, 2d Ed., Saunders, Philadelphia.

Franke, W. W., D. Mayer, E. Schmid, H. Denk, and E. Borenfreund (1981a) Differences of expression of cytoskeletal proteins in cultured rat hepatocytes and hepatoma cells. *Exp. Cell Res.* **134**:345–365.

Franke, W. W., D. L. Schiller, R. Moll, S. Winter, E. Schmid, I. Engelbrecht, and H. Denk (1981b) Diversity of cytokeratins. Differentiation specific expression of cytokeratin polypeptides in epithelial cells and tissues. *J. Mol. Biol.* **153**:933–959.

Franke, W. W., E. Schmid, C. Grund, H. Mueller, I. Engelbrecht, R. Moll, J. Stadler, and E.-D. Jarasch (1981c) Antibodies to high molecular weight polypeptides of desmosomes: Specific localization of a class of junctional proteins in cells and tissues. *Differentiation* **20**:217–241.

Franke, W. W., R. Moll, D. L. Schiller, E. Schmid, J. Kartenbeck, and H. Mueller (1982) Desmoplakins of epithelial and myocardial desmosomes are immunologically and bio-chemically related. *Differentiation* **23**:226–237.

Franke, W. W., H.-P. Kapprell, and H. Mueller (1983a) Isolation and symmetrical splitting of desmosomal structures in 9 M urea. *Eur. J. Cell Biol.* **32**:117–130.

Franke, W. W., R. Moll, H. Mueller, E. Schmid, C. Kuhn, R. Krepler, U. Artlieb, and H. Denk (1983b) Immunocytochemical identification of epithelium-derived human tumors with antibodies to desmosomal plaque proteins. *Proc. Natl. Acad. Sci. USA* **80**:543–547.

Franke, W. W., H. Mueller, S. Mittnacht, H.-P. Kapprell, and J. L. Jorcano (1983c) Significance of two desmosome plaque–associated polypeptides of molecular weights 75,000 and 83,000. *EMBO J.* **2**:2211–2215.

Franke, W. W., P. Cowin, M. Schmelz, and H.-P. Kapprell (1987a) The desmosomal plaque and the cytoskeleton. *Ciba Found. Symp.* **125**:26–44.

Franke, W. W., H.-P. Kapprell, and P. Cowin (1987b) Plakoglobin is a component of the filamentous subplasmalemmal coat of lens cells. *Eur. J. Cell Biol.* **43**:301–315.

Franke, W. W., H.-P. Kapprell, and P. Cowin (1987c) Immunolocalization of plakoglobin in endothelial junctions: Identification as a special type of zonulae adherentes. *Biol. Cell* **59**:205–218.

Franke, W. W., S. Winter, E. Schmid, P. Soellner, G. Haemmerling, and T. Achtstaetter (1987d) Monoclonal cytokeratin antibody recognizing a heterotypic complex: Immunological probing of conformational states of cytoskeletal proteins in filaments and in solution. *Exp. Cell Res.* **173**:17–37.

Franke, W. W., M. D.Goldschmidt, R. Zimbelmann, H. M. Mueller, and D. L. Schiller (1989) Molecular cloning and amino acid sequence of human plakoglobin, the common junctional plaque protein. *Proc. Natl. Acad. Sci. USA* (in press).

Geiger, B., A. H. Dutton, K. T. Tokuyasu, and S. J. Singer (1981) Immunoelectron microscope studies of membrane-microfilament interactions: Distributions of α-actinin, tropomyosin, and vinculin in intestinal epithelia brush border and in chicken gizzard smooth muscle cells. *J. Cell Biol.* **91**:614–628.

Geiger, B., E. Schmid, and W. W. Franke (1983) Spatial distribution of proteins specific for desmosomes and adherens junctions in epithelial cells demonstrated by double immuno-fluorescence microscopy.

Geiger, B., Z. Avnur, T. E. Kreis, and J. Schlessinger (1984) The dynamics of cytoskeletal organization in areas of cell contact. *Cell Muscle Motil.* **5**:195–234.

Geiger, B., Z. Avnur, T. Volberg, and T. Volk (1985a) Molecular domains of adherens junctions. In *The Cell in Contact: Adhesions and Junctions as Morphogenetic Determinants*, G. M. Edelman and J. P. Thiery, eds., pp. 461–489, Wiley, New York.

Geiger, B., T. Volk, and T. Volberg (1985b) Molecular heterogeneity of adherens junctions. *J. Cell Biol.* **101**:1523–1531.

Giudice, G. J., S. M. Cohen, N. H. Patel, and M. S. Steinberg (1984) Immunological comparison of desmosomal components from several bovine tissues. *J. Cell Biochem.* **26**:35–45.

Gorbsky, G., and M. S. Steinberg (1981) Isolation of the intercellular glycoproteins of desmosomes. *J. Cell Biol.* **90**:243–248.

Gorbsky, G., S. M. Cohen, M. Shida, G. J. Giudice, and M. S. Steinberg (1985) Isolation of the nonglycosylated proteins of desmosomes and immunolocalization of a third plaque protein: Desmoplakin III. *Proc. Natl. Acad. Sci. USA* **82**:810–814.

Gumbiner, B., and K. Simons (1986) The role of uvomorulin in the formation of epithelial occluding junctions. *Ciba Found. Symp.* **125**:168–186.

Hennings, H., D. Michael, C. Cheng, P. Steinert, K. A. Holbrook, and S. H. Yuspa (1980) Calcium regulation of growth and differentiation of mouse epidermal cells in culture. *Cell* **19**:245–254.

Hennings, H., and K. A. Holbrook (1983) Calcium regulation of cell–cell contact and differentiation of epidermal cells in culture. *Exp. Cell Res.* **143**:127–142.

Hubbard, B. D., and E. Lazarides (1979) Copurification of actin and desmin from chicken smooth muscle and their co-polymerization *in vitro* to intermediate filaments. *J. Cell Biol.* **80**:166–182.

Jones, J. C. R., and R. D. Goldman (1985) Intermediate filaments and the initiation of desmosome assembly. *J. Cell Biol.* **101**:506–517.

Kapprell, H.-P. (1987) Identifizierung und Charakterisierung von Plaque-Proteinen der interzellulären Verbindungsstrukturen. Ph.D. Thesis, Faculty of Biology, University of Heidelberg, pp. 1–227.

Kapprell, H.-P., P. Cowin, W. W. Franke, H. Ponstingl, and H. J. Opferkuch (1985) Biochemical characterization of desmosomal proteins isolated from bovine muzzle epidermis: Amino acid and carbohydrate composition. *Eur. J. Cell Biol.* **36**:217–229.

Kapprell, H.-P., P. Cowin, and W. W. Franke (1987) Biochemical characterization of the soluble form of the junctional plaque protein, plakoglobin, from different cell types. *Eur. J. Biochem.* **166**:505–517.

Kapprell, H.-P., K. Owaribe, W. W. Franke (1988) Identification of a basic protein of M_r 75,000 as an accessory desmosomal plaque protein in stratified and complex epithelia. *J. Cell Biol.* **106**:1679–1691.

Kartenbeck, J., E. Schmid, W. W. Franke, and B. Geiger (1982) Different modes of internalization of proteins associated with adherens junctions and desmosomes: Experimental separation of lateral contacts induces endocytosis of desmosomal plaque material. *EMBO J.* **1**:725–732.

Kartenbeck, J., W. W. Franke, J. G. Moser, and U. Stoffels (1983) Specific attachment of desmin filaments to desmosomal plaques in cardiac myocytes. *EMBO J.* **2**:735–742.

Kartenbeck, J., K. Schwechheimer, R. Moll, and W. W. Franke (1984) Attachment of vimentin filaments to desmosomal plaques in human meningiomal cells and arachnoidal tissue. *J. Cell Biol.* **98**:1072–1081.

Kreis, T. E., B. Geiger, and J. Schlessinger (1982) The mobility of microinjected rhodamine-actin within living chicken gizzard cells determined by fluorescence photobleaching recovery. *Cell* **29**:835–845.

Kreis, T. E., B. Geiger, E. Schmid, J. L. Jorcano, and W. W. Franke (1983) *De novo* synthesis and specific assembly of keratin filaments in non-epithelial cells after microinjection of mRNA for epidermal keratin. *Cell* **32**:1125–1137.

Magin, T. M., J. L. Jorcano, and W. W. Franke (1983) Translational products of mRNA coding for non-epidermal cytokeratins. *EMBO J.* **2**:1387–1392.

Marchesi, V. T. (1985) Stabilizing infrastructure of cell membranes. *Annu. Rev. Cell Biol.* **1**:531–561.

Mattey, D. L., and D. R. Garrod (1986a) Calcium induced desmosome formation in cultured kidney epithelial cells. *J. Cell Sci.* **85**:95–111.

Mattey, D. L., and D. R. Garrod (1986b) Splitting and internalization of the desmosomes of cultured kidney epithelial cells by reduction in calcium concentration. *J. Cell Sci.* **85**:113–124.

Miller, K., D. Mattey, H. Measures, C. Hopkins, and D. Garrod (1987) Localization of the protein and glycoprotein components of bovine nasal epithelial desmosomes by immunoelectron microscopy. *EMBO J.* **4**:885–889.

Moll, R., P. Cowin, H.-P. Kapprell, and W. W. Franke (1986) Biology of disease. Desmosomal proteins: New markers for identification and classification of tumors. *Lab. Invest.* **1**:4–25.

Mueller, H., and W. W. Franke (1983) Biochemical and immunological characterization of desmoplakins I and II, the major polypeptides of the desmosomal plaque. *J. Mol. Biol.* **163**:647–671.

Ogou, S.-I., C. Yoshida-Noro, and M. Takeichi (1983) Calcium-dependent cell–cell adhesion molecules common to hepatocytes and teratocarcinoma stem cells. *J. Cell Biol.* **97**:944–948.

O'Keefe, E., R. A. Briggaman, and B. Herman (1987) Calcium-induced assembly of adherens junctions in keratinocytes. *J. Cell Biol.* **105**:807–817.

Penn, E. J., C. Hobson, D. A. Rees, and A. I. Magee (1987) Structure and assembly of desmosome junctions: Biosynthesis, processing, and transport of the major protein and glycoprotein components in cultured epithelial cells. *J. Cell Biol.* **105**:57–68.

Peyriéras, N., F. Hyafil, D. Louvard, H. L. Ploegh, and F. Jacob (1983) Uvomorulin: A nonintegral membrane protein of early mouse embryo. *Proc. Natl. Acad. Sci. USA* **80**:6274–6277.

Pollard, T. D., and J. A. Cooper (1986) Actin and actin-binding proteins. A critical evaluation of mechanisms and functions. *Annu. Rev. Biochem.* **55**:987–1035.

Quinlan, R. A., and W. W. Franke (1982) Heteropolymer filaments of vimentin and desmin in vascular smooth muscle tissue and cultured baby hamster kidney cells demonstrated by chemical crosslinking. *Proc. Natl. Acad. Sci. USA* **79**:3452–3456.

Quinlan, R. A., and W. W. Franke (1983) Molecular interactions in intermediate-sized filaments revealed by chemical crosslinking. Heteropolymers of vimentin and glial filament protein in cultured human glioma cells. *Eur. J. Biochem.* **132**:477–484.

Rosenfeld, G. C., D. C. Hou, J. Dingus, I. Meza, and J. Bryan (1985) Isolation and partial characterization of human platelet vinculin. *J. Cell Biol.* **100**:669–676.

Schmelz, M., R. Duden, P. Cowin, and W. W. Franke (1986a) A constitutive transmembrane glycoprotein of M_r 165,000 (desmoglein) in epidermal and non-epidermal desmosomes. I. Biochemical identification of the polypeptide. *Eur. J. Cell Biol.* **42**:177–183.

Schmelz, M., R. Duden, P. Cowin, and W. W. Franke (1986b) A constitutive transmembrane glycoprotein of M_r 165,000 (desmoglein) in epidermal and non-epidermal desmosomes. II. Immunolocalization and microinjection studies. *Eur. J. Cell Biol.* **42**:184–199.

Schmid, E., W. W. Franke, C. Grund, D. L. Schiller, H. Kolb, and N. Paweletz (1983a) An epithelial cell line with fusiform myoid morphology derived from bovine mammary gland: Expression of cytokeratins and desmoplakins in abnormal arrays. *Exp. Cell Res.* **146**:309–328.

Schmid, E., D. L. Schiller, C. Grund, J. Stadler, and W. W. Franke (1983b) Tissue type–specific expression of intermediate filament proteins in a cultured epithelial cell line from bovine mammary gland. *J. Cell Biol.* **96**:37–50.

Shida, H., G. J. Giudice, M. Shida, M. S. Cohen, and M. S. Steinberg (1984) Molecular organization of the transmembrane molecules and cytoplasmic components of the desmosome. In *International Cell Biology*, S. Seno and Y. Okada, eds., p. 314, Academic, Tokyo.

Skerrow, C. J., and A. G. Matoltsy (1974a) Isolation of epidermal desmosomes. *J. Cell Biol.* **63**:515–523.

Skerrow, C. J., and A. G. Matoltsy (1974b) Chemical characterization of isolated epidermal desmosomes. *J. Cell Biol.* **63**:524–531.

Soellner, P., R. A. Quinlan, and W. W. Franke (1985) Identification of a distinct soluble subunit of an intermediate filament protein: Tetrameric vimentin from living cells. *Proc. Natl. Acad. Sci. USA* **82**:7929–7933.

Staehelin, L. A. (1974) Structure and function of intercellular junctions. *Int. Rev. Cytol.* **39**:191–283.

Steinberg, M. S., H. Shida, G. J. Giudice, M. Shida, N. H. Patel, and O. W. Blaschuk (1987) On the molecular organization, diversity, and functions of desmosomal proteins. *Ciba Found. Symp.* **125**:3–25.

Stevenson, B. R., J. D. Siliciano, M. S. Mooseker, and D. A. Goodenough (1986) Identification of ZO-1: A high molecular weight polypeptide associated with the tight junction (zonula occludens) in a variety of epithelia. *J. Cell Biol.* **103**:755–766.

Suhrbier, A., and D. Garrod (1986) An investigation of the molecular components of desmosomes in epithelial cells of five vertebrates. *J. Cell Sci.* **81**:223–242.

Tsukita, S., and S. Tsukita (1985) Desmocalmin: A calmodulin-binding high molecular weight protein isolated from desmosomes. *J. Cell Biol.* **101**:2070–2080.

Vogl, A. W., and L. J. Soucy (1985) Arrangement and possible function of actin filament bundles in ectoplasmic specializations of ground squirrel Sertoli cells. *J. Cell Biol.* **100**:814–825.

Volberg, T., B. Geiger, J. Kartenbeck, and W. W. Franke (1986) Changes in membrane–microfilament interaction in intercellular adherens junctions upon removal of extracellular Ca^{2+} ions. *J. Cell Biol.* **102**:1832–1842.

Volk, T., and B. Geiger (1984) A 135-kD membrane protein of intercellular adherens junctions. *EMBO J.* **3**:2249–2260.

Volk, T., and B. Geiger (1986a) A-CAM: A 135-kD receptor of intercellular adherens junctions. I: Immunoelectron microscopic localization and biochemical studies. *J. Cell Biol.* **103**:1441–1450.

Volk, T., and B. Geiger (1986b) A-CAM: A 135-kD receptor of intercellular adherens junctions. II. Antibody-mediated modulation of junction formation. *J. Cell Biol.* **103**:1451–1464.

Volk, T., O. Cohen, and B. Geiger (1987) Formation of heterotypic adherens-type junctions between L-CAM-containing liver cells and A-CAM-containing lens cells. *Cell* **50**:987–994.

Watts, F. M., D. L. Mattey, and D. R. Garrod (1984) Calcium-induced reorganization of desmosomal components in cultured human keratinocytes. *J. Cell Biol.* **99**:2211–2215.

Chapter 14

Desmosomes

DAVID R. GARROD
ELAINE P. PARRISH
DEREK L. MATTEY
JANE E. MARSTON
HELEN R. MEASURES
MARCELO J. VILELA

ABSTRACT

The tissue and species distribution of desmosomes is briefly described, and their role in adhesion and the structural organization of tissues is discussed. Quantitative data on the immunogold localization of desmosomal proteins (dp) and glycoproteins (dg) within bovine nasal epithelial desmosomes are presented, and a model of desmosome structure is illustrated. New biochemical data relating to the different asymmetric transmembrane distributions of dg1, dg2, and dg3 are reported. The calcium-induced formation of desmosomes by cells in tissue culture is discussed. First, evidence relating to the differential requirements of different cell types for assembly or synthesis of components in response to calcium switching is described. Second, previous work on calcium-induced desmosome assembly is summarized, and the possible roles played by calcium are discussed. Cumulative data on the staining of large numbers of certain types of human carcinomas with anti-desmosomal antibodies are presented. All 244 tumors studied possessed desmosomes. Furthermore, metastatic cell populations do not represent special cell groups that have lost the ability to form desmosomes. A possible mechanism for the generation of viable metastatic cell clumps by the breakdown of intercellular adhesion is described. New anti-desmosomal monoclonal antibodies that are reliable epithelial markers for diagnostic pathology are briefly presented.

Desmosomes are punctate, adhesive intercellular junctions, usually less than 1 μm in diameter, that occur in most types of epithelial cells (Cowin and Garrod, 1983; Cowin et al., 1984a; Garrod, 1985; Garrod and Cowin, 1986; Moll et al., 1986). They are absent, however, from chick pigmented retinal epithelial cells (Docherty et al., 1984) and the lens epithelium (Ramaekers et al., 1980). Desmosomes are not exclusively epithelial, since they are also present in a number of nonepithelial cell types: the intercalated disks of heart

315

muscle (Franke et al., 1982; Cowin and Garrod, 1983; Cowin et al., 1984a; Atherton et al., 1986), the arachnoid mater and pia mater of the meninges (Parrish et al., 1986, 1987; Alcolado et al., 1988), and the dendritic reticulum cells (DRC) of lymphoid tissues (J. C. R. Jones and D. R. Garrod, unpublished data). They are absent from skeletal and connective tissues, nervous tissue, and lymphoid cells other than DRC.

Desmosomes occur throughout the vertebrate class, as well as in a variety of different tissues. Cross-reactivity of antibodies raised against bovine desmosomal components with avian, reptilian, amphibian, and piscine tissues shows that desmosomal components are conserved, the proteins more extensively than the glycoproteins (Cowin et al., 1984a; Suhrbier and Garrod, 1986). Moreover, the adhesive recognition specificity of desmosomes is also conserved, since desmosomes have been shown to form between heterologous combinations of mammalian, avian, and amphibian cells (Overton, 1977; Mattey and Garrod, 1985). Junctions resembling desmosomes in ultrastructural appearance also occur in invertebrates, but nothing is yet known about their composition and function. These observations point to an essential role for desmosomes in the adhesive properties of various cell types.

Cells that have desmosomes invariably also possess a number of different adhesion mechanisms, perhaps as many as 10 in some cell types if the various intercellular junctions, nonjunctional adhesion molecules, and cell–matrix adhesion mechanisms are totaled. Each mechanism must be presumed to have a particular function different from that of every other adhesion mechanism. The sum of contributions from these various mechanisms represents that nebulous quantity, cellular *adhesiveness* (Garrod, 1981; Garrod and Nicol, 1981; Garrod, 1985, 1986a,b).

Desmosomes may be regarded as adhesive intercellular links between the intermediate filament cytoskeletons of adjacent cells (McNutt and Weinstein, 1973; Overton, 1974; Staehelin, 1974; Cowin et al., 1985; Garrod, 1985). Intermediate filaments attach in bundles to the cytoplasmic plaques of desmosomes, and either connect to other desmosomal plaques or extend deep into the cytoplasm to form a basketlike matrix around the nucleus. In an epithelium, desmosomes therefore act as structural links that maintain cytoskeletal continuity throughout the cell sheet or tissue (Garrod, 1985)— they are the couplings in the tissue scaffolding.

Desmosomes are much more abundant in stratified epithelia such as epidermis than in simple epithelia. In the former they are present all around the surfaces of cells in the intermediate layer of the epithelium, whereas in the latter they are confined to the lateral cell surfaces and are usually concentrated in a terminal bar region near the luminal surface (Cowin et al., 1985; Garrod, 1985, 1986a,b). It is likely, therefore, that desmosomes make a much larger contribution to cellular adhesiveness in stratified epithelia than they do in simple epithelia, where other junctions and nonjunctional adhesion mechanisms play a major role.

The abundance of desmosomes in a tissue is thought to be related to the necessity for resistance to stress and abrasion (Arnn and Staehelin, 1981). The desmosome-intermediate filament scaffolding is thus seen as a means of mechanically stabilizing a tissue. In embryonic development desmosomes' first appearance may well be associated with just such a function. In the mammal they first appear at the early blastocyst stage and are restricted to the enveloping trophoblast layer, presumably resisting its disruption by internal pressure from accumulating blastocoelic fluid (Ducibella et al., 1975). In the chick embryo, desmosomes appear in the area opaca at stage 3 of Hamburger and Hamilton, when spreading over the yolk begins, generating tension in the cell sheet. However, they are absent from the whole of the area pellucida, which at that stage is beginning the extensive cell rearrangements of gastrulation (Figure 1) (Overton, 1962; Andries et al., 1985).

It may be, however, that desmosomes play a more subtle role in some situations. The presence of desmosomes does not preclude cell rearrangements in all situations. For example, in the intestine there is a continual movement of mucosal cells, all of which possess desmosomes, from the bases of the crypts to the apexes of the villi. Also, in epidermis the continual production, movement,

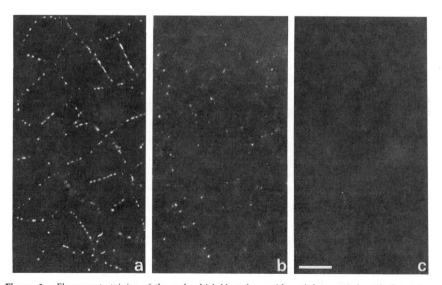

Figure 1. *Fluorescent staining of the early chick blastoderm with anti-desmosomal antibody. a:* Area opaca showing bright, punctate staining of desmosomes around the cells. *b:* Marginal zone between area opaca and area pelucida showing greatly reduced desmosomal staining. *c:* Area pelucida in region of primitive streak showing complete absence of desmosomal staining. Embryo was removed from yolk on filter paper disk, fixed in methanol, and stained with a guinea-pig polyclonal antibody to dp1 and dp2 using the indirect technique. Calibration bar = 20 μm.

and maturation of cells from the basal layer may require desmosomal breakdown and reformation (Watt et al., 1984). There is evidence for a difference between the desmosomal glycoproteins of basal and suprabasal cells (Parrish et al., 1986) that could be an important factor in the migration of cells from the basal layer.

It has been suggested that variation in desmosomal stability may contribute to the modulation of cellular adhesive properties (Mattey and Garrod, 1985, 1986a). Desmosomes of cells in tissue culture have been shown to exhibit variable stability to disruption by a reduced extracellular calcium concentration, and stability to this treatment varies with time in some cell types (Watt et al., 1984; Mattey and Garrod, 1986b; Mattey et al., 1986). Whether this variability has any physiological significance for tissue morphogenesis remains to be determined, but the potential seems to be present. Variation in desmosomal stability (as well as that of other types of junction) could also play a part in disease processes, such as the process of releasing neoplastic cells from contact with their neighbors in malignant invasion and metastasis.

LOCALIZATION OF PROTEIN AND GLYCOPROTEIN COMPONENTS WITHIN DESMOSOMES

In epidermis, desmosomes consist of eight major constituents, which we designate dp (desmosomal proteins) and dg (desmosomal glycoproteins) (Miller et al., 1987). In descending order of relative molecular weight, the proteins are dp1 (250 kD), dp2 (230 kD), dp3 (83 kD), and dp4 (75 kD), and the glycoproteins are dg1 (175 kD–164 kD), dg2 (115 kD), dg3 (107 kD), and dg4 (22 kD) (Skerrow and Matoltsy, 1974; Franke et al., 1981, 1982; Gorbsky and Steinberg, 1981; Cowin and Garrod, 1983; Garrod, 1985, 1986a,b; Miller et al., 1987). In addition to these, there are desmocalmin, a high-molecular-weight calmodulin- and cytokeratin-binding protein (Tsukita and Tsukita, 1985), and a glycoprotein of 140 kD (Jones et al., 1986). Dp1 and dp2 are biochemically and immunologically related to each other, as are dg2 and dg3, while the remaining molecules are distinct (Cohen et al., 1983; Mueller and Franke, 1983; Kapprell et al., 1985).

Miller et al. (1987) have used specific antibodies and the high-resolution technique of the immunogold labeling of ultrathin frozen sections to record the detailed ultrastructural organization of the major protein and glycoprotein components within desmosomes of bovine nasal epithelium. In frozen sections stained with uranyl acetate, desmosomes appear to consist of four distinct zones: the intercellular space (about 30 nm wide), the paired plasma membranes of the two adjacent cells (each membrane about 8 nm wide), a cytoplasmic plaque (about 17 nm wide) bearing an electron-dense lamina showing a transverse periodicity of about 2.6 nm on its inner (cytoplasmic) face, and a zone about twice as wide as the plaque, showing no structural features between the plaque and the tonofilaments. (We have referred to this zone as the satellite zone, by way of analogy with centriolar satellites, to which microtubules attach.)

We now describe the molecular distributions found, starting at the intercellular space and progressing through to the cytoplasmic region between the desmosomal plaque and the tonofilaments. The quantitative basis for these distributions is summarized in Figure 2. Labeling for dg2 and dg3 was shown to be located principally in the 30-nm-wide intercellular space, a distribution consistent with the postulated role of these molecules in desmosomal adhesion (Cowin et al., 1984b; Garrod, 1985, 1986a,b; Garrod and Cowin, 1986). Although labeling was predominantly in the intercellular space, it was unclear whether or not dg2 and dg3 are transmembrane proteins, since the antibodies

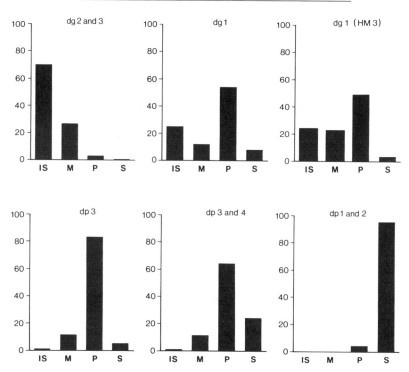

DISTRIBUTION OF GOLD PARTICLES IN LABELLED DESMOSOMES

Figure 2. *Distribution of gold particles in bovine nasal epithelial desmosomes labeled in ultrathin frozen sections with anti-desmosomal antibodies followed by protein A-gold. Vertical axis* indicates percentage distribution of gold particles between four desmosomal zones for a minimum of 50 desmosomes with each antibody. IS, intercellular space; M, plasma membrane; P, plaque; S, satellite region, a 34-nm-wide zone at the cytoplasmic face of the plaque (see Figure 3). All antibodies are guinea-pig polyclonals having specificity for desmosomal proteins and glycoproteins as indicated, except HM3, which is a rabbit polyclonal antibody. (From quantitative data given by Miller et al., 1987.)

may not react with the entire molecule. Labeling for dg1 was present both in the extracellular space and in the cytoplasmic plaque. Dg1 is therefore a transmembrane glycoprotein whose cytoplasmic portion makes a major contribution to the structure of the desmosomal plaque.

Dp3 is situated within the desmosomal plaque, with much labeling close to the cytoplasmic face of the plasma membrane (similar to that found by Gorbsky and Steinberg, 1981). Labeling with an antibody that recognized both dp3 and dp4 was similar in distribution but also extended beyond the plaque

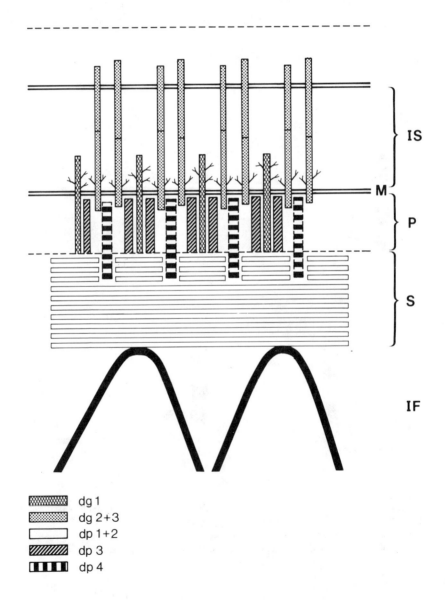

dg 1	
dg 2+3	
dp 1+2	
dp 3	
dp 4	

into the cytoplasm. Dp4 may therefore be positioned close to dp3 but distal to it from the plasma membrane, though this evidence cannot be regarded as conclusive, since a dp4-specific antibody was not available. Dp1 and dp2 were shown to be located predominantly in a region 34 nm wide between the desmosomal plaque and the tonofilaments. These proteins may therefore be involved in joining the plaque to the tonofilaments, rather than being major plaque constituents, as previously suggested by Franke et al. (1982). A diagrammatic model of desmosome structure, suggested by our results, is shown in Figure 3.

The immunogold localization of desmosomal proteins and glycoproteins in bovine nasal epithelial desmosomes has also been reported by Steinberg et al. (1986). The technique used by those authors involved the postembedding labeling of ultrathin sections, rather than the ultrathin frozen section technique used by us (Miller et al., 1987). It is rather difficult to make a precise comparison between the two sets of results because the different techniques give somewhat different impressions of desmosome structure. However, we believe the results to be broadly similar, with two apparent exceptions. First, Steinberg et al. (1986) suggest that dp1 and dp2 (which they call desmoplakin I/II) extend from a level in the cytoplasm roughly equivalent to the inner boundary of our 34-nm satellite zone all the way through the plaque to the inner leaflet of the plasma membrane. In contrast, our results indicate that dp1 and dp2 are largely confined to the satellite region and excluded from the plaque. Second, the data of Steinberg et al. (1986) suggest that the distributions of dg1, dg2, and dg3 (which they call desmogleins I and II) are identical; all three extend from the intercellular space to beyond the inner, cytoplasmic face of the plaque. In contrast, our results suggest that dg1 does indeed have this

---

Figure 3. *Diagrammatic model of desmosome structure,* based largely on immunogold labeling data given in Figure 2, but partly on trypsinization data described in text. The diagram shows one-half of a desmosome in detail: The entire desmosome would show mirror image symmetry about a line in the intercellular space (IS) midway between the two plasma membranes (M) of the adhering cells. The desmosomal glycoproteins dg1, dg2, and dg3 are depicted with different asymmetric transmembrane distributions as suggested by both gold labeling and trypsinization studies. Dg1 makes a major contribution to the structure of the desmosomal plaque (P), whereas the major portions of the dg2 and dg3 molecules (approximately 80%) are in the intercellular space. Dg2 is depicted larger than dg3. The end-to-end contact between the dg2 and dg3 molecules in the region of the midline is purely conjectural—there is no detailed evidence relating to the arrangement of molecules within the intercellular space. On the cytoplasmic side of the membrane, dp3 is depicted as being distributed throughout the plaque, although Miller et al. (1987) noted a tendency for alignment of labeling along the cytoplasmic face of the membrane. Dp1 and dp2 are depicted as being exclusively in the 34-nm-wide satellite zone (S), since there was very little labeling in the plaque region. The striped bar for dp4 is intended only to represent a range of possible positions for this molecule. The data may suggest a location toward the cytoplasmic face of the plaque, but are not conclusive. IF, intermediate filaments.

that the cytoplasmic extent of dg2 and dg3 may be considerably shorter than that of dg1.

In order to clarify the disposition of dg2 and dg3 with respect to the plasma membrane, a series of trypsinization and phosphorylation studies have been carried out with cultured Madin-Darby canine kidney (MDCK) cells. A monoclonal antibody, 52-3D, has been shown to react with a conserved epitope on dg2 and dg3, since it gave a fluorescent staining of many bovine, rat, and human epithelia. This epitope must be located either within the plasma membrane or on its cytoplasmic side, since 52-3D does not react with the surface of living cultured MDCK cells but gives peripheral staining after the cell membranes have been made permeable by acetone fixation (Figure 4). Trypsinization of MDCK cells in the absence of calcium (0.25% trypsin with 1 mM EGTA at 37°C for 40 min—no significant decrease in cell viability), followed by Western blotting with 52-3D, yields two polypeptides of 28 kD and 24 kD (Figure 5). The two undigested proteins have molecular weights of 130 kD and 115 kD. Since a monoclonal antibody recognizes one distinct epitope, it is probable that each of the two fragments comes from dg2 or dg3.

The incubation of MDCK cells with ^{32}P for 24 hours, followed by immunoprecipitation with a polyclonal anti-dg2 and anti-dg3 antibody, reveals that only dg2 is phosphorylated under these conditions (Figure 6).

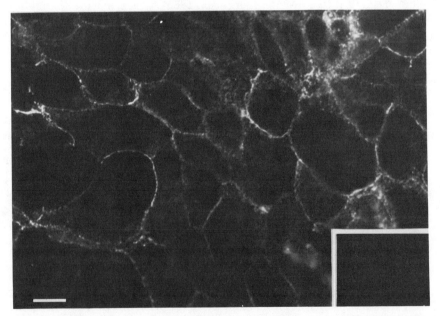

Figure 4. *Fluorescent staining of confluent MDCK cells with 52-3D, a monoclonal antibody against dp2 and dp3, after fixation of the cells with acetone.* The *inset* shows the absence of staining of the surfaces of living cells in low-calcium medium with this antibody, even though staining with polyclonal antibodies shows that these molecules are exposed on the cell surface under these conditions (see Mattey and Garrod, 1986a, Figure 3A). Calibration bar = 20 μm.

Figure 5. *Membrane-associated fragments of dg2 and dg3 in MDCK cells.* Western blot with monoclonal antibody 52-3D of an extract of whole MDCK cells (*lane A*) and trypsinized MDCK cells (*lane B*). The cells were grown to confluency in minimum essential medium (MEM) and trypsinized for 40 min at 37°C with 0.25% trypsin in 1 mM EDTA. Cell viability after this treatment was >90%.

Figure 6. *Phosphorylation of dg2 and its membrane-associated fragment.* MDCK cells were labeled for 24 hours with either [^{35}S] methionine (*lane A*) or ^{32}P (*lanes B,C*). ^{32}P-labeled cultures were solubilized either directly or after trypsinization, as described in Figure 5. Immunoprecipitation was then carried out with a polyclonal antibody to dp2 and dp3, and the immunoprecipitates were run in adjacent lanes of a polyacrylamide gel. Autoradiography was carried out for different times in order to give optimal exposure of ^{35}S- and ^{32}P-labeled bands. Note that the ^{32}P-labeled band in *lane B* has exactly the same mobility as the upper of the two ^{35}S-labeled bands. It thus corresponds to dg2.

Trypsinization of similarly treated cells leads to the immunoprecipitation of a single phosphorylated fragment of 28 kD (Figure 6), thus strengthening the view that the 28-kD and 24-kD tryptic fragments are derived, respectively, from dg2 and dg3.

We suggest that these fragments represent the membrane-associated portions of the dg2 and dg3 molecules. The sizes of the fragments appear to define maxima for the sizes of the intramembrane and cytoplasmic domains of the two molecules, and the phosphorylation of dg2 confirms that it almost certainly has a cytoplasmic domain. Furthermore, two criteria, size and phosphorylation, demonstrate heterogeneity between dg2 and dg3 in their membrane-associated domains.

   Trypsinization and immunoprecipitation experiments carried out on the dg1 molecule by Penn et al. (1987) suggest that the maximum estimate for the size of its protected domain is 92 kD. This result is consistent with the different asymmetric distributions of the dg1 compared to the dg2 and dg3 molecules suggested by our immunogold labeling data, illustrated in Figure 3.

CALCIUM-INDUCED DESMOSOME FORMATION

Assembly and Synthesis

In tissue culture media containing low calcium concentration, epithelial cells do not form desmosomes, but when the calcium concentration is raised, desmosome formation takes place (Hennings et al., 1980). We have shown that in keratinocytes and MDCK cells, desmosome formation triggered by raising calcium concentration from <0.05 mM to 1.8 mM is very rapid—approximately one hour (Watt et al., 1984; Mattey and Garrod, 1986a; Mattey et al., 1986)—whereas in MDBK (bovine kidney) cells the process takes about eight hours (Mattey and Garrod, 1986a). (1.8 mM is the calcium concentration of minimum essential medium [MEM], and is therefore used routinely for calcium switching. However, such a large rise in calcium concentration is not required for the triggering process; desmosome formation is induced by 0.1 mM calcium.) Some simple inhibitor studies suggest a possible reason for the difference in rate of desmosome formation between these cell types.

   The addition of cycloheximide to MDCK cells at 10 μg/ml, or even 100 μg/ml, while simultaneously raising the calcium concentration, has no apparent effect on desmosome formation (Figure 7a,b), even though cycloheximide at 5 μg/ml has been shown to inhibit protein synthesis by 90% in these cells (Gottlieb et al., 1986). However, if MDCK cells are pretreated for eight hours with cycloheximide (10 μg/ml) in low calcium medium (LCM) before calcium switching in the presence of cycloheximide, desmosome formation is completely inhibited. If the cycloheximide is removed at the time of calcium switching, desmosome formation occurs, but only after a delay of about three hours. Actinomycin D (1 μg/ml) also inhibits desmosome formation in MDCK cells, but only if added eight hours before the calcium switch.

   We conclude that MDCK cells in LCM possess all components necessary for desmosome formation in sufficient quantity to permit their rapid assembly into the fully formed junctions when the calcium concentration is raised. Protein synthesis is not required for this rapid assembly process. However, our results also suggest that synthesis of at least some components must occur in LCM (this has been confirmed by immunoprecipitation for dg2 and dg3 and for dp1 and dp2) and that these components are presumably subject to turnover, so that the inhibition of protein (or RNA) synthesis for eight hours results in the depletion of components and the failure of desmosome formation.

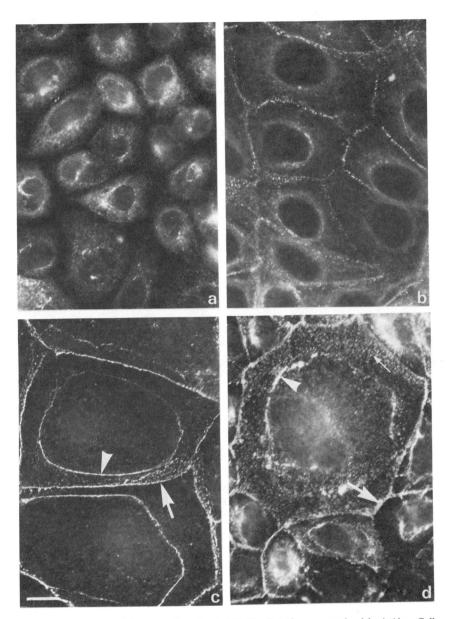

Figure 7. *Calcium-induced desmosome formation by MDCK cells in the presence of cycloheximide. a:* Cells cultured for 24 hours in low-calcium medium (LCM: calcium concentration <0.05 mM) show absence of desmosomal staining at cell periphery. *b:* Cells transferred into standard medium (SM: calcium concentration = 1.8 mM) in the presence of 100 μg/ml cycloheximide for 3 hours show peripheral staining for desmosomes. *c:* Cells cultured in SM for 24 hours, transferred to LCM for 30 min, and then back to SM for one hour. Note ring of internalizing desmosomes (*arrowhead*) and new desmosomes at cell periphery (*arrow*). *d:* Cells cultured for 24 hours in SM and then subjected to calcium switching according to the sequence SM→LCM→SM→LCM→SM over a two-hour period in the presence of 10 μg/ml cycloheximide. Ring of first internalized desmosomes (*arrowhead*), diffuse pattern of second internalized desmosomes (*small arrow*), and second set of new desmosomes (*large arrow*). Methanol fixation and staining with antibody to dp1 and dp2. Calibration bar = 20 μm.

It also seems that cells in LCM contain an excess of desmosomal components. We have shown previously that if MDCK cells that have been cultured in MEM for less than four days are placed in LCM, their desmosomes split and become internalized. If they are then returned to MEM 15–30 min later, they form new desmosomes while continuing to internalize the old ones that are not disassembled (Figure 7c) (Mattey and Garrod, 1986b; Mattey et al., 1986). If cycloheximide at 10 μg/ml is added prior to carrying out this type of experiment, the cells can undergo at least two cycles of new desmosome formation, provided that this is done within two hours of cycloheximide addition (Figure 7d). The turnover of desmosomal components once they have been incorporated into desmosomes seems to be very slow compared with the turnover in LCM, because when cells in MEM are continuously treated with cycloheximide, they show undiminished staining for desmosomes for long periods.

In contrast to MDCK cells, calcium-induced formation by MDBK cells is very slow, taking about eight hours from the time of calcium switching. Both cycloheximide (10 μg/ml) and actinomycin D (1 μg/ml) inhibit calcium-induced formation in MDBK cells, suggesting that they either do not possess all the components required to assemble desmosomes prior to the calcium switch or do not possess them in sufficient quantity, but instead need to undergo RNA and protein synthesis before they are competent to form desmosomes. In fact, the addition of actinomycin D at any time up to four or five hours after the switch inhibits desmosome formation, while the addition of cycloheximide even beyond eight hours after the switch greatly diminishes desmosome formation.

The Assembly Process

Fluorescent antibody studies of calcium-induced desmosome formation in keratinocytes and MDCK cells showed that dg2 and dg3 appeared to be evenly distributed over the surfaces of cells in LCM. Other desmosomal components were not detectable at the cell surface (Watt et al., 1984; Mattey and Garrod, 1986a). Raising the calcium concentration caused a rapid accumulation of both protein and glycoprotein components at regions of intercellular contact (first detectable at about 15 min). Subsequently, there was a gradual and progressive increase in staining for desmosomal components in regions of contact, and this was accompanied, in keratinocytes, by a loss of diffuse staining (Watt et al., 1984; Mattey and Garrod, 1986a).

Electron microscopy of calcium-switched cells showed a series of progressive stages of desmosome formation, similar to those previously described by Lentz and Trinkaus (1971) and Dembitzer et al. (1980), in which the initiating event was an accumulation of electron-dense material between adjacent cell membranes, followed by the gradual development of plaque structures. Completely formed desmosomes were found at one hour (Mattey and Garrod, 1986a). We found no evidence that completely formed half desmosomes were inserted into the cell surface and then paired between adjacent cells, as suggested for keratinocytes by Jones and Goldman (1985). The formation of desmosomes was accompanied by a rearrangement of cytokeratin to form an attachment to the cell surface at desmosomal plaques (see also Bologna et al., 1986).

Since desmosomes are adhesive structures that form only in regions of intercellular contact, we have suggested that the initiating event in desmosome formation is adhesive recognition between and patching of the dg2 and dg3 molecules. This would then be followed by the insertion and recruitment of other desmosomal components from within the cell, while the size of the desmosomal domain and therefore the strength of adhesion would be increased by the lateral recruitment of more cell surface molecules. The process might be thought of as analogous to the operation of a two-dimensional zipper (Garrod, 1985) (Figure 8).

An alternative possibility for desmosome formation is that instead of patching on the surface, adhesive recognition is initiated by targeting desmosomal adhesion molecules to desmosomal domains from within the cell. Such a mechanism may be particularly attractive in simple epithelial cells, where desmosomes are confined to the lateral cell membranes, instead of being all over the cell surface as they are in the middle layers of stratified epithelia. Adhesive interaction between desmosomes of simple and stratified epithelial cells takes place readily in tissue culture (Mattey and Garrod, 1985). It is therefore the position, and also the number, of desmosomes that

Low (Ca^{2+})

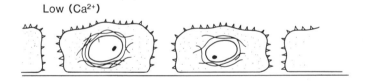

High (Ca^{2+})

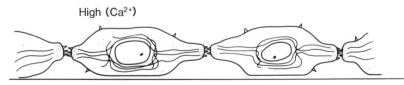

Figure 8. *Proposed redistribution of desmosomal components in keratinocytes during calcium-induced desmosome formation, as suggested by fluorescent antibody staining.* Dg2 and dg3 are indicated as *"teeth"* on the cell surface, other desmosomal components as *dots* in the cytoplasm, and the cytokeratin filaments as *lines* in the cytoplasm. A similar model would seem suitable for MDCK cells. (From Garrod, 1985.)

characterize the adhesive differences between simple and stratified epithelia, and not adhesive specificity (Garrod, 1986a,b).

The mechanism by which calcium concentration controls desmosome assembly is not understood. It seems probable that several different calcium-dependent processes may be involved. The work of Hennings et al. (1983), showing that calcium channel blockers and ionophores were without effect on desmosome formation in keratinocytes, may suggest that the role of calcium is primarily extracellular. We have also shown that a large number of inhibitors of cell function processes involving calcium are without effect on calcium-induced desmosome formation in MDCK cells. Thus, desmosome formation in MDCK cells is not inhibited by: (1) cytochalasin B, colchicine, or nocodazole, drugs that disrupt the cytoskeleton; (2) calmidazolium, chlorpromazine, or trifluoperazine, which are calmodulin antagonists; (3) TMB8, which prevents the mobilization of intracellular calcium; (4) monensin, which interferes with the transport of glycoprotein through the Golgi apparatus; (5) palmityl-D,L-carnitine, an inhibitor of protein kinase C; and (6) TPA, a tumor promoter and activator of protein kinase C. In spite of these results, it is still possible to envisage some cytoplasmic role for calcium, either by direct interaction with desmosomal components in the cytoplasm or by the modulation of some other cytoplasmic event, such as phosphorylation. Following is a series of observations that seem relevant to the control process:

1. The threshold extracellular calcium concentration required for desmosome formation is about 0.1 mM, roughly 1/20 of the concentration in most tissue culture media. The threshold appears to be similar for both the rapid, protein synthesis–independent assembly process shown by MDCK cells and the slow, protein synthesis–dependent process of MDBK cells.

2. Desmosomes of keratinocytes and MDCK cells undergo a stabilization process after their formation that makes them resistant to disruption even by chelating agents. Before stabilization, however, desmosomal adhesion may be broken simply by reducing extracellular calcium concentration to below 0.05 mM. The desmosomes of MDBK cells do not appear to undergo stabilization but always (up to two weeks in culture) remain susceptible to splitting by a reduction in extracellular calcium concentration (Mattey and Garrod, 1986b).

3. Keratinocytes may be induced to form desmosomes even in calcium concentration below 0.05 mM by incubating them with tunicamycin (1 μg/ml) for 48 hours (Mattey et al., 1986). We have now shown that this treatment results in the deglycosylation of the desmosomal glycoproteins, as described by Penn et al. (1987). Furthermore, the same effect is produced by treatment with the deglycosylating agent 2-deoxyglucose, but not with inhibitors of carbohydrate processing such as castanospermine and swainsonine (D. Mattey, unpublished data). Tunicamycin does not inhibit calcium-induced desmosome formation in keratinocytes, MDBK or MDCK cells, or chick embryonic corneal epithelial cells (Overton, 1982).

4. The major desmosomal glycoproteins (dg1, dg2, and dg3) are calcium-

binding proteins that bind calcium at micromolar concentrations or less (Mattey et al., 1986; Steinberg et al., 1986).

5. Trypsinization of isolated desmosomes in the presence of calcium yields a 42-kD glycosylated fragment that blots with some anti-dg2 and anti-dg3 antibodies and appears to be degraded by trypsin in the presence of EDTA (Mattey et al., 1986).

6. Desmosomes of bovine nasal epithelium contain desmocalmin, a protein of 240 kD that binds both calmodulin and cytokeratin. It is located in the vicinity of the desmosomal plaque (Tsukita and Tsukita, 1985).

The conclusions that we draw from these results are the following:

1. Desmosome assembly and separation (in unstabilized cells) are sensitive to changes in extracellular calcium concentration above and below a threshold of about 0.1 mM (100 μM).
2. Under physiological conditions, extracellular calcium concentration is likely always to be above that required to regulate desmosome assembly/separation.
3. Desmosomal glycoproteins lacking N-linked carbohydrates can participate in desmosome formation. Thus, N-linked carbohydrate is not required for desmosomal adhesion, confirming the suggestion of Overton (1982).
4. Although the effect of tunicamycin on keratinocytes is certainly complex, the N-linked carbohydrates of desmosomal glycoproteins may be important in regulating the calcium-induced response and also in determining desmosomal stability.
5. Desmosomal glycoproteins of keratinocytes are competent to participate in adhesive binding at calcium concentrations below 0.05 mM, at least when deglycosylated.
6. Calcium binding by desmosomal proteins in the micromolar range is unlikely to be involved in the regulatory response, which is sensitive to fluctuations in calcium concentration at a level two orders of magnitude greater than this.
7. Such calcium binding is, however, likely to be important in determining the configuration of the molecules, since EDTA treatment increases sensitivity to tryptic digestion.

DESMOSOMES AND CANCER

The reduced strength of adhesion between tumor cells may contribute to the invasive nature of malignancies and could arise through the loss of adhesive junctions. There have been varying reports on the numbers of desmosomes in tumor cells, mostly based on nonquantitative electron microscope studies.

However, Alroy et al. (1981), in a quantitative morphometric analysis, showed a correlation between reduced numbers of desmosomes and the invasiveness of carcinoma in the urinary bladder. Kocher et al. (1981) also found a decrease in the number of desmosomes in invasive carcinoma of the cervix, and basal and squamous cell carcinoma of the epidermis. In contrast, Pauli et al. (1978) found no such correlation in chemically induced bladder carcinoma in rats.

The advent of anti-desmosomal antibodies has enabled us to examine large numbers of malignant tumors for desmosomes. A survey of 244 malignant carcinomas revealed that all of these tumors stained for desmosomal antigens, regardless of their histological grade or degree of differentiation, or whether they were primary or secondary (Table 1). We conclude that all carcinomas that we have examined possessed desmosomes and that metastatic cell populations do not represent a special group of cells that have lost their ability to form desmosomal junctions.

We have examined the distribution of desmosomal staining in carcinoma of the bowel in more detail. We found that the cells of well- and moderately differentiated tumors showed an organized and polarized distribution of desmosomal staining similar to that found in normal bowel mucosa (Figure 9a,b). The punctate staining was present only at the lateral contacts between the cells, and was concentrated in the terminal bar region near the cell apices. Rather remarkably, the same polarized distribution of staining was found in liver metastases derived from these tumors (Figure 9c). This leads to the conclusion that metastatic cells derived from well- and moderately differentiated carcinoma of the bowel retain the ability to form desmosomes and to arrange them in a cellular distribution resembling that found in the primary carcinoma. Only in poorly differentiated carcinoma was the polarity of desmosomal staining disturbed (Figure 9d).

Since it seems that the absence of desmosomes does not explain the metastatic spread of colorectal carcinoma, we have attempted to determine whether the junctions of tumors are less stable (i.e., more readily disrupted)

Table 1.   Carcinomas Examined for Desmosomes with Anti-Desmosomal Antibodies

Type of Carcinoma	Primary Tumors	Secondary Tumors	Total	No. Staining Positively	Antibody[a]
Breast	77	7	84	84	064
Bowel (Colon + rectum)	47	6	53	53	11-5F
Lung (Various)	53		53	53	32-2B
Bladder	40		40	40	32-2B
Intracranial (Various)		14	14	14	11-5F

[a]064 = guinea pig anti-dp1 and -dp2 (Cowin and Garrod, 1983); 11-5F = mouse monoclonal anti-dp1 and -dp2 (Parrish et al., 1987); 32-2B = mouse monoclonal anti-dg1 (Vilela et al., 1987).

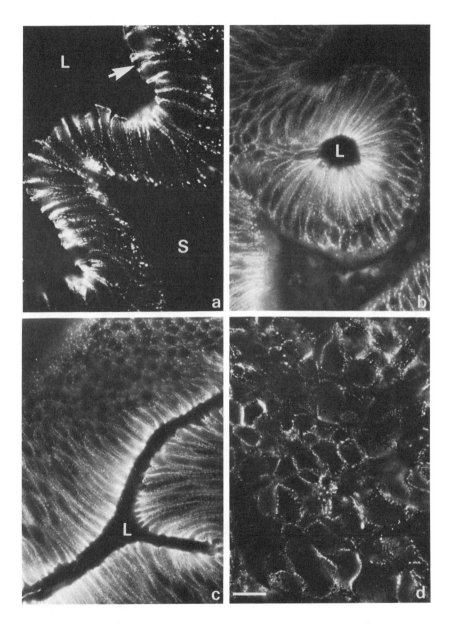

Figure 9. *Fluorescent staining of frozen sections of human colorectal carcinomas with monoclonal antibody to dp1 and dp2. a:* Uninvolved bowel mucosa showing normal distribution of desmosomal staining. The staining is strongest in the apicolateral regions of the cells (terminal bar region) and decreases toward the submucosa (S). The apical or luminal surfaces of the cells are completely devoid of staining (*arrow*). L, lumen. *b:* Moderately well differentiated primary carcinoma shows retention of polarized desmosomal staining. *c:* Metastatic deposit in liver of same tumor as shown in (*b*), again showing retention of polarized staining. *d:* Poorly differentiated carcinoma showing nonpolarized distribution of desmosomal staining. Calibration bar = 20 μm.

than those of normal tissues. Borysenko and Revel (1973) speculated that differences in the sensitivity of desmosomes in different tissues to disruption by EDTA, trypsin, or sodium deoxycholate might reflect differences in physiological stability. Watt et al. (1984) and Mattey and Garrod (1986b) showed that desmosomes of cells in tissue culture exhibit differences in stability when treated with EDTA or when the extracellular calcium concentration is simply reduced (< 0.05 mM). These differences vary between cell types, and with time in culture in a given cell type. It is important to determine whether differences of a similar nature exist between carcinoma cells and those of the normal tissue from which the tumor originates.

We have shown that cell lines derived from human bowel carcinomas exhibit calcium-dependent desmosome stability properties similar to those of MDCK cells. This prompted us to study stability in neoplastic and uninvolved bowel epithelium in response to extracellular calcium reduction, assessing the results by electron microscopy. When desmosomal junctions of normal bowel were subjected to disruption by removing extracellular calcium (4 mM EDTA), the desmosomes remained intact for up to two hours, although the non-junctional cell membranes separated and the epithelium became detached from the submucosa (Figure 10a). In tumors many intercellular contacts were disrupted within 30 min by either 4 mM EDTA or low calcium concentration (Figure 10b). However, we wish to stress that *this effect was not specific to desmosomes*, but rather involved the complete disruption of intercellular contacts.

The treatment of tumors with EDTA or low calcium concentration induced much cell degeneration, and this may have been primarily responsible for the loss of intercellular contact. Significantly, however, there were groups of cells within the degenerating regions that appeared to be viable, since their cytoplasm showed a normal density of staining and no vacuolation, while their mitochondria were also unswollen (Figure 11a). Furthermore, where contact between such cells and the surrounding degenerate ones was broken, the viable cells were found to be internalizing the half desmosomes left unpaired by the loss of contact (Figure 11b). [Desmosome internalization is a well-documented response of cells to loss of contact (Overton, 1968; Kartenbeck et al., 1982; Mattey and Garrod, 1986b).] Within these viable cell groups desmosomes remained intact, although nonjunctional membranes were separated in response to calcium reduction.

We feel that these observations may have significance in relation to the mechanism of generation of metastases. They demonstrate that cells within tumors are heterogeneous in their response to extracellular calcium concentration reduction. Such differential effects on cellular adhesion may illustrate how the loss of intercellular contact could release viable clumps of cells that may be transported to other locations.

The absolute regularity of the occurrence of desmosomes in carcinomas means that anti-desmosomal antibodies should provide reliable reagents for the diagnosis of carcinomas. With this in view, we have developed two

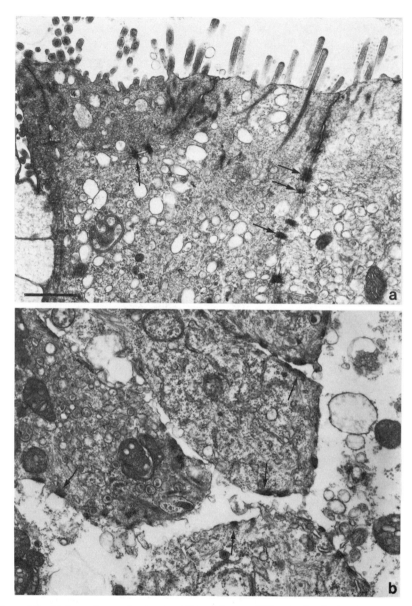

Figure 10. *Electron micrographs showing differential susceptibility of uninvolved and cancerous bowel epithelium to disruption of intercellular contacts by 30-min treatment at 30°C with 4 mM EDTA. a: Uninvolved epithelium. Desmosomes (arrows) and other intercellular junctions are intact and have an essentially normal appearance. b: Moderately well differentiated carcinoma from the same patient showing extensive disruption of intercellular contacts. Unpaired desmosomal halves remain at the cell surface (arrows). Calibration bar = 20 μm.*

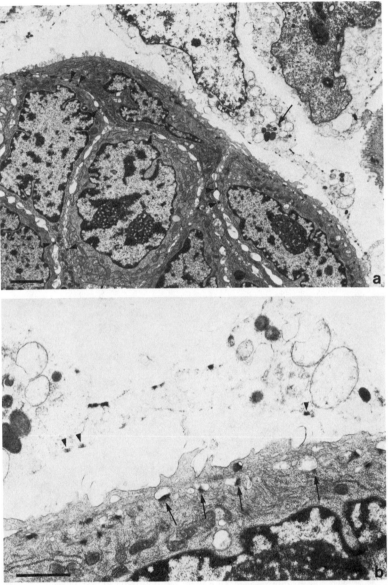

Figure 11. *Formation of apparently viable cell clumps by EDTA treatment of colorectal carcinoma.*
a: Periphery of cell clump showing loss of contact with adjacent cells, which appear degenerate
and necrotic (*arrow*). Note that within the cell clump the cell membranes have separated but
remain attached at desmosomes (*arrowheads*). Calibration bar = 2 μm. *b:* Enlargement of
portion of (*a*) showing internalization of unpaired desmosomal halves (*arrows*) at free cell
surface at periphery of clump. The unpaired desmosomal halves (*arrowheads*) in the adjacent
necrotic cell remain at the cell surface. Calibration bar = 1 μm.

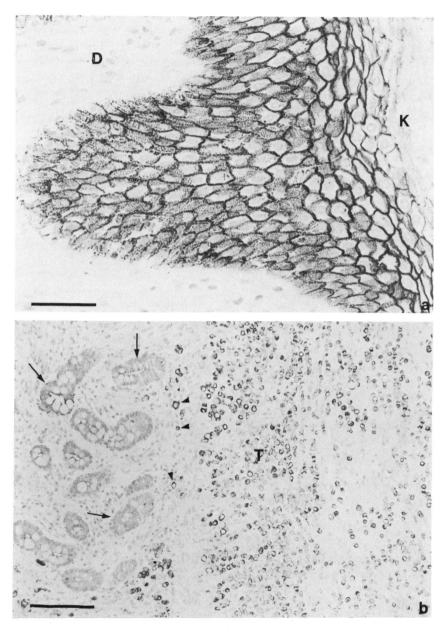

Figure 12. *Immunoperoxidase staining of formalin-fixed, paraffin-embedded pathology specimens with monoclonal antibody 32-2B against dg1. a:* Normal skin from vulva showing specificity of antibody for epidermis and punctate staining for desmosomes in the germinative, spinous, and granular layers, but not in the outer, keratinized layer. D, dermis; K, keratinized layer. Calibration bar = 20 μm. *b:* Metastatic deposit of carcinoma of breast in submucosa of small intestine. Note staining of normal bowel mucosa (*arrows*) tumor (T), illustrating the ability of the antibody to detect single, isolated tumor cells (*arrowheads*). Calibration bar = 50 μm.

monoclonal antibodies that are proving extremely reliable: 32-2B is an antibody against dg1 that reacts with carcinomas in paraffin sections as employed in routine diagnostic pathology (Vilela et al., 1987) (Figure 12), while 11-5F is an antibody against dp1 and dp2 that reacts in frozen sections and should prove useful for the intraoperative diagnosis of intracranial tumors, meningiomas, and metastatic carcinomas (Parrish et al., 1987). So far, these reagents appear to be more reliable than existing epithelial markers.

ACKNOWLEDGMENTS

Original work reported in this chapter was supported by the Cancer Research Campaign (D. R. G., E. P. P., D. L. M.), the Medical Research Council (H. R. M.), the University of Southampton (J. E. M.), Conselho Maçional de Pesquisas do Brasil, and Universidade Federal de Vicosa-MG, Brazil (M. J. V.).

REFERENCES

Alcolado, R., R. O. Weller, E. P. Parrish, and D. R. Garrod (1988) The cranial arachnoid and pia mater in man: Anatomical and ultrastructural observations. *Neuropathol. Appl. Neurobiol.* **14**:1–17.

Alroy, J., B. U. Pauli, and R. S. Weinstein (1981) Correlation between numbers of desmosomes and the aggressiveness of transitional cell carcinoma in human urinary bladder. *Cancer* **47**:104–112.

Andreis, L., F. Harrison, R. Hertseus, and L. Vakaet (1985) Cell junctions and locomotion of the blastoderm edge in gastrulating chick and quail embryos. *J. Cell Sci.* **78**:191–204.

Arnn, J., and L. A. Staehelin (1981) The structure and function of spot desmosomes. *Dermatology* **20**:330–339.

Atherton, B. T., D. M. Meyer, and D. G. Simpson (1986) Assembly and remodelling of myofibrils and intercalated discs in cultured neonatal rat heart cells. *J. Cell Sci.* **86**:233–248.

Bologna, M., R. Allen, and R. Dulbecco (1986) Organization of cytokeratin bundles by desmosomes in rat mammary epithelial cells. *J. Cell Biol.* **102**:560–567.

Borysenko, J. Z., and J.-P. Revel (1973) Experimental manipulation of desmosome structure. *J. Anat.* **137**:403–422.

Cohen, S. M., G. Gorbsky, and M. S. Steinberg (1983) Immunochemical characterization of related families of glycoproteins in desmosomes. *J. Biol. Chem.* **258**:2621–2627.

Cowin, P., and D. R. Garrod (1983) Antibodies to epithelial desmosomes show wide tissue and species cross-reactivity. *Nature* **302**:148–150.

Cowin, P., D. L. Mattey, and D. R. Garrod (1984a) Distribution of desmosomal components in the tissues of vertebrates, studied by fluorescent antibody staining. *J. Cell Sci.* **66**:119–132.

Cowin, P., D. L. Mattey, and D. R. Garrod (1984b) Identification of desmosomal surface components (desmocollins) and inhibition of desmosome formation by specific Fab'. *J. Cell Sci.* **70**:41–60.

Cowin, P., W. W. Franke, C. Grund, H.-P. Kapprell, and J. Kartenbeck (1985) The desmosome–intermediate filament complex. In *The Cell in Contact: Adhesions and Junctions as Morphogenetic Determinants,* G. M. Edelman and J.-P. Thiery, eds., pp. 427–460, Wiley, New York.

Dembitzer, H. M., F. Herz, A. Schermer, R. C. Wooley, and L. G. Koss (1980) Desmosome development in an *in vitro* model. *J. Cell Biol.* 85:695–702.

Docherty, R. J., J. G. Edwards, D. R. Garrod, and D. L. Mattey (1984) Chick embryonic pigmented retina is one of a group of epithelial tissues which lack cytokeratins and desmosomes and have intermediate filaments composed of vimentin. *J. Cell Sci.* 71:61–74.

Ducibella, T., D. E. Albertini, E. Anderson, and J. D. Biggers (1975) The preimplantation mammalian embryo: Characterization of intercellular junctions and their appearance during development. *Dev. Biol.* 45:231–250.

Franke, W. W., E. Schmid, C. Grund, H. Mueller, I. Engelbrecht, R. Moll, J. Stadler, and E. D. Jarasch (1981) Antibodies to high molecular weight polypeptides of desmosomes: Specific localisation of a class of junctional proteins in cells and tissues. *Differentiation* 20:217–241.

Franke, W. W., R. Moll, D. L. Schiller, E. Schmid, J. Kartenbeck, and H. Mueller (1982) Desmoplakins of epithelial and myocardial desmosomes are immunologically and biochemically related. *Differentiation* 23:115–127.

Garrod, D. R. (1981) Adhesive interactions of cells in development: Specificity, selectivity, and cellular adhesive potential. *Fortschr. Zool.* 26:184–195.

Garrod, D. R. (1985) The adhesions of epithelial cells. In *Cellular and Molecular Control of Direct Cell Interactions,* H.-J. Marthy, ed., NATO Adv. Study Inst. Ser. A, Life Sci. 99:43–83.

Garrod, D. R. (1986a) Formation of desmosomes in polarized and non-polarized epithelial cells: Implications for epithelial morphogenesis. *Biochem. Soc. Trans.* 14:172–175.

Garrod, D. R. (1986b) Desmosomes, cell adhesion molecules, and the adhesive properties of cells in tissues. *J. Cell Sci. (Suppl.)* 4:221–237.

Garrod, D. R., and P. Cowin (1986) Desmosome structure and function. In *Receptors in Tumour Biology,* C. M. Chadwick, ed., pp. 95–130, Cambridge Univ. Press, Cambridge, England.

Garrod, D. R., and A. Nicol (1981) Cell behaviour and molecular mechanisms of cell–cell adhesion. *Biol. Rev. Camb. Philos. Sci.* 56:199–242.

Gorbsky, G., and M. S. Steinberg (1981) Isolation of the intercellular glycoproteins of desmosomes. *J. Cell Biol.* 90:243–248.

Gottlieb, T. A., A. Gonzalez, L. Rizzolo, M. J. Rindler, M. Adesnik, and D. D. Sabatini (1986) Sorting and endocytosis of viral glycoproteins in transfected polarized epithelial cells. *J. Cell Biol.* 102:1242–1255.

Hennings, H., D. Michael, C. Cheng, P. Steinert, K. Holbrook, and S. H. Yuspa (1980) Calcium regulation of growth and differentiation of mouse epidermal cells in culture. *Cell* 19:245–254.

Hennings, H., K. A. Holbrook, and S. H. Yuspa (1983) Factors influencing calcium-induced terminal differentiation in cultured mouse epidermal cells. *J. Cell. Physiol.* 116:265–281.

Jones, J. C. R., and R. D. Goldman (1985) Intermediate filaments and the initiation of desmosome assembly. *J. Cell Biol.* 101:506–517.

Jones, J. C. R., K. M. Yokoo, and R. D. Goldman (1986) A cell surface desmosome-associated component: Identification of a tissue-specific cell adhesion molecule. *Proc. Natl. Acad. Sci. USA* 83:7282–7286.

Kapprell, H.-P., P. Cowin, W. W. Franke, H. Postingl, and H. J. Opferkuch (1985) Biochemical characterization of desmosomal proteins isolated from bovine muzzle epidermis: Amino acid and carbohydrate composition. *Eur. J. Cell Biol.* 36:217–229.

Kartenbeck, J., E. Schmid, W. W. Franke, and B. M. Geiger (1982) Different modes of internalization of proteins associated with *adhaerens* junctions and desmosomes: Experimental separation of lateral contacts induces endocytosis of desmosomal plaque material. *EMBO J.* 1:725–732.

Kocher, O., M. Amaudruz, A.-M. Schindler, and G. Gabbiani (1981) Desmosomes and gap junctions in precarcinomatous and carcinomatous conditions of squamous epithelia: An electron microscopic and morphometrical study. *J. Submicrosc. Cytol.* 13:267–281.

Lentz, T. L., and J. P. Trinkaus (1971) Differentiation of the junctional complex of surface cells in the developing *Fundulus* blastoderm. *J. Cell Biol.* **48**:455–472.

Mattey, D. L., and D. R. Garrod (1985) Mutual desmosome formation between all binary combinations of human, bovine, carmine, avian, and amphibian cells: Desmosome formation is not tissue or species specific. *J. Cell Sci.* **75**:377–399.

Mattey, D. L., and D. R. Garrod (1986a) Calcium-induced desmosome formation in cultured kidney epithelial cells. *J. Cell Sci.* **85**:95–111.

Mattey, D. L., and D. R. Garrod (1986b) Splitting and internalization of the desmosomes of cultured kidney epithelial cells by reduction in calcium concentration. *J. Cell Sci.* **85**:113–124.

Mattey, D. L., A. Suhrbier, E. Parrish, and D. R. Garrod (1986) Recognition, calcium, and the control of desmosome formation. *Ciba Found. Symp.* **125**:49–65.

McNutt, M. S., and R. S. Weinstein (1973) Membrane ultrastructure at mammalian intercellular junctions. *Prog. Biophys. Mol. Biol.* **26**:45–101.

Miller, K., D. Mattey, H. Measures, C. Hopkins, and D. R. Garrod (1987) Localisation of the protein and glycoprotein components of bovine nasal epithelial desmosomes by immunoelectron microscopy. *EMBO J.* **6**:885–889.

Moll, R., P. Cowin, H.-P. Kapprell, and W. W. Franke (1986) Desmosomal proteins: New markers for identification and classification of tumors. *Lab. Invest.* **54**:4–25.

Mueller, H., and W. W. Franke (1983) Biochemical and immunological characterization of desmoplakins I and II, the major polypeptides of the desmosomal plaque. *J. Mol. Biol.* **163**:647–671.

Overton, J. (1962) Desmosome development in normal and reassociating cells of the early chick blastoderm. *Dev. Biol.* **4**:532–548.

Overton, J. (1968) The fate of desmosomes in trypsinized tissue. *J. Exp. Zool.* **168**:203–213.

Overton, J. (1974) Cell junctions and their development. *Prog. Surf. Membr. Sci.* **8**:161–208.

Overton, J. (1977) Formation of junctions and cell sorting inaggregates of chick and mouse cells. *Dev. Biol.* **55**:103–116.

Overton, J. (1982) Inhibition of desmosome formation with tunicamycin and with lectin in corneal cell aggregates. *Dev. Biol.* **92**:66–72.

Parrish, E. P., D. R. Garrod, D. L. Mattey, L. Hand, P. Steart, and R. O. Weller (1986) Mouse antisera specific for desmosomal adhesion molecules of suprabasal skin cells, meninges, and meningioma. *Proc. Natl. Acad. Sci. USA* **83**:2657–2661.

Parrish, E. P., P. V. Steart, D. R. Garrod, and R. O. Weller (1987) Antidesmosomal monoclonal antibody in the diagnosis of intracranial tumours. *J. Pathol.* **153**:265–273.

Pauli, B. U., S. M. Cohen, J. Alroy, and J. R. Weinstein (1978) Desmosome ultrastructure and the biological behaviour of chemical carcinogen-induced urinary bladder carcinomas. *Cancer Res.* **38**:3276–3285.

Penn, E. J., C. Hobson, D. A. Rees, and A. I. Magee (1987) Structure and assembly of desmosome junctions: Biosynthesis, processing, and transport of the major protein and glycoprotein components in cultured epithelial cells. *J. Cell Biol.* **105**:57–68.

Ramaekers, F. C. S., M. Osborn, E. Schmid, K. Weber, H. Bloemendal, and W. W. Franke (1980) Identification of the cytoskeletal proteins in lens-forming cells, a special epithelioid cell type.*Exp. Cell Res.* **127**:309–327.

Skerrow, C. J., and A. G. Matoltsy (1974) Chemical characterization of epidermal desmosomes. *J. Cell Biol.* **63**:524–530.

Staehelin, L. A. (1974) Structure and function of intercellular junctions. *Int. Rev. Cytol.* **39**:191–283.

Steinberg, M. S., H. Shida, G. J. Guidice, M. Shida, N. H. Patel, and O. W. Blaschuck (1986) On the molecular organisation, diversity, and functions of desmosomal proteins. *Ciba Found. Symp.* **125**:3–25.

Suhrbier, A., and D. R. Garrod (1986) An investigation of the molecular components of desmosomes in epithelial cells of five vertebrates. *J. Cell Sci.* **81**:223–242.

Tsukita, S., and S. Tsukita (1985) Desmocalmin—A calmodulin-binding high molecular weight protein isolated from desmosomes. *J. Cell Biol.* **101**:2070–2080.

Vilela, M. J., E. P. Parrish, D. H. Wright, and D. R. Garrod (1987) Monoclonal antibody to desmosomal glycoprotein 1—A new epithelial marker for diagnostic pathology. *J. Pathol.* **153**:365–375.

Watt, F. M., D. L. Mattey, and D. R. Garrod (1984) Calcium-induced reorganization of desmosomal components in cultured human keratinocytes. *J. Cell Biol.* **99**:2211–2215.

Chapter 15

The Epithelial Tight Junction: Occluding Barrier and Fence

KAI SIMONS

ABSTRACT

The plasma membrane of epithelial cells is differentiated into two domains: the apical membrane lining the epithelial lumen, and the basolateral domain, which faces the blood supply of the tissue. The two plasma membrane domains are responsible for the vectorial functions that characterize transporting, absorptive, and secretory epithelia. The epithelial cells accomplish their polarized functions by localizing a distinct set of cell surface components to either of the two plasma domains. Not only do the apical and basolateral cell surfaces have distinct protein compositions (enzymes, transport proteins, channels, receptors, etc.), but they also contain different lipids. The boundary between the two cell surface domains is formed by a specific zone of cell–cell contacts, which encircle the top of each cell. This structure, called the tight junction or zonula occludens, connects neighboring cells and seals the spaces between the cells. The occluding barrier between the cells has been well characterized. The tight junction also has a role in maintaining the compositional differences between the two domains by forming a diffusion barrier in the plane of the membrane. This chapter summarizes our knowledge of the structure of the tight junction and its functions as an occluding barrier and as a fence for maintaining the spatial division of the plasma membrane.

The tight junction, or zonula occludens (ZO), links epithelial cells into a monolayer by forming a continuous belt of sealing contacts around the apex of each cell. In thin sections the outer leaflets of the adjoining membranes appear to be fused with one another in the contact areas (Farquhar and Palade, 1963). In freeze-fracture replicas these contact areas appear as an anastomosing network of strands (Chalcroft and Bullivant, 1970; Friend and Gilula, 1972). Early results suggested that these strands were made up of a series of individual intramembranous particles that could be cross-linked by glutaraldehyde into a continuous strand (Staehelin, 1974; van Deurs and Luft, 1979). However, quick-freezing methods demonstrated that the strands were continuous, and that the particles arose as an artifact during slow freezing (Kachar and Reese, 1982; Pinto da Silva and Kachar, 1982).

Another cellular junction immediately basal to the tight junctions is the intermediate junction, or zonula adherens (ZA) (Geiger et al., 1985). The adjoining membranes in these specialized contact areas are separated by a gap of 200 Å (Farquhar and Palade, 1963). On the cytoplasmic face there is a circumferential ring of actin filaments linked to the junction by an adhesion plaque that contains the actin-binding proteins vinculin and α-actinin (Geiger et al., 1980, 1981; Mooseker et al., 1984). Little is known about the molecular structure of these two zonular junctions in absorptive, secretory, and transporting epithelia. Although they usually are considered as separate entities, it is likely that they are interconnected both structurally and functionally (Gumbiner and Simons, 1987).

The zonular junctions form an occluding barrier that inhibits the diffusion of solutes between the cells (Diamond, 1977). This barrier is formed by the ZO strands alone, since the intercellular space in the intermediate junctions is accessible to electron-dense marker molecules (Friend and Gilula, 1972). The other important function of the junctions is to form a "fence" that divides the plasma membrane into its two domains, the apical membrane, facing the external environment, and the basolateral membrane on the serosal side of the epithelium (Diamond, 1977). This fence function has also been attributed to the ZO strands (Claude and Goodenough, 1973), although exact studies on the permeability of the intermediate junctions to membrane components have not yet been performed. This chapter briefly summarizes our present knowledge of the occluding function of the ZO (for more detailed accounts, see Diamond, 1977; Powell, 1981) and concentrates on our studies on the fence function of the tight junctions.

THE OCCLUDING BARRIER

It was long believed that the tight junctions prevented transepithelial solute fluxes from passing between the cells and forming an epithelium. Frömter and Diamond (1972) first convincingly demonstrated that this was not the case. They showed that the electrical resistance of the cell membranes of the Necturus gall bladder epithelial cells exceeded the measured transepithelial resistance by a factor of 23. Hence, the conductance through the junctions, the paracellular shunt, accounted for 22/23 of the transepithelial ion conductance. A survey of epithelial properties revealed other "leaky" epithelia in small intestine and renal proximal tubule (Diamond, 1977). Tight epithelial junctions were found in urinary bladder, salivary and sweat duct, distal and renal tubule, and stomach. The junctional resistance varied from 4 ohm.cm^2 in the leaky rat proximal tubule to over 300,000 ohm.cm^2 in the tight epithelia of the rabbit urinary bladder.

The structural basis for these variations in junctional resistance is now known to depend on the organization of the ZO (Claude and Goodenough, 1973). Several careful studies have demonstrated that the paracellular conductance is inversely proportional to the logarithm or number of ZO strands in the tight junction (Claude, 1978; Easter et al., 1983; Marcial et al., 1984; Madara and Dharmsathaphorn, 1985). Or, in other words, leaky junctions have few strands, whereas tight epithelia have an extensive network of ZO strands.

Some epithelial monolayers, such as those formed by low-resistance MDCK strain II cells, have been shown to have occluding junctions that are heterogeneous with respect to strand number (Cereijido et al., 1980). The occluding junctions do not form a homogeneous belt with uniform strand numbers, as in most native epithelia, but instead have regions with only one or two strands intervening between a network of seven or eight strands. Cereijido et al. (1980) showed by microelectrode surface scanning that there were local current sinks along the perimeter of the cells and proposed that these corresponded to the regions of lower strand number.

The selectivity of the barrier formed by the occluding junctions has been studied in numerous epithelia (Powell, 1981). In most epithelia the paracellular route is more permeable to cations than to anions. For instance, an MDCK strain II cell monolayer is three to nine times more permeable to sodium than to chloride (Cereijido et al., 1978, 1980; Rabito et al., 1978). This has been attributed to the nature of the paracellular pathway, which is thought to be lined with negative charges (carboxyl, phosphoric acid, or sulfuric acid groups) that discriminate against the passive movement of anions through the hydrated channels (Powell, 1981). However, not all epithelial tight junctions are cation selective. The paracellular shunt in the straight segment of the proximal tubule is anion selective (Schafer et al., 1984). This is reflected in the ion permeability of pig $LLC\text{-}PK_1$ cells, which have many characteristics of the proximal tubule (Rabito, 1986). The junctional permeability to sodium is only 40% of the permeability to chloride in $LLC\text{-}PK_1$ cell monolayers.

One physiologically important parameter that affects the permeability characteristics of the paracellular pathway is the proton concentration. Cation-selective junctions are relatively permeable to protons at neutral pH. However, as the pH is decreased, the junction may suddenly switch from being cation selective to becoming anion selective and thus block proton permeability (Powell, 1981). This is especially important to the function of the mucosal epithelium in the stomach. Monolayers of gastric chief cells have been shown to resist back diffusion of protons efficiently (Sanders et al., 1985). When the apical medium was acidified to pH2, the transepithelial resistance increased from 1650 to 3600 ohm.cm^2, and the monolayer maintained the 1:100,000 gradient of protons for more than 4 hours.

The pore size of the hydrated channel across the ZO has been studied in several epithelia. The junctions are essentially impermeable to molecules with radii ≥ 15 Å (Moreno and Diamond, 1975; Cereijido et al., 1978; Madara and

Dharmsathaphorn, 1985). Moreno and Diamond concluded that the radius of the permeation channel for nitrogeneous cations varies between epithelia (4.4 Å in rabbit gall bladder versus 8.1 Å in frog gall bladder) and that the selectivity of the channel can be modulated by both pH and field strength.

There is some evidence suggesting that the intermediate junctions might be involved in regulating the permeability of the occluding junctions. The microfilament-perturbing agent cytochalasin D has been shown to increase the passive transepithelial flow of ions (Meza et al., 1980, 1982) and to increase junctional permeability to mannitol (Madara et al., 1986). Cytochalasin D causes the condensation of filamentous elements associated with the contractile actin–myosin ring of the ZA that is accompanied by the contraction of the apical membrane (Madara et al., 1986). The contraction of the ZA actin filament bundle can be elicited by a mechanism involving a calcium–calmodulin-dependent myosin light chain kinase (Mooseker et al., 1984). This contraction might also be triggered by cytochalasin D and lead to alterations in occluding junction structure by shearing cross-links between strands in the anastomosing ZO network.

Another parameter that implicates the ZA in occluding junction function is the effect of calcium concentration (Gumbiner and Simons, 1987). The exposure of epithelial layers to concentrations of EGTA or EDTA causes a decrease in the transepithelial resistance by opening the occluding junctions (Martinez-Palomo et al., 1980). The restoration of calcium reseals the junctions. It was generally thought that calcium is integrally involved in the stabilization of ZO structure. However, Stevenson and Goodenough (1984) isolated junctional ribbons from hepatocytes in the presence of EGTA, and each of these ribbons contained an anastomosing network of strands similar to that seen in freeze-fractured ZOs in hepatocytes.

The calcium dependence of the occluding barrier may instead be an indirect one, mediated by uvomorulin (L-CAM), a protein located in the ZA itself and in the lateral or basolateral membrane of epithelial cells (Edelman et al., 1983; Thiery et al., 1984; Boller et al., 1985). This protein undergoes a calcium-dependent conformational change that can be monitored by specific antibodies or by trypsin sensitivity (Hyafil et al., 1981). Uvomorulin is intimately involved in establishing the epithelial occluding barrier (Gumbiner and Simons, 1987). Antibodies against the protein have been shown to block the resealing of ZOs that have been opened by calcium removal.

Gumbiner and Simons (1986) have postulated that the assembly of the ZA is mediated via uvomorulin-dependent interactions that occur in areas of cell–cell contact. As the cells polarize, the ZA becomes positioned in the apical region of the cell, perhaps through the influence of its cytoplasmic actin–myosin belt. The intercellular assembly of the ZA in turn positions the ZO network of strands above the ZA belt. These dynamic interactions between the ZA and ZO will remain speculative until the molecular components mediating the links have been identified.

FENCE FUNCTION

Proteins

The role of the ZO in sealing the epithelial layer, thereby preventing diffusion of substances from one side of the epithelium to the other, is firmly established. Its role in maintaining the spatial division of the plasma membrane of epithelial cells has been more controversial. Studies in which toad kidney A6 epithelial cell layers were probed with several fluorescent lectins showed that these lectins were essentially immobile upon binding to glycoproteins on either the apical or the basolateral surface (Dragsten et al., 1981, 1982). Thus, one would not need to postulate a diffusion barrier for maintaining the asymmetric distribution of apical and basolateral proteins.

However, this study left open the possibility that the lectins artifactually caused the immobilization of the proteins by cross-linking them to each other. This problem was overcome in a later study in which fluorescein-conjugated monovalent Fab fragments of antibodies directed against the (sodium, potassium) ATPase were used to label the plasma membrane of MDCK cell monolayers (Jesaitis and Yguerabide, 1986). The fluorescence recovery after photobleaching indicated that 50% of the enzyme was mobile. The immobile fraction was too small to account for the exclusively basolateral distribution of the (sodium, potassium) ATPase in these cells.

It seems obvious, considering the selective solute permeability of the occluding junction, that the junction must also act as a diffusion barrier for apical and basolateral proteins. It is difficult to envisage how the large exoplasmic domain of a glycoprotein could diffuse through a junctional seal that is impermeable to molecules with radii larger than 15 Å. Protein polarity does not seem to be correlated to the number of ZO strands forming the junction. A single strand may be sufficient to stop protein diffusion from one side to the other (Fuller and Simons, 1986).

The relation between protein polarity and transepithelial resistance was studied in MDCK cells. Transferrin-mediated ^{55}Fe uptake was used to measure the polarity of active transferrin receptors in filter-grown MDCK cells. The ratio of basolateral to apical receptors was 800:1 for the high-resistance strain I MDCK cells (typically >2000 ohm.cm^2), and 300:1 for the low-resistance strain II cells (<350 ohm.cm^2). This 2.6-fold difference in polarity of the total number of receptors can be expressed in terms of surface concentration of the receptors by adjusting for the measured surface areas of the two plasma membrane domains in the two cell strains. Filter-grown strain I cells have a ratio of apical to basolateral surface area of 1:7.6, so that a receptor polarity of 1:800 results in a ratio of surface concentration of 1:99, while strain II cells, with a surface ratio of 1:4 and a receptor polarity of 1:300, yield a surface concentration ratio of 1:75. These ratios are so similar that one can conclude that a low-resistance junction is just as good a fence as a high-resistance junction (Fuller and Simons, 1986).

Although basolateral proteins are efficiently excluded from the apical pole, apical proteins have been found on the basolateral side (Simons and Fuller, 1985). This does not seem to be a property of the junctional fence. Insofar as this phenomenon has been studied, it seems to reflect errors in the sorting of newly synthesized apical proteins in the Golgi complex (Pfeiffer et al., 1985). It is possible that the discriminating power of the apical sorting device is greater than that of the basolateral one. As a consequence, a small percentage of apical proteins may be missorted to the basolateral surface.

Lipids

Not only are the proteins different in the two cell surface domains of epithelial cells, but their lipids also differ in composition (Simons and Fuller, 1985). The most careful studies on the lipids of apical and basolateral membranes have been carried out with small intestinal cells. In these cells the apical brush border membrane contains about equimolar ratios of glycosphingolipids and phospholipids (Forstner and Wherrett, 1973; Kawai et al., 1974; Brasitus and Schachter, 1980). The basolateral membrane contains two to four times fewer glycosphingolipids than the apical brush borders and instead has two to four times more phosphatidylcholine. This trend is observed in other epithelial cells as well (van Meer and Simons, 1982; Hise et al., 1984; Meier et al., 1984; Carmel et al., 1985; Molitoris and Simon, 1985). The sphingolipids, mainly glycosphingolipid but also sphingomyelin, are concentrated in the apical membrane, while phosphatidylcholine is enriched in the basolateral membrane.

The role of the occluding junctions in maintaining these differences in lipid composition between the surface domains has been investigated in several systems. Dragsten et al. (1981, 1982) allowed water-soluble fluorescent lipid probes to partition into the apical membranes of A6, LLC-PK$_1$, and MDCK cells grown on tissue culture dishes. They also detached the A6 cell layer from the dish to expose the cells' basolateral membranes to the lipid probes. All the probes diffused freely in the membrane. However, some remained restricted to the domain into which they had been partitioned. These probes included 5-(N-hexadecanoyl)-aminofluorescein (AFC$_{16}$) and 1-acyl-2-(N-4-nitrobenz-2-oxa-1,3-diazole)-aminocaproylphosphatidylcholine (NBD-PC). Other probes (3,3'-dihexadecylindocarbocyanine iodide, di IC$_{16}$, di IC$_{14}$, and AFC$_{12}$) diffused over the entire cell surface at 20°C. If the temperature was 10°C or less, AFC$_{12}$ also became restricted to one domain.

Dragsten et al. (1981) explained these differences in behavior by suggesting that di IC$_{14}$, di IC$_{16}$, and AFC$_{12}$ could flip-flop from the exoplasmic leaflet to the cytoplasmic leaflet of the bilayer and that they could diffuse in the cytoplasmic leaflet to the other side of the cell. On the other hand, they proposed that AFC$_{16}$ and NBD-PC do not flip-flop and therefore remain in the exoplasmic leaflet. They postulated that the occluding junction acts as a diffusion barrier to the lipid probes in the exoplasmic leaflet only.

These results were confirmed by using another experimental approach to the question of tight junction permeability (van Meer and Simons, 1986). In these studies an experimental cell system was chosen that had been carefully studied for basolateral and apical protein polarity: MDCK cells strain I and II grown on permeable supports (Fuller et al., 1984). By growing MDCK cells on filters, conditions that closely mimicked those prevailing *in vivo* were achieved (Simons and Fuller, 1985). On filters the cells fed from the basal side and formed a monolayer of cuboidal cells that was more stable to experimental manipulations than cells grown on solid supports. Instead of partitioning water-soluble lipid analogues into the cell membranes, liposomes were fused with the apical plasma membrane domain (van Meer et al., 1985). By this method it was possible to use a water-insoluble fluorescent phospholipid, N-rhodamine-dioleoyl PE (N-Rh-PE), which was more natural in its properties than the analogues previously used.

The fluorescent lipid was introduced into both leaflets of the plasma membrane bilayer by carrying out the fusion with symmetric liposomes. Alternatively, the lipid was introduced primarily into the exoplasmic leaflet of the apical plasma membrane by fusing with liposomes in which the fluorescent lipid was enriched in the external leaflet. The behavior of the fluorescent lipid was studied at 0°C. At this temperature diffusion should occur, but vesicular transport is completely inhibited. The fusion reaction took place within 1 min at 37°C, a time too short for transcytosis to play a role in the transport to the basolateral surface. The intactness of the tight junction was assessed by measuring the transepithelial electrical resistance.

Using these methods, clear differences were observed in the behavior of N-Rh-PE implanted in the plasma membrane of MDCK cells (van Meer and Simons, 1986). When fused into the exoplasmic leaflet of the apical plasma membrane, N-Rh-PE did not distribute to the lateral plasma membrane unless the junctions were opened by calcium removal. However, when N-Rh-PE was fused into both the exoplasmic and cytoplasmic leaflets of the bilayer, a fraction of the fluorescent phospholipid immediately passed to the basolateral plasma membrane domain. Also, exogeneously implanted glycolipids have been shown to be unable to pass the tight junction in the exoplasmic bilayer leaflet (Spiegel et al., 1985).

Another approach was used to demonstrate that newly synthesized sphingo-lipids cannot diffuse across the occluding junctions after delivery to the exoplasmic leaflet of the apical and basolateral membranes. van Meere et al. (1987) used a fluorescent ceramide analogue, N-6[7-nitro-2,1,3-benzoxadiazol-4-yl]amino-caproyl-sphingosine (C6-NBD-ceramide), as a probe (Lipsky and Pagano, 1983, 1985a,b). This ceramide is readily taken up by filter-grown MDCK cells from liposomes at 0°C. After partitioning into the cell, the fluorescent probe accumulates in the Golgi complex at temperatures between 0°C and 20°C as a result of metabolic conversion into C6-NBD-sphingomyelin and C6-NBD-glucosylceramide.

Raising the temperature to 37°C for 1 hour resulted in intense plasma membrane staining and a loss of fluorescence from the Golgi complex. The addition of serum albumin to the apical medium cleared the fluorescence from the apical domain by extracting the fluorescent lipids from the membrane. No effect was seen on the fluorescence on the basolateral plasma membrane. The basolateral fluorescence could be depleted only by introducing albumin to the basal side of a monolayer of MDCK cell monolayers, grown on polycarbonate filters. The C6-NBD-sphingomyelin and C6-NBD-glucosylceramide were extracted by the albumin added through the pores of the filter. This treatment did not deplete the fluorescent probes from the apical side. The conclusion from this study was that fluorescent sphingomyelin and glucosylceramide are delivered from the Golgi complex to the exoplasmic leaflet of the plasma membrane, where they accumulate and are unable to pass the ZO in either direction (van Meer et al., 1987).

One interesting question, in light of the hexagonal lipid model of ZO strand structure (Kachar and Rees, 1982; Pinto da Silva and Kachar, 1982), is whether endogenous lipids can diffuse from one epithelial cell to another. Strain II MDCK cells possess a series of glycolipids, the globo series, which are not found in MDCK strain I cells (Hansson et al., 1986). Forssman antigen, one glycolipid of this series [GalNAc(a1–3)GalNAc(b1–3)Gal(a1–4)Gal(b1–4)Glc(b1–1)Cer], constitutes 21% of the total neutral glycosphingolipids of MDCK strain II cells. When a monolayer of these cells was labeled from the apical side with a monoclonal antibody against Forssman antigen, apical staining was observed. This glycolipid is therefore present in the exoplasmic leaflet of the apical plasma membrane.

The MDCK strain II cells were cocultured with MDCK strain I cells, which do not express Forssman antigen (van Meer et al., 1986). Tight junctions formed between the two cell types. When the mixed monolayers of MDCK I and II cells (1:1) were incubated with a monoclonal antibody against Forssman antigen, antigen was detected on the apical surface of about 50% of the cells. The boundary between stained and unstained cells was sharp. Thus, the endogenous glycolipid was unable to pass from MDCK strain II cells to the apical surface of neighboring MDCK strain I cells over a period of 96 hours of coculture at 37°C. The same results were obtained with exogenous lipid probes. N-Rh-PE and octadecyl rhodamine B, which were fused into the apical membrane of only one MDCK cell, did not diffuse to neighboring cells (van Meer et al., 1986).

These results imply that lipids in the exoplasmic leaflet cannot move from one side of the cell to the other or from one cell to the other. However, movement from one side of the cell to the other along the cytoplasmic leaflet of the plasma membrane does seem to take place. Free diffusion of lipid molecules in the cytoplasmic leaflet would lead to an identical lipid composition of the cytoplasmic leaflets of the apical and basolateral membrane domains, and has interesting consequences for the transbilayer distribution of the individual lipid classes in these membranes. If the exoplasmic leaflet of the apical domain were predominantly occupied by glycosphingolipids, as is probably the case in intestinal cells, the phospholipids of the apical domain would be mainly situated in its cytoplasmic leaflet (Barsukov et al., 1986). The phospholipid composition of the cytoplasmic leaflet of the basolateral membrane would be identical to that of the apical domain, and the distribution of the individual phospholipid classes across the basolateral membrane bilayer could be predicted from the total phospholipid composition of the basolateral membrane.

For the two major phospholipid classes, this leads to the following predictions: 65–90% of the phosphatidylethanolamine and only 10–25% of the phosphatidylcholine would be localized in the cytoplasmic leaflet. This agrees with the distribution reported for the plasma membrane of erythrocytes and blood platelets and may reflect general principles underlying phospholipid organization in mammalian plasma membranes (Bretscher and Raff, 1975; Op den Kamp, 1979).

These observations suggest that the lipid differences between the apical and basolateral membranes reside in their exoplasmic leaflets. The problem of generating the different lipid compositions of these two membrane domains would then be reduced to enriching sphingolipids, predominantly glyco-sphingolipids, and phosphatidylcholine in the exoplasmic leaflets of the apical and basolateral membranes, respectively. How are these differences generated? Several possibilities exist. One is local metabolism; in other words, the enriched lipids are synthesized in the membrane domain where they are found. This seems very unlikely, since phosphatidylcholine is synthesized mainly in the endoplasmic reticulum, and glycosphingolipids in the Golgi complex (Bell et al., 1981; Pagano and Sleight, 1985). Therefore, it is more likely that these lipids are transported to the cell surface from their sites of synthesis.

In principle, this could occur by two routes: either through the cytosol, with the help of exchange proteins, for example (Wirtz, 1974; Yaffe and Kennedy, 1983), or by vesicular transport. If vesicular transport were the mechanism for distributing the correct lipids apically and basolaterally, one would expect that lipid sorting to occur in the membrane compartment from which vesicles exit for the cell surface, namely the *trans*-most compartment of the Golgi complex, the *trans* Golgi network (Griffiths and Simons, 1986). Newly synthesized apical and basolateral proteins are also thought to be sorted from each other in the *trans* Golgi network before their transport to the cell surface (Hughson et al., 1987). Recent evidence suggests that sphingolipid sorting occurs in the Golgi complex, perhaps in the *trans* Golgi network (van Meer et al., 1987).

ROLE OF THE OCCLUDING JUNCTION IN ESTABLISHING CELL SURFACE POLARITY

Several studies in MDCK cells have addressed the question of whether the establishment of cell surface polarity requires the formation of occluding junctions. One would assume that this must be the case if the junctions form a

barrier to inhibit the intermixing of lipids and proteins. Rodriguez-Boulan et al.
(1983) reported that subconfluent MDCK cell monolayers lacking extensive
cell–cell contacts and tight junctions support polarized virus budding.
However, virus budding is not a very sensitive measure of the asymmetric
distribution of the virus membrane glycoproteins, because polar budding can
occur even when the difference in glycoprotein density is only a few–fold.
Furthermore, no quantitative analysis was performed to assess the percentage
of infected cells in which virus budding was polarized. The MDCK cells where
budding was polar may have been connected by junctions that escaped
detection. Even in MDCK cell monolayers, grown in calcium-free medium,
13% of the cell pairs analyzed had lateral membranes inaccessible to the
ruthenium red applied to the apical side (Vega-Salas et al., 1987a).

In other studies endogenous basolateral proteins were used as markers
(Balcarova-Ständer et al., 1984; Herzlinger and Ojakian, 1984). The basolateral
proteins were found on the apical side of the subconfluent MDCK cell
monolayers, and complete polarization required fully developed occluding
junctions, as measured by transepithelial resistance. In contrast to these results,
Vega-Salas et al. (1987a) found that a 184-kD apical protein became highly
polarized to the apical pole even before occluding junctions were present.
Simultaneous studies of a 63-kD basolateral protein demonstrated that this
protein was present over the entire cell surface and did not polarize prior to
junction formation.

One interpretation of these conflicting results is that epithelial cells first
develop a basal and an apical pole across which most plasma membrane
proteins and lipids can diffuse (Simons and Fuller, 1985). This bipolar
structure may provide the framework for the formation of the zonular
junctions. This in turn leads to the development of a spatial division of the
plasma membrane into two domains. When epithelial cells attach to the
substratum, plasma membrane receptors might bind to extracellular matrix
components to form a basal pole, and the free cell surface might develop an
apical pole by cytoskeletal interactions with "pole-forming" proteins (e.g., the
184-kD protein), fixing them apically in the membrane plane. Interestingly, the
surface expression of newly synthesized apical proteins is partially blocked at
this stage (Vega-Salas et al., 1987b). These proteins are trapped in intracellular
vacuoles, where they are stored until occluding junctions begin to form. At this
stage the apical proteins are delivered to the apical pole. Further studies are
necessary to elucidate the precise sequence of events leading to the establishment of cell surface polarity in epithelial cells. Although the mechanisms
involved are still unknown, the present evidence suggests that occluding
junctions play an essential role in this process.

THE MOLECULAR STRUCTURE OF THE ZONULA OCCLUDENS

Although the ZO has been well characterized from the functional point of
view, little is known about its molecular composition. Kachar et al. proposed

that the tight junction strands represent intramembranous hexagonal cylinders of lipids (Kachar and Rees, 1982; Pinto da Silva and Kachar, 1982). This model was based on the morphology and the dimensions of the strands that were found to correspond to those of lipids in the hexagonal II phase. The lipid model of the ZO postulates that the cytoplasmic leaflet of the plasma membrane is continuous from the apical to the basolateral domain. The exoplasmic leaflets of the same cell are interrupted by the hexagonal lipid cylinder, although they are thought to be continuous with those of adjacent cells (Figure 1A).

The lipid model of the tight junction makes interesting predictions for the diffusion of membrane components. Membrane-spanning proteins with an exoplasmic domain would not be able to diffuse between the apical and basolateral sides across the ZO. Passage of membrane components that are situated in only one leaflet of the plasma membrane (e.g., lipid molecules) would depend on their orientation in the bilayer. In the cytoplasmic leaflet they would freely diffuse between domains, but in the exoplasmic leaflet the tight junction would act as a diffusion barrier for movement from one domain to the other. On the other hand, lipid molecules in the external leaflet would be able to diffuse though a continuous lipid monolayer from one cell to another.

All the predictions except the last agree with the available data. Endogenous and exogenous lipids in the exoplasmic leaflet of the apical membrane do not diffuse from one epithelial cell to the next under conditions where the ZO should be intact. Another finding that is not easily explained by the lipid model is the isolation of strands from hepatocyte junctions in the presence of the

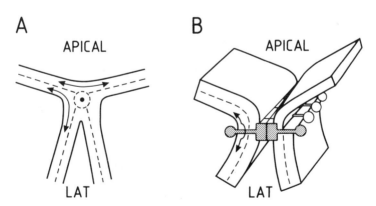

Figure 1. *Two models for tight junction structure with different predictions for lipid diffusion. A: Hexagonal lipid model. B:* Tight junction is made up of proteins spanning the bilayers and interacting with one another to restrict diffusion between the cells and in the exoplasmic leaflet. Both *A* and *B* are in agreement with the findings that free lipid diffusion occurs between the cytoplasmic but not the exoplasmic leaflets of the apical and basolateral plasma membranes. In addition, the hexagonal lipid model (*A*) predicts that lipids are free to diffuse between the exoplasmic domains of neighboring cells. Since neither endogenous nor exogenous lipids appear to be able to diffuse from one epithelial cell to the next, *B* is more likely than *A*. (Reproduced from *EMBO Journal* with permission.)

detergent deoxycholate (Stevenson and Goodenough, 1984). The detergent should solubilize to lipid strands.

Perhaps the most reasonable model of the ZO is a protein bridge that not only occludes the paracellular route to solutes with radii larger than ≥ 15 Å and forms the hydrated channel that regulates the passage of ions across but also blocks the diffusion of proteins and lipids in the exoplasmic leaflet from one domain to the other (Figure 1B). The protein model readily explains the functional data. The electron microscope findings are less consistent with those for other membrane proteins; however, the proteins of the ZO should have very unique properties.

Recently, a high-molecular-weight polypeptide of 225 kD has been identified in rat hepatocytes that is found on the cytoplasmic side of the ZO (Stevenson et al., 1986). This protein does not appear to span the membrane, and it is present in all simple epithelia investigated. The antibody that recognizes the 225-kD ZO protein also binds to endothelial and Sertoli–Sertoli cell junctions. Another peripheral protein of about 300 kD has also been identified and localized to the cytoplasmic plaque of the ZO (see Kapprell et al., this volume). Interestingly enough, this protein seems to be present only in simple epithelia and in hepatocytes but not in endothelial and Sertoli cell junctions. All attempts to identify the postulated protein components of the ZO strands have so far failed. The elucidation of the molecular structure of the ZO promises to be an exciting chapter in epithelial cell biology.

ACKNOWLEDGMENTS

I would like to thank Mark Bennett, Steve Fuller, and Angela Wandinger-Ness for critical readings of the manuscript, Marianne Remy, Rachel Wainwright, and Anne Walter for typing the manuscript, and Hilkka Virta for help with the references.

REFERENCES

Balcarova-Stånder, J., S. E. Pfeiffer, S. D. Fuller, and K. Simons (1984) Development of cell surface polarity in the epithelial Madin-Darby canine kidney (MDCK) cell line. EMBO J. 3:2687–2694.

Barsukov, L. I., L. D. Bergelson, M. Spiess, H. Hauser, and G. Semenza (1986) Phospholipid topology and flip-flop in intestinal brush-border membrane. Biochim. Biophys. Acta 862:87–99.

Bell, R. M., L. M. Ballas, and R. A. Coleman (1981) Lipid topogenesis. J. Lipid Res. 22:391–403.

Boller, K., D. Vestweber, and R. Kemler (1985) Cell-adhesion molecule uvomorulin is localized in the intermediate junctions of adult intestinal epithelial cells. J. Cell Biol. 100:327–332.

Brasitus, T. A., and D. Schachter (1980) Lipid dynamics and lipid-protein interactions in rat enterocyte basolateral and microvillus membranes. Biochemistry 19:2763–2769.

Bretscher, M. S., and M. C. Raff (1975) Mammalian plasma membranes. Nature 258:43–49.

Carmel, G., F. Rodrigus, S. Carriers, and C. Le Grimellec (1985) Composition and physical properties of lipids from plasma membranes of dog kidney. Biochim. Biophys. Acta 818:149–157.

Cereijido, M., E. S. Robbins, W. J. Dolan, C. A. Rotunno, and D. D. Sabatini (1978) Polarized monolayer formed by epithelial cells on a permeable and translucent support. *J. Cell Biol.* 77:853–880.

Cereijido, M., E. Stefani, and A. Martinez-Palomo (1980) Occluding junctions in a cultured transporting epithelium: Structural and functional heterogeneity. *J. Membr. Biol.* 53:19–32.

Chalcroft, J. P., and S. Bullivant (1970) An interpretation of liver cell membrane and junction structure based on the observation of freeze-fracture replicas of both sides of fracture. *J. Cell Biol.* 47:49–60.

Claude, P. (1978) Morphological factors influencing transepithelial permeability: A model for the resistance of the zonula occludens. *J. Membr. Biol.* 39:219–232.

Claude, P., and D. A. Goodenough (1973) Fracture faces of zonulae occludentes from "tight" and "leaky" epithelia. *J. Cell Biol.* 58:390–400.

Diamond, J. M. (1977) The epithelial junction: Bridge, gate, and fence. *Physiologist* 20:10–18.

Dragsten, P. R., R. Blumenthal, and J. S. Handler (1981) Membrane asymmetry in epithelia: Is the tight junction a barrier to diffusion in the plasma membrane? *Nature* 294:718–722.

Dragsten, P. R., J. S. Handler, and R. Blumenthal (1982) Fluorescent membrane probes and the mechanism of maintenance of cellular asymmetry in epithelia. *Fed. Proc.* 41:48–53.

Easter, D. W., J. B. Wade, and J. L. Boyer (1983) Structural integrity of hepatocyte tight junctions. *J. Cell Biol.* 96:745–749.

Edelman, G. M., W. J. Gallin, A. Delouvee, B. A. Cunningham, and J. P. Thiery (1983) Early epochal maps of two different cell adhesion molecules. *Proc. Natl. Acad. Sci. USA* 80:4384–4388.

Farquhar, M. G., and G. E. Palade (1963) Junctional complexes in various epithelia. *J. Cell Biol.* 17:375–412.

Forstner, G. G., and J. R. Wherrett (1973) Plasma membrane and mucosal glycosphingolipids in the rat intestine. *Biochim. Biophys. Acta* 306:446–459.

Friend, D. S., and N. B. Gilula (1972) Variations in tight and gap-junctions in mammalian tissues. *J. Cell Biol.* 53:758–776.

Frömter, E., and J. Diamond (1972) Route of passive ion permeation in epithelia. *Nature New Biol.* 235:9–13.

Fuller, S. D., C.-H. von Bonsdorff, and K. Simons (1984) Vesicular stomatitis virus infects and matures only through the basolateral surface of the polarized epithelial cell line, MDCK. *Cell* 38:65–77.

Fuller, S. D., and K. Simons (1986) Transferrin receptor polarity and recycling accuracy in "tight" and "leaky" strains of Madin-Darby canine kidney cells. *J. Cell Biol.* 103:1767–1779.

Geiger, B., K. T. Tokuyasu, A. H. Dutton, and S. J. Singer (1980) Vinculin, an intracellular protein localized at specialized sites where microfilament bundles terminate at cell membranes. *Proc. Natl. Acad. Sci. USA* 77:4127–4131.

Geiger, B., A. H. Dutton, K. T. Tokuyasu, and S. J. Singer (1981) Immunoelectron microscope studies of membrane–microfilament interactions: Distributions of α-actinin, tropomyosin, and vinculin in intestinal epithelial brush border and chicken gizzard smooth muscle cells. *J. Cell Biol.* 91:614–628.

Geiger, B., Z. Avnur, T. Volberg, and T. Volk (1985) Molecular domains of adherens junctions. In *The Cell in Contact: Adhesions and Junctions as Morphogenetic Determinants*, G. M. Edelman and J. P. Thiery, eds., pp. 461–490, Wiley, New York.

Griffiths, G., and K. Simons (1986) The *trans* Golgi network: Sorting at the exit site of the Golgi complex. *Science* 234:438–443.

Gumbiner, B., and K. Simons (1986) A functional assay for proteins involved in establishing an epithelial occluding barrier: Identification of a uvomorulin-like polypeptide. *J. Cell Biol.* 102:457–468.

Gumbiner, B., and K. Simons (1987) The role of uvomorulin in the formation of epithelial occluding junctions. *Ciba Found. Symp.* 125:168–180.

t specified.

354

Epithelial Tight Junctions

Hansson, G. C., K. Simons, and G. van Meer (1986) Two strains of the Madin-Darby canine kidney (MDCK) cell line have distinct glycosphingolipid compositions. EMBO J. 5:483–489.

Herzlinger, D. A., and G. K. Ojakian (1984) Studies on the development and maintenance of epithelial cell surface polarity with monoclonal antibodies. J. Cell Biol. 98:1777–1787.

Hise, M. K., W. W. Mantulin, and E. J. Weinman (1984) Fluidity and composition of brush border and basolateral membranes from rat kidney. Am. J. Physiol. 247:F434–F439.

Hughson, E., A. Wandinger-Ness, H. Gausepohl, G. Griffiths, and K. Simons (1988) The cell biology of enveloped virus infection of epithelial tissues. In The Molecular Biology and Infectious Diseases, Centenary Symposium of the Pasteur Institute, M. Schwartz, ed., pp. 75–89, Elsevier, Paris.

Hyafil, F., C. Babinet, and F. Jacob (1981) Cell–cell interactions in early embryogenesis: A molecular approach to the role of calcium. Cell 26:447–454.

Jesaitis, A. J., and J. Yguerabide (1986) The lateral mobility of the (Na^+, K^+)-dependent ATPase in Madin-Darby canine kidney cells. J. Cell Biol. 102:1256–1263.

Kachar, B., T. S. Reese (1982) Evidence for the lipidic nature of tight junction strands. Nature 296:464–466.

Kawai, K., M. Fujita, and M. Nakao (1974) Lipid components of two different regions of an intestinal epithelial cell membrane of mouse. Biochim. Biophys. Acta 369:222–233.

Lipsky, N. G., and R. E. Pagano (1983) Sphingolipid metabolism in cultured fibroblasts: Microscopic and biochemical studies employing a fluorescent ceramide analogue. Proc. Natl. Acad. Sci. USA 80:2608–2612.

Lipsky, N. G., and R. E. Pagano (1985a) Intracellular translocation of fluorescent sphingolipids in cultured fibroblasts: Endogenously synthesized sphingomyelin and glucocerebroside analogues pass through the Golgi apparatus en route to the plasma membrane. J. Cell Biol. 100:27–34.

Lipsky, N. G., and R. E. Pagano (1985b) A vital stain for the Golgi apparatus. Science 228:745–747.

Madara, J. L., and K. Dharmsathaphorn (1985) Occluding junction structure–function relationships in a cultured epithelial monolayer. J. Cell Biol. 101:2124–2133.

Madara, J. L., D. Banenberg, and S. Carlson (1986) Effects of cytochalasin D on occluding junctions of intestinal absorptive cells: Further evidence that the cytoskeleton may influence paracellular permeability and junctional charge selectivity. J. Cell Biol. 102:2125–2136.

Marcial, M. A., S. L. Carlson, and J. L. Madara (1984) Partitioning of paracellular conductance along the ileal crypt–villus axis: A hypothesis based on structural analysis with detailed consideration of tight junction structure–function relationships. J. Membr. Biol. 80:59–70.

Martinez-Palomo, A., I. Meza, G. Beaty, and M. Cereijido (1980) Experimental modulation of occluding junctions in a cultured transporting epithelium. J. Cell Biol. 87:736–745.

Meier, P. J., E. S. Sztul, A. Reuben, and J. L. Boyer (1984) Structural and functional polarity of canalicular and basolateral plasma membrane vesicles isolated in high yield from rat liver. J. Cell Biol. 98:991–1000.

Meza, I., G. Ibarra, M. Sabanero, A. Martinez-Palomo, and M. Cereijido (1980) Occluding junctions and cytoskeletal components in a cultured transporting epithelium. J. Cell Biol. 87:746–754.

Meza, I., M. Sabanero, E. Stefani, and M. Cereijido (1982) Occluding junctions in MDCK cells: Modulation of trans-epithelial permeability by the cytoskeleton. J. Cell Biochem. 18:407–421.

Molitoris, B. A., and F. R. Simon (1985) Renal cortical brush-border and basolateral membranes: Cholesterol and phospholipid composition and relative turnover. J. Membr. Biol. 83:207–215.

Mooseker, S., M. Bonder, A. Conzelman, J. Fishkind, L. Howe, and C. S. Keller (1984) Brush border cytoskeleton and integration of cellular functions. J. Cell Biol. 99:104–112.

Moreno, J. H., and J. M. Diamond (1975) Nitrogenous cations as probes of permeation channels. J. Membr. Biol. 21:197–259.

Op den Kamp, J. A. A. (1979) Lipid asymmetry in membranes. *Annu. Rev. Biochem.* **48**:47–71.

Pagano, R. E., and R. G. Sleight (1985) Emerging problems in the cell biology of lipids. *Trends Biochem. Sci.* **10**:421–425.

Pfeiffer, S., S. D. Fuller, and K. Simons (1985) Intracellular sorting and basolateral appearance of the G protein of vesicular stomatitis virus in MDCK cells. *J. Cell Biol.* **101**:470–476.

Pinto da Silva, P., and B. Kachar (1982) On tight-junction structure. *Cell* **28**:441–450.

Powell, D. W. (1981) Barrier function of epithelia. *Am. J. Physiol.* **241**:G275–G288.

Rabito, C. A. (1986) Occluding junctions in a renal cell line (LLC-PK1) with characteristics of proximal tubular cells. *Am. J. Physiol.* **250**:F734–F743.

Rabito, C. A., R. Tchao, J. Valentich, and J. Leighton (1978) Distribution and characteristics of the occluding junctions in a monolayer of a cell line (MDCK) derived from canine kidney. *J. Membr. Biol.* **43**:351–365.

Rodriguez-Boulan, E., K. T. Paskiet, and D. D. Sabatini (1983) Assembly of enveloped viruses in MDCK cells: Polarized budding from single attached cells and from clusters of cells in suspension. *J. Cell Biol.* **96**:866–874.

Sanders, M. J., A. Ayalon, M. Roll, and A. H. Soll (1985) The apical surface of canine chief cell monolayers resists H^+ back-diffusion. *Nature* **313**:52–54.

Schafer, J. A., S. L. Troutman, and T. E. Andreoli (1984) Volume reabsorption, transepithelial potential differences, and ionic permeability properties in mammalian superficial proximal straight tubules. *J. Gen. Physiol.* **64**:582–607.

Simons, K., and S. D. Fuller (1985) Cell surface polarity in epithelia. *Annu. Rev. Cell Biol.* **1**:243–288.

Spiegel, S., R. Blumenthal, P. H. Fishman, and J. S. Handler (1985) Gangliosides do not move from apical to basolateral plasma membrane in cultured epithelial cells. *Biochim. Biophys. Acta* **821**:310–318.

Staehelin, L. A. (1974) Structure and function of intercellular junctions. *Int. Rev. Cytol.* **39**:191–283.

Stevenson, B. R., and D. A. Goodenough (1984) Zonulae occludentes in junctional complex-enriched fractions from mouse liver: Preliminary morphological and biochemical characterization. *J. Cell Biol.* **98**:1209–1221.

Stevenson, B. R., J. D. Siliciano, M. S. Mooseker, and D. A. Goodenough (1986) Identification of ZO-1: A high molecular weight polypeptide associated with the tight junction (zonula occludens) in a variety of epithelia. *J. Cell Biol.* **103**:755–766.

Thiery, J.-P., A. Delouvee, W. J. Gallin, B. A. Cunningham, G. M. Edelman (1984) Ontogenetic expression of cell adhesion molecules: L-CAM is found in epithelia derived from the three preliminary germ layers. *Dev. Biol.* **102**:61–78.

van Deurs, B., and J. H. Luft (1979) Effects of glutaraldehyde fixation on the structure of tight junctions: A quantitative freeze-fracture analysis. *J. Ultrastruct. Res.* **68**:160–172.

van Meer, G., and K. Simons (1982) Viruses budding from either the apical or the basolateral plasma membrane domain of MDCK cells have unique phospholipid compositions. *EMBO J.* **1**:847–852.

van Meer, G., and K. Simons (1986) The function of tight junctions in maintaining differences in lipid composition between the apical and the basolateral cell surface domains of MDCK cells. *EMBO J.* **5**:1455–1464.

van Meer, G., J. Davoust, and K. Simons (1985) Parameters affecting low pH-mediated fusion of liposomes with the plasma membrane of cells infected with influenza virus. *Biochemistry* **24**:3593–3602.

van Meer, G., B. Gumbiner, and K. Simons (1986) The tight junction does not allow lipid molecules to diffuse from one epithelial cell to the next. *Nature* **322**:639–641.

van Meer, G., E. H. Stelzer, R. W. Wijnaendts-van-Resandt, and K. Simons (1987) Sorting of sphingolipids in epithelial (MDCK) cells. *J. Cell Biol.* **105**:1623–1635.

Vega-Salas, D. E., P. J. I. Salas, D. Gundersen, and E. Rodriguez-Boulan (1987a) Formation of the apical pole of epithelial (Madin-Darby canine kidney) cells: Polarity of an apical protein is independent of tight junctions while segregation of a basolateral marker requires cell–cell interactions. *J. Cell Biol.* **104**:905–916.

Vega-Salas, D. E., P. J. I. Salas, and E. Rodriguez-Boulan (1987b) Modulation of the expression of an apical plasma membrane protein of Madin-Darby canine kidney epithelial cells: Cell–cell interactions control the appearance of a novel intracellular storage compartment. *J. Cell Biol.* **104**:1249–1259.

Wirtz, K. W. A. (1974) Transfer of phospholipids between membranes. *Biochim. Biophys. Acta* **344**:95–117.

Yaffe, M. P., and E. P. Kennedy (1983) Intracellular phospholipid and the role of phospholipid transfer protein in animal cells. *Biochemistry* **22**:1497–1507.

Chapter 16

Gap Junctions and Intercellular Communication

DANIEL A. GOODENOUGH

ABSTRACT

Gap junctions are intercellular contacts in which protein assemblies in one cell's plasma membrane are directly joined to similar assemblies in an adjacent cell's membrane (Robertson, 1963; Revel and Karnovsky, 1967; Goodenough and Revel, 1970; Makowski et al., 1977; Unwin and Ennis, 1983, 1984). The gap junction is characterized by thin-section electron microscopy as a pair of membranes of variable area separated (in most, but not all, tissues) by a 2-nm "gap." Gap junctions have been associated with cell-to-cell diffusion of ions and molecules via a low-resistance pathway that avoids the extracellular space. In the case of ions, both the transmission of action potentials at electrical synapses (Furshpan and Potter, 1959; Weidman, 1966) between myocardial cells and neurons and the "electrotonic" transmission of experimentally evoked electrical potentials between nonexcitable cells (Bennett, 1966; Loewenstein, 1966) are thought to occur via gap junctions. In the case of molecules, a cell contact–dependent intercellular transfer of metabolites has been demonstrated in tissue culture systems, termed metabolic cooperation (Subak-Sharpe et al., 1969; Pitts and Simms, 1977). This phenomenon has been shown to be correlated with the presence of intercellular electrotonic coupling and the structural presence of gap junctions (Gilula et al., 1972).

STRUCTURAL DEFINITIONS

Gap junctions are intercellular contacts in which protein assemblies in one cell's plasma membrane are directly joined to similar assemblies in an adjacent cell's membrane (Robertson, 1963; Revel and Karnovsky, 1967; Goodenough and Revel, 1970; Makowski et al., 1977; Unwin and Ennis, 1983, 1984). The gap junction is characterized by thin-section electron microscopy as a pair of membranes of variable area separated (in most, but not all, tissues) by a 2-nm "gap." Colloidal lanthanum impregnation of the 2-nm gap reveals an array of subunits (connexons) that house a lanthanum-stainable core. In freeze-fracture replicas the structure is characterized by a macular, or plaque-shaped, differentiated region on the plasma membranes that contains the dense, sometimes crystalline, array of connexons seen as intramembrane particles on the P-fracture face and a complementary array of minute depressions or pits on the E-fracture face.

Because of the crystalline nature of the connexons in postmortem, isolated hepatocyte gap junctions, it has been possible to study some of the detailed molecular structure using Fourier techniques. A low-resolution (25 Å) structural model was developed in 1977, using a combination of X-ray diffraction and electron microscopy that has continued to be refined over the past 12 years (Makowski et al., 1977). In addition, Unwin and his colleagues (Unwin and Zampighi, 1980; Unwin and Ennis, 1984) pioneered the study of the structure of the gap junction using low-dose Fourier microscopy (Unwin and Henderson, 1975). These investigators have provided evidence that structural changes accompany changes in calcium activity; these structural changes have been interpreted as possible channel-gating mechanisms. Makowski et al. (1984) have provided data indicating that the channels in isolated gap junctions are impermeant to sucrose, a molecule small enough to traverse the junction. Thus, the average structure of the isolated junctional channels was interpreted to be in the closed state. This finding, coupled with the demonstration that amphibian intercellular communication is 10,000-fold less sensitive to calcium than to hydrogen (Spray et al., 1982), raised some questions as to what the calcium-induced structural transitions observed by Unwin might mean in physiological terms. For a review and further discussion, see Makowski (1985).

Structural studies of gap junctions imaged with cationic and anionic negative stains can be explained in terms of a fixed negative charge on the channel wall (Baker et al., 1985). This is in agreement with the evidence from permeability studies using small, charged fluorescent molecules (Flagg-Newton et al., 1979; Brink and Dewey, 1980; Flagg-Newton, 1980) that there is a fixed negative charge along the channel under physiological conditions. These observations will have to be reconciled with the finding from cDNA studies, discussed below, that the liver channel protein is highly basic, with a net +21 charge. The localization of these charges within the structure will provide interesting structural data in future studies.

FUNCTIONAL DEFINITIONS

Gap junctions have been associated with cell-to-cell diffusion of ions and molecules via a low-resistance pathway that avoids the extracellular space. In the case of ions, both the transmission of action potentials at electrical synapses (Furshpan and Potter, 1959; Weidman, 1966) between myocardial cells and neurons and the "electrotonic" transmission of experimentally evoked electrical potentials between nonexcitable cells (Bennett, 1966; Loewenstein, 1966) are thought to occur via gap junctions. In the case of cell-to-cell diffusion of molecules, a cell contact–dependent intercellular transfer of metabolites has been demonstrated in tissue culture systems, termed metabolic cooperation

(Subak-Sharpe et al., 1969; Pitts and Simms, 1977). This phenomenon has been shown to be correlated with the presence of intercellular electrotonic coupling and the structural presence of gap junctions (Gilula et al., 1972). An electrophysiological approach has shown that membrane-impermeant, fluorescent dyes can diffuse from one cell's cytoplasm to that of another in a process termed dye coupling or dye transfer (Payton et al., 1969; Flagg-Newton and Loewenstein, 1979). Studies with intracellularly injected dyes of known molecular size and charge have demonstrated that the intercellular channels between mammalian cells are about 1.5 nm in diameter and are negatively charged (Flagg-Newton et al., 1979; Brink and Dewey, 1980; Flagg-Newton, 1980), and that solutes move through the intercellular channel in hydrated forms (Brink, 1983, 1985).

SUBCELLULAR FRACTIONATION OF GAP JUNCTIONS

Procedures for the isolation of liver gap junctions were developed following Benedetti and Emmelot's (1968) observation that these intercellular junctions were insoluble in the detergent deoxycholate (Evans and Gurd, 1972; Goodenough and Stoeckenius, 1972; Henderson et al., 1979; Hertzberg and Gilula, 1979). Isolated gap junctions in thin sections retained their *in vivo* appearance. Negative-stain and freeze-fracture images of isolated junctions (Goodenough, 1975) revealed that the subunits that appeared as disordered arrays of particles in freeze fracture of whole tissues were now tightly crystallized, a sequela of the isolation procedure. Negative staining revealed the presence of a stain-penetrable core in the center of each connexon (Goodenough and Stoeckenius, 1972; Goodenough, 1974, 1976) similar to that of the lanthanum-stained specimens.

STRUCTURE–FUNCTION CORRELATIONS AND
ANTIBODY PRODUCTION

While there was initial excitement that the structural transitions of the connexons from a disordered to an ordered state might be correlated with the physiological closure (uncoupling) of the junction channels (Peracchia and Dulhunty, 1976; Peracchia, 1977, 1978), quick freezing of unfixed specimens suggested the kinetics of crystallization were too slow to be a direct effect (Raviola et al., 1980). Indeed, in recent correlated structure–function studies using different uncoupling reagents combined with quick-freezing methodology (Hanna et al., 1985; Miller and Goodenough, 1985), there were no common structural alterations in connexon packing as visualized by quick-freeze fracture that could be correlated with a block in intercellular dye transfer in the lens epithelium or in the tunicate heart, and aldehyde fixation proved to be the only way to achieve connexon crystallization on a physiologically relevant time scale.

A more direct demonstration that gap junction structures were actually mediating cell–cell communication has involved considerable effort over the past decade, requiring knowledge of the molecular components of the junction structure. The isolation protocols led to the identification of a principal polypeptide of 27 kD (called liver 27 kD) as the major protein component that copurified with the structure of the liver gap junction as assayed in the electron microscope (Henderson et al., 1979; Hertzberg and Gilula, 1979; Finbow et al., 1980). Willecke was the first to report the successful production of an anti-27-kD antiserum (Traub et al., 1982), followed soon after by Hertzberg (1984; Hertzberg and Skibbens, 1984), Warner et al. (1984), Paul (1985), and Zervos et al. (1985).

A major breakthrough was achieved in 1984 by Warner and her colleagues, who pioneered the use of antisera as specific reagents in studying the function of gap junction–mediated intercellular communication in tissue processes. These authors demonstrated that the intracellular microinjection of affinity-purified, liver anti-27-kD antiserum could block intercellular communication between *Xenopus* blastomeres, using an antiserum which must have recognized conserved structure in the junctional channel protein, and which also was able to interfere with the physiological function. Of greater interest, moreover, was how Warner showed that this interruption of communication, which did not cause cell death, could be correlated with certain developmental defects among the progeny of the injected cell. Such experiments are extremely complex, since they deal with poorly understood, multifaceted interactions between embryonic cells over long periods of time. The care and rigor of Warner's experiments nonetheless established the necessity for cells to remain networked through communication pathways in order for normal development to occur. The nature and timing of the information exchange between embryonic cells in developing tissues remains an exciting challenge for future experimentation.

With the different polyclonal antisera available in several laboratories, it has been possible to demonstrate directly that the liver 27-kD protein was a component of the electron microscope structure (Traub et al., 1982; Paul, 1985, 1986; Stevenson et al., 1986; Young et al., 1987), and that intracellularly injected anti-liver-27-kD antibodies resulted in electrical and dye uncoupling in different cell systems (Warner et al., 1984; Hertzberg et al., 1985). Indeed, these antisera permitted preliminary investigations of reconstituted gap junction channels *in vitro* (Lynch et al., 1984; Spray et al., 1986a; Young et al., 1987), showing single-channel activity similar to that in studies of single-channel measurements in isolated cell pairs (Loewenstein et al., 1978; Neyton and Trautmann, 1985; Veenstra and DeHaan, 1986). These studies are discussed further below.

In contrast to polyclonal antisera, the generation of useful monoclonal reagents has followed a slower success rate in the gap junction field. Paul

(1985) reported the first ultrastructural localization of a monoclonal by the immunocytochemical staining of isolated liver plasma membranes. While this monoclonal antibody specifically stained only gap junctions, and did not stain the nonjunctional membranes, this reagent would not recognize its antigen following SDS denaturation, so that its usefulness was very limited. Interestingly, immunohistochemical localization using this monoclonal on frozen sections of a wide variety of tissues revealed that it would recognize junctional structure only in liver, suggesting the presence of a liver-specific determinant within the junction structure.

Since then, an additional monoclonal from a rat/mouse hybridoma (R5.21) has been generated (Stevenson et al., 1986). This reagent binds to the cytoplasmic surfaces of isolated gap junctions and binds to the liver 27-kD protein in Western blots. It also binds to the protein product coded for by the liver cDNA reported below. The monoclonal recognizes gap junctions formed between AR4–2J cells, a pancreatic cell line, in addition to junctions between native exocrine pancreas, stomach, kidney tubules, and glia but does not recognize an antigen in either heart or lens. Janssen-Timmen et al. (1986) have also reported the successful generation of an anti-liver-27-kD monoclonal antibody.

Anti-liver-27-kD reagents produced in different laboratories have shown different tissue and species reactivities, some showing broad tissue reactivity (Hertzberg and Skibbens, 1984; Zervos et al., 1985), some showing broad species reactivity (Warner et al., 1984), and some showing high tissue and species specificity (Paul, 1985, 1986). Taken together, these results indicate what had been expected from both comparative biochemical studies (Goodenough et al., 1978; Goodenough, 1979; Kensler and Goodenough, 1980; Nicholson et al., 1983, 1985; Manjunath et al., 1985) and many years of physiological work (for reviews, see Bennett and Goodenough, 1978; Spray and Bennett, 1985), namely, that intercellular communication via gap junctions is mediated by a family of related proteins that show both conserved and nonconserved primary protein structure. A critical comparison of the differences in the structure and function of gap junction proteins from different tissues thus might provide clues for understanding what functions gap junctions confer on groups of nonexcitable cells, and what functions are carried out by different domains of the junctional molecules.

SINGLE-CHANNEL/RECONSTITUTION STUDIES

Along with antibody studies probing the functions of gap junction–mediated intercellular communication in tissue processes, progress has been made in the understanding of functional domains within the junctional molecules themselves. Recent physiological studies have focused on the gating properties of gap junction channels, both in cell pairs and, in the case of liver, in reconstituted systems (Neyton and Trautmann, 1985; Spray et al., 1986a,b;

Veenstra and DeHaan, 1986; Young et al., 1987). Neither liver nor heart gap junction channels are voltage sensitive (in comparison to amphibian blastomeres [Spray et al., 1984]), while both are sensitive to octanol (Spray et al., 1986b; Veenstra and DeHaan, 1986). Results different from the *in vivo* data have been obtained in reconstituted systems (Young et al., 1987), indicating exciting experimental possibilities. Both liver and myocardial gap junction channels are pH sensitive, but the apparent pK values are 6.4 for liver (Spray et al., 1986b) and 6.8 for myocardium (Reber and Weingart, 1982; White et al., 1985). Single-channel events have had conductances measuring in the range of 140–165 pS. In the reconstituted systems, the addition of a specific antiserum abolishes the channel openings.

There are data demonstrating the roles of cytoplasmic factors in the regulation and properties of communication. For example, it has been reported that cAMP-dependent phosphorylation plays a stimulatory role in the regulation of cell communication (Saez et al., 1986), while the early effects of $pp60^{src}$ and c-src protein kinase reduce communication (Atkinson et al., 1981; Azarnia and Loewenstein, 1984; see Loewenstein, 1987, for further discussion).

MOLECULAR CHARACTERIZATION

Using specific antisera and partial amino acid sequences, three laboratories have reported the successful cloning of cDNAs for liver 27-kD gap junction protein (Heynkes et al., 1986; Kumar and Gilula, 1986; Paul, 1986). In one approach an affinity-purified anti-27-kD antiserum was used to screen a liver cDNA library cloned into the expression vector λgt11 (Paul, 1986). One positive detected in 2.5×10^5 plaques was picked and cloned to purity and found to contain a 1.1-kb fragment. Since the polyclonal antibody had known contaminants, two strategies were adopted to prove that the cloned cDNA coded for the liver 27-kD protein. In the first, the cDNA from the λgt11 was excised and subcloned into the high-level expression vector pMAM-17. The resultant fusion protein, expressed in *E. coli* Y1090, was electrophoretically isolated from bug lysates and used to generate a new polyclonal antiserum. This antifusion protein antiserum stained gap junctions immunocytochemically and reacted with the 27-kD protein on Western blots.

The second strategy for confirmation was to compare the N-terminal sequence predicted from the cDNA with that sequenced by Edman degradation for the 27-kD protein (Nicholson et al., 1983). Accordingly, the 1.1-kb fragment was used to rescreen the cDNA library by hybridization. Twenty-two positives were picked, and a 1.5-kb cDNA clone was selected for complete sequencing. The cDNA sequence predicted a protein containing 283 amino acids with a calculated molecular mass of 32,007 daltons. The predicted N-terminal amino acid sequence matched exactly the N-terminal 56 amino acids determined by Nicholson et al. (1983), thereby providing a second, independent confirmation that the cloned cDNA encodes for the liver 27 kD.

Two other laboratories simultaneously and independently screened cDNA libraries using oligonucleotides synthesized according to the published amino acid N-terminal sequence, and according to internal amino acid sequences derived from peptide fragments of the 27-kD liver gap junction protein. These approaches were also fruitful and generated virtually identical sequence information from cDNAs in different species (Heynkes et al., 1986; Kumar and Gilula, 1986).

While the 50 N-terminal amino acids predicted by the liver gap junction cDNA agreed exactly with the published partial sequence of Nicholson et al. (1981), the sequence showed only a 50% homology with the N-terminal sequence of the heart gap junction protein (Nicholson et al., 1985), reported to be a 47-kD molecule (Manjunath et al., 1985). Neither the heart nor the liver junctional proteins showed any sequence homology whatsoever with the lens MP26 (Gorin et al., 1984), in keeping with some immunocytochemical data indicating that the MP26 is not a junctional molecule (Paul and Goodenough, 1983; Kistler et al., 1985; Gruijters et al., 1987). There is controversy on this point, however, since other laboratories find immunocytochemical evidence for the localization of MP26 in both junctional and nonjunctional lens fiber membranes (FitzGerald et al., 1983; Sas et al., 1985). Perhaps recent innovative reconstitution approaches will resolve this issue (Hall and Zampighi, 1985).

Northern analysis using the coding region of the liver gap junction cDNA as a probe revealed a similar 1.5-kb mRNA in the liver, brain, stomach, and kidney, in good agreement with the antibody data reported above. Also in agreement, the cDNA failed to hybridize with mRNAs from heart and lens under stringent hybridization conditions (0.1 × SSC at 65°C), but was able to hybridize to different-sized mRNAs under less stringent hybridization conditions (2.5 × SSC at 50°C). In the heart, hybridization was initially reported to a 1.3-kb mRNA (Paul, 1986); subsequent Northern analysis showed a 2.8-kb heart mRNA to be more reproducible. These observations again indicated that there were conserved and variable domains in the gap junction proteins found in different tissues.

Taking advantage of the ability of the liver cDNA to hybridize to a 2.8-kb heart message at low stringency, Drs. Eric Beyer and David Paul constructed a heart cDNA library, screened under identical low-stringency conditions, that produced the 2.8-kb Northern signal. A single positive was identified and cloned to purity. The selected cDNA was a small, 220-bp fragment that showed a 55% amino acid sequence homology to a region of the liver 27-kD protein, 37 amino acids from the N-terminal. The 220-bp cDNA was then used to rescreen the library at high stringency, and four additional clones were selected. These clones together comprised a 2.8-kb cDNA that appeared to contain a single open reading frame, currently being analyzed. The amino terminal amino acid sequence of the predicted sequence of this cDNA closely matched the published N-terminal sequence of a 29-kD polypeptide prominent in a gap junction–enriched fraction from rat myocardium reviewed above. This observation, together with regions of high-sequence homology to the liver gap junction protein, confirmed the identity of the cDNA as related to myocardial gap junctions.

To the degree that the myocardial cDNA has been analyzed, both heart and liver gap junction proteins are basic, with an estimated pI of 10.88 for liver. Counting the histidine residues, the liver has a net positive charge of $+21$. The predicted heart amino acid sequence thus far has an overall homology with that of the liver of less than 50%, and this homology is clustered into different domains of the molecules. Four major hydrophobic domains, which may correspond to transmembrane regions, and two hydrophilic domains, which may correspond to the junctional extracellular regions, show a much higher degree of conservation of amino acid sequence between the heart and liver than does a putative cytoplasmic loop on each molecule. Such data imply that junctional molecules from different tissues may be capable of interaction and junction formation at the extracellular surfaces, while at the same time retaining their unique cytoplasmic physiological regulatory domains. Indeed, asymmetric gap junctional constructs have been physiologically detected in heterologous culture systems by Flagg-Newton and Loewenstein (1980). These data, combined with the demonstration that epithelial cells can simultaneously form physiologically and structurally different gap junctions with homologous and heterologous cell neighbors, open possibilities for complex cell–cell interactions via gap junctions in whole organisms (Miller and Goodenough, 1986).

EXPRESSION OF mRNA DERIVED FROM CLONED cDNA

Kumar and Gilula (1986) have used SP6 RNA polymerase to direct mRNA synthesis from their cloned human liver cDNA. Using this mRNA to direct synthesis in a reticulocyte lysate system, these authors obtained variable results: either a 32-kD or a 27-kD transcript, depending on different batches of mRNA. The reasons for these differences are not understood. Nonetheless, it is clear that the synthesis and membrane insertion of the liver gap junction 27-kD protein is not accompanied by the cleavage of an N-terminal signal sequence, since the predicted sequence from the cDNA is identical to the N-terminal sequence derived by the Edman degradation of assembled gap junctions.

Werner et al. (1985) have shown in pioneering studies that mRNA fractions from uterus and heart induce cell–cell communication between Xenopus oocytes. In unpublished studies by Drs. Gerhard Dahl and David Paul, capped mRNA, which was transcribed in vitro from cloned liver cDNA linked to an SP6 promoter, was also injected into Xenopus oocytes, and this produced the highest increase in junctional conductance yet observed (manuscript in review). The induction of conductance showed a dose dependence on the concentration of injected mRNA over two orders of magnitude. To test that this high conductance did not mean that the oocytes had fused, coupled oocyte pairs were subjected to acidification with carbon dioxide, a treatment that reversibly uncouples embryos and hepatocytes (Turin and Warner, 1980; Spray et al., 1986b). Oocyte pairs injected with SP6-transcribed mRNA were reversibly uncoupled by acidification.

The *Xenopus* oocyte model presents a unique opportunity to study the synthesis and assembly of gap junctions directed by specific mRNAs. Asymmetric junctions between oocytes injected with different tissue mRNAs can theoretically be constructed and detected with tissue-specific antisera and electrophysiological methods. In addition, chimeric junctional molecules can be engineered from the cDNAs, with the goal of identifying specific physiological domains within the protein structure.

ACKNOWLEDGMENTS

I am indebted to Drs. Eric Beyer, Gerhard Dahl, and David Paul for their willingness to share unpublished data. The studies reported were supported by grants GM-18974 and EY-02430 to D. A. G. and GM-37751 to David Paul. Eric Beyer is a recipient of Clinician-Scientist Award 870405 from the American Heart Association.

REFERENCES

Atkinson, M. M., A. S. Menko, R. G. Johnson, J. R. Sheppard, and J. D. Sheridan (1981) Rapid and reversible reduction of junctional permeability in cells infected with a temperature-sensitive mutant of avian sarcoma virus. *J. Cell Biol.* **91**:573–578.

Azarnia, R., and W. R. Loewenstein (1984) Intercellular communication and the control of growth. X. Alteration of junctional permeability by the *src* gene. A study with temperature-sensitive mutant Rous sarcoma virus. *J. Membr. Biol.* **82**:191–205.

Baker, T. S., G. E. Sosinsky, D. L. D. Caspar, C. Gall, and D. A. Goodenough (1985) Gap junction structures. VII. Analysis of connexon images obtained with cationic and anionic negative stains. *J. Mol. Biol.* **184**:81–98.

Benedetti, E. L., and P. Emmelot (1968) Hexagonal array of subunits in tight junctions separated from isolated rat liver plasma membranes. *J. Cell Biol.* **38**:15–24.

Bennett, M. V. L. (1966) Physiology of electrotonic junctions. *Ann. N.Y. Acad. Sci.* **137**:509–539.

Bennett, M. V. L., and D. A. Goodenough (1978) Gap junctions, electrotonic coupling, and intercellular communication. *Neurosci. Res. Progr. Bull.* **16**:373–486.

Brink, P. R. (1983) Effect of deuterium oxide on junctional membrane channel permeability. *J. Membr. Biol.* **71**:79–87.

Brink, P. R. (1985) The effects of deuterium oxide on junctional membrane permeability and conductance. In *Gap Junctions*, M. V. L. Bennett and D. C. Spray, eds., pp. 123–138, Cold Spring Harbor Laboratory, Cold Spring Harbor, New York.

Brink, P. R., and M. M. Dewey (1980) Evidence for fixed charge in the nexus. *Nature* **285**:101–102.

Evans, W. H., and J. W. Gurd (1972) Preparation and properties of nexuses and lipid enriched vesicles from mouse liver plasma membranes. *Biochem. J.* **128**:691–700.

Finbow, M., S. B. Yancey, R. Johnson, and J.-P. Revel (1980) Independent lines of evidence suggesting a major gap junctional protein with a molecular weight of 26,000. *Proc. Natl. Acad. Sci. USA* **77**:970–974.

FitzGerald, P. G., D. Bok, and J. Horwitz (1983) Immunocytochemical localization of the main intrinsic polypeptide (MIP26) in ultrathin frozen sections of rat lens. *J. Cell Biol.* **97**:1491–1499.

Flagg-Newton, J. L. (1980) The permeability of the cell-to-cell membrane channel and its regulation in mammalian cell junctions. *In Vitro Cell Dev. Biol.* **16**:1043–1048.

Flagg-Newton, J. L., and W. R. Loewenstein (1979) Experimental depression of junctional membrane permeability in mammalian cell culture: A study with tracer molecules in the 300–800 Dalton range. *J. Membr. Biol.* **50**:65–100.

Flagg-Newton, J. L., and W. R. Loewenstein (1980) Asymmetrically permeable membrane channels in cell junction. *Science* **207**:771–773.

Flagg-Newton, J. L., I. Simpson, and W. R. Loewenstein (1979) Permeability of the cell-to-cell membrane channels in mammalian cell junction. *Science* **205**:404–407.

Furshpan, E. J., and D. D. Potter (1959) Transmission at the giant motor synapses of the crayfish. *J. Physiol. (Lond.)* **145**:289–325.

Gilula, N. B., O. R. Reeves, and A. Steinbach (1972) Metabolic coupling, ionic coupling, and cell contacts. *Nature* **235**:262–265.

Goodenough, D. A. (1974) Bulk isolation of mouse hepatocyte gap junctions. Characterization of the principal protein, connexin. *J. Cell Biol.* **61**:557–563.

Goodenough, D. A. (1975) The structure and permeability of isolated hepatocyte gap junctions. *Cold Spring Harbor Symp. Quant. Biol.* **40**:37–43.

Goodenough, D. A. (1976) *In vitro* formation of gap junction vesicles. *J. Cell Biol.* **68**:220–231.

Goodenough, D. A. (1979) Lens gap junctions: A structural hypothesis for nonregulated low-resistance intercellular pathways. *Invest. Ophthalmol. Vis. Sci.* **18**:1104–1122.

Goodenough, D. A., and J.-P. Revel (1970) A fine structural analysis of intercellular junctions in the mouse liver. *J. Cell Biol.* **45**:272–290.

Goodenough, D. A., and W. Stoeckenius (1972) The isolation of mouse hepatocyte gap junctions: Preliminary chemical characterization and x-ray diffraction. *J. Cell Biol.* **54**:646–656.

Goodenough, D. A., D. Paul, and K. Culbert (1978) Correlative gap junction structure. *Birth Defects* **14**:83–97.

Gorin, M. B., S. B. Yancey, J. Cline, J.-P. Revel, and J. Horwitz (1984) The major intrinsic protein (MIP) of the bovine lens fiber membrane: Characterization and structure based on cDNA cloning. *Cell* **39**:49–59.

Gruijters, W. I. M., J. Kistler, S. Bullivant, and D. A. Goodenough (1987) Immunolocalization of MP70 in lens fiber 16–17nm intercellular junctions. *J. Cell Biol.* **104**:565–572.

Hall, J. E., and G. A. Zampighi (1985) Protein from purified lens junctions induces channels in planar lipid bilayers. In *Gap Junctions*, M. V. L. Bennett and D. C. Spray, eds. pp. 177–189, Cold Spring Harbor Laboratory, Cold Spring Harbor, New York.

Hanna, R. B., R. L. Ornberg, and T. S. Reese (1985) Structural details of rapidly frozen gap junctions. In *Gap Junctions*, M. V. L. Bennett and D. C. Spray, eds., pp. 23–32, Cold Spring Harbor Laboratory, Cold Spring Harbor, New York.

Henderson, D., H. Eibl, and K. Weber (1979) Structure and biochemistry of mouse hepatic gap junctions. *J. Mol. Biol.* **132**:193–218.

Hertzberg, E. L. (1984) A detergent-independent procedure for the isolation of gap junctions from rat liver. *J. Biol. Chem.* **259**:9936–9943.

Hertzberg, E. L., and N. B. Gilula (1979) Isolation and characterization of gap junctions from rat liver. *J. Biol. Chem.* **254**:2138–2147.

Hertzberg, E. L., and R. V. Skibbens (1984) A protein homologous to the 27,000 dalton liver gap junction protein is present in a wide variety of species and tissues. *Cell* **39**:61–69.

Hertzberg, E. L., D. C. Spray, and M. V. L. Bennett (1985) Reduction of gap junctional conductance by microinjection of antibodies against the 27-kDa liver gap junction polypeptide. *Proc. Natl. Acad. Sci. USA* **82**:2412–2416.

Heynkes, R., G. Kozjek, O. Traub, and K. Willecke (1986) Identification of a rat liver cDNA and mRNA coding for the 28 kDa gap junction protein. *FEBS Lett.* **205**:56–60.

Janssen-Timmen, U., O. Traub, R. Dermeitzel, H. M. Rabes, and K. Willecke (1986) Reduced number of gap junctions in rat hepatocarcinomas detected by monoclonal antibody. *Carcinogenesis* **7**:1475–1482.

Kensler, R. W., and D. A. Goodenough (1980) Isolation of mouse myocardial gap junctions. *J. Cell Biol.* **86**:755–764.

Kistler, J., B. Kirkland, and S. Bullivant (1985) Identification of a 70,000-D protein in lens membrane junctional domains. *J. Cell Biol.* **101**:28–35.

Kumar, N. M., and N. B. Gilula (1986) Cloning and characterization of human and rat liver cDNAs coding for a gap junction protein. *J. Cell Biol.* **103**:767–776.

Loewenstein, W. R. (1966) Permeability of membrane junctions. *Ann. N.Y. Acad. Sci.* **137**:441–472.

Loewenstein, W. R. (1987) The cell-to-cell channel of gap junctions. *Cell* **8**:725–726.

Loewenstein, W. R., Y. Kanno, and S. J. Socolar (1978) Quantum jumps of conductance during formation of membrane channels at cell–cell junction. *Nature* **274**:133–136.

Lynch, E. C., A. Harris, and D. L. Paul (1984) Ion channel phenomenology reconstituted from liver gap junction preparations. *Biophys. J.* **45**:61a (Abstract).

Makowski, L. (1985) Structural domains in gap junctions: Implications for the control of intercellular communication. In *Gap Junctions*, M. V. L. Bennett and D. C. Spray, eds., pp. 5–12, Cold Spring Harbor Laboratory, Cold Spring Harbor, New York.

Makowski, L., D. L. D. Caspar, W. C. Phillips, and D. A. Goodenough (1977) Gap junction structures. II. Analysis of the x-ray diffraction data. *J. Cell Biol.* **74**:629–645.

Makowski, L., D. L. D. Caspar, W. C. Phillips, and D. A. Goodenough (1984) Gap junction structures. V. Structural chemistry inferred from x-ray diffraction measurements on sucrose accessibility and trypsin susceptibility. *J. Mol. Biol.* **174**:449–481.

Manjunath, C. K., G. E. Goings, and E. Page (1985) Proteolysis of cardiac gap junctions during their isolation from rat hearts. *J. Membr. Biol.* **85**:159–168.

Miller, T. M., and D. A. Goodenough (1985) Gap junction structures after experimental alteration of junctional channel conductance. *J. Cell Biol.* **101**:1741–1748.

Miller, T. M., and D. A. Goodenough (1986) Evidence for two physiologically distinct gap junctions expressed by the chick lens epithelial cell. *J. Cell Biol.* **102**:194–199.

Neyton, J., and A. Trautmann (1985) Single-channel currents of an intercellular junction. *Nature* **317**:331–335.

Nicholson, B. J., M. W. Hunkapiller, L. B. Grim, L. E. Hood, and J.-P. Revel (1981) Rat liver gap junction protein: Properties and partial sequence. *Proc. Natl. Acad. Sci. USA* **78**:7594–7598.

Nicholson, B. J., L. J. Takemoto, L. E. Hunkapiller, L. E. Hood, and J.-P. Revel (1983) Differences between liver gap junction protein and lens MP26 from rat: Implications for tissue specificity of gap junctions. *Cell* **32**:967–978.

Nicholson, B. J., D. B. Gros, S. B. H. Kent, L. E. Hood, and J.-P. Revel (1985) The M_r 28,000 gap junction proteins from rat heart and liver are different but related. *J. Biol. Chem.* **260**:6514–6517.

Paul, D. L. (1985) Antibody against liver gap junction 27-kD protein is tissue specific and cross-reacts with a 54-kD protein. In *Gap Junctions*, M. V. L. Bennett and D. C. Spray, eds., pp. 107–122, Cold Spring Harbor Laboratory, Cold Spring Harbor, New York.

Paul, D. L. (1986) Molecular cloning of cDNA for rat liver gap junction protein. *J. Cell Biol.* **103**:123–134.

Paul, D. L., and D. A. Goodenough (1983) Preparation, characterization, and localization of antisera against bovine MP26, an integral protein from lens fiber plasma membrane. *J. Cell Biol.* **96**:625–632.

Payton, B. W., M. V. L. Bennett, and G. D. Pappas (1969) Temperature-dependence of resistance at an electrotonic synapse. *Science* **165**:594–597.

Peracchia, C. (1977) Gap junctions: Structural changes after uncoupling procedures. *J. Cell Biol.* **72**:628–641.

Peracchia, C. (1978) Calcium effects on gap junction structure and cell coupling. *Nature* **271**:669–671.

Peracchia, C., and A. F. Dulhunty (1976) Low resistance junctions in crayfish: Structural changes with functional uncoupling. *J. Cell Biol.* **70**:419–439.

Pitts, J. D., and J. W. Simms (1977) Permeability of junctions between animal cells: Intercellular transfer of nucleotides but not macromolecules. *Exp. Cell Res.* **104**:153–163.

Raviola, E., D. A. Goodenough, and G. Raviola (1980) Structure of rapidly frozen gap junctions. *J. Cell Biol.* **87**:273–279.

Reber, W. R., and R. Weingart (1982) Ungulate cardiac Purkinje fibres: The influence of intracellular pH on the electrical cell-to-cell coupling. *J. Physiol. (Lond.)* **328**:87–104.

Revel, J.-P., and M. J. Karnovsky (1967) Hexagonal array of subunits in intercellular junctions of the mouse heart and liver. *J. Cell Biol.* **33**:C7–C12.

Robertson, J. D. (1963) The occurrence of a subunit pattern in the unit membranes of club endings in Mauthner cell synapses in goldfish brains. *J. Cell Biol.* **19**:201–221.

Saez, J. C., D. C. Spray, A. C. Nairn, E. L. Hertzberg, P. Greengard, and M. V. L. Bennett (1986) cAMP increases junctional conductance and stimulates phosphorylation of the 27-kDa principal gap junction polypeptide. *Proc. Natl. Acad. Sci. USA* **83**:2473–2477.

Sas, D. F., M. J. Sas, K. R. Johnson, A. S. Menko, and R. G. Johnson (1985) Junctions between lens fiber cells are labeled with a monoclonal antibody shown to be specific for MP26. *J. Cell Biol.* **100**:216–225.

Spray, D. C., J. H. Stern, A. L. Harris, and M. V. L. Bennett (1982) Gap junctional conductance: Comparison of sensitivities to H^+ and Ca^{++} ions. *Proc. Natl. Acad. Sci. USA* **79**:441–445.

Spray, D. C., R. L. White, A. C. Campos de Carvalho, A. L. Harris, and M. V. L. Bennett (1984) Gating of gap junction channels. *Biophys. J.* **45**:219–230.

Spray, D. C., J. H. Stern, A. L. Harris, and M. V. L. Bennett (1982) Gap junctional conductance: Comparison of sensitivities to H^+ and Ca^{++} ions. *Proc. Natl. Acad. Sci. USA* **79**:441–445.

Spray, D. C., J. C. Saez, D. Brosius, M. V. L. Bennett, and E. L. Hertzberg (1986a) Isolated liver gap junctions: Gating of transjunctional currents is similar to that in intact pairs of rat hepatocytes. *Proc. Natl. Acad. Sci. USA* **83**:5494–5497.

Spray, D. C., R. D. Ginzberg, E. A. Morales, Z. Gatmaitan, and I. M. Arias (1986b) Electrophysiological properties of gap junctions between dissociated pairs of rat hepatocytes. *J. Cell Biol.* **103**:135–144.

Stevenson, B. R., J. D. Siliciano, M. S. Mooseker, and D. A. Goodenough (1986) Identification of ZO-1: A high molecular weight polypeptide associated with the tight junction (zonula occludens) in a variety of epithelia. *J. Cell Biol.* **103**:755–766.

Subak-Sharpe, H., R. R. Burk, and J. D. Pitts (1969) Metabolic cooperation between biochemically marked mammalian cells in tissue culture. *J. Cell Sci.* **4**:353–367.

Traub, O., U. Janssen-Timmen, P. M. Druge, R. Dermietzel, and K. Willecke (1982) Immunological properties of gap junction protein from mouse liver. *J. Cell. Biochem.* **19**:27–44.

Turin, L., and A. E. Warner (1980) Intracellular pH in early *Xenopus* embryos: Its effects on current flow between blastomeres. *J. Physiol. (Lond.)* **300**:489–504.

Unwin, P. N. T., and P. D. Ennis (1983) Calcium-mediated changes in gap junction structure: Evidence from the low angle x-ray pattern. *J. Cell Biol.* **97**:1459–1466.

Unwin, P. N. T., and P. D. Ennis (1984) Two configurations of a channel-forming membrane protein. *Nature* **307**:609–613.

Unwin, P. N. T., and R. Henderson (1975) Molecular structure determination by electron microscopy of unstained crystalline specimens. *J. Mol. Biol.* **94**:425–440.

Unwin, P. N. T., and G. Zampighi (1980) Structure of the junction between communicating cells. *Nature* **283**:545–549.

Veenstra, R. D., and R. L. DeHaan (1986) Measurement of single channel currents from cardiac gap junctions. *Science* **233**:972–974.

Warner, A. E., S. C. Guthrie, and N. B. Gilula (1984) Antibodies to gap-junctional protein selectively disrupt junctional communication in the early amphibian embryo. *Nature* **311**:127–131.

Weidmann, S. (1966) The diffusion of radiopotassium across intercalated disks of mammalian cardiac muscle. *J. Physiol. (Lond.)* **187**:323–342.

Werner, R., T. Miller, R. Azarnia, and G. Dahl (1985) Translation and functional expression of cell–cell channel mRNA in *Xenopus* oocytes. *J. Membr. Biol.* **87**:253–268.

White, R. L., D. C. Spray, A. C. Campos de Carvalho, B. A. Wittenberg, and M. V. L. Bennett (1985) Some physiological and pharmacological properties of cardiac myocytes dissociated from adult rat. *Am. J. Physiol.* **249**:c447–c455.

Young, J. D.-E., Z. A. Cohn, and N. B. Gilula (1987) Functional assembly of gap junction conductance in lipid bilayers: Demonstration that the major 27 kD protein forms the junctional channel. *Cell* **48**:733–743.

Zervos, A. S., J. Hope, and W. H. Evans (1985) Preparation of a gap junction fraction from uteri of pregnant rats: The 28-kD polypeptides of uterus, liver, and heart gap junctions are homologous. *J. Cell Biol.* **101**:1363–1370.

Chapter 17

Gap Junction Protein and Its Organization into the Cell–Cell Channel

NORTON B. GILULA

ABSTRACT

The transmission of molecular information through gap junction channels that join cells has been the subject of a number of cell biological and physiological studies. On the basis of recent progress in the development of immunological and recombinant DNA probes for the major 32-kD protein of the mammalian liver gap junction, it is now possible to define the general topological relationship of this protein to the gap junctional cell–cell channel. This understanding is a direct result of being able to integrate the primary amino acid sequence for the 32-kD protein into the membrane in a topologically relevant fashion.

It has now been twenty years since the initial structural clarification was made for the cell–cell membrane specialization that is termed the gap junction. This distinction was made in a classic electron microscope analysis applying lanthanum penetration by Revel and Karnovsky (1967). The electrotonic or electrical synapse, the primary functional property of this structure, was originally described by Furshpan and Potter in 1959. Between 1959 and 1967 several important studies that provided partial clarifications of the structure of the gap junction element were published. These included: a study on the site of an electrical synapse in the fish brain by Robertson (1963); an analysis on junctions in smooth and cardiac muscle systems by Dewey and Barr, which proposed the term "nexus" for this structure (1962; Barr et al., 1965); a study by Farquhar and Palade (1963) on several junctional structures in various epithelia; and a description of the gap junction structures in subcellular fractions of hepatocyte plasma membranes by Benedetti and Emmelot (1968). The information from all of these studies was somewhat finalized by the observations reported in the paper by Revel and Karnovsky in 1967. Subsequent ultrastructural analysis employing negative stain and freeze-fracture procedures (Goodenough and Revel, 1970; McNutt and Weinstein, 1970; Friend and Gilula, 1972) provided additional criteria for distinguishing the gap junction as a unique cell surface membrane specialization distinct from other, related structures, such as tight junctions.

In 1972 a genetic and cell biological analysis that provided the first integrated evidence for the structure–function relationship of the gap junction was published (Gilula et al., 1972). This study demonstrated that the junctional structure provided a pathway for the movement of molecular information between cells. Such cell–cell transmission of information was equated with the electrical synaptic event, or electrical coupling, and a metabolic coupling event that was initially described as metabolic cooperation between cells (Subak-Sharpe et al., 1969).

During the past twenty years the study of the structure–function relationship of the gap junction has focused on electron microscope observations documenting the distribution and pleiomorphic arrangements of the gap junction structures in various tissues from both invertebrate and vertebrate organisms (for review, see Gilula, 1977); biophysical experiments documenting the permeability properties of gap junction channels (for reviews, see Loewenstein, 1981; Spray and Bennett, 1985); and finally, biochemical characterizations of enriched subcellular fractions of gap junctions, which have led to the identification of the actual chemical components of the gap junction structures (for reviews, see Hertzberg et al., 1981; Hertzberg, 1985; Revel et al., 1985). These analyses have resulted in opportunistic progress in this area over the past few years, particularly in (1) understanding the biological properties of gap junctional communication in nonexcitable systems that are engaged in developmental and differentiation processes; and (2) forming a characterization of the complete primary amino acid sequence for the major protein that exists in the mammalian liver gap junction.

The ingredients of this progress form the basis for this chapter, which considers four fundamental questions. What is the major protein of the cell–cell channel? What is the function of this protein? How is the protein organized to form the channel? How is the channel regulated?

THE MAJOR PROTEIN OF THE CELL–CELL CHANNEL

Information on the biochemical components of the gap junction has been obtained directly from the characterization of gap junction structures that can be isolated as enriched subcellular fractions from either mammalian liver (Hertzberg et al., 1981) or, more recently, mammalian heart (Manjunath and Page, 1986). Since most of the direct information on the gap junction protein has been obtained from the liver gap junctions, this chapter focuses primarily on the properties of the major protein derived from that structure.

At this writing, a number of proteins with different sizes have been reported to be associated with gap junctions. These include molecules with sizes of 16–18 kD, 21 kD, 26–28 kD, 32 kD, 47 kD, 54 kD, and 64–66 kD. Clearly, all of these molecules do *not* represent different gene products, but rather, different electrophoretic mobilities or physical states of similar or the same gene products. A reasonable consensus from several labs has been reached on the size of the major endogenous liver junction protein, a molecule of 26–28 kD (for review, see Revel et al., 1985). In addition, progress in using recombinant DNA technology has made it possible to determine the full-length sequence as deduced from rat and human cDNAs (Kumar and Gilula, 1986; Paul, 1986). These developments indicate that the full-length sequence for the major mammalian liver junction protein is actually 32 kD, rather than the 26–28-kD size determined by SDS-gel electrophoresis analysis of the native protein. From this analysis of the liver cDNA, it has been possible to demonstrate that a complete 32-kD protein can be derived from the transcript by *in vitro* translation (Kumar and Gilula, 1986). Consequently, the true size of the major human and rat liver gap junction protein is 32 kD, and this molecule can commonly associate into a dimeric form with a true size of approximately 64 kD.

The other molecules, such as the 16–18 kD reported by Finbow et al. (1983) and the 21 kD, may represent either totally independent molecules or proteolytic fragments derived from the 32-kD molecule (Zimmer et al., 1987). Clearly, the 47-kD and 54-kD sizes represent gel electrophoretic profiles of the dimeric form of the 32-kD molecule (Henderson et al., 1979; Hertzberg and Gilula, 1979). The relationships of the 26–28-kD, 32-kD, 47-kD, 54-kD, and 64–66-kD molecules can be demonstrated by various protease treatments and immunological criteria. Thus far, although an 18-kD fragment of the 32-kD molecule has been described (Zimmer et al., 1987), it is clear that there are other species of the same size (16–18 kD) that are totally unrelated to the authentic 32-kD molecule (Finbow et al., 1983, 1987). Finally, the 21-kD molecule is definitely detectable to differing extents in preparations of both mouse and rat liver gap junctions (Revel et al., 1987). The relationship of the 21-kD molecule to the intact gap junction is still uncertain. It is possible that this molecule is either an integral component or a peripherally associated component of the gap junction.

The identity and properties of the major protein present in isolated junction preparations from mammalian heart have been the subject of a number of reports (Kensler and Goodenough, 1980; Nicholson et al., 1985; Manjunath and Page, 1986). It appears from the most recent information available (Beyer, 1987) that the major protein of the heart junction preparations is a molecule of approximately 43 kD. This molecule has some relationship to the liver junction protein, shown by a comparative analysis of the amino-terminal sequence of the two proteins (Nicholson et al., 1985) and the use of antibodies (Hertzberg and Skibbens, 1984; Zervos et al., 1985).

Finally, a molecule that has been extensively studied from the mammalian eye lens, the membrane protein of 26-kD size, has been completely characterized by various approaches, including a recombinant cDNA analysis (Gorin et al., 1984). Thus far, all available biochemical, immunological, and genetic evidence shows this molecule to be totally unrelated to the major protein of the liver junction (Hertzberg et al., 1982; Revel et al., 1987). However, recent studies in the lens have focused on another membrane protein of 70-kD size, termed MP70 (Kistler et al., 1985). It is possible that the MP70 molecule may be related to the gap junction protein family.

The direct evidence that the 32-kD liver protein is an integral part of the gap junction comes primarily from an analysis of the preparations enriched for gap junction structures, as well as from direct immunocytochemical localizations. In the enriched subcellular fractions of liver junctions, the most prominent protein component consistently associated with this material has the size of the 32-kD monomer. In addition, it is very common for this monomer to be aggregated into multiple-sized species (dimers, etc.). The most direct evidence that this molecule is located in the membrane structure itself comes from using antibodies that have been generated to the junction protein. Such immuno-localization procedures have been applied on both light (Traub et al., 1982; Hertzberg and Skibbens, 1984) and electron microscope levels (Dermietzel et al., 1984; Paul, 1986; Young et al., 1987). The light microscope evidence clearly demonstrates that the junction protein, as an antigen, is highly localized on the surfaces of cells to the sites where cells come into contact with other cells. These regions are predictably the locations of gap junction structures when analyzed by electronmicroscopy.

Finally, an extension of the light microscope localization has been made to the ultrastructural level, and these include both *in vivo* and *in vitro* isolated structures. When antibodies have been applied to the isolated gap junctions, the antibodies bind directly to both sides of the cytoplasmic surfaces of the isolated bipartite structure (see Figure 1). This evidence indicates that the 32-kD protein that has been characterized as an endogenous protein, from both native material and an mRNA transcript, is an integral component of the gap junction membrane. Similar information has not yet been generated for the major protein associated with the heart gap junction fraction.

The application of the antibodies that have been generated in various laboratories has suggested that the major protein of the liver gap junction has regions of strong sequence conservation as well as regions of substantial diversity, relative to other potential gap junction proteins. For example, some antibodies show dramatic tissue specificity for detecting junction antigen (Traub et al., 1982), whereas others show broad homology when applied to tissues from various organs as well as organisms (Hertzberg and Skibbens, 1984). The reagent generated in our laboratory is able to detect homologous proteins from tissues in both vertebrate and invertebrate organisms (Warner et al., 1984; Fraser et al., 1987). Consequently, it is possible to predict that there will be some diversification to create a family of gap junction proteins. However, within this family of proteins there will be a strong conservation of amino acid sequences in certain regions, and those conserved regions will predictably be the ones that are essential for creating the structural framework for the integral membrane portion of a cell–cell channel.

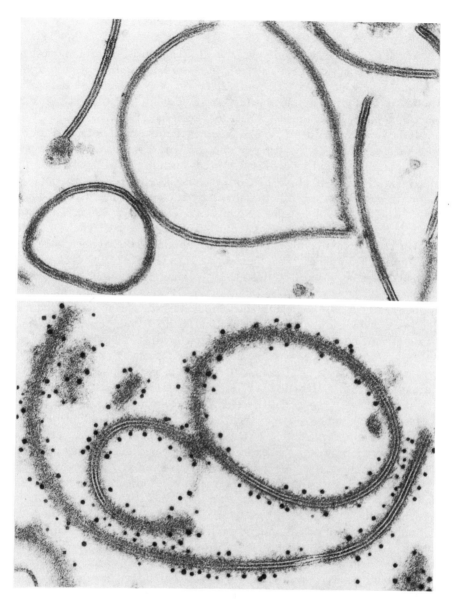

Figure 1. *Immunolocalization of preimmune IgG (top) and affinity-purified anti-32-kD junction IgG (bottom) to a subcellular fraction of isolated rat liver gap junctions.* The bound IgG is detected by a secondary reagent, a protein A–colloidal gold conjugate. Note that the electron-dense gold spheres are bound to the cytoplasmic surfaces of the anti-32-kD IgG–treated structure (*bottom*), and not to the preimmune IgG–treated material (*top*). Magnifications: × 99,000 (*top*); × 100,000 (*bottom*). (From Young et al., 1987.)

FUNCTION OF THE MAJOR GAP JUNCTION PROTEIN

The actual functional contribution of the 32-kD gap junction protein has been determined by: (1) direct analysis via *in vitro* reconstitution of the purified protein (Young et al., 1987); and (2) *in vivo* perturbation studies using antibodies that are specific for the 32-kD protein (Warner et al., 1984; Hertzberg et al., 1985; Fraser et al., 1987; Lee et al., 1987).

Studies of the function of the 32-kD protein were carried out by using an *in vitro* reconstitution of the protein into artificial lipid membrane systems (Young et al., 1987). This analysis showed that the purified protein can be integrated into phospholipid bilayer systems to generate cell–cell channels that have conductance properties strikingly similar to those that have been detected in intact cell systems (Neyton and Trautmann, 1985). The actual conductance of a single channel is about 140 pS in 0.1 M NaCl when measured in an *in vitro* system (Young et al., 1987) and about 160 pS between intact cells (Neyton and Trautmann, 1985; Veenstra and DeHaan, 1986). The relevance of the specific contribution of this 32-kD protein to channel function was determined by using the 32-kD antibodies. When these antibodies were applied to the *in vitro* reconstituted channel, the conductance of the channels decreased significantly. Such effects on channel conductance were not observed when antibodies that do not interact with the 32-kD molecule were used. Consequently, it is possible to conclude that the 32-kD protein in the presence of a bilayer lipid matrix is capable of generating a channel conductance that is, for all practical purposes, identical to the channel conductances observed *in vivo*. The channels detected in the reconstituted system presumably represent an oligomerization of the 32-kD protein in the lipid bilayer.

The determinations of the 32-kD protein function *in vivo* have been based on a perturbation strategy that uses gap junction–specific antibodies to disrupt or block the channel conductance. Initial observations demonstrated that antibodies specific for the 32-kD protein can block channel conductances as determined by both dye transfer measurements and direct electrical coupling measurements (Warner et al., 1984; Hertzberg et al., 1985). These studies were critically dependent on an antibody injection strategy, since the antibodies interact with specific antigenic determinants that are located on the cytoplasmic surface of the junction membranes. This perturbation approach was extended in three biological systems to address the issue of the relationship of cell–cell channel function to biological events of development and differentiation. In the early frog embryo it was determined that blocking communication by microinjecting an antibody into one cell of an eight-cell-stage embryo caused a serious developmental defect in the structures generated by the lineage of the initial cell that was perturbed (Warner et al., 1984). These defects were quite conspicuous at the tadpole stage; for example, strong structural asymmetries could be detected in nervous system–derived structures.

In a second analysis a similar approach was used to determine the relationship of cell–cell communication to a patterning process in the coelenterate *Hydra* (Fraser et al., 1987). In this experiment it was found that a block in communication could have a significant influence on a patterning process in this organism: in the absence of junctional communication, a secondary axis was developed. These observations are described in detail by Fraser (this volume).

Finally, a block of communication in single cells of the early mouse embryo significantly influenced the subsequent fate of cells that had been perturbed (Lee et al., 1987). In this study cells that were made communication defective by virtue of antibody microinjection subsequently decompacted and were eliminated from the remaining cells in the early multicellular preimplantation embryo.

In essence, the perturbation approach applied to the different cell systems *in vivo* has provided a direct indication that cell–cell communication via gap junction channels is an important activity during development and differentiation. In general, the three studies described above are all consistent with a role for gap junctional communication in patterning processes that occur at critical stages of development. Additional experimentation in the future may reveal what effect, if any, gap junctional communication pathways may have in other developmental processes, such as embryonic induction.

ORGANIZATION OF THE PROTEIN TO FORM A JUNCTION CHANNEL

The topological organization of the 32-kD protein in the intact junctional membrane is extremely important for understanding the precise structure–function relationship of this protein in the cell–cell channel. Furthermore, the deduction that there is a primary sequence for the protein has emphasized the importance of some direct experimental information that can be used to integrate the primary sequence into the membrane in a relevant fashion.

With this objective in mind, a topological analysis of the 32-kD protein in the intact junction membrane was carried out using selected protease digestions together with antibody immunolocalizations and calmodulin-binding data (Zimmer et al., 1987). This study considered the isolated gap junction bipartite structure as two dimensional and applied an experimental methodology for separating the two junction membranes *in vitro*. By using this multifaceted approach, a number of topologically relevant details concerning the integration of this 32-kD protein in the membrane were determined. (1) The 32-kD protein contains both exposed and buried regions as determined by protease digestions of the intact junction. (2) The most prominent region of protease accessibility is a cytoplasmic surface domain that extends from the carboxyl terminus of the molecule. (3) The most likely site for the amino terminus of the molecule is on the cytoplasmic surface. (4) Antibodies that block conductance of the channel bind to at least two different sites, both of which are exposed on the cytoplasmic surface of the junction membrane. (5) Calmodulin binding to the junction protein takes place in a region that has a potential biological

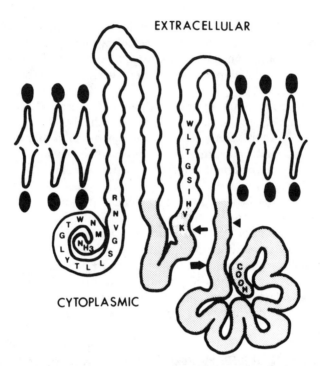

Figure 2. *Two-dimensional model of the intact 32-kD major gap junction protein from mammalian liver.* The model is based on evidence obtained from primary sequence data, antibody binding, calmodulin binding, and protease digests of intact isolated gap junctions. The *arrowhead* represents the location of a TPCK-trypsin site, the *small arrow* locates a lys-X cleavage site, and the *large arrow* denotes the approximate location of an endoproteinase-C cleavage site. (From Zimmer et al., 1987.)

influence. The sites for calmodulin binding are located on the cytoplasmic surface. (6) When the junctions are split *in vitro* into single membranes, the initial asymmetries that exist in the intact bipartite structure appear to be retained, with no detectable rearrangements for protease-sensitive sites or antibody-binding sites.

On the basis of the information presented above, a model that permits the integration of the primary sequence for the 32-kD protein into the junction membrane in a relevant fashion was generated (Figure 2). This integration of the protein indicates the location of amino- and carboxy-terminal domains on the cytoplasmic surface of the junction membrane. With a protein of 32-kD size, it is plausible that the protein will traverse the membrane at least four times; this would explain the data obtained by the "protection" approach using the protease digestions. In the proposed model two loops extend to the extracellular or "gap" surface of the membranes, and these loops could participate in the association or assembly process required to create the bipartite structure. Most of the potential sites for posttranslational modification of the junction protein, via phosphorylation in particular, exist in a suitably exposed portion of the protein; these are all situated on the extensive carboxy-terminal domain located on the cytoplasmic surface (Kumar and Gilula, 1986).

REGULATION OF THE CHANNEL

This problem area has been extensively studied, primarily by several laboratories using various biophysical and pharmacological approaches (for reviews, see Loewenstein, 1981; Sheridan and Atkinson, 1985; Spray and Bennett, 1985). In general, the regulation of channel *function* can be separated from the events surrounding the regulation of channel *formation*.

In the area of channel formation it is quite easy to summarize the available information, since very little is known about the formation process (for a review, see Sheridan, 1978). In general, it is known that channels can form very quickly between cells, and with little specificity for cell types (Michalke and Loewenstein, 1971; Epstein and Gilula, 1977). Consequently, it has been extremely difficult to study channel formation because virtually all systems have previously communicated, and it is possible that the formation process can take place in the absence of any synthesis (Epstein et al., 1977). In this regard, there has been a recent report suggesting that in a cell culture system of primary mammalian hepatocytes, the loss of expression of gap junction structures and protein can be reversed by the addition of specific components from the extracellular matrix (Spray et al., 1987). In that system a pseudo-differentiation initially occurs, which can be reversed by the addition of proteoglycan and glycosaminoglycan components from the extracellular matrix.

The regulation of channel function can be generalized to include virtually all factors that influence the homeostatic properties of cells. This general category has included ingredients such as general cellular metabolism, the production of ATP, calcium concentration, pH, and the application of various pharmacological agents (for reviews, see Loewenstein, 1981; Sheridan and Atkinson, 1985; Spray and Bennett, 1985). It is important to point out that one major difficulty in interpreting the effects of these various treatments has been that many of these treatments cause irreversible changes, with the important exception of the pH influence (Turin and Warner, 1977). Nonetheless, it will be extremely important in the near future to apply the information now available on the primary sequence of the protein to developing approaches that will have an explicit specificity for the junction protein. Such approaches will be able to reduce, if not eliminate, the complications that currently arise when trying to understand whether the effects of such agents on gap junctional channels are indirect or direct. Such studies are urgently needed, since the number of observations implicating gap junction regulatory alterations has significantly increased (Sheridan and Atkinson, 1985). Some of these alterations have been reported in important contemporary areas, such as tumor promoter treatments and transformation by viral oncogenes, areas of widespread interest. Consequently, the development of specific approaches for understanding the process of gap junction channel regulation has become imperative.

CONCLUSION

Progress on understanding the structure and function of at least one of the gap junction proteins, the 32-kD molecule from mammalian liver, has been substantial. It is now apparent that the next task will be to determine the precise molecular relationship of the 32-kD protein to the functional channel. That progress will undoubtedly come from combining a high-resolution structural analysis with molecular genetic approaches for studying the channel function *in vivo* and *in vitro*.

ACKNOWLEDGMENTS

The studies described from my laboratory have been supported by National Institutes of Health grants GM-37904 and GM-32230. I am grateful for the expert assistance of Cheryl Negus in preparing this manuscript.

REFERENCES

Barr, L., M. M. Dewey, and W. Berger (1965) Propagation of action potentials and the structure of the nexus in cardiac muscle. *J. Gen. Physiol.* **48**:797–823.

Benedetti, E. L., and P. Emmelot (1968) Hexagonal array of subunits in tight junctions separated from isolated rat liver plasma membranes. *J. Cell Biol.* **38**:15–24.

Beyer, E. C. (1987) Molecular cloning of cDNA for a rat heart gap junction. *J. Cell Biol.* **105**:263a.

Dermietzel, R., A. Leibstein, U. Frixen, U. Janssen-Timmen, O. Traub, and K. Willecke (1984) Gap junctions in several tissues share antigenic determinants with liver gap junctions. *EMBO J.* **3**:2261–2270.

Dewey, M. M., and L. Barr (1962) Intercellular connection between smooth muscle cells: The nexus. *Science* **137**:670–672.

Epstein, M. L., and N. B. Gilula (1977) A study of communication specificity between cells in culture. *J. Cell Biol.* **75**:769–787.

Epstein, M. L., J. D. Sheridan, and R. G. Johnson (1977) Formation of low resistance junctions *in vitro* in the absence of protein synthesis and ATP production. *Exp. Cell Res.* **104**:25–30.

Farquhar, M. G., and G. E. Palade (1963) Junctional complexes in various epithelia. *J. Cell Biol.* **17**:375–412.

Finbow, M. E., J. Shuttleworth, A. E. Hamilton, and J. D. Pitts (1983) Analysis of vertebrate gap junction protein. *EMBO J.* **2**:1479–1486.

Finbow, M. E., T. E. J. Buultjens, S. John, E. Kam, L. Meagher, and J. D. Pitts (1987) Molecular structure of the gap junctional channel. *Ciba Found. Symp.* **125**:92–104.

Fraser, S. E., C. R. Green, H. R. Bode, and N. B. Gilula (1987) Selective disruption of gap junctional communication interferes with a patterning process in *Hydra*. *Science* **237**:49–55.

Friend, D. S., and N. B. Gilula (1972) Variations in tight and gap junctions in mammalian tissues. *J. Cell Biol.* **53**:758–776.

Furshpan, E. J., and D. D. Potter (1959) Transmission of giant motor synapses of the crayfish. *J. Physiol.* **143**:289–325.

Gilula, N. B. (1977) Gap junctions and cell communication. In *International Cell Biology*, B. R. Brinkley and K. R. Porter, eds., pp 61–69, Rockefeller Univ. Press, New York.

Gilula, N. B., O. R. Reeves, and A. Steinbach (1972) Metabolic coupling, ionic coupling, and cell contacts. *Nature* **235**:262–265.

Goodenough, D. A., and J.-P. Revel (1970) A fine structural analysis of intercellular junctions in the mouse liver. *J. Cell Biol.* **45**:272–290.

Gorin, M. B., S. D. Yancey, J. Cline, J.-P. Revel, and J. Horwitz (1984) The major intrinsic protein (MIP) of the bovine lens fiber membrane: Characterization and structure based on cDNA cloning. *Cell* **39**:49–59.

Henderson, D., H. Eibl, and K. Weber (1979) Structure and biochemistry of mouse hepatic gap junctions. *J. Mol. Biol.* **132**:193–218.

Hertzberg, E. L. (1985) Antibody probes in the study of gap junctional communication. *Annu. Rev. Physiol.* **47**:305–318.

Hertzberg, E. L., and N. B. Gilula (1979) Isolation and characterization of gap junctions from rat liver. *J. Biol. Chem.* **254**:2138–2147.

Hertzberg, E. L., and R. V. Skibbens (1984) A protein homologous to the 27,000 dalton liver gap junction protein is present in a wide variety of species and tissues. *Cell* **39**:61–69.

Hertzberg, E. L., T. S. Lawrence, and N. B. Gilula (1981) Gap junctional communication. *Annu. Rev. Physiol.* **43**:479–491.

Hertzberg, E. L., D. J. Anderson, M. Friedlander, and N. B. Gilula (1982) Comparative analysis of the major polypeptides from liver gap junctions and lens fiber junctions. *J. Cell Biol.* **92**:53–59.

Hertzberg, A. L., D. C. Spray, and M. V. L. Bennett (1985) Reduction of gap junctional conductance by microinjection of antibodies against the 27-kDa liver gap junction polypeptide. *Proc. Natl. Acad. Sci. USA* **82**:2412–2416.

Kensler, R. W., and D. A. Goodenough (1980) Isolation of mouse myocardial gap junctions. *J. Cell Biol.* **86**:755–764.

Kistler, J., B. Kirkland, and S. Bullivant (1985) Identification of a 70,000-D protein in lens membrane junctional domains. *J. Cell Biol.* **101**:28–35.

Kumar, N. M., and N. B. Gilula (1986) Cloning and characterization of human and rat liver cDNAs coding for a gap junction protein. *J. Cell Biol.* **103**:767–776.

Lee, S., N. B. Gilula, and A. E. Warner (1987) Gap junctional communication and compaction during preimplantation stages of mouse development. *Cell* **51**:851–860.

Loewenstein, W. R. (1981) Junctional intercellular communication: The cell-to-cell membrane channel. *Physiol. Rev.* **61**:829–913.

Manjunath, C. K., and E. Page (1986) Rat heart gap junctions disulfide-bonded connexon multimers: Their depolymerization and solubilization in deoxycholate. *J. Membr. Biol.* **90**:43–57.

McNutt, N. S., and R. S. Weinstein (1970) The ultrastructure of the nexus: A correlated thin-section and freeze-cleave study. *J. Cell Biol.* **47**:666–687.

Michalke, W., and W. R. Loewenstein (1971) Communication between cells of different types. *Nature* **232**:121–122.

Neyton, J., and A. Trautmann (1985) Single-channel currents of an intercellular junction. *Nature* **317**:331–335.

Nicholson, B., D. Gros, S. Kent, L. Hood, and J.-P. Revel (1985) The M_r 28,000 gap junction proteins from rat heart and liver are different but related. *J. Biol. Chem.* **260**:6514–6517.

Paul, D. (1986) Molecular cloning of cDNA for rat liver gap junction protein. *J. Cell Biol.* **103**:123–134.

Revel, J.-P., and M. J. Karnovsky (1967) Hexagonal array of subunits in intercellular junctions of the mouse heart and liver. *J. Cell Biol.* **33**:C7–C12.

Revel, J.-P., B. J. Nicholson, and S. B. Yancey (1985) Chemistry of gap junctions. *Annu. Rev. Physiol.* **47**:263–279.

Revel, J.-P., S. B. Yancey, B. Nicholson, and J. Hoh (1987) Sequence diversity of gap junction proteins. *Ciba Found. Symp.* **125**:108–121.

Robertson, J. D. (1963) The occurrence of a subunit pattern in the unit membranes of club endings in Mauthner cell synapses in goldfish. *J. Cell Biol.* **19**:201–221.

Sheridan, J. D. (1978) Junction formation and experimental modification. In *Intercellular Junctions and Synapses*, J. Feldman, N. B. Gilula, and J. D. Pitts, eds., pp. 37–60, Chapman and Hall, London.

Sheridan, J. D., and M. M. Atkinson (1985) Physiological roles of permeable junctions: Some possibilities. *Am. Rev. Physiol.* **47**:337–353.

Spray, D. C., and M. V. L. Bennett (1985) Physiology and pharmacology of gap junctions. *Annu. Rev. Physiol.* **47**:281–303.

Spray, D., M. Fujita, J. C. Saez, H. Choi, T. Watanabe, E. Hertzberg, L. C. Rosenberg, and L. M. Reid (1987) Proteoglycans and glycosaminoglycans induce gap junction synthesis and function in primary liver cultures. *J. Cell Biol.* **105**:541–551.

Subak-Sharpe, J. H., R. R. Bürk, and J. D. Pitts (1969) Metabolic cooperation between biochemically marked mammalian cells in tissue culture. *J. Cell Sci.* **4**:353–367.

Traub, O., U. Janssen-Timmen, P. Drüge, R. Dermietzel, and K. Willecke (1982) Immunological properties of gap junction protein from mouse liver. *J. Cell Biochem.* **19**:27–44.

Turin, L., and A. Warner (1977) Carbon dioxide reversibly abolishes ionic communication between cells of the early amphibian embryo. *Nature* **270**:56–57.

Veenstra, R. D., and R. L. DeHaan (1986) Measurement of single channel currents from cardiac gap junctions. *Science* **233**:972–974.

Warner, A. E., S. C. Guthrie, and N. B. Gilula (1984) Antibodies to gap junctional protein selectively disrupt junctional communication in the early amphibian embryo. *Nature* **311**:127–131.

Young, J. D. E., Z. A. Cohn, and N. B. Gilula (1987) Functional assembly of gap junction conductance in lipid bylayers: Demonstration that the major 27 kD protein forms the junctional channel. *Cell* **48**:733–743.

Zervos, A. S., J. Hope, and W. H. Evans (1985) Preparation of a gap junction fraction from uteri of pregnant rats: The 28 kD polypeptides of uterus, liver, and heart gap junctions are homologous. *J. Cell Biol.* **101**:1363–1370.

Zimmer, D., C. R. Green, W. H. Evans, and N. B. Gilula (1987) Topological analysis of the major protein in isolated intact rat liver gap junctions and gap junction–derived single membrane structures. *J. Biol. Chem.* **262**:7751–7763.

Section 4

Function in Histogenesis and Disease

Chapter 18

Cellular and Molecular Analysis of N-CAM Expression in Skeletal Muscle

FRANK S. WALSH
GEORGE DICKSON
HILARY GOWER
C. HOWARD BARTON
PATRICK DOHERTY

ABSTRACT

The expression of N-CAM was examined in skeletal muscle cells in vitro. *Both myoblasts and myotubes express N-CAM, but myotubes have a higher level. Muscle differentiation is also accompanied by changes in N-CAM isoforms. Myoblasts express a 145-kD isoform that is down-regulated with fusion, while myotubes express isoforms of 125 kD and 155 kD. These latter isoforms are nontransmembrane and appear to be bound to the cell membrane via a phosphatidylinositol linkage. cDNA clones encoding N-CAM mRNAs from human skeletal muscle were isolated and found to correspond to the myoblast and myotube isoforms. One isolated clone encoded the 6.7-kb transmembrane isoform present in myoblasts. A second clone was found to encode the nontransmembrane myotube isoforms of 5.2 kb and 4.0 kb. A muscle-specific sequence (MSD1) was found in the myotube mRNAs. This resides in the extracellular domain of N-CAM and could dramatically change its structure. The MSD1 sequence was not present in brain or a variety of neural cell lines and is the first identified muscle-specific N-CAM sequence. The cytoplasmic domain of N-CAM was previously believed to be the major region of sequence diversity. Our results show that there is also diversity in the extracellular region. A coculture model has been developed using PC12 cells and myotubes to determine if cell–cell and/or cell–matrix interactions can modulate the biochemical responses of neuronal cells to nerve growth factor (NGF). A specific recognition event between nerve and muscle has been identified and quantitated by its ability to modulate the NGF induction of neurofilament protein and Thy-1 antigen by PC12 cells. Perturbation experiments using specific antibody reagents are under way to determine if N-CAM is involved in this early recognition event.*

Cell–cell interactions mediated by molecules at the cell surface are of crucial importance in the development of skeletal muscle. Much effort has been devoted to identifying developmentally regulated surface proteins that may be involved in events such as myoblast fusion and nerve–muscle interactions. Most of these studies have been based on the idea that specific patterns of gene expression associated with these recognition events may implicate certain glycoconjugates in these processes. Two main methods have been used to date, namely vectorial labeling methods (Walsh and Phillips, 1981) and immuno-chemical, particularly monoclonal antibody, methods (Lee and Kaufman, 1981; Walsh and Ritter, 1981).

Although a number of specific patterns of gene expression have been identified at the cell surface, little is known about the structure and function of such molecules. An alternative strategy is to test whether cell recognition molecules that have been found to be important in other developing systems are expressed in skeletal muscle. One such family of molecules that has now been extensively studied in skeletal muscle, is that of the neural cell adhesion molecules (N-CAMs). N-CAM is perhaps the best-characterized cell adhesion molecule and has been shown to be important in specific cell–cell interactions in brain and other tissues (Edelman, 1984; Rutishauser, 1984). In this chapter we review our work on N-CAM in skeletal muscle, concentrating on three areas: (1) the expression of N-CAM *in vitro,* (2) patterns of alternative splicing of the N-CAM gene in muscle, and (3) the development of coculture models with which to assess the function of N-CAM in nerve–muscle interactions.

REGULATION OF EXPRESSION OF SPECIFIC N-CAM ISOFORMS IN MUSCLE CELL CULTURE

A variety of studies have shown N-CAM to be expressed by muscle cells in culture. These include primary cell cultures of human (Moore and Walsh, 1985), rat (Covault et al., 1986), and chicken muscle (Grumet et al., 1982), and a number of permanent cell lines (Williams et al., 1985; Covault et al., 1986; Moore et al., 1987). However, since such different methods as indirect immunofluorescence, cell surface labeling, and immunoblotting have been used, it has been difficult to come to a consensus as to when N-CAM is first expressed in cell culture, and what molecular isoforms are present during myogenesis. Indirect immunofluorescence staining of the primary cell cultures of human skeletal muscle and also clonal muscle cell lines shows that N-CAM is expressed by mononucleated myoblasts and multinucleated myotubes. Figure 1 shows an example of the staining profile observed on the G8–1 muscle cell line. Myoblasts and myotubes are both positive, but myotubes appear to be more strongly stained. This analysis was confirmed by quantitating the level of N-CAM at the myoblast and myotube stage of growth. G8–1 myotube cultures expressed fourfold more N-CAM than myoblasts (Moore et al., 1987).

The factors controlling the increase in N-CAM expression during myo-genesis are not known, nor is the purpose of the increase. What is clear, however, is that there are also major changes in N-CAM molecular forms

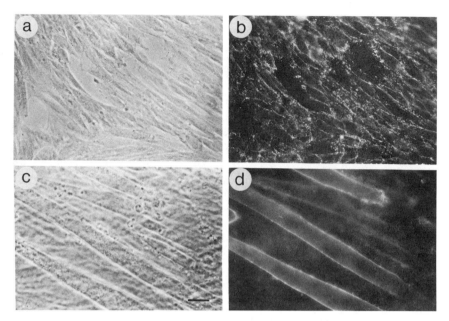

Figure 1. *Indirect immunofluorescence analysis of N-CAM expression in cell cultures of the G8–1 muscle cell line.* Cells at the myoblast or myotube stage of growth were reacted with rabbit anti-N-CAM. *a,b:* Myoblast cells. *c,d:* Myotubes. Calibration bar = 100 μm.

during myogenesis in culture. This was found by using a combination of immunoprecipitation and Western blotting methods. We showed that myoblasts mainly express a desialo N-CAM isoform of 145 kD, while myotubes express isoforms of 125 kD and 155 kD (Moore et al., 1987). The myoblast isoform is transmembrane, since it can be metabolically labeled with ^{32}P and appears to be similar to the 140-kD transmembrane isoform found in brain (Cunningham, 1985). It is also down-regulated following myoblast fusion and is not expressed by myotubes. The major myotube isoforms are not transmembrane and appear to be associated with the plasma membrane via phosphatidylinositol linkage.

This was deduced from a series of experiments where muscle cell cultures were treated with phosphatidylinositol-specific phospholipase C (PI-PLC). We found that treatment of myotubes with PI-PLC led to the release of two desialo N-CAM immunoreactive bands of 125 kD and 155 kD, whereas treatment of myoblasts did not release any N-CAM reactive bands. Control precipitations from conditioned media of G8–1 myotubes also did not show any evidence of secreted N-CAM, suggesting that the release of N-CAM induced by PI-PLC was specific. Indirect immunofluorescence analysis confirmed these observations. Muscle cultures treated with PI-PLC showed greatly reduced levels of N-CAM on the myotube membrane (Figure 2), while

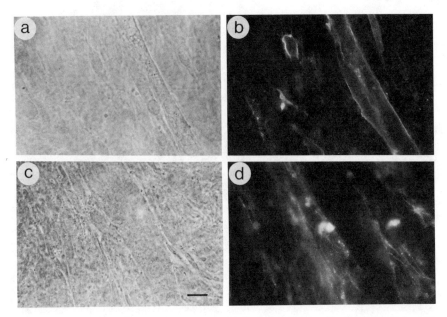

Figure 2. *Release of N-CAM from the G8–1 myotube membrane by phosphatidylinositol-specific phospholipase C.* Indirect immunofluorescence was used to monitor the release of N-CAM after treatment with the enzyme. *a,b:* Phase contrast and anti-N-CAM stain of a G8–1 myotube culture. *c,d:* A sister culture treated with the enzyme. Calibration bar = 100 μm.

myoblasts remained strongly positive. These data suggest that transmembrane N-CAM isoforms are not present in the myotube membrane at any signifiant level, and that the nontransmembrane isoforms predominate. The N-CAM isoforms present in myotubes are the same as those present in myofibers and in developing muscle (Covault et al., 1986). This suggests that these lipid-linked isoforms must be involved in the putative role of N-CAM in nerve–muscle interactions (Rutishauser et al., 1983).

An increasing number of cell membrane molecules appear to be bound to the plasma membrane via linkage through phosphatidylinositol (Low, 1987). The reason for using this type of linkage is not clear at present. It has been suggested that for N-CAM this linkage could be important in the rapid release of cell–cell bonds (He et al., 1986; Low, 1987). This has not yet been shown, nor has a specific cell surface phospholipase C been identified that could carry out the task. If this mechanism occurs, it will be interesting to determine the physiological consequences of the release of N-CAM. One possible consequence is that the second product of the reaction, namely diacylglycerol, could directly activate protein kinase C. The ability of PI-PLC to release N-CAM from the myotube membrane also offers the possibility of generating a preparation of soluble N-CAM that could be of great value in further analyzing the binding specificity of this isoform and the biochemical consequences of its binding to similar and dissimilar cell types.

ANALYSIS OF N-CAM cDNA CLONES IN SKELETAL MUSCLE

In skeletal muscle and brain a variety of transmembrane and nontrans-membrane N-CAM isoforms have been identified. It has been suggested that the size differences between these isoforms is due exclusively to the size of the COOH-terminal cytoplasmic tail, while the major extracellular regions of the molecule are believed to be similar, if not identical (Cunningham et al., 1983; Gennarini et al., 1984). This is probably an oversimplification, as there are minor size differences between N-CAM isoforms in brain and muscle. We have therefore embarked on an analysis of N-CAM cDNA clones from skeletal muscle in order to compare the similarities and differences between N-CAM in skeletal muscle and brain.

A number of recent studies have reported on the structure of N-CAM cDNA clones, and a model of the N-CAM gene is emerging (Goridis et al., 1985; Hemperly et al., 1986a,b; Murray et al., 1986a,b; Barthels et al., 1987; Dickson et al., 1987; Owens et al., 1987). While N-CAM has been shown to be a single-copy gene present on chromosome 11 in humans (Nguyen et al., 1986; Walsh et al., 1986), the most detailed analysis to date has been carried out in chickens (Owens et al., 1987). In this species there are at least 19 exons, and a model showing how alternative splicing can be used to generate the major N-CAM isoforms in brain of 180 kD (N-CAM 180), 140 kD (N-CAM 140), and 120 kD (N-CAM 120) has been presented (Owens et al., 1987). N-CAM 180 is encoded by a 7.2-kb mRNA, N-CAM 140 by a 6.7-kb mRNA, and N-CAM 120 by a 4.0-kb mRNA in the chicken; a sequence analysis of these three mRNAs has shown that diversity is generated at the COOH-terminal region of the molecule via specific alternative splicing choices and yields a large cytoplasmic domain (7.2-kb mRNA), a small cytoplasmic domain (6.7-kb mRNA), or a nontransmembrane phosphatidylinositol-linked N-CAM (4.0-kb mRNA).

In order to analyze directly the structure of human skeletal muscle N-CAM isoforms, and the transcriptional and RNA processing events controlling their expression, a series of human cDNAs were isolated and characterized from a skeletal muscle library prepared in the phage expression vector λgt11. Plaque hybridization screening with a mouse N-CAM cDNA (pM1.3) (Goridis et al., 1985), in conjunction with the anti-N-CAM immunoblot analysis of bacterial fusion proteins, allowed the isolation of 30 human muscle N-CAM clones from an initial library of $>10^6$ events. On the basis of restriction enzyme site mapping, a number of repetitive and/or overlapping cDNAs were identified.

Two clones, λ9.5 and λ4.4, which have clearly different restriction maps (Figure 3) and are likely to represent distinct mRNA isotypes, were subjected to DNA sequence analysis. Both clones contained cDNAs with long open reading frames that exhibited discrete regions of interspersed homologous and nonhomologous sequence blocks (Figure 4). The λ4.4 cDNA encodes a predicted transmembrane N-CAM isoform with extracellular, membrane-spanning, and intracellular domains, whereas the λ9.5 cDNA encodes a predicted nontransmembrane isoform with an extracellular coding sequence and a COOH-terminal hydrophobic domain only. In addition, the continuity of

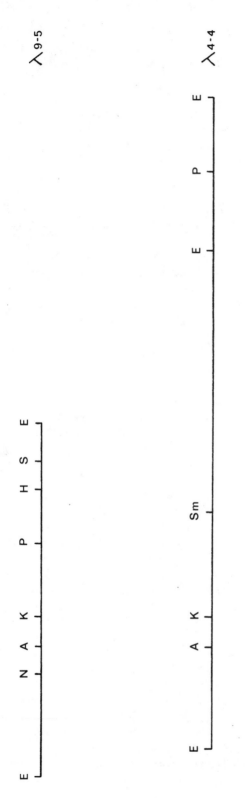

Figure 3. *Restriction enzyme site maps of human N-CAM cDNAs λ9.5 and λ4.4.* Restriction endonuclease cleavage sites are shown for ApaL1 (A), EcoR1 (E), Hind III (H), Kpn1 (K), Nhe1 (N), Pst1 (P), Sac1 (S), and Sma1 (Sm). Calibration bar = 500 bp.

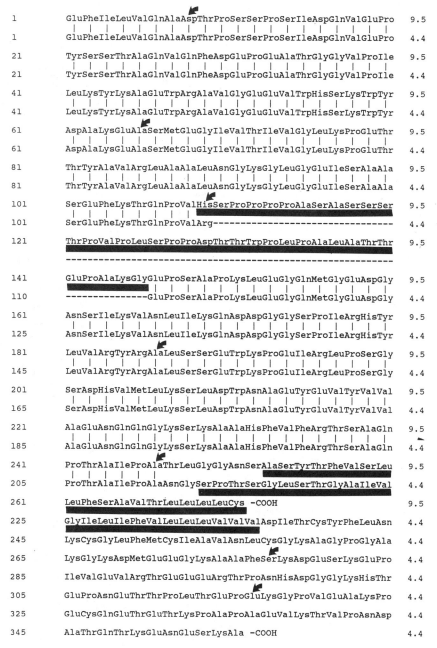

```
1     GluPheIleLeuValGlnAlaAspThrProSerSerProSerIleAspGlnValGluPro    9.5
      | | | | | | | | | | | | | | | | | | | |
1     GluPheIleLeuValGlnAlaAspThrProSerSerProSerIleAspGlnValGluPro    4.4

21    TyrSerSerThrAlaGlnValGlnPheAspGluProGluAlaThrGlyGlyValProIle    9.5
      | | | | | | | | | | | | | | | | | | | |
21    TyrSerSerThrAlaGlnValGlnPheAspGluProGluAlaThrGlyGlyValProIle    4.4

41    LeuLysTyrLysAlaGluTrpArgAlaValGlyGluGluValTrpHisSerLysTrpTyr    9.5
      | | | | | | | | | | | | | | | | | | | |
41    LeuLysTyrLysAlaGluTrpArgAlaValGlyGluGluValTrpHisSerLysTrpTyr    4.4

61    AspAlaLysGluAlaSerMetGluGlyIleValThrIleValGlyLeuLysProGluThr    9.5
      | | | | | | | | | | | | | | | | | | | |
61    AspAlaLysGluAlaSerMetGluGlyIleValThrIleValGlyLeuLysProGluThr    4.4

81    ThrTyrAlaValArgLeuAlaAlaLeuAsnGlyLysGlyLeuGlyGluIleSerAlaAla    9.5
      | | | | | | | | | | | | | | | | | | | |
81    ThrTyrAlaValArgLeuAlaAlaLeuAsnGlyLysGlyLeuGlyGluIleSerAlaAla    4.4

101   SerGluPheLysThrGlnProValHisSerProProProProAlaSerAlaSerSerSer    9.5
      | | | | | | | | |
101   SerGluPheLysThrGlnProValArg---------------------------------    4.4

121   ThrProValProLeuSerProProAspThrThrTrpProLeuProAlaLeuAlaThrThr    9.5
      -----------------------------------------------------------

141   GluProAlaLysGlyGluProSerAlaProLysLeuGluGlyGlnMetGlyGluAspGly    9.5
                                 | | | | | | | | | | | | | | |
110   --------------GluProSerAlaProLysLeuGluGlyGlnMetGlyGluAspGly    4.4

161   AsnSerIleLysValAsnLeuIleLysGlnAspAspGlyGlySerProIleArgHisTyr    9.5
      | | | | | | | | | | | | | | | | | | | |
125   AsnSerIleLysValAsnLeuIleLysGlnAspAspGlyGlySerProIleArgHisTyr    4.4

181   LeuValArgTyrArgAlaLeuSerSerGluTrpLysProGluIleArgLeuProSerGly    9.5
      | | | | | | | | | | | | | | | | | | | |
145   LeuValArgTyrArgAlaLeuSerSerGluTrpLysProGluIleArgLeuProSerGly    4.4

201   SerAspHisValMetLeuLysSerLeuAspTrpAsnAlaGluTyrGluValTyrValVal    9.5
      | | | | | | | | | | | | | | | | | | | |
165   SerAspHisValMetLeuLysSerLeuAspTrpAsnAlaGluTyrGluValTyrValVal    4.4

221   AlaGluAsnGlnGlnGlyLysSerLysAlaAlaHisPheValPheArgThrSerAlaGln    9.5
      | | | | | | | | | | | | | | | | | | | |
185   AlaGluAsnGlnGlnGlyLysSerLysAlaAlaHisPheValPheArgThrSerAlaGln    4.4

241   ProThrAlaIleProAlaThrLeuGlyGlyAsnSerAlaSerTyrThrPheValSerLeu    9.5
      | | | | | | | | | |
205   ProThrAlaIleProAlaAsnGlySerProThrSerGlyLeuSerThrGlyAlaIleVal    4.4

261   LeuPheSerAlaValThrLeuLeuLeuLeuCys  -COOH                        9.5

225   GlyIleLeuIlePheValLeuLeuLeuValValValAspIleThrCysTyrPheLeuAsn    4.4

245   LysCysGlyLeuPheMetCysIleAlaValAsnLeuCysGlyLysAlaGlyProGlyAla    4.4

265   LysGlyLysAspMetGluGluGlyLysAlaAlaPheSerLysAspGluSerLysGluPro    4.4

285   IleValGluValArgThrGluGluGluArgThrProAsnHisAspGlyGlyLysHisThr    4.4

305   GluProAsnGluThrThrProLeuThrGluProGluLysGlyProValGluAlaLysPro    4.4

325   GluCysGlnGluThrGluThrLysProAlaProAlaGluValLysThrValProAsnAsp    4.4

345   AlaThrGlnThrLysGluAsnGluSerLysAla  -COOH                        4.4
```

Figure 4. *Comparison of predicted amino acid sequences of human muscle N-CAM isoforms.* The major open reading frames in λ9.5 and λ4.4 cDNA sequences were translated and aligned. The MSD1 region of λ9.5 (residues 109 to 145 inclusive) and hydrophobic putative membrane-associated domains are *underlined.* Splice sites corresponding to those identified in the chick (Owens et al., 1987) are marked by *arrows.*

391

homology in the extracellular coding domains of the two cDNAs was discretely broken in the λ9.5 clone by the presence of a 108-bp nucleotide block (referred to as MSD1), which introduced a unique run of 37 amino acids into the predicted protein sequence of the nontransmembrane isoform.

Human cDNAs exhibited high nucleotide and predicted amino acid homologues with published chick and mouse N-CAM sequences. The λ4.4 clone corresponds to the chick cDNA pEC208 and shows continuous coding region homology with the alternatively spliced portion of this chick sequence, which encodes N-CAM 140 (Hemperly et al., 1986a; Murray et al., 1986a,b). However, an additional 18 amino acids of COOH-terminal cytoplasmic domain sequence are present in the human cDNA. In contrast to the λ4.4 cDNA, the human muscle λ9.5 clone corresponds closely to chick and mouse brain cDNAs, which encode nontransmembrane phospholipid-linked N-CAM 120 isoforms in these species (Hemperly et al., 1986b; Barthels et al., 1987). The MSD1 sequence block in the extracellular domain of the λ9.5 cDNA is not, however, expressed in these or indeed any other N-CAM sequence described to date. Furthermore, close matches are observed between the human sequences and the intron–exon boundary sequences described in the chick (Owens et al., 1987). The point of insertion of the MSD1 sequence is positioned at such a boundary (exons 12 and 13 in the chick), as is the point of divergence of λ4.4 and λ9.5 in their 3' coding regions. Thus, it is likely that alternative RNA splicing events contribute to the generation of human muscle N-CAM diversity.

In human skeletal muscle cells in culture, three N-CAM mRNAs, of 6.7 kb, 5.2 kb, and 4.0 kb, are expressed (Figure 5). Using specific subfragment probes from the 3' regions of the λ4.4 and λ9.5 cDNAs, the transmembrane and nontransmembrane (putative phospholipid-linked) isoforms they encode were shown to be the products of the 6.7-kb and the 5.2-kb and/or 4.0-kb RNAs, respectively. During myogenesis in vitro, human skeletal myoblasts express predominantly the 6.7-kb mRNA, while myotube cultures contain major 5.2-kb and 4.0-kb isotypes. Taken together with the biochemical studies outlined in the previous section, a predicted picture of N-CAM expression in skeletal muscle can be drawn. Thus, myoblasts have a 6.7-kb RNA that encodes a transmembrane N-CAM 145 isoform down-regulated with fusion, whereas myotubes express two mRNA species of 5.2 kb and 4.0 kb, whose protein products are likely to be the lipid-linked N-CAM 125 and N-CAM 155 isoforms.

Of particular significance in terms of the role of N-CAM in muscle development, further Northern analyses have shown that the MSD1 domain is present only in the 5.2-kb and 4.3-kb mRNAs expressed by myotubes. The protein sequence block it encodes is thus present specifically in the extracellular domain of the lipid-linked muscle N-CAM isoforms. Furthermore, Northern analysis of RNA from human and rodent brain tissue and from various neural cell lines clearly demonstrates that the MSD1 sequence is indeed muscle specific. The functional significance of this muscle-specific domain remains as

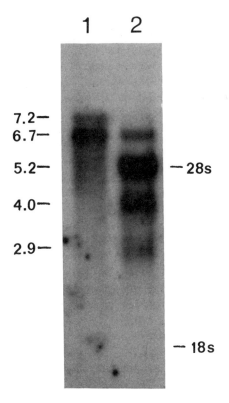

Figure 5. *N-CAM mRNAs in human tissue and muscle cells.* Poly(A)$^+$ RNA from human embryonic brain tissue (*lane 1*) and midfusion muscle cultures (*lane 2*) were subject to Northern blot analysis and probe with the λ9.5 insert. The positions of 28s and 18s mRNA standards are shown on the *right*; the size of N-CAM mRNAs is indicated on the *left* in kilobases.

yet undefined, however; we must await the outcome of cell biological studies on the properties of corresponding synthetic peptides and their antibodies.

CELL RECOGNITION EVENTS IN NERVE–MUSCLE COCULTURES

There appears to be little doubt that during embryogenesis a complex sequence of epigenetic events controls the formation of a functional synaptic contact between a motor neuron and a differentiated myotube. The spatio-temporal control of the expression of cell surface "recognition" molecules has long been regarded as a potential controlling mechanism (Jacobson, 1978; Goodman et al., 1984). By interacting with relevant cues in the extracellular matrix and/or on the surface of other cells, including the target cell, these molecules may either mediate or facilitate the transduction of information from the extracellular to the cellular compartment. Consequent inductive events, involving transcriptional, posttranscriptional, and translational changes in protein expression, may be essential prerequisite steps in the overall sequence of events that leads to the establishment of a synapse. The elucidation of the role of individual cell surface recognition molecules in the process of synapse formation, both *in vivo* and *in vitro,* will most probably require the dissection of this complex phenomenon into well-defined components.

The presence of N-CAM on both motor axons (Daniloff et al., 1986) and the target muscle tissue (Moore and Walsh, 1985) and the clear demonstration that muscle N-CAM gene expression can be regulated by innervation status (Covault and Sanes, 1985; Rieger et al., 1985; Moore and Walsh, 1986) support the contention that N-CAM may function in nerve-muscle interactions. Perturbation assays with Fab' fragments to N-CAM have implicated N-CAM in early nerve–muscle adhesive interactions *in vitro* (Rutishauser et al., 1983). However, in cocultures of embryonic-day-eight chick ciliary ganglia with chick myotubes, the same antibody reagents had little effect on the number of axonal–muscle contacts, the ability of the nerve to extend axons over myotubes, or the formation of early synaptic structures as assayed by the number of specific antigen clusters (Bixby and Reichardt, 1987). The negative results of the latter studies should be interpreted with caution. By day eight, chick ciliary ganglion neurons have reached their target tissue *in vivo,* and the authors were therefore studying events consequential to neurite regeneration rather than primary neurite induction. It is by no means certain that the cellular recognition events that underlie neurite regeneration and reinnervation reflect a recapitulation of events that control synapse formation during embryogenesis. It will be far easier to identify the role of N-CAM in nerve–muscle interactions once we understand the temporal sequence of inductive events operating between these two cell types.

We have recently developed a coculture model to study the sequence of inductive events that operate during neuritogenesis and neurite regeneration. Our model involves the addition of PC12 pheochromocytoma cells (Greene and Tischler, 1976) to monolayer cultures of nonneuronal cells. PC12 cells have previously been shown to be capable of forming functional synapses with muscle cells (Schubert et al., 1977). These cells are uniquely useful for studies on the mechanism of neurite outgrowth, as they do not require exogenous trophic support for survival and may undergo either generation (naive cells) or regeneration (primed cells) of neurites in response to nerve growth factor (NGF). An additional advantage of PC12 cells is that unlike most primary neurons that are available for culture, they may be used to study steps whereby a classical growth factor (NGF) stimulates the initiation of neurite outgrowth from cells previously unexposed to detectable levels of that factor.

In a series of "simple choice" experiments, we have determined the ability of naive PC12 cells (previously unexposed to NGF) and primed PC12 cells (exposed to maximally active concentrations of NGF for five to seven days before being subcultured to remove their neuritic network) to extend axons on a variety of nonneuronal cell substrata. We previously found an accumulation of the 155-kD subunit of neurofilament protein (as recognized by a mab coded RT97) to correlate with neuritic outgrowth (Doherty et al., 1987a,b). Our results have established the following. In the absence of NGF, both naive and primed PC12 cells fail to extend axons when grown over a three-day period on monolayer cultures of C2 and G8–1 myotubes, C6 glioma cells, and human skin fibroblasts. NGF was found to readily promote neurite outgrowth and accumulation of the neurofilament protein antigen from naive PC12 cells grown on muscle and glioma but not on fibroblast monolayers. In contrast, there was essentially no difference in NGF-induced neurite outgrowth and the accumulation of the neurofilament protein antigen when primed PC12 cells were grown on the various monolayers. An example of the modulatory effects of the cellular substratum on the induction of the neurofilament protein antigen is shown in Figure 6.

From the observations above we can conclude that cell–cell contact is not sufficient to induce neurite outgrowth. However, it can clearly modulate the responses of naive PC12 cells to NGF, but fails to do so for primed cells. Pretreatment with NGF has apparently programmed the PC12 cells in an as yet undefined manner such that neurite outgrowth becomes less dependent on specific cues within the extracellular environment. The primed cells are similar to the ciliary ganglion neurons discussed above, in that they do not readily exhibit specificity in their choice of cellular substratum for neurite regeneration.

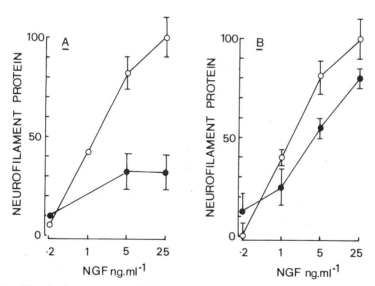

Figure 6. *Effect of substratum on neurofilament response.* Naive and primed PC12 cells were grown on monolayer cultures of C2 muscle (*open circles*) or, alternatively, human skin fibroblasts (*filled circles*). The ability of NGF to induce the expression of the 155-kD neurofilament protein subunit was determined as previously described (Doherty et al., 1987b). The results show the relative increase in expression for cultures grown in the presence as compared to the absence of NGF. Measurements were made after three days of coculture for naive PC12 cells (*A*) or after two days of coculture for primed PC12 cells (*B*). Each value is the mean of 4–6 determinations. Calibration bars show ± 1 SEM.

We are currently analyzing the sequence of inductive events operating in the coculture models with a view to characterizing the recognition molecules that modulate neurite outgrowth. We have examined NGF-induced increases in the expression of both the Thy-1 glycoprotein (Doherty et al., 1988) and N-CAM (Prentice et al., 1987). Considerable evidence exists to support the notion that these glycoproteins function as molecules involved in cell recognition and/or the transduction of biological signals across cell membranes (Edelman, 1984; Morris, 1985; Kroczek et al., 1986).

In cultures of naive PC12 cells grown on a collagen substratum, NGF induces dose-dependent increases in the cell surface expression of both Thy-1 and N-CAM, with a response readily detectable after 24 hours (Figure 7). The relative sensitivity of the responses to NGF clearly differs from the half-maximal stimulation of Thy-1 expression found at NGF concentrations that have little or no effect on N-CAM expression. The induction of Thy-1 but not N-CAM can be mimicked by the addition of low concentrations of the protein kinase C activators phorbol 12-myristate 13-acetate and phorbol 12,13-dibutyrate to PC12 cultures. The induction of Thy-1 and N-CAM can be inhibited by cordycepin (Doherty et al., 1988) and is accompanied by changes in gene transcription (Dickson et al., 1986; Prentice et al., 1987). These data suggest that NGF can activate the transcription of genes encoding cell surface glycoproteins via at least two independent pathways. Both mechanisms are

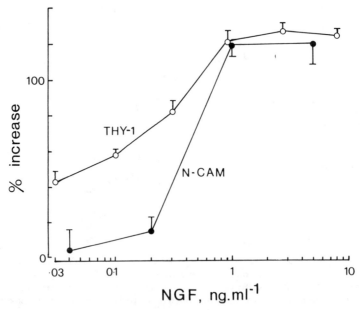

Figure 7. *Effect of NGF on Thy-1 and N-CAM expression.* Cell surface–associated Thy-1 and N-CAM antigen levels were determined as previously described (Doherty et al., 1988) for naive PC12 cultures grown over 24 hours in the presence of 0–10 ng/ml NGF. The results show the percentage increase over cultures grown in control media; each value is the mean ± 1 SEM of six independent determinations.

operative over a similar time frame with increased expression of Thy-1, but not N-CAM, and are fully accounted for by the activation of protein kinase C (Doherty et al., 1988).

The induction of Thy-1 by phorbol esters was not associated with the morphological differentiation of PC12 cells. Thus, increased expression of Thy-1 is not a consequence of neurite outgrowth, nor does it directly trigger the latter. However, the relatively high sensitivity of this response to NGF suggests that it could represent an important prerequisite step in the sequence of events that leads to stable neurite outgrowth. In support of this notion, we have recently shown receptor-mediated increases in cAMP to inhibit both NGF-induced increases in Thy-1 expression and NGF-induced neurite regeneration (Doherty and Walsh, 1987; Doherty et al., 1987a). Thus, the induction of Thy-1 is a highly attractive candidate for an early response that may be modulated by environmental cues that promote or suppress neurite outgrowth.

To test this hypothesis, we cultured naive PC12 cells on a collagen substratum or, alternatively, on monolayers of C2 myotubes and human skin fibroblasts. The results of this experiment are summarized in Figure 8. In the absence of NGF, the basal level of Thy-1 expression for PC12 cells grown in coculture with fibroblasts and cells grown on a collagen substratum was similar. In contrast, there was a twofold increase in Thy-1 expression by PC12 cells grown on the C2 monolayer. The addition of a maximally active

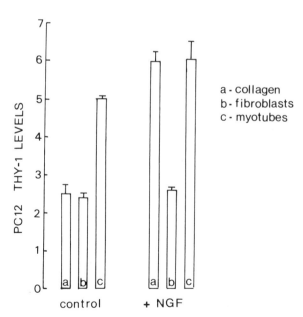

Figure 8. *Effect of substratum on Thy-1 response.* Naive PC12 cells were seeded on a collagen substratum (*a*) or monolayer cultures of C2 myotubes (*c*) and human skin fibroblasts (*b*). Cultures were grown in the presence and absence of NGF (11.5 ng/ml) for three days, and Thy-1 levels associated with PC12 cells were determined as previously described (Doherty et al., 1987b). The results show the mean ± 1 SEM of three independent determinations.

concentration of NGF to cells grown on collagen resulted in a twofold increase in Thy-1 expression, had little effect on the already elevated levels of Thy-1 in the muscle coculture, and had no effect on Thy-1 expression in the fibroblast coculture. These data suggest that a coculture with C2 myotubes can directly induce an increase in Thy-1 expression, whereas a coculture with fibroblasts can suppress NGF-induced increases in Thy-1 expression. These modulatory responses could not be mimicked by media conditioned over the various cell types, and it therefore appears probable that control is mediated via direct cell–cell and/or cell–matrix interactions.

We intend to use immunological perturbation assays to characterize the nature of the recognition events that modulate the expression of both the cell surface glycoprotein Thy-1 and the neurofilament protein antigen associated with morphological differentiation. It will be of particular interest to determine if the homophilic binding of nerve and muscle N-CAM molecules is involved in any of these inductive events.

REFERENCES

Barthels, D., M.-J. Santoni, W. Wille, C. Ruppert, J.-C. Chaix, M.-R. Hirsch, J. C. Fontecilla-Camps, and C. Goridis (1987) Isolation and nucleotide sequence of mouse N-CAM cDNA that codes for a M_r 79000 polypeptide without a membrane-spanning region. *EMBO J.* **6**:907–914.

Bixby, J. L., and L. F. Reichardt (1987) Effects of antibodies to neural cell adhesion molecule (N-CAM) on the differentiation of neuromuscular contacts between ciliary ganglion neurons and myotubes *in vitro*. *Dev. Biol.* **119**:363–372.

Covault, J., and J. R. Sanes (1985) Neural cell adhesion molecule (N-CAM) accumulates in denervated and paralyzed skeletal muscle. *Proc. Natl. Acad. Sci. USA* **82**:4544–4548.

Covault, J., J. P. Merlie, C. Goridis, and J. R. Sanes (1986) Molecular forms of N-CAM and its RNA in developing and denervated skeletal muscle. *J. Cell Biol.* **102**:731–739.

Cunningham, B. A. (1985) Structure of cell adhesion molecules. In *The Cell in Contact: Adhesions and Junctions as Morphogenetic Determinants*, G. M. Edelman and J. P. Thiery, eds., pp. 197–217, Wiley, New York.

Cunningham, B. A., S. Hoffman, U. Rutishauser, J. J. Hemperly, and G. M. Edelman (1983) Molecular topography of the neural cell adhesion molecule N-CAM: Surface orientation and location of sialic acid-rich binding regions. *Proc. Natl. Acad. Sci. USA* **80**:3116–3120.

Daniloff, J. K., G. Levi, M. Grumet, F. Rieger, and G. M. Edelman (1986) Altered expression of neuronal cell adhesion molecules induced by nerve injury and repair. *J. Cell Biol.* **103**:929–945.

Dickson, J. G., H. M. Prentice, J.-P. Julien, G. Ferrari, A. Leon, and F. S. Walsh (1986) Nerve growth factor activates Thy-1 and neurofilament gene transcription in rat PC12 cells. *EMBO J.* **5**:3449–3453.

Dickson, J. G., H. J. Gower, C. H. Barton, H. M. Prentice, V. L. Elsom, S. E. Moore, R. D. Cox, C. A. Quinn, W. Putt, and F. S. Walsh (1987) Human muscle neural cell adhesion molecule (N-CAM): Identification of a muscle-specific sequence in the extracellular domain. *Cell* **50**:1119–1130.

Doherty, P., and F. S. Walsh (1987) Control of Thy-1 glycoprotein expression in cultures of PC12 cells. *J. Neurochem.* **49**:610–616.

Doherty, P., D. A. Mann, and F. S. Walsh (1987a) Cholera toxin and dibutyryl cyclic AMP inhibit the expression of neurofilament protein induced by nerve growth factor in cultures of naive and primed PC12 cells. *J. Neurochem.* **49**:1676–1687.

Doherty, P., D. A. Mann, and F. S. Walsh (1987b) Cell–cell interactions regulate the responsiveness of PC12 cells to nerve growth factor. *Development* **101**:605–615.

Doherty, P., D. A. Mann, and F. S. Walsh (1988) Comparison of the effects of NGF, activators of protein kinase C, and a calcium ionophore on the expression of Thy-1 and N-CAM in PC12 cell cultures. *J. Cell Biol.* **107**:333–340.

Edelman, G. M. (1984) Modulation of cell adhesion during induction, histogenesis, and perinatal development of the nervous system. *Annu. Rev. Neurosci.* **7**:339–377.

Gennarini, G., G. Rougon, H. Deagostini-Bazin, M. Hirn, and C. Goridis (1984) Studies on the transmembrane disposition of the neural cell adhesion molecule N-CAM. A monoclonal antibody recognizing a cytoplasmic domain and evidence for the presence of phosphoserine residues. *Eur. J. Biochem.* **142**:57–64.

Goodman, C. S., M. J. Bastiani, C. Q. Doe, S. du Lac, S. L. Helfond, J. Y. Kuwada, and J. B. Thomas (1984) Cell recognition during neuronal development. *Science* **225**:1271–1279.

Goridis, C., M. Hirn, M.-J. Santoni, G. Gennarini, H. Deagostini-Bazin, B. R. Jordan, M. Keifer, and M. Steinmetz (1985) Isolation of mouse N-CAM related cDNA: Detection and cloning using monoclonal antibodies. *EMBO J.* **4**:631–635.

Greene, L. A., and A. S. Tischler (1976) Establishment of noradrenergic clonal line of rat adrenal pheochromocytoma cells which respond to nerve growth factor. *Proc. Natl. Acad. Sci. USA* **72**:2424–2428.

Grumet, M., U. Rutishauser, and G. M. Edelman (1982) Neural cell adhesion molecule is on embryonic muscle cells and mediates adhesion to nerve cells *in vitro*. *Nature* **295**:693–695.

He, H.-T., J. Barbet, J.-C. Chaix, and C. Goridis (1986) Phosphatidylinositol is involved in the membrane attachment of N-CAM-120, the smallest component of the neural cell adhesion molecule. *EMBO J.* **5**:2489–2494.

Hemperly, J. J., B. A. Murray, G. M. Edelman, and B. A. Cunningham (1986a) Sequence of a cDNA clone encoding the polysialic acid-rich and cytoplasmic domains of the neural cell adhesion molecule N-CAM. *Proc. Natl. Acad. Sci. USA* **83**:3037–3041.

Hemperly, J. J., G. M. Edelman, and B. A. Cunningham (1986b) cDNA clones of the neural cell adhesion molecule (N-CAM) lacking a membrane-spanning region consistent with evidence for membrane attachment via a phosphatidylinositol intermediate. *Proc. Natl. Acad. Sci. USA* **83**:9822–9826.

Jacobson, M. (1978) *Developmental Neurobiology*, Plenum, New York.

Kroczek, R. A., K. C. Gunter, R. N. Germain, and E. M. Shevach (1986) Thy-1 functions as a signal transduction molecule in T lymphocytes and transfected B lymphocytes. *Nature* **322**:181–183.

Lee, H. U., and S. J. Kaufman (1981) Use of monoclonal antibodies in the analysis of myoblast development. *Dev. Biol.* **81**:81–95.

Low, M. G. (1987) Biochemistry of the glycosylphosphatidylinositol membrane protein anchors. *Biochem. J.* **244**:1–13.

Moore, S. E., and F. S. Walsh (1985) Specific regulation of N-CAM/D2-CAM cell adhesion molecule during skeletal muscle development. *EMBO J.* **4**:623–630.

Moore, S. E., and F. S. Walsh (1986) Nerve-dependent regulation of neural cell adhesion molecule expression in skeletal muscle. *Neuroscience* **18**:499–505.

Moore, S. E., J. Thompson, V. Kirkness, J. G. Dickson, and F. S. Walsh (1987) Skeletal muscle neural cell adhesion molecule (N-CAM): Changes in protein and mRNA species during myogenesis of muscle cell lines. *J. Cell Biol.* **105**:1377–1386.

Morris, R. J. (1985) Thy-1 in developing nervous tissue. *Dev. Neurosci.* **7**:133–160.

Murray, B. A., J. J. Hemperly, E. A. Prediger, G. M. Edelman, and B. A. Cunningham (1986a) Alternatively spliced mRNAs code for different polypeptide chains of the chicken neural cell adhesion molecule (N-CAM). *J. Cell Biol.* **102**:189–193.

Murray, B. A., G. C. Owens, E. A. Prediger, K. L. Crossin, B. A. Cunningham, and G. M. Edelman

(1986b) Cell surface modulation of the neural cell adhesion molecule results from alternative mRNA splicing in a tissue specific developmental sequence. *J. Cell Biol.* **103**:1431–1439.

Nguyen, C., M.-G. Mattei, J.-F. Mattei, M.-J. Santoni, C. Goridis, and B. R. Jordan (1986) Localization of the human N-CAM gene to band q23 of chromosome 11: The third gene coding for a cell interaction molecule mapped to the distal portion of the long arm of chromosome 11. *J. Cell Biol.* **102**:711–715.

Owens, G. C., G. M. Edelman, and B. A. Cunningham (1987) Organization of the neural cell adhesion molecule (N-CAM) gene: Alternative exon usage as the basis for different membrane-associated domains. *Proc. Natl. Acad. Sci. USA* **84**:294–298.

Prentice, H. M., S. E. Moore, J. G. Dickson, P. Doherty, and F. S. Walsh (1987) Nerve growth factor-induced changes in neural cell adhesion molecule (N-CAM) in PC12 cells. *EMBO J.* **6**:1859–1863.

Rieger, F., M. Grumet, and G. M. Edelman (1985) N-CAM at the vertebrate neuromuscular junction. *J. Cell Biol.* **101**:285–293.

Rutishauser, U. (1984) Developmental biology of a neural cell adhesion molecule. *Nature* **310**:549–554.

Rutishauser, U., M. Grumet, and G. M. Edelman (1983) N-CAM mediates initial interactions between spinal cord neurons and muscle cells in culture. *J. Cell Biol.* **97**:145–152.

Schubert, D., S. Heinemann, and Y. Kidokoro (1977) Cholinergic metabolism and synapse formation by a rat nerve cell line. *Proc. Natl. Acad. Sci. USA* **74**:2579–2583.

Walsh, F. S., and E. Phillips (1981) Specific changes in cellular glycoproteins and surface proteins during myogenesis in clonal muscle cells. *Dev. Biol.* **81**:229–237.

Walsh, F. S., and M. A. Ritter (1981) Surface antigen differentiation during human myogenesis in culture. *Nature* **289**:60–64.

Walsh, F. S., W. Putt, J. G. Dickson, C. A. Quinn, R. D. Cox, M. Webb, N. Spurr, and P. N. Goodfellow (1986) Human N-CAM gene: Mapping to chromosome 11 by analysis of somatic cell hybrids with mouse and human cDNA probes. *Brain Res.* **387**:197–200.

Williams, R. K., C. Goridis, and R. Akeson (1985) Individual neural cell types express immunologically distinct N-CAM forms. *J. Cell Biol.* **101**:36–42.

Chapter 19

Regulation of Synaptogenesis in Adult Skeletal Muscle

JOSHUA R. SANES
JONATHAN COVAULT
CHRISTINE L. GATCHALIAN
DALE D. HUNTER
MICHAEL B. LASKOWSKI
JOHN P. MERLIE

ABSTRACT

Several behaviors that axons exhibit during reinnervation of adult skeletal muscle demonstrate that they are guided by cues that the muscle provides. For example, axons form synapses on denervated but not on normally innervated muscle fibers; axons selectively reinnervate original synaptic sites on denervated muscle fibers; regenerating axons become specialized for synaptic transmission only in regions where they contact muscle fibers; and motor axons prefer to reinnervate muscles derived from matching levels of the body's rostrocaudal axis. This chapter describes these phenomena and summarizes progress toward identifying the soluble, membrane-bound, and extracellular matrix molecules that underlie them.

It is clear that growing axons use a variety of extrinsic cues in deciding where and when to establish synapses. Questions of central interest to developmental neurobiologists include: What are these cues? Where are they located? How is their expression regulated? We have chosen to study the vertebrate skeletal neuromuscular junction in order to address these issues. This is because the neuromuscular junction is perhaps the simplest and certainly the best understood of all synapses, a wealth of information already being available about its structure, function, and development. Furthermore, the ready reinnervation of adult muscles following nerve damage permits synapse formation to be studied in a convenient postembryonic setting. This chapter reviews our studies of factors that guide axons reinnervating skeletal muscle, and places our work in the context of a large number of studies on similar issues from other laboratories.

PRECISE REINNERVATION OF ORIGINAL SYNAPTIC SITES

When muscle fibers are denervated, they change in many ways: they atrophy, acquire new metabolic, electrical, and contractile properties, and begin to express acetylcholine receptors (AChRs) throughout their surface. None-theless, original synaptic sites retain many of their specialized properties and are identifiable months after the nerve terminal degenerates. For example, AChRs remain concentrated at a density of $\sim 10^4/\mu m^2$ in the synaptic membrane even as new receptors are inserted (at a density of 10^2–$10^3/\mu m^2$) extrasynaptically (Salpeter and Loring, 1985). The membranous folds that invaginate the postsynaptic membrane and the aggregates of myonuclei associated with the end plate also persist. Tello (1907) was the first to report that regenerating axons preferentially form new neuromuscular junctions at these original synaptic sites. This finding has since been confirmed for mammalian, avian, and amphibian muscles reinnervated by their own or by foreign nerves (e.g., Gutmann and Young, 1944; Aitkin, 1950; Landmesser, 1972; Bennett and Pettigrew, 1975; Letinsky et al., 1976). While axons can sometimes form "ectopic" synapses on previously extrasynaptic portions of the muscle fiber surface (Figure 1; see below), precise reinnervation of original sites is very much the rule, and the topographic specificity of the process is striking. In frog muscles reinnervated following nerve crush, for example, over 90% of the original synaptic sites are reinnervated and over 95% of the nerve–

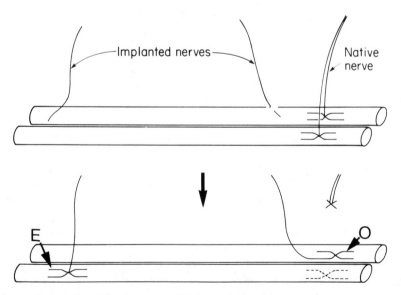

Figure 1. *Axons preferentially reinnervate original synaptic sites (O) in denervated muscle, but can form ectopic synapses (E) in some cases.* In innervated muscle, implanted axons form few if any synapses. (From Sanes and Covault, 1985.)

muscle contacts formed are at original sites, even though these sites occupy less than 0.1% of the muscle fiber surface (Sanes et al., 1978). The strong inference that can be drawn from this result is that there are factors concentrated at original sites that regenerating axons can recognize.

At least some of these factors are associated with the basal lamina (BL), which traverses the synaptic cleft at the neuromuscular junction. The experiment that led to this conclusion is sketched in Figure 2. Muscles were denervated, mechanically damaged to induce degeneration of muscle fibers, and X-irradiated to prevent regeneration of new muscle. The BL sheaths of muscle fibers survived these insults, and original synaptic sites were identifiable on the sheaths even after degeneration of all cellular elements. Axons regenerating to the region of damage contacted the BL sheaths, and over 95% of the close contacts detected electron microscopically occurred precisely at original synaptic sites. Thus, reinnervation was as topographically selective in the absence of muscle cells as in their presence (Sanes et al., 1978). Furthermore, portions of axons that contacted synaptic BL differentiated into nerve terminals, judging by a combination of morphological, electrical, and immunohistochemical criteria (Sanes et al., 1978; Glicksman and Sanes, 1983). Thus, there are components in the synaptic BL that regenerating axons recognize and to which they respond.

Molecular correlates of this functional specialization have been identified histochemically: A variety of antisera and monoclonal antibodies have been shown to stain synaptic BL far more intensely than they do extrasynaptic BL, thus defining "synaptic" antigens (Sanes and Hall, 1979; Anderson and Fambrough, 1983; Sanes and Chiu, 1983; Fallon et al., 1985). One such antigen is a heparan sulfate proteoglycan (Anderson and Fambrough, 1983); another is agrin, which induces aggregation of AChRs in the myotube membrane (Nitkin et al., 1987); a third is s-laminin, described below. In addition, studies with lectins have shown that a carbohydrate moiety that terminates in a β-N-acetylgaloctosaminyl residue is concentrated in synaptic BL, where it is associated with acetylcholinesterase and possibly with other synapse-specific mole-

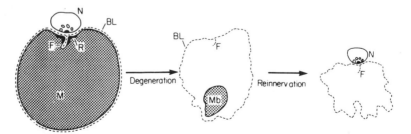

Figure 2. *Sketch of the experiment that demonstrated a role of synaptic BL in synaptic regeneration.* N, nerve terminal; M, muscle fibers; F, junction folds; R, acetylcholine receptors; BL, basal lamina; Mb, myoblastic satellite cells. See text for explanation. (From Sanes and Chiu, 1983.)

cules as well (Sanes and Cheney, 1982; Scott and Sanes, 1984; Scott et al., 1988). Finally, several monoclonal antibodies bind to extrasynaptic but not synaptic BL, thus defining epitopes that are excluded from the synaptic cleft (Sanes, 1982; Sanes and Chiu, 1983; Gatchalian et al., 1985).

In no case has any molecule concentrated in (or excluded from) synaptic BL yet been shown to serve as a recognition marker for axons *in vivo*. However, recent work has identified a novel molecule, s-laminin, that may participate in this process (Hunter et al., 1989). S-laminin was first defined by an antiserum, JS-1, that selectively stained synaptic BL (Sanes and Hall, 1979). Because JS-1 was not monospecific, we generated monoclonal antibodies that recognized JS-1-related antigens and also stained synaptic sites (Sanes and Chiu, 1983; Figure 3a–c). Using these antibodies, we identified s-laminin, an approximately 190 kD, noncollagenous, BL-associated glycoprotein (Figure 3d). The antibodies were also used to initiate molecular cloning; eventually, the complete s-laminin mRNA of approximately 5.7 kD was cloned. Sequence analysis revealed that s-laminin is homologous to the subunits of laminin (Figure 3e), a known promoter of neurite outgrowth (see below). Using native and recombinant s-laminin, we have found that this molecule is recognized by motoneurons but not by several other types of neurons. However, preliminary experiments indicate that s-laminin does not promote neurite outgrowth; instead, neurites extending on laminin cease growing when they encounter s-laminin. These results suggest a scheme whereby laminin-rich pathways provide a favorable terrain for regenerating axons to follow, whereas s-laminin at original synaptic sites might direct axons to stop growing and/or to differentiate into nerve terminals.

SUSCEPTIBILITY AND REFRACTIVENESS TO SYNAPSE FORMATION

If a motor nerve is cut and its distal end is sutured into a skeletal muscle, its axons survive and ramify locally but form few if any synapses on the innervated muscle fibers that they encounter. If, however, the muscle's own nerve is subsequently cut, the axons grow more profusely and readily form synapses. These observations, first reported by Elsberg (1917) and confirmed repeatedly since (e.g., Aitken, 1950; Jansen et al., 1973; Weinberg et al., 1981), demonstrate that muscles are able to control their susceptibility to synapse formation in accordance with their current state of innervation. Although the implanted nerve preferentially forms synapses at original synaptic sites, as discussed above, at least in mammalian muscles, axons can form new or ectopic synapses if they are implanted far from or otherwise denied access to synaptic sites. This phenomenon of ectopic synaptogenesis indicates that the refractoriness of innervated muscle fibers to synapse formation is not simply due to the physical occupancy of a single competent site, but is instead a property that extends along the length of the muscle fiber.

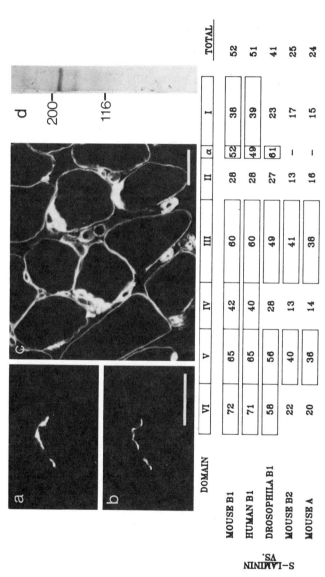

DOMAIN	VI	V	IV	III	II	α	I	TOTAL
MOUSE B1	72	65	42	60	28	52	38	52
HUMAN B1	71	65	40	60	28	49	39	51
DROSOPHILA B1	58	56	28	49	27	61	23	41
MOUSE B2	22	40	13	41	13	–	17	25
MOUSE A	20	36	14	38	16	–	15	24

Figure 3. *S-laminin is concentrated in synaptic BL. a, c:* Cryostat sections of adult rat muscle stained by an immunofluorescence method with antibodies to s-laminin (*a*) or laminin (*c*). S-laminin is highly concentrated at synaptic sites (*b* is the same field counterstained with rhodamine-α-bungarotoxin, which binds AChRs and marks synaptic sites), while laminin is present throughout muscle fiber BL (*c*). *d:* A Western blot of a BL extract reacted with anti-s-laminin. A major immunoreactive band of approximately 190 kD is apparent. *e:* The homologies (percentage identity) between amino acid sequences of s-laminin and subunits of laminin. Values of ≥35% are boxed. S-laminin sequence, references to laminin sequences, and rules for dividing these molecules into domains are given in Hunter et al. (1989). Calibration bar = 20 μm in *a–c.*

405

How does the muscle fiber inform the nerve of its susceptibility to innervation? It seems likely that innervated and denervated muscles (or active and inactive muscles) produce different amounts of some factor(s) that influence axons. A useful precedent for this idea is provided by the AChR. In normal adult muscle, AChRs are concentrated in the postsynaptic membrane, and AChR-subunit mRNAs are concentrated in synaptic areas (Merlie and Sanes, 1985; Fontaine et al., 1988). Following denervation, levels of AChR mRNA increase (Merlie et al., 1984; Goldman et al., 1985; Klarsfeld and Changeux, 1985; Covault et al., 1986), and new AChRs are synthesized and inserted throughout the muscle fiber surface. The neural control of AChR synthesis is mediated in large part by muscle activity, in that innervated but paralyzed muscles acquire extrasynaptic AChRs, while directly stimulated denervated muscles do not. Similarly, implanted axons can form ectopic synapses in innervated, paralyzed muscles but not in denervated, stimulated muscles (Jansen et al., 1973). We have proposed that the AChR-subunit genes are members of a "synapse-specific gene family" that are transcribed preferentially by synapse-associated nuclei in active muscle, and induced in nonsynaptic nuclei when muscles become inactive (Merlie and Sanes, 1985, 1986). It is tempting to speculate that genes for neuroactive factors that control a muscle's susceptibility to innervation are members of this family, and that their expression is regulated in parallel with AChRs.

One such factor might be a soluble growth factor that promotes the regeneration of motor axons. The existence of such a factor was hypothesized several years ago, based on studies of nerve growth factor (NGF), which sympathetic and sensory targets use to control their innervation. In fact, several groups have shown that extracts of muscles, or media conditioned by them, can increase the survival and differentiation of cultured motor neurons, and these extracts are more active when made from developing or denervated muscle than from innervated muscle (e.g., Henderson et al., 1983; Slack et al., 1983; Nurcombe et al., 1984; Gurney et al., 1986). Thus, increased synthesis and/or secretion of a "motor nerve growth factor" by muscle following denervation or paralysis might stimulate axonal growth and thereby promote reinnervation.

Several observations on the control of ectopic synapse formation (Sanes and Covault, 1985), as well as the obvious fact that synaptogenesis requires cell surface interactions, suggest that muscles might use cell surface molecules as well as soluble factors to regulate their susceptibility to innervation. The neural cell adhesion molecule, N-CAM, is one leading candidate for the role. Two sets of data support this suggestion. First, in vivo, levels of N-CAM are regulated in parallel with muscle fibers' susceptibility to innervation (Covault and Sanes, 1985, 1986; Rieger et al., 1985; Covault et al., 1986; Moore and Walsh, 1986; Cashman et al., 1987; see also chapters by Walsh et al. and Rieger, this volume). Embryonic myotubes, which readily accept synapses, are rich in N-CAM. N-CAM disappears perinatally, and is nearly absent from extrasynaptic portions of adult muscle fibers, which are refractory to hyperinnervation. However, N-CAM remains concentrated in and near the postsynaptic membrane in adult muscle. Following denervation, as muscles reacquire receptivity to innervation, N-CAM reappears throughout the muscle fiber, an induction that is accompanied by, and is presumably due to, an increased abundance of N-CAM mRNA (Figure 4). Paralysis also induces an accumulation of N-CAM. In denervated and paralyzed muscles, N-CAM appears not only in muscle fibers but also in interstitial areas between muscle fibers, a point discussed below. In addition, N-CAM is abundant on the growth cones of regenerating motor axons, which are therefore able, via a homophilic binding mechanism, to respond to the N-CAM that muscles provide. Finally, N-CAM disappears from extrasynaptic portions of the muscle fiber surface (and from interstitial areas) as the muscle is reinnervated and loses its susceptibility to synapse formation.

A second line of support for a role of N-CAM in mediating nerve–muscle interaction comes from antibody-blocking experiments. While no such studies have yet been reported *in vivo*, two groups have shown that anti-N-CAM can perturb nerve–muscle interactions *in vitro*. A point of disagreement between the two groups is that one (Grumet et al., 1982; Rutishauser et al., 1983) finds effects of anti-N-CAM alone, while the other (Bixby and Reichardt, 1987; Bixby et al., 1987) reports that anti-N-CAM is effective only when combined with antibodies to other adhesion systems. The latter result suggests that N-CAM is only one of several adhesion systems that participate in mediating nerve–muscle interactions, a conclusion consistent with the studies of BL described above and with the finding that an adhesive laminin–heparan sulfate proteoglycan complex that is present on the surface of innervated muscle fibers increases in level upon denervation (Chiu et al., 1986). Nonetheless, the hypothesis that N-CAM is one of the molecules that muscles use to inform

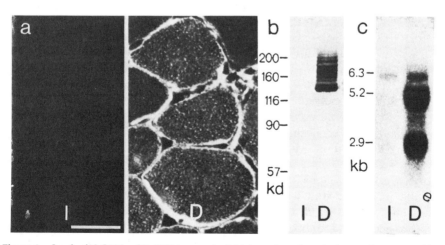

Figure 4. *Levels of N-CAM and its RNA increase in skeletal muscle as shown by immunofluorescence (a), Western blotting (b), and Northern blotting (c). In each case I indicates innervated and D indicates denervated muscle. (b and c are from Covault et al., 1986.) Calibration bar = 50 μm in a.*

nerves of their susceptibility to innervation is at the very least worth testing critically.

HOW AXONS FIND ORIGINAL SYNAPTIC SITES

Taken together, the results presented in the previous two sections give rise to a new question: If axons are capable of forming ectopic synapses, how do they find the small original synaptic sites that they preferentially reinnervate? The problem is not a trivial one, given that these sites occupy only $\sim 0.1\%$ of the muscle fiber surface; filopodia of regenerating motor axons might infrequently encounter these sites by an unguided process of random search. In fact, there are at least three potential mechanisms that may act, in conjunction with components of synaptic BL, to enhance the topographic specificity of re-innervation.

One type of guidance is provided by the connective tissue sheaths of intramuscular nerves. While distal portions of motor axons degenerate following axotomy, the supporting cells and connective tissues of the nerves survive. Regenerating axons generally grow back to the muscle fiber surface through these nerve trunks, and are thereby delivered to regions of original synaptic sites (e.g., Gutmann and Young, 1944). Axons can cross a sizable gap to enter the cut end of a nerve segment, suggesting that cells within the distal nerve stump release a soluble factor that serves as an axonal attractant (Ramón y Cajal, 1928; Kuffler, 1986, and references therein). Once within the nerve, regenerating axons grow on, and presumably adhere to, the BL of endoneurial (Schwann cell) tubes (e.g., Ide et al., 1983; Scherer and Easter, 1984). Studies of axonal extension on basal lamina–related substrata *in vitro* suggest that the active component in the BL is a laminin–heparan sulfate proteoglycan complex (Davis et al., 1985; Lander et al., 1985; Chiu et al., 1986) that is recognized by a cellular receptor (Bozyczko and Horwitz, 1986) called integrin (Hynes, 1987) in the neurite's membrane (see Sanes, 1989, for a review). Thus, nerve trunks contain both soluble and extracellular matrix molecules that orient axonal growth.

A second source of axonal guidance is suggested by immunohistochemical studies of denervated muscle. Four adhesive macromolecules accumulate in interstitial areas surrounding original synaptic sites during the first few days following denervation. These are tenascin(J1), N-CAM, fibronectin, and a heparan sulfate proteoglycan (Sanes et al., 1986). [First identified as a neuron–astrocyte adhesion molecule (Kruse et al., 1985), the 200–220 kD forms of the J1 antigen are now known to be identical to tenascin and are therefore called tenascin(J1) (Faissner et al., 1988.)] For the latter three molecules, interstitial deposits account for only a small fraction of their intramuscular contents. As noted above, fibronectin and heparan sulfate proteoglycan are normal constituents of the BL in both innervated and denervated muscle, while N-CAM appears throughout the muscle fiber surface following denervation.

Nonetheless, immunofluorescence clearly reveals that these molecules also accumulate outside of denervated muscle fibers and that the interstitial accumulations are focused near synapses.

For tenascin(J1) the situation is more dramatic: This molecule is nearly undetectable in normal muscle, and in denervated muscle it is almost entirely confined to—and thereby serves as the best available marker for—perisynaptic, interstitial areas. Because regenerating axons traverse these areas as they approach original synaptic sites, it is likely that they encounter deposits of adhesive macromolecules and possible that their growth is guided by them.

Studies on the source of the interstitial deposits of adhesive molecules suggest that they are synthesized at least in part by a population of fibroblasts that proliferate in perisynaptic areas following denervation. Fibroblasts are fairly numerous in muscle, and it has been known for some time that they are stimulated to divide by denervation (e.g., Murray and Robbins, 1982). Connor and McMahan (1987) found that these divisions—and the fibroblasts that thereby arise—are more numerous near to than far from end plates, and we have obtained similar results in rat muscle (Figure 5) (Gatchalian et al., 1989).

Electron microscope immunohistochemistry reveals that interstitial N-CAM is present on the surface of these cells, while J1 is associated with collagen fibrils that abut fibroblast processes. Using immunofluorescence and immunoprecipitation methods, we find that fibroblasts cultured from denervated muscle synthesize N-CAM, tenascin(J1), fibronectin, and heparan sulfate proteoglycan (Figure 6) (Gatchalian et al., 1989). Additional experiments show that fibroblasts from many sources can synthesize these molecules and that the synthesis of at least tenascin(J1) is up-regulated when fibroblasts are stimulated to proliferate. Together, these results suggest that the selective perisynaptic accumulation of tenascin(J1) following denervation may be part of a general program of cell activation that is induced by a mitogenic signal acting locally on ordinary fibroblasts. As axons grow toward original synaptic sites, which they preferentially reinnervate, they are likely to encounter perisynaptic fibroblasts and may be influenced by fibroblast-derived adhesive molecules.

An additional source of axonal guidance is hinted at by studies of the spatial distribution of ectopic synapses formed during the reinnervation of rat muscle. In the soleus, implanted axons form numerous ectopic synapses in distal portions of the muscle, but strikingly few synapses in an ~1-mm region surrounding the central band of original end plates (Frank et al., 1975). A similar pattern has been seen in diaphragm following neonatal denervation (Bennett and Pettigrew, 1974). It is possible that these patterns result from mechanisms already discussed: axons that grow near enough to an original synaptic site might encounter interstitial deposits of adhesive macromolecules, or fibroblasts, or nerve branches, and be stimulated to grow past perisynaptic areas to synaptic sites. An alternative, however—and one that was proposed by Bennett and Pettigrew—is that the susceptibility of muscle fibers to synapse formation is a graded phenomenon, with perisynaptic areas more refractory to

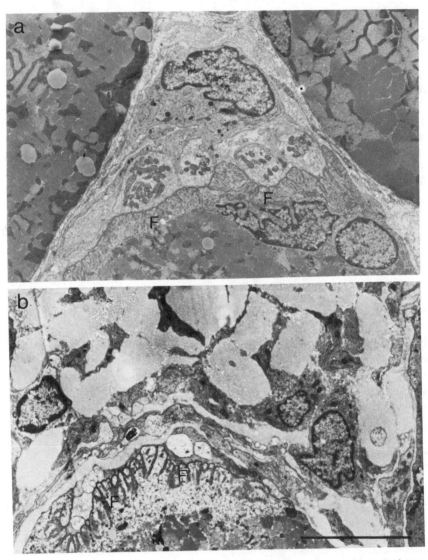

Figure 5. *Electron micrographs of synaptic sites in innervated (a) and denervated (b) rat diaphragm to show the proliferation of fibroblasts that occurs near denervated end plates.* F, junctional folds. Calibration bar = 5 μm.

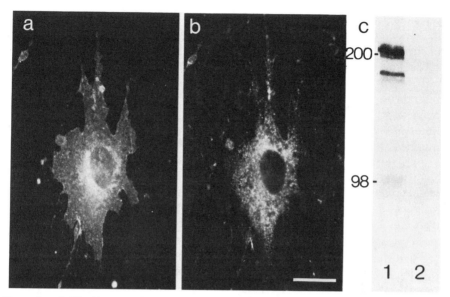

Figure 6. *a,b: Fibroblasts cultured from denervated skeletal muscle and doubly stained with anti-N-CAM (a) and fibronectin (b). c:* tenascin(J1) immunoprecipitated from a ^{35}S-methionine-labeled culture of fibroblasts from denervated muscle. *Lane 1* shows immune serum and *lane 2* preimmune serum. Calibration bar = 25 μm.

innervation than distant areas. In light of recent studies *in vitro* showing that the accumulation of acetylcholine receptors at nerve–muscle junctions is accompanied by their depletion from nearby areas (e.g., Anderson et al., 1977; Kuromi and Kidikoro, 1984), it is not difficult to imagine that other molecules important for synapse formation are in short supply in perisynaptic areas.

POSITIONALLY SELECTIVE REINNERVATION

The cues described so far are "specific" in the sense that they restrict synapse formation to particular states of muscle (denervated) or parts of muscle fibers (original synaptic sites). However, the hallmark of neural specificity is generally considered to be discrimination among potential target cells and preferential synaptogenesis among appropriately matched partners. In this classical sense, the reinnervation we have documented is not notably selective: "foreign" motor axons and even autonomic preganglionic axons can innervate denervated but not innervated muscles, and can form synapses preferentially at original synaptic sites (e.g., Landmesser, 1972; Jansen et al., 1973; Bennett and Pettigrew, 1975). In lower vertebrates, selective reinnervation of muscles and muscle fiber types by their own axons has been documented electrophysiologically (Elizalde et al., 1983; Wigston, 1986). However, most attempts to demonstrate selective reinnervation of adult mammalian muscle have been unsuccessful: native and foreign motor nerves appear to be comparably successful in reinnervating muscle, when matters of access and relative numbers of axons are taken into account (e.g., Bernstein and Guth, 1961; for references, see Wigston and Sanes, 1985). These discouraging results have been taken to mean that the (hypothetical) factors used to guide specific innervation in the embryo are poorly expressed in the adult, and that adult muscles are therefore poor subjects for studies of this sort of selectivity.

In reinvestigating this issue, we tested the idea that axons and muscles might be matched on the basis of their position in the rostrocaudal axis (Wigston and Sanes, 1982, 1985). In normal animals, muscles, which arise from segmentally arranged somites, are generally innervated by neurons from matching levels of the spinal cord. While such an arrangement presumably arises in large part from spatial relationships of growing axons and muscles in the embryo, it seemed that the segmental level might be an attractive variable to assay in seeking selective reinnervation. We therefore studied the reinnervation of intercostal muscles, each of which develops from a single somite and is normally innervated by motor neurons that lie in the corresponding segment of the spinal cord.

Pieces of intercostal muscle were transplanted from one of several levels to a common site in the neck of an adult rat, and the cervical sympathetic trunk was implanted in the muscle. This nerve was chosen because its axons, arising from several levels of the spinal cord, were known to be capable of innervating skeletal muscle, and had been shown by Purves et al. (1981) to be capable of selectively innervating sympathetic neurons. Following reinnervation, we recorded intracellularly from muscle fibers in the transplant and stimulated individual ventral roots to determine which segment(s) contained axons that innervated the muscle. The result was that muscles differed in the segmental origin of the innervation they received: Muscles from a rostral level recovered a disproportionate amount of their innervation from rostrally derived axons in the nerve, while caudal muscles were innervated by a caudally biased subset of axons. These results showed that axons can distinguish among adult mammalian muscle in a way that reflects their position of origin in the rostrocaudal axis.

We have now extended this analysis to study the innervation and reinnervation of skeletal muscles by motor axons. For this work we chose muscles that are innervated by axons that arise in more than one spinal segment but that enter the muscle through a single nerve. The serratus anterior (Figure 7a) is one such muscle. Using a combination of electrophysiological and histological methods, we found that the rostrocaudal axis of the spinal motor pool is systematically mapped onto the rostrocaudal axis of the muscle's surface (Figure 7b) (Laskowski and Sanes, 1987). Furthermore, following nerve damage, reinnervation is positionally selective: Rostral and caudal axons preferentially reinnervate rostral and caudal sectors of the muscle, respectively

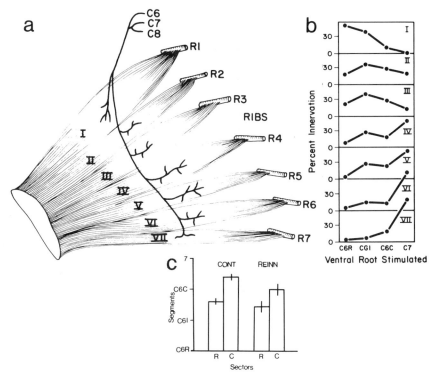

Figure 7.   *a: Diagram of the serratus anterior muscle, which is innervated through a single nerve that arises from spinal cord segments C6–8. b: Segmental origin of synaptic inputs to each sector of the serratus anterior, assessed by intracellular recording (R, I, and C indicate rostral, intermediate, and caudal rootlets of the C6 root). c: Average origin of inputs to rostral (R, II–IV) and caudal (C, V–VII) sectors of the normal and reinnervated serratus anterior demonstrating significant (p < .01) positionally selective reinnervation. Sector I was excluded from analysis because it was not completely denervated in some animals. (a and b from Laskowski and Sanes, 1987; c is redrawn from Laskowski and Sanes, 1988).*

(Figure 7c) (Laskowski and Sanes, 1988). Similar results on the topography of normal innervation and on the selectivity of reinnervation were obtained in a second muscle, the diaphragm (Laskowski and Sanes, 1987, 1988). Thus, reinnervation of adult mammalian muscle is demonstrably selective when rostrocaudal position is the variable evaluated.

In addition to providing a basis for using adult muscle to study selective synapse formation, these electrophysiological studies provide three clues about the nature of the selectivity. First, there seems to be a system of cues that matches axons and targets from corresponding levels of the rostrocaudal axis. Second, positional cues may not be limited to a single system: They can be read by either autonomic or motor axons reinnervating skeletal muscle, as well as by autonomic axons reinnervating sympathetic ganglia (Purves et al., 1981). Finally, connectivity is not rigidly specified, in that axons from virtually any part of the spinal cord can innervate muscles (or ganglia) from any level. Instead, specificity is statistical, such that the probability that innervation derives from a given spinal level changes gradually from rostral to caudal ends of the target field. This graded selectivity (noted also by Purves et al., 1981, in ganglia) is consistent with the existence of gradients of molecules that bias but do not rigidly determine synapse formation between positionally matched partners. Our aim is to apply immunological methods of the types described above to a search for molecular correlates of this positional information.

A BIOASSAY FOR CELL ADHESION MOLECULES IN DENERVATED MUSCLE

Although no antibodies or antigens have yet been described that distinguish rostral from caudal muscles, studies discussed above have identified molecules that muscles may use to inform axons of their state of innervation and proximity to synaptic sites. The distribution of several of these molecules is sketched in Figure 8. Their arrangement suggests that regenerating axons encounter increasing numbers of attractive species with decreasing distance from original synaptic sites. Thus, denervated muscles may secrete soluble growth factors that axons sense from a distance; denervated fibers within a muscle are rich in N-CAM; perisynaptic areas bear interstitial deposits of several adhesion molecules; and the BL at original synaptic sites contains molecules that provide the most highly localized—and perhaps the most attractive—cue.

We would like to use immunological blockade to test the model that this arrangement implies. Unfortunately, the two most straightforward approaches seem poorly suited to the task. Cocultures of neurons and embryonic myotubes that have been useful for a variety of studies of synaptogenesis do not display the topographically ordered distributions of molecules that we

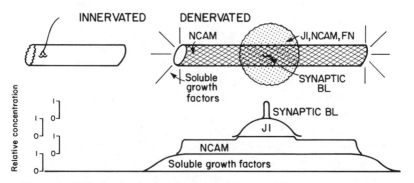

Figure 8. *Molecules that denervated muscles might use to influence regenerating axons.*

wish to investigate. Injections of antibodies *in vivo* provide a potentially more direct test, but in our opinion the difficulty of carrying out or interpreting such experiments in whole animals would be prohibitive. We have therefore assayed the behavior of neurons cultured on cryostat sections of the type already used to map the distribution of cell surface molecules. We used chick ciliary ganglion cells, because they are well-characterized neurons that are known to form cholinergic skeletal neuromuscular junctions *in vivo* and *in vitro*, but are considerably easier to isolate than spinal motoneurons. We plated these neurons on 6-μm-thick cryostat sections, incubated them in a rich medium for one to six days, then stained them with anti-N-CAM to examine their processes. Our main results (Covault et al., 1987) were the following:

1.   Neurons survived and extended processes on sections of both innervated and denervated skeletal muscle, as well as on several other tissues. In each case neurites grew preferentially along sectioned cell surfaces (Figure 9a), and several results (detailed in Covault et al., 1987) argued strongly that outgrowth was guided by specific interactions of neurites with cell surface molecules rather than by mechanical inhomogeneities in the section. Thus, the "cryoculture" system, which we adapted from a method for studying adhesion of lymphocytes to tissue elements (Stamper and Woodruff, 1976), may be useful for studying axon guidance in a number of developing and adult tissues.

2.   Neurites were, on average, longer, broader, and more highly branched on sections of denervated muscle than on sections of innervated muscle. Furthermore, outgrowth on denervated muscle was greater on sections cut from an end plate–rich area than on sections cut from an end plate–poor area. In a series of 16 cultures, average neurite length was 40 ± 3 (SEM) μm on end plate–rich innervated muscle, 56 ± 4 μm on end plate–poor denervated muscle, and 67 ± 4 μm on end plate–rich denervated muscle. While not dramatic, these differences were consistent and significant.

3.   Neurites that encountered intramuscular nerves in the sections generally confined their subsequent growth to them (Figure 9b). Neurites growing on nerve trunks or branches were considerably longer than neurites growing on muscle cell surfaces. Thus, intramuscular nerves can orient axonal growth in the cryocultures as they do *in vivo*.

4.   Neurites that reached original synaptic sites generally terminated at these points of contact (Figure 9c). These contacts were rare, owing to the shortness of the neurites and the scarcity of synaptic sites. Nonetheless, this phenomenon was sufficiently consistent to leave little doubt that axons recognize some component of the sectioned synaptic site.

Together, these results support the idea that axons can respond to molecules that are arranged on cell surfaces in denervated muscles. In addition, the cryoculture system is providing a basis for immunological tests of candidate molecules that we believe to be involved in guiding reinnervation,

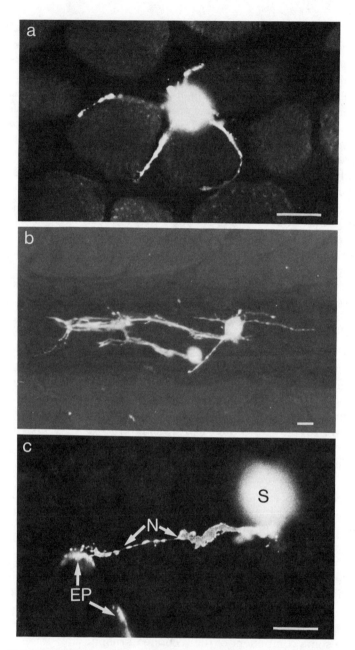

Figure 9. *N-CAM-stained neurons cultured on cryostat sections of denervated skeletal muscle.* Neurites grew along muscle fiber surfaces in *a* and *c*, but grew along an intramuscular nerve trunk in *b. c* was doubly stained with rhodamine-α-bungarotoxin as well as N-CAM to show the termination of a neurite (N) at a denervated end plate (EP). The neuronal soma (S) in *c* is out of the plane of focus. Calibration bars = 20 μm.

and may eventually be useful in the search for more subtle cues that axons use to distinguish among muscles—for example, the muscles' position in the rostrocaudal axis.

REFERENCES

Aitken, J. T. (1950) Growth of nerve implants in voluntary muscle. *J. Anat.* **84**:38–49.

Anderson, M. J., and D. M. Fambrough (1983) Aggregates of acetylcholine receptors are associated with plaques of a basal lamina heparan sulfate proteoglycan on the surface of skeletal muscle fibers. *J. Cell Biol.* **97**:1396–1411.

Anderson, M. J., M. W. Cohen, and E. Zorychta (1977) Effects of innervation on the distribution of acetylcholine receptors on cultured muscle cells. *J. Physiol. (Lond.)* **268**:731–756.

Bennett, M. R., and A. G. Pettigrew (1974) The formation of synapses in reinnervated and cross-reinnervated striated muscle during development. *J. Physiol. (Lond.)* **241**:547–573.

Bennett, M. R., and A. G. Pettigrew (1975) The formation of neuromuscular synapses. *Cold Spring Harbor Symp. Quant. Biol.* **40**:409–424.

Bernstein, J. J., and L. Guth (1961) Non-selectivity in the establishment of neuromuscular connections following nerve regeneration in the rat. *Exp. Neurol.* **4**:262–275.

Bixby, J. L., and L. F. Reichardt (1987) Effects of antibodies to neural cell adhesion molecule (N-CAM) on the differentiation of neuromuscular contacts between ciliary ganglion neurons and myotubes *in vitro. Dev. Biol.* **119**:363–372.

Bixby, J. L., R. S. Pratt, J. Lilien, and L. F. Reichardt (1987) Neurite outgrowth on muscle cell surfaces involves extracellular matrix receptors as well as calcium-dependent and independent cell adhesion molecules. *Proc. Natl. Acad. Sci. USA* **84**:2555–2558.

Bozyczko, D., and A. F. Horwitz (1986) The participation of a putative cell surface receptor for laminin and fibronectin in peripheral neurite extension. *J. Neurosci.* **6**:1241–1251.

Cashman, N. R., J. Covault, R. L. Wollman, and J. R. Sanes (1987) Neural cell adhesion molecule in normal, denervated, and myopathic muscle. *Ann. Neurol.* **21**:481–489.

Chiu, A. Y., W. D. Matthew, and P. H. Patterson (1986) A monoclonal antibody which blocks the activity of a neurite regeneration promoting factor: Studies on the binding site and its localization *in vivo. J. Cell Biol.* **102**:1383–1398.

Connor, E. A., and U. J. McMahan (1987) Cell accumulation in the junctional region of denervated muscle. *J. Cell Biol.* **104**:109–120.

Covault, J., and J. R. Sanes (1985) Neural cell adhesion molecule (N-CAM) accumulates in denervated and paralyzed skeletal muscles. *Proc. Natl. Acad. Sci. USA* **82**:4544–4548.

Covault, J., and J. R. Sanes (1986) Distribution of N-CAM in synaptic and extrasynaptic portions of developing and adult skeletal muscle. *J. Cell Biol.* **102**:716–730.

Covault, J., J. P. Merlie, C. Goridis, and J. R. Sanes (1986) Molecular forms of N-CAM and its RNA in developing and denervated skeletal muscle. *J. Cell Biol.* **102**:731–739.

Covault, J., J. M. Cunningham, and J. R. Sanes (1987) Neurite outgrowth on cryostat sections of innervated and denervated skeletal muscle. *J. Cell Biol.* **105**:2479–2488.

Davis, G. E., M. Manthorpe, E. Engvall, and S. Varon (1985) Isolation and characterization of rat Schwannoma neurite-promoting factor: Evidence that the factor contains laminin. *J. Neurosci.* **5**:2662–2671.

Elizalde, A., M. Heurta, and E. Stefani (1983) Selective reinnervation of twitch and tonic muscle fibers of the frog. *J. Physiol. (Lond.)* **340**:513–524.

Elsberg, C. A. (1917) Experiments on motor nerve regeneration and the direct neurotization of paralyzed muscles by their own and by foreign nerves. *Science* **45**:318–320.

Faissner, A., J. Kruse, R. Chiquet-Ehrismann, and E. Mackie (1988) The high molecular weight tenascin(J1) glycoproteins are immunochemically related to tenascin. *Differentiation* **37**:104–114.

Fallon, J. R., R. M. Nitkin, N. E. Reist, B. G. Wallace, and U. McMahan (1985) Acetylcholine receptor–aggregating factor is similar to molecules concentrated at neuromuscular junctions. *Nature* **315**:571–574.

Fontaine, B., D. Sassoon, M. Buckingham, and J.-P. Changeux (1988) Detection of the nicotinic acetylcholine receptor α-subunit by *in situ* hybridization at neuromuscular junctions of 15-day-old chick striated muscles. *EMBO J.* **7**:603–609.

Frank, E., J. K. S. Jansen, T. Lømo, and R. H. Westgaard (1975) The interaction between foreign and original motor nerves innervating the soleus muscle of rats. *J. Physiol. (Lond.)* **247**:725–743.

Gatchalian, C. L., M. Schachner, and J. R. Sanes (1989) Fibroblasts that proliferate near denervated synaptic sites in skeletal muscle synthesize the adhesive molecules tenascin(J1), N-CAM, fibronectin, and a heparan sulfate proteoglycan. *J. Cell Biol.* **108**:1873–1890.

Gatchalian, C., A. Y. Chiu, and J. R. Sanes (1985) Monoclonal antibodies to laminin that distinguish synaptic and extrasynaptic domains in muscle fiber basal lamina. *J. Cell Biol.* **101**:89a.

Glicksman, M., and J. R. Sanes (1983) Development of motor nerve terminals formed in the absence of muscle fibers. *J. Neurocytol.* **12**:661–671.

Goldman, D., J. Boulter, S. Heinemann, and J. Patrick (1985) Muscle denervation increases the levels of two mRNAs coding for the acetylcholine receptor α-subunit. *J. Neurosci.* **5**:2553–2558.

Grumet, M., U. Rutishauser, and G. M. Edelman (1982) Neural adhesion molecule is on embryonic muscle cells and mediates adhesion to nerve cells *in vitro*. *Nature* **295**:693–695.

Gurney, M. E., B. R. Apatoff, and S. P. Heinrich (1986) Suppression of terminal axonal sprouting at the neuromuscular junction by monoclonal antibodies against a muscle-derived antigen of 56,000 daltons. *J. Cell Biol.* **102**:2264–2272.

Gutmann, E., and J. Z. Young (1944) The re-innervation of muscle after various periods of atrophy. *J. Anat.* **78**:15–43.

Henderson, C. E., M. Huchet, and J.-P. Changeux (1983) Denervation increases a neurite-promoting activity in extracts of skeletal muscle. *Nature* **302**:609–611.

Hunter, D. D., V. Shah, J. P. Merlie, and J. R. Sanes (1989) A laminin-like adhesive protein concentrated in the synaptic cleft of the neuromuscular junction. *Nature* **338**:229–234.

Hynes, R. O. (1987) Integrins: A family of cell surface receptors. *Cell* **48**:549–554.

Ide, C., K. Tohyama, R. Yokota, T. Nitatori, and S. Onodera (1983) Schwann cell basal lamina and nerve regeneration. *Brain Res.* **288**:61–75.

Jansen, J. K. S., T. Lømo, K. Nicolaysen, and R. H. Westgaard (1973) Hyperinnervation of skeletal muscle fibers: Dependence on muscle activity. *Science* **181**:559–561.

Klarsfeld, A., and J.-P. Changeux (1985) Activity regulates the levels of acetylcholine receptor α-subunit mRNA in cultured chicken myotubes. *Proc. Natl. Acad. Sci. USA* **82**:4558–4562.

Kruse, J., G. Keilhauer, A. Faissner, R. Timpl, and M. Schachner (1985) The J1 glycoprotein—A novel nervous system cell adhesion molecule of the L2/HNK-1 family. *Nature* **316**:146–148.

Kuffler, D. P. (1986) Isolated satellite cells of a peripheral nerve direct the growth of regenerating frog axons. *J. Comp. Neurol.* **249**:57–64.

Kuromi, H., and Y. Kidikoro (1984) Nerve disperses preexisting acetylcholine receptor clusters prior to induction of receptor accumulation in *Xenopus* muscle cultures. *Dev. Biol.* **103**:53–61.

Lander, A. D., D. K. Fujii, and L. F. Reichardt (1985) Purification of a factor that promotes neurite outgrowth: Isolation of laminin and associated molecules. *J. Cell Biol.* **101**:898–913.

Landmesser, L. (1972) Pharmacological properties, cholinesterase activity, and anatomy of nerve–muscle junctions in vagus-innervated frog sartorius. *J. Physiol. (Lond.)* **220**:243–256.

Laskowski, M. B., and J. R. Sanes (1987a) Topographic mapping of motor pools onto skeletal muscles. *J. Neurosci.* 7:252–260.

Laskowski, M. B., and J. R. Sanes (1988) Topographically selective reinnervation of adult mammalian muscles. *J. Neurosci.* 8:3094–3099.

Letinsky, M. K., D. G. Fischbach, and U. J. McMahan (1976) Precision of reinnervation of original postsynaptic sites in muscle after a nerve crush. *J. Neurocytol.* 5:691–718.

Merlie, J. P., and J. R. Sanes (1985) Concentration of acetylcholine receptor mRNA in synaptic regions of adult muscle fibres. *Nature* 317:66–68.

Merlie, J. P., and J. R. Sanes (1986) Regulation of synapse-specific genes. In *Molecular Aspects of Neurobiology*, R. Levi-Montalcini, P. Calissano, E. R. Kandel, and A. Maggi, eds., pp. 000–000, Springer-Verlag, Berlin.

Merlie, J. P., K. E. Isenberg, S. D. Russell, and J. R. Sanes (1984) Denervation supersensitivity in skeletal muscle: Analysis with a cloned cDNA probe. *J. Cell Biol.* 99:332–335.

Moore, S. E., and F. S. Walsh (1986) Nerve-dependent regulation of neural cell adhesion molecule expression in skeletal muscle. *Neuroscience* 18:499–505.

Murray, M. A., and N. Robbins (1982) Cell proliferation in denervated muscle: Identity and origin of the dividing cells. *Neuroscience* 7:1823–1833.

Nitkin, R. M., M. A. Smith, C. Magill, J. R. Fallon, Y. M. Yao, B. G. Wallace, and U. J. McMahan (1987) Identification of agrin, a synaptic organizing protein from *Torpedo* electric organ. *J. Cell Biol.* 105:2471–2478.

Nurcombe, V., M. A. Hill, K. L. Eagelson, and M. R. Bennett (1984) Motor neuron survival and neuritic extension from spinal cord explants induced by factors released from skeletal muscle. *Brain Res.* 291:19–28.

Purves, D., W. Thompson, and J. W. Yip (1981) Reinnervation of ganglia transplanted to the neck from different levels of the guinea-pig sympathetic chain. *J. Physiol. (Lond.)* 313:49–63.

Ramón y Cajal, S. (1928, reprinted 1968) *Degeneration and Regeneration of the Nervous System*, Hafner, London.

Rieger, F., M. Grumet, and G. M. Edelman (1985) N-CAM at the vertebrate neuromuscular junction. *J. Cell Biol.* 101:285–293.

Rutishauser, U., M. Grumet, and G. M. Edelman (1983) Neural cell adhesion molecule mediates initial interactions between spinal cord neurons and muscle cells in culture. *J. Cell Biol.* 97:145–152.

Salpeter, M. M., and R. H. Loring (1985) Nicotinic acetylcholine receptors in vertebrate muscle: Properties, distribution, and neural control. *Prog. Neurobiol.* 25:297–325.

Sanes, J. R. (1982) Laminin, fibronectin, and collagen in synaptic and extrasynaptic portions of muscle fiber basement membrane. *J. Cell Biol.* 93:442–451.

Sanes, J. R. (1989) Extracellular matrix molecules that influence neural development. *Annu. Rev. Neurosci.* 12:491–516.

Sanes, J. R., and J. M. Cheney (1982) Lectin-binding reveals a synapse-specific carbohydrate in skeletal muscle. *Nature* 300:646–647.

Sanes, J. R., and A. Y. Chiu (1983) The basal lamina of the neuromuscular junction. *Cold Spring Harbor Symp. Quant. Biol.* 48:667–678.

Sanes, J. R., and J. Covault (1985) Axon guidance during reinnervation of skeletal muscle. *Trends Neurosci.* 8:523–528.

Sanes, J. R., and Z. W. Hall (1979) Antibodies that bind specifically to synaptic sites on muscle fiber basal lamina. *J. Cell Biol.* 83:357–370.

Sanes, J. R., L. M. Marshall, and U. J. McMahan (1978) Reinnervation of muscle fiber basal lamina after removal of myofibers. Differentiation of regenerating axons at original synaptic sites. *J. Cell Biol.* 78:176–198.

Sanes, J. R., M. Schachner, and J. Covault (1986) Expression of several adhesive macromolecules (N-CAM, L1, J1, NILE, uvomorulin, laminin, fibronectin, and a heparan sulfate proteoglycan) in embryonic, adult, and denervated adult skeletal muscles. *J. Cell Biol.* **102**:420–431.

Scherer, S. S., and S. S. Easter, Jr. (1984) Degenerative and regenerative changes in the trochlear nerve of goldfish. *J. Neurocytol.* **13**:519–565.

Scott, L. J., and J. R. Sanes (1984) A lectin that selectively stains neuromuscular junctions binds to collagen-tailed acetylcholinesterase. *Soc. Neurosci. Abstr.* **10**:546.

Scott, L. J., F. Bacou, and J. R. Sanes (1988) A synapse-specific carbohydrate at the neuromuscular junction: Association with both acetylcholinesterase and a glycolipid. *J. Neurosci.* **8**:932–944.

Slack, J. R., W. G. Hopkins, and S. Pockett (1983) Evidence for a motor nerve growth factor. *Muscle & Nerve* **6**:243–252.

Stamper, H. B., and J. J. Woodruff (1976) Lymphocyte homing into lymph nodes: *In vitro* demonstration of the selective affinity of recirculating lymphocytes for high-endothelial venules. *J. Exp. Med.* **144**:828–833.

Tello, F. (1907) Degeneration et regeneration des plaques motrices après la section des nerfs. *Trav. Lab. Recherche Biol.* **5**:117–149.

Weinberg, C. B., J. R. Sanes, and Z. W. Hall (1981) Formation of neuromuscular junctions in adult rats: Accumulation of acetylcholine receptors, acetylcholinesterase, and components of synaptic basal lamina. *Dev. Biol.* **84**:255–266.

Wigston, D. J. (1986) Selective innervation of transplanted limb muscles by regenerating motor axons in the axolotl. *J. Neurosci.* **6**:2757–2763.

Wigston, D. J., and J. R. Sanes (1982) Selective reinnervation of adult mammalian muscles by axons from different segmental levels. *Nature* **299**:464–467.

Wigston, D. J., and J. R. Sanes (1985) Selective reinnervation of rat intercostal muscles transplanted from different segmental levels to a common site. *J. Neurosci.* **5**:1208–1221.

Chapter 20

Localization and Potential Role of Cell Adhesion Molecules and Cytotactin in the Developing, Diseased, and Regenerating Neuromuscular System

FRANÇOIS RIEGER

ABSTRACT

The development, maintenance, and physiological functions of muscle depend upon specialized structures in the neuromuscular system, particularly the neuromuscular junction and nodes of Ranvier. We have identified three glycoproteins involved in cell–cell adhesion that are highly localized to these privileged sites and that vary in amount and distribution during development, during pathological changes, and in response to nerve injury: the neural cell adhesion molecule, N-CAM, the neuron–glia cell adhesion molecule, Ng-CAM, and cytotactin, an extracellular matrix protein. We have examined by histochemical and biochemical means the detailed localization of N-CAM, Ng-CAM, and cytotactin during the formation of the neuromuscular synapse and nodes of Ranvier, analyzed the distribution of these molecules in normal and dysmyelinating mouse mutant nerves, and studied the expression and addressed the question of the contribution of these molecules in regenerative events. These studies provide new insights and perspectives into the formation and maintenance of structures crucial to muscle function, as well as to their alterations in nerve–muscle diseases and dysfunction.

The neuromuscular system is a complex supracellular structure composed of a number of different cell types: sensory and motor neurons, Schwann cells, muscle cells, fibroblasts, and various support cells. Multiple interactions and reciprocal inductive events occur during development among the precursors of these cells and later among the differentiating cells. Initial connections between growing axons, Schwann cells, and muscle cells and the subsequent morphogenesis of synaptic junctions and nodes of Ranvier in the neuromuscular system require local associations of structural proteins, enzymes, ionic channels, and receptors in precise spatiotemporal sequences. The molecular nature and the steps of the interactive mechanisms responsible for the formation of these privileged sites of cell–cell interactions are not yet known.

At the motor end plate during embryonic development, the acetylcholine receptor (AChR) accumulates in the postsynaptic muscle membrane adjacent to nerve terminals (Fambrough, 1979). At later stages of development, postjunctional folds are formed, with AChR-rich crests closely apposed to the nerve terminals (Fertuck and Salpeter, 1976). In the synaptic cleft and on the surface of the muscle fiber, a basal lamina develops, connected to the underlying cytoskeleton (Hirohawa and Heuser, 1982). Growing axons arrive at developing muscle in the embryo before basal lamina is deposited and AChR aggregates appear, suggesting that AChR aggregation and basal lamina deposition are not essential for the formation of future synapses (Rieger, 1985).

In nodal development a single motile Schwann cell becomes associated with a growing axon, ensheaths an internodal segment, stops dividing, acquires a basal lamina, and begins spiral growth and myelin formation (Geren, 1954). The node of Ranvier separates two consecutive Schwann cells and is a highly differentiated annular narrowing interrupting the myelin sheath. It has an essential function in the axonal conduction of the nerve impulse: Axonal membrane depolarization leads to inward current only in nodal regions, and conduction is saltatory between them (Tasaki, 1939; Huxley and Stampfli, 1949). Nodes of Ranvier may be structurally and physiologically predetermined before they achieve complete ultrastructural and physiological differentiation (Tao-Cheng and Rosenbluth, 1982; Waxman et al., 1982).

Because the formation of both the neuromuscular junction and node of Ranvier involve the close apposition and interaction of axon and muscle cell and Schwann cell and axon, respectively, it is highly probable that molecular mechanisms of cell adhesion play a central and early role in determining subsequent cellular and molecular signals and localizations. Two neuronal cell adhesion molecules (CAMs), neural CAM (N-CAM) and the neuron–glia CAM (Ng-CAM) (see other contributions in this book for their molecular characterization), are present within the central and peripheral nervous systems (Chuong and Edelman, 1984; Thiery et al., 1985; Daniloff et al., 1986a). The properties of these CAMs suggest that they play decisive roles in developmental processes of pattern formation during early embryogenesis and histogenesis (Edelman, 1984, 1985a,b). N-CAM is a cell surface glycoprotein that mediates adhesion by a homophilic mechanism: one N-CAM molecule on one cell interacts with one N-CAM molecule on an apposing cell (Edelman, 1983, 1984). N-CAM antigenic determinants have been found on central and autonomic neurons (Thiery et al., 1982; Edelman, 1984), in the myotome (Thiery et al., 1982), and in developing skeletal muscle both in culture and in vivo (Grumet et al., 1982; Rutishauser et al., 1983; Covault and Sanes, 1985; Rieger et al., 1985). Muscle cells in culture have been shown to bind vesicles reconstituted from purified N-CAM and lipids (Rutishauser et al., 1983).

These findings suggest that N-CAM may mediate adhesion between neurons and developing muscle cells *in vivo* and raise the possibility that this molecule may play a role in the development and stabilization of nerve–muscle junctions (Rieger, 1985; Rieger et al., 1985). Ng-CAM has been found on neurons and mediates the adhesion of neurons to glial cells by a heterophilic mechanism (Grumet et al., 1984a,b). Ng-CAM is also involved in neuron–neuron adhesion such as neurite fasciculation in developing nerves (Grumet et al., 1984a; Friedlander et al., 1986). In addition to N-CAM and Ng-CAM, a substrate adhesion molecule (SAM) involved in neuron–glia interactions and synthesized by glia and not by neurons of the central and peripheral nervous systems has been identified, the extracellular matrix protein cytotactin (Grumet et al., 1985). The distribution of this molecule in nonneural sites during embryogenesis and histogenesis is particularly prominent in regions of active cell migration, and it may serve a function in allowing, restricting, or guiding cell movements (Crossin et al., 1986).

This chapter reviews and summarizes the results of our search for patterned localization of these molecules in the neuromuscular system in the normal adult, during development, in mouse neuromuscular mutants, and after experimental injury to the nerve, with special emphasis on the motor end plate and the node of Ranvier. We used immunocytochemical methods— immunofluorescence at the light microscope level and immunogold techniques at the electron microscope level—to localize all three molecules. We also used electrophoretic techniques in the presence of SDS to assess the specificity of their antibodies in the neuromuscular system (Rieger et al., 1985, 1986) and get their biochemical characterization.

N-CAM, Ng-CAM, AND CYTOTACTIN IN ADULT AND DEVELOPING MUSCLE AND NERVE

In an initial attempt to study the expression of N-CAM in muscle and nerve and establish the specificity of the antibodies used for the immunocytochemical experiments, different antibodies raised against mouse and chicken brain N-CAM were used in immunoblotting experiments. The forms of N-CAM present in muscle and nerve in mouse were resolved by 7.5% polyacrylamide gel electrophoresis in the presence of SDS, and immunoblotting was performed with a rabbit polyclonal antibody. In adult mouse diaphragm muscle and sciatic nerve, the forms of N-CAM differed; in contrast to spinal cord, which contained a major component of 120 kD (ssd chain) and minor components including those of 180 kD (ld chain) and 140 kD (sd chain) (Chuong and Edelman, 1984; Rieger et al., 1985), mouse muscle contained only one major component of about 140 kD and sciatic nerve contained two components of 140 kD and 120 kD (Rieger et al., 1985). To compare the polypeptide structures of N-CAMs from mouse muscle and neural tissues, the component of 140 kD isolated from muscle and the components of 180 kD and 140 kD isolated from brain were treated by *Staphylococcus aureus* V8 protease and peptide mapped in one-dimension gel electrophoresis experiments. Peptide fragments derived from muscle and brain showed extensive homology, demonstrating a near-identity of the N-CAM forms from brain and muscle. No such extensive characterization study has yet been conducted for Ng-CAM and cytotactin and their antibodies in muscle and nerve.

The distribution of muscle N-CAM in teased fiber preparations of adult mouse diaphragm has been studied by indirect immunofluorescence (Rieger et al., 1985). Positive staining for N-CAM was essentially observed at nerve–muscle contacts that were identified by the binding of fluorescein-conjugated α-bungarotoxin, a toxin that binds irreversibly and specifically to the AChR (Figure 1) (Rieger et al., 1985). Anti-N-CAM antibodies stained the contours of the motor end plates as delineated by the AChR-rich regions, and some presynaptic structures, essentially axon terminal branches (Figure 1). N-CAM and AChR were colocalized at adult neuromuscular junctions, in all cases including mouse (Rieger et al., 1985) and chicken (Daniloff et al., 1986b) fast and slow muscles and also frog muscle (Figure 2). In the mouse more than 90% of the neuromuscular junctions identified by α-BgTX staining were also positively stained for N-CAM (Rieger et al., 1985). In whole mount preparations of diaphragm muscles of near-term (embryonic day 19) or newborn mice, staining for N-CAM was more widespread than in adult muscle. We observed a uniform pattern with occasional foci of intense staining (Rieger et al., 1985). Axon terminal arborizations were also stained for N-CAM, and double-labeling patterns of N-CAM and of AChR were found to overlap at the developing motor end plates (Rieger et al., 1985). More than 90% of the developing motor end plates characterized by clusters of AChRs also had a quite coextensive pattern of intense staining for N-CAM. In addition to these staining patterns, the myotendinous junctions and interstitial tissues were stained for N-CAM.

Recent experiments on myotendinous junctions in mouse and chicken muscle show that N-CAM is a prominent component of the extremity of muscle fibers. In developing muscle, N-CAM is already concentrated at nerve–muscle contacts and myotendinous junctions (F. Rieger, K. L. Crossin, and G. M. Edelman, unpublished data). A more precise subcellular localization of N-CAM in muscle is now needed and awaits further ultrastructural studies, such as those using immunocytochemical methods at the electron microscope level, in order to determine a possible association of N-CAM with structures such as the pre- and postsynaptic membranes themselves, or the muscle synaptic or terminal Schwann cell basal laminae.

Cytotactin is also found in high amounts in the motor end plate region, as shown in double-staining experiments with AChR staining, using a rabbit polyclonal antibody raised against chicken brain cytotactin (Figure 3a,a′). Cytotactin seems to be preferentially associated to the terminal Schwann cell. This has been shown in double-staining experiments (Figure 4a,a′) using a

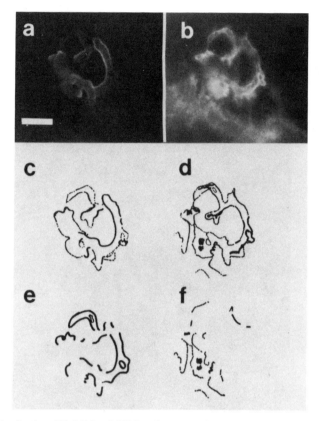

Figure 1. *Colocalization of N-CAM and AChR at the motor end plate in adult mouse diaphragm muscle.*
Micrograph of staining with FITC-α-BgTX (*a*) under optics for fluorescein. The profiles
delineate the classical contours of mature motor end plates. Micrographs of staining with anti-
N-CAM IgG (*b*) under rhodamine optics were obtained after exposure to rhodamine-labeled
goat anti-rabbit IgG. The patterns of staining of α-BgTX and N-CAM closely corresponded,
with additional regions of N-CAM staining that corresponded to the axon terminal branches.
The contours of staining for AChR (*c*) and for N-CAM (*d*) were drawn. The profiles shown in *c*
and *d* were superimiposed, and the common features (*e*) and the difference profile (*f*) were
drawn. *e* shows the characteristic features of the AChR-rich postsynaptic membrane. The
profile in *f* represents primarily axon terminal branches. Calibration bar = 10 μm.

monoclonal antibody specific for the Schwann cell, obtained after immunization with a Schwann cell–specific component (Goujet-Zalc et al., 1986; Guerci
et al., 1986). In addition, we find cytotactin at the myotendinous junction
associated to the tendinous structures bridging the muscle fiber extremity and
the bone insertions.

Ng-CAM staining has also been observed in the region of the motor end
plate in strict association with the presynaptic elements. In mouse as well as in
chicken muscle, the staining obtained after incubation with the polyclonal anti-

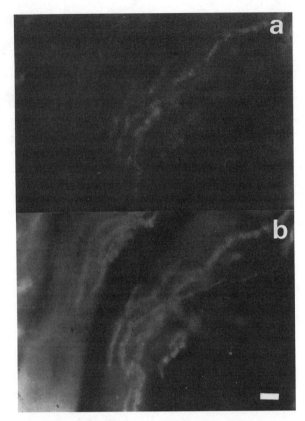

Figure 2. *Colocalization of AChR and N-CAM at the adult frog motor end plate.* Adult frog cutaneous pectoris muscles were stained with FITC-α-BgTX (*a*) and anti-N-CAM IgG (*b*) obtained after purification of *Xenopus laevii* N-CAM. The pattern of N-CAM staining closely corresponds to the pattern of α-BgTX staining. Calibration bar = 10 μm.

Ng-CAM antibodies (described in Daniloff et al., 1986a) is of rather low intensity in conditions similar to those used in N-CAM and cytotactin studies.

In adult mouse and chicken nerve we have found all three proteins, N-CAM, Ng-CAM, and cytotactin, in very specific locations. To determine the distribution of Ng-CAM and N-CAM on myelinated and unmyelinated axons, we used teased nerve fiber preparations. We performed double-staining experiments using polyclonal anti-CAM antibodies and a monoclonal antibody to the sodium channel to identify the nodal regions (Figure 5), which were also recognized morphologically by using phase contrast microscopy. We observed colocalization of Ng-CAM, N-CAM, and cytotactin with sodium channels at the nodes of Ranvier (Rieger et al., 1986). Staining, slightly above background levels, was also observed for both Ng-CAM and N-CAM in internodal regions but not for cytotactin. Small-caliber, unmyelinated fibers stained intensely for

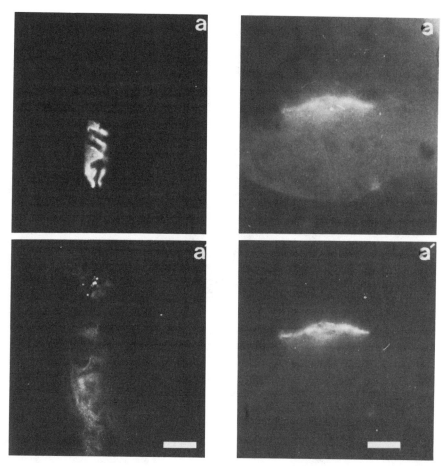

Figure 3. *Immunocytochemical staining of cyto-tactin at the adult mouse motor end plate.* Cytotactin staining (*a*) was found in the same region as α-BgTX staining (*a'*) but was not coincident with it. Calibration bar = 5 μm.

Figure 4. *Immunocytochemical staining of terminal Schwann cells in adult mouse skeletal muscle.* A monoclonal antibody, mab 224–58, that specifically labels Schwann cells stained the same structures (*a*) as the anti-cytotactin rabbit polyclonal antibody (*a'*). Calibration bar = 5 μm.

both Ng-CAM and N-CAM. Transverse sections of chicken sciatic nerve double stained for Ng-CAM and N-CAM possessed intensely stained, cross-sectioned bundles of small-caliber, unmyelinated fibers. The Ng-CAM and N-CAM staining was uniform all along the length of the unmyelinated fibers, as shown in teased fiber preparations (Rieger et al., 1986). Their cytotactin staining was negative.

The study of developing chicken peripheral nerve was then initiated to try to establish when focal accumulations of Ng-CAM and N-CAM as well as

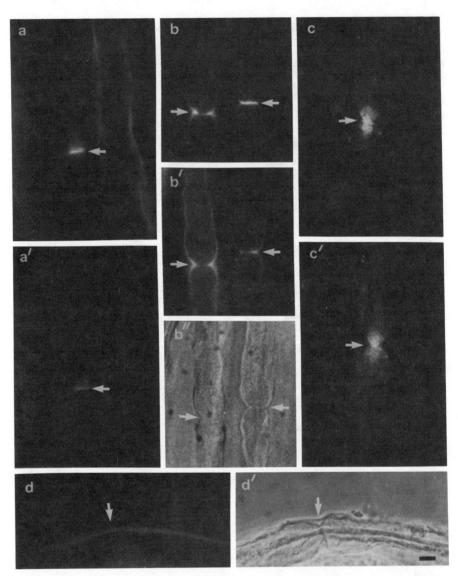

Figure 5. *Colocalization of Ng-CAM and N-CAM stainings at the node of Ranvier in teased fiber preparations of myelinated fibers of adult mouse and chicken sciatic nerves.* The distributions of both proteins (*a,b*) overlapped with accumulations of sodium channels (*a',b'*) at nodes of Ranvier (*arrows in b'*; phase contrast microscopy). All fibers shown in this figure correspond to large-caliber fibers. Intense staining for Ng-CAM and N-CAM was observed at nodal regions. Cytotactin staining was morphologically different from that of Ng-CAM and N-CAM, and extended itself partially in internodal axonal regions. Nonmyelinated fiber bundles (*arrow in a*) were also labeled by anti-Ng-CAM antibodies. In teased preparations of adult chicken sciatic nerves, intense staining for both Ng-CAM (*c*) and N-CAM (*c'*) was observed at the nodal regions (*arrow*). Anti-Thy-1 antibodies stained the entire axon in chicken sciatic nerves (*d*) and did not accumulate exclusively at identified nodes of Ranvier (*d'*, phase contrast microscopy). Calibration bar = 10 μm.

cytotactin occurred, in relation to known events of nerve differentiation, namely myelination, and to get quantitative data on Ng-CAM and N-CAM in nerve. This study brought us two important observations: (1) Nodallike accumulations of Ng-CAM and N-CAM are observed very early on small-caliber nerve fibers, at embryonic day 14 (Figure 6a,a'), that is, before myelination. The state of myelination was assessed by double-staining experiments involving Ng-CAM or N-CAM staining and a staining obtained with a monoclonal antibody directed against the P1 myelin basic protein (a gift from Dr. A. Peterson, Montreal). Later, at embryonic day 18, a few small-caliber axons with no myelin showed periodic accumulations of Ng-CAM and N-CAM. These particular nerve fibers probably correspond to nerve fibers before myelination, because later on, after hatching, no small-caliber fiber displayed focal accumulations of these proteins (Rieger et al., 1986). Small-caliber nerve fibers, intensely stained for Ng-CAM and N-CAM on their whole surface, were observed from very early (embryonic days 12–14) up until the adult stage, suggesting that unmyelinated nerve fibers are characterized by a uniform distribution of CAMs as soon as they form. (2) Cytotactin staining was

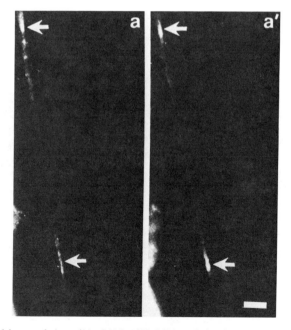

Figure 6. *Prenodal accumulations of Ng-CAM and N-CAM.* Early focal accumulations (*arrow*) of Ng-CAM (*a*) and N-CAM (*a'*) on teased, small-caliber (0.4–0.6 μm) fibers from chicken embryos at E14. These fibers are not myelinated, as indicated by the absence of P1 staining in double-staining experiments (Rieger et al., 1986). Calibration bar = 10 μm.

observed in the early steps of nerve development (embryonic days 14–20) and was localized essentially in the cytoplasm of the Schwann cell, mostly in a perinuclear location (Figure 7). This staining was transitory and vanished at about embryonic days 18–20, at the same time the first nodal accumulations of cytotactin were detected (Rieger et al., 1986). Some perineurial and endo-neurial connective tissue staining was also present in developing nerve.

On the basis of this developmental schedule of CAMs and cytotactin and their reciprocal distribution, it is tempting to hypothesize that (1) Ng-CAM and N-CAM prespecify the regions that will become the nodes of Ranvier, imposing some sort of spatial restriction for Schwann cell extension and/or movement, and (2) cytotactin, which comes later at the nodes, acts by stabilizing the nodal region and paranodal appositions of the myelinating Schwann cell and the axon.

The immunofluorescent data obtained on chicken sciatic nerve correlated well with more quantitative data obtained by immunoblotting experiments. Both Ng-CAM and N-CAM significantly decreased in quantity between embryonic day 14 and the adult stage, with most of the decrease occurring before hatching (Rieger et al., 1986). Only one major form of Ng-CAM of 135 kD was found in chicken nerves, and high molecular weight forms of N-CAM together with distinct 140-kD and 120-kD components were detected in embryonic nerve. Nerve N-CAM then lost its high molecular weight components and remained as a major 140-kD component together with a 120-kD component.

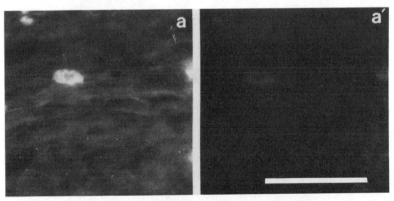

Figure 7. *Immunocytochemical staining of Schwann cells in embryonic chicken sciatic nerve.* Longi-tudinal frozen sections of sciatic nerves from embryonic day 18 were double labeled with mouse monoclonal anti-chicken cytotactin IgG (*a*) and polyclonal antibodies to S100 protein specific for Schwann cells (*a'*). Cytotactin staining is observed in the soma of the Schwann cells identified by their S100-positive staining. Calibration bar = 50 μm.

CAMs AND CYTOTACTIN EXPRESSION IN MOUSE
DYSMYELINATING MUTANT NERVES

The study of two mouse mutations known to affect primarily the peripheral nervous system, *trembler (tr)* and motor end-plate disease *(med)*, revealed other correlations between the staining patterns of the adhesion molecules and Schwann cell–axon interactions in peripheral nerve. In *trembler* the Schwann cells are the primary targets of the mutation, have lower amounts of myelin than normal (Aguayo et al., 1979), and present segmental demyelination in internodal regions as well as enhanced Schwann cell proliferation (Bray et al., 1981; Perkins et al., 1981). In *med/med* nerves nodes of Ranvier are abnormally wide as soon as myelination starts, but internodal regions are normal in length and myelin thickness (Rieger and Pinçon-Raymond, 1984).

Accumulations of Ng-CAM and N-CAM at morphologically identified nodes of Ranvier were very seldom observed in *trembler* nerves. *Trembler (+/tr)* nerve fibers were characterized by a striking abnormality consisting of large, intensely stained patches of coincident Ng-CAM and N-CAM all along the demyelinated axons (Figure 8a,a'). It is not known why Ng-CAM and N-CAM are systematically found coincident even in these pathological conditions. The distribution of cytotactin was also dramatically altered, with intense cytotactin staining of Schwann cells, similar to that of embryonic Schwann cells at the time of myelin formation, and faint staining all along the axon–Schwann cell interface (Rieger et al., 1986). *Med/med* nerve fibers had an aspect closer to normal, but most of the nodal accumulations of Ng-CAM and N-CAM were found wider than in control nerves, and corresponded to the enlarged mutant nodes (Figure 8b,b'). In some myelinated axons in *med/med* mice, paranodal accumulations of both Ng-CAM and N-CAM were found separated by a gap with decreased or no staining (Figure 8c). The cytotactin staining in *med/med* nerves was found extending in internodal regions, sometimes giving a continuous staining all along the axon.

The physicochemical nature of Ng-CAM and N-CAM in mutant nerves was assessed by SDS-gel electrophoresis and immunoblotting. Two polypeptide chains of 140 kD and 120 kD were also found in both control and mutant nerves. The only difference observed was in *trembler* nerves, which showed significant increases of Ng-CAM and N-CAM when compared with normal age-matched controls (Rieger et al., 1986), in agreement with the fluorescent data (see above). The amounts of Ng-CAM and N-CAM were similar or slightly decreased in *med/med* nerves.

Thus, our study of the dysmyelinating mutants mainly showed that defective axon–Schwann cell interactions are accompanied by important changes in the staining patterns of Ng-CAM, N-CAM, and cytotactin. These results and the observations that Schwann cells contain Ng-CAM, N-CAM, and cytotactin are consistent with (1) the idea that these molecules mediate axon–Schwann cell interactions and adhesion between glia and axons and (2)

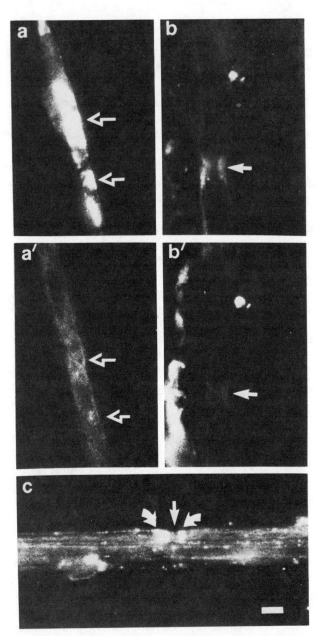

Figure 8. *Abnormalities of distributions of Ng-CAM and N-CAM in dysmyelinating mouse mutants.* In *trembler (+/tr) (a,a')*, nodal accumulations were rarely found, and Ng-CAM *(a)* and N-CAM *(a')* were spread all along the demyelinated fibers *(broken arrows)*. The staining patterns were coincident. In *med/med* sciatic nerves *(b,b',c)* nodes of Ranvier *(arrows)* were wider than normal, as expected from previous studies (Rieger and Pinçon-Raymond, 1984), and the whole nodal area stained for both Ng-CAM *(b)* and N-CAM *(b')*. In some myelinated fibers *(c)* staining was intense in paranodal areas *(curved arrows)*, with a gap of less intense staining in the nodal region. There was also more internodal staining in these fibers. Calibration bar = 10 μm.

432

the hypothesis that CAMs are important for the early establishment of the nodal pattern and also, together with cytotactin, for its maintenance at later stages.

CAMs AND CYTOTACTIN EXPRESSION DURING NERVE REGENERATION AND MUSCLE REINNERVATION

Nerve regeneration and muscle reinnervation are two phenomena probably involving extensive cell adhesion and cell–cell interactions. Two general paradigms, nerve compression and nerve transection, have been extensively used to study regeneration in vertebrates (Guth, 1956; Bennett et al., 1973; Daniloff et al., 1986b). We examined changes in the amounts, forms, and cellular distribution of Ng-CAM and N-CAM and also cytotactin in muscle, nerve, and spinal cord after transection or compression of the sciatic nerve. We compared normal, crushed, and transected sciatic nerves at low magnification to provide an overview of the changes in CAMs and cytotactin distribution that occur 10 days after nerve damage, a time intermediate between degenerative and regenerative sequences.

In longitudinal sections of normal chicken sciatic nerves, low levels of all three proteins were observed (Figure 9), with the exception of occasional nodal areas. In 10-day lesioned nerves, staining for Ng-CAM and N-CAM (Figure 9B,B',C,C') and also cytotactin was increased. In crushed nerves, the intensity of the staining was similarly increased proximal and distal to the lesion site (Figure 9B,B'). In transected nerves, very few axons had crossed the gap to enter the distal stump by 10 days (Figure 9C,C'). The intensity of staining in the proximal stump after nerve transection was intense for Ng-CAM, but the distal stump had low staining. There was also intense N-CAM staining in the proximal stump; the distal stump and the gap between the stumps expressed less N-CAM. Before the perineurium was restored, the nonneural cells in the scar region were N-CAM positive and were either low or negative for Ng-CAM. Cytotactin staining followed the same trends as Ng-CAM and N-CAM stainings in both experimental conditions and was found intense, essentially in the proliferating Schwann cells, in the bands of Büngner, and the perineurium during the active phases of regeneration. By 20 days after transection, Ng-CAM-positive fascicles of axons were present in the distal stump (Figure 10) along the bands of Büngner, composed of aligned S100-protein-positive Schwann cells, which had low, near-control Ng-CAM and N-CAM stainings and also low cytotactin staining.

Fifty days after crushing a nerve or 150 days after a cut, Ng-CAM and N-CAM approached control levels in all neural and nonneural structures of the nerve (Daniloff et al., 1986b), although in transected nerves bundles of nonmyelinated nerve, fibers were denser and larger than in control nerves.

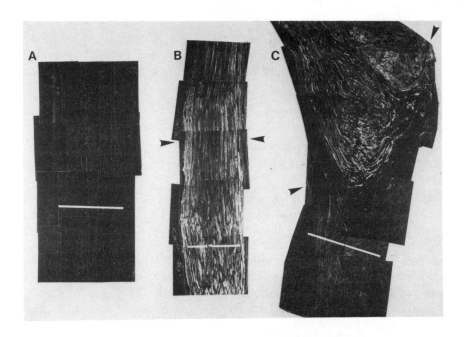

Figure 9. *Longitudinal sections of normal (A,A'), crushed (B,B') and cut (C,C') sciatic nerves 10 days after injury.* The same sections were double labeled with anti-Ng-CAM *(A,B,C)* and anti-N-CAM *(A',B',C')*. The site of the lesion is indicated by *arrowheads*. The immunocytochemical study of the sites distal to the lesions, reported in the text, was performed at the level of the *white lines*. In experimental nerves, the intensity of both Ng-CAM *(B,C)* and N-CAM *(B',C')* staining

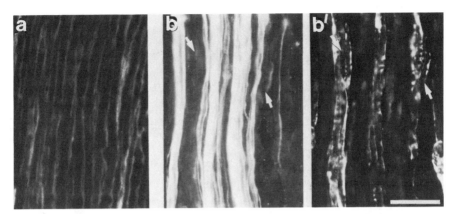

Figure 10. *Regenerated fibers in the distal stump of transected nerve.* Longitudinal sections of the distal stump at 3 and 20 days after injury were stained for anti-Ng-CAM *(a,b)* and anti S100 protein *(b')*. After 3 days there was no sprouted fiber, but endoneurial tubes were Ng-CAM positive *(a)*. After 20 days many Ng-CAM-positive fibers traversed the distal stump *(b)*. In the same section, cords of S100 protein–labeled Schwann cells *(arrows in b and b')* formed the bands of Büngner. Calibration bar = 50 μm.

To assess the changes in the target tissues after nerve compression or transection, we studied the expression of N-CAM in the gastrocnemius, soleus, and extensor digitorum longus (EDL) muscles of the mouse after various times of injury (Figure 11). At three days a significant increase in N-CAM was observed in the cytoplasm of the muscle on the myofiber surface, in interstitial spaces (Covault and Sanes, 1985; Rieger et al., 1985), and in some mononucleated cells, presumably Schwann cells. The increase evaluated by immunoblotting experiments was transitory for both crushed and cut nerves. After a crush N-CAM quickly returned to normal levels by 11 days in the fast-twitch muscles (gastrocnemius and EDL), and by three weeks in soleus, a slow-twitch muscle, as shown by immunoblotting experiments. Similar observations were made in mouse or chicken after nerve section, although control levels returned later (Daniloff et al., 1986b). Cytotactin also increased after both compression and transection and returned to normal levels with a time course similar to that of N-CAM. Areas of accumulation for cytotactin were the extracellular spaces, and particularly the regions near motor end plates (F. Rieger, K. L. Crossin, and G. M. Edelman, in preparation).

increased, compared with control levels *(A,A')*. Many Ng-CAM-positive *(B)* and N-CAM-positive *(B')* fibers traversed the compressed region to enter the distal nerve by 10 days. In contrast, very few growing fibers were observed in the distal stump of transected nerve. Calibration bar = 500 μm.

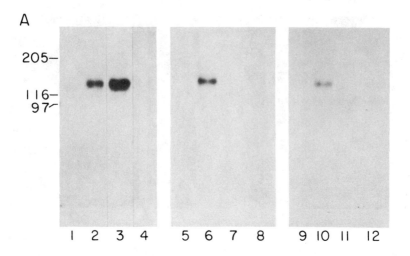

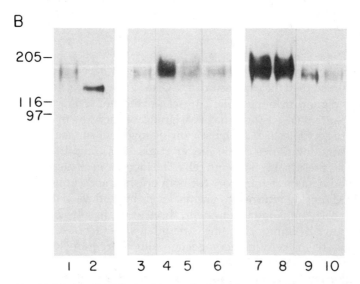

Figure 11. *Modulation of N-CAM levels in muscles from mice after crushing the sciatic nerve.* Muscles were dissected at various times after injury, extracted, resolved on SDS-PAGE (100 μg of protein), and then immunoblotted with polyclonal anti-N-CAM IgG (20 μg/ml), as described in the text. *Lanes 1–4:* soleus. *Lanes 5–8:* gastrocnemius. *Lanes 9–12:* extensor digitorum longus (EDL). Extracts from control mice (*lanes 1–5* and *9*) contained low amounts of N-CAM present as a 140-kD component. Levels of N-CAM were increased at 3 days (*lane 2*) and 7 days (*lane 3*) and decreased at 19 days (*lane 4*) in the soleus. In the gastrocnemius and EDL after a transitory increase at 3 days (*lanes 6, 10*), levels were decreased at 7 days (*lanes 7, 11*) and were back to normal at 19 days (*lanes 8, 12*).

A correlation was made between the return to normal levels of N-CAM and cytotactin and reinnervation. We performed immunoblot analyses for N-CAM and measured choline acetyltransferase (ChAT) activity as an index of the extent of reinnervation in muscle. Amounts of this enzyme correlate well with electrophysiological activity in denervated muscle and are a reliable indicator of the degree of reinnervation (A. J. Harris, F. Rieger, M. Pinçon-Raymond, and L. Houenou, unpublished data). We examined the time course of CAM expression, relating it to the extent of innervation, and also determined the levels of N-CAM in a totally denervated muscle, kept denervated by reverse suturing of the transected nerve. The changes in the levels of N-CAM in the experimental muscles were inversely related to the degree of reinnervation as assessed by ChAT activity (Table 1). We found that, after a nerve crush, the extent of reinnervation at 30 days after injury was high (65% of normal), and the amount of N-CAM was approaching control levels (1.7 times the control). Transected nerves at the same time had recovered less (40% of normal), and N-CAM levels were proportionately higher (2.5 times the control). In chronically denervated muscles the amount of N-CAM was significantly higher than that of cut nerve (4.3 times the control). These results are consistent with the idea that the amount of N-CAM expression in muscle is determined by the degree of innervation. The role of muscle activity and the precise mechanisms of regulation remain to be determined.

The changes of staining patterns in crushed or cut animals were also assessed for motor neurons of the ventral horn in spinal cord. We studied the N-CAM staining 20 days after transection on the control and lesioned side. An intense N-CAM staining was observed at the border between the cell surface of motor neurons of the ventral horn and the surrounding neuropil, forming rings around the cells. No such changes were observed in crushed animals at that time. Similar results were obtained for cytotactin (J. K. Daniloff, F. Rieger, K. L. Crossin, and G. M. Edelman, unpublished data), and double-staining experiments with anti-S100 protein antibodies suggested that Schwann cell processes were the privileged sites of cytotactin- and probably N-CAM-

Table 1. Correlation Between N-CAM Expression and Innervation in Mouse Gastrocnemius Muscles as Measured by ChAT Activity[a]

	ChAT Activity[b]	Amount of N-CAM[c]
30-day crush	1005 ± 81	100 ± 11
30-day cut	639 ± 135	177 ± 11
30-day reverse-suture	80 ± 84	427 ± 7

[a]Muscle extracts from experimental animals 30 days after injury. The two sets of data were found to be strongly correlated, using a Pearson product moment correlation coefficient.
[b]ChAT activity using a radiochemical assay (Fonnum, 1975); numbers are counts per minute.
[c]N-CAM amounts were determined by quantitative immunoblotting (Daniloff et al., 1986b). Numbers are counts per minute. Three animals were used. Standard deviations are indicated.

increased staining. There was also a general decrease in the intensity of the N-CAM fiber staining in the ventral horn.

CONCLUSIONS

The purpose of our studies has been to evaluate the expression of CAMs and cytotactin during development and during experimental or genetic perturbations of the neuromuscular system at various morphogenetic sites. The dynamic appearance and disappearance of each CAM as well as cytotactin probably reflects changes in cell adhesivity in the tissues where they appear (Edelman and Chuong, 1982; Edelman, 1983). We have detected N-CAM and Ng-CAM on both motoneurons and Schwann cells during histogenesis of the neuromuscular system. As observed in other neural systems (Thiery et al., 1985; Crossin et al., 1986), the expression of Ng-CAM and the site-restricted SAM, cytotactin, in the neuromuscular system seems to occur in association with cell or axon movements and interactive areas. In the neuromuscular system, cell–cell interactions follow a sequence of events that involve surface contacts of axons with the mesenchymal environment and peripheral glia, the Schwann cells, which are derived from the neural crest. Proliferating Schwann cells have been observed to migrate along outgrowing axons of peripheral nerves.

Thus, the restricted expression of N-CAM within specific areas of specialized cell–cell contact such as the node of Ranvier, the motor end plate, or the myotendinous junction, the previously reported involvement of Ng-CAM in neurite extension in both the CNS and PNS (Thiery et al., 1985; Daniloff et al., 1986a,b; Hoffman et al., 1986), and the presence of cytotactin in areas of cell movement such as the neural crest pathways (Crossin et al., 1986) or the Bergmann glial fibers in the developing cerebellum (Grumet et al., 1985; Fischer et al., 1986) all suggest that the modulation of CAM and cytotactin levels in the peripheral nerve and in muscle through prevalence, chemical modification, and cell polarity modulation (Edelman, 1984, 1985a) plays a critical role during histogenesis of the neuromuscular system. Such mechanisms may underlie the association of Schwann cells with axonal processes, fasciculation of small diameter axons, formation of compact myelin, maintenance of unmyelinated axolemma at the nodes of Ranvier, and synapse formation.

N-CAM, Ng-CAM, and cytotactin have several potential roles during development and regeneration. In muscle, N-CAM may participate in (1) the initial binding of motor or sensory axons to muscle fibers early in fetal development; (2) interactions with itself or other proteins to redistribute basal lamina–linked molecules on the cell surface; (3) selective strengthening of certain cell–cell bonds via the E-to-A conversion during postnatal maturation of synaptic contacts; (4) the specific reinnervation of previous synaptic sites after nerve injury; and (5) the formation of the myotendinous junctions in some association to cytotactin and maybe other molecules as well. In nerve, all three proteins alone or in combination, N-CAM, Ng-CAM, and cytotactin, may (1) modulate the migratory activity of Schwann cells during development and after injury; (2) mediate axon fasciculation and initial contacts between axons and Schwann cells and participate in axon guidance during development and regeneration; (3) specify and stabilize the areas on axons that will become nodes of Ranvier; and (4) interact with themselves or other proteins to organize or redistribute structural or functional proteins, such as the voltage-dependent sodium channel, the sodium–potassium ATPase, or the voltage-dependent potassium channel on the nodal axolemma.

ACKNOWLEDGMENTS

I am deeply indebted to my coworkers for letting me summarize part of our work in this review: Drs. K. L. Crossin, J. K. Daniloff, M. Grumet, G. Levi, M. Nicolet, M. Pinçon-Raymond, and Mrs. M. Murawsky. I wish to express my gratitude to Dr. G. M. Edelman for continuous advice and help. This work was supported by INSERM, CNRS, Fondation pour la Recherche Médicale, and A.F.M.; by United States Public Health Service grants HD-09635, HD-16550, AI-11378, and DK-04256 (to Dr. G. M. Edelman); and by an international CNRS–NSF grant (to F. R. and Dr. G. M. Edelman).

REFERENCES

Aguayo, A. J., G. M. Bray, C. M. Perkins, and D. Duncan (1979) Axon sheath cell interactions in peripheral and central system transplants. *Soc. Neurosci. Symp.* **4**:361–383.

Bennett, M. R., M. McLachlan, and R. S. Taylor (1973) Formation of synapses in reinnervated mammalian striated muscle. *J. Physiol. (Lond.)* **233**:481–500.

Bray, G. M., M. Rasminsky, and A. J. Aguayo (1981) Interactions between axons and their sheath cells. *Annu. Rev. Neurosci.* **4**:127–162.

Chuong, C.-M., and G. M. Edelman (1984) Alteration in neural cell adhesion molecules during development of different regions of the nervous system. *J. Neurosci.* **4**:2354–2368.

Covault, J., and J. Sanes (1985) Neural cell adhesion molecule (N-CAM) accumulates in denervated and paralyzed skeletal muscles. *Proc. Natl. Acad. Sci. USA* **82**:4544–4548.

Crossin, K. L., S. Hoffman, M. Grumet, J. P. Thiery, and G. M. Edelman (1986) Site-restricted expression of cytotactin during development of the chicken embryo. *J. Cell Biol.* **102**:1917–1930.

Daniloff, J. K., C.-M. Chuong, G. Levi, and G. M. Edelman (1986a) Differential distribution of cell adhesion molecules during histogenesis of the chick nervous system. *J. Neurosci.* **6**:739–758.

Daniloff, J. K., G. Levi, M. Grumet, F. Rieger, and G. M. Edelman (1986b) Altered expression of neuronal cell adhesion molecules induced by nerve injury and repair. *J. Cell Biol.* **103**:929–945.

Edelman, G. M. (1983) Cell adhesion molecules. *Science* **219**:450–457.

Edelman, G. M. (1984) Modulation of cell adhesion during induction, histogenesis, and perinatal development of the nervous system. *Annu. Rev. Neurosci.* **7**:339–377.

Edelman, G. M. (1985a) Cell adhesion and the molecular processes of morphogenesis. *Annu. Rev. Biochem.* **54**:135–169.

Edelman, G. M. (1985b) Expression of cell adhesion molecules during embryogenesis and regeneration. *Exp. Cell Res.* **161**:1–16.

Edelman, G. M., and C. M. Chuong (1982) Embryonic to adult conversion of neural cell adhesion molecules in normal and staggerer mice. *Proc. Natl. Acad. Sci. USA* **79**:7036–7040.

Fambrough, D. M. (1979) Control of acetylcholine receptors in skeletal muscle. *Physiol. Rev.* **59**:1–27.

Fertuck, H. C., and M. M. Salpeter (1976) Quantitation of junctional and extrajunctional acetylcholine receptors by electron microscope autoradiography after [^{125}I] α-bungarotoxin binding at mouse neuromuscular junctions. *J. Cell Biol.* **59**:144–158.

Fischer, G., V. Künemund, and M. Schachner (1986) Neurite outgrowth patterns in cerebellar microexplant cultures are affected by antibodies to the cell surface glycoprotein L1. *J. Neurosci.* **6**:602–612.

Fonnum, F. (1975) A rapid radiochemical method for the delineation of choline acetyltransferase. *J. Neurochem.* **24**:407–409.

Friedlander, D., M. Grumet, and G. M. Edelman (1986) Nerve growth factor enhances expression of neuron–glia cell adhesion molecule in PC12 cells. *J. Cell Biol.* **102**:413–419.

Geren, B. B. (1954) The formation from the Schwann cell surface of myelin in the peripheral nerves of chick embryos. *Exp. Cell Res.* **7**:558–562.

Goujet-Zalc, C., A. Guerci, G. Dubois, and B. Zalc (1986) Schwann cell marker defined by a monoclonal antibody (224–58) with species cross-reactivity. II. Molecular characterization of the epitope. *J. Neurochem.* **46**:435–439.

Grumet, M., U. Rutishauser, and G. M. Edelman (1982) N-CAM mediates adhesion between embryonic nerve and muscle cells *in vivo*. *Nature* **295**:693–695.

Grumet, M., S. Hoffman, C.-M. Chuong, and G. M. Edelman (1984a) Polypeptide components and binding functions of neuron–glia cell adhesion molecules. *Proc. Natl. Acad. Sci. USA* **81**:7989–7993.

Grumet, M., S. Hoffman, and G. M. Edelman (1984b) Two antigenically related cell adhesion molecules of different specificities mediate neuron–neuron and neuron–glia adhesion. *Proc. Natl. Acad. Sci. USA* **80**:267–271.

Grumet, M., S. Hoffman, K. L. Crossin, and G. M. Edelman (1985) Cytotactin, an extracellular matrix protein of neural and nonneural tissues that mediates glia–neuron interactions. *Proc. Natl. Acad. Sci. USA* **82**:8075–8079.

Guerci, A., M. Monge, A. Baron-Van-Evercoren, C. Lubetzki, S. Dancea, J. M. Boutry, C. Goujet-Zalc, and B. Zalc (1986) Schwann cell marker defined by a monoclonal antibody (224–58) with species cross-reactivity. I. Cellular localization. *J. Neurochem.* **46**:425–434.

Guth, L. (1956) Regeneration in the mammalian peripheral nervous system. *Physiol. Rev.* **36**:441–478.

Hirohawa, H., and J. E. Heuser (1982) Internal and external differentiations of the postsynaptic membranes at the neuromuscular junction. *J. Neurocytol.* **11**:487–510.

Hoffman, S., D. R. Friedlander, C.-M. Chuong, M. Grumet, and G. M. Edelman (1986) Differential contributions of Ng-CAM and N-CAM to cell adhesion in different neural regions. *J. Cell Biol.* **103**:145–158.

Huxley, A. F., and R. Stampfli (1949) Evidence for saltatory conduction in peripheral myelinated nerve fibers. *J. Physiol. (Lond.)* **108**:315–339.

Perkins, C. S., A. J. Aguayo, and G. M. Bray (1981) Schwann cell multiplication in *trembler* mice. *Neuropathol. Appl. Neurobiol.* **7**:115–126.

Rieger, F. (1985) Morphogenesis of the mouse motor end plate. In *Molecular Determinants of Animal Form*, G. M. Edelman, ed., pp. 393–421, Alan R. Liss, New York.

Rieger, F., and M. Pinçon-Raymond (1984) The node of Ranvier in neurological mutants. In *Advances in Cellular Neurobiology*, Vol. 6, S. Federoff, ed., pp. 273–310, Academic, New York.

Rieger, F., M. Grumet, and G. M. Edelman (1985) N-CAM at the vertebrate neuromuscular junction. *J. Cell Biol.* **101**:285–293.

Rieger, F., J. K. Daniloff, M. Pinçon-Raymond, K. L. Crossin, M. Grumet, and G. M. Edelman (1986) Neuronal cell adhesion molecules and cytotactin are co-localized at the node of Ranvier. *J. Cell Biol.* **103**:379–391.

Rutishauser, U., M. Grumet, and G. M. Edelman (1983) Neural cell adhesion molecule mediates initial interactions between spinal cord neurons and muscle cells in culture. *J. Cell Biol.* **97**:145–152.

Tao-Cheng, J. H., and J. Rosenbluth (1982) Development of nodal and paranodal membrane specializations in amphibian peripheral nerves. *Brain Res.* **255**:577–594.

Tasaki, I. (1939) The electro-saltatory transmission of the nerve impulse and the effect of narcosis upon the nerve fiber. *Am. J. Physiol.* **127**:211–227.

Thiery, J. P., J. L. Duband, U. Rutishauser, and G. M. Edelman (1982) Cell adhesion molecules in early embryogenesis. *Proc. Natl. Acad. Sci. USA* **79**:6737–6741.

Thiery, J. P., A. Delouvée, M. Grumet, and G. M. Edelman (1985) Initial appearance and regional distribution of the neuron–glia cell adhesion molecule in the chick embryo. *J. Cell Biol.* **100**:442–456.

Waxman, S. G., J. A. Black, and R. E. Foster (1982) Freeze-fracture heterogeneity of the axolemma of premyelinated fibers in the CNS. *Neurology* **32**:418–421.

Chapter 21

Families of Neural Cell Adhesion Molecules

MELITTA SCHACHNER
HORST ANTONICEK
THOMAS FAHRIG
ANDREAS FAISSNER
GÜNTHER FISCHER
VOLKER KÜNEMUND
RUDOLF MARTINI
ANKE MEYER
ELKE PERSOHN
ELISABETH POLLERBERG
RAINER PROBSTMEIER
KARIN SADOUL
REMY SADOUL
BERND SEILHEIMER
GAUTAM THOR

ABSTRACT

Based on the observation that out of two functionally identified cell surface glycoproteins carrying the L2/HNK-1 carbohydrate structure two were recognized as the neural cell adhesion molecules L1 and N-CAM, we formulated the hypothesis that other glycoproteins of yet unidentified number expressing this carbohydrate epitope are also involved in adhesion. In agreement with this hypothesis, we have characterized two other glycoproteins as adhesion molecules, the extracellular matrix molecule J1 and the myelin-associated glycoprotein MAG. Each of the four adhesion molecules is regulated in its expression independently from the others in time and space and mediates cell interactions between different cell types. The individual glycoproteins, however, do not always appear to act independently of each other, for example, L1 and N-CAM 180, the molecular form of N-CAM with the largest cytoplasmic domain. L1 and N-CAM 180 are associated with each other in the surface membrane; they appear late in development, and N-CAM 180 is involved in the stabilization of cell contacts, probably through its association with the cytoskeleton by means of the membrane–cytoskeleton linker protein brain spectrin. Besides the L2/HNK-1 family of adhesion molecules, another carbohydrate-based family has been found. The L3 carbohydrate epitope shared by members of this family is also expressed by some members of the L2/HNK-1 family, but not by others. The novel cell adhesion molecule on glia (AMOG)

443

belongs to the L3 family only. These observations indicate that adhesion molecules are able to express sets of carbohydrate structures in various combinations that may be important for specifying cell adhesion.

To build a complex structure such as the nervous system, a number of cellular and molecular mechanisms seem to be necessary. There is good evidence that cell surface molecules play important roles not only in specifying cell contacts between neighboring cells, but also in specifying contacts between a particular cell and its surrounding extracellular matrix. The cell surface components that mediate cell–cell interactions have operationally designed adhesion molecules, since in *in vitro* assay systems these surface components make cells adhere to each other or to a particular substrate. The questions facing us now are: How many adhesion molecules are involved in specifying cell contacts during neural development and regeneration? What are the functional and structural differences and similarities among these molecules?

One clue in our search came in the form of a monoclonal antibody designated L2, which recognizes a carbohydrate epitope that is common to several adhesion molecules (Figure 1) (Kruse et al., 1984). This carbohydrate epitope (L2 epitope) is also recognized by several other monoclonal antibodies generated against such diverse cells as human natural killer cells, neural crest cells from chicken, and crude membrane fractions from developing mouse brain (Abo and Balch, 1981; Rougon et al., 1983; Schachner et al., 1983; Kruse et al., 1984; Tucker et al., 1984).

We observed that out of two functionally characterized L2 epitope–positive molecules both L1 and N-CAM were adhesion molecules. From this we hypothesized that the other cell surface glycoproteins expressing this carbohydrate structure would also be involved in adhesion. In this chapter we describe what we have done to verify this hypothesis, further characterize the cell adhesion molecule L1, provide first hints that the cell adhesion molecules L1 and N-CAM do not act independently of each other, show that N-CAM has an "inner life" through its interaction with cytoskeletal components, summarize what we know about the structure–function relationships of the L2/HNK-1 carbohydrate epitope, characterize adhesion molecule expression in the peripheral nervous system during development and under regenerative conditions, and describe a novel cell adhesion molecule on glia (AMOG), which is a member of a newly discovered carbohydrate-based family of adhesion molecules, the L3 family.

THE MYELIN-ASSOCIATED GLYCOPROTEIN MAG: AN ADHESION MOLECULE

The myelin-associated glycoprotein MAG is a minor constituent of central and peripheral nervous system myelin sheaths (for a review, see Quarles, 1984). Because of its particular localization in periaxonal membranes, it has been

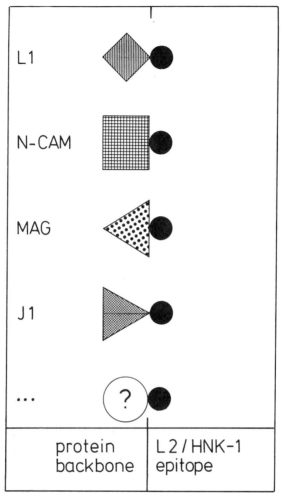

Figure 1. *Schematic representation of neural adhesion molecules' expression of the L2/HNK-1 carbohydrate structure.*

implicated in neuron–myelinating cell interactions (Sternberger et al., 1979; Trapp and Quarles, 1982). However, direct experimental evidence was missing to prove this point. On the basis of our evidence that MAG belongs to the L2/HNK-1 family, which was, at the time, found to include the cell adhesion molecules L1 and N-CAM, we suggested that MAG was also involved in cell adhesion (Kruse et al., 1984). To prove that MAG was indeed a cell adhesion molecule, it was necessary to isolate it from membranes under mild conditions, so as to obtain it in a functionally active form, and to prepare antibodies that would be useful in modifying the molecule in *in vitro* assay systems designed to probe for cell surface interactions.

Toward this goal, monoclonal antibodies were prepared in mice immunized with L2 epitope–carrying glycoproteins from chicken brain (Poltorak et al., 1987). One of the antibodies obtained recognized MAG from mouse, bovine, human, rat, chicken, and frog. The antibody reacted with the cell surface of oligodendrocytes, but not with neurons, astrocytes, or fibroblasts, as could be seen by double immunolabeling with antibodies to established cell type–specific markers. In histological sections MAG was detectable only in white matter tracts, as had been described previously, by using an antibody against MAG isolated by the lithium diiodosalicylate-phenol method. Interestingly, only approximately 30% of all MAG-positive oligodendrocytes expressed the L2/HNK-1 epitope in culture, suggesting a heterogeneity in expression of this carbohydrate moiety on MAG. This heterogeneity in L2/HNK-1 expression was observed previously for two other members of the L2/HNK-1 family, N-CAM and L1 (Kruse et al., 1984; Faissner, 1987).

To investigate whether MAG is a cell adhesion molecule, the adhesion of single-cell suspensions of small neurons from early postnatal mouse cerebellum and astrocytes and oligodendrocytes from rat cerebrum to monolayers of oligodendrocytes was measured under calcium-free conditions (Keilhauer et al., 1985) in the absence and presence of Fab fragments of monoclonal MAG antibodies. The antibodies inhibited oligodendrocyte–neuron and oligo-dendrocyte–oligodendrocyte but not oligodendrocyte–astrocyte interaction (Poltorak et al., 1987). Inhibition values were small but significantly different from control values, suggesting that oligodendrocytes display a heterogeneity in adhesion mechanisms.

Control antibodies prepared against mouse liver membranes, which reacted strongly by indirect immunofluorescence with all three cell types, did not interfere with adhesion. These experiments suggest that MAG is involved in cell adhesion among select types of neural cells. However, antibodies that bind to MAG on the cell surface may not only block MAG, but could also sterically conceal closely associated molecules that may be responsible for adhesion.

To investigate whether MAG itself is the binding ligand, MAG was isolated from nonionic detergent lysates of a crude membrane fraction of adult mouse brain by immunoaffinity purification, using the MAG monoclonal antibody column. MAG was then incorporated into phosphatidylcholine- and choles-terol-containing liposomes labeled with fluorescein (Sadoul et al., 1983). These MAG-containing liposomes were added to monolayer cultures prepared from dorsal root ganglia, spinal cord, and cerebellum. MAG liposomes specifically attached to neurites and neurite bundles, but not to the underlying Schwann cells or astrocytes. Binding of MAG liposomes to neurites was also observed in cultures of early postnatal mouse cerebellum, which consist of up to 90% of small neurons that are never contacted by myelinating oligodendro-cytes in situ. Binding was not detectable after short culture periods but was seen after 12 days, although always less pronounced than binding to spinal cord or dorsal root ganglion neurons. Binding of MAG-containing liposomes to neurites was inhibited by Fab fragments of mono- and polyclonal MAG antibodies. These experiments indicate that MAG is itself the binding ligand in adhesion and that neurons express different levels of the receptor(s) for MAG.

In addition to mediating binding between neurons and oligodendrocytes, MAG appears to be involved also in oligodendrocyte–oligodendrocyte adhesion. This indicates that MAG plays a role in the self-recognition of myelinating cell surface membranes, a striking feature of myelin formation involving the apposition of spiraling loops of cellular processes. That MAG may indeed be involved in the process of myelination is suggested by observations on the localization of MAG in the developing sciatic nerve (Martini and Schachner, 1986).

Unlike L1 and N-CAM, MAG was never found on nonmyelinating Schwann cells. On myelinating Schwann cells, MAG expression was seen when L1 and N-CAM expression ceased to be detectable and Schwann cell processes had turned approximately 1.5 to 2 loops around the axon. It was then that MAG became detectable periaxonally and was associated with the turning loops of Schwann cell processes. MAG disappeared from the turning loops once the compaction of myelin began, coinciding with the expression of myelin basic protein. In the mature myelin, MAG was never found in compact parts, but remained detectable in the uncompacted parts: outer and inner mesaxons, paranodal loops, and Schmidt-Lanterman incisures. The contention that MAG is also present in compacted myelin lamella (Webster et al., 1983) has been disputed (Trapp and Quarles, 1984) and can now be ruled out on the basis of our experiments using postembedding staining procedures that avoid antibody penetration problems.

The mechanisms by which MAG may mediate oligodendrocyte–oligo-dendrocyte, Schwann cell–Schwann cell, and myelinating cell–axon binding remain to be investigated. Whether MAG binds to itself by a self-binding mechanism that has been suggested to exist for members of the immuno-globulin superfamily, which also includes MAG and N-CAM (Williams, 1982; Hemperly et al., 1986; Arquint et al., 1987; Lai et al., 1987; Salzer et al., 1987), is at present unresolved. Also, whether MAG serves as a ligand in binding to other molecules at the cell surface will have to be investigated.

It is interesting that although MAG expresses the RGD sequence of the cell-binding domain of several matrix molecules (Ruoslahti and Pierschbacher, 1986), binding of MAG to neurites or to extracellular matrix constituents such as collagens could not be inhibited by an RGD sequence containing hexa-peptide (Fahrig et al., 1987; R. Sadoul, unpublished data). Whether an altered form of MAG in *trembler* (Inuzuka et al., 1985) and in *quaking* (Matthieu et al., 1974) mutant mice is related to a defect in cell adhesion between axon and glia or between myelinating processes remains to be investigated. Also, the function of the two forms of MAG that are developmentally regulated at different stages of cell interactions will be an important topic for further investigations. The elucidation of the cellular and molecular signals that underlie the regulation of MAG expression will yield important insights into the mechanisms of the complex process of myelination.

THE EXTRACELLULAR MATRIX ADHESION MOLECULE J1

A major glycoprotein of the L2/HNK-1 family in adult mouse brain is a 160-kD glycoprotein. To further verify our hypothesis that members of the L2/HNK-1 family are adhesion molecules, we isolated the 160-kD component and analyzed it with regard to its adhesive properties (Kruse et al., 1985). Antibodies were prepared against the 160-kD component from adult mouse brain and were isolated on L2 monoclonal antibody columns after removal of L1 and N-CAM by immunoaffinity chromatography; subsequent gel filtration yielded the J1 antigen (Kruse et al., 1985). Polyclonal antibodies prepared against this component did not react with L1, N-CAM, or MAG. During cerebellar development, J1 was detectable in Western blots of a crude membrane fraction at embryonic day 15 in the form of two faint bands with apparent molecular weights of 220 and 200 kD. With increasing age, immuno-reactivity shifted from the 220-kD and 200-kD components toward the 160-kD and a 180-kD band. In the adult mouse brain the 220-kD and 200-kD components were no longer detectable. All components of the J1 antigen are soluble in detergent-free buffers, since they were found to be secreted from cultured cells and could be recovered from soluble supernatants of a brain homogenate. Astrocytes, oligodendrocytes, and fibronectin-positive fibro-blastlike cells, but not neurons, synthesize J1. Radioimmunoassays show that the antigen is distinct from laminin, collagen type IV, fibronectin, nidogen (entactin), or a heparan sulfate proteoglycan (Kruse et al., 1984).

J1 mediates neuron–astrocyte adhesion, but not astrocyte–astrocyte or neuron–neuron adhesion. Neuron–oligodendrocyte adhesion has not been measured. J1 is also secreted by Schwann cells (Seilheimer and Schachner, 1987) and appears to mediate Schwann cell–neuron adhesion (B. Seilheimer, unpublished data). Since the J1 antigen is synthesized by astrocytes and is involved in neuron–astrocyte adhesion, it was interesting to investigate whether J1 was involved in cerebellar granule cell migration (Mugnaini and Forstrønen, 1967; Rakic, 1971). Antibodies added to an *in vitro* assay system designed to quantify the migration of cerebellar granule cell neurons (Lindner et al., 1983) did not show an inhibition of neuronal migration (Antonicek et al., 1987).

To find other functional roles of J1, we analyzed the expression of J1 by immunohistochemical methods in the central and peripheral nervous systems in the hope of being able to make inferences about its function. A striking localization was, indeed, observed in the adult rat optic nerve, where the nodes of Ranvier were preferentially labeled by J1, HNK-1, and L2 antibodies (ffrench-Constant et al., 1986). Immunoelectron microscopy showed J1 and the carbohydrate epitope to be concentrated around paranodal astrocytic processes. It is interesting that these astrocytes, designated type II (ffrench-Constant and Raff, 1986), associated with the J1 antigen and L2/HNK-1 epitope just at the site of interdigitation into the paranodal region, where the axon is not covered by myelin and lies free to be contacted by the astrocytic process. Other parts of the type II astrocyte did not show immunoreactivity, indicating a high degree of topographically selective regulation of antigen expression. Since J1 is a secreted molecule that is also expressed by oligodendrocytes, conclusions about its site of synthesis and deposition are at present only tentative. However, at least its receptor(s) should show this specialized accumulation at the interfaces between oligodendrocyte, astrocyte, and axon.

The observation that J1 and the L2/HNK-1 carbohydrate epitope are concentrated at the node of Ranvier has potentially important functional implications in that J1 is involved in neuron–astrocyte adhesion *in vitro*. Taken together, these *in vivo* and *in vitro* observations raise the possibility that J1 plays an important part in the neuron–astrocyte interactions that are presumably involved in the assembly and/or maintenance of the exquisite cytoarchitecture at the node of Ranvier.

The neuromuscular junction appears to be yet another site where J1 expression is selectively regulated, depending on the state of neuron–muscle interaction (for a review, see Sanes and Covault, 1985). To place the expression of J1 at the neuromuscular junction into perspective, a summary of the cellular and molecular events leading to synapse formation is warranted. During development, embryonic myotubes readily accept synapses, but once innervation has occurred, axons can no longer form new synapses. However, adult muscles regain their susceptibility to synapse formation if they are denervated or paralyzed. Upon innervation, these denervated muscles again become refractory to synapse formation. Thus, muscles regulate their susceptibility to synapse formation depending on their state of innervation or activity. Therefore, interest has focused on the means by which muscles inform nerves of their ability to be innervated. In addition, the denervated muscle directs the regrowth of axons to the original site of innervation, even when they are growing outside of preexisting pathways.

The expression of J1 in developing rat muscle was detectable in the form of small, discrete deposits occupying spaces between myotubes (Sanes et al., 1986). These deposits were observed in interstitial spaces between muscle cells from embryonic day 13 until birth, but were sparse at all times and disappeared completely after birth. After the denervation of adult normal muscles, however, J1 accumulated in denervated muscle. In the diaphragm, the accumulation of J1 was already quite considerable two days after denervation, the earliest time examined, and remained strongly detectable in muscles kept denervated for up to two months. J1 accumulated near denervated synaptic sites, but the extent of this accumulation varied with time after denervation. J1 was least accumulated and most widely distributed at early times after denervation (two to four days) and became focused at the original synaptic sites when denervation was maintained for one to two months. This accumulation was seen not only in the diaphragm, but also in other muscles. Accumulation of J1 in synaptic regions is thus a general response of skeletal muscle to denervation in the rat.

The induction of J1 expression was also seen, after the paralysis of lower leg muscles, by abolishing nerve activity with tetrodotoxin. Interestingly, J1 was then not obviously associated with synaptic sites. Thus, the accumulation of J1 results at least in part from the inactivity of either nerve or muscle, and does not depend on degeneration of axons. A similar, activity-dependent accumulation of acetylcholine receptors and N-CAM in denervated muscle has been observed previously (Covault and Sanes, 1985; Rieger et al., 1985). When muscles were reinnervated, J1 disappeared from the muscle soon after reinnervation was complete. Thus, axonal regeneration and/or muscle reinnervation reverses the denervation-induced accumulation of J1.

J1 in denervated muscle was found to be concentrated in interstitial spaces between muscle fibers, external to the postsynaptic membrane and basal lamina. J1 immunoreactivity was localized along large collagen fibers composed mainly of type I collagen and smaller, collagen-associated fibrils. Basal lamina and cell membranes were lightly immunoreactive. Thus, J1 is associated with the extracellular matrix and resembles in this respect other extracellular matrix–associated secreted glycoproteins, such as fibronectin, laminin, and a heparan sulfate proteoglycan (Sanes et al., 1986). Local synthesis, for instance by Schwann cells or synaptic regions of muscle fibers, could explain the restriction of J1 to synaptic areas. On the other hand, extracellular fibrils might differ in synaptic and extrasynaptic areas, resulting in J1 binding preferentially to synapse-associated components of the matrix.

The accumulation of J1 at original synaptic sites is intriguing in view of the observation that the regenerating axons preferentially reinnervate at these original synaptic sites. The accumulation of J1 is more selective than the localization of N-CAM, fibronectin, or a heparan sulfate proteoglycan and may therefore be a more useful marker for the regrowing axon. Regenerating axons may bear a neuronal receptor for J1 and interact through it with J1 during axon regrowth.

THE ADHESION MOLECULE L1

The observation that one and the same cell can express more than one cell adhesion molecule at a particular differentiable stage and at topographically important sites on the cell surface (e.g., axon versus dendrite) begs the question of why a cell permits itself the luxury of expressing more than one adhesion molecule. One could argue that by varying the steady-state levels of a particular cell adhesion molecule at the cell surface and by its strategic deposition at topographically important sites, one adhesion molecule would suffice to specify cell surface contacts. Before we describe the first hints of evidence that L1 and N-CAM may not act independently of each other within the surface membrane, thus emphasizing the principle of cooperativity, we present a short summary of our present state of knowledge of the cell adhesion molecule L1. (For N-CAM, see reviews in this volume.)

The cell surface glycoprotein L1 is, like N-CAM, involved in a calcium-independent adhesion mechanism among neural cells. During development of the mouse central nervous system, L1 appears after and is coexpressed with N-CAM on postmitotic neurons (for a previous review, see Schachner et al., 1985). L1, in contrast to N-CAM, displays a previously unrecognized restricted expression by particular neuronal cell types in that it is found, for instance, in the mouse cerebellar cortex on granule cells, but not on stellate or basket cells (Persohn and Schachner, 1987). Also, L1 is expressed only at certain topographicaly distinct subcellular compartments on the cell surface; for example, it is expressed on axons, but not on dendrites or cell bodies of Purkinje and granule cell neurons (Persohn and Schachner, 1987). L1 is expressed on fasciculating axons of granule cells and a set of neurites in the spinal cord that express L1 only in their fasciculative state, not when nonfasciculating (Holley and Schachner, 1987). Likewise, fasciculating axons in the sciatic nerve express L1, but myelinated axons do not (Martini and Schachner, 1986). It is likely that the prominent expression of L1 on fasciculating axons constitutes the molecular basis for the maintenance of the fasciculative state in adulthood.

L1 also appears to be involved in cerebellar granule cell migration, since L1 antibodies reduce the dislocation of thymidine-labeled granule cell bodies from the outer to the inner granular layer (Lindner et al., 1983). The observation that a prominent feature of granule cell migration is the apposition of the migrating cell body and its leading and trailing processes with Bergmann glial processes (Mugnaini and Forstrønen, 1967; Rakic, 1971) prompted the idea that L1 may be involved in neuron–glia adhesion. However, L1 antibodies have never been observed to interfere with neuron–glia adhesion, and only affect neuron–neuron adhesion *in vitro* (Keilhauer et al., 1985). In recent immunohistochemical experiments we have obtained evidence supporting the contention that L1 may not be involved in neuron–glia adhesion during neuronal migration in the cerebellar cortex and the telencephalic anlage (Fushiki and Schachner, 1986; Persohn and Schachner, 1987).

With regard to the migrating granule cell neuron, L1 was always found to be confined to contact sites between apposing neuronal surface membranes (Persohn and Schachner, 1987; see also Martini and Schachner, 1986) and excluded from contact sites between an L1-positive surface membrane and an apposing L1-negative one (that is, a Bergmann glial cell process). Particularly pertinent with regard to granule cell migration, therefore, is the absence of L1 at contact sites between neurons and glia and its striking presence at neuron–neuron interaction sites between granule cells (Persohn and Schachner, 1987) migrating in tandem packages along Bergmann glial processes (Altman, 1976). This raises questions about the underlying molecular mechanisms leading to the polarization of adhesion molecules in the surface membrane of one cell.

Also, in the telecephalic anlage, L1 may not be involved in the initiation of the migration of neurons on radial glial fibers, since it is not detectable at the onset of migration, but only when postmigratory neurons cluster to aggregate their cell bodies in the final stages of their migratory pathway (Fushiki and Schachner, 1986).

These observations are in contrast to those concerning Ng-CAM, which have been shown to be immunochemically identical to NILE (Friedlander et al., 1986), which, in turn, has been shown to be identical to L1 (Bock et al., 1985). Ng-CAM is thought to mediate adhesion between neurons and neurons as well as between neurons and glia (Grumet et al., 1984a,b; Hoffman et al., 1986). Thus, our present data support rather than contradict the contention that the inhibitory effect of L1 antibodies on granule cell migration *in vitro* is due to the disturbance of the tandem alignment of migrating neurons, fasciculation of granule cell axons before the cell bodies enter migration, outgrowth of growth cones on axons (Chang et al., 1987), or sorting out of postmitotic, premigratory granule cells from mitotic ones in the external granular layer. Furthermore, it is tempting to speculate that L1 on one cell may recognize and interact with L1 or N-CAM on the partner cell (see also Martini and Schachner, 1986). We have, however, not been able to show a specific and controlled interaction of purified L1 with itself, although L1 has been observed to adhere unusually tightly and in a specific antibody-inhibitable manner to itself even in the presence of higher concentrations of detergents and when inserted into liposomes (Faissner et al., 1985; Sadoul et al., 1988).

Finally, it should be mentioned that L1 is expressed in another tissue where cell migration occurs (Thor et al., 1987)—the intestinal tract, where L1 was localized in the proliferating epithelial progenitor cells of crypts, but not in the more differentiated epithelial cells of villi, showing for the first time that L1 is not restricted to only the nervous tissue in the mouse. L1 was detected in crypt cells in the molecular forms characteristic of peripheral neural cells, with apparent molecular weights of 230 kD, 180 kD, and 150 kD. The aggregation of single, enriched crypt, but not villus, cells was strongly inhibited in the presence of Fab fragments of polyclonal L1 antibodies, suggesting that L1 may play a role in the histogenesis of the intestine in the adult animal. It is noteworthy that epithelial cells in the intestine undergo a constant renewal in that they are generated in the depth of the crypts and migrate from the crypts to the villi, to be constantly shed from the tips of the villi into the intestinal lumen. It is, therefore, interesting that L1 is expressed during early phases of this epithelial cell migration, but that later, when cells prepare for shedding, expression of L1 is no longer detectable.

L1 AND N-CAM INTERDEPENDENCE IN FUNCTION

That L1 and N-CAM may not act independently of each other in adhesion has become evident in two different functional assay systems *in vitro*. When the aggregation of cell bodies in single-cell suspensions was measured at 37°C, L1 and N-CAM appeared to act synergistically with each other in the aggregation of neuroblastoma and early postnatal cerebellar cells (Faissner et al., 1984; Rathjen and Rutishauser, 1984). When mediating adhesion between single-cell suspensions and substrate-attached monolayer cells at room temperature, L1 and N-CAM antibodies blocked each other in their action, such that in the presence of the two antibodies at saturating blocking levels, their effect in the inhibition of adhesion was less than additive (Keilhauer et al., 1985). It has been suggested that this interdependence of the molecules' action within the surface membrane results from interactions between the two molecules— either between neighboring partner cells or within the plasma membrane of one cell (Schachner et al., 1985). The two possibilities were therefore investigated.

To probe the possibility that L1 may bind to N-CAM in a ligand–receptor relationship, the purified molecules were allowed to interact with each other in detergent-solubilized form in several different binding assays (R. Probstmeier and G. Thor, unpublished data). These experiments showed that L1 bound to N-CAM in an L1 and N-CAM antibody–inhibitable manner. Likewise, N-CAM bound to N-CAM and L1 bound to L1. However, when other membrane glycoproteins were introduced as control binding partners, such as the acetylcholine receptor, sodium/potassium–dependent ATPase, and calcium-dependent ATPase (which are likely not the physiological binding partners for L1 and N-CAM), binding of L1 and N-CAM, even in a specific antibody-inhibitable fashion, was also seen. These observations should caution one against making interpretations about self-binding mechanisms with presently available techniques.

To investigate cooperativity between the two adhesion molecules within the surface membrane of one cell, the association of L1 and N-CAM was assayed by antibody-induced redistribution on the surface of live cells (Thor et al., 1986). We thus used a method designed to study molecular associations between surface molecules on lymphoid cells, namely the antibody-induced patching and capping of cell surface antigens (Taylor et al., 1971). Would L1 antibody–induced patching or the capping of L1 on the cell surface redistribute N-CAM with it, or would the two molecules behave independently of each other? In other words, would a patch or cap of L1 contain N-CAM or not? When patching was carried out on cultured neuroblastoma cells with N-CAM antibodies and subsequent double-label immunofluorescence was performed with L1 antibodies, complete colocalization was seen. When patching was carried out with L1 antibodies and redistribution was assayed with N-CAM antibodies, N-CAM was seen not only in the L1 antigen–containing patches, but also outside of the patches in a more uniform cell surface labeling. Polyspecific antibodies to mouse liver membranes recognizing a 130-kD component on neuroblastoma cells (Pollerberg et al., 1986) did not co-redistribute with either of the two antibodies, indicating that not all molecules in the surface membrane are associated with one another. Interestingly, after patching with L1 antibodies and monitoring co-redistribution with monoclonal N-CAM 180 antibodies (Pollerberg et al., 1985), colocalization was seen. These observations show that there is a differential molecular association between L1 and N-CAM 180. A functional interdependence between the two molecules may depend on cooperativity in the recognition phenomenon, but also in the subsequent stabilization of initial contacts.

Before further probing into the functional properties of N-CAM 180 that distinguish it from the two other components, N-CAM 140 and N-CAM 120, we should describe some observations on the expression of L1 and N-CAM *in vivo* in two morphogenetically important events: axon fasciculation and neuron migration. In the developing spinal cord, a set of neurons displays an exquisite fine-tuning of L1 and N-CAM expression on their axons. These neurons have their cell bodies in the dorsolateral part of the spinal cord and extend their axons individually through a meshwork of nonneuronal neuro-epithelial processes. The individual axons form bundles in the ventrolateral region, where they change their direction to 90° to run up or down the spinal cord. Axons are N-CAM-positive but not L1-positive in their nonfasciculative state, but become L1 positive in the ventrolateral tract in addition to expressing N-CAM. Thus, one and the same axon can regulate the expression of L1 along the length of its surface, and the coexpression of L1 with N-CAM is correlated with fasciculation (J. A. Holley and M. Schachner, 1987). Also, in the cere-bellum, L1 and N-CAM are coexpressed on fasciculating granule cell axons (Persohn and Schachner, 1987). A functional correlate of these observations *in situ* is the involvement of L1 and N-CAM in fasciculation *in vitro*, where not only polyclonal but also monoclonal L1 and N-CAM antibodies interfere with fasciculation (Fischer et al., 1986).

Both L1 and N-CAM are also involved in cerebellar granule neuron migration. Antibodies to L1 and N-CAM could be shown to interfere with migration, although the interference by L1 antibodies was larger by a factor of three when compared to N-CAM antibodies (Lindner et al., 1986a). Interest-ingly, the effect of both antibodies together was less than additive in that the effects of the L1 antibodies were no longer apparent, and overall inhibitory activity was as low as that seen by N-CAM antibodies alone. N-CAM was always expressed together with L1 on fasciculating axons and on cell bodies of postmitotic, premigratory, and migrating granule cells at sites of neuron–neuron contact. In contrast to L1, however, N-CAM was also seen at contact sites between neuron and glia (Persohn and Schachner, 1987). While N-CAM antibodies reacting with the three major components of N-CAM showed a rather uniform labeling of all cell types during granule cell migration, antibodies to N-CAM 180 stained only the postmigratory granule cell bodies, supporting the notion that N-CAM 180 is not expressed before stable cell contacts are formed. A prevalent expression of N-CAM 180 in the internal granular layer was also found by biochemical analysis (Nagata and Schachner, 1986).

It should also be mentioned that granule cell migration does not directly depend on the conversion of the embryonic to the adult form of N-CAM (Nagata and Schachner, 1986), which would shift N-CAM from a less to a more adhesive form (Hoffman and Edelman, 1983; Sadoul et al., 1983) because of the loss of the negative charges of polysialic acid. Thus, the embryonic form of N-CAM is expressed before, during, and shortly after migration (Hekmat, unpublished data). These observations are in agreement with the finding that neuraminidase does not inhibit granule cell migration in a cerebellar explant culture system (Lindner et al., 1986a). It is possible that proteases are involved in allowing granule cell migration to occur, and it is interesting that a glia-derived protease inhibitor reduces granule cell migration (Lindner et al., 1986b). However, our hope that this protease inhibitor would block proteolysis of the higher molecular weight component of L1 (L1–200) to L1–140 has not been substantiated (J. Lindner, unpublished data). Thus, the involvement of proteases in clipping off adhesion molecules from the cell surface to mediate de-adhesion remains to be investigated. Such de-adhesive mechanisms are likely to play important roles in a dynamic process of cell interactions such as granule cell migration.

THE "INNER LIVES" OF INDIVIDUAL N-CAM COMPONENTS

The three components of N-CAM, N-CAM 180, N-CAM 140, and N-CAM 120, are supposed to have identical extracellular domains, but differ in the length of their intracellular domains (Cunningham et al., 1983; Gennarini et al., 1984b). The individual forms of N-CAM arise from apparently one gene by alternative splicing mechanisms (see Cunningham and Edelman, and Walsh et al., this volume). We are beginning to have first hints about the functional role of these individual forms inside the cell.

N-CAM 180 is expressed late in the differentiation of a particular neuron, not only in the cerebellum, but also in the telencephalon and retina (Pollerberg et al., 1985). In contrast to the other two N-CAM components, N-CAM 180 is not present in the proliferating neuronal precursors, but becomes detectable only at later stages. These observations *in situ* are supported by experiments *in vitro*, in which neuroblastoma cells, promoted to differentiate with dimethyl-sulfoxide or laminin, shifted N-CAM expression toward a predominance of N-CAM 180 (Pollerberg et al., 1985, 1986).

N-CAM 180 was also found to accumulate specifically at contact sites between neighboring cells, whereas N-CAM 140 was more uniformly distributed over the whole cell body (Pollerberg et al., 1985). The cytoskeleton membrane linker protein brain spectrin (fodrin), actin, and interestingly, L1, also showed this accumulation, whereas other cytoskeleton-associated proteins tested— neurofilament, α-actinin, filamin, vinculin, β-tubulin, ankyrin, band 4.1, and synapsin I—did not (Pollerberg et al., 1987). Brain spectrin could be shown to bind specifically to N-CAM 180, but not to N-CAM 140 or N-CAM 120 (Pollerberg et al., 1986, 1987). Growth cones of cultured neuroblastoma cells express only N-CAM 180 when in contact with a target cell. Conversely, N-CAM 140 was more apparent on growth cones when they were uncontacted and free, "searching" (Pollerberg et al., 1987). N-CAM 180 showed a reduced lateral mobility in the surface membrane and thus may be the N-CAM component ideally suited to stabilize cell contacts (Pollerberg et al., 1986).

These data raise a question concerning the extracellular or intracellular signals that underlie the accumulation of particular forms of N-CAM at particular contact sites. The first stages of the morphological differentiation of neuroblastoma cells that was rapidly induced on laminin as substrate did not coincide with the increased synthesis of N-CAM 180. Morphological differentiation per se thus does not correlate with restriction in surface mobility of N-CAM 180. With the stabilization of the differentiated phenotype, however, N-CAM 180 becomes expressed and could then mediate the association with the cytoskeleton. The stabilization of cell contacts may be achieved by conformational modifications in intracellular protein domains triggered by conformational changes in extracellular domains, or vice versa. The restriction in mobility of N-CAM 180 within the surface membrane could be induced by the prior accumulation of the cytoskeleton at contact sites between neighboring cells. Alternatively, N-CAM 180 could be accumulated by L1 antigen or other cell surface components within the surface membrane of the same or a neighboring cell or by the extracellular matrix. In any case, the restriction of lateral mobility and association with the cytoskeleton could induce the accumulation of N-CAM 180 at contact sites to stabilize cell–cell contacts selectively.

N-CAM 120, the smallest form of N-CAM lacking a transmembrane domain, was found to be anchored into the surface membrane by phosphatidylinositol (He et al., 1986; Sadoul et al., 1986). Both the adult and embryonic forms of N-CAM showed this type of membrane anchoring (Sadoul et al., 1986). These observations point to a specific release mechanism that could be used by astrocytes and neurons for the regulation of adhesion between cells. The release of N-CAM from the surface by an endogenous phosphatidylinositol phospholipase C could lead to de-adhesion, by either the removal of a surface ligand resulting in reduction of binding affinity between cells, or competition by the soluble ligand that decreases the cell-associated N-CAM-mediated binding. In this regard, it is relevant that N-CAM 120 (and not N-CAM 140 or N-CAM 180) has been observed in both a membrane-bound and a soluble form (Gennarini et al., 1984a). One consequence of the removal of N-CAM by an endogenous phosphatidylinositol-specific phospholipase C could be the production of 1,2-diacylglycerol and activation of protein kinase C, if diacylglycerol moved to the cytoplasmic side of the surface membrane (Cross, 1987; Low, 1987). Thus, N-CAM expression and intracellular metabolic processes could be coordinately regulated by this mechanism and provide yet another route to modify cell adhesion. This expected complexity in the physiological roles of N-CAM 120 remains a topic for further investigations.

REGENERATION

Of paramount importance for the restitution of neuron–target interactions in the adult mammalian nervous system is the question of the functional roles of cell adhesion molecules in neurite outgrowth. Neurons of both peripheral and central nervous system origin show the capacity for functional regrowth, but it is the unique feature of Schwann cells, continued into adulthood that supports the regrowth and regeneration of transected axons in mammals (Tello, as cited in Ramón y Cajal, 1928; Aguayo, 1985). In contrast to the peripheral glia, central nervous system glial cells do not support regrowth, while less differentiated glia do favor regrowth (Smith et al., 1986). In lower vertebrates axon regrowth in the central nervous system takes place even in adulthood. These observations have led us to search for the cellular and molecular signals that regulate adhesion molecule expression on Schwann cells during development and regeneration.

In our first study we looked at the expression of the neural cell adhesion molecules L1 and N-CAM and their shared carbohydrate epitope L2/HNK-1 during development and after the transection of sciatic nerve from adult mice. During development, L1 and N-CAM were detectable on all Schwann cells at embryonic day 17 and postnatal day 0, the earliest stages tested (Nieke and Schachner, 1985; Martini and Schachner, 1986). L1 and N-CAM showed a similar staining pattern. Both were localized on small, nonmyelinated, fasciculating axons and on axons ensheathed by nonmyelinating Schwann cells. Schwann cells were also positive for L1 and N-CAM in their nonmyelinating state (Mirsky et al., 1986; Sanes et al., 1986) and at the onset of myelination, when the Schwann cell processes had turned 1.5 to 2 loops around the axon. Thereafter, neither axon nor Schwann cell could be detected expressing the L1 antigen, whereas N-CAM was found periaxonally and, more weakly, compact myelin of myelinated fibers. Compact myelin, Schmidt-Lanterman incisures, paranodal loops, and fingerlike processes of Schwann cells at nodes of Ranvier were L1 negative. Even at nodes of Ranvier, the axolemma was also always L1-negative and N-CAM-negative. The L2/HNK-1 carbohydrate epitope coincided in its cellular and subcellular localization most closely to that observed for L1.

These studies show that axon–Schwann cell interactions are characterized by the sequential appearance of cell adhesion molecules apparently coordinated in time and space. From this sequence it may be deduced that L1 and N-CAM are involved in axon fasciculation, initial axon–Schwann cell interaction, and the onset of myelination. In contrast to L1, N-CAM may be further involved in the maintenance of compact myelin and the axon–myelin apposition of larger diameter axons.

During regeneration, a characteristic re-expression of adhesion molecules was reminiscent of development. Within two days after transection, when the first signs of degeneration were seen in the distal and also, to a lesser extent, in the distal part of the proximal nerve stumps, L1 and N-CAM appeared again on Schwann cells. Toward the end of the second week after transection, the two nerve stumps had grown together, with the distal end of the nerve still being more L1 and N-CAM positive than the proximal stump, and the Schwann cells oriented longitudinally and in parallel to form the so-called bands of Büngner. As during development, fasciculating axons and Schwann cells in contact with each other or with axons were L1 and N-CAM antigen–positive and remained so, unless myelination occurred. As in development, L1 disappeared from both axon and Schwann cell, and N-CAM was reduced when the Schwann cell loops had turned approximately 1.5 times around the axon. From there on, the temporal and spatial sequence of adhesion molecule expression, including MAG, was the same as during development.

The observation that cell adhesion molecules exhibit a certain degree of plasticity in expression during regeneration requires validation of their functional significance. It is likely that surface-mediated interactions between Schwann cells and axons and between Schwann cells themselves at the level of the mesaxon are important steps in cell recognition during axon outgrowth and myelination. During development, this interaction may be needed for the apposition of Schwann cells to axons and the initiation of myelination, while during regeneration, the L1-positive and N-CAM-positive Schwann cells may also guide the outgrowing axons. It is possible that the re-expression of the two cell adhesion molecules on Schwann cells during regeneration is a prerequisite for the successful regrowth of axons, and that the capacity for regeneration in the peripheral nervous system may be attributed to the re-expression of these adhesion molecules. Whether they indeed play an important role in the initiation of Schwann cell–axon interactions during development and regeneration will have to be studied by testing the ability of the antibodies to interfere with these contacts *in vivo* and *in vitro*.

With regard to neuron–glia interactions in the central nervous system, it is interesting that N-CAM expression in the central nervous system is restricted to a subclass of astrocytes and oligodendrocytes, many of which are developmentally immature (Keilhauer et al., 1985). Furthermore, it has been suggested that neuroepithelial end-feet guide growing axons by expressing N-CAM (Silver and Rutishauer, 1984). Whether astrocytes and oligodendrocytes are plastic with regard to N-CAM expression after the occurrence of lesions in mammals seems a worthwhile topic of study in view of the absence of regenerative potential.

The cellular and molecular signals underlying the re-expression of adhesion molecules on Schwann cells will need to be investigated to understand the remarkable plasticity of Schwann cells in rejuvenating after the incurrence of a lesion and in recapitulating development in many ways. As a first step in this direction, we studied the influence of nerve growth factor (NGF) on adhesion molecule expression by Schwann cells in culture. These experiments were instigated by the observation that upon denervation, NGF appears to be detectable in Schwann cells (Rush, 1984). Also, developing Schwann cells express NGF receptors (Rohrer and Sommer, 1983; Rohrer, 1985), and these are re-expressed *in vivo* after nerve transection and axon degeneration (Taniuchi et al., 1986). Based on these observations and the fact that PC12 pheochromocytoma cells are induced by NGF to synthesize increased levels of NILE/L1 (McGuire et al., 1978; Lee et al., 1981), it seemed pertinent to investigate whether NGF can also regulate L1 expression by Schwann cells.

Pure cultures of Schwann cells were established from early postnatal mouse sciatic nerve and could be shown to express L1 and N-CAM and L2/HNK-1. L1 and N-CAM were synthesized in molecular forms slightly different from those expressed by small cerebellar neurons or astrocytes (Seilheimer and Schachner, 1987). The expression of L1, but not N-CAM, by Schwann cells was found to be regulated by nerve growth factor. L1 expression on the cell surface was increased 1.6-fold in the presence of NGF after three days of maintenance *in vitro* and 3-fold after 16 days. Interestingly, the glia-derived neurite-promoting factor (Günther et al., 1985) increased L1 expression by a factor of 1.9 and decreased N-CAM expression by a factor of 0.4 after 3 days *in vitro*, suggesting that the Schwann cell may respond to more than one factor by a change in adhesion molecule expression. Antibodies to NGF abolished the influence of NGF on L1 expression. The addition of NGF antibodies to the Schwann cell cultures without exogenously added NGF decreased L1 expression, indicating that Schwann cells secrete NGF that may influence L1 expression by an autocrine mechanism.

Our experiments show for the first time that cell adhesion molecule expression on a nonneuronal cell, the Schwann cell, can be directly regulated by the neurotrophic factor NGF. These observations indicate that Schwann cells are able to respond to their environment with a considerable degree of plasticity in regulating cell adhesion molecule expression. Increased expression of L1 is meaningful during development and regeneration, when Schwann cells engage in cell surface contacts with neurons and promote neurite extension. An increase in L1 expression was indeed observed *in vivo* after transection of the adult sciatic nerve, thus correlating with our present observations *in vitro*. On the other hand, N-CAM was also increased after transection *in situ*, but no influence of NGF on N-CAM expression by Schwann cells could be observed *in vitro*. These results encourage our search for other regulatory signals. An understanding of these signals and the physiological role of L1 and N-CAM in the outgrowth of axons on Schwann cell surfaces will be an important step toward dissecting the cellular and molecular mechanisms of peripheral nerve regeneration.

THE L2/HNK-1 CARBOHYDRATE EPITOPE

The fact that several cell adhesion molecules express the L2/HNK-1 carbohydrate epitope begs the question of whether the epitope itself is involved in adhesion. One could argue that since it is shared by functionally important molecules, it may be important. Interestingly, the epitope is not confined to the nervous system, but is also present on subpopulations of lymphoid cells, among them natural killer cells (Abo and Balch, 1981). The carbohydrate structure is regulated during development independently of the protein backbone (Wernecke et al., 1985; Martini and Schachner, 1986). We could show that the population of N-CAM, L1, and MAG molecules is heterogeneous with respect to the expression of the L2/HNK-1 epitope (Kruse et al., 1984; Faissner, 1987; Poltorak et al., 1987). The L2/HNK-1 carbohydrate moiety is localized to the 65-kD amino-terminal fragment of N-CAM, but is not present in the amino-terminal 24-kD region of N-CAM that contains the heparin-binding domain (Cole and Schachner, 1987). Sera from patients with gammopathy and peripheral polyneuropathy also react with the L2/HNK-1 carbohydrate (for references, see Poltorak et al., 1986).

The L2/HNK-1 epitope is present not only on glycoproteins, but also on unusual glycolipids from human peripheral nerves and embryonic fetal brain (Ilyas et al., 1984; Chou et al., 1986; Noronha et al., 1986; Schwarting et al., 1987). These glycolipids are characterized as sulfate-3-glucuronylparagloboside and sulfate-3-glucuronylneolactohexasoyl ceramide (Chou et al., 1986; Ariga et al., 1987). The presence of the sulfate-3-glucuronyl moiety in the lipid is essential for antibody binding for some L2 antibodies, but not for others (Chou et al., 1986; Ilyas et al., 1986). It is not known whether the sulfate-3-glucuronyl moiety is also present on the glycoproteins expressing the epitope.

Indications that the L2/HNK-1 domain is involved in cell interactions came from studies that investigated the effect of L2 antibodies on neural cell adhesion (Keilhauer et al., 1985) or HNK-1 antibodies on neurite outgrowth (Riopelle et al., 1986). However, since antibodies do not only cover the epitope that they are directed against, but may also sterically block neighboring molecular domains from function, a more direct demonstration of the importance of the L2/HNK-1 moiety for cell interactions seemed necessary. We therefore took advantage of the possibility of using the L2 glycolipid and tetrasaccharide isolated from this glycolipid in sensitive culture systems to monitor cell–cell interactions. It could indeed be shown that the L2/HNK-1 reactive carbohydrate moiety, without the attached protein backbone, is able to interfere with not only cell to cell but, even more strikingly, also cell to substrate interactions (Künemund et al., 1988). Since the inhibitory effects in adhesion assays observed with glycolipids or tetrasaccharide were qualitatively and quantitatively amazingly similar to those observed with the L2 antibodies, the most straightforward interpretation of our findings is that the L2/HNK-1 carbohydrate moiety is itself involved as a ligand in cell interactions.

At this point it is tempting to speculate about the molecular nature of the receptors for the L2/HNK-1 carbohydrate. Whether these are adhesion molecules themselves or as yet unknown cell surface constituents remains to be resolved. An interesting hypothesis put forward by Cole, Glaser, and colleagues (Cole and Glaser, 1985; Cole et al., 1986a,b) is worth mentioning in this context. These authors showed that N-CAM has binding sites for heparin and the cell surface. They suggested that the heparin- and cell-binding domains may be identical and speculated that the L2/HNK-1 carbohydrate may well bind to the heparin-binding site. At present, however, we have no evidence that supports this notion.

The L2/HNK-1-carrying carbohydrate moiety may subserve a particular function in conjunction with others on a multifunctional glycoprotein, as has been suggested for the hormone chorionic gonadotropin (Calvo and Ryan, 1985). This hormone binds with its protein backbone to a receptor that is distinct from the binding site for the hormone's carbohydrate moiety, which is necessary for the activation of adenylyl cyclase. Enzyme activation, however, occurs only when protein backbone and carbohydrate are simultaneously bound to the cell surface, possibly inducing a cross-linking of the two receptors. Whether a similar cooperativity exists between protein backbone and carbohydrate moiety of adhesion molecules and whether the L2/HNK-1 carbohydrate plays a role in the activation of second messenger systems will have to be investigated.

THE ADHESION MOLECULE ON GLIA (AMOG) AND THE L3 FAMILY

AMOG is a novel neural cell adhesion molecule that mediates neuron–astrocyte interaction *in vitro* (Antonicek et al., 1987). It is expressed by astrocytes in the cerebellum at critical developmental stages of granule neuron migration. AMOG was not detectable on Bergmann glial cells before the onset of migration and disappeared from these cells after migration had ceased. At the end of the migratory period, AMOG became detectable on astrocytic processes in the internal granular layer and remained detectable there in adulthood. Granule neuron migration was inhibited by monoclonal AMOG antibody, probably by disturbing neuron–glia adhesion. AMOG was shown to be an integral cell surface glycoprotein of approximately 50-kD molecular mass with a carbohydrate content of approximately 30%. Thus, AMOG is yet another cell adhesion molecule involved in granule cell migration.

AMOG was not found to express the L2/HNK-1 carbohydrate epitope, but expressed a carbohydrate epitope recognized by the monoclonal antibody L3 (Kücherer et al., 1987), which is present on several glycoproteins from mouse brain, including L1 and MAG, but, interestingly, at the levels of detectability, not by J1 and N-CAM. The occurrence of this epitope thus introduces AMOG into another family of cell adhesion molecules that is characterized by a carbohydrate structure common to members within this family but also shared by some, but not all, members of another—the L2 family. The number of members in this family is yet unknown, but appears to consist of at least nine glycoproteins in adult mouse brain.

Several other features of the L3 carbohydrate epitope are reminiscent of those of the L2/HNK-1 epitope. Like L2/HNK-1, the L3 carbohydrate domain appears to be involved in cell–cell interactions, since L3 antibodies inhibit cell adhesion and cellular outgrowth patterns in explant cultures (Weber, un-published data). L3 is *N*-glycosidically linked, and its expression is regulated independently of the protein backbone of these glycoproteins. Thus, not all molecules of each member in this family express the epitope. Consistent with these findings is the observation that only 10–20% of all MAG-positive oligodendrocytes express the L3 epitope. Similarly, in cultures of early postnatal mouse cerebella, AMOG-positive astrocytes do not express the L3 epitope. Since the L3 carbohydrate epitope is expressed on both the L2-negative and L2-positive L1 molecules, a simple relationship between the L2 and L3 carbohydrate epitope–carrying variants of adhesion molecules appears unlikely.

These observations point to questions that need to be answered in the future: Is the next functionally characterized member of the L3 family also an adhesion molecule? What are the structure and function of the L3 epitope? How many families of cell adhesion molecules exist that are combined by a common family feature in the form of a distinct carbohydrate structure? And, is it possible that neural cell adhesion molecules are "presenters" of functionally important carbohydrate structures with immense combinatorial possibilities?

CONCLUSIONS AND OUTLOOK

Our studies have given evidence of the existence of carbohydrate-based families of cell adhesion molecules. First suggestions that these carbohydrate structures are themselves important in adhesion are emerging, but the molecular mechanisms of their involvement in adhesion have yet to be elucidated. The common carbohydrate structures may not be individually responsible as ligands in adhesion, but may act in concert with one another or with the protein backbone, which is endowed with probably other molecular features important for cell adhesion. Such a motif could be the Ig homologous domain of the superfamily, which is present in two out of two sequenced cell adhesion molecules, N-CAM and MAG. We would like to speculate that each of the carbohydrate structures represents a characteristic sequence, similar to the RGD sequence shared by such cell surface ligands as fibronectin, vitronectin, and thrombospondin (Ruoslahti and Pierschbacher, 1986), the specificity and avidity of which may be modified by other domains on the molecule at the carbohydrate and protein levels. Although we have only caught a glimpse of the importance of carbohydrate epitopes on neural cell adhesion molecules, it seems that trying to elucidate the structure–function relationships and possible functional symbiosis of carbohydrates with proteins is a worthwhile venture for the future.

ACKNOWLEDGMENTS

We are grateful to Deutsche Forschungsgemeinschaft, Alexander von Humboldt-Stiftung, Studienstiftung des Deutschen Volkes and Fonds der Chemischen Industrie for support.

Note added in proof: This chapter was written during the summer of 1987 and was not updated in proof.

REFERENCES

Abo, T., and C. M. Balch (1981) A differentiation antigen of human NK and K cells identified by a monoclonal antibody (HNK-1). *J. Immunol.* **127**:1024–1029.

Aguayo, A. J. (1985) Axonal regeneration from injured neurons in the adult mammalian central nervous system. In *Synaptic Plasticity*, C. W. Cotman, ed., pp. 457–483, Guilford, New York.

Altman, J. (1976) Postnatal development of the cerebellar cortex in the rat. III. Maturation of the components of the granular layer. *J. Comp. Neurol.* **145**:465–514.

Antonicek, H., E. Persohn, and M. Schachner (1987) Biochemical and functional characterization of a novel neuron–glia adhesion molecule that is involved in neuronal migration. *J. Cell Biol.* **194**:1587–1595.

Ariga, T., T. Kohriyama, L. Freddo, N. Latov, M. Saito, K. Kon, S. Ando, M. Suzuki, M. E. Hemling, K. L. Rinehart, S. Kusunoki, and R. K. Yu (1987) Characterization of sulfated glucuronic acid containing glycolipids reacting with IgM M-proteins in patients with neuropathy. *J. Biol. Chem.* **262**:848–853.

Arquint, M., J. Roder, L. S. Chia, J. Down, H. Bayley, P. Braun, and R. Dunn (1987) Molecular cloning and primary structure of myelin-associated glycoprotein. *Proc. Natl. Acad. Sci. USA* **84**:600–604.

Bock, E., R. Richter-Landsberg, A. Faissner, and M. Schachner (1985) Demonstration of immunochemical identity between the nerve growth factor–inducible large external (NILE) glycoprotein and the cell adhesion molecule L1. *EMBO J.* **4**:2765–2768.

Calvo, F. O., and R. J. Ryan (1985) Inhibition of adenylyl cyclase activity in rat corpora luteal tissue by glycopeptides of human chorionic gonadotropin and the α-subunit of human chorionic gonadotropin. *Biochemistry* **24**:1953–1959.

Chang, S., F. G. Rathjen, and J. A. Raper (1987) Extension of neurites on axons is impaired by antibodies against specific neural cell surface glycoproteins. *J. Cell Biol.* **104**:355–362.

Chou, D. K. H., A. A. Ilyas, J. E. Evans, C. Costello, R. H. Quarles, and F. B. Jungalwala (1986) Structure of sulfated glucuronyl glycolipids in the nervous system reacting with HNK-1 antibody and some IgM paraproteins in neuropathy. *J. Biol. Chem.* **261**:11717–11725.

Cole, G. J., and L. Glaser (1985) A heparin-binding domain from N-CAM is involved in neural cell–substratum adhesion. *J. Cell Biol.* **102**:403–412.

Cole, G. J., and M. Schachner (1987) Localization of the L2 monoclonal antibody binding site on N-CAM and evidence for its role in N-CAM-mediated cell adhesion. *Neurosci. Lett.* **78**:227–232.

Cole, G. J., A. Loewy, N. V. Cross, R. Akeson, and L. Glaser (1986a) Topographic localization of the heparin-binding domain of the neural cell adhesion molecule N-CAM. *J. Cell Biol.* **103**:1739–1744.

Cole, G. J., A. Loewy, and L. Glaser (1986b) Neuronal cell–cell adhesion depends on interactions of N-CAM with heparin-like molecules. *Nature* **320**:445–448.

Covault, J., and J. R. Sanes (1985) Neural cell adhesion molecule (N-CAM) accumulates in denervated and paralyzed skeletal muscle. *Proc. Natl. Acad. Sci. USA* **82**:4544–4548.

Cross, G. A. M. (1987) Eukaryotic protein modification and membrane attachment via phosphatidylinositol. *Cell* **48**:179–181.

Cunningham, B. A., S. Hoffman, U. Rutishauser, J. J. Hemperly, and G. M. Edelman (1983) Molecular topography of N-CAM: Surface orientation and the location of sialic acid–rich and binding regions. *Proc. Natl. Acad. Sci. USA* **80**:3116–3120.

Fahrig, T., C. Landa, P. Pesheva, K. Kühn, and M. Schachner (1987) Characterization of binding properties of the myelin-associated glycoprotein to extracellular matrix constituents. *EMBO J.* 6:2875–2883.

Faissner, A. (1987) Monoclonal antibody detects carbohydrate microheterogeneity on the murine cell adhesion molecule L1. *Neurosci. Lett.* 83:327–332.

Faissner, A., J. Kruse, C. Goridis, E. Bock, and M. Schachner (1984) The neural cell adhesion molecule L1 is distinct from the N-CAM related group of surface antigens BSP-2 and D2. *EMBO J.* 3:733–737.

Faissner, A., D. Teplow, D. Kübler, G. Keilhauer, V. Kinzel, and M. Schachner (1985) Biosynthesis and membrane topography of the neural cell adhesion molecule L1. *EMBO J.* 4:3105–3113.

ffrench-Constant, C., R. H. Miller, J. Kruse, M. Schachner, and M. C. Raff (1986) Molecular specialization of astrocyte processes at nodes of Ranvier in rat optic nerve. *J. Cell Biol.* 102:844–852.

ffrench-Constant, C., and M. C. Raff (1986) The oligodendrocyte-type-2-astrocyte cell lineage is specialized for myelination. *Nature* 323:335–338.

Fischer, G., V. Künemund, and M. Schachner (1986) Neurite outgrowth patterns in cerebellar microexplant cultures are affected by antibodies to the cell surface glycoprotein L1. *J. Neurosci.* 6:605–612.

Friedlander, D. R., M. Grumet, and G. M. Edelman (1986) Nerve growth factor enhances expression of neuron–glia cell adhesion molecule in PC12 cells. *J. Cell Biol.* 102:413–419.

Fushiki, S., and M. Schachner (1986) Immunocytological localization of cell adhesion molecules L1 and N-CAM and the shared carbohydrate epitope L2 during development of the mouse neocortex. *Brain Res.* 389:153–167.

Gennarini, G., M. Hirn, H. Deagostini-Bazin, and C. Goridis (1984a) Studies on the transmembrane disposition of the neural cell adhesion molecule N-CAM. The use of liposome-inserted radioiodinated N-CAM to study its transbilayer orientation. *Eur. J. Biochem.* 142:65–73.

Gennarini, G., G. Rougon, H. Deagostini-Bazin, M. Hirn, and C. Goridis (1984b) Studies on the transmembrane disposition of the neural cell adhesion molecule N-CAM. A monoclonal antibody recognizing a cytoplasmic domain and evidence for the presence of phosphoserine residues. *Eur. J. Biochem.* 142:57–64.

Grumet, M., S. Hoffman, C.-M. Chuong, and G. M. Edelman (1984a) Polypeptide components and binding functions of neuron–glia cell adhesion molecule. *Proc. Natl. Acad. Sci. USA* 81:7989–7993.

Grumet, M., S. Hoffman, and G. M. Edelman (1984b) Two antigenically related neuronal CAMs of different specificities mediate neuron–neuron and neuron–glia adhesion. *Proc. Natl. Acad. Sci. USA* 81:267–271.

Günther, J., H. Nick, and D. Monard (1985) A glia-derived neurite-promoting factor with protease inhibitory activity. *EMBO J.* 4:1963–1966.

He, H. T., J. Barbet, J. C. Chaix, and C. Goridis (1986) Phosphatidylinositol is involved in the membrane attachment of N-CAM 120, the smallest component of the neural cell adhesion molecule. *EMBO J.* 5:2489–2494.

Hemperly, J. J., B. A. Murray, G. M. Edelman, and B. A. Cunningham (1986) Sequence of a cDNA clone encoding the polysialic acid–rich cytoplasmic domains of the neural cell adhesion molecule N-CAM. *Proc. Natl. Acad. Sci. USA* 83:3037–3041.

Hoffman, S., and G. M. Edelman (1983) Kinetics of homophilic binding by embryonic and adult forms of the neural cell adhesion molecule. *Proc. Natl. Acad. Sci. USA* 80:5762–5766.

Hoffman, S., D. R. Friedlander, C.-M. Chuong, M. Grumet, and G. M. Edelman (1986) Differential contributions of Ng-CAM and N-CAM to cell adhesion in different neural regions. *J. Cell Biol.* 103:145–158.

Holley, J. A., and M. Schachner (1987) Localization of cell surface antigens N-CAM, L1, and L2 in embryonic mouse dorsal root ganglia and spinal cord (submitted).

Ilyas, A. A., R. H. Quarles, and R. O. Brady (1984) The monoclonal antibody HNK-1 reacts with a peripheral nerve ganglioside. *Biochem. Biophys. Res. Commun.* **122**:1206–1211.

Ilyas, A. A., M. C. Dalakas, R. O. Brady, and R. H. Quarles (1986) Sulfated glucuronyl glycolipids reacting with anti-myelin-associated glycoprotein monoclonal antibodies including IgM paraproteins in neuropathy: Species distribution and partial characterization of epitopes. *Brain Res.* **385**:1–9.

Inuzuka, T., R. H. Quarles, J. Heath, and B. D. Trapp (1985) Myelin-associated glycoprotein and other proteins in *trembler* mice. *J. Neurochem.* **44**:793–797.

Keilhauer, G., A. Faissner, and M. Schachner (1985) Differential inhibition of neurone–neurone, neurone–astrocyte, and astrocyte–astrocyte adhesion by L1, L2, and N-CAM antibodies. *Nature* **316**:728–730.

Kruse, J., R. Mailhammer, H. Wernecke, A. Faissner, I. Sommer, C. Goridis, and M. Schachner (1984) Neural cell adhesion molecules and myelin-associated glycoprotein share a common carbohydrate moiety recognized by monoclonal antibodies L2 and HNK-1. *Nature* **311**:153–155.

Kruse, J., G. Keilhauer, A. Faissner, R. Timpl, and M. Schachner (1985) The J1 glycoprotein—A novel nervous system cell adhesion molecule of the L2/HNK-1 family. *Nature* **316**:146–148.

Kücherer, A., A. Faissner, and M. Schachner (1987) The novel carbohydrate epitope L3 is shared by some neural cell adhesion molecules. *J. Cell Biol.* **104**:1597–1602.

Künemund, V., F. B. Jungalwala, G. Fischer, D. K. H. Chou, G. Keilhauer, and M. Schachner (1988) The L2/HNK-1 carbohydrate of neural cell adhesion molecules is involved in cell interactions. *J. Cell Biol.* **106**:213–223.

Lai, C., M. S. Brow, K.-A. Nave, A. B. Noronha, R. H. Quarles, F. E. Bloom, R. J. Milner, and J. G. Sutcliffe (1987) Two forms of 1B236/myelin-associated glycoprotein, a cell adhesion molecule for postnatal neural development, are produced by alternative splicing. *Proc. Natl. Acad. Sci. USA* **84**:1–5.

Lee, V. M., L. A. Greene, and M. L. Shelanski (1981) Identification of neural and adrenal medullary surface membrane glycoproteins recognized by antisera to cultured rat sympathetic neurons and PC12 pheochromocytoma cells. *Neuroscience* **6**:2773–2786.

Lindner, J., F. G. Rathjen, and M. Schachner (1983) L1 mono- and polyclonal antibodies modify cell migration in early postnatal mouse cerebellum. *Nature* **305**:427–430.

Lindner, J., G. Zinser, W. Werz, C. Goridis, B. Bizzini, and M. Schachner (1986a) Experimental modification of postnatal cerebellar granule cell migration *in vitro*. *Brain Res.* **377**:298–304.

Lindner, J., J. Guenther, H. Nick, G. Zinser, H. Antonicek, M. Schachner, and D. Monard (1986b) Modulation of granule cell migration by a glia-derived protein. *Proc. Natl. Acad. Sci. USA* **83**:4568–4571.

Low, M. G. (1987) Biochemistry of the glycosyl-phosphatidylinositol membrane protein anchors. *J. Biochem.* **244**:1–13.

Martini, R., and M. Schachner (1986) Immunoelectron microscopic localization of neural cell adhesion molecules (L1, N-CAM, MAG) and their shared carbohydrate epitope and myelin basic protein (MBP) in developing sciatic nerve. *J. Cell Biol.* **103**:2439–2448.

Matthieu, J. M., R. O. Brady, and R. H. Quarles (1974) Anomalies of myelin-associated glycoprotein in quaking mice. *J. Neurochem.* **22**:291–296.

McGuire, J. C., L. A. Greene, and A. V. Furano (1978) NGF stimulates incorporation of fucose or glucosamine into an external glycoprotein in cultured rat PC12 pheochromocytoma cells. *Cell* **15**:357–365.

Mirsky, R., K. R. Jessen, M. Schachner, and C. Goridis (1986) Distribution of the adhesion molecules N-CAM and L1 on peripheral neurons and glia in adult rats. *J. Neurocytol.* **15**:799–815.

Mugnaini, E., and P. F. Forstrønen (1967) Ultrastructural studies on the cerebellar histogenesis. I. Differentiation of granule cells and development of glomeruli in the chick embryo. *Zellforsch. Mikrosk. Anat.* **77**:115–143.

Nagata, I., and M. Schachner (1986) Conversion of embryonic to adult form of the neural cell adhesion molecule (N-CAM) does not correlate with pre- and postmigratory states of cerebellar granule neurons. *Neurosci. Lett.* **63**:153–158.

Nieke, J., and M. Schachner (1985) Expression of the neural cell adhesion molecules N-CAM and L1 and their common carbohydrate epitope L2/HNK-1 during development and regeneration of sciatic nerve. *Differentiation* **30**:141–151.

Noronha, A. B., A. Ilyas, H. Antonicek, M. Schachner, and R. H. Quarles (1986) Molecular specificity of L2 monoclonal antibodies that bind to carbohydrate determinants of neural cell adhesion molecules and their resemblance to other monoclonal antibodies recognizing the myelin-associated glycoprotein. *Brain Res.* **385**:237–244.

Persohn, E., and M. Schachner (1987) Immunoelectron-microscopic localization of the neural cell adhesion molecules L1 and N-CAM during postnatal development of the mouse cerebellum. *J. Cell Biol.* **105**:569–576.

Pollerberg, E. G., R. Sadoul, C. Goridis, and M. Schachner (1985) Selective expression of the 180 kD component of the neural cell adhesion molecule N-CAM during development. *J. Cell Biol.* **101**:1921–1929.

Pollerberg, G. E., M. Schachner, and J. Davoust (1986) Differentiation state–dependent surface mobilities of two forms of the neural cell adhesion molecule. *Nature* **324**:462–465.

Pollerberg, E. G., K. Burridge, K. Krebs, S. Goodman, and M. Schachner (1987) The 180 kD component of the neural cell adhesion molecule N-CAM is involved in cell–cell contacts and cytoskeleton–membrane interactions. *Cell Tissue Res.* **250**:227–236.

Poltorak, M., A. J. Steck, and M. Schachner (1986) Reactivity with neural cell adhesion molecules of the L2/HNK-1 family in sera from patients with demyelinating diseases. *Neurosci. Lett.* **65**:199–203.

Poltorak, M., R. Sadoul, G. Keilhauer, C. Landa, T. Fahrig, and M. Schachner (1987) Myelin-associated glycoprotein (MAG), a member of the L2/HNK-1 family of neural cell adhesion molecules, is involved in neuron–oligodendrocyte and oligodendrocyte–oligodendrocyte interaction. *J. Cell Biol.* **105**:1893–1899.

Quarles, R. H. (1984) Myelin-associated glycoprotein in development and disease. *Dev. Neurosci.* **6**:286–303.

Rakic, P. (1971) Neuron–glia relationship during granule cell migration in developing cerebellar cortex. A Golgi and electron microscopic study in *Macaque rhesus. J. Comp. Neurol.* **141**:283–312.

Ramón y Cajal, S. (1928) *Degeneration and Regeneration of the Nervous System,* Vol. 2, R. M. May, trans. and ed., pp. 397–769, Oxford Univ. Press, London.

Rathjen, F. G., and U. Rutishauser (1984) Comparison of two cell surface molecules involved in neural cell adhesion. *EMBO J.* **3**:1–10.

Rieger, F., M. Grumet, and G. M. Edelman (1985) N-CAM at the vertebrate neuromuscular junction. *J. Cell Biol.* **101**:285–293.

Riopelle, R. J., R. C. McGarry, and J. C. Roder (1986) Adhesion properties of a neuronal epitope recognized by the monoclonal antibody HNK-1. *Brain Res.* **367**:20–25.

Rohrer, H. (1985) Nonneuronal cells from chick sympathetic and dorsal root sensory ganglia express catecholamine uptake and receptors for nerve growth factor during development. *Dev. Biol.* **111**:95–107.

Rohrer, H., and I. Sommer (1983) Simultaneous expression of neuronal and glial properties by chick ciliary ganglion cells during development. *J. Neurosci.* **3**:1683–1693.

Rougon, G., M. R. Hirsch, M. Hirn, J. L. Guenet, C. Goridis (1983) Monoclonal antibody to neural cell surface protein: Identification of a glycoprotein family of restricted cellular localization. *Neuroscience* **10**:511–520.

Ruoslahti, E., and M. D. Pierschbacher (1986) Arg-Gly-Asp: A versatile cell recognition signal. *Cell* **44**:517–518.

Rush, R. A. (1984) Immunohistochemical localization of endogenous nerve growth factor. *Nature* **312**:364–367.

Sadoul, R., M. Hirn, H. Deagostini-Bazin, G. Rougon, and C. Goridis (1983) Adult and embryonic forms of a mouse neural cell adhesion molecule have different binding properties. **304**:347–348.

Sadoul, K., A. Meyer, M. G. Low, and M. Schachner (1986) Release of the 120 kD component of the neural cell adhesion molecule N-CAM from cell surfaces by phosphatidylinositol-specific phospholipase C. *Neurosci. Lett.* **72**:341–346.

Sadoul, K., R. Sadoul, A. Faissner, and M. Schachner (1988) Biochemical characterization of different molecular forms of the neural adhesion molecule L1. *J. Neurochem.* **50**:510–521.

Salzer, J. L., W. P. Holmes, and D. R. Colman (1987) The amino acid sequences of the myelin-associated glycoproteins: Homology to the immunoglobulin gene superfamily. *J. Cell Biol.* **104**:957–965.

Sanes, J. R., and J. Covault (1985) Axon guidance during reinnervation of skeletal muscle. *Trends Neurosci.* **8**:423–428.

Sanes, J. R., M. Schachner, and J. Covault (1986) Expression of several adhesive macromolecules (N-CAM, L1, J1, NILE, uvomorulin, laminin, fibronectin, and a heparan sulfate proteoglycan) in embryonic, adult, and denervated adult skeletal muscle. *J. Cell Biol.* **102**:420–431.

Schachner, M., A. Faissner, J. Kruse, J. Lindner, D. H. Meier, F. G. Rathjen, and H. Wernecke (1983) Cell-type specificity and developmental expression of neural cell-surface components involved in cell interactions and of structurally related molecules. *Cold Spring Harbor Symp. Quant. Biol.* **48**:557–568.

Schachner, M., A. Faissner, G. Fischer, G. Keilhauer, J. Kruse, V. Künemund, J. Lindner, and H. Wernecke (1985) Functional and structural aspects of the cell surface in mammalian nervous system development. In *The Cell in Contact: Adhesions and Junctions as Morphogenetic Determinants*, G. M. Edelman and J. P. Thiery, eds., pp. 257–275, Wiley, New York.

Schwarting, G. A., F. B. Jungalwala, D. K. H. Chou, A. M. Boyer, and M. Yamamoto (1987) Sulfated glucuronic acid containing glycoconjugates are temporally and spatially regulated antigens in the developing mammalian nervous system. *Dev. Biol.* **20**:65–76.

Seilheimer, B., and M. Schachner (1987) Regulation of neural cell adhesion molecule expression on cultured mouse Schwann cells by nerve growth factor. *EMBO J.* **6**:1611–1616.

Silver, J., and U. Rutishauser (1984) Guidance of optic axons *in vivo* by a preformed adhesive pathway on neuroepithelial end feet. *Dev. Biol.* **106**:485–499.

Smith, G. W., R. H. Miller, and J. Silver (1986) Changing role of forebrain astrocytes during development, regenerative failure, and induced regeneration upon transplantation. *J. Comp. Neurol.* **251**:23–43.

Sternberger, N. H., R. H. Quarles, Y. Itoyama, and H. deF. Webster (1979) Myelin-associated glycoprotein demonstrated immunocytochemically in myelin and myelin-forming cells of developing rat. *Proc. Natl. Acad. Sci. USA* **76**:1510–1514.

Taniuchi, M., H. B. Clark, and E. M. Johnson, Jr. (1986) Induction of nerve growth factor receptor in Schwann cells after axotomy. *Proc. Natl. Acad. Sci. USA* **83**:4094–4098.

Taylor, R. B., W. P. H. Duffus, M. C. Raff, and S. de Petris (1971) Redistribution and pinocytosis of lymphocyte surface immunoglobulin molecules induced by anti-immunoglobulin antibody. *Nature* **233**:225–229.

Thor, G., E. Pollerberg, and M. Schachner (1986) Molecular association of two neural cell adhesion molecules, L1 antigen and the 180 kD component of N-CAM, within the surface membrane of cultured neuroblastoma cells. *Neurosci. Lett.* **66**:121–126.

Thor, G., R. Probstmeier, and M. Schachner (1987) Characterization of the cell adhesion molecules L1, N-CAM, and J1 in the mouse intestine. *EMBO J.* **6**:2581–2586.

Trapp, B. D., and R. H. Quarles (1982) Presence of the myelin-associated glycoprotein correlates with alterations in the periodicity of the peripheral myelin. *J. Cell Biol.* **92**:877–882.

Trapp, B. D., and R. H. Quarles (1984) Immunocytochemical localization of the myelin-associated glycoprotein. Fact or artifact? *J. Neuroimmunol.* **6**:231–249.

Tucker, G. C., H. Aoyama, M. Lipinski, T. Tursz, and J. P. Thiery (1984) Identical reactivity of monoclonal antibodies HNK-1 and NC-1: Conservation in vertebrates on cells derived from the neural tube and on some leukocytes. *Cell Differ.* **14**:223–230.

Webster, H. deF., C. G. Palkovits, G. L. Stoner, J. T. Favilla, D. E. Frail, and P. E. Braun (1983) Myelin-associated glycoprotein: Electron microscopic immunocytochemical localization in compact developing and adult central nervous system myelin. *J. Neurochem.* **41**:1469–1479.

Wernecke, H., J. Lindner, and M. Schachner (1985) Cell type specificity and developmental expression of the L2/HNK-1 epitope in mouse cerebellum. *J. Neuroimmunol.* **9**:115–130.

Williams, A. F. (1982) Surface molecules and cell interactions. *J. Theoret. Biol.* **98**:221–234.

Chapter 22

Glia-Guided Neuronal Migration *In Vitro*

MARY E. HATTEN

ABSTRACT

To study the mechanism of glia-guided neuronal migration, we have developed an in vitro *model system in which the migration of cerebellar granule neurons along astroglial processes can be observed in real time. As viewed by video-enhanced differential interference contrast microscopy, the migrating neuron forms an extensive cell–cell contact with the glial fiber along the region of neuronal cell soma and extends a leading process in the direction of migration (Edmondson and Hatten, 1987). The leading process is closely apposed to the glial fiber along its entire length and has features characteristic of the neuronal growth cone and of migrating fibroblasts. In the neurological mutant mouse* weaver, *granule cell migration fails in concert with abnormalities in the form and disposition of Bergmann glial processes, processes that guide the migration of granule cells in normal animals (Rakic and Sidman, 1973; Sotelo and Changeux, 1974). The* in vitro *recombination of granule cells and astroglia purified from* weaver *and normal littermates demonstrates that the granule cell is a primary site of action of the* weaver *gene and that the* weaver *granule cell fails to form the neuron–glia contact typical of migrating neurons (Hatten et al., 1986). To identify antigens involved in neuron–glia interactions, a microculture system was used as a functional assay for immune activities that block the formation of neuron–glia contacts. A polyclonal antiserum raised in rabbits against early postnatal mouse cerebellar cells that blocks neuron–glia interactions is described (Edmondson et al., 1988). The prominent antigen recognized by this immune activity, which we have named anti-astrotactin, is a glycoprotein with an apparent molecular weight of 100 kD. This antigen appears to be distinct from either N-CAM (Thiery et al., 1977) (BSP-2; Hirn et al., 1981) or the NILE glycoprotein (McGuire et al., 1978) (L1, Ng-CAM; Grumet et al., 1984; Rathjen and Schachner, 1984), neither of which influences neuron–glia contacts in this culture system. By immunoprecipitation of Triton extracts of [^{3}H] fucose-labeled P7 cerebellar cells, the band at 100 kD is missing or defective in granule cells from the* weaver *mouse (Edmondson et al., 1988).*

One of the most striking cell–cell interactions in the developing mammalian brain is that between young migrating neurons and the radial processes of astroglial cells (Ramón y Cajal, 1911). Rakic and his colleagues have shown glial fibers to be a primary guidance system for migrating neurons in many cortical areas, ushering postmitotic neurons from the proliferative zones, where they are generated, out into more superficial zones of the cortex, where they form synaptic connections (Rakic, 1971, 1972; Rakic and Sidman, 1973; Sidman and Rakic, 1973; Rakic et al., 1974).

By reconstructing serial electron micrographs of migrating neurons, Rakic has shown that the neurons assume a characteristic posture on the glial process (Rakic et al., 1974; Rakic, 1985). The migrating neuron closely apposes its cell soma against the glial arm, extends a thickened leading process in the direction of migration, and scales the glial process by winding its way around it. Once in the proper layer of the cortex, the neuron retracts the leading process, detaches its cell body from the radial glial process, and begins to form synaptic contacts.

Several *in vitro* systems have been developed to study neuronal migration. Trenkner has analyzed granule cell migration along fascicles of glial and neuronal fibers in microwell cultures of mouse cerebellar cells (Trenkner and Sidman, 1977) and has suggested that the timing of migration is governed by an internal clock in the neuron (Trenkner et al., 1984). Explant cultures of cerebellar tissue have been used to monitor the disposition of [^{3}H]thymidine-labeled cells after treatment of the explant with antibodies or with proteases and their inhibitors (Moonen et al., 1982; Lindner and Schachner, 1983; Lindner et al., 1986; Antonicek et al., 1987).

NEURON–GLIA INTERACTIONS IN MICROCULTURES

To analyze the neuron–glia relationship of migrating granule neurons and to provide an assay system for the molecules requisite for migration, we developed a monolayer microculture system for mouse cerebellar cells (Hatten and Liem, 1981). We chose the mouse cerebellum because the cerebellum presents a paradigm of glial guidance of neuronal migration, that of the granule cell along Bergmann glial processes, and because of the existence of the neurological mutant mouse *weaver*, an animal that suffers a failure of granule cell migration in concert with abnormalities in the form and alignment of the Bergmann glial cells (Rezai and Yoon, 1972; Hirano and Dembitzer, 1973; Rakic and Sidman, 1973; Sotelo and Changeux, 1974).

When cells are harvested from the early postnatal mouse cerebellum in the period of granule cell migration *in vivo*, neuron–glia interactions can be visualized *in vitro* by immunostaining with cellular antigen markers. The most useful of these has been the glial filament protein (GFP), because it highlights astroglial processes and facilitates quantitation of the disposition of neurons vis-à-vis these stained glial fibers (Hatten and Liem, 1981). When the cells are harvested from the first postnatal week, the astroglial cells provide a template for the positioning of granule cells in the culture (Hatten and Liem, 1981).

Several features of the microculture system should be emphasized. First, the small volume of the cultures (10–50 μl) promotes the rapid recovery of the cells from the procedures used to dissociate the tissue. Generally, glial process extension commences within 5–15 min of the time of plating. Second, maintaining the cells in horse serum promotes astroglial differentiation and selects against oligodendroglia. Very few oligodendroglial cells survive in horse serum. Third, treating the substratum with low concentrations of polylysine (25–100 μg/ml) promotes neuron–glia interactions over neuronal interactions with the culture surface. Finally, recently described methods for the recombination of purified populations of granule neurons and astroglial cells (Hatten, 1985) allow us to examine the interactions of defined cell types.

Two forms of astroglial cells are seen in the microcultures (Hatten et al., 1984a): a stellate form that accounts for more than 80% of the GFP-stained cells in the culture, and a highly elongated astroglial cell that resembles the Bergmann glia seen *in vivo* (Figure 1). A key question we wished to address with these cultures was whether the neurons would migrate on glial fibers *in vitro*. To analyze the behavior of granule cells on these two astroglial forms, we studied the movement of the neurons with time-lapse, phase-contrast video microscopy. These studies showed that neuronal migration occurs on the fibers of highly elongated astroglial cells (Hatten et al., 1984a) and that migrating neurons *in vitro* assume a posture on the glial arm that closely resembles that described by Rakic for migrating neurons *in vivo* (Figure 1b) (Rakic et al., 1974; Rakic, 1985).

An underlying theme of neuronal migration along astroglial cells, such as radial glia or the Bergmann glia, is how the glial cell attains the elongated shape needed to support migration. Consequently, glial morphological differentiation has become an important derivative issue for our studies on migration. In particular, we wanted to determine whether astroglial cells attain their shape by process extension. To analyze this, we studied the time course of the development of glial form in microcultures by correlated immunocyto-chemistry and video and electron microscopy. The emergence of both of the principal astroglial forms seen in these cultures occurs via similar means (Mason et al., 1988): the outgrowth of a broad process, tipped by a motile ending that resembles a growth cone (Figure 2a–c). The central core of the emerging glial process is filled with motile mitochondria and masses of glial filaments. Its borders are undulating lamellae fringed by microspikes. Process extension commences within 30 min of plating and is virtually complete for stellate astroglial forms within 8 hours. The second population of glial cells, the elongated forms that support migration, continue to extend processes for 24 hours, generating cells with processes 100–200 μm in length. Thus, both stellate and Bergmann glial cells attain their complex shapes by process outgrowth.

A unique feature of the emerging astroglial process is its rapid interactions with neurons (Figure 2d), an event that results in a stable binding of neurons to astroglia. This, in turn, is a prerequisite for the formation of glial processes (Hatten, 1985, 1987). As will be discussed later, both glial forms require cell–cell contacts with neurons for process outgrowth to commence.

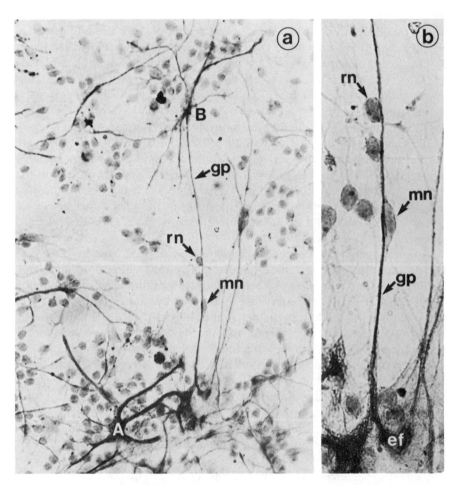

Figure 1. *Two forms of astroglial cells in microcultures of mouse cerebellar cells harvested on postnatal day 7, cultured for 48 hours and immunostained with anti-GFP antisera to visualize astroglial cells (a).* The predominant form (A) has a stellate shape, and numerous neurons are seen among its arms. More elongated glial forms (B) are also seen, and these appear to support migration. A neuron with the posture of a migrating cell (mn) seen along the glial process (gp) in *a* is shown at higher magnification in *b*. (From Hatten et al., 1984a.)

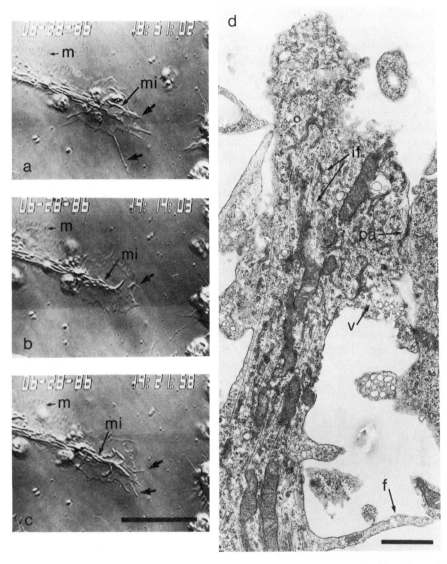

Figure 2. *Features of the astroglial growth cone.* By video microscopy (*a–c*), filopodia and microspikes extend and withdraw, and mitochondria move in and out of the central core of organelles over the 30-min observation period. Peripheral lamellae border the emerging glial process. By electron microscopy, the cytology of an elongated glial process is seen five hours after the cells are plated. Microtubules extend through the core of the process, and intermediate filaments (if) extend to the borders. A system of smooth ER is evident. v, vesicles; F, filopodium. A punctate density (pa) is seen at the apposition of the glial growth cone with a neuron. Calibration bar in $c = 20\mu$m; in $d = 1$ μm. (From Mason et al., 1988.)

NEURONAL MIGRATION *IN VITRO*

To provide a detailed view of the living, migrating neuron and to resolve the cytology of the leading process and its relationship to the glial fiber, we applied the techniques developed by Allen, Inouye, and their colleagues (Allen et al., 1981; Inouye, 1986), video-enhanced differential interference contrast microscopy, to cerebellar cells *in vitro* (Edmondson and Hatten, 1987). The features of neuronal migration, seen *in vitro* in real time, provide a striking image of the relationship between the migrating neuron and the glial guide (Figure 3).

In the culture setting, neuronal migration is bidirectional, generally at speeds of 10–60 μm per hour (Hatten et al., 1984a; Edmondson and Hatten, 1987). Migration appears to be discontinuous, the cells often pausing for periods of minutes before bursting forward again (Figure 4). Two features of the migrating neuron are of special interest. First, the neuron extends a "leading process" in the direction of migration. The surface of the leading process is highly ruffled, and both lamellopodia and filopodia are numerous from the tip of the leading edge back to the cell soma. Filopodial extensions are short, generally 5 μm in length, and lamellopodial extensions generally wrap around the glial arm. Second, the leading process closely apposes the astroglial process along its entire length, and an extensive region of cell–cell contact is maintained during migration between the cell soma of the neuron and the astroglial fiber (Figure 3).

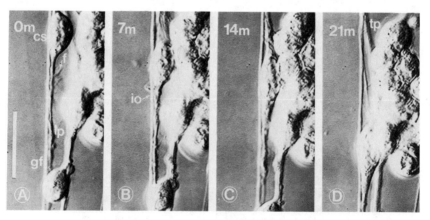

Figure 3. *The cytology of the migrating neuron as viewed by video-enhanced differential interference contrast microscopy. A–D: The leading process (lp) extends forward from the neuronal cell soma (cs) in the direction of migration in close apposition to the glial fiber (gf). (A trailing axonal process [tp] extends away from the glial fiber.) Within the leading process the movements of intracellular organelles (io) can be seen. The surface of the leading process displays extensive ruffling, lamellopodia, and short filopodia. The time interval (minutes) is indicated at the top of each frame. Calibration bar = 20 μm. (From Edmondson and Hatten, 1987.)*

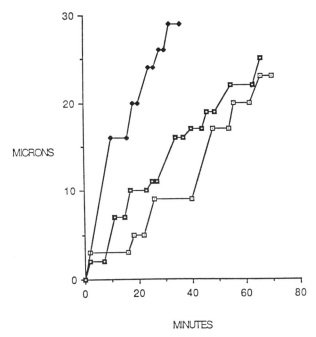

Figure 4. *Pattern of movement of three granule cells migrating along glial fibers* in vitro. The position of each neuron was recorded at 2-min intervals over a period of one hour. At the start of movement, the rate was approximately 60 μm/hour. Movement was interrupted by periods of rest, seen as *horizontal* portions of the curve. The average rate of migration was approximately 30 μm/hour. (From Edmondson and Hatten, 1987.)

The leading process of the migrating neuron appears to share cytological features with both neuronal growth cones and the leading edge of fibroblasts. A distinction between the neuronal leading process and either the growth cone or fibroblast, as studied in tissue culture, is that the leading process restricts its motions to the curved surface of the glial arm, a cylindrical structure only several microns in diameter, whereas both neuronal growth cones and fibroblasts survey an expansive, flat substratum *in vitro*.

The form of the leading tip of the migrating neuron, especially the presence and behavior of filopodia as they extend, rotate backward, and are then resorbed into the process, is similar to that of growth cones *in vitro*. However, whereas filopodia of neuronal growth cones often extend as much as 100 μm, those of the leading process of migrating neurons rarely exceed 5 μm. One interpretation of this finding is that the length of filopodia corresponds to the distance between sites of adhesion and that these sites are closely spaced along the glial arm. This view is consistent with a "guidepost" sensing role for filopodia, relating filopodial extension to the distance between sites of adhesion (Bastiani and Goodman, 1984; Bentley and Toroian-Raymond, 1984; Hammarback and Letourneau, 1986).

Like the neuronal growth cone and the leading edge of a fibroblast, the surface of the leading process of the migrating neuron displays extensive ruffling of the surface membrane and undulation of lamellae along its borders (Bunge, 1977; Abercrombie, 1980; Letourneau, 1982; Wessells, 1982). In addition, the rate of movement of the migrating cell is within the general range of rates seen for neuronal growth cones and fibroblasts, 10–60 μm/hour. A number of cytological features of the leading process are similar to the neuronal growth cone. These include longitudinally oriented microtubules extending from the cell body into the tip of the process, filopodia containing bundles of microfilaments, and clumps of smooth vesicles after glutaraldehyde fixation (Ludueña and Wessells, 1973; Rees and Reese, 1981; Cheng and Reese, 1985; Gregory et al., 1987a).

Features that distinguish the leading process from the neuronal growth cone are the morphology of the tip and the pattern of contact with the substratum. The leading process of the migrating granule cell is a tapered expansion of cytoplasm, extending from the somata out to the tip of the process, and it is closely apposed to the glial fiber along its entire length. In contrast, a neuronal growth cone is an expansive motile ending on a thin neuritic shaft, and the neurite forms adhesions only at the growing tip.

Two models have been proposed for growth cone locomotion. In Bray's model (Bray and Chapman, 1985; Bray, 1987), filopodia pull the growth cone forward, by the attachment of the tip of the filopodium to the substratum and the subsequent contraction of microfilaments. In an opposing view, obtained from studies on *Aplysia* axon formation, Goldberg has proposed that growth cones advance during periods of lamellopodial extension and that filopodia extend between, not during, periods of advance (Goldberg and Burmeister, 1986; Aletta and Greene, 1988).

Neither of these models for growth cone locomotion adequately describes the motion of the migrating granule cell. Instead, the fibroblast appears to provide a more useful model for the movement of the migrating neuron. Like the fibroblast, the granule neuron migrates in an irregular, saltatory motion, short intervals of movement punctuated by regular pauses (Figure 4). The mechanisms of fibroblast movement have been summarized by Abercrombie (Abercrombie et al., 1970; Abercrombie, 1980). The cell moves by four steps: the extension of a broad, leading lamella, the formation of adhesion sites between this lamella and the substratum, the contraction of intracellular bundles of microfilaments that connect the nucleus and cell–substratum adhesion sites, and finally, the movement of the nucleus toward the site of adhesion.

To apply this model to the migrating neuron, one has to consider the geometrical constraint imposed by the glial process, a thin cylindrical structure generally 1–5 μm in diameter. The question as to how the curvature of the substratum affects fibroblast movement has been analyzed by Dunn and Heath

(1976), who showed that whereas fibroblasts can move either circumferentially or longitudinally on fibers with a diameter larger than 200 μm, movement is restricted to the longitudinal direction when the fiber diameter is decreased to below 200 μm. An important aspect of fibroblast orientation on fibers is the alignment of microfilaments parallel to the longitudinal axis of the fiber (Dunn and Heath, 1976). Thus, the longitudinal orientation of granule cell movement along the glial arm probably relates to the thin radius of the glial fiber, a dense distribution of sites for adhesive contacts along the glial process, and the alignment of microfilaments (presumably relative to focal adhesions) in the direction of migration.

To analyze the junction formed between migrating neurons and the astroglial process, we combined video microscopy with electron microscopy and examined the cell–cell contacts of neurons shown to be migrating by time-lapse video microscopy (Gregory et al., 1988). These studies indicate that "interstitial densities" are seen between the somata of the migrating neuron and the astroglial process. The interstitial density region is slightly dilated to an intercellular distance of 20 μm and is filled with filamentous material (Figure 5). Some of the fibrils present in interstitial densities appear to be either contiguous with or extensions of cytoskeletal elements, suggesting a possible role in cell locomotion (Gregory et al., 1988).

Experiments are in progress to localize cell adhesion molecules thought to play a role in cell contacts and in cell movements (Gregory et al., 1987). It will be important to correlate the distribution of these molecules with sites of cell contact and with the organization of cytoskeletal elements thought to propel the cell forward along the glial process.

THE *WEAVER* MOUSE

The neurological mutant mouse *weaver* has been studied extensively *in vivo* and *in vitro* as a test of the hypothesis that Bergmann fibers guide the migration of granule neurons. In *weaver*, an autosomal recessive mutation that leads to the failure of granule cell migration along Bergmann glia, granule cells fail to form close appositions with glial processes that are abnormally aligned in the *weaver* cortex (Rezai and Yoon, 1972; Hirano and Dembitzer, 1973). Previous studies on the *weaver* mouse fueled a debate over which cell was the regulatory agent in the migration—the neuron or the glial cell. In one view, abnormalities in glial form precluded neuronal migration (Rakic and Sidman, 1973); in the other, the failure of migration was caused by the action of the *weaver* gene on the neuron (Sotelo and Changeux, 1974; Goldowitz and Mullen, 1982).

When *weaver* cerebellar cells are placed in microcultures, the neurons fail to form characteristic cell–cell interactions with astroglial cells; the glia, in turn, have stunted, abnormal processes (Hatten et al., 1984a), and granule neurons fail to extend neurites (Willinger and Margolis, 1985). To assess the contribution of intrinsic genetic information to the formation of neuron–glia contacts, we

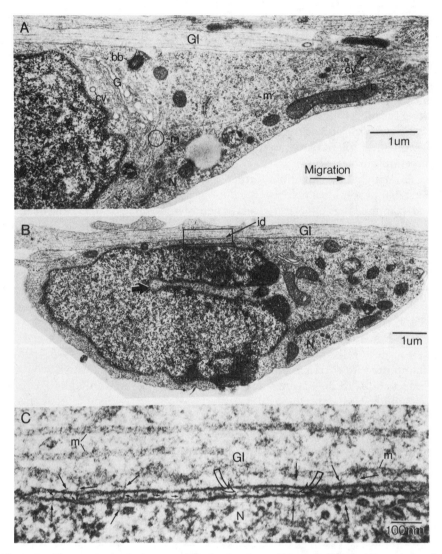

Figure 5. *Electron micrograph of a neuron (N) migrating along a glial process (G1). In A the rostral end of the soma contains Golgi (G), basal body (bb), coated vesicles (cv), and ribosomes. Microtubules are seen around the nucleus and extend into the leading process (lp). A different section through the same neuron shows the characteristic caudal placement of the nucleus during migration, with a nuclear indentation on the forward surface (arrow). A region of pronounced interstitial density (id) is seen between the migrating neuron and adjoining glial process (G1). A higher magnification of the interstitial density seen in B shows fibrillar material spanning the density between the cells. Submembranous filamentous structures are present in the process and neuron (larger, solid arrows). Some submembranous cytoskeletal elements (C) are contiguous with, or transmembranous extensions of, interstitial fibrils (coaligned smaller and larger solid arrows). (From Gregory et al., 1987a.)*

developed a technique for the rapid purification of granule neurons and astroglial cells (Hatten, 1985). This allowed us to analyze the behavior of *weaver* granule cells on normal astroglial processes and vice versa. When cocultured with normal astroglial cells, *weaver* granule neurons failed to form neuron–glia contacts typical of migrating neurons, the close apposition of the cell soma to the glial arm and the extension of a leading process out along the glial process (Hatten et al., 1986). In contrast, normal granule cells formed tight appositions with *weaver* astroglial processes, apposition that resembled those of normal neurons cultured with normal astroglial cells (Figure 6).

These studies underscore the importance of the cell–cell contact between neurons and glial cells for neuronal migration and suggest, in agreement with the studies of Sotelo and Changeux (1974) and Goldowitz and Mullen (1982), that the granule neuron is a primary site of action of the *weaver* gene.

Thus, a series of cell contact relationships is required for neuronal migration. The granule neuron has to form cell–cell contacts with astroglia for the glial cells to extend the highly elongated processes needed to support migration. For cell movement to follow, the granule cell has to assume a particular posture on the glial process, to form a specialized junction at the level of the cell soma, and to extend a leading process in the direction of migration.

NEURONAL REGULATION OF ASTROGLIAL DIFFERENTIATION

To analyze directly the influence of neurons on astroglial growth and differentiation, we developed methods to purify and then recombine granule cells and astroglia rapidly prior to plating them in microcultures. In the absence of neurons, astroglial cells proliferated rapidly and had a flattened morphology instead of the highly differentiated shapes normally seen in the microcultures. When neurons were added back to the astroglial cells, glial cell DNA synthesis ceased within 6 hours, and astroglial process extension commenced (Hatten, 1985).

The transformation of astroglial cells from flat, proliferating cells into the complex shapes seen *in vivo* depended on the presence of a neuron–glia ratio of at least 4:1. Below that ratio, many of the glial cells remained free of neuronal contacts and continued to proliferate. Above that ratio, the proportion of highly elongated glial forms increased dramatically, and extensive migration occurred along these elongated glial arms. This latter finding has provided a convenient technique for producing cultures with large numbers of migrating neurons (Edmondson and Hatten, 1987).

Recent experiments indicate that neuronal regulation of glial proliferation is membrane mediated (Hatten, 1987). Granule cell membranes arrest glial growth in a dose-dependent manner. In contrast, membrane material purified from PC12 neurons, 3T3 cells, or PTK cells do not influence glial DNA

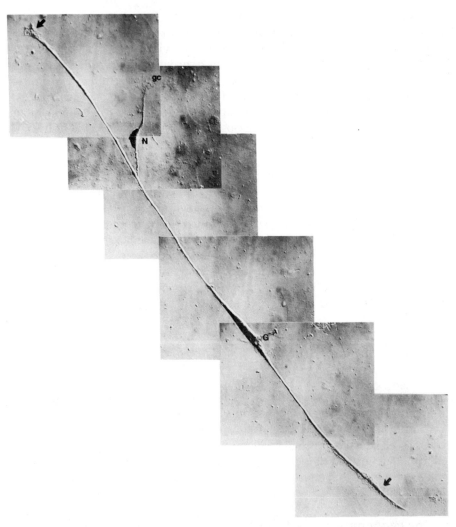

Figure 6. Weaver *granule neurons fail to form a migration apposition on wild-type astroglial cells.* Granule neurons were purified from P4 *weaver* B6CBA-A^w-J-*wv (wv/wv)* cerebellum and cocultured with astroglial cells purified from a normal (+/+) littermate. After 48 hours in culture, the cells were immunostained with AbGFP to visualize astroglial cells and their processes. *Weaver* neurons (N) failed to form cell–cell contacts between the neuronal somata and the glial process and attached only at the tips of short neurites. N, granule cell; gc, growth cone; G, glial cell body; *arrow*, broadened end foot typical of astroglia in coculture with *weaver* neurons. Video-enhanced differential interference Nomarski light microscopy (Allen et al., 1981) × 445 was used. (From Hatten et al., 1986.)

synthesis. This finding argues for a specific regulation of astroglial growth by CNS neurons and raises the possibility that membrane elements involved in neuron–glia contacts regulate glial growth (Hatten, 1987).

ASTROTACTIN, A NOVEL NEURONAL ANTIGEN INVOLVED IN NEURON–GLIA INTERACTIONS

To study the molecular mechanism of neuron–glia interactions, we used the microculture system to identify antigens that block neuron–glia interactions *in vitro*. Neuron–glia interactions were quantitated by measuring the distance of each neuron from glial processes immunostained with anti-glial filament protein antibodies and then plotting the distribution of the neurons relative to glial processes in the culture (Hatten and Liem, 1981). Immune sera, raised by immunizing rabbits with intact cerebellar cells maintained in culture for two days prior to injection, were added to cultures at the time the cells were plated, and neuron–glial contacts were quantitated 24 hours later. In the absence of antibodies, more than 90% of the neurons were located within 20 μm of a glial process.

In the presence of one of the immune sera we tested, potent blocking activity was seen. When Fab fragments (0.5 mg/ml) of this antiserum were added to the culture medium at the time the cells were plated, a random distribution of the neurons was seen 24 hours later; that is, fewer than 50% of the neurons were within 20 μm of a stained glial process (Figure 7). In addition, the form of the astroglial cells was affected. In the presence of immune serum or Fab fragments of the serum, glial process outgrowth was severely impaired. We named this blocking activity anti-astrotactin.

To distinguish whether the anti-astrotactin Fab fragments disrupted the formation of neuron–glia contacts, an event that other experiments indicate is necessary for glial process extension, we added Fab fragments (0.5 mg/ml) to the cultures at the time of plating and used high-resolution video microscopy to monitor their effects on the establishment of cell–cell contacts. Whereas in control cultures, stable contacts formed between neurons and glia within the first minute after the initial encounter between filopodia on the glial process and the granule neuron, stable contacts did not form in the presence of anti-astrotactin antibodies. Instead, the filopodia repeatedly withdrew without forming stable cell–cell contacts (Figures 8, 9).

Partial purification of the astrotactin-blocking antibodies was obtained by cellular absorption with PC12 cells, a clonal cell line that expresses both N-CAM (Thiery et al., 1977; Rutishauser, 1983) (BSP-2; Hirn et al., 1981) and NILE (McGuire et al., 1978; Salton et al., 1983; Sajovic et al., 1986) (Ng-CAM, L1; Grumet et al., 1984; Rathjen and Schachner, 1984) antigens, after which the blocking activity remained. Subsequent absorption with purified granule neurons, but not with astroglial cells, removed the blocking activity, suggesting that the antigens bound by the blocking antibodies are neuronal.

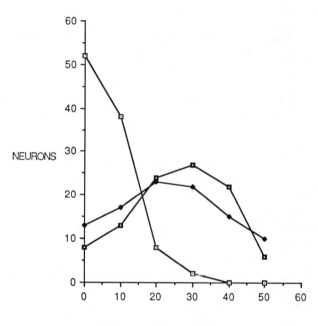

60 ┐

50

40

NEURONS 30

20

10

0

0   10   20   30   40   50   60

MICRONS TO NEAREST GLIAL PROCESS

Figure 7. *Anti-astrotactin antibodies block the formation of specific associations between neurons and astroglia in microcultures.* Distribution of the distances of individual neurons from AbGFP-stained glial processes in the presence (*filled squares*) or absence (*dotted squares*) of anti-astrotactin Fab fragments (0.5 mg/ml), added at plating. After 24 hours in untreated cultures, more than 90% of the neurons are within 20 μm of a glial process. In antibody-treated cultures, the distribution is bell shaped, resembling a computer-generated random distribution (*open squares*) of the cells. (From Edmondson et al., 1988.)

Figure 8. *(Top, right) Video-enhanced differential interference contrast, time-lapse video microscopy of astroglial process outgrowth and initial neuron–glial interactions in untreated normal cultures (a,b) and anti-astrotactin Fab fragment-treated (0.5 mg/ml) cultures (c,d).* In *a* and *b* stable cell–cell contacts form rapidly between the filopodia of the growing glial process (gc) and the granule cell (gn). In contrast, in *c* and *d* the filopodia withdraw after making contact and do not form a stable binding. Time is shown in minutes at the top of the fields. Calibration bar = 20 μm. (From Edmondson et al., 1988.)

Figure 9. *(Bottom, right) Video-enhanced differential interference contrast, time-lapse microscopy of astroglial process outgrowth and the formation of neuron–glial contacts in untreated cultures (a,b) and cultures treated with Fab fragments (0.5 mg/ml) of anti-astrotactin antibodies (c,d).* Time in minutes at the top of the fields is a continuation of Figure 8. Calibration bar = 20 μm. (From Edmondson et al., 1988).

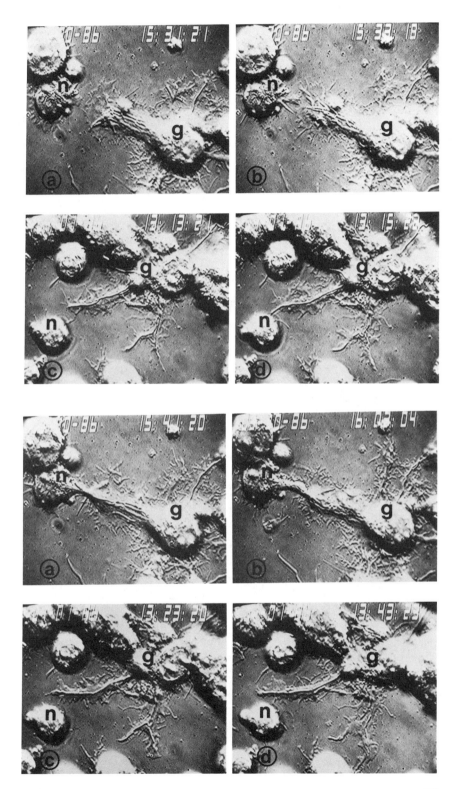

Immunoprecipitation of [^{35}S] methionine-radiolabeled or [^{3}H] fucose-radiolabeled Triton extracts of early postnatal mouse cells showed that the starting immune serum bound a large number of bands (Figure 10), including those with apparent molecular weights of N-CAM (180 kD and 140 kD) (Thiery et al., 1977) and NILE (230 kD) (McGuire et al., 1978). After absorption with PC12 cells, these bands were removed, and the bound bands were reduced to a prominent band at 100 kD and a smear of material between 80 ar

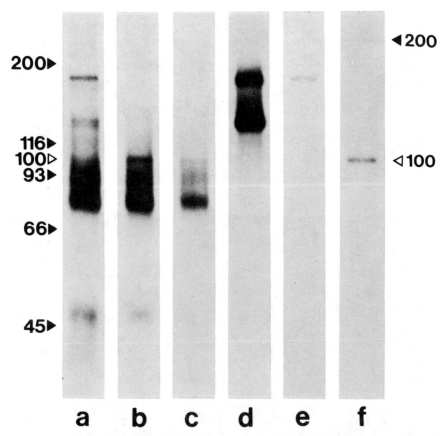

Figure 10. *Immunoprecipitation of [^{3}H] fucose-labeled material solubilized from P7 mouse cerebellar cells grown in microcultures for 5 days.* In *lane a* the starting serum binds a large number of bands. In *lane b* PC12 cell-absorbed antibodies recognize a major band at 100 kD and a diffuse smear of material at 80–90 kD. Subsequent absorption with purified granule cells (*lane c*) removes the band at 100 kD and the blocking activity in the culture assay. Immunoprecipitation of duplicate samples with antibodies against BSP-2 (*lane d*) and NILE (*lane e*) shows that these antigens are recognized by the starting serum, but are removed by PC12 cell absorption. *Lane f* shows Triton extracts of [^{3}H] fucose-labeled rat cerebellar cells immunoprecipitated with PC12 cell-absorbed serum. A single prominent band is seen at 100 kD. The molecular weight markers for lanes *a–e* are shown at *left*, those for *lane f* at *right*. (From Edmondson et al., 1988.)

90 kD. The band at 100 kD was removed by absorption with granule cells, a step that also removed the blocking activity in the culture assay. Immunoprecipitation of [³H] fucose-labeled rat cerebellar cells showed a single band at 100 kD, supporting the interpretation that this band is the prominent activity in the anti-astrotactin serum (Figure 10).

Studies on cells from the neurological mutant mouse *weaver* suggested that the astrotactin activity is missing or defective on *weaver* granule cells. When anti-astrotactin Fab fragments were absorbed with *weaver* granule neurons, the blocking activity was not removed. In contrast, cells from the wild-type littermate did remove the blocking activity. In addition, immunoprecipitation of [³H] fucose-radiolabeled Triton extracts of *weaver* cells showed that the intensity of the band at 100 kD was reduced by 95% in *weaver* cells (Edmondson et al., 1988).

To test directly whether antibodies against the cell adhesion molecules BSP-2 (N-CAM) or NILE (L1, Ng-CAM) would block neuron–glia interactions *in vitro*, we added Fab fragments of anti-BSP-2, kindly provided by Dr. C. Goridis (Marseilles), anti-L1, kindly provided by Dr. M. Schachner (Heidelberg), or anti-NILE, kindly provided by Dr. L. A. Greene of our department, either at the time the cells were plated or during the course of migration *in vitro*. None of the antisera that we tested interfered with neuron–glia contacts or inhibited neuronal migration along glial processes (Edmondson et al., 1988).

These findings differ from those of Lindner and Schachner (1983), who showed that anti-L1 antibodies altered the distribution of [³H] thymidine-labeled cells in explants of cerebellar tissue. Our experiments suggest that the perturbations seen in granule cell migration in slice preparations do not involve the cell–cell contact between the neuron and the astroglial process.

Here it is important to recall that the migration of the granule neuron involves not only the translocation of the cell soma down the arm of the Bergmann cell, but also the extension of parallel fibers, the outgrowth of which begins just prior to migration (Ramón y Cajal, 1911). As the granule cell leaves the external granular layer, it extends a leading process forward of the cell soma down the glial process and two parallel fibers into the molecular layer in the plane perpendicular to migration (Rakic, 1971). One explanation for the finding that anti-L1 antibodies block neuronal movements in tissue slices but do not block neuron–glia interactions in the microculture system is that these antibodies block axon–axon interactions needed to align the parallel fibers, but do not block neuron–glia contacts. This interpretation is consistent with experiments by Keilhauer et al. (1985), who showed that anti-L1 antibodies block neuron–neuron but not neuron–astroglial cell adhesion, and by Persohn and Schachner (1987), who showed by immunoelectron microscope localization that the L1 antigen is present on parallel fibers but not at neuron–glia appositions in developing mouse cerebellar cortex.

An advantage of the microwell culture system is that it allows analysis of the behavior of single migrating granule cells on a glial fiber in real time. A major task for the future will be to study the interactions and interrelationships of N-

CAM, NILE, astrotactin, the recently described adhesion molecule on glia (AMOG) (Antonicek et al., 1987), and other molecules thought to influence migration, including the extracellular matrix glycoprotein laminin (Liesi, 1985), plasminogen activator (Moonen et al., 1982), and a glia-derived protein that is an inhibitor of serine proteases (Lindner et al., 1986). One issue will be whether blocking activity will be seen in the cultures when the antibodies are combined or added in sequence, or whether indeed anti-NILE-blocking antibodies exist, but are very rare. In addition, it will be of interest to localize these antigens on the surfaces of living, migrating neurons (Gregory et al., 1987).

Although the microculture system provides a sensitive assay for whether a given antibody disrupts neuron–glia relationships, it is not well suited for monitoring the biochemical purification of astrotactin, because of the time needed to prepare, immunostain, and analyze the cultures. Recently we have developed a rapid assay for this purpose (Stitt and Hatten, 1987), an assay that stems from our finding that the kinetics of the binding of granule cell membranes to a number of astrocytoma cell lines, including the G26–24 line obtained originally from the C57Bl mouse, approximates that of granule cell membranes to primary cerebellar astroglial cells (Hatten and Shelanski, 1988).

The binding of [^{35}S] Met-radiolabeled rat granule cell membrane material to a monolayer of G26–24 cells is maximal after 30–60 min. In the presence of anti-astrotactin Fab fragments, binding is inhibited 70%. Thus, the rate of binding of granule cell membranes to glioma cells is a convenient, rapid assay for activities that block neuron–glia interactions.

SUMMARY

Neuronal migration is a complex event that depends on the exquisite timing of the developmental stage of the neuron and the glial cell, and on the appropriate differentiation of both cells prior to the contacts involved in the actual movements of the neuron. The critical cytological features of glia-guided migration appear to be the formation of an interstitial density between the cell soma of the migrating neuron and the glial process and the extension of a leading process out in the direction of migration. Initial evidence suggests that the immune activity we have named astrotactin disrupts neuron–glia interactions. Experiments are in progress to purify astrotactin and to assess its role in granule cell migration.

ACKNOWLEDGMENTS

I am grateful to my colleagues Drs. Jim Edmondson, Bill Gregory, Ron Liem, Joan Kuster, Carol Mason, Mike Shelanski, and Trevor Stitt, who collaborated with me on portions of this work. The research was supported by National Institutes of Health grants NS-15429 and NS-21097.

REFERENCES

Abercrombie, M. (1980) The crawling movement of metazoan cells. *Proc. R. Soc. Lond. [Biol.]* **207**:129–147.

Abercrombie, M., J. E. M. Heaysman, and S. M. Pegrimm (1970) The locomotion of fibroblasts in culture. I. Movements of the leading edge. *Exp. Cell Res.* **59**:393–398.

Aletta, J. M., and L. A. Greene (1988) Growth cone configuration and advance: A time-lapse study using video-enhanced differential interference contrast microscopy. *J. Neurosci.* **8**:1425–1435.

Allen, R. D., N. S. Allen, and J. L. Travis (1981) Video-enhanced differential interference contrast (AVEC-DIC) microscopy: A new method capable of analyzing related motility in the reticulopodial network of *Allogromia laticollaris. Cell Motil.* **1**:291–302.

Antonicek, H., E. Persohn, and M. Schachner (1987) Biochemical and functional characterization of a novel neuron–glia adhesion molecule that is involved in neuronal migration. *J. Cell Biol.* **104**:1587–1595.

Bastiani, M. J., and C. S. Goodman (1984) Neuronal growth specific interaction mediated by filopodial insertion and induction of coated vesicles. *Proc. Natl. Acad. Sci. USA* **81**:1849–1853.

Bentley, D., and A. Toroian-Raymond (1984) Disoriented pathfinding by pioneer growth cones deprived of filopodia by cytochalasin treatment. *Nature* **323**:712–715.

Bray, D. (1987) Growth cones: Do they pull or are they pushed? *Trends Neurosci.* **10**:431–434.

Bray, D., and K. Chapman (1985) Analysis of microspike movements on the neuronal growth cone. *J. Neurosci.* **5**:3204–3213.

Bunge, M. B. (1977) Initial endocytosis of peroxidase or ferritin by growth cones of cultured nerve cells. *J. Neurocytol.* **6**:407–439.

Cheng, T. P. O., and T. S. Reese (1985) Polarized compartmentalization of organelles in growth cones from developing optic tectum. *J. Cell Biol.* **101**:1473–1478.

Dunn, G. A., and J. P. Heath (1976) A new hypothesis for contact guidance in tissue cells. *Exp. Cell Res.* **101**:1–14.

Edmondson, J. C., and M. E. Hatten (1987) Glial-guided neuronal migration *in vitro:* A high resolution time-lapse video microscopic study. *J. Neurosci.* **7**:1928–1934.

Edmondson, J. C., R. K. H. Liem, J. E. Kuster, and M. E. Hatten (1988) Astrotactin: A novel neuronal cell surface antigen that mediates neuron–astroglial interactions in cerebellar microcultures. *J. Cell Biol.* **106**:505–517.

Goldberg, D. J. and D. W. Burmeister (1986) Stages in axon formation. Observations of growth of *Aplysia* axons in culture using video-enhanced contrast differential interference contrast microscopy. *J. Cell Biol.* **103**:1921–1931.

Goldowitz, D., and R. J. Mullen (1982) Granule cell as a site of gene action in the *weaver* mouse cerebellum: Evidence from heterozygous mutant chimeras. *J. Neurosci.* **2**:1474–1485.

Gregory, W. A., J. C. Edmondson, M. E. Hatten, and C. A. Mason (1988) Cytology and neuron–glial apposition of migrating cerebellar granule cells *in vitro. J. Neurosci.* **8**:1728–1738.

Gregory, W. A., C. A. Mason, and M. E. Hatten (1987) Immunocytochemical studies of cell surface molecules of migrating neurons *in vitro. Soc. Neurosci. Abstr.* **13**:1636.

Grumet, M., S. Hoffman, and G. M. Edelman (1984) Two antigenically related neuronal adhesion molecules of different specificities mediate neuron–neuron and neuron–glia adhesion. *Proc. Natl. Acad. Sci. USA* **81**:267–271.

Hammarback, J. A., and P. C. Letourneau (1986) Neurite extension across regions of low cell–substratum adhesivity: Implications for the guidepost hypothesis of axonal pathfinding. *Dev. Biol.* **117**:655–662.

Hatten, M. E. (1985) Neuronal regulation of astroglial morphology and proliferation *in vitro. J. Cell Biol.* **100**:384–396.

Hatten, M. E. (1987) Neuronal inhibition of astroglial cell proliferation is membrane mediated. *J. Cell Biol.* **104**:1353–1360.

Hatten, M. E., and R. K. H. Liem (1981) Astroglial cells provide a template for the positioning of developing cerebellar neurons *in vitro. J. Cell Biol.* **90**:622–630.

Hatten, M. E., R. K. H. Liem, and C. A. Mason (1984a) Two forms of cerebellar glial cells interact differently with neurons *in vitro. J. Cell Biol.* **98**:193–204.

Hatten, M. E., R. K. H. Liem, and C. A. Mason (1984b) Defects in specific associations between astroglia and neurons occur in microcultures of *weaver* mouse cerebellar cells. *J. Neurosci.* **4**:1163–1172.

Hatten, M. E., R. K. H. Liem, and C. A. Mason (1986) *Weaver* cerebellar cells fail to migrate on wild-type astroglial processes *in vitro. J. Neurosci.* **9**:2676–2683.

Hatten, M. E., and M. L. Shelanski (1988) Cerebellar granule neurons arrest the proliferation of human and rodent astrocytoma cell lines. *J. Neurosci.* **8**:1447–1453.

Hirano, A., and H. M. Dembitzer (1973) Cerebellar alterations in the *weaver* mouse. *J. Cell Biol.* **56**:478–486.

Hirn, M., M. Pierres, H. Deagostini-Bazin, M. Hirsch, and C. Goridis (1981) Monoclonal antibody against cell surface glycoprotein of neurons. *Brain Res.* **214**:433–439.

Inouye, S. (1986) *Video Microscopy,* Plenum, New York.

Keilhauer, G., A. Faissner, and M. Schachner (1985) Differential inhibition of neurone–neurone, neurone–astrocyte, and astrocyte–astrocyte adhesion by L1, L2, and N-CAM antibodies. *Nature* **316**:728–730.

Letourneau, P. C. (1982) Nerve fiber growth and its regulation by extrinsic factors. In *Neuronal Development,* N. C. Spitzer, ed., pp. 213–254, Plenum, New York.

Liesi, P. (1985) Do neurons in the CNS migrate on laminin? *EMBO J.* **4**:1163–1170.

Lindner, J., and M. Schachner (1983) L1-mono and polyclonal antibodies modify cell migration in early postnatal cerebellum. *Nature* **305**:427–429.

Lindner, J., J. Guenthner, H. Nick, G. Zinser, H. Antonicek, M. Schachner, and D. Monard (1986) Modulation of granule cell migration by a glia-derived protein. *Proc. Natl. Acad. Sci. USA* **83**:4568–4571.

Ludueña, M. A., and N. K. Wessells (1973) Cell locomotion, nerve elongation, and microfilaments. *Dev. Biol.* **30**:427–440.

Mason, C. A., J. C. Edmondson, and M. E. Hatten (1988) The extending astroglial process: Development of glial shape, the growing tip, and interactions with neurons. *J. Neurosci.* **8**:3124–3134.

McGuire, J. C., L. A. Greene, and A. V. Furano (1978) NGF stimulated incorporation of fucose or glucosamine into an external glycoprotein in cultured rat PC12 pheochromocytoma cells. *Cell* **15**:357–365.

Moonen, G., M. P. Grau-Wagemans, and I. Selak (1982) Plasminogen activator-plasmin system and neuronal migration. *Nature* **298**:753–755.

Persohn, E., and M. Schachner (1987) Immunoelectron microscopic localization of the neural cell adhesion molecules L1 and N-CAM during postnatal development of the mouse cerebellum. *J. Cell Biol.* **105**:569–576.

Rakic, P. (1971) Neuron–glia relationship during granule migration in *Macacus rhesus. J. Comp. Neurol.* **141**:312.

Rakic, P. (1972) Mode of migration of the superficial layers of the fetal monkey neocortex. *J. Comp. Neurol.* **145**:85.

Rakic, P. (1985) Contact regulation of neuronal migration. In *The Cell in Contact: Adhesions and Junctions as Morphogenetic Determinants,* G. M. Edelman and J. P. Thiery, eds., pp. 67–91, Wiley, New York.

Rakic, P., and R. L. Sidman (1973) *Weaver* mutant mouse cerebellum: Defective neuronal migration secondary to abnormality of Bergmann glia. *Proc. Natl. Acad. Sci. USA* **70**:240–244.

Rakic, P., L. J. Stensaas, E. P. Sayre, and R. L. Sidman (1974) Computer-aided three-dimensional reconstruction and quantitative analysis of serial electron microscopic montages of fetal monkey brain. *Nature* **250**:31–34.

Ramón Y Cajal, S. (1911) *Histologie du système nerveux de l'homme et des vertébrés*, Maloine, Paris. Reprinted Consejo Superior de Investigaciónes Científicas, Madrid, 1955, Vol. 2.

Rathjen, F. G., and M. Schachner (1984) Immunocytochemical and biochemical characterization of a new neuronal surface component (L1 antigen) which is involved in cell adhesion. *EMBO J.* **3**:1–10.

Rees, R. P., and T. S. Reese (1981) New structural features of freeze-substituted neuritic growth cones. *Neuroscience* **6**:247–254.

Rezai, Z., and C. H. Yoon (1972) Abnormal rate of granule cell migration in the cerebellum of *weaver* mutant mice. *Dev. Biol.* **29**:17–26.

Rutishauser, U. (1983) Molecular biological properties of a neural cell adhesion molecule. *Cold Spring Harbor Symp. Quant. Biol.* **48**:501–514.

Sajovic, P., E. Kouvelas, and E. Trenkner (1986) Probable identity of NILE glycoprotein and the high molecular weight component of L1 antigen. *J. Neurochem.* **47**:541–546.

Salton, S. R. J., C. Richter-Landsberg, L. A. Greene, and M. L. Shelanski (1983) Nerve-growth factor inducible large external (NILE) glycoprotein: Studies of a central and peripheral neuronal marker. *J. Neurosci.* **3**:441–454.

Sidman, R. L., and P. Rakic (1973) Neuronal migration with special reference to human brain: A review. *Brain Res.* **62**:1–35.

Sotelo, C. E., and J.-P. Changeux (1974) Bergmann fibers and granule cell migration in the cerebellum of the homozygous *weaver* mutant mouse. *Brain Res.* **77**:484–491.

Stitt, T. N., and M. E. Hatten (1987) Development of a rapid biological assay for the cell-surface ligand astrotactin. *Soc. Neurosci. Abstr.* **13**:193.

Thiery, J. P., R. Brackenbury, U. Rutishauser, and G. M. Edelman (1977) Adhesion among neural cells of the chick embryo. II. Purification and characterization of a cell adhesion molecule from neural retina. *J. Biol. Chem.* **252**:6841–6845.

Trenkner, E., and R. L. Sidman (1977) Histogenesis of mouse cerebellum in microwell cultures: Cell reaggregation migration, fiber and synapse formation. *J. Cell Biol.* **75**:915–940.

Trenkner, E., D. Smith, and N. Siegel (1984) Is granule cell maturation controlled by an internal clock? *J. Neurosci.* **4**:2850–2855.

Wessells, N. K. (1982) Axon elongation: A special case of cell locomotion. In *Cell Behavior: A Tribute to Michael Abercrombie*, R. Bellairs, ed., pp. 225–246, Cambridge Univ. Press, New York.

Willinger, M., and D. M. Margolis (1985) Effect of the *weaver (wv)* mutation on cerebellar neuron differentiation. I. Qualitative observations of neuron behavior in culture. *Dev. Biol.* **107**:156–172.

Chapter 23

The Nerve Growth Factor–Inducible Large External (NILE) Glycoprotein: Biochemistry and Regulation of Synthesis

LLOYD A. GREENE
MICHAEL L. SHELANSKI

ABSTRACT

In this chapter we review the biochemistry and regulation of synthesis of the nerve growth factor–inducible large external glycoprotein (NILE GP). Originally described as an NGF-inducible surface glycoprotein in PC12 pheochromocytoma cells, this molecule is identical to components recognized by antisera to glycoproteins B2, Ng-CAM, and L1. The NILE GP appears to be present on all mammalian neurons and on Schwann cells, but not on most other cell types. It has an apparent molecular weight by SDS-PAGE of 200–230 kD and is both phosphorylated and sulfated. The size heterogeneity of NILE GP obtained from different cell types appears to be due, at least in large part, to differences in complex oligosaccharide structure and, in particular, to the variable presence of poly(N-acetyllactosaminyl) oligosaccharides. The nonglycosylated core peptide has an apparent molecular weight by SDS-PAGE of 150–160 kD. Although NILE GP is an intrinsic membrane component, it is also released from the cell surface in a slightly smaller, soluble form (s-NILE GP). The hydrodynamic properties of the molecule indicate that it is rod shaped. Recent molecular cloning of cDNAs for NILE GP reveal additional structural insight into the molecule, including its membership in the IgG superfamily. Studies of the regulation of NILE GP levels in PC12 cells indicate no effect of a variety of agents, with the exception of NGF and fibroblast growth factor. Continued examination of NILE GP biochemistry and synthesis regulation will be highly relevant to understanding its functional role in the nervous system.

AIMS AND BACKGROUND

The aim of this chapter is to review knowledge regarding the biochemistry and regulation of synthesis of what has been termed by our laboratory the nerve growth factor (NGF)–inducible large external glycoprotein (NILE GP) (McGuire et al., 1978). It should be noted that this chapter was largely written in the spring of 1987, with a degree of updating having been carried out at the proof stage.

As described below, NILE GP bears a relationship to components that have been subsequently described by a variety of other terminologies, including B2, L1, and Ng-CAM. For the purposes of this chapter, however, the NILE GP designation will be retained.

The NILE GP was described and so named as a result of studies designed to detect the effects of NGF treatment on the protein composition of cultured rat PC12 pheochromocytoma cells (McGuire et al., 1978). PC12 cells respond to NGF by undergoing a change from a phenotype that resembles dividing chromaffin/pheochromoblast cells to a nondividing, neurite-bearing pheno-type that resembles sympathetic neurons (Greene and Tischler, 1976). One- and two-dimensional electrophoretic analyses were used to resolve and characterize PC12 cell proteins labeled with [^{14}C]amino acids or [^{3}H]fucose either before or after various times of NGF exposure. Although few changes were observed in overall protein composition, a substantial NGF-dependent increase was noted in the relative incorporation of fucose into a GP of apparent molecular weight of 230 kD. The proteolysis of this component that occurs when intact PC12 cells are exposed to trypsin indicated that it is surface exposed. The characteristic responsiveness to NGF, high apparent molecular weight, surface localization, and fucose content of this protein led to the acronym NILE GP. As discussed in greater detail below, all neurons and, to a lesser extent, Schwann cells bear NILE GP, whereas most other cell types do not appear to do so.

RELATIONSHIP OF NILE GP TO OTHER IDENTIFIED GLYCOPROTEINS

The study of the NILE GP has been aided by its high antigenicity (Lee et al., 1977). Polyclonal antisera have been generated from PC12 cell material that specifically recognize this component in PC12 cell extracts (Salton et al., 1983a; Stallcup et al., 1983). Such antisera have been quite useful for establishing the relationship of the NILE GP to glycoproteins described by others. Sweadner (1983a) has described a glycoprotein (B1), its precursor (P1), and its modified derivatives (B2 and S2) that are present in cultures of rat sympathetic neurons and that are in the range of 210–230 kD. These were all found to cross-react with antiserum (Salton et al., 1983a) against PC12 cell NILE GP; likewise, a monoclonal antibody against sympathetic neuron material that recognizes B1 and B2 also specifically recognized PC12 cell NILE GP (Sweadner, 1983a).

Monoclonal antisera and polyclonal antibodies have been raised against material, originally isolated from chicken tissue, that has been designated as Ng-CAM (for reviews, see Edelman, 1985, 1986; Cunningham and Edelman, this volume). Several types of evidence (Friedlander et al., 1986) indicate that in PC12 cells these antibodies recognize a single NGF-regulated surface com-ponent of apparent molecular weight of 230 kD that appears to be identical to the component recognized by an anti–NILE GP antiserum (Stallcup et al., 1983). A single component of about this size was also recognized in mouse brain (Friedlander et al., 1986). Such findings have led to the conclusion that the Ng-CAM is the avian equivalent to NILE GP.

Another antiserum has been prepared against what has been designated the L1 antigen (Rathjen and Schachner, 1984; for a review, see Schachner et al., this volume). This antiserum recognizes glycoproteins of 140 kD and 200 kD in adult mouse brain. Two studies (Bock et al., 1985; Sajovic et al., 1986) have shown that anti–NILE GP antiserum (Salton et al., 1983a) and anti-L1 (Rathjen and Schachner, 1984) both recognize the same 230-kD glycoprotein in extracts of PC12 cells. Thus, NILE GP and the larger component of L1 antigen present in brain also appear to be equivalent. This conclusion has recently been confirmed by cloning studies (Moos et al., 1988; Prince et al., 1989).

RECOGNITION OF NILE GP BY ANTISERA

As originally defined, and for the purposes of this chapter, NILE GP is a molecule of apparent molecular weight in the range of 200–230 kD. The high antigenicity of the NILE GP has led to the preparation of a variety of immunological reagents that recognize this component. These have been of major use in characterizing the biochemical properties of NILE GP and, as reviewed above, in establishing its relationship to other glycoproteins. Moreover, much interest in this glycoprotein has recently arisen from the observations that antibodies that recognize NILE GP/Ng-CAM/L1 are reported to show specific patterns of staining in the nervous system (cf. Faissner et al., 1984; Edelman, 1985, 1986; Stallcup et al., 1985; Cunningham and Edelman, this volume; Schachner et al., this volume) and to interfere with neuron–glia interaction (for reviews, see Edelman, 1985, 1986; Cunningham and Edelman, this volume) and neurite–neurite interaction (Edelman, 1985, 1986; Stallcup and Beasley, 1985b; Cunningham and Edelman, this volume; Schachner et al., this volume) as well as with developmental neuronal migration (Lindner et al., 1983; Edelman, 1985, 1986; Cunningham and Edelman, this volume; Schachner et al., this volume). It is therefore of importance to consider precisely what molecules such antibodies are reported to recognize in various neural tissues. This will, in turn, aid in evaluating the certitude that the 230-kD species is the only species localized in the stained structures and whether it is unambigously involved in the tested functional behaviors.

Polyclonal antisera have been prepared against whole PC12 cells and membrane fractions thereof (Lee et al., 1977, 1981; Stallcup et al., 1983). These recognize NILE GP as well as several other glycoproteins and thus are of only limited use per se in studying localization and function. An antiserum was also prepared against NILE GP isolated from PC12 cells by lectin-affinity chromatography and SDS-PAGE (Salton et al., 1983a). With this antiserum, a single 230-kD band was specifically immunoprecipitated from PC12 cells labeled with methionine (Salton et al., 1983a) or fucose (Salton et al., 1983a; Bock et al., 1985; Sajovic et al., 1986). Similar experiments with fucose-labeled neurons revealed a major component corresponding to the NILE GP (see below for details) and several very minor components of 140–180 kD (Salton et al., 1983a). It was speculated that the latter might be degradation products. Since this antiserum also recognizes the nonglycosylated core peptide of NILE GP (see below for discussion), at least part of the activity in this material is directed against the protein rather than carbohydrate portion of the molecule.

The same antiserum was also tested by immunoblotting crude membrane fractions of adult mouse cerebellum (Bock et al., 1985). Along with the NILE GP band, there was also prominent recognition of components of 140 kD and 80 kD. It was hypothesized that the latter were proteolytic degradation products. A second, independent set of PC12 cell anti–NILE GP antisera has been raised and characterized (Stallcup et al., 1983; Stallcup and Beasley, 1985a). In this case antisera were raised against NGF-treated PC12 cells, and IgG fractions thereof were absorbed exhaustively with several nonneuronal cell lines and with rat adrenal tissue. These antisera specifically immuno-precipitated fucose-labeled NILE GP from PC12 cells and from a variety of cell lines. In addition, it was reported that this antisera can immunoprecipitate a very minor glycoprotein of 150–170 kD (Stallcup et al., 1985). Degradation was again invoked as a possible source of such material. At least part of the activity of this antiserum was also found to be against the peptide portion of the NILE GP molecule.

Several monoclonal antibodies that appear to recognize the NILE GP have been reported. One, designated ASCS4, was prepared against sympathetic neurons and immunoprecipitates NILE GP from fucose-labeled PC12 cells and cultured rat sympathetic neurons (Sweadner, 1983a). Another, prepared against cultured rat cerebellar cells and designated 69-A1, immunoprecipitates two prominent components from lysates of iodinated cell surface molecules of cultured rat cerebellar cells (Pigott and Kelly, 1986). One appears to be NILE GP, and the other component is 150 kD.

As noted above, anti-L1 antiserum recognizes NILE GP in PC12 cells and in brain. In the latter, an additional component of 140 kD is also recognized by this antiserum (cf. Bock et al., 1985). Of the two studies using anti-L1 antiserum to immunoprecipitate fucose-labeled material from PC12 cells, one (Bock et al., 1985) reported that a single band corresponding to NILE GP was observed, while the other (Sajovic et al., 1986) noted that a small amount of an additional band of 140–150 kD was also specifically immunoprecipitated. The latter study did not observe such material when immunoprecipitation was with anti–NILE GP antiserum.

Finally, as discussed above, antibodies against chicken Ng-CAM recognize PC12 cell NILE GP (Friedlander et al., 1986). Although such antibodies recognize three components in chicken brain, by both the blotting of mouse brain cell membrane extracts and the immunoprecipitation of surface ^{125}I-labeled rat brain, dorsal root ganglia, and PC12 cells, only a single component corresponding to NILE GP was reported.

In summary, at least several of the antisera that recognize NILE/L1/Ng-CAM also cross-react to various degrees with additional components. Such components have been suggested to be degradation products of the parent molecule. In support of this, there is evidence that NILE/L1/Ng-CAM can be proteolytically cleaved into fragments of apparent molecular weights similar to those of the cross-reactive species (Salton et al., 1983b; Edelman, 1985, 1986; Faissner et al., 1985; Sadoul et al., 1988; Prince et al., 1989). However, the presence of cross-reacting species nevertheless suggests caution in the use of immunological tools to attribute specific functional or localization properties to the NILE GP.

TISSUE LOCALIZATION OF NILE GP

The availability of antisera that recognize NILE GP has enabled its direct localization in mammalian tissues. By immunoprecipitation analysis, NILE GP has been found in a variety of types of PNS and CNS neurons (Lee et al., 1981; Margolis et al., 1983; Salton et al., 1983a; Stallcup et al., 1983; Friedlander et al., 1986). Western blotting analysis has also confirmed the presence of NILE GP in the CNS (Bock et al., 1985; Friedlander et al., 1986). Immunostaining has confirmed this and has indicated that the NILE GP is present on virtually all types of mammalian PNS and CNS neurons (cf. Salton et al., 1983a; Faissner et al., 1984; Stallcup et al., 1985). Immunoprecipitation of metabolically labeled glycoproteins has also directly demonstrated that cultured Schwann cells can synthesize the NILE GP (Salton et al., 1983a; Stallcup et al., 1983). Aside from neurons and Schwann cells, however, the synthesis or presence of NILE GP or of NILE GP cross-reactive proteins has not been detected in a wide variety of other tissues surveyed (cf. Lee et al., 1981; Salton et al., 1983a; Stallcup et al., 1983; Edelman, 1985, 1986; Cunningham and Edelman, this volume; Schachner et al., this volume). Immunochemical data, however, indicate the presence of NILE GP in crypt cells of the intestine (Thor et al., 1987).

A number of antisera and antibodies that recognize NILE GP have been used for the cellular localization of NILE GP cross-reactive activity (cf. Faissner et al., 1984; Edelman, 1985, 1986; Pigott and Kelly, 1986; Rieger et al., 1986; Sajovic et al., 1986; Sanes et al., 1986; Beasley and Stallcup, 1987; and chapters in this volume by Cunningham and Edelman, Rieger, and Schachner et al.). For the details of these findings, the reader is referred to the chapters cited above.

HETEROGENEITY OF THE APPARENT MOLECULAR WEIGHT OF NILE GP ON DIFFERENT CELL TYPES

As originally described on both NGF-treated and NGF-untreated PC12 cells (McGuire et al., 1978), the apparent molecular weight of NILE GP as estimated by SDS-PAGE is 230 kD. It was noted that the estimated electrophoretic mobility of NILE GP present on cultured rat sympathetic and sensory neurons

(Lee et al., 1981; Salton et al., 1983a; Stallcup et al., 1983; Sweadner, 1983a) as well as on cultured Schwann cells (Salton et al., 1983a; Stallcup et al., 1983), is consistently slightly greater than that on PC12 cells. The apparent molecular weight of such material is 220–225 kD. NILE GP immunoprecipitated from the N18 murine neuroblastoma and B35 neuronal cell lines (Stallcup et al., 1983; Stallcup and Beasley, 1985a) and from a human neuroblastoma cell line (Salton et al., 1983a) also migrates with an apparent molecular weight of approximately 220–225 kD. Analysis of NILE GP in the central nervous system by SDS-PAGE has revealed this material to be of even greater mobility, with an apparent molecular weight of 200–215 kD. This is so for material immunoprecipitated from cultured rodent cerebellum, cerebral cortex, and spinal cord (Salton et al., 1983a; Stallcup et al., 1983; Stallcup and Beasley, 1985a; Bock et al., 1986) or from glucosamine-labeled brains of adult rats (Salton et al., 1983a) or recognized by immunoblots of adult mouse brain (Bock et al., 1986; Friedlander et al., 1986). Such findings suggest that the forms of NILE GP present in the PNS and CNS are slightly different from one another and that these in turn differ slightly from that present on PC12 cells. The likely source of these differences will be discussed below. Another point that arises from such studies is that no differences have yet been detected between NILE GP in the developing and adult nervous systems.

SYNTHESIS OF NILE GP

To examine the nonglycosylated core peptide of the NILE GP, cultured neuronal cells have been exposed to tunicamycin, an inhibitor of N-linked glycosylation. As detected immunologically, and analyzed by SDS-PAGE, the core protein has an apparent molecular weight of 150–160 kD (Salton et al., 1983b; Faissner et al., 1985; Stallcup and Beasley, 1985a). Similar results were achieved with both NGF-treated and NGF-untreated PC12 cells and cultures of dissociated cerebellum and cerebral cortex, although the possibility remains that a very slight difference could be detected in the electrophoretic mobilities of the apoproteins of PC12 cells and cultured CNS neurons (Salton et al., 1983b). Treatment of NILE GP from various sources with neuraminidase followed by endoglycosidase F also produced a component with an apparent molecular weight near 160 kD (Stallcup and Beasley, 1985a; see also Faissner et al., 1985). However, there were slight differences in the electrophoretic mobilities of products obtained by these means from different cell types, raising the suggestion that there may be multiple forms of the nonglycosylated core protein.

Pulse-labeling experiments have been employed to examine sequential processing during synthesis of NILE GP in PC12 cells (Stallcup and Beasley, 1985a) and cultured rat sympathetic neurons (Sweadner, 1983a). In the former study a 190-kD neuraminidase-insensitive precursor was detected that was then processed, presumably in the Golgi, to the mature neuraminidase-sensitive 230-kD molecule. The latter study provided evidence for the initial appearance of a precursor of 210 kD that is not surface exposed, and is poor or lacking in fucose, rich in mannose, and nonsialylated. Within an hour, this precursor is sequentially converted to surface-exposed mature forms of 230 kD and 215 kD, presumably by the removal of most of the mannose and the formation of complex oligosaccharide units containing fucose and sialic acid.

THE OLIGOSACCHARIDE STRUCTURE OF NILE GP AND ITS CONTRIBUTION TO THE APPARENT MOLECULAR HETEROGENEITY OF NILE GP FROM DIFFERENT CELL TYPES

Complex Oligosaccharide Structure

The oligosaccharide structures of NILE GP on PC12 cells and in rat brain have been analyzed and compared (Margolis et al., 1983, 1986). Tri- and tetra-antennary complex oligosaccharides account for 80–90% of the total carbohydrate in NILE GP; 2–3% consists of high-mannose oligosaccharides, and the remainder of bi-antennary complex oligosaccharides. A susceptibility to alkali treatment and the complete inhibition of glycosylation by tunicamycin (Salton et al., 1983b; Stallcup and Beasley, 1985a) suggest the absence of O-glycosidically linked oligosaccharides in the NILE GP. For the PC12 cell NILE GP, a moderate degree of microheterogeneity was found among the predominant tri- and tetra-antennary oligosaccharide units with respect to the presence of core fucose, outer galactose, and sialic acid residues, and the substitution positions on the α-linked mannose residues. However, there was no apparent significant effect of NGF treatment on the PC12 cell NILE GP oligosaccharide structure. Furthermore, analysis of brain and PC12 cell NILE GP glycopeptides revealed that the difference in apparent molecular weight between NILE GP from these two sources is apparently due to the greater size, rather than number, of the PC12 cell tri- and tetra-antennary complex oligosaccharides.

Poly(N-Acetyllactosaminyl) Oligosaccharides

As one aspect of investigating the nature of the size differences between the oligosaccharides present on PC12 cell and rat brain NILE GP, glycopeptides derived from each NILE GP were treated with endo-β-galactosidase (Margolis et al., 1986). Susceptibility to the latter enzyme is considered to be diagnostic for the presence of poly(N-acetyllactosaminyl) oligosaccharides. Although this had no effect on CNS NILE GP glycopeptide fractions, there was a significant effect on comparable fractions from PC12 cells. Moreover, the enzyme-treated PC12 material was the same size as that derived from brain. These findings indicate the presence of poly(N-acetyllactosaminyl) oligosaccharides on the PC12 cell, but not rat brain, NILE GP and suggest that such groups account for much of the difference in apparent size between the two forms of the molecule. To further characterize PC12 cell glycoproteins containing poly(N-acetyl-lactosaminyl) oligosaccharides, cultures were treated sequentially with endo-β-galactosidase and galactosyltransferase + UDP-[^{14}C]galactose (Margolis et al., 1986). This revealed that at least 10–20 PC12 cell glycoproteins, including (as anticipated) the NILE GP, possess such oligosaccharides and that NGF treatment did not alter the relative number of such groups on the NILE GP.

Although NILE GP on peripheral neurons has not yet been examined for the presence of poly(N-acetyllactosaminyl) oligosaccharides, one may presently hypothesize that such groups could at least partly account for the difference between the apparent sizes of the PNS and CNS forms of the molecule. The possible functional role of such modifications is not yet clear. In at least several tissues the presence and fine structure of poly(N-acetyllactosaminyl) oligo-saccharides appear to be related to specific stages of cell and tissue dif-ferentiation (cf. Fukada, 1985).

Sialic Acid Residues

The neuraminidase sensitivity of NILE GP isolated from various cellular sources has been characterized (Stallcup and Beasley, 1985a). Material from PC12 cells, murine neuroblastoma cells, B35 cells, and P5 cerebellar cultures all showed an increase in electrophoretic mobility after treatment with neur-aminidase. The treated PC12 material had an apparent molecular weight of 215 kD, while both the degree of shift in electrophoretic mobility and apparent molecular weight of that from the other cell types was somewhat smaller. It was concluded that NILE GP from different sources vary in their contents of sialic acid and that this contributes to their differences in electrophoretic mobility. The analysis of PC12 cell and rat brain oligosaccharides (Margolis et al., 1983) is consistent with this conclusion. Finally, although the NILE GP contains sialic acid, there is no evidence of polysialation of the molecule.

PHOSPHORYLATION OF NILE GP

Immunoprecipitation (with anti–NILE GP antiserum) of extracts of PC12 cells or cultured neonatal rat sympathetic neurons metabolically labeled by exposure to [^{32}P]orthophosphate has established that the NILE GP is a phos-phoprotein (Salton et al., 1983b; see also Faisnner et al., 1985; Sadoul et al., 1988; Prince et al., 1989). The phosphate-labeled PC12 cell NILE GP shows the same pattern by two-dimensional IEF × SDS-PAGE as does NILE GP labeled with [^{3}H]fucose; this indicates that PC12 cells are unlikely to possess an unphosphorylated form of this GP. Although several PC12 cell proteins show rapid changes in levels of phosphorylation after short-term (i.e., 0.25–2 hours) exposure of cultures to NGF, little or no change in NILE GP phosphorylation is apparent within 2–24 hours of treatment with the factor. Scintillation counting and autoradiography of immunoprecipitated material indicate that longer times of NGF exposure also do not alter the relative level of phosphate incorporation into NILE GP.

Alkaline treatment of gels containing immunoprecipitated phosphate-labeled NILE GP suggests that most of the phosphate is serine linked, with a small proportion possibly linked to threonine or tyrosine residues. Incubation of anti–NILE GP immunoprecipitates with γ-labeled [^{32}P]ATP showed no detectable catalysis of the phosphorylation of either immunoglobulin or of NILE GP itself, thus indicating that this GP is not itself active as a phospho-transferase. There is as yet no information as to the identity of the kinase(s) responsible for phosphorylation of NILE GP.

When intact, phosphate-labeled PC12 cells are subjected to mild trypsini-zation followed by immunoprecipitation with anti-NILE antiserum, a prominent membrane-associated phosphorylated fragment of approximately 90 kD is detected, along with several less abundant phosphorylated membrane-bound fragments of approximately 40 kD, 32 kD, and 28 kD (Salton et al., 1983; see also Sadoul, 1988; Prince et al., 1989). Of these, only the approximately 90-kD fragment contains incorporated fucose. After more extensive trypsinization, immunoprecipitation yields only a single phosphate-labeled membrane-associated fragment of approximately 28 kD. This fragment, however, cannot be observed if labeling is with [^{3}H] fucose. These observations suggest: (1) The phosphate is confined to a 28-kD domain of the protein that does not possess carbohydrates and that is likely to contain the transmembrane region(s) of the molecule; and (2) starting from the membrane-bound terminus of the molecule, the carbohydrates begin to be present somewhere between ap-proximately the 40th and 90th residues. With regard to these observations, it is of interest that immunoprecipitation of phosphate-labeled proteins from rat sympathetic neurons with anti–NILE GP antiserum yields, along with the characteristic NILE GP band at 210 kD, a somewhat less prominent band of 90 kD and several minor components of 116–200 kD. Though of less relative abundance, phosphorylated bands of similar size can also be detected in immunoprecipitates from NGF-treated PC12 cells. This raises the possibility that sympathetic neurons and PC12 cells may normally possess small amounts of proteolytically cleaved NILE GP. This, however, cannot be presently distinguished from the possible occurrence of NILE GP cross-reactive phosphoproteins.

SULFATION OF NILE GP

To test whether NILE GP may contain sulfate groups, PC12 cells have been metabolically labeled with [^{35}S]SO$_4$ and cell extracts immunoprecipitated with anti–NILE GP antiserum (P. Sajovic, L. A. Greene, R. U. Margolis, and R. K. Margolis, unpublished data). Analysis of the immunoprecipitate by SDS-

PAGE and autoradiography reveals a single sulfate-labeled band in the position of NILE GP (230 kD). The relative degree of incorporation of the sulfate does not appear to be significantly altered after treatment of the cells with NGF. Preliminary experiments carried out with tunicamycin-treated cultures indicate that most or all of the incorporated sulfate is attached to carbohydrates. Experiments carried out with SO_4-labeled rat brain (Bock et al., 1986) have demonstrated that the L1 antigen also contains sulfate groups. It is not yet clear whether sulfation alters the physical and/or functional properties of the NILE GP, nor has it been established whether the entire cellular population of NILE GP is sulfated.

RELEASE OF SOLUBLE NILE GP FROM CELLS

Several studies have demonstrated the release of NILE GP in a soluble form from PC12 cells (Salton et al., 1983b; Stallcup et al., 1983; Richter-Landsberg et al., 1984) and other cell lines (Prince et al., 1989), as well as by cultured neonatal rat sympathetic neurons (Sweadner, 1983a; Richter-Landsberg et al., 1984) and central nervous system neurons (Sadoul et al., 1988). Spontaneous release of soluble NILE GP (s-NILE GP) along with several other glycoproteins was estimated to constitute about 2% per hour of the total cellular complement (Richter-Landsberg et al., 1984). Release could be greatly accelerated by exposure of the cultured neurons to a variety of agents, including black widow spider venom, elevated potassium, veratridine, ionophore A23187, and monensin (Sweadner, 1983a). Such evoked release requires the presence of extracellular calcium.

A number of different criteria indicate that s-NILE GP released both spontaneously and by evocation is derived from cell surface NILE GP rather than from an intracellular site. By SDS-PAGE, s-NILE GP has a slightly faster mobility than NILE GP extracted from membranes (Salton et al., 1983b; Stallcup et al., 1983; Sweadner, 1983a; Richter-Landsberg et al., 1984). This suggests that release is accompanied, and most likely caused, by modification. One possible mechanism for this is proteolytic cleavage. However, a wide variety of protease inhibitors were found to be ineffective in blocking the black widow spider venom–evoked release of s-NILE GP from sympathetic neurons (Sweadner, 1983a). Another possibility is release via cleavage of a phosphatidyl-inositol (PI) linkage. But treatment of PC12 cells with a PI-phospholipase C that is specific for such types of linkages does not evoke release of s-NILE GP (L. A. Greene and M. L. Shelanski, unpublished data). Thus, the mechanism whereby NILE GP as well as several other GPs are released from neurons is unclear. Moreover, although a variety of attractive hypothesis may be constructed in this regard, there is presently no information as to the potential functional role of s-NILE GP.

PHYSICAL PROPERTIES

As noted above, NILE GP is surface exposed on cells. The extraction properties of the molecule indicate that it is an integral membrane component (Salton et al., 1983b). By analysis with two-dimensional isoelectric focusing × SDS-PAGE, PC12 cell NILE GP appears as a streak of several spots with pI ranging between 6.4 and 6.6 (McGuire et al., 1978; Salton et al., 1983b). NILE GP in cultured rat sympathetic neurons also focuses as a streak with components of slightly different apparent molecular weight; the pI of this material has been reported to be in the approximate range of 5.3–5.8 (Sweadner, 1983a). These properties thus indicate that mature NILE GP is slightly acidic. In view of the findings that NILE GP is glycosylated, phosphorylated, and sulfated, it would seem likely that the apparent microheterogeneity observed by two-dimensional electrophoresis reflects variable degrees of posttranslational modification of the molecule.

The hydrodynamic properties of both membrane-bound and released NILE GP have been investigated in detail (Sweadner et al., 1983b). Both forms were determined to have large Stokes radii (76 and 68 Å, respectively) and small sedimentation coefficients (about 5 and 5.6, respectively), indicating that they have large frictional coefficients (calculated to be 2.1 and 1.9, respectively). A likely interpretation of these data is that NILE and s-NILE GPs are rod shaped rather than globular and that s-NILE is monomeric in solution.

CLONING OF THE NILE/L1 GLYCOPROTEIN

Moos et al. (1988) have reported the isolation of cDNA clones for mouse L1 and the use of these to deduce the complete amino acid sequence of the molecule. This analysis showed it to contain 1241 amino acids with an apoprotein molecular weight of 138,531. The deduced sequence data revealed several interesting features. First, there appears to be a single membrane-spanning domain near the carboxy terminus of the protein. Second, there are six immunoglobulinlike domains (of the C subgroup), thus placing the molecule in the immunoglobulin superfamily. Other members of this family that are expressed in the mammalian nervous system include thy-1, MAG, and N-CAM. Third, there are two RDG sequences, which in other proteins are involved in binding to the extracellular matrix (see Hynes et al., this volume). Fourth, there are three domains that bear close homology to the "repeating unit type III" of fibronectin. Such domains may be involved in cell–cell and cell–substrate binding. Blot analyses performed by Tacke et al. (1987) indicated that L1 is encoded by a single gene which is transcribed as a single 6-kb mRNA. Prince et al. (1989) recently presented sequence data for a partial cDNA clone for the rat PC12 cell NILE GP. The clone contained an open reading frame coding for 79 amino acids, of which only two differed from that determined from the homologous portion of mouse L1. This further confirms the identity of L1 and the NILE GP.

REGULATION OF NILE GP LEVELS

Given the importance of understanding the means by which the expression of cell surface glycoproteins is regulated, several studies have been directed toward this subject in the case of NILE GP (McGuire et al., 1978; Salton et al., 1983a; Friedlander et al., 1986). As discussed above, NILE GP was originally the subject of particular interest because of the observation that its apparent levels (as monitored by labeling with [^{3}H]fucose) are specifically increased in PC12 cells in response to NGF (McGuire et al., 1978). Subsequent quantification by immunochemical means verified that the induction is due to specific increases (by three- to fivefold) in NILE GP levels relative to total cell protein (Salton et al., 1983a; Friedlander et al., 1986). This contrasts with the finding that most other PC12 cell proteins and glycoproteins do not show changes in relative levels after treatment with NGF. Exposure to an inhibitor of transcription suggests that regulation of NILE GP levels by NGF requires a new synthesis of mRNA. Consistent with this idea, the NGF-dependent increase begins only after a lag of at least one to two days, is progressive with time of treatment, and generally follows a time course similar to that of neurite outgrowth (McGuire et al., 1978; Salton et al., 1983a). However, the regulation of NILE GP levels by NGF does not appear to be a consequence of neurite outgrowth; induction can occur even when the cells are exposed to NGF under conditions in which they are unattached to a substrate and cannot extend processes.

A number of agents in addition to NGF have been screened for their capacities to affect NILE GP levels in PC12 cells. Exposure to epidermal growth factor, permeant cAMP derivatives, forskolin, cholera toxin, phorbol esters, insulin, dexamethasone, and depolarizing concentrations of potassium, among other treatments, has had no effect. It is of interest that PC12 cells do show some type of response to each of these agents, but that none is capable of promoting PC12 cell neuronal differentiation (for a review, see Greene and Tischler, 1982). The one agent thus far found to cause enhancement of PC12 cell NILE GP levels is fibroblast growth factor (FGF). Exposure of PC12 cells to FGF promotes neurite outgrowth as well as a number of other responses similar to those evoked by NGF (Togari et al., 1985; Rydel and Greene, 1987). Among such responses is an induction of NILE GP equal in magnitude to that found with NGF (Rydel and Greene, 1987).

The observation that both NGF and FGF can modulate NILE GP levels is of potential biological importance. NGF clearly plays a major role in regulating the differentiation and development of sympathetic and sensory neurons and also affects at least several populations of CNS neurons. Recent evidence indicates that FGF can, in addition to its actions as a mitogen on many cell types, act as a neurotrophic factor on certain CNS neurons (cf. Morrison et al., 1986). Thus, the possibility is raised that the expression of NILE GP may be under the influence of neurotrophic agents during and after nervous system development. While effects thus far have been limited only to NGF and FGF, it is quite possible that additional neurotrophic agents will be found to influence NILE GP levels on other types of neurons.

An additional point of interest is the contrast between the regulation of NILE GP and that of another neuronal cell surface GP, thy-1. Thy-1 GP levels are also enhanced by the exposure of PC12 cells to NGF (Richter-Landsberg et al., 1985; Drexler and Greene, 1986) or FGF (Rydel and Greene, 1987). However, in contrast to NILE GP, the induction of thy-1 begins within a day of factor exposure and, after reaching a maximum within three days, returns to pretreatment levels after one to two weeks of treatment. In addition, unlike NILE GP, thy-1 can be induced by culture in the presence of elevated levels of potassium or phorbol ester (Drexler and Greene, 1986). These observations thus suggest the absence of a unitary mechanism whereby neurotrophic factors regulate cell surface glycoprotein levels.

CONCLUSIONS AND PERSPECTIVES

A good deal of information has been accumulated over the past 10 years regarding the regulation and biochemistry of the NILE GP. This has been complemented by a number of exciting studies on the neural localization and function of NILE GP/L1/Ng-CAM cross-reactive components. Recent cloning studies have begun to provide further insight as to how the biochemical properties of the NILE GP may be translated into its functional activities. Future biochemical work will include determining whether and how modifications of the protein may affect its function, and identifying the nature and molecular mechanism of action of the cellular signals that regulate NILE GP synthesis and modification.

ACKNOWLEDGMENT

The studies described here from our laboratories were supported by National Institutes of Health grant NS-16839.

REFERENCES

Beasley, L., and W. B. Stallcup (1987) The nerve growth factor–inducible large external (NILE) glycoprotein and neural cell adhesion molecule (N-CAM) have distinct patterns of expression in the developing rat central nervous system. *J. Neurosci.* 7:708–715.

Bock, E., C. Richter-Landsberg, A. Faissner, and M. Schachner (1985) Demonstration of immunochemical identity between the nerve growth factor–inducible large external (NILE) glycoprotein and the cell adhesion molecule L1. *EMBO J.* **4**:2765–2768.

Bock, E., D. Linnemann, O. Nybroe, and H. Rohde (1986) Biosynthesis of the cell adhesion molecules L1 and N-CAM. *Eur. Soc. Neurosci. Abstr.* **6**:81.

Drexler, S. A., and L. A. Greene (1986) Regulation of thy-1 mRNA in PC12 cells by NGF and elevated potassium. *Soc. Neurosci. Abstr.* **12**:215.

Edelman, G. M. (1985) Cell adhesion and the molecular process of morphogenesis. *Annu. Rev. Biochem.* **54**:135–169.

Edelman, G. M. (1986) Cell adhesion molecules in the regulation of animal form and tissue pattern. *Annu. Rev. Cell Biol.* **2**:81–116.

Faissner, A., J. Kruse, J. Nieke, and M. Schachner (1984) Expression of neural cell adhesion molecule L1 during development in neurological mutants and in the peripheral nervous system. *Dev. Brain Res.* **15**:69–82.

Faissner, A., D. Teplow, D. Kubler, G. Keilhauer, V. Kinzel, and M. Schachner (1985) Biosynthesis and membrane topography of the neural adhesion molecule L1. *EMBO J.* **4**:3105–3113.

Friedlander, D. R., M. Grumet, and G. M. Edelman (1986) Nerve growth factor enhances expression of neuron–glia cell adhesion molecule in PC12 cells. *J. Cell Biol.* **102**:413–419.

Fukada, M. (1985) Cell surface glycoconjugates as onco-differentiation markers in hemopoietic cells. *Biochim. Biophys. Acta* **780**:119–150.

Greene, L. A., and A. S. Tischler (1976) Establishment of a noradrenergic clonal line of rat adrenal pheochromocytoma cells which respond to nerve growth factor. *Proc. Natl. Acad. Sci. USA* **73**:2424–2428.

Greene, L. A., and A. S. Tischler (1982) PC12 pheochromocytoma cultures in neurobiological research. In *Advances in Cellular Neurobiology*, Vol. 3, S. Federoff and L. Hertz, eds., pp. 373–414, Academic, New York.

Lee, V., M. L. Shelanski, and L. A. Greene (1977) Specific neural and adrenal medullary antigens detected by antisera to clonal PC12 pheochromocytoma cells. *Proc. Natl. Acad. Sci. USA* **74**:5021–5025.

Lee, V., L. A. Greene, and M. L. Shelanski (1981) Identification of neural and adrenal medullary surface membrane glycoproteins recognized by antisera to cultured rat sympathetic neurons and PC12 pheochromocytoma cells. *Neuroscience* **6**:2773–2786.

Leonard, D. G. B., L. A. Greene, and E. B. Ziff (1987) Identification and characterization of mRNAs regulated by nerve growth factor in PC12 cells. *Mol. Cell Biol.* **7**:3156–3167.

Lindner, J., F. G. Rathjen, and M. Schachner (1983) L1 mono- and polyclonal antibodies modify cell migration in early postnatal mouse cerebellum. *Nature* **305**:427–429.

Margolis, R. K., S. R. Salton, and R. U. Margolis (1983) Structural features of the nerve growth factor inducible large external glycoprotein of PC12 pheochromocytoma cells and brain. *J. Neurochem.* **41**:1635–1640.

Margolis, R. K., L. A. Greene, and R. U. Margolis (1986) Poly(N-acetyllactosaminyl) oligosaccharides in glycoproteins of PC12 pheochromocytoma cells and sympathetic neurons. *Biochemistry* **25**:3463–3468.

McGuire, J. C., L. A. Greene, and A. V. Fuvano (1978) NGF stimulates incorporation of fucose or glucosamine into an external glycoprotein in cultured rat PC12 pheochromocytoma cells. *Cell* **15**:357–365.

Moos, M., R. Tacke, H. Scherer, D. Teplow, K. Früh, and M. Schachner (1988) Neural adhesion molecule L1 as a member of the immunoglobulin superfamily with binding domains similar to fibronectin. *Nature* **334**:701–703.

Morrison, R. S., A. Sharma, J. De Vellis, and R. A. Bradshaw (1986) Basic fibroblast growth factor supports the survival of cerebral cortical neurons in primary culture. *Proc. Natl. Acad. Sci. USA* **83**:7537–7541.

Pigott, R., and J. S. Kelly (1986) Immunocytochemical and biochemical studies with the monoclonal antibody 69A1: Similarities of the antigen with cell adhesion molecules L1, NILE, and Ng-CAM. *Brain Res.* **394**:111–222.

Prince, J. T., N. Milona, and W. B. Stallcup (1989) Characterization of a partial cDNA clone for the NILE glycoprotein and identification of the encoded polypeptide domain. *J. Neurosci.* **9**:876–883.

Rathjen, F. G., and M. Schachner (1984) Immunocytological and biochemical characterization of a new neuronal cell surface component (L1 antigen) which is involved in cell adhesion. *EMBO J.* **3**:1–10.

Richter-Landsberg, C., V. M. Lee, S. R. Salton, M. L. Shelanski, and L. A. Greene (1984) Release of the NILE and other glycoproteins from cultured PC12 rat pheochromocytoma cells and sympathetic neurons. *J. Neurochem.* **43**:841–848.

Richter-Landsberg, C., L. A. Greene, and M. L. Shelanski (1985) Cell surface thy-1-cross-reactive glycoprotein in cultured PC12 cells: Modulation by nerve growth factor and association with the cytoskeleton. *J. Neurosci.* **5**:468–476.

Rieger, F., J. K. Daniloff, M. Pincon-Raymond, K. L. Crossin, M. Grumet, and G. M. Edelman (1986) Neuronal adhesion molecules and cytotactin are colocalized at the node of Ranvier. *J. Cell Biol.* **103**:379–391.

Rydel, R. E., and L. A. Greene (1987) Acidic and basic fibroblast growth factors promote stable neurite outgrowth and neuronal differentiation in cultures of PC12 cells. *J. Neurosci.* **7**:3639–3653.

Sadoul, K., R. Sadoul, A. Faissner, and M. Schachner (1988) Biochemical characterization of different molecular forms of the neural adhesion molecule L1. *J. Neurochem.* **50**:510–521.

Sajovic, P., E. Kouvelas, and E. Trenkner (1986) Probable identity of NILE glycoprotein and the high-molecular-weight component of L1 antigen. *J. Neurochem.* **47**:541–546.

Salton, S. R., C. Richter-Landsberg, L. A. Greene, and M. L. Shelanski (1983a) Nerve growth factor–inducible large external (NILE) glycoprotein: Studies of a central and peripheral neuronal marker. *J. Neurosci.* **3**:441–454.

Salton, S. R., M. L. Shelanski, and L. A. Greene (1983b) Biochemical properties of the nerve growth factor–inducible large external (NILE) glycoprotein. *J. Neurosci.* **3**:2420–2430.

Sanes, J. R., M. Schachner, and J. Covault (1986) Expression of several adhesive macromolecules (N-CAM, L1, J1, NILE, uvomorulin, laminin, fibronectin, and a heparan sulfate proteoglycan) in embryonic, adult, and denervated adult skeletal muscle. *J. Cell Biol.* **102**:420–431.

Stallcup, W. B., and L. Beasley (1985a) Polymorphism among NILE-related glycoproteins from different types of neurons. *Brain Res.* **346**:287–293.

Stallcup, W. B., and L. Beasley (1985b) Involvement of the nerve growth factor–inducible large external glycoprotein (NILE) in neurite fasciculation in primary cultures of rat brain. *Proc. Natl. Acad. Sci. USA* **82**:1276–1280.

Stallcup, W. B., L. S. Arner, and J. M. Levine (1983) An antiserum against the PC12 cell line defines cell surface antigens specific for neurons and Schwann cells. *J. Neurosci.* **3**:53–68.

Stallcup, W. B., L. Beasley, and J. M. Levine (1985) Antibody against nerve growth factor–inducible large external (NILE) glycoprotein labels nerve fiber tracts in the developing rat nervous system. *J. Neurosci.* **5**:1090–1101.

Sweadner, K. J. (1983a) Post-translational modification and evoked release of two large surface proteins of sympathetic neurons. *J. Neurosci.* **3**:2504–2517.

Sweadner, K. J. (1983b) Size, shape, and solubility of a class of releasable cell surface proteins of sympathetic neurons. *J. Neurosci.* **3**:2518–2524.

Tacke, R., M. Moos, D. B. Teplow, K. Früh, H. Scherer, A. Bach, and M. Schachner (1987) Identification of cDNA clones of the mouse neural cell adhesion molecule L1. *Neurosci. Lett.* **82**:89–94.

Thor, G., R. Probstmeier, and M. Schachner (1987) Characterization of the cell adhesion molecules L1, N-CAM, and J1 in the mouse intestine. *EMBO J.* **6**:2581–2586.

Togari, A., G. Dickens, H. Kuzuya, and G. Guroff (1985) The effect of fibroblast growth factor on PC12 cells. *J. Neurosci.* **5**:307–316.

Chapter 24

Soluble N-CAM

ELISABETH BOCK
KLAUS EDVARDSEN
ANN GIBSON
DORTE LINNEMANN
JOAN M. LYLES
OLE NYBROE

ABSTRACT

Soluble forms of N-CAM have been demonstrated in cerebrospinal fluid, serum, amniotic fluid, and media from neural cell cultures. Studies on N-CAM levels in cerebrospinal fluid have shown that measurement may be of diagnostic value in normal pressure hydrocephalus. Furthermore, a strong increase in cerebrospinal fluid N-CAM has been demonstrated in multiple sclerosis patients two to three weeks after an acute exacerbation of the disease. The analysis of N-CAM in amniotic fluid is of diagnostic value in fetal neural tube defects and other malformations. At early developmental stages brain contains a soluble N-CAM polypeptide of 170 kD, whereas in adult brain a soluble N-CAM polypeptide of 115 kD is predominant. The latter is probably identical to the 115-kD membrane-associated N-CAM polypeptide C. Soluble N-CAM constitutes approximately 1–2% of total N-CAM in whole brain. In medium from cultured neuronal cells, the 170-kD form is found, whereas medium from astroglial cell culture contains the 115-kD form. The soluble N-CAM forms in body fluids, brain supernatant, and cell culture media are immunochemically identical to the membrane-associated forms. The rate of appearance of soluble N-CAM in medium is very rapid, the soluble form appearing in the medium within 15–25 min after the start of synthesis, indicating that it appears by secretion and not as a degradation product of the membrane-associated forms.

The neural cell adhesion molecule (N-CAM) was originally described as a rat neuronal membrane antigen (Jørgensen and Bock, 1974). Nearly all biochemical studies on N-CAM have been performed on the membrane-associated form. However, soluble forms of N-CAM have been observed and quantified in human cerebrospinal fluid, amniotic fluid, and serum (Jørgensen and Bock, 1975; Jørgensen and Nørgaard-Pedersen, 1981; Ibsen et al., 1983b). In cell culture supernatants the presence of N-CAM was indicated by

Rutishauser et al. (1978), who produced antibodies against N-CAM using chick retinal cell culture supernatants as a source of antigen, and by Thiery et al. (1977), who were able to block the inhibitory effect of N-CAM antibodies on cell–cell adhesion by means of soluble antigens from chick retinal cell culture supernatant. Schubert et al. (1983) showed that embryonic chick retinal cells release macromolecular aggregates, termed adherons, into their culture medium. Adherons promote cell–substratum attachment, and a polypeptide of an apparent molecular weight of 170 kD, immunochemically identical to N-CAM, was demonstrated to be the cell–substratum adhesion molecule of the adherons (Cole and Glaser, 1986).

N-CAM-mediated cell–cell adhesion occurs by a homophilic binding mechanism (Rutishauser et al., 1982), in which N-CAM on the surface of one cell binds to N-CAM on another cell. Cole et al. (1986) have shown that although N-CAM-mediated adhesion involves homophilic binding, the binding of cell surface heparan sulfate to N-CAM is also required. The membrane-associated forms of N-CAM consist of three polypeptides which in SDS-polyacrylamide gels migrate with molecular weights of approximately 190 kD, 135 kD, and 115 kD (hereafter referred to as polypeptides A, B, and C). The main difference between N-CAM A and B is the size of their cytoplasmic domains (Gennarini et al., 1984; Nybroe et al., 1985; Murray et al., 1986). In contrast, N-CAM C lacks a transmembrane domain (Nybroe et al., 1985) and is anchored in the membrane not by a stretch of hydrophobic amino acids, but by a phospholipid (He et al., 1986). Even though the existence of soluble forms of N-CAM in body fluids and culture supernatants has long been recognized, little is known about the origin and structure of these components.

N-CAM IN CEREBROSPINAL FLUID, AMNIOTIC FLUID, AND SERUM

Because of the inaccessibility of the living human brain, considerable efforts have been made to demonstrate and quantify brain-derived proteins in cerebrospinal fluid, blood, urine, and amniotic fluid. N-CAM was originally demonstrated in cerebrospinal fluid by means of an immunoprecipitation-in-gel technique (Jørgensen and Bock, 1975). In the first study on N-CAM in cerebrospinal fluid, patients suffering from a variety of neurological and psychiatric disorders (multiple sclerosis, inflammatory–degenerative neurological diseases, brain hemorrhage or thrombosis, manic–depressive psychosis, and schizophrenia) were examined. However, no statistically significant changes were observed (Jørgensen and Bock, 1975).

Later, Jørgensen et al. (1977) studied N-CAM levels in cerebrospinal fluid from manic-depressive patients. It was found that the concentration of N-CAM increased with the age of the patients until about 35 years of age. No differences were found when N-CAM levels from a control group were compared with different manic-depressive subgroups. N-CAM concentration in cerebrospinal fluids collected from patients during depression or mania was also compared with values in cerebrospinal fluid collected from the same patients when their mood was normalized. In the case of depressed patients, it was found that N-CAM concentration increased slightly, but statistically significantly, when the mood was normalized. The biological basis for the rise in N-CAM from endogenously depressed patients during periods of recovery can only be speculated upon at present. The changes may be related to synaptic re-modeling. Alternatively, the rise in N-CAM levels may reflect the cleaning-up process of the breakdown and removal of damaged neural cells.

The immunoprecipitation procedure employed for these first studies demanded concentration of the cerebrospinal fluid and was rather com-plicated to perform. Therefore a simpler and more sensitive assay suited for routine analysis was needed. An enzyme-linked immunosorbent assay (ELISA) was established using purified human brain N-CAM and rabbit anti–human brain N-CAM antibodies (Ibsen et al., 1983a). The N-CAM ELISA worked in the range 50–2000 ng/ml. Results of the N-CAM determinations obtained by this procedure are presented in Table 1 and Figure 1. The N-CAM concen-tration in cerebrospinal fluid was found to be significantly lower in normal-pressure hydrocephalus patients than in patients suffering from primary degenerative dementia ($p < .001$) and in controls ($p < .01$) (Sørensen et al., 1983). The subnormal N-CAM concentration in cerebrospinal fluid in normal pressure hydrocephalus seems to be a unique finding.

A major problem in adult-communicating hydrocephalus has been to identify those patients who might benefit from shunting procedures. In particular, it has been difficult to distinguish between normal-pressure hydrocephalus patients with cortical atrophy and patients with enlarged ventricles associated with degenerative brain disease. It is unknown whether the decreased N-CAM concentration in normal-pressure hydrocephalus reflects a defect in the brain of patients who might develop this disease, or is a result of the hydrocephalic state itself. However, the results indicate that measurement of N-CAM in cerebrospinal fluid provides an additional diagnostic tool for discrimination between patients with normal-pressure hydrocephalus and patients with hydrocephalus as part of a diffuse atrophy of the brain.

Table 1.   N-CAM Concentrations in Cerebrospinal Fluid (Means ± SEM)

Diagnosis	n	ng N-CAM/ml
Controls	24	641 ± 45
Primary degenerative dementia	14	658 ± 50
Normal pressure hydrocephalus	13	299 ± 48
Multiple sclerosis (acute phase)	20	589 ± 74
Multiple sclerosis (nonacute phase)	10	497 ± 95

Source: Sørensen et al. (1983); Massaro et al. (1987).

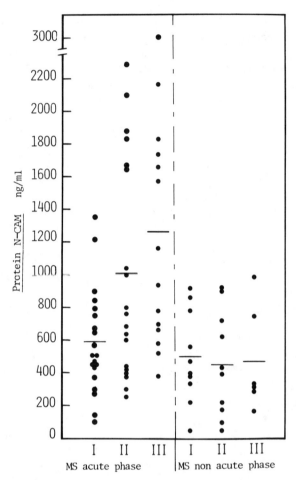

Figure 1. *Cerebrospinal fluid concentration of N-CAM in acute and nonacute phase multiple sclerosis patients.* For each group, values obtained from first (I), second (II), and third (III) lumbar punctures are shown. (From Massaro et al., 1987.)

N-CAM levels in cerebrospinal fluid have recently been studied in patients with multiple sclerosis (Massaro et al., 1987). All multiple sclerosis patients had two or three lumbar punctures performed at about one-week intervals. No differences in N-CAM concentrations in cerebrospinal fluids obtained by the first lumbar puncture could be demonstrated between acute and nonacute multiple sclerosis patients and controls. However, an increase of N-CAM was shown when the values of first (I), second (II), and third (III) lumbar puncture samples from the acute phase multiple sclerosis patients were compared (Figure 1). The mean N-CAM concentration of first (589 ng/ml), second (1006 ng/ml), and third (1065 ng/ml) lumbar puncture samples differed statistically significantly from each other ($p < .02$ comparing I and II; $p < .005$ comparing I and III; $p < .01$ comparing II and III). On the other hand, no significant changes could be shown in the follow-up of the nonacute patients, indicating a role for N-CAM in the processes involved in acute demyelination and possibly in the following evolution of the sclerotic plaque (Massaro et al., 1987).

Jørgensen and Nørgaard-Pedersen (1981) demonstrated the correlation between raised N-CAM concentration in amniotic fluid and neural tube defects. These authors found a 100% specificity of the screening procedure. Ibsen et al. (1983b) performed a more extensive study on amniotic fluids from pregnancies with neural tube defects and other malformations and found that when a cutoff level of 150 ng N-CAM/ml amniotic fluid was used, no false positive values were obtained. The sensitivity for detecting anencephaly was 100%, whereas the sensitivity for detecting spina bifida only was 75%. Ibsen et al. (1983b) also observed raised N-CAM levels in amniotic fluids from fetuses with other defects, such as omphalocele, Meckel's syndrome, multiple malformations, and intrauterine death. In normal pregnancies levels below 50 ng/ml were found.

From Table 1 it appears that the level of N-CAM in human cerebrospinal fluid from adult individuals is approximately 600 ng/ml. Surprisingly, N-CAM in serum is in the range of 1000–3000 ng/ml (Ibsen et al., 1983b), indicating that sources other than the brain may contribute N-CAM to the blood. No correlation between N-CAM levels in blood and any pathological conditions has been demonstrated so far.

STRUCTURE AND ORIGIN OF SOLUBLE N-CAM

Immunochemical Comparisons Between Soluble and Membrane-Bound Forms of N-CAM

In order to determine the immunochemical relationship between soluble and membrane-bound forms of N-CAM, Ibsen et al. (1983a) performed ELISA determinations of N-CAM on serial dilutions of adult human brain, serum, cerebrospinal fluid, and amniotic fluid. All dilutions resulted in plots that paralleled the curve obtained from the N-CAM standard (derived from human fetal brain), implying an immunochemical identity between the different N-CAM antigens with regard to the employed antibody. Bock et al. (1987) collected conditioned medium from fetal rat brain neuronal cultures, detergent extract of rat brain membranes, and supernatant from rat brain homogenates at postnatal days 0 and 40. Serial dilutions of these five preparations all resulted in parallel dilution curves. Thus the soluble forms of N-CAM in brain, body fluids, and medium from neuronal cultures carry the same epitopes as the membrane-bound forms. Using a specific polyclonal antibody, this means that approximately 5–10 epitopes, each corresponding to approximately 6–8 amino acids, are identical in the various forms of N-CAM.

Amount of Soluble N-CAM in Rat Brain

Soluble N-CAM is a minor fraction of the proteins in brain. Bock et al. (1987) determined the amount in brains from rats at embryonic day 17 and postnatal days 4, 15, and 40 (Table 2). The results are expressed relative to brain total protein or to total amount of N-CAM. It appears from the table that the amount of soluble N-CAM in rat brain varies from 85 to 194 ng/gm total protein, with the lowest value at embryonic day 17 and the peak value at postnatal day 15. For comparison, the amount of total N-CAM per gram total protein was determined, and soluble N-CAM was expressed at the percentage of total N-CAM.

It can be seen that soluble N-CAM constitutes a steadily increasing fraction during development from 0.7% at embryonic day 17 to 2.3% at postnatal day 40 (Table 2). If the amount of soluble N-CAM was correlated to wet weight of the brain, approximately 5 μg soluble N-CAM per gram wet weight was found at embryonic day 17, the amount increasing to approximately 18 μg in the adult brain. If we assume that all soluble N-CAM is present in the extracellular space, it can be calculated that the concentration here during development varies from approximately 25 to 100 μg/gm, assuming that the volume of the extracellular space is 15–20% of the tissue. This means that during development the concentration in the extracellular space is approximately 5–10% of the total N-CAM concentration, which varies from 420 μg at embryonic day 17 to the maximum 1250 μg per gram wet weight at postnatal day 15 (Linnemann et al., 1985). The extracellular N-CAM concentration is considerably higher than the concentration in cerebrospinal fluid, which is approximately 0.6 μg/ml (Sørensen et al., 1983).

Polypeptide Composition of Soluble N-CAM in Brain

The polypeptide composition of soluble N-CAM in brain has been studied by the radioiodination of supernatants obtained from homogenates of rat brains collected at postnatal days 0 and 40. After labeling, N-CAM was isolated by immunoprecipitation and submitted to SDS-polyacrylamide gel electrophoresis (Bock et al., 1987). The polypeptide composition of the soluble form

Table 2.   Concentration of Soluble N-CAM in Rat Brain Homogenate at Various Ages (Means $\pm$ SEM, $n = 4$)

Age	ng/gm Total Protein	% of Total N-CAM
Embryonic day 17	85 $\pm$ 8	0.7 $\pm$ 0.04
Postnatal day 4	109 $\pm$ 3	0.8 $\pm$ 0.02
Postnatal day 15	194 $\pm$ 6	1.5 $\pm$ 0.06
Postnatal day 40	158 $\pm$ 4	2.3 $\pm$ 0.11

Source: Bock et al. (1987).

differed from that of membrane-associated N-CAM. At postnatal day 0, soluble N-CAM is composed of a polypeptide of 170 kD, whereas at postnatal day 40, soluble N-CAM is composed of a polypeptide of 115 kD (see Figure 2). In the figure, membrane-bound N-CAM at the corresponding ages also is shown.

It can be seen that the 170-kD soluble N-CAM polypeptide has a migration distinctly different from the membrane-bound forms, whereas the 115-kD soluble N-CAM form in adult brain supernatant has a migration identical to the membrane-associated form of the C polypeptide. Since it has been shown that the membrane-associated form of N-CAM of 190 kD, polypeptide A, is an integral membrane protein with a cytoplasmic domain of approximately 30 kD (Nybroe et al., 1985; Hemperly et al., 1986), it is unlikely that the soluble 170-kD N-CAM is derived from this molecule. On the other hand, the soluble 115-kD form presumably is identical to polypeptide C, which *in vitro* has been shown to be synthesized as a soluble protein (Nybroe et al., 1985).

Polypeptide Composition of Soluble N-CAM in Media from Neuronal and Glial Cell Cultures

A 170-kD N-CAM form has been demonstrated in the so-called adheron fraction of media from chick retinal cells in culture (Cole and Glaser, 1986). Saermark et al. (1988) prepared adheron fractions from media from fetal rat brain neuronal cells in culture. They found that both the adheron fraction and the postadheron supernatant contained a 170-kD N-CAM polypeptide. Bock et al. (1987) studied the polypeptide composition of soluble N-CAM in media

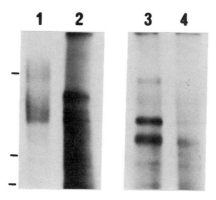

Figure 2. *Soluble N-CAM polypeptides in brain extract.* Rat brains from postnatal days 0 and 40 were homogenized in a physiological buffer, and soluble proteins and membranes were obtained by centrifugation. The obtained fractions were radioiodinated and [^{125}I]N-CAM was immuno-isolated and analyzed by SDS-polyacrylamide gel electrophoresis and autoradiography. *Lanes 1* and *2* show membrane-associated N-CAM and soluble N-CAM at postnatal day 0. *Lanes 3* and *4* show membrane-associated N-CAM and soluble N-CAM at postnatal day 40. The positions of molecular weight standard proteins are indicated in the *left margin*; from *top* to *bottom*: 200 kD, 92 kD, and 69 kD. (From Bock et al., 1987.)

from fetal neuronal cells and astroglial cells in culture. In Figure 3 it can be seen that N-CAM in centrifuged, conditioned medium from neuronal cells consists of a polypeptide of 170 kD, whereas N-CAM in centrifuged conditioned medium from astroglial cells is composed of a polypeptide of 115 kD. The electrophoretic migration of N-CAM in neuronal medium was identical to that of the soluble N-CAM polypeptide demonstrated in newborn brain in Figure 1. The molecular weight of the glial medium N-CAM of 115 kD is identical to the molecular weight of the soluble form found at postnatal day 40 in brain supernatant and to the membrane-associated polypeptide C.

These data indicate that the 170-kD form in brain originates from neurons, whereas the 115 kD in brain is synthesized by cells that have the capacity for polypeptide C expression, that is, astroglial cells (Noble et al., 1985), and to a lesser extent, neurons (Nybroe et al., 1986). Polypeptide C is a major fraction of the total N-CAM in brain, constituting approximately 10% of total N-CAM at early developmental stages and approximately 30% in adult brain (Linnemann and Bock, 1986). Thus, it can be calculated from the data shown in Table 2 that only 5–10% of all polypeptide C present in adult brain is in a soluble form.

Biosynthesis of Soluble N-CAM in Neuronal Cell Cultures

The rate of appearance of N-CAM in media from fetal rat brain neurons in culture was examined by Bock et al. (1987). Fetal rat brain neurons were labeled with [^{35}S] methionine for 15 min, rinsed, and incubated with complete medium. Chase medium was harvested sequentially from the same culture after 10, 20, and 50 min. N-CAM was immunoisolated, and dpm incorporated into N-CAM was determined. In Figure 4 the amount of N-CAM dpm released into the culture medium per minute in each of the three periods is indicated.

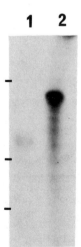

Figure 3. *Soluble N-CAM in centrifuged media (100,000 gm × 3 hours) obtained from glial and neuronal cell cultures. Lane 1, N-CAM in astroglial cell culture medium; lane 2, N-CAM in neuronal cell culture medium. The positions of molecular weight standards are indicated in the left margin; from top to bottom: 200 kD, 92 kD, and 69 kD. (From Bock et al., 1987.)*

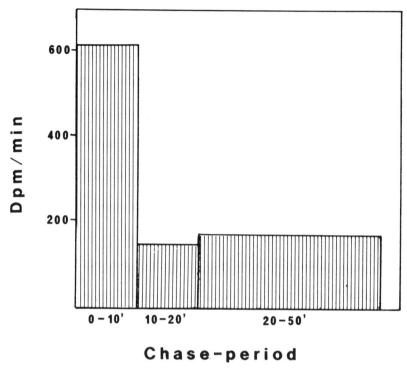

Chase-period

Figure 4. *Rates of secretion of soluble N-CAM in neuronal culture.* A cell culture was labeled for 15 min with [^{35}S]methionine. New medium was added and recollected after 10, 20, and 50 min, respectively. N-CAM was immunoisolated from each of the media, and the amount of N-CAM dpm secreted per min in each period was determined. (From Bock et al., 1987.)

The experiment shows that nearly 50% of all N-CAM secreted in the observation period appears within 15–25 min after the start of synthesis, indicating a very rapid secretion. The incorporation of labeled amino acid into soluble N-CAM relative to the total protein incorporation 15–25 min after synthesis was found to be 6.7%, whereas the incorporation percentages 25–35 min and 35–65 min after synthesis were found to be 1.7% and 1.4%, respectively. This indicates that N-CAM appears extracellularly earlier than the majority of secreted proteins.

Previously, Lyles et al. (1984) showed that newly synthesized membrane-bound N-CAM reaches the cell surface within 35 min after the start of synthesis. It is generally assumed that more time is required for the post-translational modification and transport of membrane-bound proteins to the cell surface than is required in the case of secreted proteins. Detection of soluble N-CAM in the medium soon after the start of synthesis and before the accumulation of N-CAM on the surface of the membrane indicates that the appearance of soluble 170-kD N-CAM is an active secretory process rather than a passive release of membrane-associated N-CAM or degradation products of the same.

CONCLUSION

The soluble forms of N-CAM demonstrated in brain and body fluids are probably actively secreted by neurons and astroglial cells. Soluble N-CAM is composed of at least two polypeptides of 170 kD and 115 kD, the latter presumably representing a soluble subpopulation of the membrane-associated polypeptide C. We suggest that the functional role of extracellular N-CAM in soluble form may be the regulation of the adhesivity of neighboring cells by binding N-CAM molecules to the cell surface, thereby changing their adhesive function. A consequence of such binding may be a changed migratory capacity. The function of extracellular matrix-bound N-CAM, so-called adheron N-CAM, on the other hand, may be to act as an insolubilized chemotactic factor that guides migrating cells in specified directions. Thus a possible target cell may secrete soluble N-CAM, which subsequently is bound to the extracellular matrix, and thereby create an insolubilized gradient of N-CAM molecules around the secreting target cells. According to this theory, N-CAM's role in morphogenesis is expanded. Not only may N-CAM participate in cell–cell adhesion, but the secreted forms may modulate cell migration and create migratory pathways for N-CAM-expressing cells.

ACKNOWLEDGMENTS

The financial support of the Danish Cancer Society (grants 84–104, 86–064, 86–078, 86–030) and the Danish Medical Research Council (grant 12–6517) is gratefully acknowledged.

REFERENCES

Bock, E., K. Edvardsen, A. Gibson, D. Linnemann, J. M. Lyles, and O. Nybroe (1987) Characterization of soluble forms of N-CAM. *FEBS Lett.* **225**:33–36.

Cole, G. J., and L. Glaser (1986) A heparin binding domain from N-CAM is involved in neural cell–substratum adhesion. *J. Cell Biol.* **102**:403–412.

Cole, G. J., A. Loewy, and L. Glaser (1986) Neuronal cell–cell adhesion depends on interactions of N-CAM with heparinlike molecules. *Nature* **320**:445–447.

Gennarini, G., M. Hirn, H. Deagostini-Bazin, and C. Goridis (1984) Studies on the transmembrane disposition of the neural cell adhesion molecule N-CAM. *Eur. J. Biochem.* **142**:65–73.

He, H.-T., J. Barbet, J.-C. Chaix, and C. Goridis (1986) Phosphatidylinositol is involved in the membrane attachment of NCAM-120, the smallest component of the neural cell adhesion molecule. *EMBO J.* **5**:2489–2494.

Hemperly, J. J., B. A. Murray, G. M. Edelman, and B. A. Cunningham (1986) Sequence of a cDNA clone encoding the polysialic acid-rich and cytoplasmic domains of the neural cell adhesion molecule N-CAM. *Proc. Natl. Acad. Sci. USA* **83**:3037–3041.

Ibsen, S., V. Berezin, B. Nørgaard-Pedersen, and E. Bock (1983a) Enzyme-linked immunosorbent assay of the D2-glycoprotein. *J. Neurochem.* **41**:356–362.

Ibsen, S., V. Berezin, B. Nørgaard-Pedersen, and E. Bock (1983b) Quantification of D2 glyco-protein in amniotic fluid and serum from pregnancies with neural tube defects. *J. Neurochem.* **41**:363–366.

Jørgensen, O. S., and E. Bock (1974) Brain specific synaptosomal membrane proteins demonstrated by crossed immunoelectrophoresis. *J. Neurochem.* **23**:879–880.

Jørgensen, O. S., and E. Bock (1975) Synaptic membrane antigen D2 measured in human cerebrospinal fluid by rocket line immunoelectrophoresis: Determination in psychiatric and neurological patients. *Scand. J. Immunol.* **1**:3–16.

Jørgensen, O. S., and B. Nørgaard-Pedersen (1981) The synaptic membrane D2-protein in amniotic fluid from pregnancies with fetal neural tube defects. *Prenat. Diagn.* **1**:3–6.

Jørgensen, O. S., E. Bock, P. Bech, and O. J. Rafaelsen (1977) Synaptic membrane protein D2 in the cerebrospinal fluid of manic-melancholic patients. *Acta Psychiatr. Scand.* **56**:50–56.

Linnemann, D., and E. Bock (1986) A developmental study of detergent solubility and polypeptide composition of the neural cell adhesion molecule. *Dev. Neurosci.* **8**:24–30.

Linneman, D., J. M. Lyles, and E. Bock (1985) A developmental study of the biosynthesis of the neural cell adhesion molecule. *Dev. Neurosci.* **7**:230–238.

Lyles, J. M., D. Linnemann, and E. Bock (1984) Biosynthesis of the D2-cell adhesion molecule: Post-translational modifications, intracellular transport, and developmental changes. *J. Cell Biol.* **99**:2082–2091.

Massaro, A. R., M. Albrechtsen, and E. Bock (1987) N-CAM in cerebrospinal fluid: A marker of synaptic remodeling after acute phases of multiple sclerosis. *Ital. J. Neurol. Sci. (Suppl.)* **6**:85–88.

Murray, B. A., G. C. Owens, E. A. Prediger, K. L. Crossin, B. A. Cunningham, and G. M. Edelman (1986) Cell surface modulation of the neural cell adhesion molecule resulting from alternative mRNA splicing in a tissue-specific developmental sequence. *J. Cell Biol.* **103**:1431–1439.

Noble, M., M. Albrechtsen, C. Møller, J. Lyles, E. Bock, C. Goridis, M. Watanabe, and U. Rutishauser (1985) Purified astrocytes express N-CAM/D2-CAM-like molecules *in vitro*. *Nature* **316**:725–728.

Nybroe, O., M. Albrechtsen, J. Dahlin, D. Linnemann, J. M. Lyles, C. J. Møller, and E. Bock (1985) Biosynthesis of the neural cell adhesion molecule: Characterization of polypeptide C. *J. Cell Biol.* **101**:2310–2315.

Nybroe, O., A. Gibson, C. J. Møller, H. Rohde, J. Dahlin, and E. Bock (1986) Expression of N-CAM polypeptides in neurons. *Neurochem. Int.* **9**:539–544.

Rutishauser, U., W. E. Gall, and G. M. Edelman (1978) Adhesion among neural cells of the chick embryo. IV. Role of the cell surface molecule CAM in the formation of neurite bundles in cultures of spinal ganglia. *J. Cell Biol.* **79**:382–393.

Rutishauser, U., S. Hoffman, and G. M. Edelman (1982) Binding properties of a cell adhesion molecule from neural tissue. *Proc. Natl. Acad. Sci. USA* **79**:685–689.

Saermark, T., O. Nybroe, A. Gibson, and E. Bock (1988) Endocytosis of the neural cell adhesion molecule (N-CAM) (submitted).

Schubert, D., M. LaCorbiere, F. G. Klier, and C. Birdwell (1983) A role for adherons in neural retina cell adhesion. *J. Cell Biol.* **96**:990–998.

Sørensen, P. S., F. Gjerris, S. Ibsen, and E. Bock (1983) Low cerebrospinal fluid concentration of brain specific protein D2 in patients with normal pressure hydrocephalus. *J. Neurol. Sci.* **62**:59–65.

Thiery, J.-P., R. Brackenbury, U. Rutishauser, and G. M. Edelman (1977) Adhesion among neural cells of the chick embryo. II. Purification and characterization of a cell adhesion molecule from neural retina. *J. Biol. Chem.* **252**:6841–6845.

Section 5

Morphology and Development

Chapter 25

Gap Junctions and Intercellular Communication in Developmental Patterning

SCOTT E. FRASER

ABSTRACT

Gap junctions can mediate the direct exchange of small molecules between the cytoplasms of neighboring cells. This ability has prompted the suggestion that communication through gap junctions may play an important role during the establishment of patterning in developing tissues. Several approaches, ranging from correlational studies to perturbation experiments, have been used to test this proposal. The evidence accumulated to date supports this role in some cases but appears to disprove it in others.

Gap junctions, first described on the basis of their morphology in thin sectioned material viewed in the electron microscope (Revel and Karnovsky, 1967), are now known to provide a direct pathway for the exchange of small molecules between the cytoplasms of neighboring cells (for reviews, see Gilula, 1980; Caveney, 1985). Several different physiological, structural, and chemical approaches have been applied to the study of the gap junction in order to better understand how it accomplishes its role in intercellular communication. Physiological experiments demonstrate that the gap junction can mediate a significant exchange of molecules up to about 1 kD (Simpson et al., 1977), indicating that it forms a channel of about 10 Å diameter. Structural experiments suggest that the active channel is made up of a hexamer of integral proteins in one cell membrane aligned with a similar hexamer in a neighboring cell to produce a central pore linking the interiors of the two cells (Caspar et al., 1977; Unwin and Ennis, 1984). Biochemical analyses have implicated a major protein with an apparent molecular weight of 27 kD as the channel-forming protein (see Hertzberg and Gilula, 1979), but there is considerable debate concerning the involvement of other related proteins in the gap junction (Finbow et al., 1983; Revel et al., 1986). These findings provide only an incomplete picture of the gap junction, yet they are exciting to developmental biologists because they suggest that the gap junction can mediate a direct pathway for intercellular communication.

The nature and extent of interaction between neighboring cells in the embryo have been the focus of a great body of experimental work (for a review, see Trinkaus, 1984). Cell interactions appear to play a central role in the patterning of developing tissues in systems ranging from neural induction in vertebrates to the imaginal disks of *Drosophila*. For example, the patterning of the developing or regenerating vertebrate limb has been the focus of several lines of experimental embryology, resulting in several hotly debated interpretations of the underlying mechanisms (cf. Holder, 1984; Javois, 1984; Stocum and Fallon, 1984). Although very different in detail, each proposed mechanism requires that neighboring cells communicate with one another to establish the final pattern of the limb. Relatively few mechanisms are available to mediate these intercellular interactions; obvious possibilities are the direct contact of either cell surface adhesion molecules or other cell surface receptors and the more indirect exchange of diffusible signals through either the extracellular space or gap junctions. Evidence can be garnered in favor of each of these possibilities, but until recently this evidence has been limited to correlations and consistency arguments. For example, gap junctions have been identified in thin sections of developing tissues, and simple calculations show that the time required for patterning within an embryonic field is approximately that expected for the diffusion of a small molecule through the cells (Crick, 1970).

In the past decade tremendous advances have been made in the molecular and cell biology of some aspects of cell interactions. For example, the adhesive protein N-CAM has been isolated and reconstituted into lipid vesicles to establish its ability to mediate adhesive interactions (Hoffman and Edelman, 1983). Furthermore, the protein sequence and gene structure of some cell adhesion molecules and receptors for extracellular matrix components are now known in some detail (see Cunningham and Edelman, Hynes et al., and Kemler et al., this volume). Such advances make it possible to perform test experiments that are both feasible and interpretable. Knowledge of the protein sequences of some extracellular matrix (ECM) components and their cell surface receptors has permitted a detailed molecular analysis of their interactions. Tests of their role(s) are now made possible by specific antibodies and peptides that inhibit these cell surface receptor–ECM interactions (see Horwitz et al., and Ruoslahti et al., this volume).

Molecular approaches to the study of the gap junction have been fruitfully employed, resulting in a knowledge of the complete primary protein sequence for the liver gap junction in both rats and humans (Kumar and Gilula, 1986; Paul, 1986). The use of site-specific antibodies has yielded some details on the structural disposition of the protein (see Goodenough, this volume, and Gilula, this volume). Some antibodies against the gap junction protein have been shown to disrupt gap junctional communication between some cell types (Warner et al., 1984; Hertzberg et al., 1985), making perturbation experiments possible. However, because the gap junction is proposed to be a mediator of cell–cell interactions rather than an effector of the interactions, detailed analyses of the function of gap junctions in developmental patterning have been difficult. Nevertheless, through a combined correlation and perturbation approach, evidence concerning the role of gap junctions in developmental patterning is now being assembled. In this chapter I present examples of correlation and perturbation studies that have been used in several laboratories in an attempt to build a case for a role of gap junctions in developmental patterning, and end with a description of a combined approach used to examine the role of gap junctions in the patterning of *Hydra*.

A CORRELATIONAL APPROACH

A popular method for exploring the possible role of gap junctional communication in developing tissues has been to compare the presence of developmental boundaries with the patterns of dye coupling in the tissue. The hope has been to demonstrate strong gap junction–mediated coupling in a group of cells that are participating in some patterning interaction, and an absence of coupling in nonparticipating cells. Qualitative support has been obtained in systems ranging from the squid embryo (Potter et al., 1966) to the mouse embryo (Lo and Gilula, 1979), where gap junction–mediated coupling has been found to be progressively restricted as the embryo matures and the developmental fates of the cells diverge. More recent experiments have shown a restriction of dye coupling at the segmental borders of the insect integument (Warner and Lawrence, 1982; Blennerhassett and Caveney, 1984). Cells within the segment are much more strongly dye coupled to one another than they are to cells in the neighboring segment. This pattern is especially intriguing because the segmental boundaries are lineage restriction boundaries; clones of marked cells grow to the segmental boundary but do not cross it (Lawrence, 1973). Furthermore, grafting operations have indicated that the cells possess some form of positional information about their position within a segment, and this positional information is repeated from segment to segment (Locke, 1959). The segmental boundaries therefore separate the extreme rostral positional value of one segment from the extreme caudal positional values of the next segment. If gap junctions were involved in the establishment of the positional information within the segment, such a discontinuity in coupling might well be expected.

Correlations between developmental boundaries and cell coupling can fuel suspicion that gap junctions play a role in a developmental process but cannot provide more than circumstantial evidence. In addition, caution must be applied to the interpretation of correlational studies because of incomplete knowledge of the phenomena with which the data are correlated. For example, different models of cell patterning and compartment formation make distinct predictions about the nature of intercellular communication. Some require the abolition of cell coupling between different sets of cells in order for the cells to maintain their developmental differences, while others require the maintenance of cell communications for the cells to remain different (Meinhardt, 1982). If taken to the extreme, these two classes of models would permit any pattern of dye coupling to be taken as supporting the idea of an important role for gap junctions. In the first case, coherence between developmental and gap junctional boundaries would demonstrate an important role for gap junctions; in the second, lack of coherence would demonstrate an equally important role.

Another limitation of correlational studies is that the demonstration of exact coherence between gap junctional "communication compartments" and developmental compartments may be difficult. Any discontinuity might appear to align in a significant way. For example, dye passage in the *Drosophila* imaginal wing disk shows discontinuities that have been asserted to form communication compartments (Weir and Lo, 1982, 1984) that align with the lineage restriction or compartment boundaries of the disk (cf. Garcia-Bellido et al., 1973). In the most complete analysis, claims have been made for many more communication compartment boundaries than the number of lineage compartment borders for which there is even weak evidence (Weir and Lo, 1984). In fact, with a large number of communication compartment boundaries, no dye-filled cell could be farther than a few cell diameters away from at least one, and all discontinuities might appear to coincide with one of the boundaries. If one considers that the most compelling data for lineage restrictions are available for only the dorsoventral and anteroposterior lineage restriction boundaries, and attempts to compare these with the positions of the dye coupling "boundaries" in slightly more detail, no compelling correlation emerges (Fraser and Bryant, 1985).

A variant on the usual correlation experiment is to attempt to force a test of the correlation experimentally. The experimental tissue is altered in some way, and the effects on the developing tissue and on cell coupling are then followed. We have recently employed this approach on the *Drosophila* imaginal disk (Bryant and Fraser, 1988), a system in which the cells appear to interact and in which gap junctions are prevalent. A quarter of an imaginal wing disk was surgically ablated, the cut edges were marked by microinjecting fluorescent dextran (Gimlich and Braun, 1985) into the cells neighboring the cut, and the disk was then cultured in the abdomen of an adult fly. At different times after the ablation, the disk was excised, the degree of new tissue growth was determined by measuring the distance between the dextran-labeled cells, and the extent of dye coupling was assayed by injecting fluorescein complexon into a nearby cell. During the first day of culture, the cut edges healed together, and cell division began (O'Brochta and Bryant, 1987; Bryant and Fraser, 1988), yet there was no detectable cell coupling into the regenerating region as assayed by dye injections. Cell coupling only weakly reappeared between the original and regenerating tissues on the second day of culture, and became nearly as strong as in the original tissue on the third day, when tissue growth largely ceased. In this experimentally perturbed preparation, therefore, many events take place before the reestablishment of significant gap junctional communication. Thus these findings do not support the proposal that gap junctional communication between the cells allows them to determine that tissue has been deleted, although the findings may be consistent with the idea that the loss of coupling serves as a stimulus to begin tissue growth.

A PERTURBATION APPROACH

An alternative approach that circumvents some of the potential pitfalls of a correlational analysis of gap junction function is to perform perturbation analyses. Through the examination of the developmental effects of disrupting gap junctional communication, a more direct analysis of the processes that require gap junction function may be possible. While it is clearly possible to overinterpret the results of perturbation experiments, the approach has great promise. At the very least, combining a perturbation experiment with more conventional methods should permit us to progress closer to understanding what gap junctions can do in development instead of merely predicting what role(s) they might play.

A perturbation approach has been made possible by the availability of antibodies against the major gap junction protein (Hertzberg, 1984; Warner et al., 1984). Some of these antibodies disrupt communication through gap junctions in embryos when microinjected intracellularly (Warner et al., 1984). In one perturbation experiment antibodies were injected into the blastomere of an eight-cell-stage *Xenopus* embryo whose descendants give rise to much of the neural plate on one side. The antibodies greatly diminished or abolished dye coupling into or out of the injected cell and its immediate descendants. When allowed to develop to hatching stages, the embryos were found to have unilateral defects in their neural structures on the injected side (Warner et al., 1984). The defects varied in their severity, from a slight asymmetry to a nearly complete absence of neural structures, perhaps suggesting that the antibodies interfered with neural induction or the cell's response to induction. Additional experiments will clearly be required to determine what processes the antibody interferes with to produce the observed defects.

A recent perturbation experiment offers evidence of a cell interaction that does not require the presence of functional cell coupling. Warner and Gurdon (1987) examined the expression of a muscle-specific actin that can be experimentally induced by grafting together blastomeres of the animal and vegetal tiers of the early *Xenopus* embryo. Previous coculture experiments had demonstrated that this induction required cell contact; coculture of the cells without contact failed to induce expression (Gurdon et al., 1985). Embryos injected with the anti–gap junction antibody and confirmed to lack dye coupling between the cells were still able to participate in the induction of the muscle-specific gene expression (Warner and Gurdon, 1987). The clear demonstration of a cell interaction that is not blocked by the antibody-produced abolition of gap junctional communication increases confidence in results that demonstrate a defect following antibody treatment.

COMMUNICATION AND PATTERNING IN *HYDRA*

Many perturbation experiments are limited by incomplete access to the phenomenon or the cells under study. In order to test both the sufficiency and the necessity of gap junctional communication in a tissue-patterning process, we chose a more accessible system, *Hydra*. This freshwater coelenterate has been studied extensively to gain insight into the interactions important in pattern formation (for a review, see Bode and Bode, 1984). Detailed developmental studies have been possible because of the animal's simple body plan, with a head and a foot at opposite ends of a hollow, two-layered epithelial body column. This simple body plan greatly simplifies the design, performance, and interpretation of experiments.

Hydra tissue has the ability to repattern following experimental insult: A small fragment excised from any part of the *Hydra* body column is capable of reorganizing into a complete miniature version of the animal. The more apical end of the fragment always forms the head in the repatterned animal, implying that head regeneration is controlled by some process dependent on the original positional identity of the isolated tissue. By grafting together fragments of *Hydra* tissue in novel arrangements, two developmental gradients have been shown to play a role in establishing the position of a regenerating head (for reviews, see Bode and Bode, 1984; Bode et al., 1987). The first gradient, head activation, is relatively stable and distributed monotonically, with the tissue at the upper end of the body column possessing a greater ability to form a head than the tissue farther down the body column (MacWilliams, 1983b). The second, the head inhibition gradient, is a labile monotonic property that appears to originate from an established head, travel down the body column, and inhibit the tissue from forming another head (MacWilliams, 1983a). Head inhibition is commonly assumed to be due to the steady-state production of a diffusible "head inhibitor" by the head and its degradation in the tissues through which it diffuses. This proposal has been given some support by the ability of models incorporating it to fit much of the experimental data (Meinhardt and Gierer, 1974; MacWilliams, 1982; Meinhardt, 1982).

Head inhibition is easily understood from the results of simple grafting operations. A small ring of tissue, excised from the apical end of the body column of a donor *Hydra*, can be grafted midway between the head and the bud of a host animal. In such a graft in normal animals, the host suppresses the graft from the formation of a secondary axis (head and tentacles) in a majority (about 90%) of the cases, and the grafted tissue is eventually resorbed into the column. This inhibition is dependent on the head of the host animal. If the head of the host is removed, the implant is no longer inhibited from developing a head and will form a secondary axis in the majority of the cases (70–90%). The complete formation of the secondary axis requires three or four days, but the commitment to form or not form a second head takes place within the first day; about 50% of the implants form secondary axes if the head is removed and replaced about 12 hours later (Bode et al., 1987).

To determine if the head inhibition gradient of *Hydra* is transmitted by gap junctions, we began by building a case on circumstantial evidence. Head inhibition appears to be a process mediated by the diffusional exchange of an inhibitor substance through the epithelium. Electron microscopy of the epithelial cells of the *Hydra* body column demonstrates that they are extensively linked to one another by gap junctions (Filshie and Flower, 1977; Wood, 1977). Experiments with intracellular microelectrodes indicate that the epithelial cells are both dye and electrically coupled, demonstrating the presence of a functional communication pathway between the cells (Fraser and Bode, 1981; Fraser et al., 1987). Finally, an activity has been isolated from *Hydra* with some of the qualities and behaviors expected of a head inhibitor, and appears to be sufficiently small and hydrophilic to pass through gap junctions (Berking, 1979).

As a further test, we determined whether microinjected dyes could move at a sufficient rate and reform rapidly enough after a grafting operation to fit with the known kinetics of head inhibition. A simple measure of the rate of dye passage was made by determining the time required for the dye to spread to a point 1 mm away from the injection site. Such measurements yielded an estimated diffusion constant of 3×10^{-6} cm^2/sec, consistent with the rate of spread of head inhibition predicted from grafting experiments by MacWilliams (1983a). The time required for dye coupling to return following a grafting operation was determined by grafting together two rings of body column tissue and testing the dye passage between them at various times using a double-blind protocol. Dye coupling was first detectable at 40–50 min after a grafting operation, with all grafts demonstrating coupling by one hour. Thus a strong correlational case can be built for a role of gap junctions in the head inhibition gradient: they are in the right place, provide a sufficiently rapid path for the exchange of small molecules, and reform rapidly between grafted tissue fragments.

A more direct test of the role of gap junctions requires a perturbation approach. The perturbation was made possible by two affinity-purified rabbit polyclonal antibodies (Warner et al., 1984), generated by immunizing rabbits with the 27-kD protein electroeluted from SDS-polyacrylamide gels of base-extracted gap junctions (Hertzberg, 1984). The affinity-purified antibodies bound to the cytoplasmic surfaces of intact rat liver gap junctions (Young et al., 1987), and recognized homologous antigens (27-kD and 54-kD proteins on 12.5% polyacrylamide gels) in Western blots of rat liver and *Hydra* (Fraser et al., 1987; Young et al., 1987). Indirect immunofluorescence, performed on whole mounts of fixed *Hydra*, demonstrated the presence of the antigen along the lateral membranes of the epithelial cells, consistent with electron microscope studies (Filshie and Flower, 1977; Wood, 1977).

The antibodies against the 27-kD rat liver protein were initially found to block gap junctional communication in *Hydra* by microinjecting them into single cells and testing dye coupling one hour later. Extracellular application of the antibodies had no detectable effects, so a means to introduce the antibody to the interior of all the cells of *Hydra* was developed. The epithelial cells of *Hydra* were loaded with an antibody using the permeabilizing agent dimethyl sulfoxide (5% DMSO for 30 min at 4°C). Following this treatment, the *Hydra* remained healthy, but those treated with immune antibodies no longer showed dye passage, whereas those treated with preimmune antibodies showed normal dye passage. By our criterion of dye passage—movement of the dye more than three cell diameters—the uncoupling effect of the antibodies lasted for at least seven hours (Fraser et al., 1987).

The antibody was used to test whether decreasing the communication through gap junctions reduced the ability of the host head to inhibit secondary axis formation by a graft of untreated tissue. This experimental design has the distinct advantage that the perturbation (blockage of the inhibition signal) should lead to a positive result (the formation of a second head) by untreated tissue, thereby minimizing concerns about side effects of the reagents. More than 800 grafts were performed, and the percentage of the animals in which the implant developed into a full head (hypostome and two to five tentacles, termed a secondary axis) was determined. Grafts to untreated animals resulted in the formation of secondary axes in only 11% of the cases, while 80% of the grafts to permanently decapitated animals formed secondary axes. Preimmune treatment of the host animals resulted in a small but insignificant decrease in the fraction of secondary axes to 7% ($p = .1$, by a test of differences between proportions; Freund et al., 1960). Immune treatment of the host animals with either of the immune reagents increased the incidence of secondary axes to 22%, which is statistically significant when compared to the grafts performed on both the preimmune treated ($p < .002$) and untreated ($p < .01$) hosts.

CONCLUSION

Both correlation perturbation experiments can offer some insights into the importance of gap junctions in the patterning of developing tissues. Through a combination of the two approaches, the final set of experiments attempts to demonstrate that gap junctions are both sufficient and necessary for the patterning of *Hydra* tissue. The experiments measuring the rate of dye passage and the rate of return of dye coupling after a graft demonstrate that gap junctions are a sufficient pathway for the passage of head inhibition down the body column. The antibody treatment resulted in a significant increase in the percentage of secondary axes formed (7% vs. 22%), indicating that gap junction communication is required for some aspect of head inhibition. However, the increase in secondary axes did not reach the 50% of temporarily decapitated animals for 8–12 hours, about the same amount of time as the antibody is effective. This may result from any of several factors. One possibility is suggested by evidence that the endoderm may play an important role in the transmission of head inhibition (Nishimaya et al., 1986). The means of applying the antibody we employed predominantly exposed the outer ectodermal layer of the *Hydra* to the antibody. If the inner endoderm layer had not been as blocked by the antibody, part of the pathway for the passage of head inhibition might have remained intact. Future work will attempt to refine the delivery of the antibody to determine if the elimination of gap junction communication can have as profound an effect as temporary decapitation.

A perturbation experiment has several advantages but cannot supplant a correlational approach; in fact, the two approaches may be most fruitfully performed in parallel. Correlational experiments can help to focus attention on systems that will later be examined with a perturbation experiment. Furthermore, the limitations of the two approaches are quite different, and concerns of one approach can be partially dealt with by the strengths of the other. Some of the limitations of correlational experiments were presented briefly earlier in this chapter.

Perturbation experiments also have an important set of limitations. Any perturbation can have its effects through avenues not obvious to the experimenter, and it can be quite difficult to determine the true causes of any defect. In addition, the results of many perturbation experiments have been the loss of some structure, making control experiments difficult to design and interpret. In the *Hydra* experiments presented here, a successful perturbation resulted in the addition of a second head to the *Hydra*, consistent with the blockade of the passage of head inhibition. It remains for future work to show that the effect is only through a blockade of communication and not due to some other factor such as decreased production of head inhibitor in the presence of the antibody. The unique advantages of *Hydra* should permit the experimental manipulations necessary to test alternative interpretations and better determine the role of gap junctions in the patterning of this animal.

ACKNOWLEDGMENTS

Much of the work described here was performed in collaboration with H. R. Bode, P. M. Bode, P. J. Bryant, N. B. Gilula, and C. R. Green, and I gratefully acknowledge their roles in the design, performance, and interpretation of the experiments. The work performed in my laboratory was supported by National Science Foundation grants DBC-8510891 and BNS-8406307.

REFERENCES

Berking, S. (1979) Analysis of head and foot formation in *Hydra* by means of an endogenous inhibitor. *Wilhelm Roux's Arch. Dev. Biol.* **186**:189–210.

Blennerhassett, M. G., and S. Caveney (1984) Separation of developmental compartments by a cell type with reduced junctional permeability. *Nature* **309**:361–364.

Bode, P. M., and H. R. Bode (1984) Patterning in *Hydra*. In *Pattern Formation: A Primer in Developmental Biology*, G. M. Malacinski and S. V. Bryant, eds., pp. 213–241, Macmillan, New York.

Bode, H. R., S. E. Fraser, C. R. Green, P. M. Bode, and N. B. Gilula (1987) Gap junctions are involved in a patterning process in *Hydra*. In *Genetic Regulation of Invertebrate Development*, W. Loomis, ed., Alan R. Liss, New York.

Bryant, P. J., and S. E. Fraser (1988) Wound healing, cell communication, and DNA synthesis during imaginal disc regeneration in *Drosophilia*. *Dev. Biol.* **127**:197–207.

Caspar, D. L. D., D. A. Goodenough, L. Makowski, and W. C. Phillips (1977) Gap junction structures. I. Correlated electron microscopy and X-ray diffraction. *J. Cell Biol.* **92**:213–220.

Caveney, S. (1985) The role of gap junctions in development. *Annu. Rev. Physiol.* **47**:305–318.

Crick, F. (1970) Diffusion in embryogenesis. *Nature* **225**:420–422.

Filshie, B. K., and N. E. Flower (1977) Junctional structures in *Hydra*. *J. Cell Sci.* **23**:151–172.

Finbow, M. E., J. Shuttleworth, A. E. Hamilton, and J. D. Pitts (1983) Analysis of vertebrate gap junction protein. *EMBO J.* **2**:1479–1486.

Fraser, S. E., and H. R. Bode (1981) Epithelial cells of *Hydra* are dye-coupled. *Nature* **294**:356–358.

Fraser, S. E., and P. J. Bryant (1985) Patterns of dye coupling in the imaginal wing disc. *Nature* **317**:533–536.

Fraser, S. E., C. R. Green, H. R. Bode, and N. B. Gilula (1987) Selective disruption of gap junctional communication interferes with a patterning process in *Hydra*. *Science* **237**:49–55.

Freund, J. E., P. E. Livermore, and I. Miller (1960) *Manual of Experimental Statistics*, Prentice-Hall, Englewood Cliffs, New Jersey.

Garcia-Bellido, A., P. Ripoll, and G. Morata (1973) Developmental compartmentalization of the wing disc of *Drosophila*. *Nature* **245**:251–253.

Gilula, N. B. (1980) Cell-to-cell communication and development. In *The Cell Surface: Mediator of Developmental Processes*, S. Subtelny and N. K. Wessels, eds., pp. 23–41, Academic, New York.

Gimlich, R. L., and J. Braun (1985) Improved fluorescent compounds for tracing cell lineage. *Dev. Biol.* **109**:509–514.

Gurdon, J. B., S. Fairman, T. J. Mohun, and S. Brennan (1985) Activation of muscle-specific actin genes in *Xenopus* development by an induction between animal and vegetal cells of a blastula. *Cell* **41**:913–922.

Hertzberg, E. (1984) A detergent-independent procedure for the isolation of gap junctions from rat liver. *J. Biol. Chem.* **259**:9936–9943.

Hertzberg, E., and N. B. Gilula (1979) Isolation and characterization of gap junctions from rat liver. *J. Biol. Chem.* **254**:2138–2147.

Hertzberg, E., D. C. Spray, and M. V. L. Bennett (1985) Antibodies to gap junctions block junctional conductance. *Proc. Natl. Acad. Sci. USA* **82**:2412–2416.

Hoffman, S., and G. M. Edelman (1983) Kinetics of homophilic binding by E and A forms of the neural cell adhesion molecule. *Proc. Natl. Acad. Sci. USA* **80**:5762–5766.

Holder, N. (1984) Regeneration of the axolotl limb patterns and polar coordinates. In *Pattern Formation: A Primer in Developmental Biology*, G. M. Malacinski and S. V. Bryant, eds., pp. 521–537, Macmillan, New York.

Javois, L. C. (1984) Pattern specification in the developing chick limb. In *Pattern Formation: A Primer in Developmental Biology*, G. M. Malacinski and S. V. Bryant, eds., pp. 557–579, Macmillan, New York.

Kumar, N., and N. B. Gilula (1986) Cloning and characterization of human and rat liver cDNAs coding for a gap junction protein. *J. Cell Biol.* **103**:767–776.

Lawrence, P. A. (1973) A clonal analysis of segment development in *Oncopeltus* (Hemiptera). *J. Embryol. Exp. Morphol.* **30**:681–699.

Lo, C. W., and N. B. Gilula (1979) Gap junctional communication in the postimplantation mouse embryo. *Cell* **18**:411–422.

Locke, M. (1959) The cuticular pattern in an insect, *Rhodnius prolixus* (Stål). *J. Exp. Biol.* **36**:459–477.

MacWilliams, H. (1982) Numerical simulation of *Hydra* head regeneration using a proportion regulating version of the Gierer–Meinhardt model. *J. Theoret. Biol.* **99**:681–703.

MacWilliams, H. (1983a) *Hydra* transplantation phenomena and the mechanism of hydra head regeneration. I. Properties of head inhibition. *Dev. Biol.* **96**:217–238.

MacWilliams, H. (1983b) *Hydra* transplantation phenomena and the mechanism of hydra head regeneration. II. Properties of head activation. *Dev. Biol.* **96**:239–272.

Meinhardt, H. (1982) *Models of Biological Pattern Formation*, Academic, New York.

Meinhardt, H., and A. Gierer (1974) Applications of a theory of biological pattern formation based on lateral inhibition. *J. Cell Sci.* **15**:321–326.

Nishimaya, C., N. Wanek, and T. Sugiyama (1986) Genetic analysis of the developmental mechanisms of *Hydra*. XIV. Identification of the cell lineages responsible for the altered developmental gradients in a mutant strain, reg-16. *Dev. Biol.* **115**:469–478.

O'Brochta, D. A., and P. J. Bryant (1987) Distribution of S-phase cells during the regeneration of *Drosophila* imaginal wing discs. *Dev. Biol.* **119**:137–142.

Paul, D. (1986) Molecular cloning of cDNA for rat liver gap junction protein. *J. Cell Biol.* **103**:123–134.

Potter, D. D., E. J. Furshpan, and E. S. Lennox (1966) Connections between cells of the developing squid as revealed by electrophysiological methods. *Proc. Natl. Acad. Sci. USA* **55**:328–336.

Revel, J.-P., and M. J. Karnovsky (1967) Hexagonal array of subunits in intercellular junctions of mouse heart and liver. *J. Cell Biol.* **33**:C7–C12.

Revel, J.-P., B. Nicholson, and S. B. Yancey (1986) Molecular structure of gap junctions. *Proc. Int. Congr. Electron Micros.* **11**:1857–1860.

Simpson, I., B. Rose, and W. R. Loewenstein (1977) Size limit of molecules permeating the junctional membrane channels. *Science* **195**:294–296.

Stocum, D. L., and J. F. Fallon (1984) Mechanisms of polarization and pattern formation in urolele limb ontogeny: A polarizing zone model. In *Pattern Formulation: A Primer in Developmental Biology*, G. M. Malacinski and S. V. Bryant, eds., pp. 507–520, Macmillan, New York.

Trinkaus, J. P. (1984) *Cells Into Organs: The Forces That Shape the Embryo*, Prentice-Hall, Englewood Cliffs, New Jersey.

Unwin, P. N. T., and P. D. Ennis (1984) Two configurations of a channel-forming membrane protein. *Nature* **307**:609–613.

Warner, A. E., and J. B. Gurdon (1987) Functional gap junctions are not required for muscle gene activation by induction in *Xenopus* embryos. *J. Cell Biol.* **104**:557–564.

Warner, A. E., and P. A. Lawrence (1982) Permeability of gap junctions at the segmental border in insect epidermis. *Cell* **28**:243–252.

Warner, A. E., S. C. Guthrie, and N. B. Gilula (1984) Antibodies to gap-junctional protein selectively disrupt communication in the early amphibian embryo. *Nature* **311**:127–131.

Weir, M. P., and C. W. Lo (1982) Gap junctional communication compartments in the *Drosophila* wing disk. *Proc. Natl. Acad. Sci. USA* **79**:3232–3235.

Weir, M. P., and C. W. Lo (1984) Gap junctional communication compartments in the *Drosophila* wing imaginal disk. *Dev. Biol.* **102**:130–146.

Wood, R. L. (1977) The cell junctions of *Hydra* as viewed by freeze-fracture replication. *J. Ultrastruct. Res.* **58**:299–315.

Young, J. D.-E., Z. A. Cohen, and N. B. Gilula (1987) Functional assembly of gap junction conductance in lipid bilayers: Demonstration that the major 27 kD protein forms the junction channel. *Cell* **48**:733–743.

Chapter 26

Sequential Changes in Adhesion During Morphogenesis

DAVID R. McCLAY
GREGORY A. WRAY
MARK C. ALLIEGRO
CHARLES A. ETTENSOHN

ABSTRACT

A rapid series of morphogenetic movements characterizes the early development of the sea urchin embryo. Many of these movements have been studied to determine the molecular complexity of a morphogenetic sequence. This chapter describes the cellular and some of the molecular changes contributing to the morphogenesis of primary mesenchyme cells (PMC) as they make the skeleton of the embryonic larva. At ingression there are three independent changes in PMC affinity for the three substrates used up until ingression. The PMCs lose an affinity for the hyaline layer, they lose an affinity for other cells, and they gain an affinity for the basal lamina. When the components of the basal lamina were tested, PMCs increased their adhesive affinity for fibronectin, a sea urchin matrix protein. PMCs were injected into the blastocoel cavity of the embryo to examine other adhesive properties as the cells differentiated into skeleton-synthesizing cells. It was shown that the PMCs alone of all the cells injected moved toward the vegetal plate, demonstrating adhesive recognition for a specific region of the blastocoel. Experimentally, that recognition was shown to be time specific as well as cell-type specific. The extracellular matrix lining the vegetal plate was the substrate to which the PMCs migrated and adhered before they later began migrating away from that area. Other experiments demonstrated that the PMCs would not migrate up the sides of the blastocoel until the embryo reached a certain age. Initially, the PMCs migrate individually in the blastocoel without adhering to other PMCs. After several hours of migration, the PMCs begin to adhere to one another, and then to fuse to form a syncytium. Finally, as the cells begin to make spicules, experiments show that a quantitative property released by PMCs prevents secondary mesenchyme cells (SMC) from converting and becoming PMCs. If PMCs are removed from the blastocoel, the SMCs convert to become PMCs. Thus, over a short time span of only about 10 hours, the cells pass through a number of morphogenetic changes, many of which involve cell adhesion or cell recognition, in order to acomplish a variety of functions. The molecular basis of several of those interactions is described.

533

At gastrulation a sequence of cell rearrangements establishes the three germ layers. In the transparent sea urchin embryo this process has been observed in detail, using a number of probes (Gustafson and Wolpert, 1967; Ettensohn, 1985; McClay and Ettensohn, 1987). It is clear from those and many other studies that gastrulation is not one morphogenetic movement; rather, it is a series of coordinated movements by cells that are in the process of differentiating into many cell types. To study details of morphogenesis in gastrulation, it is necessary to focus on single events against a background of rapid change. This is a difficult task, but in the sea urchin a number of experimental approaches have been developed that allow such a focus.

This chapter is limited to a description of molecular events associated with cell adhesion during morphogenetic movements of the primary mesenchyme cells. In limiting the study to cell–cell and/or cell–substrate interactions, we describe only one aspect of a morphogenetic movement, the complete picture of which involves an understanding of adhesion, motility, the genetic mechanisms that control these phenomena, plus a variety of mechanisms associated with pattern formation.

Confining the discussion to cell adhesion of a single cell type during morphogenesis still does not narrow the focus adequately. There may be many adhesion-related functions being performed by a given cell population at any time in development. When Sperry (1963) considered the neuronal pathways followed in organizing the brain, it occurred to him that cell adhesion, direction of movement, and synaptogenesis must require coded positional information to reduce the complexity that might otherwise be required if each cell had its own independent morphogenetic information. Since then, a number of models have attempted to describe how relatively few molecules might be necessary to provide adequate morphogenetic information for most complex cell rearrangements (Gustafson and Wolpert, 1967; Bell, 1978; Fraser, 1980; Edelman, 1984). The number of adhesion molecules that might be present is only one question in the story of morphogenetic complexity. Many cell adhesion molecules are discussed elsewhere in this volume, yet there is still little experimental information as to how these molecules provide morphogenetic landmarks (if in fact they do), and there are only guesses as to the precise function for many of these. Our approach has been to examine individual morphogenetic movements in an attempt to determine their molecular complexity. To illustrate that approach, we describe the events surrounding morphogenetic movements of primary mesenchyme cells in the sea urchin, beginning with ingression into the blastocoel and ending several hours later with the synthesis by these cells of the embryonic skeleton. We show that adhesion participates in morphogenesis in a number of ways. The expression of pattern in such a framework appears to utilize a collection of adhesive processes rather than one mechanism.

PRIMARY MESENCHYME CELL INGRESSION

An unequal cell division at the 16-cell stage produces 4 small cells at the vegetal plate that are called micromeres. The micromeres divide several times prior to the mesenchyme blastula stage, giving rise to a population of about 60 cells. At 10 hours of development (the time varies with species and with the temperature of the culture) the micromere progeny, now called primary mesenchyme cells, ingress into the blastocoel (Figure 1). Ingression lasts for little more than an hour and results in the entire population of primary mesenchyme cells assuming a new relationship with the basal lamina lining the blastocoel.

During ingression the PMCs lose contact with the extraembryonic matrix called hyaline, and with other cells. They migrate through and adhere to the inside of the basal lamina. Experimentally, it has been shown that PMCs simultaneously lose an affinity for hyaline and for other cells, while they gain an affinity for the basal lamina (Figure 2) (McClay and Fink, 1982; Fink and McClay, 1985). Thus, for a single morphogenetic movement lasting an hour, the cells undergo at least three independent cell adhesive changes at the same time (and coincidentally the cells become motile).

Several of these interactions have been partially characterized. For example, the cell–hyaline interaction occurs only on the apical surface of the cells and is lost (only from the PMC population) at the mesenchyme blastula stage. Prior to that time, the cells bind to hyaline. Work on the site of binding has included purification of hyaline, production of a proteolytic fragment still recognized by the cells, and isolation of a monoclonal antibody to the fragment that blocks cell-to-hyaline interactions (Adelson and Humphreys, 1988).

The increase in adhesion to the basal lamina was examined further by asking whether the PMCs increased their affinity for one or more than one of the substrate materials known to be present in the sea urchin basal lamina (Solursh

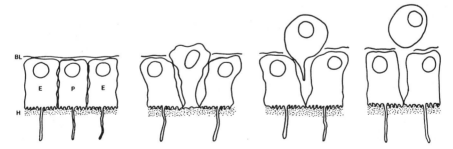

Figure 1. *Sequence of cell shape changes at ingression.* The presumptive primary mesenchyme cells (P) detach from the hyaline layer (H) and from other cells of the blastula cell wall (E), and they migrate through the basal lamina (BL). This morphogenetic sequence lasts for about one hour.

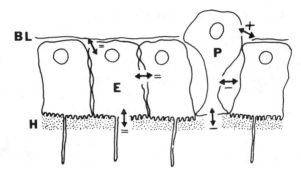

Figure 2. *Affinity changes that occur at the mesenchyme blastula stage.* Primary mesenchyme cells (P) lose affinity for hyaline and for other cells (−), and they gain an affinity for the basal lamina (+). Presumptive ectoderm and endoderm cells (E) retain the same affinities for the three substrates during this interval (=).

and Katow, 1982; Wessel et al., 1984). It was found that of all the materials tested, the PMCs demonstrated an increase in affinity only for fibronectin during the time of ingression (Fink and McClay, 1985). The substrate used for the test was vertebrate fibronectin. The sea urchin equivalent has yet to be characterized, though there are several reports suggesting that a related molecule exists (Spiegel et al., 1980; Wessel et al., 1984; DeSimone et al., 1985; Iwata and Nakano, 1985). In another approach, a sea urchin protein that serves as a better substrate for cell binding than mammalian fibronectin was isolated and purified (Alliegro et al., 1988). This molecule, called echinonectin, is quite different from fibronectin (though it is possible that the functional parts of the molecule are conserved).

Aside from the fibronectin relationship, little else is known about the cell–basal laminar interactions that appear to change at the mesenchyme blastula stage. The behavioral changes could be simple quantitative changes in the substrate, or new materials known to be deposited might enable the PMCs to escape the vegetal pole area. Whatever the case, the system will allow a molecular dissection of these changes.

Micromeres in culture contain all the information necessary to pass through development and differentiate in the absence of the remainder of the embryo (Okazaki, 1975). This means that micromere–PMC differentiation and many of the later interactions seen between PMCs and other cells or substrates are not required for PMC morphogenesis. Figure 3 shows a comparison between the events that can be monitored between micromeres grown *in vitro* and *in vivo*. On plastic they display the same sequence of behaviors that are observed *in vivo*: the cells adhere to one another until the mesenchyme blastula stage, when they lose cell–cell contacts. At the same time, the cells become motile and begin to move on the substrate. Later, the cells in culture reassociate with one another and make spicules, again in parallel to the normal functions of these cells *in vivo* (Figure 3). In order for the micromeres to grow in culture, it was

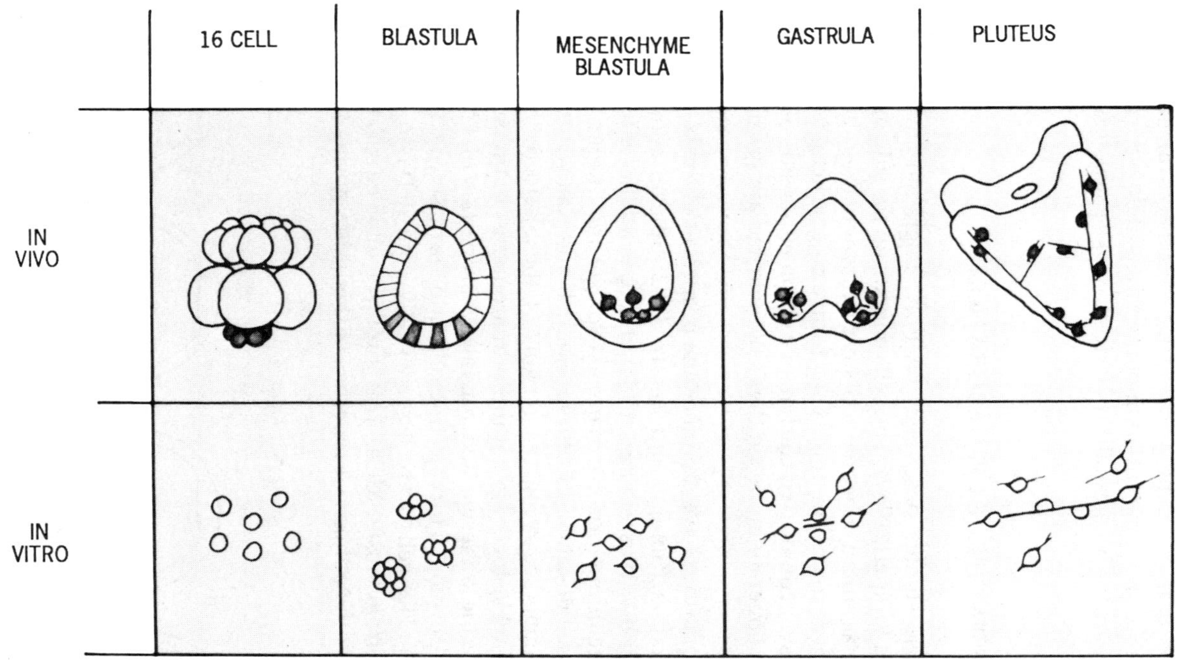

Figure 3. *Comparison between the different behaviors of micromeres in vivo and the behaviors that can be observed with micromeres in culture. The cells in culture lose contact with other PMCs, migrate, reassociate, and makes spicules in a behavioral sequence that mimics those behaviors in vivo.*

537

necessary for Okazaki to supply horse serum. If fibronectin is depleted from the horse serum, there is a reduction in the ability of micromeres to differentiate on their own *in vitro*. Thus, it would appear that the interaction with fibronectin, or in the urchin with the fibronectin equivalent, is a necessary component for further differentiation of the PMCs. Below we use the culture system to compare the innate ability of PMCs to make spicules with the pattern produced *in vivo*.

PRIMARY MESENCHYME CELL MIGRATION

Early time-lapse studies described the PMC migration in the blastocoel (Gustafson and Wolpert, 1967). The cells migrate up the side of the blastocoel and eventually assume a subequatorial position, where they again acquire the capacity to adhere to one another. They form a ring structure (Figure 4) and fuse to form a syncytium. Two clusters of cells on either side of the blastocoel then begin to cooperate in the synthesis of the triradiate spicule that is made from $CaCO_3$.

Several laboratories have examined the properties of the blastocoel during the PMC migration (Solursh and Katow, 1982; Spiegel and Burger, 1982; Spiegel et al., 1983; Wessel et al., 1984). There is a differential synthesis of wheat germ agglutinin–positive material (Spiegel and Burger, 1982), there is deposition of sulfate-labeled material (Solursh and Katow, 1982), and some monoclonal antibodies recognize regions of the blastocoel (Wessel et al., 1984). Thus, the blastocoel is not uniform around its entire surface.

To examine the adhesive and migratory properties of PMCs *in vivo*, a technique was developed for injecting cells into the blastocoel (Ettensohn and McClay, 1986). The cells were tagged vitally with RITC so they could be distinguished from host PMCs. This protocol allowed us to ask several questions about the adhesive properties of the PMCs during their migratory

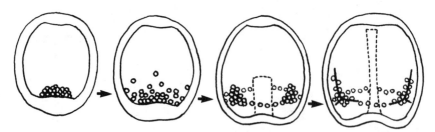

Figure 4. *Migratory behavior of PMCs during the mesenchyme blastula stage.* The cells leave the vegetal plate and migrate individually up the wall of the blastocoel. After several hours the cells begin to associate with one another to form a ring around the invaginating archenteron (*dotted line*). The cells in the ring fuse to form a syncytium, and two clusters of cells called the ventrolateral clusters are the sites of initiation of spicule formation.

behavior. When PMCs were injected into any site in the blastocoel, the cells immediately began to move toward the vegetal side of the blastocoel. Tracings of time-lapse films suggested that some of the cells moved to the vegetal plate in a straight line (Ettensohn and McClay, 1986). One hypothesis that explains this behavior is an adhesive gradient recognized by cells and directing them to the vegetal plate area. Alternatively, the cells may extend long filopodia that explore the blastocoel surface at random. If there were an adhesive surface at the vegetal plate recognized by the filopodia and the cells were to follow the filopodia to that surface, then the straight migration by the cell soma could be in response to a random capture of the filopodia at the adhesive vegetal plate. In either case, the vegetal plate is the only site to which the cells move. Further, that movement is a property of the PMCs only. Other cell types injected into the blastocoel do not migrate to the vegetal plate (Figure 5).

Since the basal plate has a special adhesive property recognized by PMCs, we next asked when during development the vegetal plate acquired this property (Ettensohn and McClay, 1986). By the injection of PMCs into young blastulas, it was learned that the adhesive area was present long before

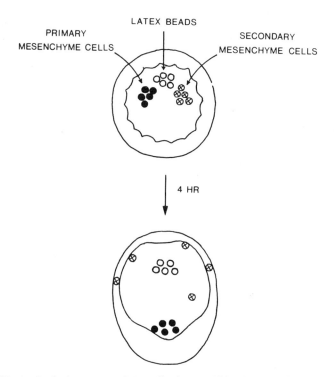

Figure 5. *Movement of primary mesenchyme cells after transplantation into a blastocoel of a recipient embryo.* PMCs move to the vegetal plate. Transplanted secondary mesenchyme cells move in the blastocoel at random, and latex beads do not move when inserted into the same region.

ingression of the host PMCs. However, the cells then exhibited unexpected behavior. In the control situation where PMCs were transplanted into embryos of the same age, the cells migrated to the vegetal pole and immediately began to participate in the morphogenetic movements leading to ring formation. In the transplants into younger embryos, the older PMCs moved to the vegetal plate, but they then did not move any further until the host embryo reached the mesenchyme blastula stage. The older PMCs, along with the host PMCs, then joined together in the morphogenetic movements, carrying the cells away from the vegetal plate. These data show that the areas lateral to the vegetal plate must change before they can serve as a substrate for the migration of PMCs.

At ring formation, the PMCs first associate with one another before fusing to form a syncytium. This ability of cells to associate with one another was lost at the mesenchyme blastula stage (Frink and McClay, 1985), and the reacquisition of adhesiveness demonstrates yet another adhesive change in these cells in the short time since ingression began six hours earlier. In experiments so far, the primary mesenchyme cells demonstrate about eight different cell–cell and cell–substrate adhesive changes in about six hours. Additional adhesive events may be required to control direction of movement, and if adhesion is necessary for feedback control over differentiation, there may be more adhesive interactions present during this time.

REGULATION OF PRIMARY MESENCHYME CELL NUMBER

In the transplantation experiments cells were added to or removed from the blastocoel, yet the experimental embryos always seemed to go on to form perfect skeletons. This was somewhat curious, since the number of PMCs is normally tightly regulated at 60 ± 2 in the species we study most commonly (*Lytechinus variegatus*). It appeared as though the PMCs somehow had the ability to regulate to different numbers and form a normal skeleton.

To study this phenomenon, we began by adding or subtracting PMCs from the blastocoel in a systematic study (Ettensohn and McClay, 1988). When the number of cells was doubled, there still was a normal skeleton produced. Only two sites of spicule growth initiation were present, and the growth of the spicule was regulated to a normal morphology. In culture, the spicule size and morphology are chaotic. Thus, it appeared that PMCs are somehow programmed to produce spicules and that the correct pattern is imposed on the cells by the blastocoelic environment. Of particular significance in this morphogenesis is the fact that there appear to be only two specific sites where the triradiate spicule rudiments initiate growth of the elongated spicules.

When PMCs were removed from the blastocoel, the embryo retained the capacity to make spicules. Even when all the primary mesenchyme cells were removed, the embryo still made normal skeletons. This was quite surprising, since normally only PMCs contribute to the spicules, as can be seen by staining the cells surrounding spicules with a monoclonal antibody specific for PMCs (McClay et al., 1983). If we removed half the PMCs, allowed the embryo to make a skeleton, then stained the embryo for PMCs, it was observed that there were about 60 cells. This too was puzzling, since the PMCs normally do not divide for a long period of time during spiculogenesis. Thus, removal of some PMCs either stimulated the remaining PMCs to divide or stimulated a separate cell population to become PMCs as seen by expression of the PMC marker antigens.

To distinguish between these possibilities, a new experimental strategy was adapted. PMCs were entirely eliminated from host embryos. Fluorescently (red) labeled donor PMCs were then added to the blastocoel (Figure 6). Before

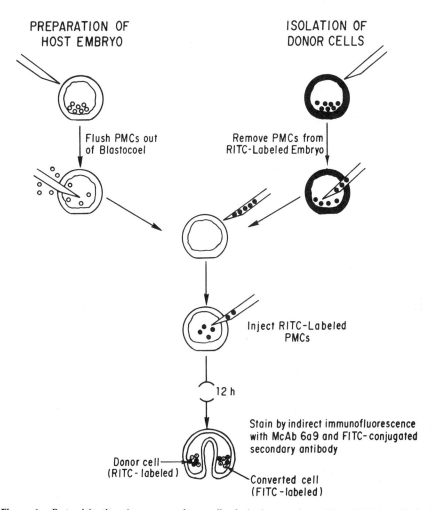

Figure 6. *Protocol for the primary mesenchyme cell substitution experiment.* Host PMCs are flushed from the blastocoel. RITC-labeled PMCs are then inserted into the blastocoel. Later, the embryos are stained with a PMC-specific monoclonal antibody carrying an FITC label. In this way one can determine which cells are PMC (red and green), and which are converted secondary mesenchyme cells (green only).

testing the hypotheses to explain the deletion, it was necessary to determine whether the red PMCs would support normal skeletogenesis. If 60 red PMCs were injected into the empty blastocoels, and the embryos were then allowed to grow, it was observed that normal skeletons resulted and the injected PMCs (red) were the only PMCs contributing to the skeleton (as confirmed by staining with the antibody to PMCs using green fluorescence).

The next experiment was to add a certain number of red PMCs to the blastocoel. Again normal skeletons were produced, and about 60 cells were stained with the antibody. However, the number of red cells was the same in each case as the number injected. This meant the balance of the cells had to have been recruited from the host. Since all PMCs originally had been removed from the host embryo, this result indicated that cells not normally of the PMC lineage had to have been converted to become PMCs. Further investigation showed that cells usually designated secondary mesenchyme cells are the cells that undergo the conversion to express the PMC markers and participate in the growth of spicules (Ettensohn and McClay, 1988).

There are several important implications of these results. First, though SMCs retain the ability to be converted and express the PMC phenotype, they normally do not. The presence of PMCs somehow prevents the conversion. Second, the spicules always have about 60 PMCs supporting their growth. After the conversion of secondary mesenchyme to primary, there again appear to be 60 cells that make skeletons. This suggests that the cells can somehow regulate to produce the correct number of PMCs. The experiments show that SMCs somehow recognize quantitatively how many PMCs are removed and are able to replace approximately that number.

The next task was to determine how the conversion is accomplished. What is missing that is recognized by the secondary mesenchyme cells, and how do SMCs know quantitatively how many to convert? Though the answer to these questions is not known at present, there are several possibilities. First, the cells could convert in response to the absence of a PMC-synthesized substrate. It is known that PMCs make substrate components, so perhaps one of those molecules must be present to prevent SMC conversion. There is also a possibility that a diffusible substance is missing from the blastocoelic environment, thereby permitting SMC conversion. If an interaction with the substrate is the critical factor in the control of the conversion, then yet another recognition phenomenon is present in the mesenchyme system. This recognition phenomenon would be important to morphogenesis not just for its adhesive properties but for the ability to regulate gene expression in the phenotypic conversion.

Yet another observation contributes to the phenotypic conversion phenomenon. The SMC cell lineage consists of several populations of cells that normally make pigmented cells, smooth muscle, primitive nerves, and the coelomic pouches. If one follows the experimental embryos during conversion, it is clear that the earliest cells to leave the lead end of the archenteron do not convert to become the PMCs. These cells constitute the subpopulation of

SMCs that become pigmented cells. A later population of SMCs then leaves the archenteron, and these cells are the ones converted. Thus, not all SMCs are capable of the phenotypic conversion. This means that the SMCs leaving the archenteron later retain more of their stem cell properties than the population leaving early.

PRIMARY MESENCHYME CELL HETEROCHRONY

One way of bringing about evolutionary change in morphogenesis is simply to shift the timing of when a particular event occurs. If the timing is changed appropriately relative to other morphogenetic events, then structural differences can occur. This evolutionary process is called heterochrony, and it has the potential of bringing about phenotypic differences without changing the structural genes in the pathway. In the course of our studies on PMCs, we have come across a naturally occurring example of heterochrony that has a striking resemblance to the experiments on PMC removal.

The *Cideroids* are a group of echinoids (considered to be evolutionary relics) that do not have primary mesenchyme cells (Schroeder, 1981), yet still produce skeletons. In *Eucidaris tribuloides* a time-lapse study has shown that a few SMCs migrate to an area on either side of the vegetal plate to synthesize the spicules (Wray and McClay, 1988). If the same phenomenon is examined with the PMC-recognizing monoclonal antibody marker, the cells in the skeletogenic population each express the PMC antigen. When traced to the point of first expression, skeleton-producing cells first become positive for the antibody marker as they leave the tip of the archenteron and move to the wall of the blastocoel. This behavior exactly parallels the conversion phenomenon in the modern species that have PMCs.

In lineage-tracing studies asking which cells give rise to the skeletogenic population in the ancient species, it is observed that micromeres are the only cells that participate in the production of skeletons (Wray and McClay, 1988). This was determined by adding RITC-labeled macromeres, mesomeres, or micromeres to unstained embryos. It was observed that only the red micromeres were included in skeleton-producing cells.

Developmentally, these experiments can be interpreted to suggest that the PMC population in the *Echinoids* is brought about by a precocious ingression of the skeletogenic cell population. The *Cideroid* species retain the primitive pattern and ingress later. In both cases the larvae have similar skeletons. It is difficult to guess the selective pressure that leads to the precocious ingression, though one possibility could be that the early ingression might speed the time required to reach the feeding larval stage. The *Echinoids* studied reach this stage in two days, whereas *Eucidaris* requires at least a week at the same temperature of culture.

PATTERN FORMATION

Over the last 20 years there has been great interest in the mechanisms by which pattern is expressed in embryos. Wolpert provided the intellectual stimulus for this interest, and models for positional information and expression of pattern have been described and tested (Turing, 1954; Crick, 1970; Wolpert et al., 1971; French et al., 1976). The models have two features in common. They each postulate the existence of a simplified code understood by cells that provides positional information. They further suggest a feedback loop between the environment and the cell that enables the cell to communicate with the environment and the environment with the cell. This section describes the information available to the PMC population and describes the response of the PMCs to that information.

The experiments on PMCs show a variety of adhesions and movements over a relatively short time span. They show an ordered sequential series of changes in the cell–cell and cell–substrate adhesions. They also show a feedback relationship between the cell surface and a change in cell behavior. Examples of this feedback are seen in the PMC response to being in a young embryonic environment. In control embryos, transplanted PMCs (of the same age as the host) continue to move and form the ring pattern, a response that involves a sequence of cell behaviors. In a young embryo the older donor PMCs are delayed in passing through that sequence but eventually do when the host substrate reaches the mesenchyme blastula stage. The delay in performing a behavioral sequence suggests that there is a certain substrate required for the PMC morphogenetic program to be expressed normally. Until that substrate becomes available, the competent PMCs seem suspended in their activities. Not all the activities are suspended, however. The older PMCs begin to make spicule rudiments earlier than the host embryo would normally begin to make spicules. Even so, the collaboration between old PMCs and a young blastula continues to produce a normal skeleton.

Another example is the conversion of the SMCs. These cells normally differentiate into a variety of cell types that do not include a skeleton. If the environment is such that the SMCs detect the absence of PMCs, some of the SMCs convert and express a PMC developmental program. The signal causing this feedback must be detected in a quantitative manner, since the final number of PMCs is about 60, with the converted SMCs filling any void left by PMCs. Here the feedback is negative, with the presence of PMCs preventing SMC conversion.

At the beginning of spiculogenesis a triradiate spicule is produced in only two locations in the embryo. Occasionally an extra triradiate rudiment is produced, and often the rudiments are produced away from the ventrolateral area they later occupy. Eventually, however, only two rudiments associate with the two ventrolateral regions, the rudiments assume the correct orientation, and they produce a skeleton (Okazaki, 1975). In culture, spicules can begin to grow as well, but usually there is not a triradiate beginning point, nor is there a restriction in spicule number or in the size of the rods that elongate from the rudiment (Okazaki, 1975). Thus, the embryonic environment supplies information that is somehow read by the PMCs, which produce the correct skeleton product.

The distribution of cells in the ring structure is another example of cells following rules of pattern formation. It has been shown that normally 70% of the PMCs enter into the ventrolateral clusters, 22% form the dorsal ring segment, and about 8% form the ventral ring segment (Ettensohn and McClay, 1986). When some PMCs are deleted (but before SMC conversion), the PMCs remaining form a ring structure with the same relative distribution of cells as in control embryos. Thus, the cells are somehow able to space themselves around the ring regardless of total cell number. The spacing is not due to a predisposition to a certain spot in the ring, since randomly transplanted cells accommodate to form roughly the same proportions as in control embryos.

In unraveling the pattern and molecular basis of spiculogenesis, it is remarkable how events are coordinated. Clearly from the *in vitro* studies the PMCs are capable of making spicules on their own. In the embryo the PMCs are responsive to the embryonic environment. The intrinsic developmental program can be delayed if the PMCs are put into a younger embryo. It can also be adjusted in terms of timing or in terms of experimentally altered cell number. The PMCs also have a role in the regulation of pattern expression of other cells by inhibition of SMC conversion. Through spicule growth, they have an effect on the overall shape of the embryo.

Thus, this example of one cell type in the sea urchin embryo shows that pattern expression is a patchwork assembly of a number of properties each of which contributes to structure. Collectively these properties are called morphogenesis. The isolation of any single molecule and its characterization is a useful undertaking, but it should be realized that complex morphogenetic behaviors probably involve more than one molecule at any given step.

REFERENCES

Adelson, D., and T. Humphreys (1988) Aspects of morphogenesis and gene expression during sea urchin embryonic development are perturbed by a monoclonal antibody specific for hyalin. *Development* **104**:391–402.

Alliegro, M. A., C. A. Burdsal, D. R. McClay, and C. A. Ettensohn (1988) Echinonectin: A new extracellular matrix protein with adhesive function in sea urchin development. *J. Cell Biol.* **107**:2319–2327.

Bell, G. I. (1978) Models for the specific adhesion of cells to cells. *Science* **200**:618–627.

Crick, F. (1970) Diffusion in embryogenesis. *Nature* **225**:420–422.

DeSimone, D. W., E. Spiegel, and M. Spiegel (1985) The biochemical identification of fibronectin in the sea urchin embryo. *Biochem. Biophys. Res. Commun.* **133**:183–188.

Edelman, G. M. (1984) Cell adhesion and morphogenesis: The regulator hypothesis. *Proc. Natl. Acad. Sci. USA* **81**:1460–1464.

Ettensohn, C. A. (1985) Gastrulation in the sea urchin embryo is accompanied by the rearrangement of invaginating epithelial cells. *Dev. Biol.* **112**:383–390.

Ettensohn, C. A., and D. R. McClay (1986) The regulation of primary mesenchyme cell migration in the sea urchin embryo: Transplantations of cells and latex beads. *Dev. Biol.* **117**:380–391.

Ettensohn, C. A., and D. R. McClay (1988) Cell lineage conversion in the sea urchin embryo. *Dev. Biol.* **125**:396–409.

Fink, R. D., and D. R. McClay (1985) Three cell recognition changes accompany the ingression of sea urchin primary mesenchyme cells. *Dev. Biol.* **107**:66–74.

Fraser, S. E. (1980) A differential adhesion approach to the patterning of nerve connections. *Dev. Biol.* **79**:453–464.

French, V., P. J. Bryant, and S. V. Bryant (1976) Pattern regulation in epimorphic fields. *Science* **193**:969–981.

Gustafson, T., and L. Wolpert (1967) Cellular movement and contact in sea urchin morphogenesis. *Biol. Rev. Camb. Philos. Soc.* **42**:442–498.

Iwata, M., and E. Nakano (1985) Fibronectin-binding acid polysaccharide in the sea urchin embryo. *Wilhelm Roux's Arch. Dev. Biol.* **194**:377–384.

McClay, D. R., and C. A. Ettensohn (1987) Cell recognition during sea urchin gastrulation. In *Genetic Regulation of Invertebrate Development*, W. Loomis, ed., pp. 111–128, Alan R. Liss, New York.

McClay, D. R., and R. D. Fink (1982) The role of hyalin in early sea urchin development. *Dev. Biol.* **92**:285–293.

McClay, D. R., G. W. Cannon, G. M. Wessel, R. D. Fink, and R. B. Marchase (1983) Patterns of antigenic expression in early sea urchin development. In *Time, Space, and Pattern in Embryonic Development*, W. R. Jeffreys and R. A. Raff, eds., pp. 157–169, Alan R. Liss, New York.

Okazaki, K. (1975) Spicule formation by isolated micromeres of the sea urchin embryo. *Am. Zool.* **15**:567–581.

Schroeder, T. (1981) Development of a "primitive" sea urchin (*Eucidaris tribuloides*): Irregularities in the hyaline layer, micromeres, and primary mesenchyme. *Biol. Bull.* **161**:141–151.

Solursh, M., and H. Katow (1982) Initial characterization of sulfated macromolecules in the blastocoels of mesenchyme blastulae of *Strongylocentrotus purpuratus* and *Lytechinus pictus*. *Dev. Biol.* **94**:326–336.

Sperry, R. W. (1963) Chemoaffinity in the orderly growth of nerve fiber patterns and connections. *Proc. Natl. Acad. Sci. USA* **50**:703–710.

Spiegel, E., M. Burger, and M. Spiegel (1980) Fibronectin in the developing sea urchin embryo. *J. Cell Biol.* **87**:308–313.

Spiegel, E., M. Burger, and M. Spiegel (1983) Fibronectin and laminin in the extracellular matrix and basement membrane of sea urchin embryos. *Exp. Cell Res.* **144**:47–55.

Spiegel, M., and M. Burger (1982) Cell adhesion during gastrulation. *Exp. Cell Res.* **139**:377–382.

Turing, A. M. (1954) The chemical basis of morphogenesis. *Philos. Trans. R. Soc. Lond. [Biol.]* **237**:37–72.

Wessel, G. M., R. B. Marchase, and D. R. McClay (1984) Ontogeny of the basal lamina in the sea urchin embryo. *Dev. Biol.* **103**:235–245.

Wolpert, L., J. Hicklin, and A. Hornbruch (1971) Positional information and pattern regulation in regeneration of *Hydra*. *Symp. Soc. Exp. Biol.* **25**:391–415.

Wray, G. A., and D. R. McClay (1988) The development of spicule-forming cells in a "primitive" sea urchin, *Eucidaris tribuloides*. *Development* **103**:305–315.

Chapter 27

Cell–Cell and Cell–Matrix Adhesion Receptors: Family Relationships and Roles in Migratory and Invasive Systems

CAROLINE H. DAMSKY
MARGARET J. WHEELOCK
KEVIN J. TOMASELLI
ANN SUTHERLAND
EILEEN CROWLEY
SUSAN J. FISHER

ABSTRACT

Biochemical characterization and gene cloning studies have shown that many cell–cell and cell–matrix adhesion receptors exist in families, providing a mechanism for groups of cells in a similar environment to distinguish themselves from one another by virtue of the subsets of the related adhesion molecules they express. Through efforts to determine the mechanisms of action of the calcium-dependent family of cell–cell adhesion molecules, it is shown that they participate directly in cell adhesion. Their roles in the organization of junctional complexes and in the maintenance of a noninvasive phenotype are discussed. Studies presented on the role of the integrin superfamily of cell–matrix adhesion receptors show that complexes of the β_1-subfamily of integrinlike cell–matrix receptors in mammalian cells, which are distinct from both the fibronectin receptor integrin complex and the 68-kD laminin receptor, can act as receptors for laminin and collagen type IV. This is another illustration that cells have multiple mechanisms to attach to particular matrix molecules. Elucidation of the different kinds of signals transmitted to cells via these contact-mediated interactions is a difficult challenge for the future. Trophoblast cells and the placenta are described as a model biological system that will be useful for the study of both cell–cell and cell–matrix receptors, as well as the regulation of migratory and invasive behavior.

The concept that cells use selected representatives of several families of adhesion receptor molecules to interact with an environment of great complexity has become a powerful one in recent years. It has the appeal of economy and efficiency as well as fulfilling the necessity for diversity and flexibility. The two families of adhesion receptors that have been studied most carefully to date are the family of calcium-dependent cell–cell adhesion molecules, called cadherins by Takeichi and coworkers (Yoshido-Noro et al., 1984), and the superfamily of heterodimeric cell adhesion receptors, recently coined integrins (Tamkun et al., 1986) or cytoadhesins (Plow et al., 1986). Most of the members of this latter family are cell–matrix adhesion receptors, although one subgroup of the family is involved in heterophilic cell–cell adhesion interactions (Kishimoto et al., 1987). These two families have in common the fact that at least some of their representatives are present on cells at very early times in development and are widely distributed in the early embryo. Studies in the nervous system (Grumet and Edelman, 1984; Rathjen and Schachner, 1984; Edelman, 1986; Bastiani et al., 1987) and in the liver (Öbrink et al., 1985) have shown that later in development, during histogenesis and organogenesis, additional adhesion molecules with more restricted tissue distributions are expressed. Some of these newly expressed molecules are members of the two families indicated above, while others are unrelated and in some cases are thought to be members of additional, distinct families of adhesion receptors.

Results such as these provide at least a suggestion that there is a hierarchy of expression of adhesion receptors that starts with a few widely distributed molecules in early development and, as morphogenesis and histogenesis proceed, allows for the expression of additional adhesion receptors with a more restricted tissue distribution. The expression of the original molecules at these later times often becomes less global and/or is modulated with respect to cell surface distribution, posttranslational modification, or prevalence. During particular stages in development, one group of cells may display one, two, or more members of a particular family as well as representatives of other families. The total repertoire of cell–cell or cell–substratum adhesion receptor molecules expressed by particular groups of neighboring cells at any particular time would then determine the behavior of that group of cells in response to a given environmental signal and may serve to segregate particular groups of cells from one another during development, preventing inappropriate migration or mixing of cell populations. During the course of development, the expression of many cell adhesion receptors is transient. In fact, it is this transience and modulation, which seem to occur at times of significant morphogenetic events, that suggest that adhesion molecules are important participants in the repeating cycle of induction and response that is the hallmark of development: signals are given that influence the expression of adhesion molecules, and their expression in turn provides new signals for further response (for a review, see Edelman, 1986).

THE CALCIUM-DEPENDENT CAM FAMILY

Over the years we have studied representatives of two adhesion receptor families. Cell-CAM 120/80, originally isolated from human mammary tumor epithelial cells (Damsky et al., 1983), has turned out to be the same molecule as uvomorulin (Hyafil et al., 1981) or E-cadherin (Yoshido-Noro et al., 1984), isolated from mouse teratocarcinoma cells; ARC-1 (Imhof et al., 1983), identified on MDCK cells; and L-CAM (Gallin et al., 1983), isolated from chick liver. This molecule is a representative of the calcium-dependent cadherin family of cell adhesion molecules. It is present on all cells of the expanded mouse blastocyst, but is restricted to epithelial cells in fetal and adult tissues (Damsky et al., 1985a; Damjanov et al., 1986). The first cells to lose expression of this molecule are the extra-embryonic parietal endoderm, a highly migratory cell type, the embryonic mesoderm as it detaches from the ectoderm during gastrulation, and the neurectoderm at the time of neural fold development (Damjanov et al., 1986). Although expression of cell-CAM 120/80 is turned off, the latter two embryonic cell types acquire a different member of the calcium-dependent family, N-cadherin (Hatta and Takeichi, 1986).

Calcium-Dependent CAMs Participate Directly in Cell–Cell Adhesion

In most work to date, molecules have been designated as being cell–cell adhesion molecules because specific antibodies against them disrupt a particular adhesion event. For example, in the case of cell-CAM 120/80, we and others have shown that antibodies to the 80-kD soluble fragment of cell-CAM 120/80 disrupt compaction of the 8–16-cell mouse embryo and also prevent the segregation of a primitive endoderm from the inner cell mass (Damsky et al., 1983; Richa et al., 1985). The question frequently arises as to whether these candidate adhesion molecules act directly to mediate cell adhesion or whether they act indirectly by modulating some global signal, such as ion transport or cell surface proteolysis, which in turn affects adhesion receptors. Wheelock et al. (1987) have been able to show that the large, trypsin-stable, soluble 80-kD fragment of cell-CAM 120/80 is itself able to disrupt cell–cell adhesion when added to the culture medium (Figure 1), indicating that cell-CAM 120/80, and by analogy the other calcium-dependent CAMs, are direct participants in cell–cell adhesion. This question can now also be addressed using molecular biology approaches. The genes for chick L-CAM (Gallin et al., 1987; Cunningham and Edelman, this volume) and for mouse E-cadherin (Nagafuchi et al., 1987) and uvomorulin (Kemler et al., this volume), all of which are homologues of the human cell-CAM 120/80, have been cloned. Using transfection techniques, the gene for E-cadherin has been expressed in mouse L-cells, which normally have a fibroblastic morphology and do not express E-cadherin. The transfected cells display an epithelioid morphology and express the E-cadherin molecule on the cell surface at regions of cell–cell contact (Nagafuchi et al., 1987). The transfection experiment also suggests, but does not prove, that the E-cadherin mediates cell–cell adhesion in a homophilic manner, since only the E-cadherin was transfected and an epithelioid morphology was achieved.

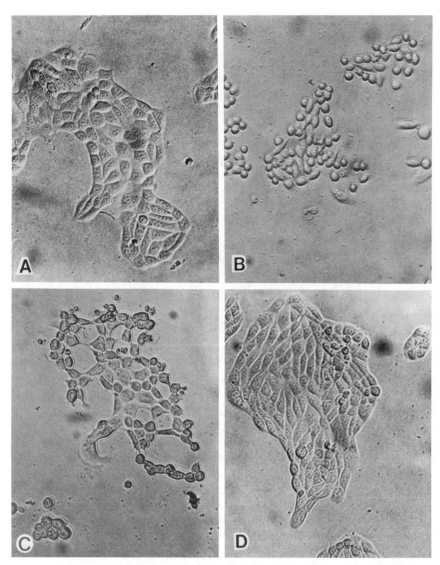

Figure 1. *The 80-kD fragment of cell-CAM 120/80 inhibits cell–cell adhesion. A:* Mammary tumor epithelial cells plated for 16 hours in culture medium containing buffer that has gone through the protocol for purifying GP80. Islands of tightly adhering cells are formed. *B:* Cells plated in culture medium containing antibody against the 80-kD fragment of cell-CAM 120/80. As reported previously, cells in the presence of this antibody form much looser islands, within which cells are not tightly adhered. *C:* Cells exposed to culture medium containing purified GP80. The purified fragment disrupts the tightly associated cells. *D:* Cells treated with medium containing a mixture of anti-GP80 and purified GP80. The fragment and the antibody neutralize each other's effects, and the cells appear as in *A*. (From Wheelock et al., 1987.) × 100.

Biological Role of Calcium-Dependent CAMs

Two other reported attributes of the members of the calcium-dependent CAM family are of interest in trying to understand the real functions of its members in cell–cell adhesion. The first relates to their distribution on the basolateral domain of the cell surface. In many differentiated fetal and adult tissues, cell-CAM 120/80 is enriched in an apical zonular band on the lateral surface (Figure 2), a location compatible with its being involved in the organization or function of the junctional complex. Boller et al. (1985) have localized uvomorulin in the adult mouse intestine and found it to be enriched at the zonula adherens regions of the junctional complex just below the tight junction. In this location calcium-dependent CAMs are in an excellent position to interact with the actin-based cytoskeleton and to mediate morphogenetically significant cell shape changes during, for instance, somitogenesis or neurulation (Duband et al., 1987; Thiery et al., this volume).

On the precompaction mouse embryo (up to the 8–16 cells), prior to the presence of organized intercellular junctions, the distribution of cell-CAM 120/80 is quite diffuse along the region of cell–cell contact. The embryo must form a functional tight junctional seal throughout its trophectoderm in order to transport fluid during blastulation. Kemler et al. (this volume) have shown that uvomorulin is redistributed to the apical region of the basolateral surface in trophectoderm prior to blastulation. Gumbiner and Simons (1986) have shown that an antibody to uvomorulin can inhibit the reestablishment of high-resistance tight junctions in EGTA-treated MDCK cells, following the re-introduction of calcium. Thus, in using both morphological and physiological approaches, uvomorulin is implicated in the organization and function of the junctional complex.

The second feature of cell-CAM 120/80 that might shed light on its function relates to its possible role in maintaining a noninvasive phenotype for epithelial cells. Cell-CAM 120/80 is present in almost all normal epithelia and in many primary carcinomas and tumor cell lines of epithelial origin (Damsky et al., 1985a). It is particularly instructive to look at the invading front of an actively metastasizing tumor. In sections showing the invasive front, cell-CAM 120/80 is still associated with groups of invading cells at their lateral borders, although it is not detected at the free edges of cells that are interacting with the surrounding environment. This would suggest that modulation of uvomorulin does not play an important role in maintaining a noninvasive phenotype. One drawback of this approach is that it is difficult to compare the amounts of an antigen present in immunohistochemically stained sections of normal and tumor tissue. Thus, it is still possible that there has been a functionally meaningful reduction in the amount of cell-CAM 120/80, or a change in its distribution along the lateral border in the metastasizing cells.

The idea that reduced expression of uvomorulin might be part of the metastatic phenotype is suggested by the observations of Behrens and Birchmeier that MDCK cells treated with anti-uvomorulin are able to invade

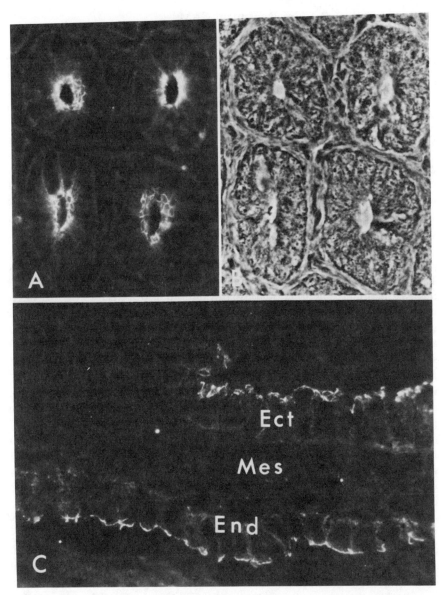

Figure 2. *Cell-CAM 120/80 is enriched at the apical zone of the basolateral cell surface of many adult and embryonic tissues. A:* Mouse adult kidney tubules in cross section stained with anti-GP80. Staining is most prominent as a zonular apical band around the individual epithelial cells of the tubule. *B:* Phase-contrast micrograph of the same field. *C:* Embryonic region of a 7.5-day mouse embryo stained with anti-GP80. Staining is prominent at the apical region of cell–cell boundaries in both the ectoderm (Ect) and endoderm (End), but is absent from the mesoderm (Mes). × 480.

collagen gels or chick heart explants, whereas untreated MDCK cells are not invasive. In addition, MDCK cells can be transformed by various oncogenic viruses to produce a series of cell lines with varying amounts of surface uvomorulin. Those cell lines with low or no expression of surface uvomorulin are more invasive than those with high levels of uvomorulin expression (Birchmeier et al., 1987). These studies do not demonstrate the mechanism by which the regulation of uvomorulin might be related to invasiveness. For example, it is possible that uvomorulin expression is reduced simply as a consequence of enhanced protease production by metastatic cells. However, the data are consistent with the idea that the presence of a certain level of uvomorulin is relevant to the maintenance of the integrity of differentiated epithelial tissues. Additional studies are clearly required to explore this interesting question.

Although recent progress has been rapid, several questions remain about the biology and regulation of calcium-dependent CAMs. The mechanisms by which the expression of different members of the calcium-dependent family is regulated are unknown. Also unknown is the nature of the cell contact–mediated signal transmitted by these CAMs. From biochemical and sequence information obtained recently, it is known that these CAMs are transmembrane, serine and threonine phosphorylated, calcium-binding glycoproteins. The location in the uvomorulin molecule of the putative transmembrane domain suggests that its cytoplasmic domain is about 35–40 kD. This gives ample provision for the interaction of uvomorulin with cytoskeletal and other cytoplasmic components in response to contact-related environmental signals. In attempts to determine which proteins might associate with cell-CAM 120/80, Peyrieras et al. (1983) and Wheelock (M. J. Wheelock, unpublished data) have shown that there is a prominent nonglycosylated protein of about 95–100 kD that consistently immunoprecipitates along with cell-CAM 120/80, but that does not react in the immunoblot procedure with antibodies against cell-CAM 120/80 (or uvomorulin). This is a candidate for a molecule that may interact with the cytoplasmic domain of cell-CAM 120/80.

Also germane to the issue of signal transmission is the ability of uvomorulin to interact with itself and/or with other cell surface molecules, within the plane of the membrane. The fact that the distribution of cell-CAM 120/80 and apparently N-cadherin as well (Duband et al., 1987) is in some cases diffuse on the lateral borders of neighboring cells, and in other cases is highly enriched at the apical junctional complex, suggests that its surface density can be spatially regulated, perhaps by interacting with other junction-associated membrane or cytoskeletal molecules. Transfection experiments such as those described by Nagafuchi et al. (1987), using altered forms of uvomorulin, should be a powerful approach to determining the functional roles of the different domains of the uvomorulin molecule.

THE SUPERFAMILY OF CELL–MATRIX ADHESION RECEPTORS

Members of the large integrin (Tamkun et al., 1987) or cytoadhesin (Plow et al., 1986) cell–matrix adhesion receptor family have been identified by two different approaches. In the first, molecules were purified from cell extracts and tested for their ability to neutralize the effects of adhesion-disrupting antisera or monoclonal antibodies. The best-characterized adhesion receptor identified by this approach is avian integrin, a complex of at least three glycoproteins in the molecular weight range of 110–165 kD (nonreduced: Greve and Gottlieb, 1982; Neff et al., 1982; Horwitz et al., 1985; Knudsen et al., 1985; Akiyama et al., 1986). The monoclonal antibodies that recognize this complex (CSAT or JG22) inhibit cell attachment to many matrix molecules, including fibronectin (FN), laminin (LN), vitronectin (VN), and collagen type IV (Col IV: Horwitz et al., 1985; Hall et al., 1987).

The second major approach to identifying adhesion receptors has been to isolate them by ligand-affinity chromatography. This approach has been successful in isolating the FN, VN, and fibrinogen receptor heterodimeric complexes (Pytela et al., 1985a,b, 1986).[1] The FN and VN receptor α- and β-chain heterodimers have relatively stringent receptor activities, recognizing primarily FN and VN, respectively. In contrast, the fibrinogen receptor complex GP IIb/IIIa can recognize FN, VN, von Willebrand factor, and thrombospondin in addition to fibrinogen (Ruoslahti and Pierschbacher, 1986). The heterodimeric FN receptor and the avian integrin complex are related, in that the human FN receptor β-chain and the 110-kD chain of the avian integrin are about 90% homologous. Thus, a heterodimeric FN receptor is likely to be a component within the avian integrin complex. The 110-kD component of the avian integrin (henceforth called the β-chain of avian integrin) is also related to the β-chain of the human VLA family of heterodimers (Hemler et al., 1987; Takada et al., 1987). Studies using the VLA β-chain have shown that it can associate with up to six α-chains. The β-chains of antibodies referred to as anti-GP140 (Brown and Juliano, 1986) or anti-ECMr (Knudsen et al., 1981; Tomaselli et al., 1987; Sutherland et al., 1988) are also related to the avian integrin β-subunit (Tomaselli et al., 1987). These are all members of one subfamily (the integrin β_1-subfamily) of a superfamily of adhesion receptor heterodimers called either integrins (Hynes, 1987) or cytoadhesins (Plow et al., 1986). The β-subunits of the VN and fibrinogen receptors are not the same as those of the FN, VLA, ECMr, or avian integrin receptor complexes, but are related to one another and make up a distinct integrin β_3-subfamily of the integrin or cytoadhesin superfamily (see Hynes et al., and Ruoslahti et al., this volume).

Evidence for an Integrin-Related Laminin Receptor

All of the adhesion ligands identified thus far that are recognized by any of the members of the integrin subfamilies described above have been shown to contain the Arg-Gly-Asp sequence, originally determined as the smallest region of the fibronectin molecule that can inhibit the attachment of some cell types to intact fibronectin (Pierschbacher and Ruoslahti, 1984). Because intact LN has not been shown to have a functional Arg-Gly-Asp sequence and because a different LN-binding protein of 68 kD with a much higher affinity for LN has already been described by several laboratories (for a review, see Liotta et al., 1986), the idea of an integrin-related complex as an LN receptor has been controversial. However, there are several pieces of evidence from our laboratories and others that suggest that the integrin complex is involved in an important way in mediating cell attachment to LN.

First, antibodies against avian integrin and integrinlike receptors in mammalian cells have been shown to inhibit the attachment to LN of a wide variety of cell types, including retinal neurons (CSAT monoclonal antibody; Hall et al., 1987), the neuronlike, NGF-stimulated PC12 cell line (anti-ECMr and anti-GP140; Tomaselli et al., 1987), many kinds of fibroblasts (CSAT, anti-ECMr; Decker et al., 1984; Tomaselli et al., 1987), mouse and human trophoblasts (anti-ECMr; Sutherland et al., 1988; C. Damsky and E. Crowley, unpublished data), and murine mammary epithelial cells and melanoma cells (anti-ECMr; Kramer et al., 1989). The PC12 cell line attaches very well to LN and very poorly to FN, suggesting that the interference by anti-ECMr with PC12 attachment to LN is not an indirect result of its interaction with a specific FN receptor (Tomaselli et al., 1987).

Second, murine B-16 melanoma cells have been shown to express the 68-kD high-affinity LN-binding protein (Liotta et al., 1986) as well as a 140-kD ECMr glycoprotein complex (Kramer et al., 1989). Anti-ECMr inhibits attachment of these cells to LN (and to FN as well). Thus, when both LN recognition activities are present on the same cell, an antibody against the 140-kD complex that does not recognize the 68-kD LN-binding protein can, by itself, inhibit cell attachment to LN.

Third, direct binding between LN and the purified avian β_1-integrin complex has been observed by equilibrium gel filtration. This interaction can be inhibited by Arg-Gly-Asp-containing peptides (Horwitz et al., 1985).[1]

Fourth, when normal fibroblasts are spread on LN and stained with a CSAT monoclonal antibody that recognizes the avian integrin β_1-subunit (Buck et al., 1986), the staining pattern, at the light microscope level, is similar to that observed when they are spread on FN (Figure 3 and Damsky et al., 1985b), suggesting that the β_1-subunit, in conjunction with the appropriate α-chain partner, can respond to LN or to FN in like manner. Similarly, the plating of transformed chick fibroblasts on either LN or FN can cause the temporary reorganization into a control cell–staining pattern, of both microfilament bundles and the β_1-integrin complex (Damsky et al., 1985c).

Correlation of Adhesive Behaviors and the Presence of Integrin-Related Polypeptides

Analysis of the polypeptide composition of immunoprecipitates, using cells with distinct adhesive properties and antibodies with differing adhesion

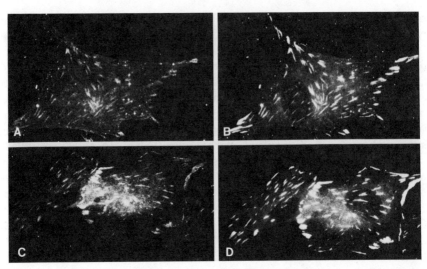

Figure 3. *Embryonic chick fibroblasts plated on LN or FN and stained with CSAT monoclonal antibody and antibodies against talin and vinculin. A:* Cardiac fibroblast plated on a LN substrate for three hours and stained with CSAT monoclonal antibody. *B:* Same cell stained with rabbit anti-vinculin. *C:* Tendon fibroblast plated on a FN substrate and stained with CSAT monoclonal antibody. *D:* Same cell stained with rabbit anti-talin. Both substrates promote the formation of well-organized adhesion plaques. $\times$ 596.

inhibitory activities, can be useful in sorting out which adhesion components are active in inhibiting cell attachment to LN and Col IV as opposed to FN. In recent studies reported by Tomaselli et al. (1987), we have examined the polypeptides recognized by anti-ECMr on cell lines with different adhesion properties (Table 1). PC12 cells adhere very poorly to FN and very well to both Col IV and LN. BHK (baby hamster kidney) cells adhere very well to FN, less well to LN, and very poorly to Col IV. NRK (normal rat kidney) cells adhere well to all three substrates. All these cell lines were immunoprecipitated with anti-ECMr and with anti-GP140 (Brown and Juliano, 1986). Figure 4 shows that PC12 cells contain material at 180 kD and 135 kD in addition to a 120-kD component. The 120-kD component of the PC12 precipitate cross-reacts with the avian integrin β_1-subunit (Tomaselli et al., 1987), and is assumed to represent the β_1-subunit in mammalian cells. BHK cells contain the 120-kD band and a broad, poorly resolved band spanning the 135–150-kD region, but no material at 180 kD. NRK cells contain components at 120 kD, 135 kD, 150 kD, and 180 kD.

These experiments show that a band at 150 kD is associated with cells that adhere well to FN. This band together with the 120-kD component presumably constitute the FN receptor heterodimer isolated by Pytela et al. (1985a), using FN-affinity chromatography. Materials at 135 kD and 180 kD, in conjunction with the 120-kD component, are therefore candidates for involvement in adhesion to Col IV and LN. In comparing the adhesive behavior on Col IV and

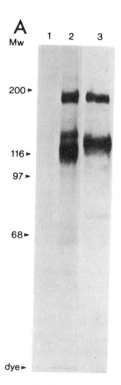

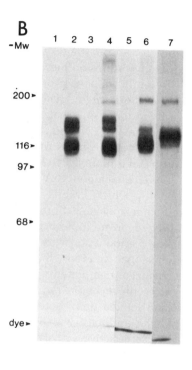

Figure 4. *SDS-PAGE analysis of immunoprecipitates using anti-ECMr and anti-GP140 (gift of Brown and Juliano) on rat cells with different adhesion properties. A:* Immunoprecipitation of PC12 cells using anti-ECMr. *Lane 1:* Immunoprecipitation of surface-iodinated PC12 cells using preimmune goat serum. *Lane 2:* Immunoprecipitation of surface-iodinated PC12 cells using anti-ECMr serum and analyzed under nonreducing conditions. Material at 120 kD, 135 kD, and 180 kD is recognized. *Lane 3:* As in *lane 2,* but analyzed under reducing conditions. As is typical for the integrin β_1-family, the β-subunit increases in apparent molecular weight under reducing conditions (Knudsen et al., 1985) and is not resolved from the 135-kD component. *B:* Immunoprecipitation of cell lines with anti-GP140. *Lanes 1, 3, 5:* Immunoprecipitation of surface-iodinated BHK, NRK, and PC12 cells, respectively, with preimmune goat serum. *Lanes 2, 4, 6:* Immunoprecipitation of surface-iodinated BHK, NRK, and PC12 cells with anti-GP140 and analysis under nonreducing conditions. *Lane 7:* PC12 cells immunoprecipitated with anti-GP140 and run under reducing conditions. The precipitates by anti-ECMr and anti-GP140 of PC12 cells are indistinguishable. When compared with NRK cells, the BHK cells lack a 180-kD component, and the PC12 cells lack a 150-kD component. The facts that PC12 cells, which do not adhere well to FN, are lacking the 150-kD component and that BHK cells, which do not adhere well to Col TIV, lack the 180-kD component are consistent with the suggestion (but do not prove it) that dimers of 120/150 kD, 120/135 kD, and 120/180 kD are related to cell attachment to FN, LN, and Col IV, respectively. (From Tomaselli et al., 1987.)

LN of BHK and NRK with the polypeptide pattern of material immuno-precipitated from each cell line using anti-ECMr and anti-GP140 (see Table 1), we tentatively suggest that the 180-kD band is associated with adhesion to Col IV, while the 135-kD band is associated with adhesion to LN. It is possible that the 120-kD component could serve as a companion β-subunit for both. Direct proof for these suggestions is lacking, and they must be confirmed by correlative adhesion perturbation experiments.[2]

Recently we have produced a panel of monoclonal antibodies against the human choriocarcinoma (trophoblast) cell line BeWo (Patillo and Gey, 1968). All hybridoma supernatants were screened for their ability to inhibit BeWo attachment to LN and FN. Two hybridomas were found that inhibited attachment to FN but not to Col IV or LN. Two hybridoma supernatants were found that inhibited attachment to FN, LN, and Col IV. None of the antibodies inhibited attachment to VN, distinguishing them from the CSAT monoclonal antibody, which inhibits attachment of avian cells to VN as well as to FN, LN, and some collagens. Both of the antibodies against the trophoblast cells that inhibited on multiple ligands gave the same immunoprecipitation pattern: major components were seen at 120 kD, 135–150 kD, and 190 kD. The monoclonal antibody that inhibited trophoblast attachment to FN alone recognized components at 120 kD and 135–150 kD, as expected for a fibronectin receptor, but not at 190 kD. Of interest is that the 135–150-kD region looked quite heterogeneous. Sequential immunoprecipitation using first the antibody with FN receptor activity, and then the antibody with multiple receptor activity (Figure 5), demonstrates that components in the 150-kD and 120-kD regions can be depleted from the extract, leaving behind material at 190 kD, 135 kD, and 120 kD that is recognized by the antibodies with multiple inhibitory activity. These results are consistent with those of anti-ECMr and PC12 cells described above: namely, that bands of 135 kD and 190 kD are associated with attachment to LN and Col IV. The ability of an antibody that inhibits attachment only to FN to deplete one set of polypeptides from a cell extract (and leave behind a distinct set of polypeptides that is still recognized by antibodies that also inhibit cell attachment on LN and Col IV) is strong evidence that the effects of these antibodies on LN and Col TIV adhesion are specific and do not act indirectly as a consequence of binding to the FN receptor heterodimer.

Table 1.   Correlation of Adhesive Behavior with Composition of Immunoprecipitates Using Anti-ECMr[+] IgG or Anti-BeWo* Monoclonal Antibodies

	FN	LN	Col IV	Ippt. Composition (kD)
PC12[+]	+/−	++	++	120, 130, —, 180
NRK[+]	++	++	++	120, 130, 150, 180
BHK[+]	++	+	+/−	120, 130–150, —
BeWo*	++	++	++	120, 130–150, 190

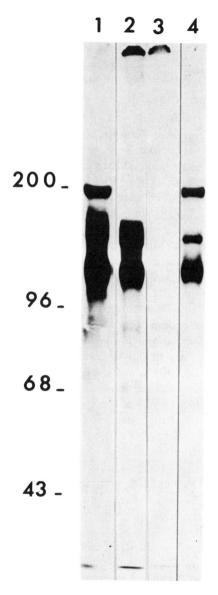

Figure 5. *SDS-PAGE analysis, under nonreducing conditions, of immunoprecipitates of [³H]glucosamine-labeled human choriocarcinoma cells with monoclonal antibodies that inhibit cell attachment to LN, Col IV, and/or FN. Lane 1:* Extract precipitated with antibody that inhibits on all three substrates. Material at 120 kD, 135–150 kD, and a sharp band at 190 kD are recognized. *Lanes 2, 3:* Cell extract precipitated for the first time *(2)* and the third time *(3)* with an antibody that inhibits only FN attachment activity. This antibody recognizes material at 120 kD and in the 135–150-kD region. After three successive precipitations, the extract is depleted of material recognized by this antibody. *Lane 4:* Depleted extract precipitated with the antibody used in *lane 1,* which inhibits cell attachment on all three substrates. Material at 120 kD and sharp bands at 135 kD and 190 kD are present. These are candidate polypeptides for involvement in attachment to LN and Col IV.

Taken together, these data suggest that there are distinct integrinlike receptors of the β_1-subfamily type for Col IV and LN as well as for FN.[1,2]

Why Do Cells Have Multiple Receptors for ECM Ligands?

Although the integrinlike receptors affect adhesion to a wide variety of substrates, each of these substrates has additional mechanisms by which it can interact with the cell surface. For example, LN has a 68-kD cell surface receptor that is unrelated to the integrinlike LN receptor. A distinctive feature of the 68-kD receptor is that it is up-regulated on tumor cells (Liotta et al., 1986). The integrinlike LN receptor activity appears to be present in similar amounts on tumor and normal cells, although its cell surface distribution changes (Damsky et al., 1985c; Chen et al., 1986; but see Plantefaber and Hynes, 1989). The question of why cells should have multiple receptor classes for particular ECM ligands raises the concept that the cell surface receptors that have been identified as adhesion molecules over the past several years are transmitters of important and distinct regulatory signals from the environment to the cell interior. In some cases they may only incidentally be adhesion molecules by virtue of the fact that the environment from which they transmit information and with which they must come in contact is an immobilized ECM. They have been identified as adhesion molecules because they have been discovered in adhesion assays. The nature of signals transmitted as a consequence of cell contact with ECM components is not known definitively in any case.

However, a recent series of experiments by Dedhar et al. (1987) presents an interesting model system for trying to reconstruct the nature of such a signal in the case of one particular cell type. In these experiments MG-63 osteosarcoma cells were treated with increasing concentrations of an Arg-Gly-Asp-containing peptide with cell detachment activity in order to isolate a cell line that remains adherent in the presence of large amounts of this peptide. Not unexpectedly, this treatment resulted in a cell line with an amplified production of the cell surface FN receptor. Of even greater interest is the fact that this cell line also produced substantial amounts of bone matrix, something the original un-differentiated osteosarcoma cell line was unable to do. Thus, amplifying expression of an adhesion receptor resulted in a more differentiated phenotype for this tumor cell line. Studies such as these, which try to understand the role of cell contact in the regulation of cell behavior and differentiation, as well as in the regulation of cell shape, cell location, and cell migration, are in their infancy, but are indicative of an important new wave of experimentation.

TROPHOBLAST CELLS AND PLACENTA AS A SYSTEM FOR STUDYING THE BIOLOGY OF CELL ADHESION MOLECULES

The choriocarcinoma cell line, BeWo, used in the set of experiments described in Figure 5, is a malignant derivative of the human placental cytotrophoblast cell. The trophoblast comprises the fetal portion of the placenta and is derived from the trophectoderm layer of the hatched mammalian blastocyst. The placenta has a well-defined lifetime (21 days in the mouse, 9 months in humans). During that time the trophoblast cells exhibit several types of temporally and spatially regulated adhesive and invasive behaviors. For example, during the first trimester of gestation they are highly proliferative and invasive. At about day 5 in the mouse, or day 6 in the human, the trophectoderm cell layer of the blastocyst is highly adhesive and makes contact with the uterine epithelium in the first step of embryo implantation. As implantation proceeds, the trophoblast cells penetrate the uterine epithelium and migrate through the uterine stroma, which has been extensively modified in the decidualization reaction. These trophoblast cells also encounter and invade maternal blood vessels in the establishment of the hemochorial placenta, in which fetal trophoblast cells are in direct contact with the maternal circulation. This type of placentation is common to both humans and mice. Thus, the trophoblast exhibits a number of interesting adhesive and invasive behaviors that are amenable to study in the mouse, using isolated embryos and implantation sites, and in humans, using the permanent trophoblast-derived cell lines and/or with primary cytotrophoblast cells isolated from placental tissue of different gestational ages. We are in the process of studying the adhesion receptors of both the calcium-dependent cell–cell adhesion family and the integrin cell–matrix adhesion receptor families in trophoblast cells from both mouse and human sources.

Blastocyst ECM Interactions in the Mouse

We have studied the initial phases of implantation, using as an *in vitro* model mouse blastocyst interactions with defined and complex extracellular matrices of the type that are likely to be important for blastocyst–uterine interaction *in vivo*. Wewer et al. (1986) have shown in the decidualization reaction of the 6–7-day pregnant mouse uterus that stromal cells deposit a pericellular basement membrane–like matrix consisting of laminin, entactin, collagen TIV, and heparan sulfate proteoglycan as well as fibronectin. These components are deposited around individual stromal cells, forming an array of basal lamina–like structures encountered by the trophoblast during blastocyst implantation. These basal lamina–clad cells may serve as stepping stones for the implanting embryo. When blastocysts first hatch from the zona pellucida *in vitro*, they are nonadhesive. By 16–24 hours after hatching, the cells of the trophectoderm display extensive blebbing, and the blastocysts become attachment competent at about 24 hours posthatching when cultured in Nutridoma-supplemented medium without serum.

We (Sutherland et al., 1988) and Armant et al. (1986) have shown that blastocysts can attach to and grow out on monolayers of FN, LN, VN, and Col IV, and on a complex basement membrane laid down by bovine corneal endothelial cells (BCE). Outgrowth on all these matrices is inhibited by anti-

ECMr. The trophoblast cells of the 48-hour outgrowth display components on their surfaces of 120–130 kD and 150 kD (Sutherland et al., 1988). This is similar, although not identical in complexity, to the pattern displayed by the human BeWo choriocarcinoma cells (Figure 5) as well as other rodent cells, such as NRK cells (Figure 4), that attach to the broad range of substrates to which the mouse trophoblast outgrowth can attach. This suggests that there is not a distinctive form of this attachment complex on embryonic trophoblast that is grossly different from that found on established cell lines.

The trophoblast cells of the outgrowth can be stained to determine the distribution of the ECMr complex, using anti-ECMr and antibodies against the FN receptor, and to determine the distribution of cytoskeletal components known to be present in cell matrix contact sites of other cells. Figure 6 shows a phase micrograph of a blastocyst that has attached and grown out on an FN substrate for 48 hours. Although it is difficult to localize antigens very precisely in the rounded inner cell mass, it is possible to observe antigen distribution clearly in the flat trophoblast outgrowth region. The highly flattened cells with large nuclei that have migrated farthest display considerable diffuse staining with both anti-ECMr and an antibody against the human fibronectin receptor (anti-FN-R). However, a distinct pattern of plaquelike structures can also be observed that is reminiscent of the distribution of CSAT monoclonal antibody on avian fibroblastic cells. This pattern codistributes with that of vinculin, a cytoskeletal molecule known to be enriched in adhesion plaques of cultured fibroblasts and at the periphery of islands of cultured epithelial cells. The antibody inhibition and immunofluoresence experiments described above indicate that mouse blastocysts *in vitro* utilize the β_1-subfamily for attachment to and outgrowth on the same kinds of molecules that they encounter in the uterine decidua *in vivo*.

Adhesive and Invasive Behavior of Human Trophoblast

We have also been studying adhesive and invasive behavior of trophoblast cells, using the first-trimester human placenta as a source of cytotrophoblasts. The anatomy of the first-trimester placenta, including both maternal and fetal components, is diagramed in Figure 7. The structural unit of the placenta is the chorionic villus. These villi are of two types. The floating villi consist of a surface syncytium or syncytiotrophoblast layer that is formed from the fusion of underlying mononucleated cytotrophoblasts. These cytotrophoblasts constitute a proliferative stem cell population capable of supporting the renewal and expansion of the syncytiotrophoblast layer during the rapid growth of the placenta that takes place during early gestation. The floating villi are suspended in maternal blood, and the syncytiotrophoblast layer is responsible for the nutrient and waste exchange activities of the placenta.

In addition to floating villi, there are anchoring villi. These are formed when columns of cytotrophoblasts break through the syncytium, invade the uterine

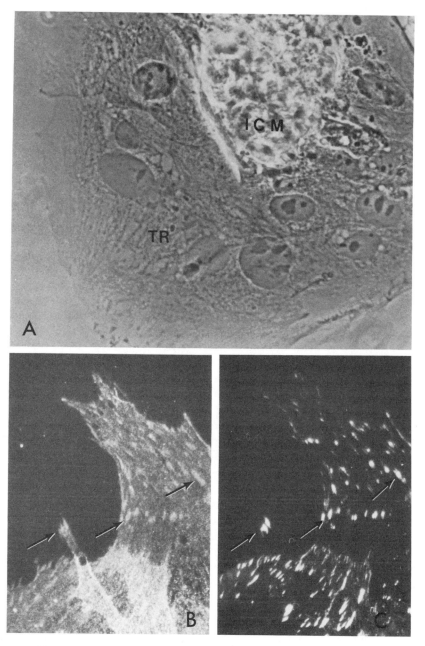

Figure 6. *Mouse embryo 48-hour outgrowth cultured in Nutridoma-containing serum-free medium on a FN-coated (10 µg/ml) glass coverslip. A:* Phase contrast of outgrowth showing the rounded central inner cell mass area and the flat area of outgrowth. *B:* Cells at periphery of outgrowth stained with anti-ECMr. *C:* Same field stained with anti-vinculin. (From Sutherland et al., 1988.)

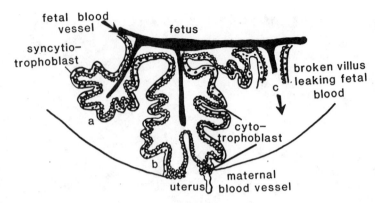

Figure 7. *Diagram of first-trimester placenta anatomy.* The structural units or villi are of two types. Floating villi consist of a cellular cytotrophoblast layer underlying an outer syncytial layer and surrounding a stromal core that contains fetal blood vessels. Anchoring villi are formed when columns of cytotrophoblasts break through the syncytium and penetrate through the maternal decidua and uterine spiral arteries, ultimately replacing the maternal vessel endothelium.

stroma, penetrate the maternal spiral arteries of the endometrium, and line these arteries, displacing the endothelium. These villi anchor the placenta, and therefore the embryo, to the uterus. The cytotrophoblasts that participate in the formation of anchoring villi are clearly highly invasive cells, sharing with malignant tumor cells the capacity to migrate through the extracellular matrix and to penetrate the basal laminae of blood vessels. In contrast to malignant tumor cells, however, cytotrophoblast invasive behavior is tightly regulated in both space and time. The cytotrophoblast invasion stops at the myometrial boundary. Furthermore, invasive activity peaks at about the twelfth week of pregnancy, declining rapidly during the second trimester (Enders, 1968). Thus, cytotrophoblast cells in the human are a highly specialized epithelium capable of rapid proliferation as well as active migratory and invasive behavior.

As an epithelium, mononucleated cytotrophoblast cells would be expected to express cell-CAM 120/80 (E-cadherin, uvomorulin, etc.). Figure 8 shows an oblique section of a first-trimester human placental floating villus stained with a monoclonal antibody against cell-CAM 120/80 (Wheelock et al., 1987). In the fetal portion of the placenta, cell-CAM 120/80 stains only the cytotrophoblast layer. Other studies have shown that in the uterus, cell-CAM 120/80 is found in the uterine epithelium of both the pregnant and nonpregnant uterus, but not in the decidua of the pregnant uterus (Damjanov et al., 1986). Nose and Takeichi (1986) have reported that in the mouse the trophoblast also expresses P-cadherin, an additional member of the calcium-dependent CAM family. In contrast to cell-CAM 120/80 (E-cadherin), which is expressed during pregnancy only on the fetal side of the mouse placenta, P-cadherin is expressed on the maternal decidua as well as on the fetal trophoblast, although not on

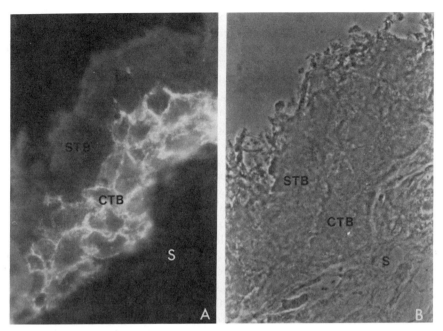

Figure 8. *An oblique section through a portion of a first-trimester placental villus. A:* Stained with antibody against the 80-kD fragment of cell-CAM 120/80 (anti-GP80). Only the cytotrophoblast cells (CTB) are positive for this antigen. The syncytiotrophoblast layer (STB) and the stromal core (S) are negative. × 355.

uterine epithelium at any time (Nose and Takeichi, 1986). This suggests that P-cadherin could play an important role in calcium-dependent, homophilic cell interactions between maternal and fetal cells during implantation.

We also looked at the distribution of cell-CAM 120/80 in placental bed biopsies taken in the first trimester. These biopsies contain portions of the maternal–fetal interface, including both cytotrophoblasts from fetal-anchoring villi and maternal decidua. These cytotrophoblasts have therefore successfully invaded the uterus. In preliminary studies this tissue has been examined microscopically and stained with antibodies to cytokeratins, which are present in cytotrophoblasts but not uterine decidual cells. Many cytokeratin-positive cytotrophoblast cells are present, along with decidual cells, in the placental bed. In many cases the cytotrophoblasts appear as rows or strings of connected cells. When this tissue is stained for cell-CAM 120/80, these rows of cells are positive for the antigen. Some apparently single cells with morphology similar to the strings of attached cells also stain for cell-CAM 120/80. The larger, pale decidual cells are negative for both the cytokeratin and cell-CAM 120/80 (Fisher et al., 1989). These results have a bearing on the previous suggestion that uvomorulin may be down-regulated at the invasive front of malignant tumors. Further examination of the placental bed with antibodies to cell-CAM 120/80, other CAMs, cytokeratins, and additional cell-type-specific markers should clarify whether actively invading cytotrophoblast cells express calcium-dependent CAMs on their surfaces. This question is also being addressed using the *in vitro* model for human placentation, described below.

In an effort to study the adhesive and invasive behavior of human trophoblast cells, an *in vitro* model has been established in which explanted fragments of first-trimester placental floating villi are cultured on basal laminae previously laid down by endothelial or teratocarcinoma cell monolayers (Fisher et al., 1985). As in the mouse decidualization reaction described above, human uterine decidual cells also surround themselves with a pericellular basal lamina (Liotta et al., 1986), suggesting that human trophoblast cells encounter and perhaps migrate on maternal basal lamina structures during the implantation process. The explanted first-trimester placental villi attach to the isolated extracellular matrices, and cells grow out as sheets from the explants. These cells are positive for cell-CAM 120/80. Of great interest is that these cells are able to penetrate and clear the layer of complex ECM on which they were plated. Identical attachment and displacement behavior is observed if villi are plated onto monolayers of endothelial cells (Figure 9), suggesting that some features of the displacement by trophoblasts of maternal blood vessel endothelium *in vivo* are retained *in vitro*. The matrix-clearing activity of the outgrowth is stage specific: villi explanted from second-trimester placenta adhere and cells grow out, but the matrix is not cleared.

In this model it has been assumed that the cell type responsible for the matrix-clearing activity is the cytotrophoblast. In order to demonstrate this conclusively and reduce the complexity of the model, we have recently modified methods used to isolate cytotrophoblasts from term placenta (Kliman et al., 1985), in order to isolate purified populations of first-trimester cytotrophoblasts. Since within the placenta, cell-CAM 120/80 is present only on cytotrophoblasts, it can be used as a marker for the purity of cytotrophoblast

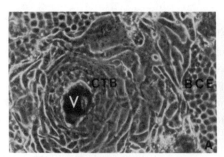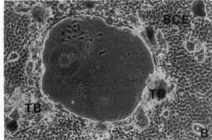

Figure 9. *Explanted first-trimester placental villus plated on a monolayer of bovine corneal endothelial cells (BCE). A:* 16 hours. *B:* 48 hours. The outgrowth of cytotrophoblast cells (CTB) from the villus (V) is able to clear the underlying BCE cell layer. TB, trophoblast. × 50.

preparations. Figure 10 shows a monolayer of cells isolated by sequential collagenase-hyaluronidase and mild trypsin treatment of first-trimester placental villi. These cells stain strongly with a monoclonal antibody against cell-CAM 120/80 and are active in matrix clearing in a stage-specific manner, as shown in Figure 11. Cytotrophoblasts isolated from second-trimester placental villi do not clear the underlying matrix. These results indicate that the model retains some of the regulated features displayed by trophoblast cells *in vivo*. These experiments suggest as well that cytotrophoblasts with invasive capability are present in tissue that contains primarily floating villi. It should be possible, therefore, to study carefully whether the cytotrophoblasts that are actively invading the matrix down-regulate the expression of cell-CAM 120/80.[3]

In summary, there are several unique advantages to using the placenta and trophoblast cells as a system for the study of adhesive, migratory, and invasive behavior. First, the placenta is a readily available normal tissue that can be obtained from human as well as animal sources. Second, its specialized

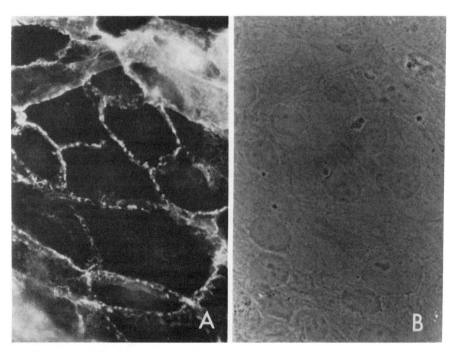

Figure 10. *First-trimester cytotrophoblasts isolated from placenta using sequential hyaluronidase/ collagenase and trypsin treatment followed by Percoll gradient centrifugation. A:* Cells stained with antibody against the 80-kD fragment of cell-CAM 120/80, which is found only in the cytotrophoblast layer of the placenta. All cells are stained, indicating that the purified cell population consists of cytotrophoblasts. *B:* Corresponding phase-contrast micrograph. × 365. (From Fisher et al., 1989.)

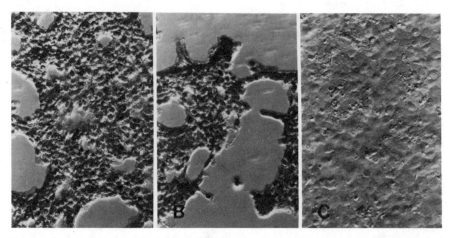

Figure 11. *Cytotrophoblast cells plated as a confluent monolayer on ECM previously laid down by a PF HR-9 teratocarcinoma cell monolayer. A:* First-trimester cytotrophoblasts 24 hours after plating. Holes have appeared, indicating clearing of the matrix. *B:* First-trimester cytotrophoblasts 48 hours after plating. Matrix clearing has increased. *C:* Late second-trimester cytotrophoblasts 72 hours after plating. The cells have formed a monolayer, but have not cleared the underlying matrix. $\times$ 71. (From Fisher et al., 1989.)

epithelial component, the trophoblast, transiently displays several types of behavior in common with metastatic carcinoma cells. However, in contrast to that of malignant cells, the invasive behavior of trophoblast cells is regulated in both time and space. This offers the possibility of studying the regulation of expression of cell–cell and cell–matrix adhesion molecules within the context of a changing pattern of invasive behavior. Third, an *in vitro* model is available in which the interaction of cytotrophoblast cells from different gestational ages with extracellular matrices and cell monolayers can be studied. This model retains important features of cytotrophoblast behavior *in vivo*. Fourth, placental bed biopsies can be examined in order to establish the relevance of findings using the *in vitro* model to the properties of trophoblasts at the maternal–fetal interface. Finally, permanent cell lines can be and have been isolated from malignant choriocarcinomas, offering the possibility of comparing the adhesive and invasive behavior of the normal early-gestation invasive trophoblast cell with its truly malignant counterpart.[3]

SUMMARY

The last seven years have seen an explosion of information concerning the identity, biochemical characterization, and cloning of cell surface adhesion receptors. Among the most interesting concepts resulting from these activities are that many adhesion receptors exist as members of large families and that cells have multiple mechanisms for adhering to particular adhesion ligands, suggesting that adhesion may not be the only important result of cell contact. These discoveries will lay the foundation for major research efforts for the next five years. Among the questions to be answered are: How is the expression of the members of an adhesion family regulated over time during development and in space in the differentiated organism? What environmental signals are initiated and transmitted by cell–cell and cell–matrix contact? How are the transmitted signals read by the cell? What kinds of cell behavior are under contact-mediated mechanisms of regulation?[4]

ACKNOWLEDGMENTS

We thank the following for their gifts of reagents: P. Brown and R. Juliano for anti-GP140 against the hamster FN receptor; M. Pierschbacher and E. Ruoslahti for human anti-FN receptor and purified vitronectin; M. Beckerle and K. Burridge for chicken anti-talin. The work was supported by the following grants: CA-42032, HD-22210, HD-22593, and NS-19090 from the National Institutes of Health, and DCB8502686 from the National Science Foundation. K. J. T. was supported by a predoctoral Chancellors Foundation Award from the University of California, San Francisco. M. J. W. was supported by N.R.S.A. grant CA-07572.

Notes added in proof: New relevant data have been published by us and others since the submission of this article in July, 1987.

[1]Affinity chromatography has been used successfully to isolate β_1 integrin heterodimers that are receptors for Col (Kramer and Marks, 1989) and LN (α200 kD/β_1: Ignatius and Reichardt, 1988; α135 kD/β_1: Gehlsen et al., 1988).

[2]Using an expanded panel of adhesion-perturbing monoclonal antibodies, we have determined that α_1/β_1 and α_6/β_1 heterodimers mediate human choriocarcinoma cell attachments to distinct sites on LN. The α_1/β_1 recognizes the cross-region of LN (elastase E1–4 domain), while the α_6/β_1 recognizes both the long arm (E8) and the E1–4 domains of LN. These studies also reveal that the α_1/β_1 heterodimer is a receptor for Col IV. [D. Hall, L. Reichardt, B. Holley, H. Moezzi, A. Sonnenberg, and C. H. Damsky (1989) α_1/β_1 and α_6/β_1 integrin heterodimers recognize LN at distinct sites. *J. Cell Biol.,* submitted.]

[3]Using the *in vitro* model system described, we have gathered data indicating that cytotrophoblasts regulate the expression of the α_6/β_1 integrin heterodimer (L. Zhang and C. H. Damsky, unpublished experiments) and the glycosylation of the α_5/β_1 FN receptor during gestation (Moss et al., 1988). They also regulate the expression of matrix-degrading metalloendoproteinases (Fisher et al., 1989). Thus this model can be used to study several aspects of cytotrophoblast adhesion and invasion.

[4]One possibility is that the state of adhesion receptor interaction with extracellular matrix components can regulate matrix remodeling. This is suggested by experiments showing that synovial fibroblasts, which do not express interstitial collagenase or stromelysin when plated on intact FN, are induced to express the genes for these matrix-degrading proteases if they are plated on substrates coated with an adhesion-disrupting monoclonal antibody against the α_5/β_1 FN receptor or on fragments of FN (Werb et al., 1989).

REFERENCES

Akiyama, S., S. Yamada, and K. Yamada (1986) Characterization of the 140 kD avian cell surface antigen as a fibronectin binding molecule. *J. Cell Biol.* **102**:442–449.

Armant, R., H. Kaplan, H. Mover, and W. Lennarz (1986) The effect of hexapeptides on attachment and outgrowth of mouse blastocysts cultured *in vitro*: Evidence for the involvement of the cell recognition tripeptide, ARG-GLY-ASP. *Proc. Natl. Acad. Sci. USA* **83**:6751–6755.

Bastiani, M., A. Harrelson, P. Snow, and C. Goodman (1987) Expression of fasciclin I and II glycoproteins on subsets of axon pathways during neuronal development in the grasshopper. *Cell* **48**:745–755.

Birchmeier, W., I. Wieland, J. Behrens, M. Kluge, U. Frixen, and P. Cledon (1987) Molecules involved in tumor invasion and metastasis. In *International Congress on Cancer Metastasis: Biological and Biochemical Mechanisms*, p. 53 (abstract), Bologna.

Boller, K., D. Vestweber, and R. Kemler (1985) Cell-adhesion molecule uvomorulin is localized in the intermediate junctions of adult intestinal epithelial cells. *J. Cell Biol.* **100**:327–332.

Brown, P., and R. Juliano (1986) Expression and function of a putative cell surface receptor for fibronectin in hamster and human cell lines. *J. Cell Biol.* **103**:1595–1603.

Buck, C., E. Shea, K. Duggan, and A. Horwitz (1986) Integrin (the CSAT antigen): Functionality requires oligomeric integrity. *J. Cell Biol.* **103**:2421–2428.

Chen, W.-T., J. Wang, T. Hasagawa, S. Yamada, and K. Yamada (1986) Regulation of fibronectin receptor distribution by transformation, exogenous fibronectin, and synthetic peptides. *J. Cell Biol.* **103**:1649–1660.

Damjanov, I., A. Damjanov, and C. H. Damsky (1986) Developmentally regulated expression of the cell–cell adhesion glycoprotein cell-CAM 120/80 in peri-implantation mouse embryos and extraembryonic membranes. *Dev. Biol.* **116**:194–202.

Damsky, C. H., J. Richa, D. Solter, K. Knudsen, and C. Buck (1983) Identification and purification of a cell surface glycoprotein mediating intercellular adhesion in embryonic and adult tissue. *Cell* **34**:455–466.

Damsky, C. H., J. Richa, M. Wheelock, I. Damjanov, and C. Buck (1985a) Characterization of cell-CAM 120/80 and the role of surface membrane glycoproteins in early events in mouse embryo morphogenesis. In *The Cell in Contact: Adhesions and Junctions as Morphogenetic Determinants*, G. M. Edelman and J. P. Thiery, eds., pp. 233–257, Wiley, New York.

Damsky, C. H., K. Knudsen, D. Bradley, C. Buck, and A. Horwitz (1985b) Distribution of the cell substratum attachment (CSAT) antigen on myogenic and fibroblastic cells in culture. *J. Cell Biol.* **100**:1528–1539.

Damsky, C. H., K. Knudsen, A. Horwitz, M. Wheelock, P. Gruber, and C. A. Buck (1985c) Integral membrane adhesion glycoproteins: What is their role in metastasis? In *Biochemistry and Molecular Genetics of Cancer Metastasis*, K. Lapis, L. Liotta, and A. Rabson, eds., pp. 25–42, Martinus Nijhoff, Boston.

Decker, C., R. Greggs, K. Duggan, J. Stubbs, and A. Horwitz (1984) Adhesive multiplicity in the interactions of fibroblasts and myoblasts with extracellular matrices. *J. Cell Biol.* **99**:1398–1404.

Dedhar, S., W. S. Argraves, S. Suzuki, E. Ruoslahti, and M. D. Pierschbacher (1987) Human osteosarcoma cells resistant to detachment by an Arg-Gly-Asp-containing peptide over-produce the fibronectin receptor. *J. Cell Biol.* **105**:1175–1182.

Duband, J.-L., S. Dufour, K. Hatta, M. Takeichi, G. M. Edelman, and J. P. Thiery (1987) Adhesion molecules during somitogenesis in the avian embryo. *J. Cell Biol.* **104**:1361–1374.

Edelman, G. M. (1986) Cell adhesion molecules in the regulation of animal form and tissue pattern. *Annu. Rev. Cell Biol.* **2**:81–116.

Enders, A. (1968) Fine structure of anchoring villi of the human placenta. *Am. J. Anat.* **122**:419–452.

Fisher, S. J., M. Leitch, M. Cantor, C. Basbaum, and R. Kramer (1985) Degradation of extracellular matrix by the trophoblastic cells of first trimester placentas. *J. Cell. Biochem.* **27**:31–42.

Fisher, S. J., T.-Y. Cui, L. Zhang, L. Hartman, K. Grahl, G.-Y. Zhang, J. Tarpey, and C. H. Damsky (1989) Adhesive and degradative properties of human placental cytotrophoblast cells *in vitro*. *J. Cell Biol.* **109** (in press).

Gallin, W. J., G. M. Edelman, and B. A. Cunningham (1983) Characterization of L-CAM, a major cell adhesion molecule from embryonic liver cells. *Proc. Natl. Acad. Sci. USA* **80**:1038–1042.

Gallin, W. J., B. C. Sorkin, G. M. Edelman, and B. A. Cunningham (1987) Sequence analysis of a cDNA clone encoding the liver cell adhesion molecule, L-CAM. *Proc. Natl. Acad. Sci. USA* **84**:2808–2812.

Gehlsen, K., E. Engvall, and E. Ruoslahti (1988) The human laminin receptor is a member of the integrin family of adhesion receptors. *Science* **241**:1228–1229.

Greve, J. M., and D. I. Gottlieb (1982) Monoclonal antibodies which alter the morphology of cultured chick myogenic cells. *J. Cell. Biochem.* **18**:221–230.

Grumet, M., and G. M. Edelman (1984) Heterotypic binding between neuronal membrane vesicles and glial cells is mediated by a specific neuron–glial cell adhesion molecule. *J. Cell Biol.* **98**:1746–1756.

Gumbiner, B., and K. Simons (1986) A functional assay for proteins involved in establishing an epithelial occluding barrier: Identification of an uvomorulin-like polypeptide. *J. Cell Biol.* **102**:457–467.

Hall, D., K. Neugebauer, and L. Reichardt (1987) Embryonic neural retina cell response to extracellular matrix proteins: Developmental changes and effects of the cell substratum attachment antibody, CSAT. *J. Cell Biol.* **104**:623–634.

Hatta, K., and M. Takeichi (1986) Expression of N-cadherin adhesion molecules associated with early morphogenetic events in chick development. *Nature* **320**:447–449.

Hemler, M., C. Huang, and L. Schwartz (1987) The VLA protein family: Characterization of five distinct cell surface heterodimers each with a common 130,000 Mr subunit. *J. Biol. Chem.* **262**:3300–3309.

Horwitz, A., K. Duggan, R. Greggs, C. Decker, and C. Buck (1985) The cell substrate attachment (CSAT) antigen has properties of a receptor for laminin and fibronectin. *J. Cell Biol.* **101**:2134–2144.

Hyafil, F., C. Babinet, and F. Jacob (1981) Cell–cell interactions in early embryogenesis: A molecular approach to the role of calcium. *Cell* **26**:447–454.

Hynes, R. O. (1987) Integrins: A family of cell surface receptors. *Cell* **48**:549–554.

Ignatius, M., and L. F. Reichardt (1988) Identification of a neuronal laminin receptor: An Mr 200/120 k integrin heterodimer that binds laminin in a divalent cation-dependent manner. *Neuron* **1**:713–725.

Imhof, B., H. Vollmers, S. Goodman, and W. Birchmeier (1983) Cell–cell interaction and polarity of epithelial cells: Specific perturbation. *Cell* **35**:667–675.

Kishimoto, T., K. O'Connor, A. Lee, T. Roberts, and T. Springer (1987) Cloning of the beta subunit of the leukocyte adhesion proteins: Homology to an extracellular matrix receptor defines a novel supergene family. *Cell* **48**:681–690.

Kliman, H., J. Nestler, E. Sermasi, J. Sanger, and J. Strauss (1985) Purification, characterization, and *in vitro* differentiation of cytotrophoblasts from human term placenta. *Endocrinology* **118**:1567–1582.

Knudsen, K., P. Rao, C. H. Damsky, and C. Buck (1981) Membrane glycoproteins involved in cell–substratum adhesion. *Proc. Natl. Acad. Sci. USA* **78**:6071–6075.

Knudsen, K., A. Horwitz, and C. Buck (1985) A monoclonal antibody identifies a glycoprotein complex involved in cell–substratum adhesion. *Exp. Cell Res.* **157**:218–226.

Kramer, R. H., and N. Marks (1989) Identification of collagen receptors on human melanoma cells. *J. Biol. Chem.* **264**:4684–4688.

Kramer, R. H., K. A. McDonald, E. Crowley, D. Ramos, and C. H. Damsky (1989) Melanoma cell adhesion to basement membrane mediated by integrin-related complexes. *Cancer Res.* **49**:393–402.

Liotta, L. A., N. C. Rao, and U. M. Wewer (1986) Biochemical interaction of tumor cells with basement membrane. *Annu. Rev. Biochem.* **55**:1037–1057.

Moss, L., S. J. Fisher, and C. H. Damsky (1988) Stage and tissue specificity in the glycosylation of human trophoblast integrins. *J. Cell Biol.* **107**:153A.

Nagafuchi, A., Y. Shirayoshi, K. Okazaki, K. Yasuda, and M. Takeichi (1987) Transformation of cell adhesion properties by exogenously introduced E-cadherin cDNA. *Nature* **329**:341–343.

Neff, N., C. Lowry, C. Decker, A. Tovar, C. H. Damsky, C. Buck, and A. Horwitz (1982) A monoclonal antibody detaches embryonic skeletal muscle from extracellular matrices. *J. Cell Biol.* **95**:654–666.

Nose, A., and M. Takeichi (1986) A novel cadherin cell adhesion molecule: Its expression patterns associated with implantation and organogenesis. *J. Cell Biol.* **103**:2649–2658.

Öbrink, B., C. Ocklind, P. Odin, and K. Rubin (1985) Adhesion reactions of rat hepatocytes: Cell surface molecules related to cell–cell and cell–collagen adhesion. In *The Cell in Contact: Adhesions and Junctions as Morphogenetic Determinants*, G. M. Edelman and J. P. Thiery, eds., pp. 227–300, Wiley, New York.

Patillo, R., and G. Gey (1968) The establishment of a cell line of human hormone synthesizing trophoblastic cells *in vitro*. *Cancer Res.* **28**:1231–1236.

Peyrieras, N., F. Hyafil, D. Louvard, H. Ploegh, and F. Jacob (1983) Uvomorulin, a non-integral membrane of early mouse embryos. *Proc. Natl. Acad. Sci. USA* **80**:6274–6277.

Pierschbacher, M., and E. Ruoslahti (1984) Variants of the cell recognition site of fibronectin that retains attachment promoting activity. *Proc. Natl. Acad. Sci. USA* **81**:5985–5988.

Plantefaber, L., and R. O. Hynes (1989) Changes in integrin receptors on oncogenically transformed cells. *Cell* **56**:281–290.

Plow, E., J. Loftus, E. Levin, D. Fair, D. Dixon, J. Forsyth, and M. Ginsburg (1986) Immunologic relationship between platelet membrane glycoprotein GPIIb/IIIa and cell surface molecules expressed by a variety of cells. *Proc. Natl. Acad. Sci. USA* **83**:6002–6006.

Pytela, R., M. Pierschbacher, and E. Ruoslahti (1985a) Identification and isolation of 140 kD cell surface glycoprotein with properties of a fibronectin receptor. *Cell* **40**:191–198.

Pytela, R., M. Pierschbacher, and E. Ruoslahti (1985b) A 125/115 kD cell surface receptor specific for vitronectin interacts with an arginine, glycine, aspartic acid adhesion sequence derived from fibronectin. *Proc. Natl. Acad. Sci. USA* **82**:5766–5770.

Pytela, R., M. Pierschbacher, M. Ginsburg, E. Plow, and E. Ruoslahti (1986) Platelet membrane glycoprotein IIb/IIIa: A member of a family of ARG-GLY-ASP specific adhesion receptors. *Science* **231**:1559–1567.

Rathjen, F. G., and M. Schachner (1984) Immunocytochemical and biochemical characterization of a new neuronal cell surface component (L1 antigen), which is an adhesion molecule. *EMBO J.* **3**:1–10.

Richa, J., C. H. Damsky, C. Buck, B. Knowles, and D. Solter (1985) Cell surface glycoproteins mediate compaction, trophoblast attachment, and endoderm formation during early mouse development. *Dev. Biol.* **108**:513–521.

Ruoslahti, E., and M. Pierschbacher (1986) Arg-Gly-Asp: A versatile recognition signal. *Cell* **44**:517–518.

Sutherland, A., P. Calarco, and C. H. Damsky (1988) Expression and function of cell surface extracellular matrix receptors in mouse blastocyst attachment and outgrowth. *J. Cell Biol.* **106**:1331–1348.

Takada, H., J. Strominger, and M. Hemler (1987) The very late antigen family of heterodimers is part of a superfamily of molecules involved in adhesion and embryogenesis. *Proc. Natl. Acad. Sci. USA* **84**:3239–3243.

Tamkun, J., D. W. DeSimone, D. Fonda, R. Patel, C. A. Buck, A. F. Horwitz, and R. O. Hynes (1987) Structure of integrin, a glycoprotein involved in the transmembrane linkage between fibronectin and actin. *Cell* **46**:271–282.

Tomaselli, K., C. H. Damsky, and L. Reichardt (1987) Interactions of a neuronal cell line (PC12) with the extracellular matrix proteins fibronectin, laminin, and collagen type IV: Identification of cell surface glycoproteins involved in attachment and outgrowth. *J. Cell Biol.* **105**:2347–2358.

Werb, Z., P. Tremble, E. Crowley, O. Berentsen, and C. H. Damsky (1989) Signal transduction through the fibronectin receptor induces collagenase and stromelysin gene expression. *J. Cell Biol.* **109** (in press).

Wewer, U., A. Damjanov, J. Weiss, L. Liotta, and I. Damjanov (1986) Mouse endometrial stromal cells produce basement membrane components. *Differentiation* **32**:49–58.

Wheelock, M., C. Buck, K. Bechtol, and C. H. Damsky (1987) The 80 kD soluble fragment of cell-CAM 120/80 inhibits cell–cell adhesion. *J. Cell. Biochem.* **34**:187–202.

Yoshida-Noro, C., N. Suzuki, and M. Takeichi (1984) Molecular nature of the calcium-dependent cell–cell adhesion system in mouse teratocarcinoma and embryonic cells studied with a monoclonal antibody. *Dev. Biol.* **101**:19–27.

Chapter 28

Cell-CAM 105: Molecular Properties, Tissue Prevalence, and Dynamics in Fetal and Regenerating Liver

BJÖRN ÖBRINK
PER ODIN
ANDERS TINGSTRÖM
MAGNUS HANSSON
PETER SVALANDER

ABSTRACT

Cell-CAM 105 is a cell surface glycoprotein that is involved in reaggregation of isolated rat hepatocytes in vitro. *The protein has been purified to homogeneity from rat liver membranes, and chemical characterization has demonstrated that it contains two structurally similar N-glycosylated peptide chains that can be phosphorylated on serine residues. Experiments with cell-CAM 105 incorporated into liposomes indicate that it has the potential of binding to itself in a homophilic manner. Immunohistochemical investigations have demonstrated that cell-CAM 105 is expressed not only in the liver but also in a variety of both stratified and simple epithelia, as well as in megakaryocytes, platelets, and granulocytes. The expression of cell-CAM 105 in hepatocellular carcinomas is significantly altered. During early embryonic development cell-CAM 105 is transiently expressed in trophoblast cells. In fetal development cell-CAM 105 is first observed in liver, where it appears in the hepatocytes on day 16 of gestation and then increases in concentration up to three weeks after birth, when the level characteristic of mature liver is reached. A transient decrease of the concentration of cell-CAM 105 is observed during liver regeneration. The cell surface location of cell-CAM 105 varies in different cell types. In the hepatocytes of the mature liver, cell-CAM 105 is located in the bile canalicular domain, but during critical time periods it occurs on all faces of the hepatocytes both in fetal and in regenerating hepatocytes. In stratified epithelia, cell-CAM 105 is located in the cell–cell borders of the suprabasal cells. In several brush border–containing epithelia, cell-CAM 105 is highly concentrated in the microvilli of the brush borders. In platelets and in granulocytes, cell-CAM 105 seems to occur primarily in intracellular sites. In platelets, cell-CAM 105 becomes expressed on the surface of the cells when they are induced to aggregate. The present knowledge about cell-CAM 105 indicates that the protein might have several functions, where a common denominator could be membrane–membrane binding. Such intermembranous binding could be involved either in cell–cell adhesion, as for example in stratified epithelia, or in direct interactions between microvilli of brush borders and bile canaliculi, respectively.*

It is generally believed that specific and dynamic spatiotemporal interactions between various types of cells are key events and fundamental driving forces in the formation of tissues and organs, both in embryonic development and in tissue regeneration. Central to these interactions are selective recognition mechanisms between, and perhaps also within, the surfaces of the participating cells. Such recognition events result in changes in cell aggregation, grouping and regrouping of cells, dissociation of existing cellular organizations, and changes in cellular motility and migration, and influence cellular differentiation. However, for the creation of functioning tissues and organs it is not enough that the cells become organized in correct three-dimensional patterns; cell type–specific shapes, polarization of cell surface components into different domains, and highly specialized structural organizations of the individual cell surfaces are also required. Examples of specialized cell surface organizations include microvilli, which can be relatively unorganized or regularly packed as in brush borders, cilia, axopods, other types of specific cell surface protrusions, fenestrations, intercellular junctions, and so on. Such structures are generally formed after the cells have reached their final positions within the forming tissues and may be regarded as the final stage of morphogenesis.

Experimental efforts in several laboratories have resulted in the disclosure of a class of cell surface molecules that seem to be involved in recognition events influencing the intercellular organizations of cells in developing and regenerating tissues (Edelman, 1986; Öbrink, 1986a). They are known as cell adhesion molecules (CAMs), and the combined work during the last few years has resulted in a dramatic increase in knowledge about these molecules and the processes they are involved in. However, our knowledge about mechanisms that regulate the terminal morphogenesis, that is, the cell surface polarization and formation and the organization of the specialized cell surface structures, is rudimentary.

Several CAMs have been identified and characterized (Öbrink, 1986a,b). Although some of these molecules have been discovered independently in different laboratories and in different cellular systems, we know enough today to conclude that many distinct CAMs exist. These molecules occur in different cell types and appear and disappear in specific spatiotemporal patterns during the course of embryonic development (Edelman, 1985, 1986). However, it has also been demonstrated that most cell types have more than one adhesive system in which distinct CAMs are operating (Öbrink, 1986a). An obvious corollary of this fact is that the function(s) of CAMs cannot only be to bind cells together simply and passively. If that were the case, one would think that it would be enough to have one adhesion system with one CAM. Thus, at least some of the CAMs must have more sophisticated functions. Experimental evidence supporting this hypothesis has recently been provided in a study of the development of the cerebellum (Chuong et al., 1987).

Another fact, important to remember when the functions of CAMs are considered, is that most of the CAMs have been identified by indirect approaches in which antibodies have been shown to perturb cellular associations (Öbrink, 1986a). Accordingly, although the available data point to important functions for CAMs in direct cellular interactions, future investigations may well reveal that CAM interactions influence cellular behavior via other mechanisms than by merely affecting physical adhesion.

In the original search for CAMs most groups looked for adhesion-related components in embryonic or tumor cells. This resulted in the discovery of N-CAM (Thiery et al., 1977), L-CAM (Gallin et al., 1983), the cadherins (Takeichi et al., 1986), Ng-CAM/L1 (Grumet and Edelman, 1984; Rathjen and Schachner, 1984; Bock et al., 1985; Friedlander et al., 1986), and other molecules that appear during embryogenesis and seem to be involved in various morphogenetic events. These molecules are discussed in other parts of this volume. We chose a slightly different approach, since we were interested in recognition events and cell surface interactions occurring in terminally differentiated cells that might be involved in the acquisition and/or maintenance of the differentiated state and the behavior of terminally differentiated cells. Therefore, we looked for adhesion-related molecules in mature adult cells.

We used hepatocytes isolated from young adult rats as a model system. By using an immunological approach to study the molecular basis for hepatocyte reaggregation, we succeeded in identifying a cell surface glycoprotein that is involved in the intercellular adhesion of these cells *in vitro* (Ocklind and Öbrink, 1982). This molecule, named cell-CAM 105, was purified to homogeneity (Odin et al., 1986) and has now been characterized extensively both at the molecular level (Odin et al., 1986) and the cellular and tissue levels (Odin and Öbrink, 1986, 1987, 1988; Odin, 1987; Odin et al., 1988). In this chapter we review the present knowledge of the properties, expression, cell surface location, and tentative functions of cell-CAM 105. Our results indicate that one of the functions of cell-CAM 105 might be to influence the final stage of morphogenesis via formation of specialized cell surface domains.

MOLECULAR PROPERTIES OF CELL-CAM 105

Cell-CAM 105 was purified from rat liver membranes by a combination of immunoaffinity chromatography and FPLC chromatography on a gel exclusion column and an ion exchange column (Odin et al., 1986). Biochemical analyses demonstrated that the protein consists of two peptide chains with apparent molecular weights of 105 kD and 110 kD in SDS-PAGE, respectively. Peptide mapping indicated that the two peptides are structurally closely related. Amino acid analysis demonstrated that the protein contains almost no tryptophan, but rather high amounts of glycine, serine, threonine, and acidic amino acids. Both peptide chains are highly glycosylated, seemingly only with N-linked oligosaccharides. One-third of the molecular mass is carbohydrate consisting of mannose, galactose, fucose, N-acetylglucosamine, and sialic acid. Treatment with neuraminidase shifted the apparent molecular weight from 105 kD to 90 kD in SDS-PAGE, and complete deglycosylation by digestion with endoglycosidase F resulted in the appearance of two peptide chains with apparent molecular weights of 54 kD and 58 kD, respectively. Isolated cell-CAM 105 contained phosphate, and the glycoprotein could be phosphorylated in hepatocytes cultured *in vitro*. Phosphoamino acid analysis revealed that serine residues, but not threonine or tyrosine residues, became phosphorylated under these conditions.

FUNCTIONAL AND BINDING PROPERTIES OF CELL-CAM 105

Cell-CAM 105 was originally identified as a CAM by an immunological functional assay (Ocklind and Öbrink, 1982). It proved to be the component that neutralized the inhibition of hepatocyte reaggregation exerted by mono-valent polyspecific rabbit antibodies raised against liver plasma membranes. We could then demonstrate that Fab fragments of both monospecific, polyclonal antibodies and monoclonal antibodies (Öbrink et al., 1986) against cell-CAM 105 effectively inhibited hepatocyte reaggregation *in vitro*. However, these data demonstrated only that cell-CAM 105 in some way is involved in hepatocyte intercellular adhesion; they did not prove that the glycoprotein participates directly in the binding reaction.

One piece of evidence indicating that cell-CAM 105 is directly involved in cell binding was provided in experiments where the purified molecule was added to isolated hepatocytes. In these experiments we found that the aggregation of the hepatocytes was significantly inhibited at low concentrations of cell-CAM 105 (100 ng/ml to 10 μg/ml) (Odin et al., 1986).

The next step in our strategy to investigate the mode of function of cell-CAM 105 at the molecular level was to incorporate the glycoprotein into liposomes and to study their properties and behavior (Öbrink et al., 1986; Tingström and Öbrink, in preparation). Cell-CAM 105 was reconstituted into phosphatidylcholine liposomes by dialysis of an octylglucoside-containing mixture of cell-CAM 105 and phosphatidylcholine. Under these conditions all of the protein became incorporated into lipid vesicles, which gives further support to our initial results indicating that cell-CAM 105 is an integral membrane glycoprotein. The cell-CAM-containing liposomes were studied with respect to their ability to bind to hepatocytes and to self-aggregate.

Liposomes containing cell-CAM 105 were able to bind to freshly isolated hepatocytes (Figure 1). The extent of binding increased with increasing concentration of the protein in the lipid vesicles. The binding was specifically inhibited by Fab fragments of monospecific antibodies against cell-CAM 105. Furthermore, the binding could be significantly inhibited if the hepatocytes were first preincubated with antibodies against cell-CAM 105 and washed before the cell-CAM-containing liposomes were added (Figure 2). These

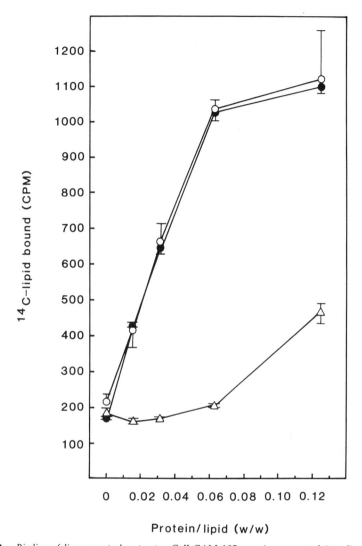

Figure 1. *Binding of liposomes to hepatocytes.* Cell-CAM 105 was incorporated into liposomes by dialysis of solutions containing cell-CAM 105, dioleylphosphatidylcholine, [^{14}C]phosphatidylcholine, and octylglucoside. The protein/lipid ratio was varied by adding different amounts of cell-CAM 105 to a constant amount of phospholipids. Aliquots of the reconstituted liposomes were incubated with hepatocytes for 60 min at 37°C under gyratory shaking. Unbound liposomes were separated from the cells by centrifugation through a cushion of Percoll. The cellular pellet was dissolved and analyzed by liquid scintillation counting. The incubation of the cells and liposomes was performed in the presence or absence of Fab fragments of antibodies against cell-CAM 105 or preimmune IgG, respectively. *Open circles:* Incubation in the absence of Fab fragments. *Filled circles:* Incubation in the presence of 100 µg/ml of preimmune Fab fragments. *Triangles:* Incubation in the presence of 100 µg/ml of anti-cell-CAM Fab fragments. The symbols represent the means and ranges of duplicate experiments.

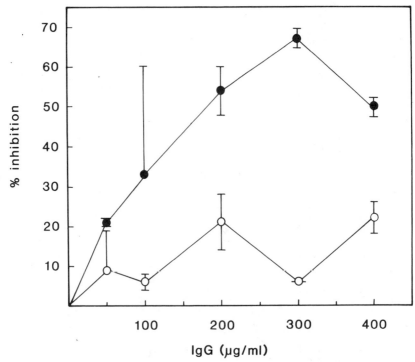

Figure 2. *Inhibition of liposome binding.* Hepatocytes were preincubated with various concentrations of anti-cell-CAM IgG or preimmune IgG for two hours at 4°C. The cells were then
washed twice and incubated with [^{14}C] liposomes containing cell-CAM 105 (protein/lipid
ratio 0.06) as described in Figure 1. *Filled circles:* Preincubation with anti-cell-CAM IgG. *Open
circles:* Preincubation with preimmune IgG. The symbols represent the means and ranges of
duplicate experiments.

results suggest that (1) cell-CAM 105 has properties that permit binding to the
cell surface of hepatocytes; and (2) the binding is homophilic in nature,
meaning that cell-CAM 105 can bind to itself.

   With increasing amounts of cell-CAM 105 incorporated into the lipid
vesicles, a spontaneous aggregation of the liposomes occurred (Figure 3). This
aggregation, which was calcium independent, could be reversibly dissociated
by elevating the pH. Upon neutralization the liposomes aggregated again.
Furthermore, Fab fragments of antibodies against cell-CAM 105 effectively
and specifically blocked the self-aggregation of the cell-CAM-containing
liposomes (Figure 3). These results thus give further support to the hypothesis
that cell-CAM 105 can participate in a homophilic binding reaction.

   Another important task was to study the location of cell-CAM 105 in
reaggregating hepatocytes, to see if it was compatible with cell-CAM 105's
being a ligand in cell–cell binding. This was done by immunofluorescence
microscopy employing affinity-purified antibodies on freshly isolated hepato-

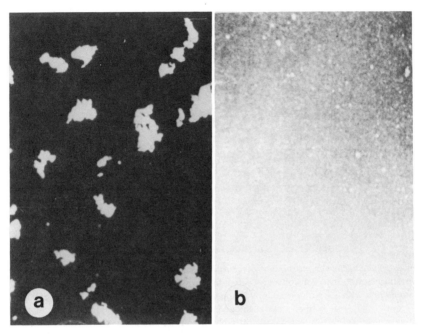

Figure 3. *Aggregation of liposomes containing cell-CAM 105.* Cell-CAM 105 was incorporated into dioleylphosphatidylcholine liposomes (protein/lipid ratio 0.1), in the presence of carboxy-fluorescein, as described in Figure 1. This resulted in fluorescent liposomes that could be observed by fluorescence microscopy. Under these conditions the cell-CAM-containing liposomes became highly aggregated, whereas liposomes without cell-CAM 105 did not aggregate. Addition of Fab fragments of anti-cell-CAM antibodies rapidly dissociated the aggregates. Fab fragments of preimmune IgG had no effect. *a:* Liposomes containing cell-CAM 105, observed by fluorescence microscopy 10 min after addition of preimmune Fab fragments (0.5 mg/ml). *b:* Liposomes containing cell-CAM 105, observed by fluorescence microscopy 10 min after addition of anti-cell-CAM Fab fragments (0.5 mg/ml).

cytes that had been seeded on collagen-coated substrata. Immediately after the isolation, cell-CAM-specific staining was seen on all faces, all around the cells. However, after a few hours in culture, when the hepatocytes had formed confluent sheets of closely associated cells, cell-CAM 105 became concentrated in the cell–cell borders where the cells were in contact with one another (Figure 4). At this stage, surfaces that were not in contact with other cells exhibited no or very weak staining for cell-CAM 105. The location of cell-CAM 105 in hepatocytes *in vitro* is thus compatible with a role in cell–cell binding.

We can thus conclude that cell-CAM 105 both appears in cell–cell contact areas *in vitro* and has binding properties that would allow the molecule to mediate cell–cell binding. The next question we wanted to address was what the functional and biological role(s) of cell-CAM-mediated cell–cell binding might be. As a first attempt to tackle this question, we investigated the occurrence and amounts of cell-CAM 105 in fetal and adult tissues in the rat.

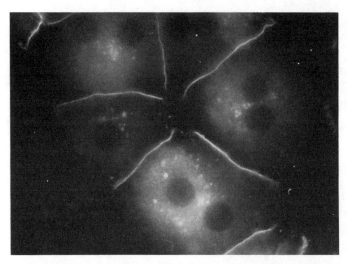

Figure 4. *Immunofluorescence of cultured rat hepatocytes.* Hepatocytes were isolated by collagenase perfusion and seeded on collagen-coated dishes. Five hours after seeding, the cells were fixed for 30 min with 2% paraformaldehyde and 0.05% glutaraldehyde and permeabilized by 0.1% Triton X-100. The fixed and permeabilized cells were stained with rabbit anti-cell-CAM IgG followed by FITC-conjugated goat anti-rabbit immunoglobulins, and viewed in a fluorescence microscope. The picture shows a colony of reaggregated hepatocytes. The cell–cell borders are specifically stained, whereas the free cell surfaces lack staining for cell-CAM 105.

OCCURRENCE OF CELL-CAM 105 IN ADULT TISSUES

The occurrence of cell-CAM 105 was determined in all types of rat organs and tissues, both quantitatively by a radioimmunoassay (Odin and Öbrink, 1987) and qualitatively (Odin et al., 1988) by immunoblotting and immunohisto-chemical techniques. Cell-CAM 105 was found in a variety of organs. The highest concentrations of the glycoprotein were found in the small intestinal mucosa, liver, salivary glands, and vagina. In these organs cell-CAM 105 occurred primarily in epithelial cells. In addition, the glycoprotein was found in vessel endothelia of capillaries, arterioles, and venules, and in platelets, megakaryocytes, granulocytes, and myelopoietic cells in the bone marrow. Nervous tissue, in both the central and peripheral nervous systems, and all kinds of muscle and connective tissues were cell-CAM negative. Thus, in the adult organism cell-CAM 105 seems to exert its major function(s) primarily in epithelia, vessel endothelia, and peripheral blood cells.

The biochemical properties of cell-CAM 105 in the various organs and cells differed somewhat (Odin et al., 1988). This was investigated by polyacrylamide gel electrophoresis in SDS. In all the organs a specific component with macromolecular properties similar to those of liver cell-CAM was observed (Figure 5). In some of the organs, such as the small intestine and kidney, and in platelets components were found that were slightly larger than liver cell-CAM

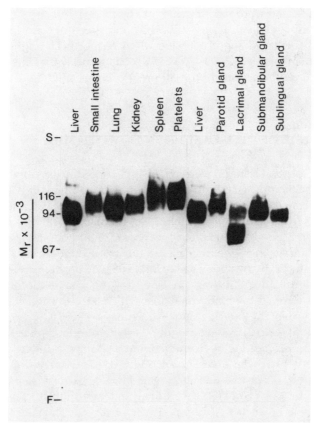

Figure 5. *Immunoblotting of various rat tissues.* Organs and cells were solubilized in 1% Triton X-100, preincubated with fixed staphylococci (*S. aureus,* Cowan I), electrophoresed in SDS on polyacrylamide gels under nonreducing conditions, and transferred to nitrocellulose sheets. The nitrocellulose sheets were incubated sequentially with affinity-purified anti-cell-CAM antibodies and ^{125}I-labeled protein A, and developed by autoradiography. The apparent molecular weight of cell-CAM 105 from the different sources varied slightly. The molecular weights of standard proteins are indicated to the *left.* S and F indicate the start and front, respectively, of the electrophoresis.

under nonreducing conditions. In the lacrimal gland a molecule smaller than liver cell-CAM occurred.

Furthermore, cell-CAM 105 of the small intestinal mucosa behaved differently from liver cell-CAM 105 upon reductive heating (Hansson et al., 1989). The two chains of liver cell-CAM increased in size from a mean apparent molecular weight of 90 kD to 105 kD when reduced. The mean apparent molecular weight of intestinal cell-CAM 105, on the other hand, decreased from 115 kD to 90 kD when heated and/or reduced. This behavior suggests that intestinal cell-CAM 105 occurs in a complex with some other component that is dissociated upon heating and/or reduction. Still another pattern was observed in platelet cell-CAM 105, which consisted of four distinct peptide chains that increased slightly in size when reduced. The differences in the biochemical properties of cell-CAM 105 might indicate that this molecule has slightly different functions in the different cell types.

OCCURRENCE OF CELL-CAM 105 IN FETAL TISSUES

Another important issue concerning the biological function of cell-CAM 105 is its appearance and behavior during embryonic and fetal development. This was investigated both in preimplantation rat embryos (Svalander et al., 1988) and in rat fetuses (Odin and Öbrink, 1986, 1988; Odin, 1987). In preimplantation embryos, cell-CAM 105 first appeared at the blastocyst stage, where it was expressed in the trophoblast cells of the trophectoderm. At the onset of implantation the distribution of cell-CAM 105 became highly polarized on the trophectoderm. Thus, it was present only on the embryonal pole, and disappeared from the abembryonal pole, which is the portion that first adheres to and invades the uterine mucosa (Alden, 1948; Chavez, 1986). The expression of cell-CAM 105 in the early embryonic development after implantation has not yet been studied, but its behavior in late fetal development has recently been investigated (Odin, 1987; Odin and Öbrink, 1988).

The first fetal stage that was investigated was day 12 of gestation. Sections of all parts of the 12-day-old fetus were analyzed immunohistochemically, but in no parts could any cell-CAM reactivity be observed. Thus, it seems that cell-CAM 105 is not expressed in any type of tissue during a certain period of embryonic development. The first time cell-CAM 105 was observed again by immunohistochemical analysis was at day 13 of gestation. It then appeared in the fetal liver, which is a blood-forming organ. The only cell-CAM-positive cells that were present at day 13 were megakaryocytes. At that time no cell-CAM-positive staining was seen in the hepatocytes of the liver parenchyme. The liver parenchyme showed cell-CAM reactivity for the first time on day 16 of gestation. This was also the first time cell-CAM 105 could be detected by quantitative RIA analysis in the fetal liver (Odin and Öbrink, 1986). The amount of cell-CAM 105 then increased linearly. At birth the concentration was one-third of that found in the mature liver. The level characteristic of mature liver was reached three weeks after birth.

At about the same time, several other tissues started to express cell-CAM 105 (Odin and Öbrink, 1988). Thus, the epithelial cells of both the proximal kidney tubules and the small intestinal mucosa showed cell-CAM-positive staining for the first time on day 17 of gestation. These results demonstrate that except for the blastocyst trophectoderm, cell-CAM 105 appears late in development, that is, in terminally differentiating cells.

OCCURRENCE OF CELL-CAM 105 IN REGENERATING LIVER

Tissue regeneration and restoration after injury are characterized by proliferation, migration, reorganization, and differentiation of a variety of cell types. It might accordingly be expected that dynamic changes in the prevalence and location of CAMs occur in regenerating tissues. To that end, we considered it to be of interest to investigate the fate of cell-CAM 105 during liver regeneration.

Rats were subjected to partial hepatectomy, in which two-thirds of the liver was surgically removed. After such an operation the remaining liver starts an active proliferation, and after 10–15 days, when it has reached the size of the original liver, the growth ceases. The hepatocytes are the first cells to divide, and somewhat later the cells of the sinusoidal walls also proliferate (Kovacs and Lapis, 1984). The major peak of mitotic division of the hepatocytes occurs about 28 hours after the partial hepatectomy. About 24 hours later a second, less pronounced round of hepatocyte division is observed.

Quantitative determination of cell-CAM 105 during liver regeneration demonstrated that this molecule started to decrease in amount about 12–24 hours posthepatectomy (Odin and Öbrink, 1986). A minimum was reached about 2–3 days after the operation, when the concentration of cell-CAM 105 in the plasma membranes of the hepatocytes was one-third of that in normal mature liver. The concentration of cell-CAM 105 then increased linearly and reached the normal level 10–15 days posthepatectomy.

Immunocytochemical analyses of cell-CAM 105 during liver regeneration demonstrated that its cell surface location also changed dynamically (Odin, 1987; Odin and Öbrink, 1988). Twenty-four hours after the operation the bile canaliculi became dilated and irregularly shaped. The next few days cell-CAM 105 then spread out of the canalicular area and exhibited a more or less unpolarized location; it was present on all faces of the surfaces of the hepatocytes (Figure 6). About 15 days posthepatectomy, when the growth of the liver had ceased, the cell surface location of cell-CAM 105 had normalized again, and the protein was seen primarily in the bile canalicular areas. Thus, the most dramatic changes in both the concentration and the cell surface location of cell-CAM 105 occurred when the reorganization of the hepatocytes was initiated (Ogawa et al., 1979).

CELL-CAM 105 IN HEPATOCELLULAR CARCINOMAS

In transplantable hepatocellular carcinomas (THC), both qualitative and quantitative changes in the cell surface glycoproteins of the hepatocytes occur (Hixson et al., 1985). A number of different THCs were investigated for the expression of nine different surface glycoproteins (Hixson et al., 1985). It was found that all THCs exhibited different patterns of expression of these glycoproteins. However, a common feature of all THCs was that two proteins

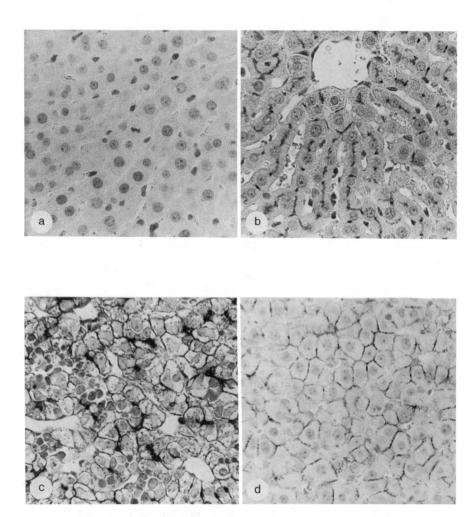

Figure 6. *Immunohistochemical staining of rat liver.* The livers were fixed with 4% paraformaldehyde, embedded in paraffin, sectioned, and stained with affinity-purified anti-cell-CAM
antibodies (or preimmune IgG) followed by biotinylated swine anti-rabbit immunoglobulin
and avidin plus biotinylated horseradish peroxidase (ABComplex). The peroxidase was
visualized by incubation with 3-amino-9-ethylcarbamazide, which gives a red staining
product. The tissues were counterstained with hematoxylin. *a:* Mature, adult liver stained with
preimmune IgG. Neither cell borders nor bile canaliculi are stained. *b:* Mature, adult liver
stained with anti-cell-CAM IgG. Note the specific staining of the bile canaliculi, which appear
both cross-sectioned and longitudinally sectioned. *c:* Fetal liver at day 21 of gestation, stained
with anti-cell-CAM IgG. Note that both the irregularly shaped bile canaliculi and the
contiguous faces of the hepatocytes bordering on adjacent hepatocytes are specifically stained
for cell-CAM 105. The hematopoietic cells do not show any cell-CAM staining. *d:* Regenerating
liver four days posthepatectomy, stained with anti-cell-CAM IgG. Note that all faces of the
hepatocyte cell surfaces are stained. No bile canalicular staining pattern similar to that of
normal liver is seen.

were missing. Immunochemical analyses demonstrated that these two proteins were the A-chain and the B-chain of cell-CAM 105. Further investigations have shown that not all THCs completely lacked cell-CAM 105, but those that expressed it made significantly lower amounts and the glycoprotein was chemically different. Thus, all hepatocellular carcinomas that have been investigated so far exhibit an altered expression of cell-CAM 105.

THE CELL SURFACE LOCATION OF CELL-CAM 105

A molecule that is involved in recognition and adhesion between cells should be expressed on the surface of the cells it resides in. This is clearly the case for cell-CAM 105, but its subcellular distribution and cell surface location vary in different cell types. The cell surface location of cell-CAM 105 is not only different in different cell types, but also varies in the same cell type according to the state of differentiation or differences in the environment.

We have observed essentially three different patterns of subcellular distribution and cell surface location of cell-CAM 105 *in vivo* (Figures 6–8)

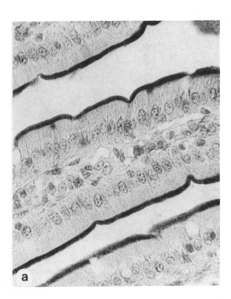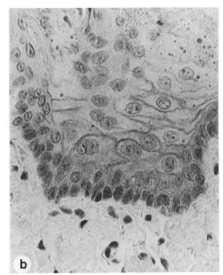

Figure 7. *Immunohistochemical staining of rat intestinal and tongue epithelia.* The tissues were treated and stained as described in Figure 6. *a:* Small intestinal mucosa stained with anti-cell-CAM IgG. Note the specific staining of the brush border region of the epithelial cells. No staining is observed in the subepithelial tissue. *b:* Tongue epithelium stained with anti-cell-CAM IgG. Note the specific staining of the cell–cell borders of the suprabasal cells of the stratified epithelium. No staining is observed in the subepithelial tissue.

(Odin, 1987; Odin et al., 1988). In stratified epithelia, cell-CAM 105 is located in the cell–cell borders of the suprabasal cell layers. In some simple epithelia, including the liver, in some vessel endothelia, and in the trophectoderm of preimplantating blastocysts, cell-CAM 105 is located at the apical surfaces of the cells. Finally, in platelets and in granulocytes, cell-CAM 105 is located primarily in intracellular compartments. The varying cell surface location of cell-CAM 105 as a function of the state of differentiation or of different environmental factors is exemplified by its dynamic expression in fetal, regenerating, and adult hepatocytes *in vivo* and *in vitro*.

The expression of cell-CAM 105 in the cell–cell borders of the suprabasal cells of stratified epithelia is compatible with a function in cell–cell binding and requires no further discussion. The expression at the apical surface of simple epithelia and in the bile canaliculi of hepatocytes (Figures 6–8) may at first sight be unexpected for a cell adhesion molecule. However, a closer examination of the morphology may give a hint to the role of cell-CAM 105 in these cells.

The simple epithelia that express cell-CAM 105 also carry a brush border on their apical surfaces. Brush borders consist of closely packed, regularly sized microvilli. For example, in the brush border of the small intestine, the microvilli are very uniform in size, and the average distances between the

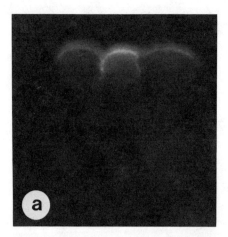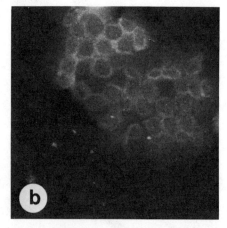

Figure 8. *Immunofluorescence staining of intestinal epithelial cells.* Sheets of small intestinal epithelial cells were isolated by treatment of the intestine with a citrate-containing balanced salt solution, and fixed with 4% paraformaldehyde. The cells were stained with rabbit anti-cell-CAM IgG followed by FITC-conjugated goat anti-rabbit immunoglobulins, and viewed in a fluorescence microscope. *a:* A group of cells seen in lateral projection, perpendicular to the long axis of the cells. The specific staining for cell-CAM 105 is located in the top portions of the cells, corresponding to the brush border region and possibly to the lateral cell borders. *b:* A group of cells seen in top view projection, parallel to the long axis of the cells. The specific staining for cell-CAM 105, which appears in a honeycomblike pattern, is concentrated at the circumference of the cells, where they border on each other.

outer surfaces of the microvillar membranes are of the order 10–20 nm (see Figures 2–3, 27–8, and 27–10 in Bloom and Fawcett, 1975). Thus, it is possible that cell-CAM 105 might be involved in linking adjacent microvilli to each other (Öbrink et al., 1986).

The hepatocyte membranes that make up the bile canaliculi also contain a large number of microvilli. The diameter of the lumen of the bile canaliculi varies according to the state of bile secretion (Bloom and Fawcett, 1975, p. 710). In the nonsecreting stage the lumen is collapsed and the bile canaliculus is obliterated by numerous microvilli in close contact with each other. Thus, the microvilli might interact via cell-CAM 105–mediated bonds in the bile canaliculi as well (Öbrink et al., 1986).

Furthermore, both in the brush border of the small intestine and in the bile canaliculi of mature liver we have observed that the immunocytochemical staining for cell-CAM 105 is not homogeneously distributed (Odin et al., 1988). In both tissues there is a higher staining intensity in the margins of cells and bile canaliculi, respectively (Figures 8, 9). This corresponds to those areas where adjacent cells border on each other. Thus, a higher concentration of cell-CAM 105 also occurs in the cell–cell borders in the brush border region of simple epithelia and in the bile canaliculi of hepatocytes. Accordingly, it might

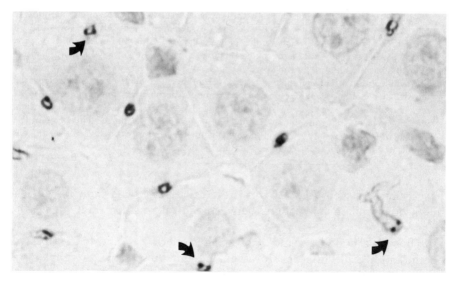

Figure 9. *Heterogeneity of bile canalicular staining of rat liver.* A mature rat liver was treated and stained for cell-CAM 105 as described in Figure 6. The positive cell-CAM staining of all bile canaliculi is heterogeneous, and often appears as punctate areas that are more intensely stained than the remainder of the bile canaliculi. The punctate areas often appear in pairs; such pairs are marked by *arrows*.

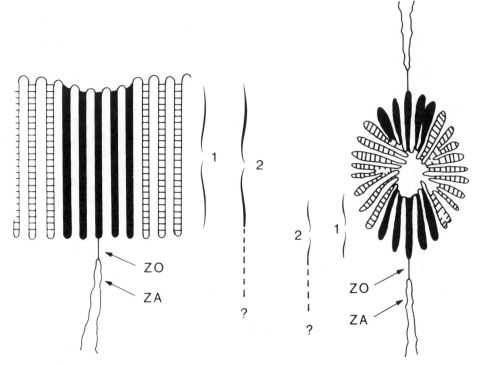

Figure 10. *A schematic illustration of the microvillar and junctional regions of intestinal epithelial cells and hepatocytes.* The *left-hand portion* illustrates the border between two adjacent intestinal epithelial cells. The *right-hand portion* illustrates a bile canaliculus formed between two adjacent hepatocytes. Immunolocalization studies indicate that cell-CAM 105 occurs on the apical microvilli of these cells. The heterogeneous staining shown in Figures 8 and 9 is interpreted as a more intense staining of the microvilli at the cell–cell borders (*filled areas* between the microvilli), and a weaker staining of the remainder of the microvilli (*bars* perpendicular to the microvillar membranes). While the immunostaining experiments strongly indicate that the region indicated by bracket (1) is stained for cell-CAM 105, the resolution at the light microscope level is not enough to determine if some part of the junctional complex might also be stained. This uncertainty in the confinement of the staining toward the basal sides of the cells is indicated by bracket (2). ZO, zonula occludens; ZA, zonula adherens.

be possible that cell-CAM 105 in these cells is involved both in cell–cell interactions and in interactions between the microvilli of individual cells (Figure 10).

In the vessel endothelia and in the trophectoderm, cell-CAM 105 is located on the apical cell surfaces. However, in both these types of cellular organizations there is a higher concentration of cell-CAM 105 in the cell–cell borders, again suggesting that cell-CAM 105 might be involved in cell–cell interactions in these tissues.

Our immunohistochemical investigations have yielded results suggesting that cell-CAM 105 in megakaryocytes, in unactivated platelets, and in granulocytes occurs primarily in intracellular compartments (Odin et al., 1988). When platelets were activated to aggregate, however, the entire surface of the cells in the aggregates became strongly cell-CAM 105 positive. Thus, it is possible that the intracellular locations of cell-CAM 105 in the blood cells enable it to serve as storage pools that are utilized in aggregation phenomena or other types of cellular interactions, which are important functions of both platelets and granulocytes.

The cell surface location of cell-CAM 105 in hepatocytes does not seem to be static, but varies as a function of the state of both the cellular differentiation and the environment of the cells. In the hepatocytes of the fetal liver, cell-CAM 105 first appeared in structures we interpreted as being immature bile canaliculi (Odin and Öbrink, 1988). However, around birth the molecule also became expressed on the contiguous faces of the hepatocytes (Figure 6), which at that time condensed and became closely associated with each other. The expression of cell-CAM 105 on these domains of the hepatocytes then gradually disappeared, and in the mature liver it was primarily the bile canalicular domains that contained this glycoprotein (Figure 6).

Also, in the hepatocytes of regenerating liver a changing surface distribution of cell-CAM 105 was observed, as mentioned above. During a time period of two to five days posthepatectomy, the cell surface distribution of cell-CAM 105 became unpolarized, and the protein appeared on all faces, including the contiguous faces, of the cells (Figure 6) (Odin and Öbrink, 1988). Fifteen days posthepatectomy, the surface distribution of cell-CAM 105 was almost normalized again.

In hepatocytes cultured *in vitro,* the cell surface distribution of cell-CAM 105 became more like that of fetal and regenerating cells than like that of hepatocytes organized in mature liver. Immediately after seeding, cell-CAM 105 was present on all faces all around the cell surface. During the next few hours, when the cells associated with each other and formed large sheets of confluent cells, cell-CAM 105 became concentrated into the domains where the cells were in contact with each other, whereas free surfaces not in contact with other hepatocytes became devoid of cell-CAM 105 (Figure 4). When the hepatocytes were cultured for a few days, bile canaliculi were often reformed. Cell-CAM 105 was observed both in the bile canalicular membranes and on the apical surfaces of the cultured hepatocytes. The only face that did not contain this glycoprotein was the undersurface of the cells, facing the solid substratum to which they were attached.

SPECULATIONS ABOUT POTENTIAL FUNCTIONS OF CELL-CAM 105

While cell-CAM 105 was identified as a result of our efforts to find cell surface components that are involved in intercellular adhesion, we clearly lack knowledge about the cell-biological function of this molecule. However, our studies of the molecular properties of cell-CAM 105 have demonstrated that the molecule can interact with itself in a way that is compatible with a role in cell–cell adhesion. Furthermore, our studies of isolated hepatocytes indicate that this molecule is indeed involved in intercellular adhesion, at least *in vitro*. Some aspects of the functional role of cell-CAM 105 have been mentioned in the preceding sections, but we find it appropriate to present a more general discussion about its potential functions at this stage.

There are three properties of cell-CAM 105 that have to be considered when its *in vivo* functions are discussed. (1) Several forms of the molecule with slightly different molecular properties seem to exist. (2) The cell surface location is different in different tissues and cell types. (3) Cell-CAM 105 appears late in embryonic development and becomes expressed primarily in cells that are undergoing terminal differentiation, with the exception of its expression in trophoblast cells. The different molecular forms of cell-CAM 105 indicate that the function of the molecule can be modulated in different tissues. The varying cell surface locations indicate that cell–cell binding may not be the only function of cell-CAM 105. The late appearance of cell-CAM 105 suggests that this molecule is important for functions specific to the differentiated state.

However, even if our observations suggest that cell-CAM 105 may play other roles in addition to cell–cell binding, a common denominator in its functions could be membrane–membrane interactions occurring on the cell surface. In stratified epithelia these interactions can occur between adjacent cells. This might also be the case in brush border–containing simple epithelial cells, but in addition, cell-CAM 105 could play a role in mediating interactions between adjacent microvilli of the apical surfaces of these cells (Figure 10). Such a cross-linking of microvilli, which might be dynamic and changing according to the functional stage of the cells, could be of importance in creating an extracellular compartment consisting of narrow channels, which might have a role in ion and fluid secretion, for example. Such a mechanism for the promotion of isosmotic secretion has been suggested in the proximal part of the Malpighian tubules of the insect *Rhodnius prolixus* (Bradley, 1983). Specific molecular interactions between the microvilli could also participate in the regulation of microvillar and cellular motility.

The apical location of cell-CAM 105 in some vessel endothelia might also be important for the organization of the cell surface in these cells. Cell-CAM 105 is not expressed in all kinds of vessel endothelia, but occurs primarily in capillaries and in arterioles, not in large arteries or veins (Odin, 1987; Odin et al., 1988). A functional difference between these vessels is that the arterioles and capillaries can reduce their lumina considerably by active contraction, which is not a feature of the larger vessels. When an arteriole or a capillary contracts, it would thus be appropriate, for hemodynamic reasons, if the apical luminal surfaces of the endothelial cells became organized in some way, for example, by a regular pleating. Membrane–membrane interactions in such hypothetical pleats might be mediated by cell-CAM 105. Another role for cell-CAM 105 on the apical endothelial surfaces might be to trap platelet aggregates by binding to cell-CAM 105 that becomes exposed on the surface of aggregates during aggregation. Binding of activated platelets to endothelial surfaces, especially at the endothelial cell–cell borders, has been observed *in vitro* (Solberg et al., 1985).

The intracellular location of cell-CAM 105 in platelets and granulocytes may at first sight seem incompatible with a role in cell surface membrane–membrane interactions. However, the intracellular location of cell-CAM 105 may serve as a storage pool that can be released at the cell surface when the cells are activated to participate in various cellular interactions. That this might be the case is indicated by our observations that the surface of aggregating platelets becomes strongly cell-CAM positive.

We have not been able to inhibit platelet aggregation with antibodies against cell-CAM 105. Thus, cell-CAM 105 does not seem to participate in the prima⁗y processes of platelet aggregation, but may well be involved in secondary platelet–platelet interactions within the aggregates. It may, for example, play a role in clot retraction. As mentioned above, it is also possible that cell-CAM 105 that becomes exposed on the surface of platelet aggregates participates in the binding of platelet aggregates to the endothelial surfaces of capillaries and small arteries and veins.

In conclusion, our data on cell-CAM 105 indicate that this molecule is multifunctional. However, a common denominator in its functions might be membrane–membrane binding at the cell surface. In all cell types and in all situations where cell-CAM 105 has been observed, it seems to appear on the cell surface after the establishment of primary cell–cell associations. This is the case both in solid tissues and in aggregating platelets.

When cell-CAM-mediated membrane–membrane interactions occur within a single type of cell, for example, intestinal epithelial cells, specialized cell surface structures such as domains of specifically organized microvilli may appear. In such a situation cell-CAM 105 will contribute to the final stage of morphogenesis mentioned at the beginning of this chapter. Cell-CAM-mediated membrane interactions between adjacent cells may also influence other properties that are characteristic of terminally differentiated cells. One possibility is that cell-CAM interactions contribute to the establishment of the sophisticated intercellular architecture characteristic of mature tissues by fine-tuning the cell–cell interactions during perinatal development. It is striking that in both perinatal and regenerating liver, cell-CAM 105 appears on all faces of the hepatocyte surfaces during the period when the hepatocytes reorganize and become more closely associated with each other. Thus, it might be possible that interactions mediated by cell-CAM 105 contribute to the achievement of the precise cellular architecture characteristic of mature liver.

Another property of mature cells to which cell-CAM-mediated interactions might contribute is their stationary status. A down-regulation or a change in the molecular properties of cell-CAM 105 might thus lead to an increased invasive potential of the cells. There is indeed an inverse correlation between the degree of expression of cell-CAM 105 and the invasive potential of trophoblast cells (Svalander et al., 1988), hepatoma cells (Hixson et al., 1985), and fetal and regenerating hepatocytes (Odin and Öbrink, 1986). In the context of blastocyst implantation, we have recently observed that cell-CAM 105 is down-regulated not only on the trophoblast cells but also on the lining epithelium of the uterine mucosa at the time of implantation (Svalander et al., 1989). Thus, it might be possible that a down-regulation of cell-CAM 105 both in the trophectoderm and in the uterine epithelium might facilitate the invasion of the trophoblasts into the endometrium.

Although our experimental observations are compatible with the suggested functions of cell-CAM 105, it should be stressed that at present we can only speculate about these things. Much more work, at both the molecular and cellular levels, is needed in order to elucidate the biological functions of this molecule.

CONCLUSIONS

In this chapter we have summarized current knowledge about the structure, prevalence, and function of cell-CAM 105. The available data indicate that cell-CAM 105 appears primarily in cells that either are terminally differentiated or are undergoing terminal differentiation. This is in contrast to several other CAMs, such as N-CAM, L-CAM, and the cadherins, which become expressed very early in embryonic development. CAMs that are expressed early have been termed primary CAMs, whereas CAMs appearing at later stages and having more restricted tissue distributions have been named secondary CAMs (Edelman, 1985). By this nomenclature, cell-CAM 105 belongs to the group of secondary CAMs.

As demonstrated in the preceding sections, cell-CAM 105 seems to become expressed at a stage when the cells expressing it already have formed contacts by other mechanisms. This suggests that cell-CAM 105 may not be a primary driving force for the creation of specific intercellular patterns during early embryogenesis and morphogenesis. Rather, its functions are associated with the final stages of morphogenesis and/or the regulation of cellular functions that are characteristic of mature, differentiated cells.
that are characteristic of mature, differentiated cells.

In more general terms, this means that the molecules known as cell adhesion molecules constitute a heterogeneous group of components. The common characteristics of these molecules are that they, at least under experimental conditions, are involved in various cell surface recognition phenomena. However, the cellular responses to these recognition events may vary considerably. Whereas the major functions for some CAMs may be to mediate stable cell–cell bonds, others may work by influencing cellular behavior and biosynthetic and metabolic functions in various ways. The adhesion mediated by CAMs *in vitro* may thus be an *in vitro* epiphenomenon reflecting diverse cell surface recognition reactions.

ACKNOWLEDGMENTS

This work was supported by grants from the Swedish Medical Research Council (project no. 05200), the Swedish Cancer Foundation (project no. 1389), Konung Gustaf V's 80-årsfond, and Magn. Bergvalls Stiftelse.

REFERENCES

Alden, R. H. (1948) Implantation of the rat egg. III. Origin and development of primary trophoblast giant cells. *Am. J. Anat.* **83**:143–182.

Bloom, W., and D. W. Fawcett (1975) *A Textbook of Histology*, 10th Ed., W. B. Saunders, Philadelphia.

Bock, E., C. Richter-Landsberg, A. Faissner, and M. Schachner (1985) Demonstration of immunochemical identity between the nerve growth factor–inducible large external (NILE) glycoprotein and the cell adhesion molecule L1. *EMBO J.* **4**:2765–2768.

Bradley, T. J. (1983) Functional design of microvilli in the Malpighian tubules of the insect *Rhodnius prolixus. J. Cell Sci.* **60**:117–135.

Chavez, D. J. (1986) Cell surface of mouse blastocysts at the trophectoderm–uterine interface during the adhesive stage of implantation. *Am. J. Anat.* **176**:153–158.

Chuong, C.-M., K. L. Crossin, and G. M. Edelman (1987) Sequential expression and differential function of multiple adhesion molecules during the formation of cerebellar cortical layers. *J. Cell Biol.* **104**:331–342.

Edelman, G. M. (1985) Expression of cell adhesion molecules during embryogenesis and regeneration. *Exp. Cell Res.* **161**:1–16.

Edelman, G. M. (1986) Cell adhesion molecules in the regulation of animal form and tissue pattern. *Annu. Rev. Cell Biol.* **2**:81–116.

Friedlander, D. R., M. Grumet, and G. M. Edelman (1986) Nerve growth factor enhances expression of neuron–glia cell adhesion molecule in PC12 cells. *J. Cell Biol.* **102**:413–419.

Gallin, W. J., G. M. Edelman, and B. A. Cunningham (1983) Characterization of L-CAM, a major cell adhesion molecule from embryonic liver cells. *Proc. Natl. Acad. Sci. USA* **80**:1038–1042.

Grumet, M., and G. M. Edelman (1984) Heterotypic binding between neuronal membrane vesicles and glial cells is mediated by a specific cell adhesion molecule. *J. Cell Biol.* **98**:1746–1756.

Hansson, M., I. Blikstad, and B. Öbrink (1989) Cell-surface location and molecular properties of cell-CAM 105 in intestinal epithelial cells. *Exp. Cell Res.* **181**:63–74.

Hixson, D. C., K. D. McEntire, and B. Öbrink (1985) Alterations in the expression of a hepatocyte cell adhesion molecule by transplantable rat hepatocellular carcinomas. *Cancer Res.* **45**:3742–3749.

Kovacs, L., and K. Lapis (1984) Cell proliferation in the liver. In *Regulation and Control of Cell Proliferation*, K. Lapis and A. Jeney, eds., pp. 229–249, Akademiai Kiado, Budapest.

Öbrink, B. (1986a) Epithelial cell adhesion molecules. *Exp. Cell Res.* **163**:1–21.

Öbrink, B. (1986b) Cell adhesion molecules. *Front. Matrix Biol.* **11**:123–138.

Öbrink, B., P. Odin, A. Tingström, M. Hansson, K. Rubin, and I. Blikstad (1986) Cell adhesion molecules involved in cell–cell adhesion phenomena. Structure and function of cell-CAM 105. In *Biology and Pathology of Platelet-Vessel Wall Interactions*, G. Jolles, Y. Legrand, and A. T. Nurden, eds., pp. 161–178, Academic, London.

Ocklind, C., and B. Öbrink (1982) Intercellular adhesion of rat hepatocytes: Identification of a cell surface glycoprotein involved in the initial adhesion process. *J. Biol. Chem.* **257**:6788–6795.

Odin, P. (1987) Chemical, immunochemical, and immunohistochemical characterization of the cell adhesion molecule cell-CAM 105. Acta Universitatis Upsaliensis. Comprehensive Summaries of Uppsala Dissertations from the Faculty of Medicine.

Odin, P., and B. Öbrink (1986) Dynamic expression of the cell adhesion molecule cell-CAM 105 in fetal and regenerating rat liver. *Exp. Cell Res.* **164**:103–114.

Odin, P., and B. Öbrink (1987) Quantitative determination of the organ distribution of the cell adhesion molecule cell-CAM 105 by radioimmunoassay. *Exp. Cell Res.* **171**:1–15.

Odin, P., and B. Öbrink (1988) The cell surface expression of the cell adhesion molecule cell-CAM 105 in rat fetal tissues and regenerating rat liver *Exp. Cell Res.* **179**:89–103.

Odin, P. A. Tingström, and B. Öbrink (1986) Chemical characterization of cell-CAM 105, a cell-adhesion molecule isolated from rat liver membranes. *Biochem. J.* **236**:559–568.

Odin, P., M. Asplund, C. Busch, and B. Öbrink (1988) Immunocytochemical localization of cell-CAM 105 in rat tissues: Appearance in epithelia, platelets, and granulocytes. *J. Histochem. Cytochem.* **36**:729–739.

Ogawa, K., A. Medline, and E. Farber (1979) Sequential analysis of hepatic carcinogenesis: The comparative architecture of preneoplastic, malignant, prenatal, postnatal, and regenerating liver. *Br. J. Cancer* **40**:782–790.

Rathjen, F. G., and M. Schachner (1984) Immunocytochemical and biochemical characterization of a new neuronal cell surface component (L1 antigen) which is involved in cell adhesion. *EMBO J.* **3**:1–10.

Solberg, S., T. Larsen, and L. Jorgensen (1985) Differences in reactivity of confluent and nonconfluent cultures of human endothelial cells toward thrombin-stimulated platelets or heparinized salt solution. *In Vitro Cell Dev. Biol.* **21**:612–616.

Svalander, P., P. Odin, B. O. Nilsson, and B. Öbrink (1988) Trophectoderm surface expression of the cell adhesion molecule cell-CAM 105 on rat blastocysts. *Development* **100**:652–660.

Svalander, P., P. Odin, B. O. Nilsson, and B. Öbrink (1989) Expression of cell-CAM 105 in the optical surface of rat uterine epithelium is controlled by ovarian steroid hormones. (submitted.)

Takeichi, M., Y. Shirayoshi, K. Hatta, and A. Nose (1986) Cadherins: Their morphogenetic role in animal development. *Prog. Clin. Biol. Res.* **217**:17–27.

Thiery, J. P., R. Brackenbury, U. Rutishauser, and G. M. Edelman (1977) Adhesion among neural cells of the chick embryo. II. Purification and characterization of a cell adhesion molecule from neural retina. *J. Biol. Chem.* **252**:6841–6845.

Chapter 29

Adhesion Systems in Morphogenesis and Cell Migration During Avian Embryogenesis

JEAN PAUL THIERY
JEAN-LOUP DUBAND
SYLVIE DUFOUR

ABSTRACT

In this chapter we analyze different modes of adhesion during early avian embryogenesis, and use somitogenesis to illustrate the mechanisms involved in epithelium–mesenchyme interconversion. In this system, extracellular matrix molecules, including fibronectins, laminin, collagens, cytotactin or tenascin, and their receptors, are expressed during certain stages of somitogenesis, but no direct role in the process of assembly and remodeling of the somites can be established at the present time. In contrast, the cell adhesion molecules N-CAM and N-cadherin are found to act as ligands between somitic cells. In vitro and in vivo, there is a good correlation between the spatiotemporal distribution of N-cadherin and N-CAM and the critical phases of epithelialization and subsequent reorganization of the somites. In addition, antibodies to N-cadherin and to a lesser extent to N-CAM are able to disrupt already established structures. The neural crest is also discussed since, in this model system, the conversion from neural epithelium to mesenchymal cells is followed by extensive migration before clustering at defined sites.

There is an excellent correlation between the different periods of development of the neural crest and modes of adhesion. First, the acquisition of the mesenchymal state is accompanied by the loss of L-CAM, then N-cadherin, and finally N-CAM; second, migratory mesenchymal cells express a high specificity for fibronectins; and third, the cell adhesion molecules N-CAM and N-cadherin are re-expressed by cells regrouping into aggregates according to a schedule characteristic of each type of ganglion. The role of fibronectins is further demonstrated by a variety of experiments, including perturbations using specific antibodies and peptides derived from cell-binding sites. Cell surface modulation of fibronectin receptors is clearly associated with the mechanisms that control the transition from the motile to the stationary state. This surface modulation is accompanied by a drastic reorganization of cytoskeletal elements directly implicated in cell locomotion and focal adhesion. Hemopoietic cells at the origin of T-cells are analyzed to demonstrate that in a quite different migratory system mediated by chemotaxis, the same substrate adhesion systems control critical phases during invasion of the thymus. These experiments should lead to a deeper molecular understanding of the control mechanisms implicated in the activation of specific modes of adhesion during morphogenesis and histogenesis.

With continuing advances in studies of the various processes of cell adhesion, some of the mechanisms operating during morphogenesis and histogenesis can now be approached. In this chapter we analyze the contributions of several modes of adhesion to the control of cell migration and pattern formation in the avian embryo. There is now a large body of data indicating that adhesion molecules can be classified with respect to their primary functions as either controlling intercellular adhesion or mediating interactions with the extracellular matrix. However, each of these molecules may carry multiple binding sites, allowing involvement in both cell–cell and cell–extracellular matrix interactions.

Another important concept stems from the discovery that several different types of adhesion molecules are expressed very early in development. In vertebrates a blastodermal cell may carry as many as six different adhesion systems (Figure 1). The cell–cell adhesion systems detected include N-CAM (neural cell adhesion molecule) and L-CAM (liver cell adhesion molecule), which have been designated as primary cell adhesion molecules (CAMs; see Cunningham and Edelman, this volume). N-cadherin (Hatta and Takeichi, 1986), very similar if not identical to A-CAM (adherens junction–specific cell adhesion molecule; Volk and Geiger, 1986a,b), is also found at an early stage, although it is expressed slightly later than N-CAM and L-CAM. There are at least three different families of receptors for the extracellular matrix, which are also expressed at that time. Fibronectin receptors are the best characterized so far, and their expression and functions are fairly well documented (Hynes, 1987; see also Humphries and Yamada, Horwitz et al., Hynes et al., and Ruoslahti et al., this volume). Since laminin and collagens are associated with cells of the blastoderm, one must consider that their corresponding receptors are also expressed.

Developmental studies of these adhesion systems have shown that CAMs are modulated during epithelium–mesenchyme interconversion. This phenomenon is particularly obvious at the onset of the transformation of epithelia into dispersing mesenchymes, when CAMs are down-regulated and the fibronectin adhesion system becomes predominant (Thiery et al., 1982a,b, 1984; Duband et al., 1986, 1988a; Hatta et al., 1987). The new cell assemblies that arise after tissue remodeling or at the cessation of migration express a subset of the primary CAMs as well as other CAMs designated as secondary CAMs. For instance, in the nervous system, N-CAM, N-cadherin (or A-CAM) as well as Ng-CAM (neuron–glia cell adhesion molecule) are found at the surface of neurons and on some glia (Duband et al., 1985, 1988a; Thiery et al., 1985; Hatta et al., 1987). Conversely, L-CAM becomes predominant in newly formed epithelial tissues and in those derived from the ectoderm and the endoderm (Thiery et al., 1984). It should be noted, however, that N-CAM and N-cadherin (or A-CAM) are also expressed transiently in restricted territories of epithelial tissues derived from the three primary germ layers (Thiery et al., 1982a; Chuong and Edelman, 1985a,b; Hatta et al., 1987; Richardson et al., 1987; Duband et al., 1988a). Other surface-associated molecules, such as those

ADHESION SYSTEMS

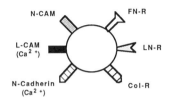

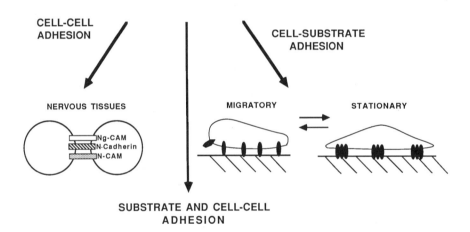

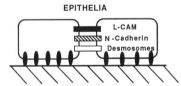

Figure 1. *Embryonic cells are able to express many adhesive molecules that play key roles in embryonic processes.* Some of them, fibronectins, laminin, collagens, and their cell surface receptors (FN-R, LN-R, and Col-R), called SAMs, are involved in cell–substrate adhesion. Other adhesive molecules such as N-CAM, N-cadherin, Ng-CAM, L-CAM, and desmosome components, called CAMs, are involved in cell–cell adhesion. Differential expression at the cell surface of these adhesive molecules controls cell behavior such as cell migration, cellular assembly into epithelia, and the formation and topological organization of the nervous system.

confined to tight junctions (Stevenson et al., 1986) and to desmosomes (see Garrod et al., and Kapprell et al., this volume) are detected in these epithelia. Extracellular matrix receptors are also maintained in epithelial cells, which express high amounts of laminin, collagens, and fibronectin receptors (Chen et al., 1985a; Koda et al., 1985; Rapraeger et al., 1986; Rubin et al., 1986; Dehdar et al., 1987).

The coordinated regulation of the expression and functions of adhesion molecules has come to be of major importance for a better understanding of the mechanisms implicated in morphogenesis. In this chapter we describe three model systems that should help to clarify the basic principles underlying this formidable problem.

SOMITOGENESIS

In vertebrates the axial mesoderm formed during gastrulation progressively organizes into metameric structures, the somites. This morphogenetic process involves several phases during which a loose mesenchymal structure is transformed into an epithelium, the latter subsequently undergoing a profound remodeling. Immunohistological analyses using antibodies to CAMs have shown that both N-CAM and N-cadherin are expressed at the surface of mesenchymal cells of the axial mesoderm as soon as it is formed (Duband et al., 1987a, 1988a; Hatta et al., 1987). Levels of N-CAM and N-cadherin progressively increase during the compaction and epithelialization of the mesenchyme. However, whereas N-CAM is distributed quite homogeneously at the surface of somitic cells, N-cadherin molecules become enriched in the junctional area of the lateral surfaces proximal to the lumen of the somites (Figure 2). L-CAM is never detected at any stage of somitogenesis, in contrast to the situation found in the kidney, another epithelial derivative of the mesoderm (Thiery et al., 1984).

Substratum adhesion molecules (SAMs) are also present during somitogenesis. Laminin is expressed rather late, at the basal surface of the newly formed somites. Fibronectins are barely detectable in the loose mesenchyme but progressively accumulate at the basal surface of the developing epithelium (Duband et al., 1987a).

These observations suggest that both N-CAM and N-cadherin may be directly involved in the different phases of somitogenesis. Single cells were prepared from both segmental plates and epithelial somites using either high-trypsin calcium (TC) or low-trypsin EDTA (LTE) treatment in order to maintain the calcium-dependent and calcium-independent adhesion mechanisms, respectively. A subsequent short-term quantitative aggregation assay in the presence of subsequent antibodies clearly showed that N-CAM is involved in the aggregation of LTE-treated cells, where N-cadherin mediates the adhesion of TC-treated cells (Table 1). No significant differences in the rate of aggregation could be detected between the two mechanisms of adhesion.

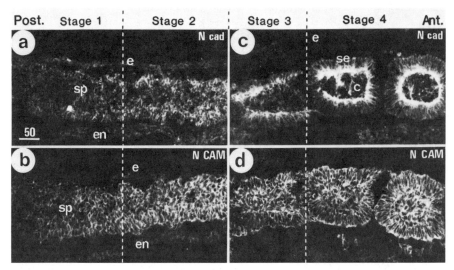

Figure 2. *Distribution of adhesion molecules during somite formation in a chicken embryo at Hamburger and Hamilton stage 15 studied by* in situ *immunofluorescence. a and b show the segmental plate and c and d show nascent somites and the last somites of the same embryo. a and c and b and d show two consecutive sagittal sections stained for N-cadherin and N-CAM, respectively. Four different stages can be distinguished in the distribution of the adhesion molecules and organization of the tissues; these stages are separated by dotted lines on the pictures. While stage 1 is characterized by low levels of the two cell adhesion molecules, stage 2 is characterized by increases in N-cadherin and N-CAM concomitantly with an increase in cell–cell adhesion. At stage 3 the nascent somite starts separating from the rest of the segmental plate and N-cadherin becomes polarized at the apical surface of the epithelial cells. Somite formation is complete by stage 4. Post. and ant. indicate the caudal and cranial sides of the sections; c, core (of the somite); e, ectoderm; en, endoderm; se, somitic epithelium; sp, segmental plate. Calibration bar = 50 μm.*

Another interesting feature of somites is their rapid remodeling into dermamyotomes and sclerotomes. Indeed, somites partially disorganize at their medioventral surfaces, and cells from this region, together with core cells that have not acquired an epithelial structure, all contribute to the formation of the sclerotome. By an as yet unknown process, the dorsal part of the somite becomes a double-layered epithelium, of which the more dorsal layer soon dissociates to form the dermis while the lower layer forms the myotome. N-CAM and N-cadherin are found to be maintained in these structures, particularly in the dermamyotome and later in the myotome. In contrast, N-cadherin (or A-CAM) disappears from the medioventral portion of the somites before the appearance of the sclerotome, while N-CAM remains present on the sclerotomal cells (Duband et al., 1987a, 1988).

In vitro, somites spontaneously dissociate on fibronectin-coated substrata. As observed *in vivo* during the early stage of remodeling, N-cadherin disappears from the surface of newly emigrating cells, while N-CAM levels

Table 1. Specificity of Aggregation of Segmental Plate and Somitic Cells[a]

Condition of Dissociation	Condition of Aggregation	Segmental Plate Cells % Aggregation	Somite Cells % Aggregation
Trypsin Ca^{++}	$-Ca^{++} - Ab$	7 ± 5	21 ± 2
	$-Ca^{++} +$ anti-N-cadherin	13 ± 8	6 ± 5
	$-Ca^{++} +$ anti-N-CAM	20 ± 3	16 ± 4
	$+Ca^{++} - Ab$	92 ± 2	73 ± 7
Trypsin Ca^{++}	$+Ca^{++} +$ anti-N-cadherin	54 ± 3	2 ± 2
	$+Ca^{++} +$ anti-N-CAM	84 ± 2	50 ± 1
	$-Ca^{++} - Ab$	72 ± 1	76 ± 2
Light trypsin EDTA	$-Ca^{++} +$ anti-N-cadherin	73 ± 2	73 ± 4
	$-Ca^{++} +$ anti-N-CAM	46 ± 5	59 ± 5
	$+Ca^{++} - Ab$	83 ± 1	61 ± 4
Light trypsin EDTA	$+Ca^{++} +$ anti-N-cadherin	65 ± 3	61 ± 3
	$+Ca^{++} +$ anti-N-CAM	47 ± 2	47 ± 2

[a]Cells were dissociated in the presence of either 0.001% trypsin and 1 mM EDTA (light trypsin EDTA) or 0.01% trypsin and 1 mM $CaCl_2$ (trypsin Ca^{++}) in order to prepare cells that retain either calcium-independent or calcium-dependent adhesion systems. Cells were then allowed to reaggregate in wells containing PBS with calcium ($+Ca^{++}$) or without calcium ($-Ca^{++}$) and in the presence of no antibodies ($-Ab$), of antibodies to N-cadherin ($+$ anti-N-cadherin), or of antibodies to N-CAM ($+$ anti-N-CAM). The total number of particles in each well was counted. The degree of aggregation of cells was expressed as the percentage of aggregation: ($1 -$ number of particles after the aggregation assay/initial particle number) $\times$ 100. The inhibition of cell aggregation by antibodies was expressed as the percentage of inhibition: ($1 - \%$ aggregation with antibodies/% aggregation without antibodies) $\times$ 100.

decrease very progressively. Cells maintained for three to four days in culture form new intercellular contacts at which N-cadherin but not N-CAM becomes concentrated (Duband et al., 1987a). It is not yet known how closely the *in vitro* dissociation of somitic cells and their subsequent reassociation mimic some of the processes observed *in vivo*; however, the very striking modulation of the expression of N-cadherin is reminiscent of the formation of sclerotomal cells. Since, *in vitro*, the explants are deprived of the inductive signals from the neural tube and the notochord (Teillet and Le Douarin, 1983), they fail to differentiate into the three components derived from the somites, and somitic cells may therefore reaggregate using N-cadherin.

The respective roles of the different adhesion systems have been analyzed *in vitro* on explants of segmental plates and on individual somites (Duband et al., 1987a). Monovalent antibodies to fibronectins and to laminin do not modify the structure of the explant over at least 15 hours. In contrast, monovalent antibodies and monoclonal antibodies to N-CAM and monoclonal antibodies to N-cadherin show a clear dissociating effect (Figures 3, 4). The most obvious effect is observed on segmental plates, which are readily dissociated within a few hours by anti-N-cadherin antibodies. Antibodies to N-CAM have a less pronounced effect on segmental plates, and do not disrupt somites (Figure 4). Consistent with the activity of N-cadherin antibodies, the removal of calcium in the culture medium is very efficient in disrupting the segmental plates.

These experiments suggest that while segmental plate cells and somitic cells share several adhesive mechanisms, N-cadherin may predominate in some key steps of somitogenesis. However, the possibility that other adhesion molecules contribute synergistically or independently to the process of reorganization of the axial mesoderm cannot be excluded. There are several

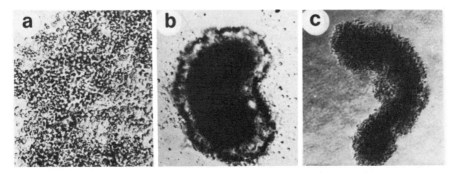

Figure 3. *Effect of antibodies to adhesion molecules on explants of segmental plates.* Explants were incubated in the presence of monoclonal antibodies to N-cadherin (*a*), control monoclonal Ig (*b*) (both at 50 μg/ml), and monovalent antibodies to N-CAM (*c*) at 1 mg/ml. The dissociating effect of antibodies to N-cadherin can be clearly seen within 3 hours and is complete within 15 hours (*a*). Control antibodies do not alter the cohesion of cells within the explant (*b*). Anti-N-CAM antibodies only induce the dissociation of cells located in the periphery of the explant (*c*).

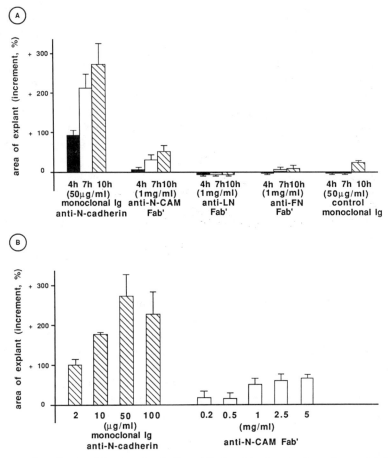

Figure 4. *Quantification of the effect of antibodies to adhesion molecules on explants of segmental plates as a function of time (A) and as a function of the concentration of the antibodies after 10 hours of incubation (B).* Explants were incubated in the presence of monoclonal antibodies to N-cadherin at concentrations ranging from 2 to 100 μg/ml, control monoclonal Ig at 50 μg/ml, monovalent antibodies to N-CAM at concentrations between 0.2 and 5 mg/ml, and monovalent antibodies to FN and LN at 1 mg/ml. The areas of the explants were recorded at various times, and the results are expressed as the increase or decrease of these areas with respect to the initial value.

indications that this may be the case: On the one hand, cellular fibronectins and the cell-binding site peptides can induce premature condensation, of the most rostral part of segmental plates into somites (Lash et al., 1984; Lash and Yamada, 1986); on the other hand, at least *in vitro*, N-CAM mediates the calcium-independent adhesion process. Other adhesion molecules, such as those associated with desmosomes, may participate in the construction of the epithelial somites. Finally, cytotactin or tenascin may be involved in some aspects of the remodeling of somites, and in particular in the formation of the sclerotome (Crossin et al., 1986; Mackie et al., 1988).

THE NEURAL CREST

In vertebrates the neural crest is a transient embryonic structure at the origin of multiple tissues including most of the peripheral nervous system, melanocytes, and craniofacial structures (for a review, see Le Douarin, 1982). Neural crest cells can be identified with certainty only at the stage of emergence from the dorsal border of the neural tube. In avian embryos most of the migration pathways have been established (Duband and Thiery, 1982; Thiery et al., 1982b; Le Douarin et al., 1984a; Vincent and Thiery, 1984; Rickmann et al., 1985; Bronner-Fraser, 1986; Thiery and Duband, 1986; Teillet et al., 1987). Neural crest cells are often encountered as a dense population that mingles with the extracellular matrix, forming transient pathways between epithelial structures. These pathways are specific to each axial level and contain primarily glycosaminoglycans (Derby, 1978), collagen type I (Duband and Thiery, 1987), cytotactin or tenascin (Crossin et al., 1986; Tan et al., 1987; Mackie et al., 1988), and fibronectins (Newgreen and Thiery, 1980; Duband and Thiery, 1982; Thiery et al., 1982b; Krotoski et al., 1986). Type IV collagen and laminin are found only as basal lamina components delimiting transient epithelial structures such as the ectoderm, the neural tube, and the somites (Sternberg and Kimber, 1986; Duband and Thiery, 1987).

Immunohistological analyses have shown that L-CAM disappears from the surface of neural epithelial cells in a ventrodorsal gradient, while N-CAM becomes progressively more abundant (Thiery et al., 1984; Crossin et al., 1986). A similar result was obtained by *in situ* hybridization using N-CAM cDNA probes in *Xenopus* neural plate: An N-CAM message appeared in a ventrodorsal gradient (Kintner and Melton, 1987). Thus, neural crest cells individualizing at the boundary between the presumptive epidermis and the neural tube have already lost L-CAM at the time of their emigration.

Early migrating crest cells rapidly become devoid of N-cadherin (or A-CAM), while N-CAM diminishes more progressively from their surfaces (Figure 5), (Duband et al., 1985, 1988a; Hatta et al., 1987). Neural crest cells cultured *in vitro* bind preferentially to fibronectins or to fibronectin-containing extracellular matrices deposited by a variety of mesenchymal cells (Newgreen et al., 1982; Rovasio et al., 1983). In contrast, neural crest cells do not bind to collagen type I (Rovasio et al., 1983), glycosaminoglycans (Newgreen et al., 1982; Tucker and Erickson, 1984), or cytotactin (Tan et al., 1987; Mackie et al., 1988); attachment of neural crest cells to laminin is significant, but their spreading on this substratum is very poor (Rovasio et al., 1983). In prolonged cultures, however, neural crest cells become as adhesive to laminin as to fibronectins (Rovasio et al., 1983). These cells may have undergone differentiation into glialike cells that are known to secrete and interact with laminin (Liesi et al., 1983).

Neural crest cells dissociated by either the LTE or TC procedure are able to aggregate in a short-term assay (Aoyama et al., 1985). Calcium-independent adhesion can be inhibited by anti-N-CAM antibodies. It remains to be

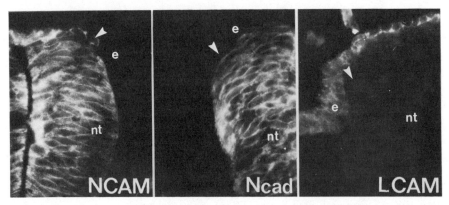

Figure 5. *Distribution of adhesion molecules during truncal neural crest cell emigration studied by im-munofluorescence for N-CAM (left), N-cadherin (center), and L-CAM (right).* Prior to their emigration, neural crest cells that express N-CAM and N-cadherin but not L-CAM are integrated into the neural epithelium. Conversely, the ectoderm is stained for L-CAM but not for N-CAM and N-cadherin. At the onset of their migration, crest cells (*arrowheads*) lose N-cadherin but not N-CAM; the latter will disappear from their surface later during cell migration. e, ectoderm; nt, neural tube.

established whether the calcium-dependent mechanism is mediated by N-cadherin, although the *de novo* expression of this molecule *in vitro* and perturbation of aggregation by antibodies to N-cadherin are consistent with such a role (Duband et al., in preparation). *In vivo,* neural crest cells give rise to sensory ganglia after accumulating between the neural tube and the derma-myotome, above the sclerotome. These ganglion rudiments express N-cadherin first and then N-CAM at a relatively late stage, and the first postmitotic sensory neurons start to synthesize Ng-CAM, which accumulates mostly at the surface of neurites (Thiery et al., 1982a, 1985; Duband et al., 1985, 1988a; Hatta et al., 1987). Neural crest cells also give rise to autonomic ganglia and to aortic and enteric plexuses. Cells reaching these sites express N-cadherin (or A-CAM) and N-CAM before forming clusters (Duband et al., 1985, 1988a; Hatta et al., 1987). These observations could suggest that the timing of CAM-gene expression is modulated according to the local environment and to the topological constraints put upon neural crest cells.

The data accumulated so far are summarized in Figure 6. As was pointed out, there are differences in the timing or reexpression of N-CAM and N-cadherin (or A-CAM) in the autonomic and sensory precursors. These observations suggest that CAMs may play different roles in different ganglia. The relative ability to bind laminin and fibronectins has not been precisely determined at all stages of the differentiation of neural crest cells into glial, sensory, or sympathetic neurons. These precursors produce laminin and fibronectins; in addition, they continue to express fibronectin receptors, at least as detected by antibodies to the β-subunit (Duband et al., 1986).

EXPRESSION OF ADHESION MOLECULES

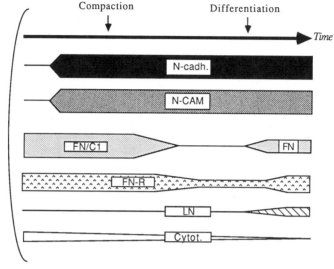

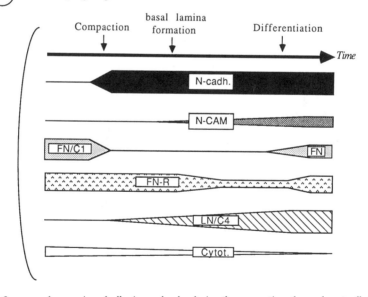

Figure 6. *Sequence of expression of adhesion molecules during the aggregation of neural crest cells into autonomic (a) and sensory (b) ganglia.* Expression of N-cadherin (N-cadh.) and N-CAM, and the reorganization of fibronectins (FN), type I collagen (C1), laminin (LN), type IV collagen (C4), cytotactin (Cytot.), and the fibronectin receptor complex (FN-R) are indicated in relation to the main morphological events occurring during cell aggregation. The thicknesses of individual lines indicate the approximate relative amounts of each molecule.

In vitro, attempts were made to analyze further the mechanisms involved in neural crest cell motility and in the acquisition of the stationary state (Tucker et al., 1985; Duband et al., 1986, 1988b,c). Migrating neural crest cells have a poorly organized cytoskeletal network. Microfilaments are found only in the cortex, whereas vinculin and talin are detected almost exclusively throughout the cytoplasm. Fibronectin receptors remain diffuse at the surface of migrating crest cells (Figure 7). Focal contacts could be detected only at the tips of filopodia, using either interference–reflection microscopy or immunofluorescence labeling of fibronectin receptors and vinculin. Interestingly, talin is not accumulated in these focal contacts. When cells are dislodged, rosette structures called podosomes (Chen et al., 1985b; Tarone et al., 1985) remain attached to the substratum. The same analyses performed on transformed cells result in similar observations; these cells have a disorganized cytoskeleton and a diffuse distribution of fibronectin receptors (Chen et al., 1986).

In contrast, when crest cells reach the stationary state, focal contacts become detectable; cells spread more on the substratum and their cytoskeleton is fully organized. A more pronounced effect can be observed during the transition from mobile to stationary states of somitic cells (Figure 7). In these cells well-developed focal contacts form in areas at which accumulations of talin and vinculin are detected, at the tips of the microfilament bundles. In addition, a significant percentage of the fibronectin receptors are clustered at these adhesive plaques (Figure 8).

The fluorescence recovery after the photobleaching technique was applied to neural crest cells as well as to other embryonic cells (Duband et al., 1988b). Rhodamine-labeled monoclonal antibodies to fibronectin receptors were applied to the cultures. Measurements of the rate of fluorescence recovery in bleached areas were made in different locations on the cell surface, mostly on the basal side. Approximately 70–90% of the fibronectin receptors were mobile in migratory cells, as opposed to less than 20% in stationary cells. In both cases, the diffusion coefficient of the mobile receptors is found to be in the range of those measured for any transmembrane glycoproteins ($1-2 \times 10^{-10}$ cm^2/sec). These studies strongly suggest that most receptors localized in focal adhesion plaques are anchored to cytoskeletal elements internally and to fibronectins externally. Indeed, fibronectin receptors have been shown to interact directly with talin, which in turn may interact with other cytoskeletal proteins, including vinculin (see Horwitz et al., this volume). It is interesting to note that while migratory neural crest cells cultured on silicon rubber can locomote on this substratum without any deformation, stationary cells are firmly anchored and induce wrinkles in the silicon-rubber sheet (Tucker et al., 1985).

We may hypothesize that, in contrast to the moderate affinity expressed by individual receptors for fibronectins (see Humphries and Yamada, this volume), clustered immobile receptors may express a much higher affinity for the substratum. Interestingly, attachment of cells to substrate-linked Arg-Gly-

**MIGRATING
NEURAL CREST CELLS**

**STATIONARY
SOMITIC CELLS**

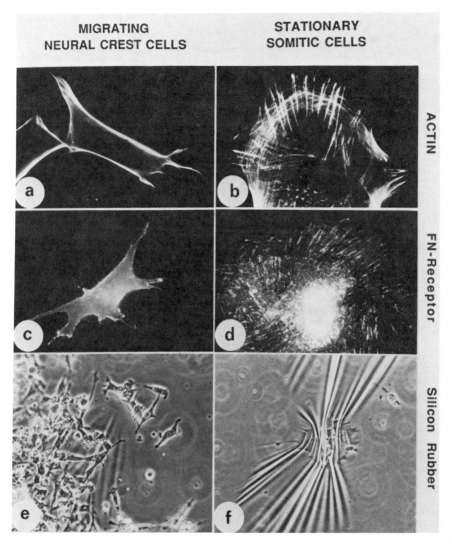

ACTIN

FN-Receptor

Silicon Rubber

Figure 7. *Properties of neural crest cells (a,c,e) compared to nonmotile somitic cells (b,d,f). a,b:* Localization of actin. In neural crest cells actin is mostly distributed in the cellular cortex, while in somitic cells it is highly organized into complex bundles. *c,d:* Localization of fibronectin receptors. The receptor is diffusely organized in the membrane of neural crest cells, in contrast to somitic cells, in which it is concentrated in the cell–substratum adhesion sites. *e,f:* Behavior on silicon rubber. Somitic cells induce wrinkles in the silicon rubber, while neural crest cells do not. *e* and *f* are by courtesy of R. P. Tucker and C. A. Erickson.

609

MIGRATORY CELL

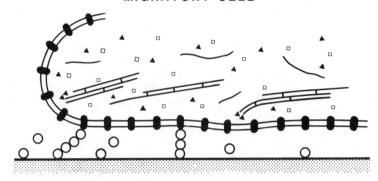

STATIONARY CELL

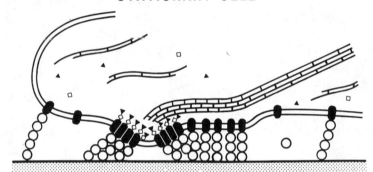

Figure 8. *Speculative diagrams comparing possible modes of cell–substratum adhesion of migratory and stationary cells. Top:* Migratory cells are characterized by poorly organized microfilaments and numerous FN receptors arranged diffusely in the membrane. The receptors are mobile within the plane of the membrane, and a number of them are not bound to FN. Neither vinculin nor talin is associated with the receptors or the cytoskeleton. *Bottom:* In stationary cells FN receptors are concentrated in clusters close to focal contact sites and microfilament bundles. They are linked to the cytoskeleton through talin and vinculin. These receptors have low membrane mobility and most are firmly anchored to FN fibers. *Open circles,* fibronectin; *filled ovals,* FN receptor; *diamonds,* talin; *triangles,* vinculin; *long lines,* actin; *cross-link between lines,* α-actinin.

Asp-Ser-containing peptides induces them to form early focal contact sites at which fibronectin receptors accumulate (Singer et al., 1987). Alternatively, the transduction of the adhesive signal to the cytoskeleton may be more efficient through clustered receptors, resulting in the development of internal tension, which in turn induces deformation of the substratum.

To evaluate further this hypothesis, neural crest cells were allowed to emigrate from neural tube explants over antibodies to the fibronectin receptor coated onto the substratum (Duband et al., 1986). The first neural crest cells reaching the substratum spread extensively and form a limited halo around the neural tube; time-lapse video cinematography revealed that these cells become rapidly paralyzed and assemble into an epitheliallike structure. These experiments suggest that an increase in the affinity for the substratum will prevent translocation. The simplest explanation may be that the binding of the fibronectin receptors to antibodies is almost irreversible, in contrast to the binding to fibronectins, during which receptors can detach and bind to other fibronectin molecules during filopodial activity. Such labile binding between fibronectin receptors and fibronectins does not result from phosphorylation of tyrosine residues of the receptors (Duband et al., 1988c), but instead from rapid internalization and possibly recycling of these receptors (Duband et al., 1988d). In addition, local proteolysis mediated in part by surface metallo-proteases may contribute to the destabilization of the binding to fibronectins that are highly sensitive to proteases, as described in transformed cells (Chen et al., 1984; Chen and Chen, 1987).

Early migrating neural crest cells express a very specific binding for pure fibronectins or for fibronectin-containing substrata. Antibodies to fibronectins and to fibronectin receptors inhibit the attachment and spreading of neural crest cells on fibronectin-coated substrata. These antibodies also arrest moving cells and induce rounding up (Rovasio et al., 1983; Duband et al., 1986). Similar observations were made with synthetic peptides containing the Arg-Gly-Asp-Ser adhesive recognition signal of fibronectins (Figure 9; Boucaut et al., 1984). These experiments indicate that migrating neural crest cells interact with fibronectin molecules through an Arg-Gly-Asp-Ser-dependent mechanism.

An important discovery has been that, while short fragments of fibronectins containing the Arg-Gly-Asp-Ser sequence retain some adhesive functions, their estimated affinity for the receptor is decreased by one or two orders of magnitude if they are below a certain size. Using wild type and mutated fusion proteins of different sizes in the Arg-Gly-Asp-Ser region, it was possible to map a new adhesion site located approximately 200 amino acids toward the amino terminus from the Arg-Gly-Asp-Ser sequence (Obara et al., 1988). This site, in synergy with the Arg-Gly-Asp-Ser site, provides full affinity for fibronectin receptors. Using the same approach, it was found that a fusion protein containing Arg-Gly-Asp-Ser but not extending to this synergistic site does not permit attachment of neural crest cells, while a longer fragment containing the two sites allows neural crest cells to adhere and spread efficiently and also to migrate, but only moderately (Dufour et al., 1988). These findings indicate that neural crest cell motility on fibronectins requires direct interaction with other adhesion sites on fibronectin molecules.

In a series of studies using variants of the adhesive signal peptides, it has been found that melanomas, highly metastatic tumors derived from neural crest cells, do not attach to fibronectins through the Arg-Gly-Asp-Ser sequence but use another binding site instead; this site contains the Arg-Glu-

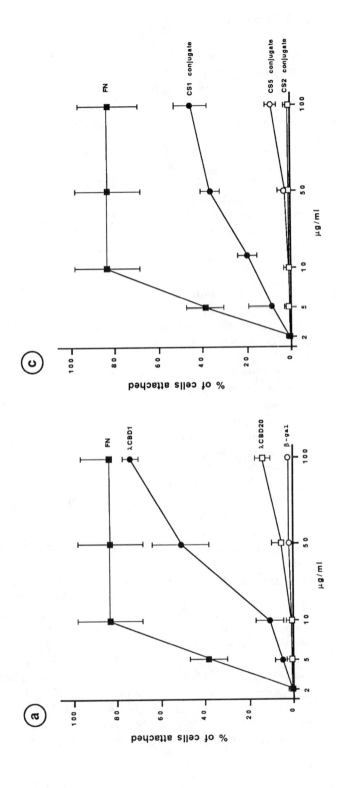

612

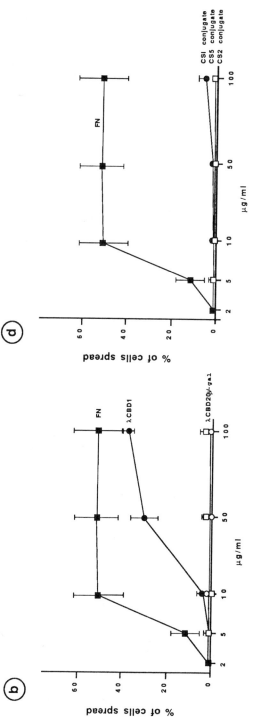

Figure 9. *Attachment (a,c) and spreading (b,d) of neural crest cells on substrates composed of pure fibronectins, β-galactosidase (β-gal), fusion proteins covering either the Arg-Gly-Asp-Ser binding site plus the synergistic adhesion site (λCBD1) or the Arg-Gly-Asp-Ser site alone (λCBD20), or on various peptides in the IIICS region conjugated to IgGs (CS1, CS2, and CS5 conjugates, respectively).* Only fusion proteins covering both the Arg-Gly-Asp-Ser and the synergistic sites and CS1 conjugates can support attachment of neural crest cells. However, CS1 conjugates do not allow cell spreading.

Asp-Val sequence and is located in a particular region of fibronectin molecules called the type III homology–connecting segment (IIICS; Humphries et al., 1986). Further investigation of different peptides covering the entire IIICS region revealed that another 25-amino-acid peptide located in IIICS (termed CS1) contains a high-affinity binding site for melanoma cells (Humphries et al., 1987; see Humphries and Yamada, this volume). Together these two sequences could mediate in an additive manner the spreading of melanoma cells.

Similar studies using neural crest cells indicate that these cells can also recognize the IIICS region (Figure 9; Dufour et al., 1988). First, the binding and spreading of crest cells can be partially inhibited in the presence of Gly-Arg-Gly-*Glu*-Ser peptides; these peptides specifically inhibit the IIICS adhesive sequences. Second, IgG-conjugated CS1 peptides or fusion proteins containing this site were found to promote attachment and little migration of these cells, but not their spreading. The motility of neural crest cells was further studied in the presence of peptides and/or antibodies known to interfere either with the Arg-Gly-Asp-Ser-containing site or with the IIICS adhesive sequences. It was found that both the Arg-Gly-Asp-Ser domain and the CS1 sequence are required in association, each with functional specificity, to permit effective locomotion of neural crest cells (Dufour et al., 1988). We conclude from these studies that, in agreement with the important findings on new adhesive sites (Donaldson et al., 1985; Humphries et al., 1986, 1987; McCarthy et al., 1986; Obara et al., 1988; Furcht et al., personal communication), neural crest cells can interact with different regions of fibronectin molecules and respond by expressing different cellular behaviors (Figure 10).

As detected by immunofluorescence labeling using polyclonal antibodies, fibronectins are localized at many different sites in the embryo. At the time of neural crest cell migration, fibronectins are as abundant in areas not invaded by the crest cells as they are in the normal pathways (Thiery et al., 1982b). Thus far there has been no satisfactory explanation for the pathfinding mechanisms. One intriguing possibility is that different types of fibronectins may be synthesized and deposited at critical periods and at specific sites during embryogenesis. Since the IIICS domain is spliced with high frequency, it is conceivable that the absence of the CS1 sequence in fibronectins located in potential pathways may prevent crest cells from migrating in these areas. Programmed fluctuations in the splicing of the IIICS region may therefore play a crucial regulatory role in pathfinding mechanisms. However, using *in situ* hybridization techniques, it has not yet been possible to detect any specific sites of mRNA production of the different variants (ffrench-Constant and Hynes, 1988). Interestingly, our preliminary studies reveal that a fibronectin variant that contains an alternatively spliced type III homology called extradomain I (EDI) is not uniformly distributed in extraembryonic spaces in the early avian embryo and is particularly abundant along the migratory routes of some neural crest subpopulations (Duband et al., in preparation).

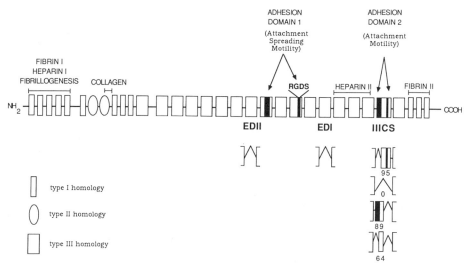

Figure 10. *Primary structure of fibronectins.* Only one chain of the dimer is shown. The three types of internal repeats, called homologies, and the various functional domains are indicated. The two major adhesion domains mapped so far have been called domains 1 and 2 and are represented by *dark boxes*. These domains are composed of separate sites that function either in synergy (domain 1) or in an additive manner (domain 2). The function of each domain in neural crest cell attachment, spreading, and motility is indicated. Variation of fibronectins is generated by alternative splicing occurring in three different regions of fibronectin pre-messengers, EDI, EDII, and IIICS. EDI and EDII can be spliced out entirely (together with the adjacent introns), or can remain unspliced. In humans, five variants of the IIICS region can be generated by differential splicing of this domain. The numbers under each variant indicate the length of these variants in amino acids.

COLONIZATION OF THE EMBRYONIC THYMUS BY HEMOPOIETIC PRECURSORS

In the thymus of birds and mammals, there is now ample evidence for an extrinsic origin of precursors for T-lymphocytes, dendritic cells, and macrophages (Le Douarin et al., 1984b). Early in embryogenesis, precursor cells circulate in blood vessels from which they extravasate in the vicinity of the thymic organ.

In birds these precursors are attracted by the thymus at defined periods. *In vitro* studies have shown that a subset of bone marrow cells is attracted by chemotactic peptides secreted by the thymic epithelium (Ben Slimane et al., 1983; Champion et al., 1986). The chemotactic mechanism involves the transient expression of specific receptors at the surface of hemopoietic precursor cells. Once stimulated, the responding cells adhere to the endothelia of the jugular vein and small blood vessels that are located in the vicinity of the thymus. Hemopoietic cells are also found in the mesenchyme and adhering to the basement membrane of the epithelial thymus (Figure 11; Savagner et al., 1986).

Soon after the onset of the attractive period, hemopoietic cells penetrate into the thymus and become inserted between epithelial cells. Immuno-fluorescence labeling of the sites traversed by hemopoietic cells has shown that laminin is restricted to the basement membrane of the thymus. Fibronectins are associated with the basement membrane and are also randomly distributed in the mesenchyme (Savagner et al., 1986). All hemopoietic cells, with the exception of mature erythrocytes, express fibronectin receptors at their surfaces.

The human amniotic basement membrane was used as a model to study the ability of a subset of these cells to invade a basement membrane containing fibronectins and laminin. Some bone marrow cells deposited in the upper well of the Boyden chamber containing the amniotic basement membrane are able to traverse the basement membrane when the lower compartment contains a partially purified chemotactic peptide solution. Antibodies to the β-subunit of the fibronectin receptor, antibodies to fibronectins, and peptides containing the Arg-Gly-Asp-Ser sequence all prevent cells from crossing the amniotic basement membrane. Anti-laminin antibodies have similar effects. These results indicate that the adhesion of hemopoietic precursors to extracellular matrix components is one of the steps required for the attachment and the subsequent invasion of the basement membrane of the thymus (Figure 12). It is interesting to note that the invasive mechanism is induced by the

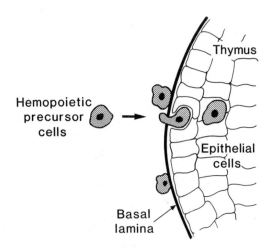

Figure 11. *Schematic representation of the attachment and invasion by hemopoietic cells of the embryonic thymus. The basal lamina contains laminin and is associated with a dense meshwork of FN. Hemopoietic cells interact transiently with the basement membrane before inserting themselves between the thymic epithelial cells.*

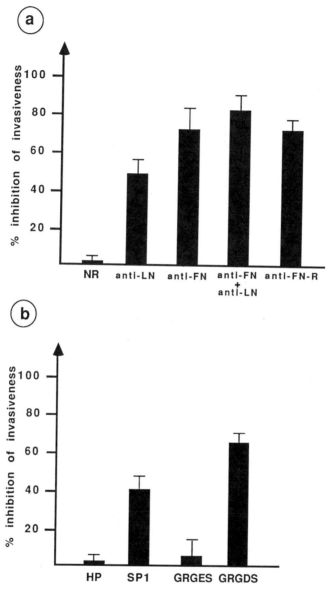

Figure 12. *Inhibition of* in vitro *invasiveness by antibodies and synthetic peptides. a:* Monospecific polyclonal antibodies against laminin (0.1 mg/ml) or fibronectins (0.1 mg/ml) or fibronectin receptors (0.1 mg/ml) and control rabbit antibodies (0.1 mg/ml) were preincubated with amniotic basement membrane alone for 20 hours before loading cells into the Boyden chambers. Cell migration was allowed to proceed in the presence of bound antibodies for a further 20 hours in the chambers. *b:* Cells were incubated with 2 mg/ml of each peptide for 20 hours in invasion chambers. As a control, the hexapeptide Leu-Trp-Met-Arg-Phe-Ala, with an amino acid sequence unrelated to fibronectin, was used. After migration, cells that crossed the basement membrane were fixed and counted. Inhibition is calculated in all cases as % of inhibition = 100 × (1 − the ratio between cell number found in the presence of antibodies or peptides and cell number found in the absence of antibodies or peptides).

chemotactic peptides, possibly as a result of induction of the locomotory machinery of cells as well as of the secretion of proteolytic activities. A similar mechanism has already been proposed for polymorphonuclear leukocytes (Russo et al., 1981).

CONCLUSIONS

The adhesive mechanisms displayed by embryonic cells are highly regulated in both time and space. These adhesive mechanisms contribute to the establishment of cell collectives, such as those forming stable epithelia. In these tissues, junctions defined at the ultrastructural level including desmosomes, tight junctions, and adherens junctions are now receiving much attention, in the hope of defining the supramolecular assemblies directly involved in the intercellular adhesion process. However, these junctions have not been extensively described and localized in early embryogenesis. It is therefore difficult to assess their specific roles in morphogenetic processes, particularly in transient epithelia.

In contrast, CAMs and SAMs have been clearly implicated in all the morphogenetic and histogenetic processes analyzed thus far. In this short review the respective roles of CAMs and SAMs have been analyzed in epithelial–mesenchymal cell interconversion and in migratory processes of a quite different nature. Other systems, including compaction and implantation in mammals (see Kemler et al. and Damsky et al., this volume), epiboly and gastrulation in amphibians, fishes, and birds (Duband et al., 1987b), migration of primordial germ cells in amphibians (Heasman et al., 1981) and mammals (Alvarez-Buylla and Merchand-Larios, 1986), histogenesis of feathers, skin, and lungs in birds (Chuong and Edelman, 1985a,b; Gallin et al., 1986; Hirai et al., 1989a,b) and of the central nervous system in amphibians (Fraser et al., 1984), in birds (Rathjen et al., 1987), and in mammals (Chuong et al., 1987; see also Cunningham and Edelman, and Schachner et al., this volume), the auditory systems in birds (Richardson et al., 1987) have all provided evidence for the independent or conjugated roles of CAMs and SAMs.

Much work is being performed to analyze systems such as the hemopoietic tissues. Indeed, there is ample evidence for specific interactions between different classes of T-cells and between lymphoid and myeloid cells (Springer et al., 1987). In addition, both lymphoid and myeloid cells are also known to interact with the stroma of hemopoietic tissues (Witte, personal communication) and with endothelial cells (Dustin et al., 1986; Gallatin et al., 1986; Streeter et al., 1988; Bevilacqua et al., 1989; Siegelman et al., 1989).

It is now becoming clear that adhesion molecules can be grouped into a limited number of families. This is already the case for N-CAM (Edelman, 1987), Ng-CAM/L1 (Moos et al., 1988), myelin-associated glycoprotein (Arquint et al., 1987; Salzer et al., 1987), I-CAM (Springer et al., 1987), and fasciclin II (Harrelson and Goodman, 1988), which belong to the superfamily

MODES OF ADHESION OF AVIAN
NEURAL CREST CELLS

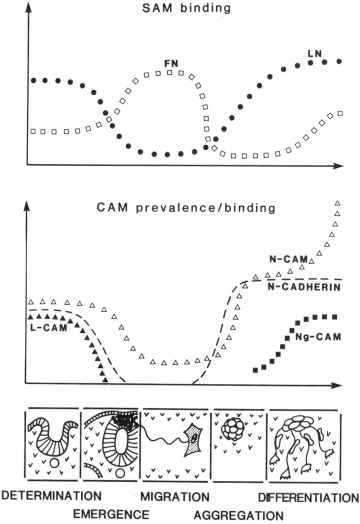

Figure 13. *Correlation between adhesive properties of crest cells, expression of the cell adhesion molecules, and the presence of fibronectins and laminin.* Five periods were chosen, corresponding to the main changes in the behavior of this population: determination, emergence, migration, aggregation, and differentiation into ganglia. The loss of L-CAM occurs during the stage of determination, while N-cadherin disappears from neural crest cells at the time of their emigration. During migration, N-CAM levels decrease, while adhesion to fibronectins in the extracellular environment increases to a plateau. At the same time, laminin disappears from the extracellular environment of crest cells. These expression levels are reversed during aggregation: the binding to fibronectins is much reduced; these molecules disappear from the ganglion rudiment, in contrast to laminin, which reappears. N-CAM and N-cadherin are re-expressed during the aggregation of neural crest cells. N-CAM is likely to increase at the surface of differentiating neurons and may contribute to the early formation of neurite fascicles in association with Ng-CAM.

of immunoglobulin genes (see Cunningham and Edelman, this volume), and for receptors recognizing the Arg-Gly-Asp adhesive signals of extra-cellular matrix proteins (see Ruoslahti et al., this volume). The calcium-dependent adhesion molecules N-cadherin (A-CAM), E-cadherin, L-CAM, and P-cadherin share sequence identities, notably in the amino- and carboxy-terminal regions, and have been regrouped in a family called cadherins (Takeichi, 1988).

It is also conceivable that members of the SAM superfamily also act as CAMs and vice versa. For instance, LFA-1 expressed by hemopoietic cells is a heterodimer related to the fibronectin receptor that also binds to I-CAM on target cells (Springer et al., 1987). Similarly, one of the N-CAM isoforms can be secreted into the extracellular milieu and may act as a SAM (see Sanes et al., this volume).

One of the major issues to be addressed (as revealed by neural crest cells throughout their natural history) is the possible coregulation of these two modes of adhesion (Figure 13). Indeed, we have found that the inhibition of neural crest cell motility either by RGDS peptides or by local removal of the fibronectin substrate induces A-CAM-mediated cell aggregation (Duband et al., in preparation). Also of particular interest is the search for natural inducers or repressors of the functional expression of these two modes of adhesion. NGF, FGF, and TGF-β may be candidates for such a role: On the one hand, NGF is the only known inducer of Ng-CAM or NILE protein (Friedlander et al., 1986; see Greene and Shelanski, this volume); on the other hand, FGF and TGF-β have together been shown to be a primary inducer in amphibians (Kimelman and Kirchner, 1987; Slack et al., 1987; Rosa et al., 1988), and TGF-β itself is known to increase the level of integrins and extracellular matrix molecules (Ignotz and Massagué, 1987). These growth factors may exert pleiotropic effects during early embryogenesis through their inductive capacity to commit cells to new adhesive and differentiation properties. Finally, scatter factors may also be important in the control of epithelial to mesenchymal cell interconversion (Stoker et al., 1987).

ACKNOWLEDGMENTS

We would like to thank our colleagues Drs. Gerald Edelman, Bruce Cunningham, Masatoshi Takeichi, Benjamin Geiger, and Kenneth Yamada for their fruitful collaborations, and Dr. Chris Henderson for critical reading of the manuscript. This work was supported by grants from the Centre National de la Recherche Scientifique, the Institut National de la Santé et de la Recherche Médicale, the Association pour la Recherche sur le Cancer, the Ligue Nationale Française contre le Cancer, the Fondation pour la Recherche sur la Myopathie, and the March of Dimes/Birth Defects.

REFERENCES

Alvarez-Buylla, A., and H. Merchant-Larios (1986) Mouse primordial germ cells use fibronectin as a substrate for migration. *Exp. Cell Res.* **165**:362–368.

Aoyama, H., A. Delouvée, and J. P. Thiery (1985) Cell adhesion mechanisms in gangliogenesis studied in avian embryo and in a model system. *Cell Differ.* **17**:247–260.

Arquint, M., J. Roder, L.-S. Chia, J. Down, D. Wilkinson, H. Bayley, P. Braun, and R. Dunn (1987) Molecular cloning and primary structure of myelin-associated glycoprotein. *Proc. Natl. Acad. Sci. USA* **84**:600–604.

Ben Slimane, S., F. Houllier, G. C. Tucker, and J. P. Thiery (1983) *In vitro* migration of avian hemopoietic cells to the thymus: Preliminary characterization of a chemotactic mechanism. *Cell Differ.* **13**:1–24.

Bevilacqua, M. P., S. Stengelin, M. A. Gimbrone, and B. Seed, Jr. (1989) Endothelial leukocyte adhesion molecule 1: An inducible receptor for neutrophils related to complement regulatory proteins and lectins. *Science* **243**:1160–1165.

Bronner-Fraser, M. (1986) Analysis of the early stages of trunk neural crest migration in avian embryos using monoclonal antibody HNK-1. *Dev. Biol.* **115**:44–55.

Champion, S., B. Imhof, P. Savagner, and J. P. Thiery (1986) The embryonic thymus produces chemotactic peptides involved in the homing of hemopoietic precursors. *Cell* **44**:781–790.

Chen, J.-M., and W.-T. Chen (1987) Fibronectin-degrading proteases from the membranes of transformed cells. *Cell* **48**:193–203.

Chen, W.-T., K. Olden, B. A. Bernard, and F. F. Chu (1984) Expression of transformation-associated protease(s) that degrade fibronectin at cell contact sites. *J. Cell Biol.* **98**:1546–1555.

Chen, W.-T., J. M. Greve, D. I. Gottlieb, and S. J. Singer (1985a) Immunocytological localization of 140-kD cell adhesion molecules in cultured chicken fibroblasts, and in chicken smooth muscle and intestinal epithelial tissues. *J. Histochem. Cytochem.* **33**:576–586.

Chen, W.-T., J.-M. Chen, S. J. Parsons, and J. T. Parsons (1985b) Local degradation of fibronectin at sites of expression of the transforming gene product pp60[src]. *Nature* **316**:156–158.

Chen, W.-T., J. Wang, T. Hasegawa, S. S. Yamada, and K. M. Yamada (1986) Regulation of fibronectin receptor distribution by transformation, exogenous fibronectin, and synthetic peptides. *J. Cell Biol.* **103**:1649–1661.

Chuong, C.-M., and G. M. Edelman (1985a) Expression of cell adhesion molecules in embryonic induction. I. Morphogenesis of nestling feathers. *J. Cell Biol.* **101**:1009–1026.

Chuong, C.-M., and G. M. Edelman (1985b) Expression of cell adhesion molecules in embryonic induction. II. Morphogenesis of adult feathers. *J. Cell Biol.* **101**:1027–1043.

Chuong, C.-M., K. L. Crossin, and G. M. Edelman (1987) Sequential expression and differential function of multiple adhesion molecules during the formation of cerebellar cortical layers. *J. Cell Biol.* **104**:331–342.

Crossin, K. L., S. Hoffman, M. Grumet, J. P. Thiery, and G. M. Edelman (1986) Site-restricted expression of cytotactin during development of the chicken embryo. *J. Cell Biol.* **102**:1917–1930.

Dedhar, S., E. Ruoslahti, and M. D. Pierschbacher (1987) A cell surface receptor complex for collagen type I recognizes the Arg-Gly-Asp sequence. *J. Cell Biol.* **104**:585–593.

Derby, M. A. (1978) Analysis of glycosaminoglycans within the extracellular environment encountered by migrating neural crest cells. *Dev. Biol.* **66**:321–336.

Donaldson, D. J. P., J. I. Mahan, D. L. Hasty, J. B. McCarthy, and L. T. Furcht (1985) Location of a fibronectin domain involved in newt epidermal cell migration. *J. Cell Biol.* **101**:73–78.

Duband, J.-L., and J. P. Thiery (1982) Distribution of fibronectin in the early phase of avian cephalic neural crest cell migration. *Dev. Biol.* **93**:308–323.

Duband, J.-L., and J. P. Thiery (1987) Distribution of laminin and collagens during avian neural crest development. *Development* **101**:461–478.

Duband, J.-L., G. C. Tucker, T. J. Poole, M. Vincent, H. Aoyama, and J. P. Thiery (1985) How do the migratory and adhesive properties of the neural crest govern ganglia formation in the avian peripheral nervous system? *J. Cell. Biochem.* **27**:189–203.

Duband, J.-L., S. Rocher, W.-T. Chen, K. M. Yamada, and J. P. Thiery (1986) Cell adhesion and migration in the early vertebrate embryo: Location and possible role of the putative fibronectin–receptor complex. *J. Cell Biol.* **102**:160–178.

Duband, J.-L., S. Dufour, K. Hatta, M. Takeichi, G. M. Edelman, and J. P. Thiery (1987a) Adhesion molecules during somitogenesis in the avian embryo. *J. Cell Biol.* **104**:1361–1374.

Duband, J.-L., T. Darribère, J.-C. Boucaut, H. Boulekbache, and J. P. Thiery (1987b) Regulation of development by the extracellular matrix. In *Cell Membranes: Methods and Reviews,* Vol. 3, E. L. Elson, W. A. Frazier, and L. Glaser, eds., pp. 1–53, Plenum, New York.

Duband, J.-L., T. Volberg, I. Sabanay, J. P. Thiery, and B. Geiger (1988a) Spatial and temporal distribution of the' adherens junction–specific adhesion molecule A-CAM during avian embryogenesis. *Development* **103**:325–344.

Duband, J.-L., G. H. Nickolls, A. Ishihara, T. Hasegawa, K. M. Yamada, J. P. Thiery, and K. Jacobson (1988b) Fibronectin receptor exhibits high lateral mobility in embryonic locomoting cells but is immobile in focal contacts and fibrillar streaks in stationary cells. *J. Cell Biol.* **107**:1385–1396.

Duband, J.-L., S. Dufour, K. M. Yamada, and J. P. Thiery (1988c) The migratory behavior of avian embryonic cells does not require phosphorylation of the fibronectin–receptor complex. *FEBS Lett.* **230**:181–185.

Duband, J.-L., S. Dufour, and J. P. Thiery (1988d) Extracellular matrix–cytoskeleton interactions in locomoting embryonic cells. *Protoplasma* **145**:112–119.

Dufour, S., J.-L. Duband, M. Humphries, M. Obara, K. M. Yamada, and J. P. Thiery (1988) Attachment, spreading, and motility of avian neural crest cells are mediated by multiple adhesion sites on fibronectin molecules. *EMBO J.* **7**:2661–2671.

Dustin, M. L., R. Rothlein, A. K. Bhan, C. A. Dinarello, and T. A. Springer (1986) A natural adherence molecule (ICAM-1): Induction by IL 1 and interferon-gamma, tissue distribution, biochemistry, and function. *J. Immunol.* **137**:245–254.

Edelman, G. M. (1987) CAMs and Igs: Cell adhesion and the evolutionary origins of immunity. *Immunol. Rev.* **1**:11–45.

Fraser, S. E., B. A. Murray, C.-M. Chuong, and G. M. Edelman (1984) Alteration of the retinotectal map in *Xenopus* by antibodies to neural cell adhesion molecules. *Proc. Natl. Acad. Sci. USA* **81**:4222–4226.

ffrench-Constant, C., and R. O. Hynes (1988) Patterns of fibronectin gene expression and splicing during cell migration in chicken embryos. *Development* **104**:369–382.

Friedlander, D. R., M. Grumet, and G. M. Edelman (1986) Nerve growth factor enhances expression of neuron–glia cell adhesion molecule in PC12 cells. *J. Cell Biol.* **102**:413–419.

Gallatin, W. M., T. P. S. John, M. Siegelman, R. Reichert, E. C. Butcher, and I. L. Weissman (1986) Lymphocyte homing receptors. *Cell* **44**:673–680.

Gallin, W. J., C.-M. Chuong, L. H. Finkel, and G. M. Edelman (1986) Antibodies to liver cell adhesion molecule perturb inductive interactions and alter feather pattern and structure. *Proc. Natl. Acad. Sci. USA* **83**:8325–8329.

Harrelson, A. L., and C. S. Goodman (1988) Growth cone guidance in insects: Fasciclin II is a member of the immunoglobulin superfamily. *Science* **242**:700–707.

Hatta, K., and M. Takeichi (1986) Expression of N-cadherin adhesion molecules associated with early morphogenetic events in chick development. *Nature* **320**:447–449.

Hatta, K., S. Takagi, H. Fujisawa, and M. Takeichi (1987) Spatial and temporal expression pattern of N-cadherin cell adhesion molecules correlated with morphogenetic processes of chicken embryos. *Dev. Biol.* **120**:215–227.

Heasman, J., R. O. Hynes, M. Swan, V. Thomas, and C. C. Wylie (1981) Primordial germ cells of Xenopus embryos: The role of fibronectin in their adhesion during migration. *Cell* **27**:437–447.

Hirai, Y., A. Nose, S. Kobayashi, and M. Takeichi (1989a) Expression and role of E- and P-cadherin adhesion molecules in embryonic histogenesis. I. Lung epithelial morphogenesis. *Development* **105**:263–270.

Hirai, Y., A. Nose, S. Kobayashi, and M. Takeichi (1989b) Expression and role of E- and P-cadherin adhesion molecules in embryonic histogenesis. II. Skin morphogenesis. *Development* **105**:271–277.

Humphries, M. J., S. K. Akiyama, A. Komoriya, K. Olden, and K. M. Yamada (1986) Identification of an alternatively-spliced site in human plasma fibronectin that mediates cell type–specific adhesion. *J. Cell Biol.* **103**:2637–2647.

Humphries, M. J., A. Komoriya, S. K. Akiyama, K. Olden, and K. M. Yamada (1987) Identification of two distinct regions of the type III connecting segment of human plasma fibronectin that promote cell type-specific adhesion. *J. Biol. Chem.* **262**:6886–6892.

Hynes, R. O. (1987) Integrins: A family of cell surface receptors. *Cell* **48**:549–554.

Ignotz, R. A., and J. Massagué (1987) Cell adhesion protein receptors as target for transforming growth factor-β action. *Cell* **51**:189–197.

Kimelman, D., and M. Kirschner (1987) Synergistic induction of mesoderm by FGF and TGF-β and the identification of an mRNA coding for FGF in the early Xenopus embryo. *Cell* **51**:869–877.

Kintner, C. R., and D. A. Melton (1987) Expression of Xenopus N-CAM RNA in ectoderm is an early response to neural induction. *Development* **99**:311–325.

Koda, J. E., A. Rapraeger, and M. Bernfield (1985) Heparan-sulfate proteoglycans from mouse mammary epithelial cells: Cell-surface proteoglycan as a receptor for interstitial collagens. *J. Biol. Chem.* **260**:8157–8162.

Krotoski, D. M., C. Domingo, and M. Bronner-Fraser (1986) Distribution of a putative cell surface receptor for fibronectin and laminin in the avian embryo. *J. Cell Biol.* **103**:1061–1071.

Lash, J. W., and K. M. Yamada (1986) The adhesion recognition signal of fibronectin: A possible trigger mechanism for compaction during somitogenesis. In *Somites in Developing Embryos*, R. Bellairs, D. A. Ede, and J. W. Lash, eds., pp. 201–208, Plenum, New York.

Lash, J. W., A. W. Seitz, C. M. Cheney, and D. Ostrovsky (1984) On the role of fibronectin during the compaction stage of somitogenesis in the chick embryo. *J. Exp. Zool.* **232**:197–206.

Le Douarin, N. M., P. Cochard, M. Vincent, J.-L. Duband, G. C. Tucker, M.-A. Teillet, and J. P. Thiery (1984a) Nuclear, cytoplasmic, and membrane markers to follow neural crest cell migration: A comparative study. In *The Role of Extracellular Matrix in Development*, R. L. Treslad, ed., pp. 373–398, Alan R. Liss, New York.

Le Douarin, N. M., F. Dieterlen-Lièvre, and P. D. Oliver (1984b) Ontogeny of primary lymphoid organs and lymphoid stem cells. *Am. J. Anat.* **170**:261–299.

Liesi, P., D. Dahl, and A. Vaheri (1983) Laminin is produced by early rat astrocytes in primary rat culture. *J. Cell Biol.* **96**:920–924.

Mackie, E. J., R. P. Tucker, W. Halfter, R. Chiquet-Ehrismann, and H. H. Epperlein (1988) The distribution of tenascin coincides with pathways of neural crest cell migration. *Development* **102**:237–250.

McCarthy, J. B., S. T. Hagen, and L. T. Furcht (1986) Human fibronectin contains distinct adhesion- and motility-promoting domains for metastatic melanoma cells. *J. Cell Biol.* **102**:179–188.

Moos, M., R. Tacke, H. Scherer, D. Teplow, K. Früh, and M. Schachner (1988) Neural adhesion molecule L1 as a member of the immunoglobulin superfamily with binding domains similar to fibronectins. *Nature* **334**:701–703.

Newgreen, D. F., and J. P. Thiery (1980) Fibronectin in early avian embryos: Synthesis and distribution along the migration pathways of neural crest cells. *Cell Tissue Res.* **211**:269–291.

Newgreen, D. F., I. L. Gibbins, J. Sauter, B. Wallenfels, and R. Wütz (1982) Ultrastructural and tissue-culture studies on the role of fibronectin, collagen and glycosaminoglycans in the migration of neural crest cells in the fowl embryo. *Cell Tissue Res.* **221**:521–549.

Rapraeger, A., M. Jalkanen, and M. Bernfield (1986) Cell surface proteoglycan associates with the cytoskeleton at the basolateral cell surface of mouse mammary epithelial cells. *J. Cell Biol.* **103**:2683–2696.

Rathjen, F. G., J. M. Wolff, R. Frank, F. Bonhoeffer, and U. Rutishauser (1987) Membrane glycoproteins involved in neurite fasciculation. *J. Cell Biol.* **104**:343–353.

Richardson, G., K. L. Crossin, C.-M. Chuong, and G. M. Edelman (1987) Expression of cell adhesion molecules during embryonic induction. III. Development of the otic placode. *Dev. Biol.* **119**:217–230.

Rickmann, M., J. W. Fawcett, and R. J. Keynes (1985) The migration of neural crest cells and the growth of motor axons through the rostral half of the chick somite. *J. Embryol. Exp. Morphol.* **90**:437–455.

Rosa, F., A. Robets, D. Danielpour, L. Dart, M. Sporn, and I. David (1988) Mesoderm induction in amphibians: The role of TGF-β2-like factors. *Science* **239**:783–785.

Rovasio, R. A., A. Delouvée, K. M. Yamada, R. Timpl, and J. P. Thiery (1983) Neural crest cell migration: Requirement for exogenous fibronectin and high cell density. *J. Cell Biol.* **96**:462–473.

Rubin, K., D. Gullberg, T. K. Borg, and B. Öbrink (1986) Hepatocyte adhesion to collagen: Isolation of membrane glycoproteins involved in adhesion to collagen. *Exp. Cell Res.* **164**:127–138.

Russo, R. G., L. A. Liotta, U. Thorgeirsson, R. Brundage, and E. Schiffmann (1981) Polymorphonuclear leukocyte migration through human amnion membrane. *J. Cell Biol.* **91**:459–467.

Salzer, J. L., W. P. Holmes, and D. R. Colman (1987) The amino-acid sequences of the myelin-associated glycoproteins: Homology to the immunoglobulin gene superfamily. *J. Cell Biol.* **104**:957–965.

Savagner, P., B. A. Imhof, K. M. Yamada, and J. P. Thiery (1986) Homing of hemopoietic precursor cells to the embryonic thymus: Characterization of an invasive mechanism induced by chemotactic peptides. *J. Cell Biol.* **103**:2715–2727.

Siegelman, M. H., M. van de Rijn, and I. L. Weissman (1989) Mouse lymph node homing receptor cDNA clone encodes a glycoprotein revealing tandem interaction domains. *Science* **243**:1165–1172.

Singer, I. I., D. W. Kawka, S. Scott, R. A. Mumford, and M. W. Lark (1987) The fibronectin cell attachment sequence Arg-Gly-Asp-Ser promotes focal contact formation during early fibroblast attachment and spreading. *J. Cell Biol.* **104**:573–584.

Slack, J. M. W., B. G. Darlington, J. K. Heath, and S. F. Godsave (1987) Mesoderm induction in early *Xenopus* embryos by heparin-binding growth factors. *Nature* **326**:197–200.

Springer, T. A., M. L. Dustin, T. K. Kishimoto, and S. D. Marlin (1987) The lymphocyte function-associated LFA-1, CD2, and LFA-3 molecules: Cell adhesion receptors of the immune system. *Annu. Rev. Immunol.* **5**:223–252.

Stevenson, B. R., J. D. Siliciano, M. S. Mooseker, and D. A. Goodenough (1986) Identification of ZO-1: A high molecular weight polypeptide associated with the tight junction (zonula occludens) in a variety of epithelia. *J. Cell Biol.* **103**:755–766.

Stoker, M., E. Gherardi, M. Perryman, and J. Gray (1987) Scatter factor is a fibroblast-derived modulator of epithelial cell mobility. *Nature* **327**:239–242.

Streeter, P. R., B. T. N. Rouse, and E. C. Butcher (1988) Immunohistologic and functional characterization of a vascular addressin involved in lymphocyte homing into peripheral lymph nodes. *J. Cell Biol.* **107**:1853–1862.

Takeichi, M. (1988) The cadherins: Cell–cell adhesion molecules controlling animal morphogenesis. *Development* **102**:639–655.

Tarone, G., D. Cirillo, P. G. Giancotti, P. M. Comoglio, and P.-C. Marchisio (1985) Rous sarcoma virus-transformed fibroblasts adhere primarily at discrete protrusions of the ventral membrane called podosomes. *Exp. Cell Res.* **159**:141–157.

Teillet, M.-A., and N. M. Le Douarin (1983) Consequences of neural tube and notochord excision on the development of the peripheral nervous system in the chick embryo. *Dev. Biol.* **98**:192–211.

Teillet, M.-A., C. Kalcheim, and N. M. Le Douarin (1987) Formation of the dorsal root ganglia in the avian embryo: Segmental origin and migratory behavior of neural crest progenitor cells. *Dev. Biol.* **120**:329–347.

Thiery, J. P., and J.-L. Duband (1986) Role of tissue environment and fibronectin in the patterning of neural crest derivatives. *Trends Neurosci.* **9**:565–570.

Thiery, J. P., J.-L. Duband, U. Rutishauser, and G. M. Edelman (1982a) Cell adhesion molecules in early chicken embryogenesis. *Proc. Natl. Acad. Sci. USA* **79**:6737–6741.

Thiery, J. P., J.-L. Duband, and A. Delouvée (1982b) Pathways and mechanism of avian trunk neural crest cell migration and localization. *Dev. Biol.* **93**:324–343.

Thiery, J. P., A. Delouvée, W. Gallin, B. A. Cunningham, and G. M. Edelman (1984) Ontogenetic expression of cell adhesion molecules: L-CAM is found in epithelia derived from the three primary germ layers. *Dev. Biol.* **102**:61–78.

Thiery, J. P., A. Delouvée, M. Grumet, and G. M. Edelman (1985) Appearance and regional distribution of the neuron–glia cell adhesion molecule (Ng-CAM) in the chick embryo. *J. Cell Biol.* **100**:442–456.

Tucker, R. P., and C. A. Erickson (1984) Morphology and behavior of quail neural crest cells in artificial three dimensional extracellular matrices. *Dev. Biol.* **104**:390–405.

Tucker, R. P., B. F. Edwards, and C. A. Erickson (1985) Tension in the culture dish: Microfilament organization and migratory behavior of quail neural crest cells. *Cell Motil.* **5**:225–237.

Vincent, M., and J. P. Thiery (1984) A cell surface marker for neural crest and placodal cells: Further evolution in peripheral and central nervous system. *Dev. Biol.* **103**:468–481.

Volk, D., and B. Geiger (1986a) A-CAM: A 135-kD receptor of intercellular adherens junctions. I. Immunoelectron microscopic localization and biochemical studies. *J. Cell Biol.* **103**:1441–1450.

Volk, D., and B. Geiger (1986b) A-CAM: A 135-kD receptor of intercellular adherens junctions. II. Antibody-mediated modulation of junction formation. *J. Cell Biol.* **103**:1451–1464.

Contributors

Mark C. Alliegro
Department of Zoology
Duke University
Durham, North Carolina 27706

Horst Antonicek
Department of Neurobiology
University of Heidelberg
6900 Heidelberg
Federal Republic of Germany

W. Scott Argraves
Cancer Research Center
La Jolla Cancer Research Foundation
La Jolla, California 92037

C. Howard Barton
Institute of Neurology
London WC1
United Kingdom

Elisabeth Bock
The Protein Laboratory
University of Copenhagen
Copenhagen, Denmark

Donna Bozyczko
Department of Biochemistry
University of Pennsylvania School of
    Medicine
Philadelphia, Pennsylvania 19104

Clayton A. Buck
The Wistar Institute of Anatomy
    & Biology
Philadelphia, Pennsylvania 19104

Jonathan Covault
Department of Physiology & Biophysics
Washington University Medical Center
St. Louis, Missouri 63110

Kathryn L. Crossin
The Rockefeller University
New York, New York 10021

Eileen Crowley
Departments of Anatomy & Oral Biology
University of California/San Francisco
San Francisco, California 94143

Bruce A. Cunningham
The Rockefeller University
New York, New York 10021

Caroline H. Damsky
Departments of Anatomy & Oral Biology
University of California/San Francisco
San Francisco, California 94143

Douglas W. DeSimone
Department of Anatomy & Cell Biology
University of Virginia School of Medicine
Charlottesville, Virginia 22908

Peter D'Eustachio
Department of Biochemistry
New York University Medical Center
New York, New York 10016

George Dickson
Institute of Neurology
London WC1
United Kingdom

Patrick Doherty
Institute of Neurology
London WC1
United Kingdom

Jean-Loup Duband
Laboratory of Developmental Pathology
Ecole Normale Supérieure, CNRS
75230 Paris
France

Rainer Duden
Institute of Cell & Tumor Biology
German Cancer Research Center
6900 Heidelberg
Federal Republic of Germany

Sylvie Dufour
Laboratory of Developmental Pathology
Ecole Normale Supérieure, CNRS
75230 Paris
France

Gerald M. Edelman
The Rockefeller University
New York, New York 10021

Klaus Edvardsen
The Protein Laboratory
University of Copenhagen
Copenhagen, Denmark

Jürgen Engel
Biozentrum
Universität Basel
Switzerland

Charles A. Ettensohn
Department of Biological Sciences
Carnegie-Mellon University
Pittsburgh, Pennsylvania 15213

Thomas Fahrig
Department of Neurobiology
University of Heidelberg
6900 Heidelberg
Federal Republic of Germany

Andreas Faissner
Department of Neurobiology
University of Heidelberg
6900 Heidelberg
Federal Republic of Germany

Jukka Finne
Department of Medical Biochemistry
University of Turku
SF-20520 Turku
Finland

Günther Fischer
Department of Neurobiology
University of Heidelberg
6900 Heidelberg
Federal Republic of Germany

Susan J. Fisher
Departments of Anatomy & Oral Biology
University of California/San Francisco
San Francisco, California 94143

Werner W. Franke
Institute of Cell & Tumor Biology
German Cancer Research Center
6900 Heidelberg
Federal Republic of Germany

Scott E. Fraser
Departments of Physiology & Biophysics
University of California/Irvine
Irvine, California 92717

James Gailit
Cancer Research Center
La Jolla Cancer Research Foundation
La Jolla, California 92037

David R. Garrod
Faculty of Medicine
University of Southampton
Southampton SO9 4XY
United Kingdom

Christine L. Gatchalian
Department of Physiology & Biophysics
Washington University Medical Center
St. Louis, Missouri 63110

Kurt R. Gehlsen
Cancer Research Center
La Jolla Cancer Research Foundation
La Jolla, California 92037

Benjamin Geiger
Department of Chemical Immunology
The Weizmann Institute of Science
Rehovot 76100
Israel

Ann Gibson
The Protein Laboratory
University of Copenhagen
Copenhagen, Denmark

Norton B. Gilula
Department of Cell Biology
Baylor College of Medicine
Houston, Texas 77030

Daniel A. Goodenough
Department of Anatomy
Harvard University School of Medicine
Boston, Massachusetts 02115

Achim Gossler
Friedrich-Miescher-Laboratorium
Max Planck Institute
7400 Tübingen
Federal Republic of Germany

Hilary Gower
Institute of Neurology
London WC1
United Kingdom

Jeannette Graf
Developmental Biology & Anomalies
National Institutes of Health
Bethesda, Maryland 20205

Lloyd A. Greene
Department of Pathology
Columbia University
  College of Physicians & Surgeons
New York, New York 10032

Magnus Hansson
Department of Medical & Physiological
  Chemistry
University of Uppsala
S-751 23 Uppsala
Sweden

Mary E. Hatten
Department of Pharmacology
New York University Medical Center
New York, New York 10016

Stanley Hoffman
The Rockefeller University
New York, New York 10021

Brigid L. M. Hogan
Department of Cell Biology
Vanderbilt University School of Medicine
Nashville, Tennessee 37232

Peter W. H. Holland
National Institute for Medical Research
London NW7 1AA
United Kingdom

Alan F. Horwitz
Department of Anatomical Sciences
University of Illinois/Urbana-Champaign
Urbana, Illinois 61801

Martin J. Humphriés
Department of Biochemistry &
  Molecular Biology
University of Manchester
Manchester
United Kingdom

Dale D. Hunter
Department of Physiology & Biophysics
Washington University Medical Center
St. Louis, Missouri 63110

Richard O. Hynes
Center for Cancer Research
Massachusetts Institute of Technology
Boston, Massachusetts 02139

Yukihide Iwamato
Developmental Biology & Anomalies
National Institutes of Health
Bethesda, Maryland 20205

Hans-Peter Kapprell
Institute of Cell & Tumor Biology
German Cancer Research Center
6900 Heidelberg
Federal Republic of Germany

Seishi Kato
Sagami Chemical Research Center
Sagami, Kanagawa, Japan

Rolf Kemler
Max Planck Institute for Immunology
D-7800 Freiburg-Zähringen
Federal Republic of Germany

Hynda K. Kleinman
Department of Biochemistry
Oita Medical School
Oita, Japan

Kimitoshi Kohno
Department of Biochemistry
Oita Medical School
Oita, Japan

Volker Künemund
Department of Neurobiology
University of Heidelberg
6900 Heidelberg
Federal Republic of Germany

Michael B. Laskowski
Department of Physiology & Biophysics
Washington University Medical Center
St. Louis, Missouri 63110

Jack J. Lawler
Department of Pathology
Harvard University School of Medicine
Cambridge, Massachusetts 02115

Dorte Linnemann
The Protein Laboratory
University of Copenhagen
Copenhagen, Denmark

Joan M. Lyles
The Protein Laboratory
University of Copenhagen
Copenhagen, Denmark

Ahmed Mansouri
Friedrich-Miescher-Laboratorium
Max Planck Institute
7400 Tübingen
Federal Republic of Germany

Eugene E. Marcantonio
Center for Cancer Research
Massachusetts Institute of Technology
Boston, Massachusetts 02139

Jane E. Marston
Department of Surgery
University of Southampton
Southampton SO9 4XY
United Kingdom

George R. Martin
Developmental Biology & Anomalies
National Institutes of Health
Bethesda, Maryland 20205

Rudolf Martini
Department of Neurobiology
University of Heidelberg
6900 Heidelberg
Federal Republic of Germany

Derek L. Mattey
Faculty of Medicine
University of Southampton
Southampton SO9 4XY
United Kingdom

David R. McClay
Department of Zoology
Duke University
Durham, North Carolina 27706

Helen R. Measures
NERC Institute of Virology
Oxford OX135R
England

John P. Merlie
Department of Pharmacology
Washington University School of
    Medicine
St. Louis, Missouri 63110

Anke Meyer
Department of Neurobiology
University of Heidelberg
6900 Heidelberg
Federal Republic of Germany

Pamela A. Norton
Center for Cancer Research
Massachusetts Institute of Technology
Boston, Massachusetts 02139

Ole Nybroe
The Protein Laboratory
University of Copenhagen
Copenhagen, Denmark

Björn Öbrink
Department of Medical & Physiological
    Chemistry
University of Uppsala
S-751 23 Uppsala
Sweden

Erich Odermatt
COBI
Walliserwerke Lonza A.G.
3930 Visp
Switzerland

Per Odin
Department of Medical & Physiological
    Chemistry
University of Uppsala
S-751 23 Uppsala
Sweden

Kohei Ogawa
Developmental Biology & Anomalies
National Institutes of Health
Bethesda, Maryland 20205

Katsushi Owaribe
Institute of Cell & Tumor Biology
German Cancer Research Center
6900 Heidelberg
Federal Republic of Germany

Elaine P. Parrish
The Rockefeller University
New York, New York 10021

Ramila S. Patel
Center for Cancer Research
Massachusetts Institute of Technology
Boston, Massachusetts 02139

Jeremy I. Paul
Department of Biochemistry
    & Molecular Biology
University of Chicago
Chicago, Illinois 60637

Elke Persohn
Department of Neurobiology
University of Heidelberg
6900 Heidelberg
Federal Republic of Germany

Michael D. Pierschbacher
Cancer Research Center
La Jolla Cancer Research Foundation
La Jolla, California 92037

Elisabeth Pollerberg
Department of Neurobiology
University of Heidelberg
6900 Heidelberg
Federal Republic of Germany

Rainer Probstmeier
Department of Neurobiology
University of Heidelberg
6900 Heidelberg
Federal Republic of Germany

François Rieger
Neurobiology & Fundamental
    Neuropathology Laboratory
INSERM U153
75005 Paris
France

Erkki Ruoslahti
Cancer Research Center
La Jolla Cancer Research Foundation
La Jolla, California 92037

Ilana Sabanay
Department of Chemical Immunology
The Weizmann Institute of Science
Rehovot 76100
Israel

Karin Sadoul
Department of Neurobiology
University of Heidelberg
6900 Heidelberg
Federal Republic of Germany

Remy Sadoul
Department of Neurobiology
University of Heidelberg
6900 Heidelberg
Federal Republic of Germany

Joshua R. Sanes
Department of Physiology & Biophysics
Washington University Medical Center
St. Louis, Missouri 63110

Makoto Sasaki
Developmental Biology & Anomalies
National Institutes of Health
Bethesda, Maryland 20205

Melitta Schachner
Department of Neurobiology
University of Heidelberg
6900 Heidelberg
Federal Republic of Germany

Monika Schmelz
Institute of Cell & Tumor Biology
German Cancer Research Center
6900 Heidelberg
Federal Republic of Germany

Jean E. Schwarzbauer
Department of Biology
Princeton University
Princeton, New Jersey 08544

Bernd Seilheimer
Department of Neurobiology
University of Heidelberg
6900 Heidelberg
Federal Republic of Germany

Michael L. Shelanski
Department of Pathology
Columbia University College of
    Physicians & Surgeons
New York, New York 10032

Kai Simons
European Molecular Biology
    Laboratory
6900 Heidelberg
Federal Republic of Germany

Mary Ann Stepp
Eye Research Institute
Boston, Massachusetts 02114

Ann Sutherland
Department of Anatomy
University of California/San Francisco
San Francisco, California 94143

Peter Svalander
Department of Anatomy
University of Uppsala
S-751 23 Uppsala
Sweden

John W. Tamkun
Department of Molecular, Cellular
    & Developmental Biology
University of Colorado
Boulder, Colorado 80309

Jean Paul Thiery
Laboratory of Developmental Pathology
Ecole Normale Supérieure, CNRS
75230 Paris
France

Gautam Thor
Department of Neurobiology
University of Heidelberg
6900 Heidelberg
Federal Republic of Germany

Anders Tingström
Department of Medical & Physiological
    Chemistry
University of Uppsala
S-751 23 Uppsala
Sweden

Kevin J. Tomaselli
Department of Physiology
University of California/San Francisco
    School of Medicine
San Francisco, California 94143

Dietmar Vestweber
Friedrich-Miescher-Laboratorium
Max Planck Institute
7400 Tübingen
Federal Republic of Germany

Marcelo J. Vilela
Faculty of Medicine
University of Southampton
Southampton SO9 4XY
United Kingdom

Tova Volberg
Department of Chemical Immunology
The Weizmann Institute of Science
Rehovot 76100
Israel

Talila Volk
Department of Chemical Immunology
The Weizmann Institute of Science
Rehovot 76100
Israel

Frank S. Walsh
Institute of Neurology
London WC1
United Kingdom

Margaret J. Wheelock
The Wistar Institute of Anatomy
    & Biology
Philadelphia, Pennsylvania 19104

Gregory A. Wray
Department of Zoology
Duke University
Durham, North Carolina 27706

Kenneth M. Yamada
Laboratory of Molecular Biology
National Cancer Institute
Bethesda, Maryland 20205

Yoshihiko Yamada
Developmental Biology & Anomalies
National Institutes of Health
Bethesda, Maryland 20205

Index